Soils, Plants and Clay Minerals

Bruce Velde · Pierre Barré

Soils, Plants and Clay Minerals

Mineral and Biologic Interactions

Springer

Prof. Dr. Bruce Velde
Laboratoire de Géologie
École Normale Supérieure
24 rue Lhomond
75231 Paris
France

Dr. Pierre Barré
Laboratoire de Géologie
École Normale Supérieure
24 rue Lhomond
75231 Paris
France

ISBN 978-3-642-03498-5 e-ISBN 978-3-642-03499-2
DOI 10.1007/978-3-642-03499-2
Springer Heidelberg Dordrecht London New York

Library of Congress Control Number: 2009935566

Cover design: deblik, Berlin

Printed on acid-free paper

Springer is part of Springer Science+Business Media (www.springer.com)

Preface

This book represents a rather complicated history of encounters, changes in research interest and some very interesting results. Initially it is the very fruitful interaction of Ecology and Geology. The point of view of ecologists is extremely refreshing for hard science people. Interaction and inter-relationships are the focus of Ecology whereas the traditional sciences, such as Geology, have tried to isolate the natural phenomena so that thye could be studied in a more rigorous manner.

The traditional sciences were of course natural science – based since the world to be observed was at the door step of everyone, mountains, weather patterns, plants and so forth. Chemistry and Physics were defined after Mathematics in order to establish more precise and viable principles of the behavior of the materials that formed the world around mankind. It became quite clear that the observation of the natural world was too complicated to consider all of the possible variables which could affect an observed process or situation. The systems were simplified and taken into the laboratory in order to better master the phenomena observed. Physics concerned itself with non-reacting materials, subjected to essentially mechanical forces. Chemistry was given the task of understanding the interactions and transmutations of matter according to the affinities of atoms and ions. As it turned out the interactions of electrons was the key to the chemical problem. In these endeavors, the systems were simplified to better understand the fundamental principles. However nature remains a multidimensional problem and although great progress has been made using the discoveries of Chemistry and Physics to understand the underlying principles of phenomena in natural systems, much escapes still.

One of the major problems today in the study of the earth's surface: contact between rock, plant and atmosphere, is the understanding of the interactions and influences possible in this zone of intersection of phenomenological forces. For example, most mineralogists concerned with clays, those silicates present and active in the surface zone, feel that the influence of the bio-zone is only negative, one of destruction and dissolution, being part of the general degradation of silicate materials by surface interactions. Plants and bio-agents are generally considered to be classified in the alteration category such as that encountered in the contact of rain water with rocks. However, numerous, little heard authors have insisted for quite some time, over a hundred years, that the interactions can be important and transfer of materials from depth to the surface can condition the soil, interaction zone.

Plants, minerals, bio-agents such as micorrhize and bacteria all interact to form a living community, in the Ecological sense.

The interaction of the sphere of living materials (plants and bio-agents) with the slowly reacting and stable silicate (rock) material under the influence of climate (rain) are the subject of this new approach to science at the surface of the earth.

We would like to thank our many collaborators for their enthusiastic help, from various parts of China, to the Unites States and Canada and parts of Europe. It is clear that Science is a world venture, and a pleasure to be part of.

Closer to us, our colleagues in our laboratory and the technical staff without which we would be more or less helpless, have been a great inspiration during the preparation of this book and the scientific studies that have led up to its expression. No one is alone in this complex modern world. We wish to thank all of those who have contributed to oru efforts.

Paris

Bruce Velde
Pierre Barré

Contents

Chapter 1
Introduction: Objectives and Background

Contents

B. Velde, P. Barré, *Soils, Plants and Clay Minerals*,
DOI 10.1007/978-3-642-03499-2_1, © Springer-Verlag Berlin Heidelberg 2010

1.1 The Need to Understand the Plant – Earth Interface in the Future World of Pollution, Increased Exploitation of Natural Resources and the Quest of a Bio-Friendly Society

In this, the beginning of a new millennium and apparently a new era, more and more questions are asked about the nature of nature. We want bio-friendly processes in farming, in the means of obtaining heat energy and the means of propelling our automobiles. We want this to be done with minimum risk at the same time to ourselves and our children as well as to the biological community which co-inhabits the surface of the world with us. Perhaps this is a too big a problem with many different answers, sometime contradictory, but it is a true challenge to the young people of our world. As it turns out, the world of science is rather poorly equipped to deal with society's new concerns. This is not due to negligence but to a historical process.

First we must put things into perspective. Until very recently and since the philosophical and material triumph of Science in the post-war period of the latter half of the twentieth century, society has come to question some of the *tenants* of this era of material progress. In the Cold War era each of the protagonists, Communist and Democracy (Free Enterprise) were bent on out producing each other in goods and services. The well being of humanity could be measured by its GNP or some other measure of the productivity of a society divided by the number of its inhabitants or participants. Materialism was on both sides of the curtain. A half century of such intense competition entrenched the major themes and objectives of the system. With the fall of the separation of the two organizations, it was seen that the communist countries produced their material wealth often at the expense of the health of the producers. This gave satisfaction to the opponents of this economic system but it turns out that the race to material well – being has taken and is still taking its toll in the western countries also. This is and has been perceived by the populations of the western nations as well as the so-called eastern block. No longer can one equate Science with Progress with Well-being. Change is not necessarily beneficial in itself. Most students today share this view and hence they turn away from the classical fields of Science. It is time to attempt to resolve some of the problems of materialistic society and its over production in a reasonable manner. To do this, many people propose a more biologically friendly approach to sustained systems of development. Few people can argue this today. However there are major problems in such a quest.

First one needs to understand the bio-systems and their ecology (interaction and development) in relation to their environment. Man has altered the surface of the earth to such an extent that it is hardly recognisable compared to its state 4,000 years ago. First the quest for a stable and high energy diet lead to the introduction of grazing and then farming at different levels of intensity in different places on the earth. Gradually the methods of development of the biosphere came to converge into a system of intense agriculture spread over large sections of the inhabitable globe. For several thousand years the agriculture practiced by different communities was bent on conserving the natural resources of the land, which is in a sense sustainable agricultural. Written texts of the ancient Greeks attest this quite clearly. Crop rotation and land use were highly recommended. This traditional farming was practiced in many areas, Europe, the circum-pacific region, Africa and the Asian continent which modified the use of the land to different degrees. However the limits of energy and monetary or resource input kept the objectives much the same: maintenance of soil fertility from one year to the other. With the advent of the mechanisation and above all the use of factory produced chemicals applied massively in the post War period the concept of soil fertility seemed to have become less important. The use of external stimulants was considered to be sufficient to overcome any mechanical or biological problems which would be encountered that would hinder increases production. Such practices have led to the situation which is present today where chemicals used in agriculture seem to be all pervasive and their long term effects as yet only poorly known. Basically many people would like to return to a more "natural" system of food production and use of the earth's surface for the ever increasing needs of humanity.

However, in order to do this it is necessary to understand the basic functioning of the biological – soil systems. Such work has essentially stopped in the last 50 years or so to the benefit of plant development and mechanical – chemical manipulation of the soil. We know how to grow large, productive plants with the aid of fertilizer and phyto-sanitary aids and how to move very large machines that plant, treat and harvest them at an incredibly rapid rate. However, there is little recent information on what is happening in the soil – plant interaction. Modern methods of investigation of the clay minerals for example have not been used to observe the changes brought about by modern farming on the mineral substrate nor the way that organic matter is bound to the clays in the soils nor the reciprocal effects of such relations. What are the major functions of the plant – soil interaction? How can one observe them and control them to benefit a long term and stable interaction? Of course we do not wish to return to the farming methods of our great grandparents nor the significantly lower productivity of these and more distant times. There is a need to feed an ever increasing population. However, this being the case, there are ultimate limits to the use of land from which it can possibly not recover. Soils will be more and more called upon to produce the needs of modern society by the development of bio fuels for example.

Our task here is to outline the structure of the plant – soil interface and to structure the relations which are known. Given this introduction we hope to suggest a number of avenues for research in the future which could possibly aid in the development of our natural resources. In order to do this we propose to use the structure of present knowledge of plant – soil interactions according to the classical methods of investigation. With some new insight it is possible to have a different view of the natural system and perhaps using this basis progress in the understanding of the soil – plant ecosystem.

The framework of interest here is the world of plants and rocks. This is the basis for life on earth. However, one must not forget water, which is the *sina qua non* for life, and also that for the transformation of rocks into soils, i.e. clay minerals.

The two basic ingredients are water and rock, whereupon life can exist. The transformation of hard, un-yielding materials into soft, water conserving materials such as soils, is accomplished by a dissolution – recrystallization or by a dissolution – transformation process in an aqueous medium. Hence in deserts, for lack of water most of the time, little reaction can occur and little life can subsist. In the extreme cold of the arctic, water does not exist most of the time (it is ice) and hence mineral transformations and life do not exist in abundance. The best place for life is where rocks become clays and where the plants can use this substrate to form their life sustaining substrate. This is why temperate climates are called temperate; they foster the abundance of living things. Out objective is to understand the relations between water, rock and plants. Herein lies the secret of the inter-relations of plants and minerals. Each individual actor uses the other to form his life – sustaining needs. Plants and eventually animals are the manipulators of the chemical systems found at the surface of the earth, the bio-zone.

1.2 Understanding How and Where the Plants We Know Today Occur and Interact with the Mineral World: The Setting that we Live in Today and its History

The world we live in today seems very old to many people, geologists and planetologists for example. Mountains, forests and other features appear to have been around for quite some time. The major factor of change, recent to say the least, is the transformation of the flora and fauna due to the use of the earth's surface by man. Some recent events, such as global warming appear to be a major threat to the stability of the system also, but put into geological perspectives they appear to be part of the normal course of fluctuating events. One must remember that in the past million years or so, a minor part of the history of the earth since the appearance of oceans and continents which is some thousands of times older, the earth has experienced 20 or more periods of glacial extension and severe climate change affecting vegetation patterns (Potts and Behrensmeyer, 1992, p. 461) and 10 or so in the last million years (Shackelton, 1995). These cycles have greatly affected the disposition of plant and faunal regimes in the continents.

However in the long history of the earth, continents have supported plant life for some 400 million years or so, and these plants were quite different from the majority of those we know now. In fact what we see today at the surface of the earth is relatively recent and some of the major geologic events seem to have played a role in the configuration of the climate and as a result the plants which occur on the earth's surface. We insist here on the plants as a major if not the major conditioning factor of life in that animals can migrate relatively rapidly, given a specific land mass to move about in, to adapt to changes in habitat but plants are more bound to local conditions, or at least some plants adapt to these changes in a more static way. It is important to put the present day configuration of the surface of the earth, mineral and biological, into perspective. The changes wrought by man's activity are enormous, to be sure, but what are they changing? How long have things been as they were before the advent of agriculture and forestry? This is the perspective that one should take before considering the interaction of plants, minerals and the use of these relations by human kind.

1.2.1 Configuration of the Surface Environment of the Earth: Mountains and Climate

The surface of the planet is irregular in that land masses arise out of the oceans at irregular distances. The present day land masses are above all occur in the northern hemisphere, an asymmetry which one should not forget. Also a continent is present at the South Pole but it is covered with ice, unlike the situation at the North Pole where no continent is present but land masses are not far away; North America, Greenland, and the Eurasia. This situation of asymmetry is important to the circulation of ocean water and to a lesser degree the air masses that dominate in climate formation. The sailing contests of "around the world" carefully skirt the Antarctic and are not bothered by detours around bothersome large continents. Their trajectory is essentially north–south. It is a lot shorter in distance and more predictable weatherwise.

Climate on the continents, which is our interest here, is dictated by their situation on the earth (latitude) and their geographic – geomorphological configuration. If continents were not present, Aikin (1991, Fig. 1-5) indicates the general wind and pressure belts which might exist if the earth had no land masses. The inter-tropical convergence zone of low pressure is where the tropics would be found and where rainfall is high as well as are temperatures. To either side (latitude) one finds cellular high pressure zones at 30° latitudes followed by more irregular and smaller high pressure zones near 60° latitude. These air mass movements are both vertical and horizontal due to the rotation of the earth and thermal effects driven by the albedo of the water masses. The salient feature in such an analysis is that there are steep pressure gradients around the equator which are lessened in further latitudes to the north and south. The end result is that one finds a strong rainy zone around the equator and dry climates directly adjacent. As the pressure gradients lessen one finds more clement climates, the temperate ones with more sporadic rainfall. The major movement of air masses is west to east due to the rotation of the earth under the air.

If one considers the oceans only as we see them today, the weather pattern is roughly regular, apparently dictated by air circulations centered on the equator. High rainfall and high temperatures seem to be typical of there areas. Typically in the present day configuration of continents and oceans on the average rainfall drops off to the north and south while temperatures fall less rapidly (Fig. 1.1).

Somewhere in mid-latitude rainfall increases as temperature falls gradually until the polar regions are reached where temperatures are low and rainfall is also. One can follow this by looking at the daily weather map of the world where it is seen to

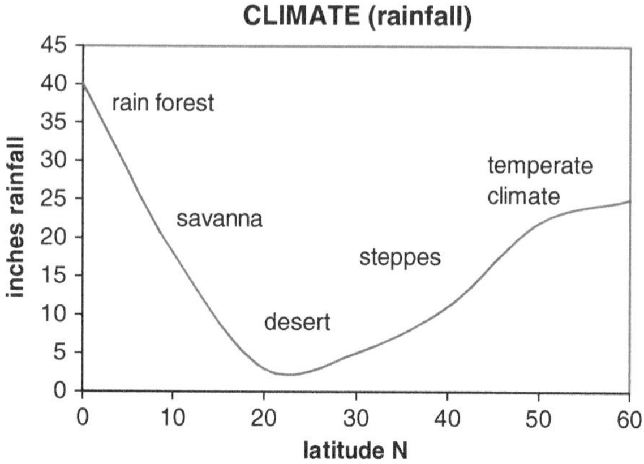

Fig. 1.1 Relations between rainfall and latitude, generalized from climatic maps (data taken from different maps in the Encyclopaedia Britannica, 1969) The different, general, climate types are indicated on the diagram. On either side of the low rainfall zone, desert, one finds a grassland (steppe and savanna). The decrease in rainfall from tropical forest to desert is much greater than that toward the temperate climates from the desert

rain at or near much of the equator and just to either side one finds no rain. Almost daily rain is centered on the equator in the summer and shifts somewhat to the south in the winter. Deserts adjacent to the tropics have little rain and the temperate, mid latitude climates have variable weather patterns.

However the continents change the simple, ocean dominated, weather pattern to various degrees and in different ways. Basically, the tropical rain belt is nourished by ocean evaporation. These wet air masses come into contact with continents. The Americas are typified by high, north–south mountain chains on their western boundaries with some much lower, and less consequential, massifs on the eastern edge in North America. The high western reliefs, where the weather comes from, act as a barrier to air movement onto the continent and the effect on climate is striking (Fig. 1.2). In North America, especially in the United States, the mountain barrier creates a rain shadow to the east which is gradually decreased by mid continent. This is true in the southern part of South America but lack of significantly high relief in Central America and the north of South America allows the humid equatorial air to move across the land masses to the east into the north of South America where it releases its humidity creating the Amazon forest basin and spilling further south from the equator. Gradually the rainfall diminishes some in that it is not renewed from oceanic evaporative resources.

By contrast Africa, essentially a low relief continent, is strongly affected by the oceanic climate pattern directly with a rain forest at the equator and deserts to the north and south. The equatorial rainfall decreases by the time the air masses reach the eastern edge of the continent where high temperatures are found in the presence of arid climates. Here when the resources of humidity are not renewed enough and the rainfall decreases greatly until there are near desert climatic conditions in the eastern portion of equatorial Africa (Fig. 1.3). The regular weather

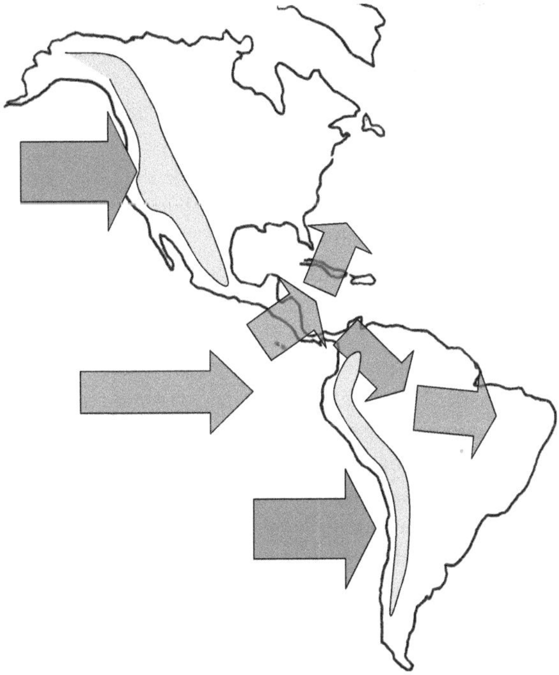

Fig. 1.2 Schematic map showing air movements bringing humid air over continents. For the Americas, the mountain barriers north and south, largely prevent the entry of these air masses onto the land masses, creating a rainfall gradient west to east. Nevertheless these sources of rainfall make their ways north and above all south to create the large, wet areas of forest growth, especially in the Amazon basin region

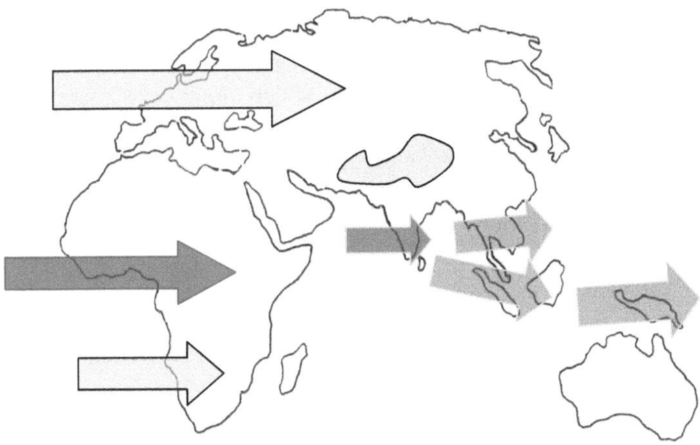

Fig. 1.3 The normal airflow of air masses over continents with high mountain barriers in their interiors, especially visible in the Himalayan region, is illustrated by the Eurasian land masses. The humidity laden air stream loses its force over Africa towards the eastern edge of the continent, but picks up moisture over the Indian Ocean to create humid climates further to the east over Southeast Asia. The air masses which cross Europe, eventually lose their humidity and leave the increasingly drier regions of the Eurasian continent where the system is aggravated still further by the Himalayan massif which blocks any movement to the north creating the great Gobi desert

pattern of ocean dominated movement picks up again as the air masses gain humidity over the Indian ocean and enter onto the Indian subcontinent. The western edge is subject to high rainfall, especially on the southern part. Here the imprint of global air mass circulations is highly visible (general configuration adapted from Trewartha, 1954).

The south Pacific area is dominated by the equatorial rain climate in that the islands do not make a formidable barrier to the air mass movements and hence the tropics of the equator are wet. Australia, to the south, follows the pattern of desert below tropical rain forests and some change to higher rainfall areas with lower temperatures in its southern areas.

The European area follows the general ocean dominated climate pattern in its western and north, north-eastern areas being only slightly disturbed by the alpine age massifs. However the Eurasian continent is most strongly affected by the highest and most massive of the mountain chains, the Himalaya. Here the equatorial air masses are deviated to the south somewhat, giving high intensity rainfall in Southeast Asia. North of the great mountain chain one finds the greatest deserts present today eventually changing into cold and low rainfall areas below the arctic areas to the north and gradually into steppes and deciduous forest climates to the northeast. This pattern is particularly important to China, where the great desert is followed by the steppes of Manchuria and succeeded eventually by the deciduous forests of Russia on the edge of the Pacific.

Hence it is clear that the configuration of high mountain chains affect the climate to a high degree. These mountain chains, present today, are for the most part geologically recent, that is being formed gradually and cyclically from about 30 million years ago (Lefort, 1996; Duff, 1993; King, 1962). Thus the configuration of the earth, continents and mountain chains which affect the climate patterns due to air mass circulation, has been roughly the same for the last 30 million years. Prior periods showed climates at times significantly warmer than today, as much as 20° at the poles and 5°C at the equator (Duff, 1993, p. 462). One can expect that the climatic configuration created by these 30 million year old events has been more or less constant. However, superposed upon this general topography dominated climatic pattern, the occurrence of glacial periods (occurring over the last million years) have strongly affected the pattern, shifting the boundaries of the climatic zones squeezing them towards the equator in cold periods and away from it depending on the cycle of ice accumulation. These shifts have been strongly felt in most areas except perhaps the very central equatorial zones.

One remarkable fact is that for the most part, excepting northern Southeast Asia, the tropical climatic areas are of low relief and here the bedrocks are old, not having been renewed by orogenesis, and the soils are also very old. This is particularly true of Africa and the Amazon region. By contrast in the Southeast Asia area, especially the island chains, the relief can be locally high and the bedrocks recent, being of volcanic origin. Hence the tropical areas can consist of strongly contrasting geology and topography in the different regions of the globe.

The effects of climate on soils is of course primordial. The climate inductive to prairie – type soils, mollisol in the vocabulary of pedologists, is shown in

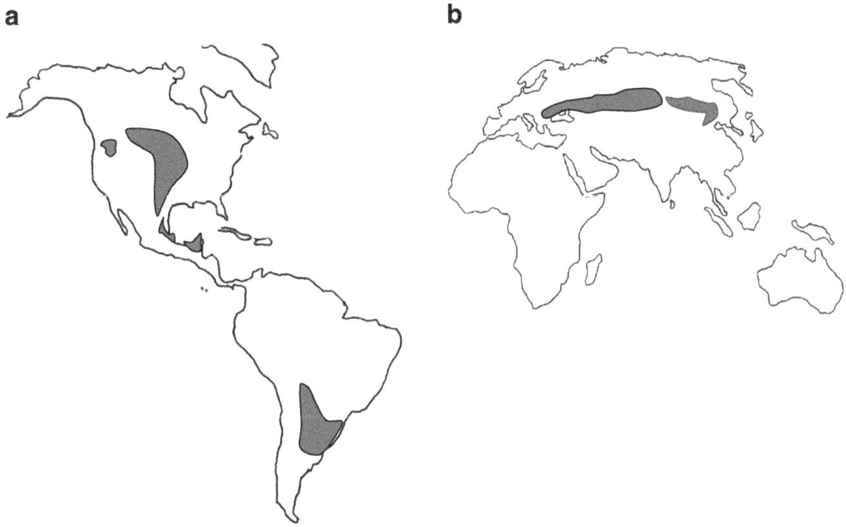

Fig. 1.4 Consequences of the global air mass movements as it affects the soil types. The grass lands, due to lower rainfall under moderate temperature conditions, create the Mollisol soil facies, rich in organic matter and fertile for crops based upon grassland plants such as wheat, oats and lentils. These distributions follow the climatic boundaries defined by topography and air mass movement, north south in the Americas (**a**) and west east over the Eurasian continents (**b**).

Fig. 1.4. The strong tendency for a north–south trend in the Americas compared to the east–west trend on the Eurasian continent is evident. The mountain chains of the west coast in the Americas re-orient the climate according to this physical barrier whereas the lack of such a barrier allows normal east–west climatic influences to reign on the Eurasian continent.

1.2.2 Vegetation and Climate Today and Yesterday

The largest part of today's plants have originated since the last 20 million years (families) and all genera since 10 Ma (Behrensmeyer, et al. 1992, p. 428), with some strong changes toward grasses for example during more recent periods, the last 13 million years up to 5 Ma. Essentially the dicotyledans are of this more recent era, with the formation of tree species as we know them today and slightly more recently the grasses. Hence the vegetal configuration we know today was created or developed when the mountain chains were formed in the geologically recent past. As vegetal life on land has been abundant for more than 400 million years this suggests that the "recent" geological events have changed things greatly. Particularly the grasses were rather late comers to the vegetal panoply but their impact on the biomass landscape was certainly decisive. The grasses allowed the development of the large grazing animals of the steppes and savannas which became the first motors of human existence in the sense that they were these animals provided the first stages of traction power stronger than that of man himself. These large quadrupeds allowed the development of nomadic hunting cultures and above all agriculture, multiplying the energy available for the needs of mankind. Grasses were of course the basis for a large portion of farmed crops also and still remain primordial for a large part of humanity. The widespread new occurrence of grasses occurred in the last 10 Ma years or so (Adams, 2007), with a peak in ecological diversity near 12 Ma. The spread of steppes propitious to their growth appears to be correlated with the rise of recent mountain chains (Behrensmeyer et al., 1992, p. 430).

Hence these recent changes, recent for the age of the earth but very old for the span of time that human or humanoids have been present which is only a couple of million years, have shaped the earth's living surface as we know it. It seems that the stage was set for the world as we know it in the rather recent geological past. Terrestrial plants were more than common in the Carboniferous period, some 300 Ma years ago. However the ecotones in this distant past were quite different from those known commonly today.

The stage was set, but climatic fluctuations occurred and especially in the last one million years or so. During this period glaciation has created displacements of the vegetal sequences in numerous cycles, up to twenty, (Potts and Behrensmeyer, 1992, p. 461). The cold periods squeezed the temperate climates towards the equator and warming released them making way for greater areas of steppes and then forests which we have known in the period of human occupation during the apprenticeship of agricultural life (see Bell and Walker, 1992, Figs. 4.12 and 4.13 for example). Thus in a given geographical place, the climate and vegetation has changed periodically over the last million years in numerous cycles. However these changes or displacements seem not to have affected the number or types of plant species to any large extent. Nevertheless they are part, and an important one, of present day preoccupations with climate change.

The overall change from megathermal vegetation, broad leaved and evergreen, dominating in the late Cretaceous period gradually changed to one dominated by deciduous trees in the 100–50 Ma period (Stewart and Rothwell, 1993, p. 496) and eventually large sectors of the continents were covered by grasses, depending upon the glacially controlled climate favourable to these plants.

1.2.3 Present Day Context of Plant Biomes

Plants appear to dominate the surface features of soils. In fact soil scientists implicitly indicate that soils exist when plants are present. The upper most horizon of an alteration sequence in soil is one where organic matter due to plant growth is present. This is normal in that the input of vegetal matter is primordial in the development of this surface zone which structures and conditions the alteration sequence beneath (see Sects. 2.3 and 5.1). Different types of plants produce different types of organic matter which attract different forms of microbial life producing the complex chemistry of soil organic matter. The organic matter deposited on the surface, and that exuded by roots or the decayed roots themselves is quickly modified by bacterial and other actions of living matter in the soils zone. The soil zone is defined, in general, where the plants have the greatest influence and where the organic matter being decomposed gives a high organic carbon content and controls many other chemical variables. Elements in soil solutions as ionic species and hydrogen ions (pH) are among the major factors of soil solution chemistry controlled by the plant derived organic matter which affects the minerals in the soils.

The major types of vegetal cover have been identified and essentially defined according to the ratio of trees compared to grasses and other low cover. In a very general way one can divide the great vegetal associations into several types (Morley, 2000, p. 37) with forests dominated by broad leaved evergreens in the tropics, deciduous forests in temperate zones, needle leaved forests in areas of abundant to moderate rainfall and cooler temperatures and, where rainfall is more limiting, one finds areas dominated by grasses, the steppes and eventually shrub lands. The types of forest present are determined by a combination of temperature and range in temperature during the year (Morley, 2000, p. 51). World maps giving the present day geographic distribution of these different types of vegetation are found in numerous texts such as the Encyclopaedia Britannica (1969), Trewartha (1954) or Aiken (1991, Fig. 6-1). In this context, the smaller changes in temperature (from hot to cold) favour tropical broad leaved evergreen rain forests, needle leaved coniferous forests and tundra under abundant to normal rainfall. All of these forests produce acid soils. Braque (1987, p. 53) indicates that the deciduous characteristic of a given plant type can depend upon the climate it has adapted to. Desert and tropical humid climates favour non-deciduous behaviour or trees and shrubs, whereas temperate climate conditions favour deciduous behaviour. Depending upon the average temperature, the major control of soil biological activity, the leaf litter is more or less actively decomposed. In cold, wet climates the humic organic matter becomes a dominant feature in the surface horizons often forming high organic deposits. In hot, wet climates the biological activity greatly reduces the presence of humic matter in the A, soil horizon.

Drier climates and those with contrasted seasons favour grasslands. Intermediate rainfall conditions favor the deciduous forests. Prairie soils are neutral to perhaps basic in pH, rich in humic matter and in general clay rich. The deciduous forest soils are slightly more acid, but tend to have less clay in the surface horizons. Coniferous forests, found in colder and wetter climates, tend to produce acid soil pH. The presence of these general ecotone zones is due to climate, a combination of total rainfall, temperature and variations of both. Neither temperature nor total rainfall are enough to define a climate and the impact it has in creating the vegetation present.

In a very general way the basic regions of the world according to general vegetation type as a function of rainfall and latitude. As mentioned above, the effects of glaciation and glacial retreat determine the importance of the different zones which will shift according to the overall polar temperatures which impact the interpolar space of the earth. At present, in a warming climate, the desert zones expand pushing the frontiers of the grasslands towards the forest cover. However for the moment, based upon studies of flora in the past (late Tertiary and Plio-pleistocene) there seems to be no specific, climate driven, change in the number or types of species present. It is more of a displacement phenomena seen during the many glacial events in the past one million years. Overall today, 31% of the continents are covered in forest and savanna, 24% pasture, i.e. grazed agriculturally, 11% under agricultural practice and planted with only 34% under other conditions, those basically not used by man (Solbrig, 1991).

The biomass production is variable according to the climate affecting the type of vegetation. For example tropical forests produce 20 t/ha, forests in warm temperate climates 17 t/ha, temperate climate forests 13 t/ha, taiga forests 10 t/ha down to 4 in cold climates whereas prairies under temperate conditions (Normandy in France) produce but 9 t/ha, most of in the root zone (Braque, 1987, p. 53).

However climate has a very strong influence on the growth of plants in that they have more or less adapted to a certain general climate. This effect can be illustrated by the data in Fig. 1.5 where the production of corn in the United States in 1919 is shown and climatic variables superimposed on the mapped distribution (Jenny, 1994 and Cox, 1925). These data indicates the natural adaptation and growing efficiency of the corn plant before artificial irrigation methods were used which influenced the growth patterns of this crop.

The soils of the Central United states are based largely upon similar materials, either loess deposits or sedimentary rocks where the clay assemblages resemble one another greatly. Thus the major parameters are temperature and rainfall, which seem to form a zone of preference for corn production. There is an apparent need for between 750 and 1,000 mm of rain fall and a thermal optimum of between 12 and 8°C average temperature. The drop off in corn production to the east is due to the

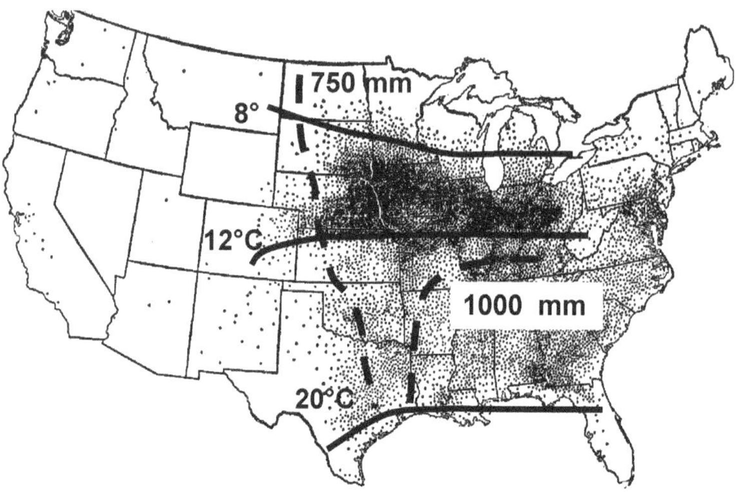

Fig. 1.5 Composite map based upon a compilation of corn productivity in the United States in 1916 (Cox, 1925) and climatic averages given by Jenny (1994). One sees the optimum conditions for corn productivity, indicated by the amount of culture of this crop, as being conditioned by rainfall (greater than 200 mm per year, and average temperature between 12 and 8°C). This indicates the importance of climate concerning plant growth and prosperity

presence of the Appalachian Mountains which change the climate characteristics and patterns of course.

This example illustrates the inter-relations between climate and plant growth where, as is well known, plants have adapted to certain climates over the course of time. In the case of corn (*zea mais*) the intervention of man in selecting varieties adapted to his needs has not overcome the natural adaptation of the plant before selection. Corn was planted where it grew best in the Central part of the Midwestern United States, familiarly known as the Corn Belt.

1.2.4 Local Variations in the Chemical Context (Geology) and Climatic Variables of Plant – Soil Interaction

The continents are variably inhabited by plants. Twenty eight percent has no vegetation (Braque, 1987, p. 53) and the rest is dominated, as far as biomass is concerned, by forests which produce 80% of the vegetative cover in mass but not in surface area. Thirty one percent of the surface area is in forest and savanna, 24 in pasture 24 in agriculture (Solbrig, 1991, p. 101). These numbers indicate that much of the continent areas are without vegetation, forests dominate in producing biomass but almost half is of use or possible use for agricultural purposes.

This very general pattern of climate and biotome are nuanced locally by two major factors: geology (the composition of the source material for soils) and topography which changes the climate locally. These two factors are of greatest importance in the field, where they often dominate concerning the plants present and the soils that are formed.

1.2.4.1 Geology

Geology as such is a vast subject of investigation, at least to geologists. In the realm of plant – soil interaction geology means chemistry (See Jenny, 1994, Chap. 4). In fact the geology of the surface concerns the rocks that are at the air – rock interface. Sediments, sedimentary rocks, metamorphic rocks and plutonic rocks are the vocabulary of the geologist in this sense. Each type of rock which is given a geological name in fact designates a specific chemical composition, combined with some other criteria of crystal size and relations of texture, species of mineral present and so forth. But what concerns the biosphere is the ratio of chemical elements present, their susceptibility to be destroyed in water interaction and the tendency to create small particles during the process of alteration. Thus chemistry controls such factors as pH, potassium availability, calcium and magnesium availability and phosphorous availability, among other elements. As is well known certain plants have adapted to granitic or acid soils such as broom grass, and calcareous soils which favour the easy growth of grasses and, by extension, common grain crops. Overall, if one looks at the great grain producing areas of the world one looks at the geological map of sediment outcrops (the sedimentary basins) of North America, Western and Eastern Europe and China. Very often these sedimentary basins, where sedimentary rocks outcrop, are a composed of series of flat lying strata, which means nice flat land surfaces amenable to reasonable drainage but little erosion and hence deeper soils (the zones of fine grained material developed by plant – silicate interaction).

However, under the same climate different biomes can be present due ot the geologic substratum. An example is given by Blackford (1993) for Ireland where pine trees were the general forest tree in the period 7500 BC except for the limestone plateau in the north where oak was present. This is but one of many examples of an influence of geological substrate on plant species and growth.

More striking, according to Richards (1976, p. 221) bedrock can change the biome from rain forest in the tropics to scrub land vegetation or even savanna. The influence of the chemistry of the substrate changes the vegetation as much as climate could. However even though the tropical forest climate savanna is primarily grass, the grasses will not be the same as in a contrasted season savanna.

A fact or great importance in this effect is the type or form of mineral present in the rocks of the soil basement.

Sedimentary rocks are just that, compacted sediments and hence compacted accumulations of soils which are slightly recrystallized to form minerals similar to those the initial sedimentary material which hardens the material to form a rock. Soils are the major contributors to sedimentation, either in continental (i.e. non-marine) or ocean accumulations. As such they condition the minerals present which will be gradually changed into new minerals as the sediments are buried deeper under others. Depth indicates higher temperatures and hence eventually mineral change. Gharrabi et al. (1998) indicate the conditions necessary to radically change the minerals found in sediments (soil clays) into new, metamorphic minerals. These changes, re-crystallizations, are a function of time and temperature in that the transformation of soil clays into metamorphic minerals is a long process and rather slow in the context of sediment burial in sedimentary basins. It takes as much as 200 Ma at 150°C to produce a decent low grade metamorphic assemblage approaching that of muscovite and chlorite (see Velde and Vasseur, 1992). The transformation of soil minerals into new phases occurs well before the stage of metamorphism however. Virtually all of the minerals in a sediment buried at several kilometres depth for several million years are different in all characteristics from surface minerals. This is essentially determined by the loss of expanding (smectite) minerals from the mixed layer mineral assemblages. However new minerals can form before this point, those normally called metamorphic. Chlorite is one such phase. Velde et al. (1991) indicate the compositional changes which occur in the new mineral formed upon change in temperature due to sediment burial. Again, the deeper the sediment, the higher the temperature affecting the soil clay minerals and the faster the reaction will occur. Initially chlorites form mineral grains of differing composition which become more homogeneous as temperature and time factors increase. Hence if one has a soil based upon a young or little diagenetically transformed sedimentary rock, one can expect that the chlorites found in it will be of rather different compositions. This will give secondary clays with significantly different compositions and properties, such as layer charge. If the material of the rock has been subjected to stronger metamorphic conditions the phyllosilicates will be more homogeneous even though they can have significant variations in compositions due to successive metamorphic events (Saliot and Velde, 1982). Thus one should not be surprised to find that the clays developed from the same initial mineral in a rock have variable compositions, and the point which concerns us the most, different layer charges due to these compositional variations. This means that the same material can form the 2:1 clay minerals with different interlayer behaviours, an observation frequently made which creates variable behaviour of mixed layer minerals and segregation under chemical treatments such as potassium saturation.

Metamorphic rocks are sediments where the minerals have been very strongly changed from those of soils. In most sedimentary rocks one finds quartz, the most refractory of minerals which remains largely inert during weathering and soil development. Hence quartz – rich rocks (sandstones) do not make for good plant growth. They are barren to plant sustainers depending upon the amount of clays that were concentrated with the sands. Sediments containing more clays form metamorphic rocks which are most often dominated by micas, and chlorites which are phyllosilicates very similar to soil clays, 2:1 minerals which have a larger grain size than clays. Other minerals can be found depending upon the temperature which the rock has experienced. Higher temperatures tend to destroy chlorite and eventually muscovite mica. However the largest portion of metamorphic rocks contain mica, especially muscovite, the potassic, aluminous phyllosilicte mineral.

Sediments containing carbonates do not change much with metamorphism, maintaining their chemical and mineralogical characteristics.

Igneous rocks form a range in compositions from those similar in bulk composition to argillaceous sedimentary rocks (granites) to those completely different from sedimentary materials (basalts). For the most part, igneous rocks do not form ideal sites for plant growth. In temperate latitudes the plant cover tends to be one of small shrubs with dispersed grasses. Extremes of pH accompany the igneous rocks of granite and basalt composition rendering them out of range for normal forest or grass growth.

The differences in vegetation determined by bedrock are largely effaced under conditions of warm tropical climates where the alteration and soil layer are sufficiently transformed so as to form a sort of convergence in their effect on vegetation. Thus under extreme climatic conditions, the climate becomes more important than geological (bedrock) factors, under extreme tundras, deserts and warm tropical conditions.

Loess is a very important component of soils in the northern hemisphere where the effects of glaciation have produced a cover of fine grained wind worn material of varying depths. This is especially apparent on the north American continent where a large part of the agricultural United States is based upon loess based soils. The same is true for Europe. These loess deposits were formed from wind – born sedimentation at the end of the last glacial period, upon glacial retreat. Much of the loessic material forming today in China has most likely the same origins but since the initial deposition of the glacial material is on the arid high plateaus north of the Himalaya chain, it is still susceptible to wind erosion (see Sect. 3.1). The loessic sediment is most often a chlorite – muscovite material, susceptible to rapid alteration at the surface by either water chemical attack or by plant transformations. Other loess material has already been transformed by surface alteration and is part of a very fertile material for plant growth.

1.2.4.2 Topography

The local morphology of the land surface can change the effective climate (temperature and rainfall) within a given climate or geographic "region" (see Jenny, 1994, Chap. 5 and Birkeland et al., 2003). Hills and valleys give different vegetal cover, especially in grassland landscapes. Mountains of course produce very strong differences in climate and plant cover.

Figure 1.6 shows some possibilities of these local climatic effects. In Fig. 1.6 the situation in temperate climate prairies where grasses are on the plateaus and woods in the bottom lands can be found in south West England where the acidic granitic soils favored in the period 7700 BC grasses on the uplands and oak – alder woods in the bottom lands. The difference in altitude was 100 m, not enough to change the major climatic factors (Casteldine and Hutton, 1993). Hence one must conclude that local drainage patterns can determine the type of vegetation present.

Sequences of soils developed upon largely the same material but under different topographic conditions (Catena) are an important study in Soil Science. The changes in vegetation are largely controlled by the conditions of water availability in many

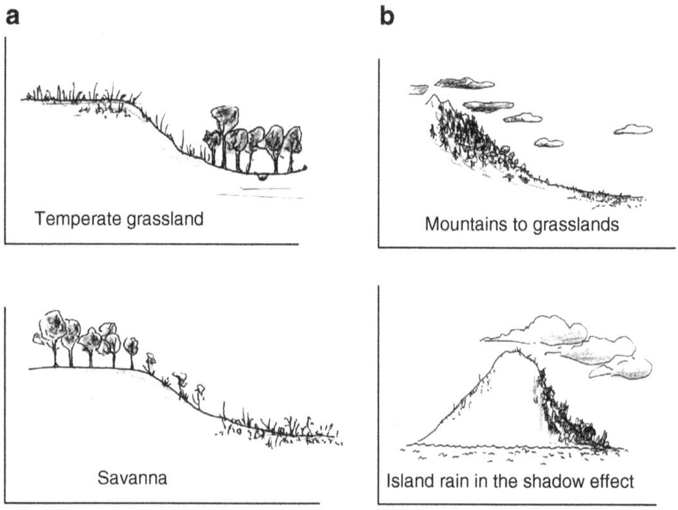

Fig. 1.6 Schematic representation of the effects of microclimate on biomes. (**a**) In temperate grass lands the valley *bottoms* are places where trees row preferentially. In savanna areas, towards the tropical climates, the trees tend to be on the plateaus. (**b**) In zones of high mountains, especially those of the Americas, the mountain heights have a higher rainfall than the lower plains found in the "rain shadow" to the east. Isolated island in ocean areas frequently are arid to the windward, source of humid air movement, and wet to the leeward, on the protected side of the elevation. Here the lee side is the wet side

instances and temperature when confronted by high mountains. Along with water, soil materials can move also. We will treat this situation in greater length later in Sect. 3.1.

However, even changes in orientation with respect to the sun (north – facing and south – facing) can create differences in vegetation and hence soils. Birkeland (1999, p. 233) shows such an instance for soils in Ohio, a temperate climate. On one side of the river valley one finds mixed oak forests, with a typical forest soil and alteration profile and on the other side of the valley one more typical of grasses. Grasses show higher pH values, more organic matter in the soil horizon and more exchangeable ions, indicating different types of clay minerals present. One side (that dominated by grasses) would be much better for farming than the other (dominated by forest).

The effect of morphology on soils is driven by water resources which affect plant regimes. The effect of altitude is to change the average temperature and hence the plant regime; These changes in vegetation produce different characteristics in the soil properties such as pH, organic matter content, base exchange capacity (exchange ions) which derive from the presence of 2:1 clay minerals and to a lesser extent organic matter. Alterability of the bedrock can be affected also. The topography determines the water regime, and thus the amount of dissolution which will occur as well as the activity of plants. Creek bottoms on the Great Plains of the United States have trees while the uplands have grasses present. The depth of organic matter, clay content and pH are quite different. These factors have been important in the development of the agricultural conquest by man of the land in which he has lived.

In Fig. 1.6 there are two types of landscapes where grasses or prairies are found, the temperate climate grasslands with prairies on the uplands and trees in the bottomlands and savanna grasslands formed in hot contrasted climate zones towards the equator where trees are found on the uplands and grasses in the expansive lower plains.

The importance of major mountain relief is very great in determining vegetation types and soils (see Birkeland et al., 2003). The presence of mountain chains on the west coast of the American continents provides a barrier to normal wind currents with the result that in middle latitudes the moisture laden air masses do not penetrate beyond the mountains, creating drier areas behind them which gradually change into more moisture – rich regimes as one progresses to the east. Thus the Americas have a high mountain wet forest followed by semi-desert and gradual change to prairie and then deciduous forest further east. This succession, desert to prairie (steppe) to deciduous leaved forest is south to north on the European continent. The Himalayas change the direction of the air masses also, with the result that the desert is to their north, followed by prairie in the Manchuria, north eastern China, and then deciduous forests in Russia on the edge of the Asian continent above Korea. Here the succession is on a slight diagonal from southwest to northeast.

Another, well known topographic effect is that of mountain islands in oceans. Here typically the air mass does not bring moisture until it reaches the lee side of the island, the reverse of continental mountain chains. The tropical valley vegetation is found on the slopes facing away from the major air mass movement. The islands of Hawaii are a fine example of this effect.

1.2.5 Summary

These "planetary" factors are in their turn modified by more local topography, plateaus, river canyons, island mountains and so forth which modify the vegetation and the soils which are related to them. The geological materials forming the surface further modify the vegetation which can adapt to different basic constraints of chemistry, mainly acid or basic pH values which strongly affect the plant regimes. Acid bedrocks favour forests while basic pH bedrocks favour grasses given equivalent temperate rainfall resources for example. The vegetation present is not only a factor of climate on a large scale, but on a smaller scale and to a lesser extent. The chemistry of the soil forming material can also change the ecotome, favoring trees with to respect to grasses for example under the same climatic conditions. However at the extremes of climate, tropical or tundra, the major factor is one of rainfall and temperature, which largely affect and modify the effects of bedrock and to a certain extent local topography.

It is important to note that the situation of the climate, topography of the continental surface is relatively recent, 20–30 million years old, and the plants we see today have developed largely since this period. The world we live in is recent, geologically speaking, and has evolved relatively rapidly, compared to the relatively small changes over the long periods of initial plant colonization of the continents.

This is the setting of surface plant – rock interaction where the ultimate chemical changes to form soils (defined as the zone where plants interact with silicate and other matter) dominates.

1.3 An Unorthodox but Operational Definition Mineral Types and Functions for Phyllosilicate Clay Minerals

The alteration processes at the surface, either in the water – rock or plant – silicate zone produce fine grained material called clay minerals. Any interpretation of the interaction between plants and their substrate must use clay identification procedures in order to understand the chemical changes and equilibria that occur. As it turns out plants strongly affect the chemistry of the soil in which they evolve and in doing so change the chemistry of the clays present. In order to understand the dynamics of the plant systems one must be able to identify the clay minerals and hence estimate their chemistry.

In order to put the reader in a functional framework, it is useful to simplify the nomenclature of clay minerals to give a working understanding of the majority of the materials present at the surface involved in plant – silicate interfaces. Soil clays are a field in themselves. Initially the concept of clay minerals is one based upon physical size. This definition, an anomaly in crystallography, is founded upon the history of scientific development. Much of mineral classification was accomplished in the later part of the 19th century. The basic tool for such an endeavour was the optical microscope. Manipulations of crystals submitted to variations in polarization of light transversing the crystal was used to help identify and classify minerals. Optical petroscopy was the basic tool of investigation. Crystal structures and atomic alignments were deduced from such manipulations. These methods were continued well into the 20th century. As with all things, the method of investigation, the microscope, has a limit of resolution. In this case it is the dimension of 0.002 mm, or 2 μm. No crystallographic determination was possible below this limit. However, mineralogists could extract and purify materials for chemical identification below this limit of observation. Since it was clear that mineral matter, classsifiable by its chemistry was present at scales below microscopic identification, these minerals, identified using their chemical characteristics, were called clay minerals, essentially, those species not amenable to normal optical identification. Use of X-ray diffraction, practiced in mid 20th century allowed a more correct definition of the crystallographic parameters of this crystalline matter. However clays remained clays, those objects which could not be visually differentiated, having a size below 2 μm. Thus clay minerals can be crystallographically determined, by X-ray diffraction, but they remain a mixed crystallographic and mineralogical group. In fact any fine grained mineral can be found in the clay fraction, especially quartz and calcite for example. But the largest part of fine grained materials found in nature have a similar crystallographic structure, that of a phyllosilicate. These minerals, found in rocks also where they can be identified optically, are essentially a two dimensional structure where the crystallographic directions *a* and *b* are much greater than the *c* crystallographic direction in the structure.

1.3.1 Network Structures

We will initially divide the most frequently found clay minerals, phyllosilicates, into two groups. First we must consider the fundamental structural differences among minerals found in soils which are based upon crystallo – chemical criteria. Clays minerals, whose structure is continuous or almost so in two crystallographic directions and limited in a third., are composed of basic chemical components of which oxygen composes about 50% of the atoms present, and a large portion of the remaining ions are silicon. Hence the name silicate mineral is applied to these crystals. The name is more than simply a name because the basic chemical characteristic of phyllosilicates, as most silicates, is that of the Si-O bond where the atoms are interlinked through sharing of electrons into a strongly co-valent structure. The strongest chemical bonding is between Si and O ions. At times an aluminium ion is substituted for silicon but usually in a minority of silicon sites in the structure. The resultant Si-O polymer of shared atoms is the dominant factor in the structure but it extends in a sheet like array, with less strong bonding between layers of Si-O atoms. The geometry of these arrays of oxygen and silicon atoms is hexagonal where at intervals a silicon atom is missing and a "hole" is present in the sheet. The silicon-oxygen ratio is such that a silicon atom is linked to four oxygens, and hence is called a silicon-oxygen tetrahedron. The dimensions of the atoms in these structures are such that the oxygens dominate in volume and other cations, such as Si, are found in the interstitial points within close packed oxygen atom arrays. The silicon-oxygen array of atoms is always linked, covalently, to one layer of other cation through shared oxygens, a second but lower energy covalent bonding.

This second variation of oxygen – cation linkages in phyllosilicates is one where atoms are bonded with six oxygens instead of four. The co-linked cations surrounded by six oxygens are called octahedrally coordinated ions. In clay mineral structures there is always a silicon-oxygen linkage with shared oxygens linking to the octahedrally coordinated linkage. This type of inter-linkage is called a tetrahedral – octahedral structure, TO (Fig. 1.7).

Thus one can have a tetrahedral and octahedral structure alternating with shared oxygen ions. This is commonly referred to as a 1:1 structure recalling the ratio of silica (tetrahedrally coordinated) to octahedrally coordinated cations. In the octahedrally coordination layer some oxygen ions are linked to hydrogen cations (OH units). The OH units are not co-linked to silicon ions. The presence of hydrogen ions gives a hydrated dimension to the mineral. Upon heating, to temperatures above 400°C or so, these hydroxyl groups leave the structure as water molecules and hence one refers to the clays as hydrated minerals. However there is no water present in the structure as such (H_2O molecules) but water appears when the mineral looses coherence due to heating.

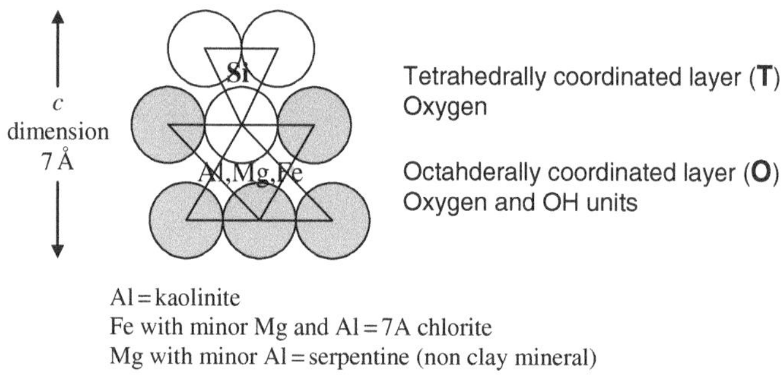

Fig. 1.7 Representation of the two cation layer structure of a tetrahedrally coordinated ion layer with an octahedrally coordinated layer of cations. Cations are Si (tetrahedral coordination) and Al, Mg or Fe (octahedrally coordination) with oxygen anions and/or hydroxyl (oxygen with a hydrogen cation) cations (*grey* in the figure). The structure can be designated as a TO structure for tetrahedral – octahedral sequence or a 1:1 structure with two layers of cations. The *c* crystallographic dimension is indicated, perpendicular to the layers of ion is the structure which is the determining measurement for X-ray diffraction identification. In the TO structure the *c* dimension is close to 7 Å

Two variations of site occupation are possible for the octahedrally coordinated site, one where three divalent ions are present, making a cationic charge of six for the unit cell, or one where two trivalent ions are present in the octahedrally coordinated layer sites leaving one site vacant. The first, completely filled, is called tri-octahedral and the second di-octahedral. In soil systems, the most common ionic occupancy in clays is by aluminium and hence most soil clay minerals are dioctahedral in structure. However a non-negligible amount of material can at times be of the tri-octahedral type. The number of sites in the octahedral sites in phyllosilicate clay minerals varies then from 2 to 3. More precisely, the occupancy is between 3 and 2.5 for the so-called tri-octahedral minerals and very close to two ions for the di-octahedral minerals.

As it turns out there is a clay mineral with the TO structure, two layers of coordinated ions, which is called kaolinite. It is almost strictly of an Al-Si composition and hence dioctahedral in site occupation. Very few ions are found to substitute for these two cations. This is unusual for clay minerals which usually show ionic substitutions of one cation for another. The small variation of the kaolinite structure makes its properties quite predictable.

In a more complex, three layer, structure one can find a silicon-oxygen layer interlinked with an aluminium-oxygen, hydroxyl layer which in its turn is linked to another silicon-oxygen layer. This gives a triple cation layered structure, tetrahedral – octahedral – tetrahedral, or TOT. Most often it is referred to as a 2:1 structure recalling the ratio of silica to octahedrally coordinated ions (Fig. 1.8).

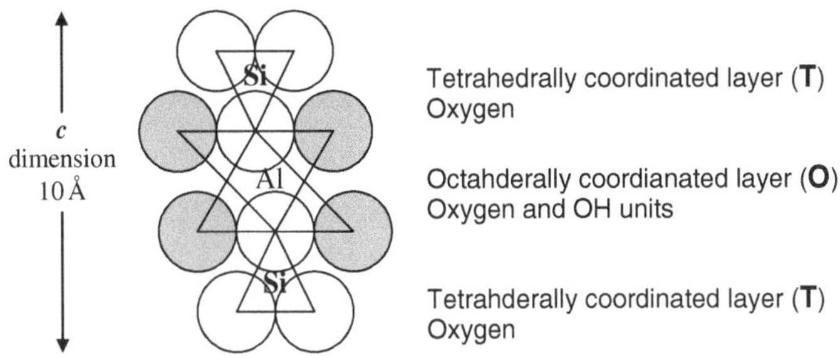

Fig. 1.8 Representation of a three cation layer structure where two tetrahedrally coordinated cation layers are present (essentially Si cations coordinated with oxygen anions) and one octahedrally coordinated cation layer. Again the octahedrally coordinated cations are related to OH ion units where the oxygen anion is coordianted to with a hydrogen cation. In soil clays the tetrahedrally coordinated ions are predominantly Si and the octahedrally coordinated ions are predominantly Al, but can be Mg ions also. Fe is probably more rare in that the oxidation state of iron would be divalent, more rare in the oxidizing environment of most soils. Trivalent iron ions are relatively rare in phyllosilicate structures. These structures are called TOT (tetrahedral-octahedral-tetrahedral) or 2:1 types because of the three cation layers coordinated to form a completed mineral structure. The *c* crystallographic dimension or the TOT layer is near 10 Å

1.3.2 Substitutions and Layer Charge

Substitutions can occur in the cation positions in the Si-O networks in 2:1 (TOT) minerals. One type is an iso-charge exchange with Mg^{2+} for Fe^{2+} ions for instance, typical of the tri-octahedral type of mineral. Also it is possible to have those of different charges, Al for Si for instance where the substitution is in the tetrahedral site. Such non-equivalent substitutions create a charge imbalance, where a trivalent ion (Al^{3+}) is put in the place of a quadrivalent ion (Si^{4+}). Substitutions which are not charge neutral necessitate the presence of another ion inserted into the 2:1 network structure to satisfy electronic neutrality. These ions can be present in the network structure itself, for example a substitution of Al^{3+} for Si^{4+} with a net negative charge could be compensated by a substitution of Al^{3+} for Mg^{2+} in the octahedral site in the network giving a positive charge locally which compensates for the negative charge at the other site, in the tetrehadral position. As a result the 2:1 layer is charge neutral due to internal compensations of different ionic substitutions.

Another possibility of charge compensation can occur as substitutions on the surfaces of the 2:1 structure between two adjacent units. The ionic substitutions creating charge dislocations (compensation from one site to another) are extremely important in the development of basic clay chemical properties. For example when there is a strong charge substitution not compensated within the network itself, one positive charge for 22 negative charges, the compensating substitution is usually

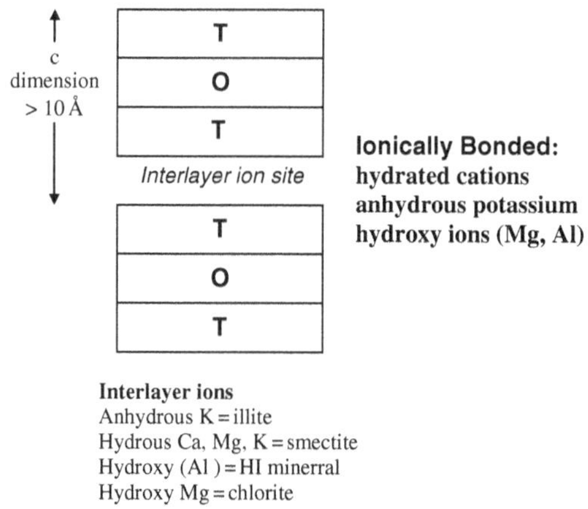

Fig. 1.9 Assemblage of TOT (2:1) layer unitts which for crystallites of soil clays. Most often one finds that there is a cation occupation between the units which compensates surface charges on the aggregate TOT layers engendered by substitution of cations within the structure of different charges. These residual charges (negative) attract interlayer ions which are situated between the TOT layers (2:1). The cation occupancy, with or without water or hydroxyl anions, determines the interlayer spacing (c dimension) allowing identification by X-ray diffraction methods. These interlayer ions, determining the X-ray diffraction characteristics, give rise to a nomenclature of clay minerals based upon the occupancy of the interlayer ion sites. The *c* crystallograpic dimension or the TOT layer is near 10 Å

made by potassium inserted into the surface oxygen network between the TOT units, forming a mica. The chemical formulation is:
$\mathbf{K}Al_2Si_3\mathbf{Al}O_{10}\,(OH)_2$ instead of the charge neutral 2:1 structure of $Al_2Si_4O_{10}(OH)_2$ Where K and Al are the substituted ions in the manner of $Si^{4+} = Al^{3+}K^+$

The position in of the potassium ion in the structure which is substituting to compensate for the charge imbalance within the 2:1 unit is in between the silicon ion coordinated layers of the structure in what is called the "interlayer" crystallographic site (Fig. 1.9)

1.3.3 Interlayer Ion Occupancy

This interlayer site can be occupied by different ions. Usually, the maximum charge compensated in this site is one per a total of 22 negative charges of the oxygen – hydroxyl network. This is a high charge structure and usually designated as a mica. The only ion present in high charge structures at the surface (in soils) is potassium. However, if the average total substitution on the structure is less than one per unit cell (22 oxygens), one develops what is called a low charge structure. Here a variety of compensating ions can be present in the interlayer site. Their species is largely determined by the chemistry of the soil fluids. In many cases, these ions can be exchanged as the fluid composition changes. These are called exchangeable ions. The amount of exchangeable ions present per mass of clay is called the cation exchange capacity or CEC.

Basically three types of cation occupations are present in the interlayer sites:

1. those which are associated with hydroxyl ions, which gives the mineral the name of hydroxyl – interlayered (HI) mineral. The basic interlayer spacing of the 2:1 structure in the *c* crystallographic direction is 14.2 Å.
2. those which are accompanied by water molecules, which gives them the name of hydrated interlayer ions. The basic interlayer spacing in the c crystallographic direction is 15.2 Å. Under conditions of higher temperature than normal or very low humidity the exchange cation layer can contain only one water layer giving a 12.5 Å spacing. This spacing tends to be present more in mono – valent cation substitutions such as Na or K ions.
3. those which are not accompanied by other ions (anhydrous), where the name given is mica or illite, since the only ion of this sort is potassium. The basic interlayer spacing in the *c* crystallographic direction is 10 Å.

The intensity with which the interlayer ions are held is a factor of total charge and charge site, the charge in the tetrahedral site appears to be relatively stronger than that in the octahedral site (Malla and Douglas, 1987). The question of mineral identification lies in the identification of the type of interlayer ion present between the layers of the TOT structure. These interlayer ion sites are the key to determining mineral species and the function of the clays in the soil environment.

1.3.4 Identification of Phyllosilicate Clay Types

Our identification of the minerals present and their relative abundance is based upon a curve decomposition method (Lanson, 1997) which allows one to identify and estimate the relative abundance of the different components of a complex diffractogram. Peak position and width depends upon the composition of the mineral present and the relative number of diffracting layers in the crystallites. Smaller diffracting domains widen the peak and shift it to larger peak positions (Moore and Reynolds, 1997). Complex spectra can give large, unidentified peaks using visual examination, whereas the components can be precisely identified using decomposition methods. The use of this very important tool of interpretation fo X-ray diffractograms, the only viable method of clay mineral identification, is given in more detail in the Annex at the end of the book.

In order to interpret the diffractograms one must have some idea of clay mineral structures for those phases found in soils.

1.3.4.1 The 1:1 Minerals (TO)

The TO minerals have a basal spacing (crystallographic direction perpendicular to the basic layer structure) of near 7.2 Å. This indicates the presence of one silica layer and, in most soils, the presence of one alumina layer with hydroxyl ions at the surface. This is the kaolinite mineral. The chemical composition is almost constant, to within several percent of the cations present. In some cases, alteration regimes that are very water – rich, a hydrated mineral (H_2O ions present) can occur. Here the spacing is closer to 10 Å, but the mineral is unstable upon heating to about 200°C. Some mixtures of kaolinite and halloysite can be found. Kaolinite is present under regimes of high rainfall but also is part of the normal rock – water interaction reactions which occur in most climates. The water in halloysite is in a way similar to that founding zeolites, loosely held playing little structural role. However the shape of the kaolinite crystals is dramatically changed from a sheet structure to one of a tubular form. This mineral is called halloysite. Thus the soil clays with a TO or 1:1 structure (one tetrahedral unit and one octahedral unit) can be defined as an anhydrous and a hydrous variety.

1.3.4.2 The 2:1 Minerals (TOT)

The interlayer ion occupancy of these minerals affects the intensities of the different basic X-ray diffraction reflections of the structure. The most often observed for mineral identification are the reflections dependent on the *c* crystallographic spacing, the *(00l)* spacings (*see* Moore and Reynolds, 1997 for a definitive discussion of the problem). When hydrated or hydroxylated ions are present in the interlayer sites between the 2:1 units or when a less than maximum or ideal number of cations are present, the 001 spacing is by far the most intense of the *(00l)* series (see Meunier, 2007). This is especially frequent for the case for hydroxy interlayer minerals found in soils. This is also the case for hydrated ion structures, i.e. cations linked to water molecules. However, when no OH or water is present and the occupation of the interlayer sites is near maximum, one cation for 12 oxygen plus hydroxyl anions for instance, the intensity of the first order peak 001 is often less than that of other lower order peaks 002, 003. etc. Hence a high intensity first order peak in the range of 15–10 Å indicates either incomplete occupation or the presence of water molecules.

Smectites

Cations associated with water molecules filling the interlayered sites give rise to the smectite minerals. Normally these ions are easily exchanged for others when introduced to an aqueous solution containing another dominant cation. Smectites are characterised by exchangeable ions. By changing the amount of water present in the surrounding of a smectite, one can change the water content which in turn changes the dimension of the clay mineral. Highly hydrated cations have a two water layer structure (15.2 Å spacing), increasing temperature or decreasing water vapour pressure reduce the amount of water to a one layer structure (12.5 Å) and eventurally an anhydrous state can be reached under conditions not expected under normal earth surface conditions giving a 10 Å spacing. This property of water loss or gain gives rise to the descriptive term of expandable mineral. If another medium is present (organic molecules in liquid or gaseous form) the water is exchanged for the other and hence the expandable nature of smectites is again expressed. molecule changing the interlayer spacing due to the differences in molecular dimension. Smectites are characterised by the ability to expand or contract.

HI minerals

In the case of a complete occupancy in interlayer sites by MgAlOH complexes, the 001 (14.2 Å) peak can be more or less intense than the 002 peak (7 Å). In the case of stoichiometric structures (fully occupied interlayer sites) the relative intensities of the *00l* peak series are a function of the cations in the interlayer site and in the octahedral layer of the 2:1 mineral. Hydroxy interlayered minerals can be either dioctahedral or trioctahedral in character. All of these minerals can be called chlorites. Further, simple elemental substitution, Fe for Mg, or Al for Mg for example, changes the relative intensities of the different peaks of the *(00l)* series. In Fig. 1.10 the occupation of the octahedral site in the 2:1 structure is either aluminium (a) or magnesium (b) in a di-octahedral configuration or a tri-octahedral configuration (calculations of these spectra were made by Eric Ferrage, Univ Poitiers using the ASN program). The structures contain all Mg or Al ions in both 2:1 and interlayer sites. If we mix the ion types from octahedral sites to interlayer sites the effect is important also. Basically, when Al or Mg are present in the interlayer, site the first order reflection is at 14.2 Å. Hence interlayer ion composition can affect the relative intensities of the X-ray diffraction patterns of the clays as well as can a low occupancy of the interlayer sites.

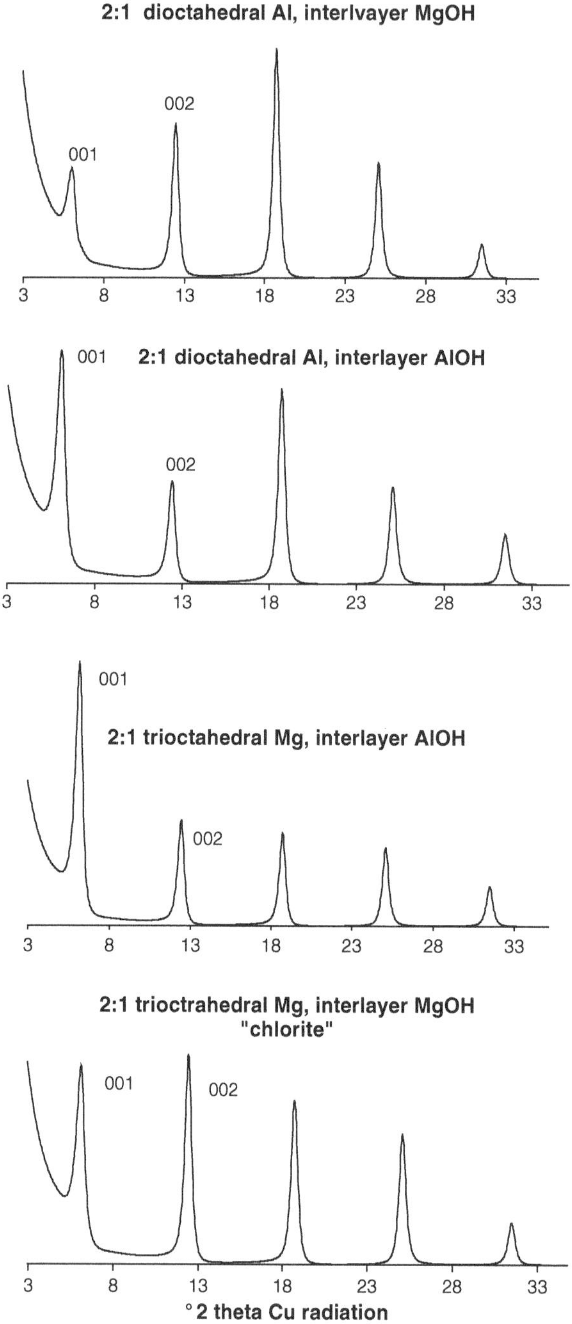

Fig. 1.10 (continued)

Micas

Minerals with no interlayer water or hydroxyl elements present will be anhydrous, and have an interlayer spacing of 10 Å.

1.3.5 Classification of 2:1 Minerals

Using the above criteria one can construct a classification scheme which focuses on the interlayer ion occupation of the 2:1 (TOT) structure of the mineral. Traditional mineral identification of clays is based upon X-ray diffraction which in fact indicates the interlayer ion population. However mineralogists usually stop when they have identified the interlayer ion site ions with a set of laboratory treatments of the clay assemblages. The different laboratory tests are used to estimate the occupation and stability of the interlayer ion sites is designed to test either the exchange of the interlayer ions with other elements in aqueous solution or the stability of the hydrous state of the interlayer ions, one or two water layers. If the elements in the interlayer ion site are easily exchanged, a treatment with a moderately concentrated solution of $CaCl_2$ or $MgCl_2$ will displace the different ions leaving a structure with a more or less homogeneous hydrated interlayer ion structure in the air dried state. Usually divalent cations are used for this laboratory test, Ca and Mg are the favourites. In this case one has a 15.2 Å structure which gradually de-hydrates, collapse to 10 Å, at temperatures between 300 and 400°C.

Treatment with ethylene glycol or glycerol vapour increases the spacing of the hydrous ion species of clays to 17 Å due to exchange of the organic molecule with water molecules. The micaceous and HI minerals are un-affected by such laboratory treatments. The change in basal spacing upon glycol or glycerol treatment, produces and expansion of the mineral, hence the nomenclature of "expanding mineral" which is roughly synonymous with "smectite" (see pertinent chapters in Dixon and Weed, 1989)

Minerals which remain in the anhydrous state after exchange cation saturation have a spacing of 10 Å are considered to be micaceous in nature and called illite in soil clays. Normally potassium is the interlayer cation. Minerals which retain a 14.2 Å spacing, and do not dehydrate easily nor exchange with divalent cations

Fig. 1.10 HI minerals. Example of the importance of cation types (atomic mass and resulting electronic density) on the relative intensities of the basal spacing reflections where 001 is the first order, 002 the second, 003 the third and so forth. Peak intensity rations can give an idea of the cation layer compositions. In the case of HI minerals with complete occupation of the interlayer site with hydroxyl coordinated cations (chlorites), the octahedrally coordinated ions can vary significantly. In some cases the TOT, 2:1 structure contains a dioctahedral ion complex (Al ions) and in others a trioctahedral (Mg dominated) complex. Depending upon the interlayer site occupancy and the 2:1 layer site occupancy the ratios of 001–002 intensities change as do those of the other basal spacing bands. For example, with a dioctahedral octahedral layer in the 2:1 structure the presence of Mg or Al in the interlayer site changes the 001/002 intensity ratio completely compared to the 003 peak intensity which is relatively constant. The spectra are obtained by calculations using the *ASN* program developed recently by Lanson and Ferrage in aqueous solution are considered to be HI or hydroxyl ion complex minerals or chlorites. Heating to temperatures above 400°C usually collapses the HI minerals, at least partially, but does not affect the chlorites.

The usual identification procedure using X-ray diffraction methods, done in the Ca, Mg – saturated air dried state gives:

10 Å = illite
15.2 = hydrous cation expandable mineral under room temperature and humidity conditions
14.2 Å = HI hydroxyl mineral

Such a state is reasonably close to the situation in a soil environment.

1.3.6 Identification of Smectites and Illites

In soils one often finds an intimate mixture of 10 and 15.2 Å structures, illite and smectite, apparently in the same crystallite. Examples of the end members frequently present in soils are presented in Fig. 1.11. One can see the typical diffraction maxima of these different types of soil clay minerals. The spectra are decomposed into their components showing individual peaks representing a mineral type and the overall envelope of the spectrum. In Fig. 1.11 we have represented typical high smectite content clay minerals (a) and a highly smectitic mixed layer clay accompanied by an illite. The smectite (minerals of high smectite component can be mixed layered with a 10 Å component, illite, within the same mineral grain structure). These are mixed layer clays. A peak positin near 15.2 Å indicates a nearly pure smectites mineral. Positions nearer to 10 Å indicate a mica component. The major smectite peaks show a peak width at half height of greater than 1° 2 theta, indicating a small grain size and/or an interlayered mineral structure. Small diffracting domain and interlayering tend to broaden the diffraction peak (Moore and Reynolds, 1997). For the illite (lower spectrum in figure) usually two peaks are present. The more narrow 10 Å peak indicates the presence of fully anhydrous interlayer ions, which are called illites (WCI or well crystallized illite). The wider peak at 10.2–10.6 Å in general represents the small grained illite material (PCI or poorly crystallized illite), usually of less than 6 unit layers thickness (see Lanson, 1997 for a more detailed explanation of the X-ray diffraction phenomena).

The smectitic phase X-ray diffraction peaks will be displaced to near 17 Å upon treatment with glycol or glycerol, typical laboratory treatments used to indicate the possibility of swelling, or incorporation of organic molecules in the place of the typical two water layers around the exchange cations, Ca or Mg. The illite peak is not displaced upon treatment with organic molecules and it remains at near 10 Å.

1.3.7 Mixed Layer Illite/Smectite Crystals

In Fig. 1.12 an example of mixed layer illite/smectite clays is presented. In many soils, the mixed layer minerals form twodifferent groups, one smectites rich, S/I,

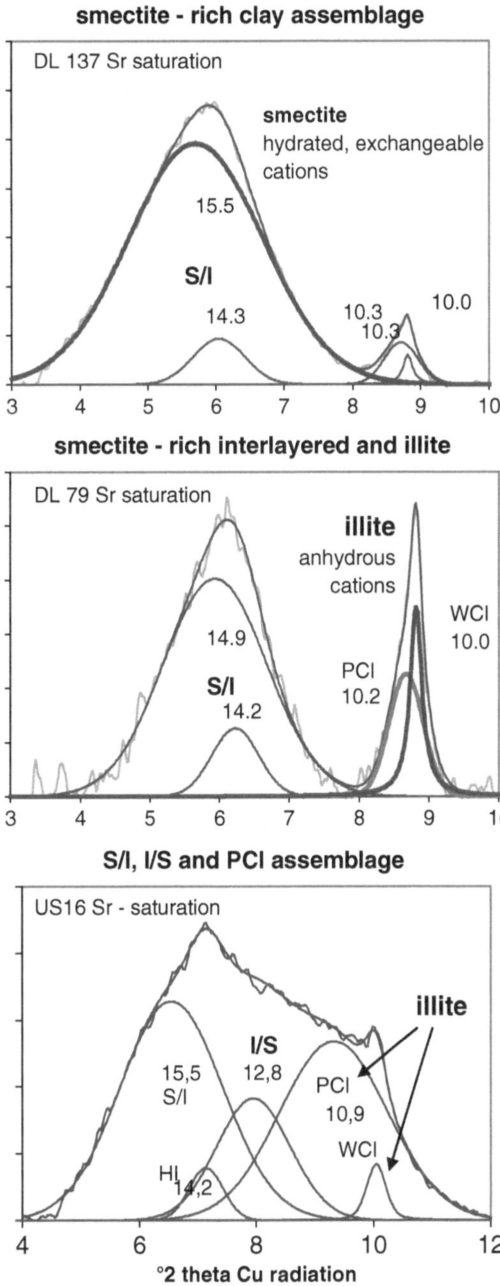

Fig. 1.11 Examples of two soil clay assemblages from central southern Iowa. Background – subtracted X-ray diffractogram treated using decomposition methods. The *upper* figure (backround subtracted spectrum) shows an assemblage dominated by a 15.5 Å mineral, close to the smectites end member. Curve decomposition shows the presence of a small 10 Å peak indicating a minor illite component and a 14.3 Å peak indicating an HI mineral.. The *lower* spectrum shows a large (in surface area) peak of slightly less than full smectite composition, (14.9 Å instead of 15.2 Å for smectite) indicating a small interlayering of illite with another narrow peak at 10 Å indicting the presence of a significant illite mineral (anhydrous potassium interlayer ions)

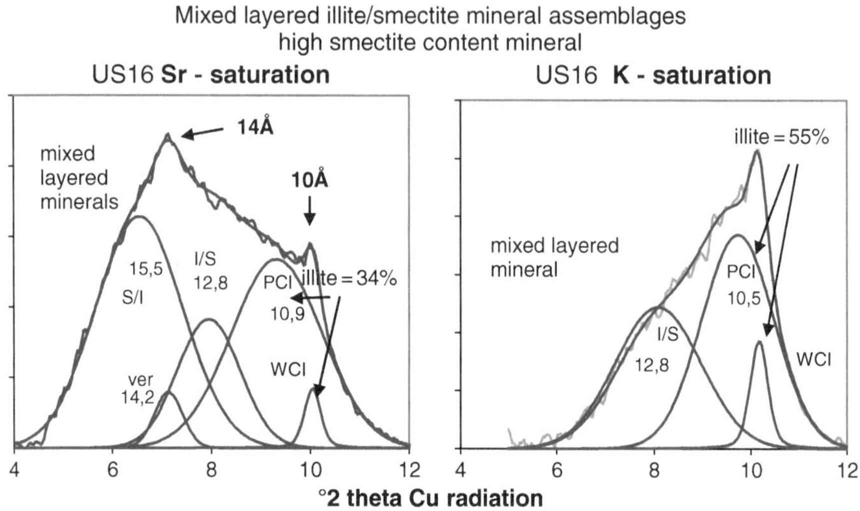

Fig. 1.12 Example of a complex X-ray diffraction spectrum (background subtracted with component curves indicated for the phases present). The peak envelope shows two maxima, at 14.2 and 10 Å. However the decomposition of the overall curve shows the presence of five components, where the sharp peaks are the effects of only minor components as far as peak surface area is concerned. Two illite-smectite bands are present, one representing a smectite-rich phase (S/I) and the other less smectitic (I/S). The illite component is dominated by the poorly crystallized component (PCI, of small diffracting domain size) and a more minor well crystallized illite (WCI). Potassium saturation shifts the illite-smectite peaks to a more illitic position (nearer 10 Å) and increases the relative surface area of the illite peak. The total illite component changes from 34% of the peak area to 55% upon potassium saturation

with a peak near 15.5 Å and the other illite rich, I/S, with a peak nearer to 10 Å. In the example presented the second illite-smectite mineral has a spacing of 12.8 Å indicating an intermediate composition between illite and smectite. A small HI mineral peak is also present, near 14.2 Å. Normally these minerals are accompanied by illite in soils. In the example of Fig. 1.12 the poorly crystallized illite (PCI) is of greater abundance than the well crystallized illite (WCI). In the overall spectrum one sees peaks (sharp intensities, at 14.2 and 10 Å). However the peak decomposition of the spectrum shows that these intensities are of relatively small surface area, and hence represent only a small portion of the material present. Without using this method of determination, one would miss the importance of most of the material and miss identify it as an assemblage of illite and HI minerals. The S/I, I/S and PCI components are dominant but do not give sharp maxima in the combined spectrum.

Here it is imperative to use curve decomposition methods. Otherwise one would not suspect the presence of two types of illite-smectie mixed layered minerals nor the relative importance of the poorly crystallized, fine grained illite compared to the better crystallized illite component.

1.3.8 Identification of Soil HI Minerals ("Vermiculite")

In Fig. 1.13 one finds what is generally called soil vermiculite or an HI (hydroxyl interlayered) mineral and a small amount of illite, where the major peak is rather narrow, about 0.8° 2 theta, and is found near 14.2 Å. It is not affected by divalent ion cation exchange as are smectites, nor heating to temperatures below 200°C nor to glycol treatment. Very often the soil HI mineral is accompanied by mixed layer minerals, of the type smectite – HI with peak positions between 15.2 and 14.2 Å which show expansion to positions closer to 17 Å upon glycol treatment) or HI – mica mixed layer minerals, with peak positions between 14.2 and 10 Å unaffected by glycol treatment.

Thus we can identify the major types of traditional 2:1 clay minerals using simple laboratory techniques. Three end member types can be identified using X-ray diffraction; illite (10 Å) smectites (15.2 Å) and vermiculite, HI (14.2 Å) after treatment with dilute solutions of divalent exchange cations (Ca or Mg).

Further treatments give more details of the minerals present.

1.3.9 Potassium Saturation and Identification of 2:1 Clay Types

1.3.9.1 Smectites

Potassium saturation can often change minerals from apparent HI or hydrous cation minerals to those with an anhydrous interlayer ion population producing an new illite type mineral (April et al., 1986; Fichter et al., 1998; Carnicelli et al., 1997 for example). It has been noted that the relative proportion of illite 001 peak area for clay assemblages increases from saturation with low concentration (0.01 M KCl) to a maximum near 0.5 M solutions (Barré et al., 2007). This treatment changes the basal spacings from 15.2 to 10 Å in X-ray diffraction spectra. The anhydrous layer ions present in the interlayer ion sites of the 2:1 structures do not respond to glycol, glycerol or mild heat treatments (see examples in Velde and Meunier, 2008, Chap. 2 for example). This operation indicates that the type of exchange cation present in the interlayer site can modify the behaviour of the clays to a significant extent. Further the anhydrous ion population of these clays can be extracted by strong cation exchange treatment (1 M SrCl2 for example). In Fig. 1.12 the behaviour of illite-smectite interlayer mineral assemblages is shown under conditions of dilute Sr saturation and under potassium saturation at dilute concentration. Both treatments use dilute chloride solutions (about 0.01 M). In the smectites rich mixed layer mineral the peak position shifts to lower d-spacings, nearer to 10 Å, or the illite position. Two peaks are visible in the first example (a) indicating an inhomogeneous behaviour of the different grains present. In the second example not only do the mixed layer smectites-illite peaks shift but the illite peaks become relatively more important, changing from 34% of peak surface area to 55%. Such a behaviour is frequent in soil clays. Some of the smectites can become illite in appearance upon light treatment with KCl. These are "high charge" components of the smectites

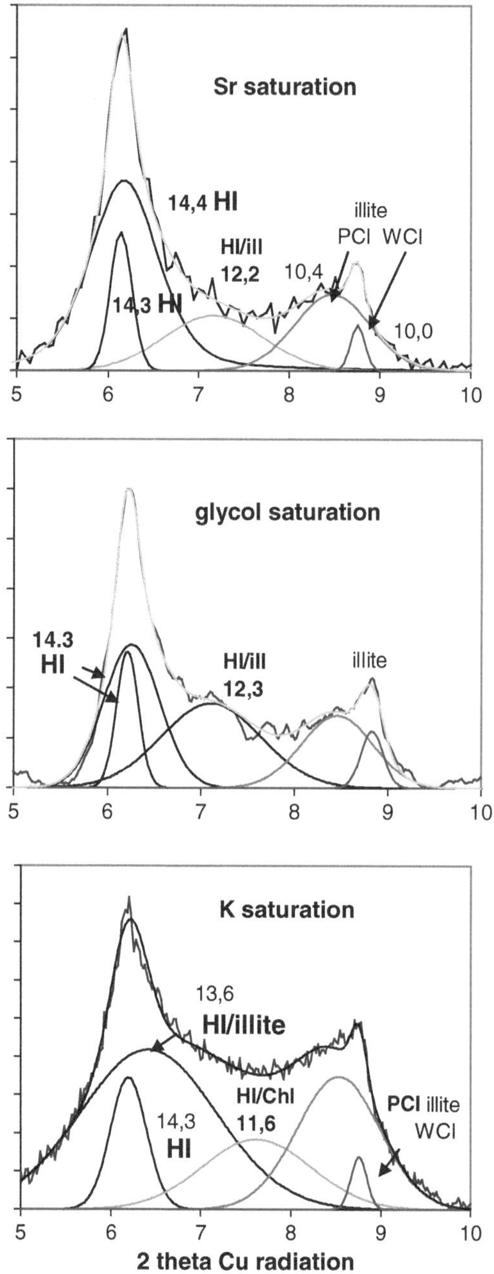

Fig. 1.13 Example of a soil clay assemblage containing a major component of HI minerals. Smaller components of mixed layer minerals between illite (10 Å) and HI (14.2 Å) minerals are seen (peak positions between 14.2 and 10 Å). Glycol treatment does not change the peak positions significantly indicating that there is no or little smectites present. Potassium saturation decreases the relative intensity of the HI mineral and increases that of the illite peak area indicating that some of the hydroxyl interlayered ions are exchangeable with potassium

minerals which select potassium as an inter layer ion in the anhydrous state. Smectites can be of a high charge variety or low charge type. High charge layers contract to he anhydrous state with potassium saturation and low charge smectites remain in a hydrous state of either two water layers (15.2 Å) or one layer (12.5 Å).

1.3.9.2 HI minerals

HI minerals, those with an initial 14.2 Å spacing after treatment with dilute divalent cation solutions, can respond to glycol treatment in increasing the basal spacing beyond 14.2 Å. These are smectite type minerals where the interlayer hydroxyl cations are loosely bound. They are HI low charge smectites.

Numerous published studies on soils containing HI minerals show a contraction and formation of "illite" upon saturation with potassium (Carnicelli et al., 1997; Ildefonse et al., 1979; Fichter et al., 1998; April et al.,1986; Zanelli et al., 2006, among many others). This capacity to fix potassium in an anhydrous state forming an illite – like mineral is very important for the interpretation of soil clay mineralogy as it is obviously important for the nutrition of plants. If the clay is affected by an exchange ion such as potassium, a portion of the hydroxyl interlayer complex is exchangeable, at least to the potassium ion. This can be called the high charge vermiculite material which in the natural state is hydroxyl interlayered, with a narrow 14.2 Å peak.

The characteristic then of most soil HI minerals is the partial exchange of the hydroxyl interlayer ions which give a 14.2 Å peak with potassium giving an illite peak position (10 Å). Typically the mixed layer HI/illite minerals shift to lower spacings as more illite layers are formed by potassium saturation of some interlayer sites. In the Fig. 1.13 such a behaviour is demonstrated where not all of the HI minerals are potassium saturated, some remaining at their typical 14.2 Å spacing while others either become totally potassium saturated forming the more intense illite peak at 10 Å, and also forming interlayered HI/illlite phases.

The material which remains in the 14.2 Å state will be called the HI or soil chlorite phase. Both normally have a strong 14.2 Å peak compared to the 7 Å peak. Such behaviour is seen in Fig. 1.13 where a soil hydroxyl ion mineral is affected by potassium treatment, with a portion of the material forming illite, and mixed layer minerals while some of the 14.2 Å material remains intact after the saturation.

A special case, and perhaps not all that special, of potassium saturation of interlayer sites in HI minerals occurs in soils derived from loessic materials found in the Po river delta region of Pavia, Italy (data provided by F Terrible, Univ Frederico II Agronomy dept, Napoli). The clay mineralogy of the soils in the Lombardi region near Pavia, shows an apparent illite – chlorite mineralogy. The diffraction peaks are relatively narrow, 0.33–0.15° two theta at peak half height for all minerals, indicating well formed and probably metamorphic minerals having been little subjected to alteration processes at the surface. However, the soils are quite fertile and respond well to potassium fertilization. One wonders how the soil minerals can retain potassium for growth throughout the growing season when there are no apparent exchange sites in minerals such as smectites or mixed layered illite smectites. When the soil clays are subjected to potassium saturation, the peak intensities change, with the first order chlorite peak, at 14 Å, decreasing in intensity and becoming slightly less wide (change from 0.33 to 0.15° 2 theta width at half height). The relative intensity of the 14 Å peak compared to the second order peak at 7 Å changes from 0.66 in the normal state (Mg saturation) to 0.26 in the potassium saturated state

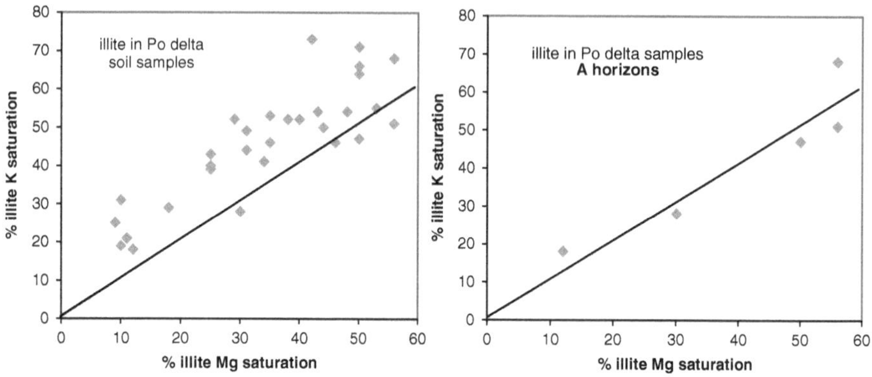

Fig. 1.14 Plot of peak intensities of illite compared to chlorite (002 peak at 7.2 Å) in Po valley illite – chlorite mineralogy agricultural soils. The saturation of potassium increases the relative intensity of illite in most cases but when one looks at the A horizon soils mineralogy, one sees that the potassium saturation does not significantly change peak intensity indicating that the agricultural practice of potassium fertilization has filled the potential illite interlayer sites which were not exchanged by magnesium saturation

(Fig. 1.14.) This indicates that the mineral initially seen as chlorite approaches the normal peak ratio for a magnesian chlorite (see Fig. 1.10). The illite peak changes from a relative intensity (14, 10 and 7 Å peak intensities) of 28–36%. Thus the illite peak increases in intensity and the chlorite peaks change relative intensity. This indicates that the initial clay assemblage (Mg saturated is in fact one of chlorite – vermiculite plus illite) which is capable of fixing significant amounts of potassium in the interlayer space of the vermiculitic minerals. At first inspection, one would assume that the clay assemblage is near that of a metamorphic rock assemblage, illite plus chlorite. However, the soil clay mineralogy has been altered to produce a potassium fixing mineralogy, one of a highly ordered soil vermiculite with little disorder nor alumina substitutions in the interlayer site (see Meunier, 2007).

In Fig. 1.14 we indicate the average increase in illite peak area for five soil profiles in the Po delta soils mentioned above. Again it is clear that the potassium saturation increases the "illite" component, i.e. the 2:1 minerals with an anhydrous potassium cation interlayer occupation. It is interesting to note that the illite percentages seem almost the same for the different saturation treatments for A (agricultural) horizon samples. This suggests that the fertilizer treatments are more than sufficient to fill the sites in clays susceptible to fix them as anhydrous interlayer ions to form illite.

1.3.10 Summary

Very briefly, one can identify four basic types of soil clay phyllosilicate minerals by X-ray diffraction. The basic state is in the Ca or Mg saturated state and observed at room temperature and humidity conditions.

TO structure (1;1): *Kaolinite* and *halloysite* which are 1:1 minerals. Heating to 110°C or so will contract the halloysite 10 Å peak to 7.2 Å, otherwise a 7.2 Å indicates the presence of kaolinite.

TOT structure (2:1): Three types of phases are possible based upon the 2:1 structure where ionic substitution occurs in interlayer ion site, between the 2:1 layers;

1. 10 Å minerals where anhydrous potassium is present. In this state the mineral is called *illite*.
2. 15.2 Å minerals where the interlayer spacing changes with glycol or glycerol treatment to a spacing near 17 Å. These are *smectites*. Formation of a 10 Å mineral on potassium saturation indicates a *high charge smectite*. Those layers remaining at interlayer spacings above 10 Å are *low charge smectites*.
3. 14.2 Å spacing are *HI minerals*. There are essentially three behaviours and we will give three names to them.

a. One is a stable 14.2 Å spacing after potassium saturation or glycol treatment. This HI mineral can be called a *soil chlorite*. The interlayer ion sites are completely filled with non exchangeable Al-Mg hydroxy complexes.
b. 17 Å spacing upon glycol or glycerol treatment. This is a *soil, low charge HI mineral*
c. 10 Å peak spacing upon potassium saturation. This is a *soil, high charge HI mineral*

One then can use the terms illite, smectite and HI mineral on first inspection of X-ray diffraction diagrams in the Ca-Mg treated state under room atmospheric conditions. The HI minerals can be subdivided into soil chlorites, high or low charge vermiculites upon further treatment.

Although the above is not what one is accustomed to find in text books on clay minerals, the shema outlined for 2:1 mineral types is clearly illustrated by Birkeland (1999, p. 101) following the outline given by Jackson (1964) concerning the alteration or formation of clay minerals in soils from pre-existing phyllosilicates. The nomenclature in these articles is similar to that proposed above and the mineral behaviour is the same. Hence the functional description and mineral nomenclature proposed here is not new, but only unusual.

1.4 Recommendations

Therefore we suggest that the normal laboratory determinations and nomenclature commonly used to identify soil clay minerals should be simplified to a large extent. The existence of a diffraction peak position in the moderately treated (Ca, Mg, Sr chloride concentrations of 0.1 M) at between 10 and 15.2 Å indicates the presence of 2:1 smectite minerals. Those with a peak at near 14.2 Å that do not change position upon chemical or moderate heat treatment (below 400°C) are of the hydroxyl interlayer ion type similar to chlorite, a high temperature mineral with a tri-octahedral occupancy in both 2:1 layer units and interlayer sites. Those that do not change peak position from 10 Å upon chemical and heat treatment are illites, of mica like behaviour. Material that is affected by chemical treatments, exchange ions is exchangeable in character. The extent of exchangeability between different ions under different ionic concentrations determines the behaviour of these clays in their soil environment.

A second treatment deals with structures of high intensity charge but that allow cation exchange under certain circumstances. 0.5 M KCl solutions will close the interlayer site with anhydrous potassium ions creating an illite – like mineral behaviour. These minerals can be either of dioctahedral (two ions per unit cell in the octahedral layer, usually aluminum) or trioctahedral (three ions in the octahedral site, usually magnesium) in nature. The increase in illite peak intensity (10 Å) can be at the expense of smectite peak intensity (15.2 Å) under divalent ion saturation or at the expense of the 14.2 Å HI minerals.

Peak positions at near 7.2 Å can indicate the presence of kaolinite. HI minerals (hydroxyl interlayered), have peak positions smaller than those of kaolinite The difference between kaolinite and second order peaks of 14 Å minerals (HI minerals or the high temperature mineral chlorite) is often difficult to ascertain but one can be identify the different minerals by looking at the peak positions at near 3.5 Å where kaolinite is found at near 3.59 Å and HI minerals at smaller spacings, nearer to 3.54 Å.

1.5 K Saturation (1 M Solution)

Formation of 10 Å mineral peaks from either HI minerals or smectites layers indicates the presence of high charge vermiculites or high charge smectites.

One problem with clay mineral identification in soils is the difficulty in identifying high temperature phyllosilicates (2:1 minerals) which are inherited from sedimentary rocks or metamorphic rocks. Chlorites are similar to HI chlorites in basal spacings. However, the 001 14 Å peak is less intense than that found at 7.1 Å (the second order 002 peak). Also, in general it is less wider having a peak width at half height of around 0.3° 2 theta while the HI minerals are slightly wider, near or above 0.5° 2 theta on X-ray diffractograms. Mica and the clay illite, with 10 Å are more difficult to differentiate. The relations are given in Table 1.1.

Table 1.1 Ca or Mg saturation (0.1 M solutions)

001 Peak position Å	Mineral name
10	Illite
14.2	HI
15.2	Smectite
7.21	Kaolinite

Smectite, illite type and HI minerals can be either dioctahedral or trioctahedral in character (Millot and Camez, 1963). However, it appears that there is often a gradual change from tri to di-octahedral site occupancy in many clays (Righi et al., 1999; Churchman, 1980; Churchman and Bruce, 1988). The general tendency is to increase the alumina content of soil clays at the expense of magnesium – iron types.

1.5.1 Summary

The clay minerals found in soils, are for the most part phyllosilicates, i.e. formed of silica complex structures which have limited dimensions in one direction and much less limited dimensions in two others thus forming a sheet – like structure. Phyllosilicates exist in sedimentary, metamorphic and to a lesser extent igneous rocks also. Hence there is a possible continuum in transformations at the surface between high temperature minerals and those stable in the biosphere soil zone or water – rock interaction zone.

Essentially two types of soil phyllosilicate structures can be considered: the TO or 1:1 tetrahedral – octahedral coordination of oxygen with cations. Silica is dominant in the T sites and alumina in the O, octahedral sites. By far the most common 1:1 mineral in soils is kaolinite. Some occurrence of a hydrous form of this mineral (halloysite) are found. The second and much more frequent mineral type is a TOT, three layer tetrahedral – octahedral – tetrahedral mineral. Identification of different types of these minerals is made by identifying the cations found between the 2:1 layers of the structure. The interlayer site ionic occupancy is determined by two factors: the excess layer charge on the 2:1 layers engendered by substitution of ions of different valence in either tetrahedral or octahedral sites and the chemistry of the soil solutions. Hence these clay minerals respond to their chemical environment. This point is extremely important to the biosphere – mineral interactions.

Thus one has two types of phyllosilicates in soils, 1:1 and 2:1 minerals. The names of the 2:1 minerals are determined by an estimation of interlayer ion occupancy and the possibility to change ionic occupation by cation exchange. Laboratory determinations of soil clays should take these factors into account in order to characterise 2:1 clay minerals as a function of their function in soil systems.

What needs to be done

Operational definition clay minerals

Coherent system of clay mineral identification according to function

More observations of field examples to validate recommmendations

References

Adams J (2007) Vegetation – Climate Interaction, Springer, 232 pp

Aikin W E (1991) Global Patterns; Climate, Vegetation and Soils, Univ Oklahoma Press, 370 pp

April R, Hulchy M and Newton R (1986) The nature of vermiculite in Adirondack soils and till. Clays Clay Min 43: 549–556

Barré P, Velde B, Catel N and Abbadie L (2007) Soil-potassium transfer: Impact of plant activity on clay minerals as seen from X-ray diffraction. Plant Soil 292: 137–146.
Behrensmeyer A, Damuth J, DiMichele W, Potts R, Sues H-D and Wing S (1992) Terrestrial Ecosystems Throughout Time, Univ Chicago Press, p. 568
Bell M and Walker M J C (1992) Late Quaternary Environmental Change, Longmans, p. 274
Birkeland P (1999) Soils and Geomorphology, Oxford Univ Press, p. 430.
Birkeland P, Shroba R, Burns S, Price A and Tonkin P (2003) Integrating soils and geomorphology in mountains – an example from the Front Range of Colorado. Geomorphology 55: 329–344.
Blackford J (1993) Peat bogs as sources of proxy climate data: Past approaches and future research, 47–65. In E Chambers (ed.), Climate Change and Human Impact on the Landscape, Chapman Hall, p. 303.
Braque R (1987) Biogeographie des Continents, Masson, p. 470.
Carnicelli S, Mirabella A, Cecchini G and Sanesi G (1997) Weathering of chlorite to a low charge expandable mineral in a spodosol on the Apennine mountains, Italy. Clays Clay Miner 45: 28–41.
Casteldine C and Hutton J (1993) The development of high mooreland on Dartmoor: Fire and the influence of Mesolithic activity o vegetation change, 117–131. In E Chambers (ed.), Climate Change and Human Impact on the Landscape, Chapman Hall, p. 303.
Chruchman G and Bruce J (1988) Relationships between loess deposition and mineral weathering in some soils in Southland, New Zealand, 11–31. In D Eden and R Furkert(eds.), Loess Balkema Rotterdam, 383 pp
Churchman G (1980) Clay minerals formed from micas and chlorites in some New Zealand soils. Clay Miner 15: 59–78.
Cox J (1925) Crop Production and Management, Wiley, p. 517.
Dixon J and Weed S (1989) Minerals in Soil Environments, Soil Sci Soc Ame Publication, 1244 pp
Duff P (1993) Holmes' Principles of Physical Geology, Chapman and Hall, p. 791.
Encyclopaedia Britannica (1969) William Benton, Chicago, USA
Fichter J, Turpault M-P, Dambrine E and Ranger J (1998) Mineral evolution of a acid forest soil in the Strengbach catchment (Vosges mountains, NE France). Geoderma 82: 315–340.
Gharrabi M, Velde B and Sagon J-P. (1998), The transformation of illite to muscovite in pelitic rocks: Constraints from X-ray diffraction. Clays Clay Miner. 46: 79–88.
Ildefonse P, Copin E and Velde B (1979) A soil vermiculite formed from a meta – gabbro, Moire Atlantique France. Clay Miner 14: 201–210.
Jackson M (1964) Chemical composition of soils, 71–141. In F Bear (ed.), Chemistry of the Soil, Reinhold, p. 515.
Jenny H (1994) Factors of Soil Formation, Dover, 281 pp
Lanson B (1997) Decomposition of experimental X-ray diffraction patterns (profile fitting) a convenient way to study clay minerals. Clays Clay Miner 40: 40–52.
King L (1962) The Morphology of the Earth, Oliver and Boyd, p. 699.
Lefort O (1996) Evolution of the Himalayas, 95–109. In A Yin and M Harisson (eds.), The Tectonic Evolution of Asia, Cambridge Press, p. 666.
Malla P and Douglas L (1987) Layer charge properties in smectites and vermiculites: Tetrahedral vs. octahedral. Soil Sci Soc Ame J 51: 1362–1378.
Meunier A (2007) Soil hydroxyl-interlayered minerals: A re-interpretation of their crystallochemical properties. Clays Clay Miner 55: 380–388.
Millot G and Camez T (1963) Genesis of vermiculite and mixed – layered vermiculite in the evolution of the soils of France, 90–96. In A Swineford (ed.), Clays Clay Minerals Monograph 12 Earth Sciences, Pergamon Press, 509 pp.
Moore D and Reynolds R (1997) X-Ray Diffraction and the Identification and Analysis of Clay Minerals, Oxford Univ Press, p. 378.
Morley RJ (2000) Origin and Evolution of Tropical Rain Forests, John Wiley, p. 362

Potts R and Behrensmeyer A (1992) Late Cenozoic Terrestrial Ecosytems, 491–520. In A Behrensmeyer, J Damuth, W DiMichele, R Potts, H-D Sues and S Wing (eds.), Terrestrial Ecosystems Throughout Time, Univ Chicago Press, p. 568.

Richards P (1976) The Tropical Rain Forest: An Ecological Study, Cambridge Univ Press, p. 450.

Righi D, Huber K and Keller C (1999) Clay formation and podzol development from postglacial moraines in Switzerland. Clay Miner 34: 319–332.

Saliot P and Velde B (1982) Phengite compositions and post-nappe high-pressure metamorphism in the Pennine zone of the French Alps. Earth Planet Sci Lett 57: 133–138.

Shackelton N (1995) New data on the evolution of Pliocene climatic variability, 242–249. In E Ubra, K Denton, T Partridge and L Burchal (eds.), Paleoclimate and Evolution with Emphasis on Human Origins, Yale Univ Press, 546 pp.

Solbrig OT (1991) Ecosystem and global environmental change, 95–108. In RW Corell and P A Anderson (eds.), Global Environmental Change, Springer, p. 264.

Stewart WN and Rothwell GW (1993) Paleobotany and the Evolution of Plants, Cambridge Univ Press, p. 515.

Trewartha G (1954) An Introduction to Climate, McGraw Hill, p. 402.

Velde B, El Moutaouakkil N and Iijima A (1991) Compositional homogeneity in low-temperature chlorites. Cont Mineral Petrol 107: 21–26.

Velde B and Vasseur G (1992) Estimation of the diagenetic smectite to illite transformation in time-temperature space. Ame Mineral 77, 967–976.

Velde B and Meunier A (2008) The Origin of Clay Minerals in soils and Weathered Rocks, Springer 406 pp.

Zanelli R, Egli M, Mirabella A, Abdelmoula M, Plötze M and Nôtsli M (2006) “Black”soils in the Southern Alps: clay mineral formation and transformation, X-ray amorphous Al phases and Fe forms. Clays Clay Miner 54: 703–720.

Chapter 2
The Soil Profile: The Structure of Plant – Mineral Interaction Space

Contents

2.1 Origins of Alteration Clay Mineral Formation

Overall the transformation of rock into altered material is one that changes the chemical associations of the materials found in rocks. Rocks are formed at temperatures above those of the surface of the earth, by burial and thermal "hot spots" due to movement of molten materials coming from great depths. When these mineral assemblages are exposed to abundant water resources at low temperatures, they become unstable in a thermodynamic and chemical sense. New minerals are formed. If one looks at the overall result of such events, it appears that some elements are taken from the mineral phase while others are left within the solid materials. Li

B. Velde, P. Barré, *Soils, Plants and Clay Minerals*,
DOI 10.1007/978-3-642-03499-2_2, © Springer-Verlag Berlin Heidelberg 2010

and Schoonmaker (2004) indicate that the composition of average shales, which represent the soils of the earth moved into sedimentary basins, are much the same as average crustal rocks. The upper crust on continents is dominated by aluminous materials (called pelitic rocks) which have relatively abundant alkali an alkaline earth contents. However the shales, sedimentation products due to soil movement, contain much the same Al and Si ratios, but greatly different alkali and alkaline earth contents. The biggest loss is that of Ca and Na from the silicate materials. Strong loss of sodium (some 70%) and Ca (50%) show that chemical segregation is effected by the weathering process. The Ca ends up in carbonate concentrations thanks to biological activity and the sea becomes salty (NaCl) as a residue of the alteration process. The chemical process is of course accompanied by dissolution – recrystallization and other physical processes.

The interaction of water, plants and rock forms small sized silicate and oxide minerals called clay minerals. The most frequent definition, by mineralogists of this fine grained material is a particle with less than 2 μm in its largest dimension. Most alteration minerals, newly formed at the surface, are found in this size category. The definition is arbitrary, being based upon the limits of microscopic resolution which were and still are to a large extent, that of 2 μm. Thus clays are minerals which cannot be identified by optical methods. Traditionally, since the latter half of the 20th century, identification has been done using X-ray diffraction methods.

The origin of alteration mineralogy is due to the inherent instability of rocks in the presence of surface chemical forces and reactions in aqueous solution. This includes those caused strictly by mineral chemistry which we will call water – rock interaction as opposed to those engendered by bio-reactions which control aqueous solution chemistry. In the case of water – rock interaction the agent of change is the slightly acidic rain water in contact with the minerals in rocks. This is a situation of thermodynamic and chemical instability. Rock mineralogy has the imprint of compression and temperatures greater that those found at the surface of the earth. This is due to the nature and origin of rocks. They are hard and durable because they have been formed at higher pressures and temperatures than those of the surface. The mineralogy has responded to these physical conditions by producing phases that correspond to physical constraints other than those of one atmosphere pressure and ambient surface temperatures. Since they have re-adjusted to their physical environment to form hard rock, they will not be stable at the new conditions of the surface. First the temperature is lower than those under which rocks formed and second the ratio of solids to water is quite different, there being more water present under conditions of alteration at the surface than those of rock formation. In fact as sedimented materials are buried and become rocks they tend to lose most of the ambient water that formed the sedimentary material. Those rocks having passed through the high temperature conditions of a molten state of course are still less hydrous. Temperature and the availability of water in relatively large quantities are the hallmark of surface mineral alteration. Basically the rock minerals tend to come to equilibrium with the surrounding water by dissolving in it, either integrally or partially (incongruent dissolution). The materials which do not enter into solution are almost always more hydrous than the initial material. Water – rock interaction is dissolution and hydration. The formation of new minerals at the surface is due to mineral chemical processes and equilibria.

Plants and other living organisms, notably bacteria, can hasten the process of conversion from high temperature, low hydrous state to low temperature hydrous minerals. The processes can be complex but in the end the result is nearly the same as that of simple water – rock interaction. The chemistry of formation of the new surface minerals, usually called clays because of their small grain size, is the same.

2.2 Some Fundamental Principles in the Alteration of a Rock are as Follows:

2.2.1 Thermal expansion

First and foremost, rocks are susceptible to effects of differential thermal expansion. Most minerals are anisotropic in their expansion (Clark, 1966) which leads to tension and unresolved stress in a rock where minerals are oriented in various and un-ordered directions. Figure 2.1 indicates such a situation in a multi-mineral rock. Thermal expansion, due to daily heating causes stress vectors which oppose each other across mineral grain boundaries. In cases of parallel contacts, the minerals can have opposing expansion – contraction vectors which create shearing stresses. In cases where a mineral is confronted with another but with opposing expansion

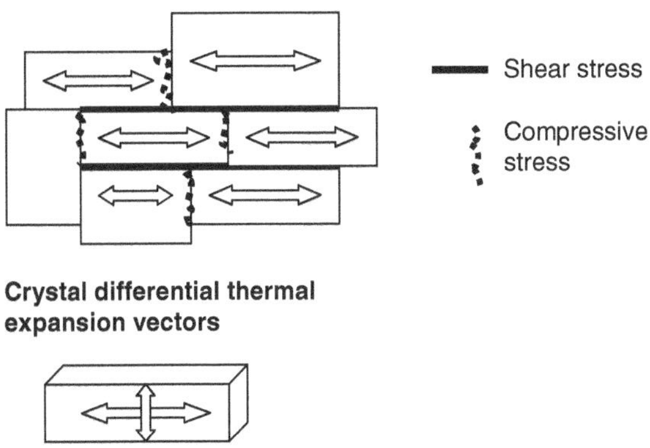

Fig. 2.1 Diagram representing the importance of differential mineral thermal expansion on rock coherence. Most silicate minerals have strongly anisotropic thermal expansion characteristics which engender opposing forces in a crystal aggregate such as a rock. Double arrows indicate the strongest direction of thermal expansion. Crystal contacts parallel to these expansion – contraction movements create shear stress forces while those at the junction of crystals, pushing on one another, create compressive stress. These forces create tensions and rupture between the different crystals making a type of sand by physical forces alone

fronted into the crystal, the stress would be one of compression or dilatation. Since the differential thermal dilatation coefficients can be on the order of six fold for a given mineral, according to the crystallographic orientation, one can imagine that the stability of a mineral assemblage at the surface of the earth where temperatures can change on the order of tens of hours up to tens of degrees centigrade, would create forces which would overcome those of normal crystallographic cohesion. Such effects lead to breaking of the interface surfaces between minerals or possibly the rupture of a grain itself. A long term experiment carried out at the experimental station at Versailles (Millot, 1964) where broken granite was exposed to atmospheric conditions for 20 years resulted in the formation of much smaller grain size particles even though no new clays or residual minerals were detected. The abundant rain and its quick evacuation leached the dissolved products of mineral degradation into solution with its transportation out of the system, leaving no residue. However the thermal effects on the rock debris were clearly seen in the production of sand and silt sized particles, approaching the mineral grains size of the granite. The fact of breaking a rock into near mineral grain size increases the surface of the minerals and their chemical reactivity. This process should not be overlooked in analysis of chemical reaction at rock – water interfaces. The importance of fracture passages in rocks which are open to water flow and their effect on alteration mechanisms and rates is discussed at length in Velde and Meunier (2008, Chap. 4).

2.2.2 Chemical Controls

Several chemical forces operate at the earth's surface which were not or little apparent under a rock formation regime. The conditions of contact with air, in general a force of oxidation and with rain water, a dissolution and hydration force are the most evident.

2.2.2.1 Congruent Dissolution

The first reaction of a rock in contact with water is to attempt to attain a solid – solution equilibrium by dissolution of a part of the solid phase in the aqueous solution. This action is fundamental to understanding alteration processes. The most important aspect of an altered rock is its decrease in density due to the dissolution of material. The more frequent the rainfall, the more material is dissolved and the more the influence of dissolution is apparent. Density can decrease by 20% or so due to increase in porosity (see discussion in Velde and Meunier, 2008, p. 119) or a loss of material. Of course non-silicates such as carbonates dissolve integrally with mineral reaction or formation of new phases such as oceans where the majority of carbonates form anew. The Versailles experiment mentioned in the preceding paragraph demonstrates this principle in that no new phases were observed in the essentially free flow experiment of rain water running over crushed granite. Dissolution in this experimental situation was the major chemical effect in the alteration process.

2.2.2.2 Oxidation

The major chemical principles of surface alteration mineral formation are change in oxidation state of iron and substitution of hydrogen ions for loosely bonded (ionic bonding) cations. The higher the temperature of formation for a mineral, the less stable it is, in general. Some minerals are recalcitrant to alteration, above all those without or few ionically bound cations or without iron. The most striking example is quartz, almost pure SiO_2, which is present in most soils even though it should in many cases have been dissolved to meet the activity requirements of silica in solution.

Oxidation is important in that most iron ions in silicates are present in the divalent state. Oxidation is then evident under surface conditions where the normal atmosphere is conducive to oxidation with the resulting change in electronic balance in the mineral which makes most iron – bearing minerals formed at high temperature unstable and induces mineral change. However, when iron becomes trivalent it tends to form an oxide or hydroxyl oxide. In other words it forms its own phase. Thus a large part of new silicate minerals formed through weathering are of relatively low iron content. Iron separates from the silicate minerals, especially when there is enough aluminium to combine with silica to form a clay mineral. This leaves an "iron stain" in the alteration zone (iron oxide), where the more minor amounts of iron take on a strong visual impact, making the altered zone reddish, tan or yellow depending upon the type of iron oxide present. The minerals and their colors are indicated in Table 2.1.

In cases where iron is in great abundance, it can combine in its trivalent state to form a new clay mineral. Thus olivines, pyroxenes and other high temperature minerals are relatively rapidly transformed by weathering into iron bearing clay minerals. Low alumina content of the reacting minerals dictates the formation of a ferric mineral.

2.2.2.3 Incongruent Dissolution: Hydration and Ionic Substitution

The second action is one of ionic substitution of hydrogen from the slightly acidic rain water in equilibrium with atmospheric CO_2 for cations in the silicate min-

Table 2.1 Color of Fe oxides and hydrosulfates (simplified from Scheinost and Schwertmann, 1999)

Fe oxide	Most frequent color	Composition
Hematite	Moderate reddish brown	α-Fe_2O_3
Goethite	Strong yellowish brown	α-FeOOH
Lepidocrocite	Moderate orange	γ-FeOOH
Ferrihydrite	Brownish orange	$Fe_5HO_8 \cdot 4H_2O$
Akaganeite	Strong brown	ß-FeOOH
Feroxyhite	Strong brown	δ-FeOOH
Maghemite	Dark yellowish brown	γ-Fe_2O_3

erals. The cations Na, Ca and K are extracted from such igneous minerals as feldspars, micas and various minerals of metamorphic origin such as amphiboles. The chemical reaction is as follows

$$H^+ = M^+ \text{ or } 2H^+ = M^{2+}$$

Where M^+ is the large ionically bonded cation in the initial silicate mineral.

The residuum is enriched in alumina as well as elements such as iron, titanium or manganese with in general a loss of silica. This residual material is the core of clay minerals and oxide phases found in alteration products. Sodium is largely lost to altering solutions while some Ca and much K is retained in the clays. Very often silica is lost to solution in the initial conversion of high temperature silicates to clay minerals. The elements which remain form clay silicate phases. Magnesium is partially lost in many instances, some being retained in the clay minerals and much being taken into solution. This is due to the fact that Mg is not accepted in large amounts in alumino-silicate clays. When enough magnesium is present in the altering minerals, the new clay minerals will be of relatively low alumina content and high in Mg and often in ferric iron. The general scheme of change in the hydration, ion – exchange process is to form low alkali and alkaline earth phases with either aluminous or magnesio-ferric minerals.

2.2.2.4 Destabilization of Clays and Silica Loss

The last step in the alteration process is the dissolution or destabilization of the newly formed clays themselves to produce, in the long term, oxy-hydroxide phases of iron and alumina. Loss of silica is the key to this change. It is a step wise process where relatively siliceous clays (Si > Al ionic content) are replaced by the intermediate mineral kaolinite (Si = Al) and then the aluminium hydroxide gibbsite (see Pedro, 1966). In this process the presence of kaolinite and gibbsite are considered to be indices of alteration intensity. The assumed sequence of clay mineral alteration is 2:1 mineral (vermiculite, illite, smectites) to kaolinite to gibbsite. These mineral names are elucidated in Sect. 1.4. The important observation is the destabilisation of the clay minerals initially produced by weathering processes which in their turn become unstable chemically and are transformed to a lesser degree than the destruction of high temperature silicate minerals which form clays in the initial water – rock interactions.

Overall one can schematically write the hydrogen ion exchange reaction and that of silica loss as:

Hydrogen ion exchange for cation in high temperature minerals

$$\underset{\text{feldspar}}{3KAlSi_3O_8} + 2H^+ \text{ -> } \underset{\text{muscovite–like clay mica}}{K_{0.9}Al_{2.9}Si_{3.1}O_{10}(OH)_2} + 2\,K + 6SiO_2aq$$

then hydrogen ion exchange in the clays clays

$$\underset{\text{muscovite clay mica}}{2KAl_3Si_3O_{10}(OH)_2} + 2H^+ = \underset{\text{kaolinite clay}}{3Al_2Si_2O_5(OH)_4} + 2\,K^+$$

and eventually hydration of clays after silica loss

$$\underset{\text{kaolinite clay}}{Al_2Si_2O_5(OH)_4} + H_2O = \underset{\text{gibbsite}}{2Al(OH)_3} + 2SiO_2\ \text{aq}$$

2.2.2.5 Scale of Alteration and Localization of Clay Minerals in Rock – Water Interaction

The silicate minerals in rocks, those that make clays, are affected to differing degrees by the passage of aqueous fluids (rain water) as it moves through gravity flow across a rock. It is evident that the initial contact of this water with silicate minerals in the rock is through fractures and fissures in the rock itself. The larger discontinuities are present due to the relaxation of pressures exerted on the rock during its trajectory from depth to its present day position near the surface of the earth. Tectonic forces tend to deform the rock under pressure and create discontinuities. Also the simple fact of releasing several kilobars of overburden pressure allows differential expansion of the rock creating fractures and fissures. These are the initial passages of water into the un-altered rock. Further, thermal expansion and retraction of the minerals in the rock subjected to diurnal thermal variations creates further ruptures between minerals and within minerals. Thus one finds passage ways of varying size and continuity in the rock itself.

As water enters the rock it will flow at differential rates depending upon the possibility of flow, i.e. width of the passagways. The larger fractures will carry more water, which will be more dilute containing little dissolved material while water moving more slowly through smaller openings will contain more dissolved material due to a more prolonged contact time with the unstable silicate minerals. The result is a hierarchy of flow systems which to a certain extent imitate the effect of rainfall of different intensities on a rock surface. With more rainfall there is more alteration by dissolution but the solutions are more dilute in dissolved components. In the rock itself, alteration intensity by dissolution is greater in large fractures while the smaller fissures and small cracks show less interaction, imitating climates of less rainfall. Clay mineral genesis is favored by slow flowing subsystems where rock – water interaction occurs with solutions carrying a large amount of dissolved materials. Under these conditions, new phases (clays) will form as the solutions become saturated with the elements which make up the clays. Usually these elements are iron, alumina and silica. The end result is that in each alteration system one has a series of chemical sub-systems which produce different minerals depending upon the relation of residence time and the amount of the aqueous fluids present in cracks of different size and the resulting chemical controls of element concentration in solution (Fig. 2.2).

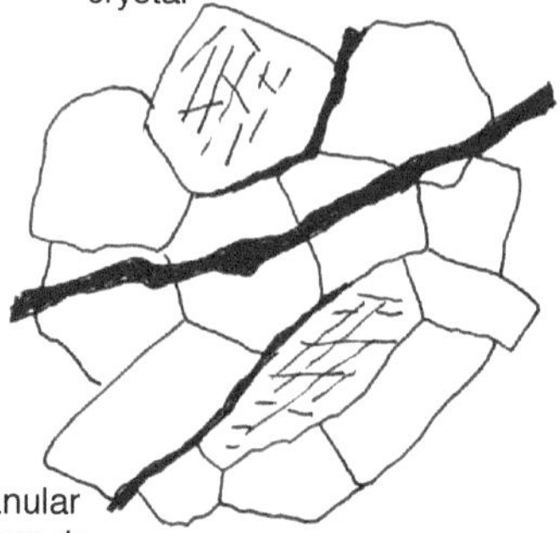

Fig. 2.2 Different types of cracks and fractures common in rocks due to their thermal and tectonic history. (1) Internal fractures within a crystal (2) inter granular fractures (3) trans-crystal fractures. All lead to the introduction of water into the different passageways in the alteration process of water – rock contact at the surface

The consequences of these factors of scale in the interaction of aqueous solution and silicate minerals is a difference in the chemical controls of the alteration process; In a very general way, the larger the scale of the system, the fewer the number of minerals present in a given site. This is a very well known maxim of phase equilibrium analysis (see Denbigh, 1987, Chap. 5 for example). The greater the number of chemical variables imposed from the exterior of a given system (mineral grain for example) the smaller the number of phases present at equilibrium. In general, the more water flow through a rock or soil system, the more intercommunication of the different parts and the more chemical constraints which will be imposed from outside of a given part of the system.

Concerning alteration mineralogy, if the scale is sufficiently restricted, through the limitation of fluid flow, each high temperature mineral crystal will form a new mineral or set of new minerals depending upon its specific chemistry. As the system is more open to fluid flow with an increasing communication through the formation of dissolution cracks, the diffusing species of elements in the fluids increases (larger and more continuous cracks and fissures) and the number of minerals formed decreases, being subjected to the chemical constraints of other altering minerals in different parts of the soil profile. In a way, the increased number of actors in the alteration process dilutes that importance of each individual one and a more general system is installed. Chemical activity of the different participating elements becomes influenced by adjacent phases and eventually by a large portion of the rock which is altering. The larger the fissure or crack, the fewer the number of minerals present and the less variable their chemistry. In extreme cases where water is plentiful and the rock never comes in equilibrium with the altering aqueous solution the number of phases present is zero, all is dissolved. One can find a more detailed discussion of these effects in Chap. 5 of Velde and Meunier (2008).

One can use the example of a potassium feldspar grain in an altering granite. Initially the altering fluids will be under saturated in potassium compared to feldspar which is the most potassium – rich mineral in the rock. As the feldspar is destabilized, some silica is lost to the solution since it is the most silica – rich mineral, other than the chemically inert mineral quartz. This is shown in Fig. 2.3, which is a chemiographic projection of mineral compositional relations. The overall composition of the feldspar system shifts down from the potassium apex and slightly toward the alumina side of the diagram. The phase assemblage instead of being one of only feldspar (single phase) becomes illite – and feldspar which is a two phase system, shown as Step 1 in the alteration process. The new micaceous phase occurs on the edges of the feldspar. In fact then the earliest new minerals formed are illites, a potassic soil clay mineral. This is the initiation of the alteration process is the initial mineral reaction. The new mineral, illite, takes up less space than the feldspar, creating small voids which inevitably let more fluid into the system.

Following this opening of the mineral grain network to new mineral formation and fluid flow, the feldspar starts to alter internally creating a *primary plasma*. This step (2) is in fact the formation of a mineral assemblage of new low temperature phases. In the case of potassium feldspar, one finds illite, as before and smectites along with fragments of feldspar. The overall composition of the old feldspar volume is now further depleted in potassium and to a small extent in silica and it becomes a three phase system.

More fluid flow induced by further decrease in the volume of solids and the creation of voids leads to a new set of equilibria, the *secondary plasma*, Step 3 in Fig. 2.3 which involves fluids originating from the alteration of other minerals in the rock. In the case of the old feldspar volume, containing three phases, one finds the formation of a new illite, which is less potassic, and kaolinite. Here the number of

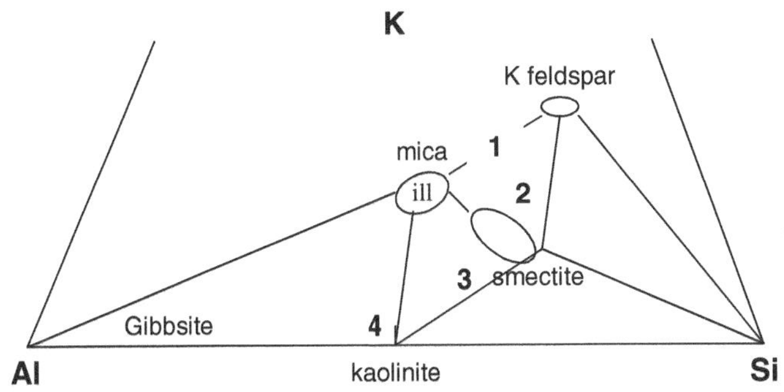

Fig. 2.3 Diagram representing the chemical relations of minerals common in granite potassium (K) feldspar, mi (muscovite) and quartz. Alteration minerals are illite (near the composition of muscovite) smectite, kaolinite and gibbsite. Initial phase stabilities are between muscite and feldspar (1), which change under water – rock alteration to include smectites (illite – K feldspar – smectites (2) and upon further alteration (more contatct with dilute aqueous solutions) the assemblage illite – smectie – kaolinite forms (4). The alteration process is one of loss of potassium and silica, in going from the assemblage 1–4

alteration minerals decreases as potassium and silica are lost creating a two phase system.

In and along the walls of large fractures which are conduits for rapid water flow to the groundwater table, one finds only kaolinite present. This is the *fissural microsytem*. Only one mineral is present due to the still larger amount of chemical intercommunication. A testimony of such interaction can be seen in the data compiled by Valeton (1972, p. 45) for water compositions in arkosic sediments (those containing quartz and feldspars). The flowing pore water is in chemical equilibria with kaoliite and not with potassium mica nor feldspar according to the equilibrium values given by Garrels and Christ (1965). This is Step 4 in the alteration sequence where the minimum number of phases is seen.

The result of the initial water – rock alteration processes in rocks, which occurs at different scales, is the formation of minerals which have responded to very local chemical potentials creating a large number of heterogeneous assemblages in specific areas or volumes which are not ultimately stable together in a larger system. As alteration of the rock continues, the amount of new clays increases until it dominates the alterite zone above the rock – water alteration interface. The number of clay types decreases. The new mineral assemblages are less dense than the old minerals in the rock due to element moss through the dissolution process. The new mineral assemblages tend to contain fewer phases. Table 2.2.

Some material escapes the system. The spatial distribution of the new phases determined by the old high temperature phases is maintained in the initial grain contact and primary plasma stage of alteration. Due to lower density and the creation of void space within such a system, water can circulate more freely and the chemical potentials imposed by the new clays as well as the incoming unsaturated rainwater become more homogeneous throughout the rock – alteration zone. The clays react to these new chemical constraints forming a more stable, general assemblage with fewer phases present. This process is maintained up into the root soil or plant – soil interaction zone.

In the primary plasma stage of alteration, different old minerals can form the same clay mineral phase, for example smectites, but the new alteration minerals can have different compositions dictated by the old mineral grain chemical composition (see Proust and Velde, 1978). One finds new minerals being formed in old grains reflecting the chemistry of each mineral as it comes into contact with slow moving fluids and their water chemistry. Gradually the chemical potential of a more large system is imposed upon the alteration minerals formed and forming and a new, more

Table 2.2 Stages of weathering of a potassium feldspar in a granite

1 Fluid flow at mineral interface	illite
2 Fluid saturation within the crystal *primary plasma*	Illite + smectite + feldspar
3 Fluid flow in cracks and fractures *secondary plasma*	smectite + kaolinite
4 High fluid flow in large fractures *Fissural microsystem*	kaolinite

general chemistry produces different minerals of a more simplified number. This new chemistry is one of lower alkali and alkaline earth ion activity. Loss of K, Na and Ca as dissolved species in solution which is exported from the system decreases the chemical variability of the phases present. In the initial stages of alteration, the activity of these elements is imposed by the altering phases, and is relatively high. As more communication is made with flowing water out of the system, the chemical activity of the more soluble elements decreases, reducing the number of new alteration phases present. Eventually the number of phases in highly altered rocks is two, hydrous iron and aluminium oxides. This is the ultimate stage of water – rock interaction.

This progression can be seen, in a very general way, as alteration clay assemblages become more simplified, essentially losing alkalis and silica. In going from temperate climate multi assemblages of different 2:1 minerals plus kaolinite, increased rainfall in more tropical climates produces more kaolinite and less 2:1 minerals, eventually forming kaolinite alone and eventually gibbsite and hematite, the aluminium and iron oxy – hydroxide assemblages where not only all alkalis are lost to altering solutions but most of the active silica also. Within a single altering rock under temperate climate conditions one can find the various stages of climate controlled soil clay mineralogy, going from alkali – containing clays to kaolinite. Hence the driving force of alteration is the relative amount of water compared to the reacting silicate solids. In fractures with free flowing soil solutions the mineralogy reflects chemical controls of undersaturated solutions and the formation of kaolinite present under conditions of temperate climate. The same conditions are found when rainfall is very intense, under tropical conditions, and the whole soil system responds with a kaolinite mineralogy and under very intense conditions only gibbsite and iron oxides.

Under conditions of temperate climate alteration, the new minerals present are multiple, depending upon the local conditions of water flow and chemical activity. As rainfall increases the clays become more dominated by kaolinite and eventually gibbsite. The temperate climate alteration mineralogy is dominated by 2:1 clay minerals.

A review of mineral alteration products by Velde and Meunier (2008, Chap. 4) indicates that the new minerals formed (i.e. those crystallized into recognizable and stale minerals) are most often 2:1 phases. Table 2.3 indicates the types of minerals formed in the initial stages of mineral alteration, within the original high temperature mineral crystal. The hydroxyl phase gibbsite is indicated whereas the oxy and hydroxyl forms of iron oxide are not in that they are almost always formed in the process of alteration of iron-bearing minerals.

The 2:1 alteration minerals are either high charge (identified as vermiculites), or low charge minerals (identified as smectites). Potassium feldspar is frequently associated with illite, a lower charge and more silica – rich form of muscovite. It is clear from these observations that the major, principal alteration mineral in the initial stages of alteration by water – rock interaction is a 2:1 mineral. These are the most silica – rich of the surface clays. The presence of 2:1 minerals reflects the normal ratio of silica compared to Al or Mg in minerals which favors silica and hence the silcia – rich secondary phases found in alterites and soils.

Table 2.3 Clay mineral types formed from high temperature silicate minerals

Mineral	2:1	1:1	gibbsite
Potassic feldspar	x	x	–
Plagioclase	x	x	–
Muscovite	x	x	
Biotite	x	x	
Chlorite	x		
Clinopyroxene	x	–	
Orthopyroxene	x		
Olivine	x		
Amphibole	x		

An interesting detail, perhaps not all that minor in importance, is the difference in alteration products of potassium feldspars, as detailed above, and the alteration products of muscovite in the same granite. Velde and Meunier (2008, p. 173) indicate that the new minerals formed in muscovites subjected to water – rock alteration processes, is an assemblage of vermiculite (HI mineral) and kaolinite. In general one can consider the presence of Al-OH interlayer ions in a 2:1 structure indicates a high charge, somewhere above 0.6 per unit cell whereas the smectites, described in potassium feldspar alteration indicate a lower charge, somewhere in the vicinity of 0.3 charges. This suggests that the smectite compositional field from low to high charge is represented in these water – rock alteration processes, where the muscovite alters to a high charge mineral and the feldspar, in the initial stages to an illite, potassic high charge mineral, and smectites of lower charge. Loss of potassium from muscovite is compensated by $Al(OH)_2^+$ ions whereas the higher potassium content of the feldspar solutions fixes this element in the 2:1 neo-formed minerals. Thus the new 2:1 minerals formed in granite alteration are high charge forms, forming illite under conditions of high potassium availability due to feldspar alteration, and forming an HI mineral under conditions of lower potassium availability.

2.2.3 *Biological Degradation of Rocks and Minerals*

Beyond the initial and fundamental interaction of water and rock to form new minerals stable at the earth's surface it is most important to consider the potential action of living organisms. In forest experiments where the land was clear cut (Bormann and Likens, 1994, p. 155), the loss of potassium to stream water was 10 kg/ha in the initial stages or forest re-growth decreasing to half that after four years. The uptake by the growing biomass was initially 2 kg/ha increasing to 31 kg after four years. This indicates that the water – rock interaction process (stream water loss at the starting point of the experiment) is relatively small when compared to the uptake by vigorous plant growth. Hence on can suspect that in the biosphere, living organisms will dominate in the transfer elements and mineral changes which one sees in the transformation of rock to soil. Biological activity is the most important factor in the stability of rock minerals.

Dissolution of silicate minerals has often been attributed to the presence of organic acids (see Barker et al., 1997 for a review). Organic acids are seen in laboratory experiments to hasten the dissolution of different minerals, especially feldspars. These are the actors of the chemical reactions where acidity and chelating effects contribute to the transformation of silicates in to soluble materials (Ullman and Welch, 2002). Silica release rates due to organic acid interaction are 2–5 times higher than those produced by mineral acids. Thus we see that biological controls and interactions can be much more important than those of mineral origin. Often it is assumed that the origin of much organic matter in the form of aggressive acids is due to the degradation of plant material transformed by bacterial action. However there may be a more direct and more efficient interaction guided directly by living organisms towards the transformation of high temperature silicate minerals found in source rocks in soils (Ullman et al., 1997). Raulund-Rasmussen et al. (1998) feel that the interaction between organic molecules and silicates is more a matter of concentration of the acids or other molecules than their specific types. This would suggest that the importance is in the concentration of the molecules and this is controlled locally by plant or other living organisms mor than ambient pore solution concentrations.

The initial stages of rock alteration in the biosphere are affected by the more simple or smaller forms of life such as bacteria, fungi and eventually lichens. Terrestrial life before land plants was most likely algal in nature (Wellman, 2003) where algal mats covered land surfaces forming the initial stages of soils. We have only partial information on the effects and importance of these agents of mineral change at present but it seems that they can be significant in the soil clay mineral formation processes. Intuitively one would think that a rooted plant would need some fine grained and clay type material to prosper though different plants can be seen today to grow, apparently, from rock surfaces where man has tried to present a clean surface of building material. Many pictures of plants growing from a cathedral wall or other "timeless" edifice have been made to show the ephemeral qualities of the efforts of man, the great manipulators of nature.

Bacterial action as an alteration agent is best known at present, in a general way, and lichen or fungal action is less well documented, at least from the stand point of the importance in the formation of clay minerals. In fact all of the forms of life discussed are now know to have and impact on the chemical stability of silicate minerals but in all instances there is little information on the clays or other residual minerals formed in the interaction process. We will consider the information on bacteria, the most simple form of life interacting and that of lichens, a rather complex but fundamental life form in the initial stages of rock – plant interaction in the early stages of alteration.

2.2.3.1 Bacteria

Bacterial interaction for silicates and sulfide minerals seems to have two major impulses: one respiratory action (Blake et al., 1994 or Newmann (2001) for example) where the bacteria are taking energy from minerals by changing the oxidation state of iron. This is called *autotrophic* interaction (Berthelin, 1983). Here the action is primary and immediate. Iron is oxidized or reduced with a consequent change in the mineral present. Normally the silicates which are involved are dissolved and a portion recrystalized (see Banfield and Nealson, 1997 for examples). In some cases even clay minerals, formed at the surface are de-stabilized and dissolved (Kostka et al., 1999 for example). This interaction can be called, here in any event, primary where the energy benefit for the bacteria is immediate. This *primary interaction* can occur on a bare rock which is in contact with bacteria. All that is needed is a bit of water. It is one of the first stages of rock alteration.

A second type of interaction, *heterotrophic* (Berthelin, 1983) appears to be that operated by silicate – dissolving bacteria which to a large extent integrally dissolve the mineral, as far as one has investigated the question at present. The bacteria obtain sustenance from organic molecules in ambient aqueous solutions, often exuded by plant roots and other plant materials (Berthelin, 1983). This effect can be called secondary interaction in that the bacteria are dependent on the organic molecule sustenance for their prosperity which is provided plants, either as root exudates or the decomposition of litter root matter, in order to get vital nutrients from rock minerals in abundance. There is some indication (Banfield and Nealson, 1997; Kawano and Tomita, 2002) that clays or proto-clays can be formed in these processes. The interaction between these bacteria and silicate minerals seems to be encouraged or probably sustained by plant action, specifically root exudation of complex organic compounds used by the bacteria as a source of sustenance. Badr et al. (2006) describe the results of such action in laboratory and hot house experiments which indicate the benefit of this interaction is to the plant and the bacteria. In the case reported, potassium and phosphorous are extracted by several bacterial strains to the benefit of the plants. The relation seems to be symbiotic.

Numerous studies have been carried out to better understand the interaction between plants and silicate dissolution by bacteria. Welch and Ullman (1999) indicate that the ambient temperature affects the bacterial action, but not always in simple ways. A lower temperature (5°C) induces low pH in solutions which increases silica dissolution from feldspars 20 fold while at 35°C the effect is such that the dissolution of the minerals is not distinguishable from abiotic experiments. At 20°C the increase in silica dissolution was 2 fold. Similar results were obtained by Barker et al. (1998) for room temperature experiments. Welch et al. (1999) indicate that the type of organic matter available to bacterial can have a significant effect on dissolution rates, ranging from an increase ranging from 5 to 100 fold or no effect at all. Barker et al. (1997) summarize numerous publications concerning bacterial or organic products of bacterial action relating to silicate mineral dissolution rates which indicate that CO_2 and temperature would be controlling factors in general and that biologic enhancement of dissolution can be two to three orders of magnitude that of normal abiotic water – rock interaction.

By contrast, Wu et al. (2008, p. 412) indicate that the overall effects of bacterial interaction on granite in laboratory experiments are only moderate. This could well be due to the fact that bacterial action is mineral selective in a granite where much material, such as quartz, is little reactive to bacterial attack. Conversely, bacterial action would very well be mineral selective, yielding certain elements useful to the ecosystem as a whole. Ullman and Welch (2002) give detailed considerations of the mechanisms of silicate (feldspar in this case) dissolution. Augusto et al. (2000) found that the dissolution rate of plagioclase feldspars in field experiments where forest vegetation prevailed showed little difference with abiotic laboratory dissolution rates. In their observations the dissolution of feldspar was essentially concentrated in the upper 10 cm of the soil profile and strongly pH dependent, which was controlled by the species of tree present.

Thus there is need of proximity of root matter or decomposing plant debris in order for the organic matter to be available to the bacteria and the dissolved ionic species to be available to the plant. Calvaruso et al. (2006) have calculated a chemical budget for this interaction in the case of trees. Augusto et al. (2000) indicate that different species of tree have very different effects on mineral dissolution, assuming that the intermediary between plant and silicate mineral is that of microbial dissolution. Taking the process one step further, seed inoculation experiments with bacteria indicate a strong effect due to plant – bacteria interaction in the dissolution of ambient detrital minerals (Badr et al., 2006; Sheng, 2005 for example).

2.2.3.2 Mycorrhizal Fungi

Cochran and Berner (1996) observed that silicate rock alteration was much higher under higher plants, i.e. more complex than bacteria, mosses and lichens. This is perhaps due to the greater needs of the bigger, more developed plants or due to a more efficient system of mineral extraction. One of the most important factors in root interaction with silicate minerals is through micorrizal contact with the silicate minerals. This has been operating for along period of time, perhaps since plants made their way onto the continents (Johnson and Gehring, 2007) and Taylor et al. (2009) summarize mycorrhizal relations in ecosystems. Basically they are everywhere where there plant roots. Further they have been around since plants became terrestrial inhabitants, more than 400 million years ago. Mycorrhize are fungal extensions of the root – plant structure below ground. They have been around for as long as plants were present on the continents. Their abundance increases the root mass by a factor of about 60 in many cases. Mycorrhize gain their sustenance from plant exudates, based upon photosynthetic energy which they exchange for mineral element extraction from silicates in the soil. Relationships are complex, depending upon plant species, climate and soil chemical characteristics. In general plants with thicker roots have more micorrhizal associations than those with strongly divided systems. Hence trees depend more upon micorrhize than grasses for example. The precise actions between minerals, especially silicates, and the fungi are not yet fully explored but it appears that they have a strong influence upon mineral stability. Not that they destroy higher temperature silicate minerals but that they modify them to the advantage of the plants seeking mineral elements.

Arocena et al. (1999) and Glowa et al. (2004) report the change of phyllosilicate minerals in soils under the influence of fungi which are closely tied to tree root physiological action. It is apparent that the high temperature phyllosilicates in the soils, micas and chlorites, are affected by a subtraction of interlayer ions, potassium and magnesium, to the benefit of the associated root systems. In exchange the fungi receive chlorophyll – produced substances for their metabolism. The modification of the high temperature phyllosilicate minerals is different depending upon the fungal species present, producing a more or less high charge vermiculitic type of mineral as are seen depending upon the type of tree involved. These differences indicate variations in the chemical activity of the fungi and perhaps differences in the general chemical context of the host tree detritus which conditions the soil. Here we see that the phyllosilicates are retained but modified into ion exchangeable clays.

In the laboratory studies of microbial action on silicate stability, generally two types of minerals have been used, phyllosilicates such as muscovite (an aluminous, potassic silicate mineral) and biotite (an iron – magnesium potassic silicate mineral) or feldspars (plagioclase N-Ca, and orthoclase K tectosilicates). Reaction rates have not really been compared directly. However Smits et al. (2005) indicate that the amount of material dissolved or released to the soils through feldspar weathering by tunnelling of fungal hyphae would be quite small, even after 1,650 years, less than 1% of mineral weathering. It is assumed that the faster reacting mineral would be biotite where iron oxidation can occur as well as dissolution and extraction of potassium. Muscovite is generally assumed to be less reactive, holding potassium in the structure, as observed by April and Keller (1990) for forest soil root system related mineral changes. Feldspars appear to dissolve integrally, perhaps forming other clay minerals wich have not been detected. However one does not see why micorrhize dissolve feldspars while being content to just extract K from a phyllosilicate structure. A partial answer is found in the work of Glowa et al. (2003) where detailed chemical analyses were made on mycorrhize themselves and it was found that he major mineral constituents of K, Mg, were found within the organisms whereas Ca appears to be fixed as an oxalate on the peripheral parts of the structures. It is clear that a portion of the mineral elements are part of the metabolism of the mycorrhize themselves. However their extraction powers exceed their needs and are part of the plant – mycorrhize system of interaction. We do not have a good idea of what mechanism would do this in muscovite where charge balance cannot be differentially affected as in biotite where higher ionic change in the structure by oxidation of iron would force out a potassium ion. The extraction mechanism of potassium from feldspar is more obscure. There are questions that remain.

Fungal Interactions in Partial Modification of Minerals

Fungi are the type symbiotic material. Fungi survive by using energy, chemical, sources supplied by other organisms in "exchange" for transformed materials useful for their host. In soil systems roots are frequently, or perhaps always, associated with mycorrhizal materials which "extend" the chemical realm explored by roots. Arocena et al. (2003) give evidence of fungal penetration from crustose lichens into chlorite interlayer sites indicating that overall dissolution is not the initial bio-reaction with chlorites but one of selective extraction of elements.

Studies by Arocena et al. (1999) and Glowa et al. 2003 give more precise information concerning fungal interaction with chlorites. Both studies involved natural soils under coniferous vegetation. The clay minerals produced by ectomycorrhizosphere interaction associated with roots of Abies lasiocarpa (subalpine fir) produce an expandable clay (Ca saturation and glycol or glycerol saturation) which largely collapses upon potassium saturation. Ca – saturation in the air dried state shows a 14.7 Å spacing, instead of the normal hydrated (two water layer) state of 15.2 Å. The soil materials not associated with the ectomycorhiza did not show significant expandable clay material but a strong chlorite peak. This indicates the formation of a relatively high charge smectite from the chlorites present. It is interesting to note the strong contraction to an anhydrous mica like mineral with potassium saturation.

Glowa et al. (2003) take such a study further in using two different types of ectomycorrhizal fungi in till based soils under spruce (Picea glauca) forest cover. In this study, depending on the dominant fungal material (ECS – A or ECS – B) the chlorite dominated phyllosilicate mineralogy changes to an expandable mineral, to a greater or lesser degree depending on the species present, from an initial chlorite rich soil. Here again potassium saturation largely collapses the interlayer spacing due to the formation of an anhydrous potassium ion layer. Ca saturation in the air dried state leaves a 14.7 Å spacing. The intensity of alteration or transformation seems to be related to the type of ectomycorrhizal fungus present.

From these two studies, albeit based upon coniferous forest soils, it appears that chlorite is not destroyed, at least not to a great extent, but is transformed into a high charge 2:1 mineral which is susceptible to capture large amounts of potassium. The expandable mineral is most likely of high charge as shown by a smaller spacing in the air dried Ca Saturated state, 14.7 instead of 15.2 Å for a bi-hydrate cation structure.

However, Khademi and Arocena (2008) show that the clays palygorskite and sepiolite (non-phyllosilicates, i.e. sheet structure minerals) are completely altered from their initial chain silicate structures to form a kaolinite (TO) mineral under micorrhize action. Thus one sees that plant bio-agents do not appear to destroy sheet silicates (especially 2:1, TOT minerals) but they tend to preserve the basic structure modifying the interlayer ion portion fo the minerals.

To summarize, bacterial have two types of interaction, one primary (autotrophic) and the other a secondary effect (heterotrophic) conditioned by interaction with higher level plant species, via roots. The primary autotrophic interaction, based upon oxidation – reduction of iron, is independent of other life forms. The secondary type of interaction depends on plants for sustenance of the bacteria in that they apparently do not benefit from the destabilization of the silicate minerals. One can deduce that in the primary interactions, bacteria are specialized in interaction with iron – bearing minerals and hence the water – rock interaction processes will dominate in the initial stages of alteration. Bacteria and micorrhize tend to destabilize non-phyllosilicates entirely whereas they allow phyllosilicates to remain largely intact, modifying the interlayer sites in a 2:1 structure to extract elements or eventually store material in these sites. This appears to be especially true of micorrhize actions.

2.2.3.3 Lichens

These plants are a hybrid between chlorophyll species (algae) and fungi. This is a condensation of sorts of the plant root – fungi symbiotic relationships described briefly above. Less data is available for these systems concerning the weathering of silicate minerals than for those of bacteria or fungi. However the information available (to a large part summarized in Chen et al. (2000)) indicates that the same relations hold as those in the microbiotic systems.

Biotite minerals have been observed to be transformed into vermiculite type clay minerals preserving the 2:1 clay framework (Wierzchos and Ascaso, 1996). However ferric clays have been seen to be dissolved as well as plagioclase feldspars which most likely form kaolinite minerals observed at times. However one report (Adamo and Violante, 2000) shows the formation of a mixed layered mica/smectite phase and kaolinite formed on a non-phyllosilicate bearing basaltic rock, mount Vesuvius. In this review paper, the rare reported case of the formation of a 2:1 mineral from a non-phyllosilicate mineral base is very striking. Here the creation of a soil – like clay is apparently the only indication at present of the formation of 2:1 soil clay minerals from non-phyllosilicate minerals. The destabilization of feldspars to kaolinite is most likely a wide spread process, according to this review, but the formation of a 2:1 new soil clay mineral indicates a different relationship in the chain of silicate reactions. If a 2:1 mineral can be formed in the first stages of mineral transformation from non-phyllosilicate minerals, then such a reaction chain could well be responsible for a portion of soil clays. The importance of such a creation is dealt with at greater length in Chap. 4.

Overall, it appears that lichens follow the trends of bacterial and fungal interaction where phyllosilicates are preserved and other silicates are destroyed to form kaolinite for example, such as is the case for plagioclase feldspar. We have less information on mineral formation from other silicate minerals which seem, in a first instance to form amorphous silica rich materials and iron oxides (Chen et al., 2000).

One observation made recently by Balogh-Brunstad et al. (2008) indicates the very important interaction of bio-extractors and plants. It was shown by laboratory experimentation that the installation of the micorrhize system of communication with plant roots decreases the amount of potassium in fluid percolating solutions in a soil. Initially the soil fluids contained potassium and other ions but very rapidly the soil solutions showed strongly reduced K contents. This indicates that the alteration process was dominated by the micorrhize root association allowing little potassium to escape into the pore fluids. The plants (Scots pine in this case) continued to grow with extracted potassium. This study indicates that the symbiosis of the micorrhize – root couple is extremely efficient in obtaining needed nutrients and the destabilisation and formation of new minerals in soils occurs with reduced loss of needed nutrient elements. Thus the measure of soil solution nutrients is not a good indicator of the processes of mineral element transfer between high temperature rock minerals and plants systems.

April and Keller (1990) indicate the complexity of chemical reactions at the bio-mineral interface. They find various products of interaction present at the

Fig. 2.4 Indication of the importance of lichen action. The photograph shows a 16th century church wall made of granite covered by the plant community. (Sainte Nonna, Pen March, Finnistere, France) Photo DV

root – clay interface such as amorphous silica, hydroxy – alumina materials and mica with fractured and fragmented portions adjacent to the root surfaces. Fragmentation of the high temperature minerals muscovite and biotite result in fin grained, clay sized, material which is characteristic of soil clays. The transformation of these and other materials is surely complex, but it appears that the phyllosilicates remain for the most part intact while the other silicate minerals are dissolved to form other phases or have the dissolved portions transported out of the soil system in percolating pore waters.

Figure 2.4 indicates the potential impact of the initial stages of bio-rock interaction. Here a venerable church wall (Sainte Nonna, Pen March, 16th century), made of granite has been colonized by lichens of different types (crustose, foliose and fructose) all combining in a concerted effort of bringing down the wall by reducing it to sand and clay.

A more recent event, 20th century in the church yard of Saint Nonna, is shown on a grave stone where crustose lichens have begun their work (Fig. 2.5).

2.2.3.4 Mosses

Carter and Arocena (2000) present data which shows the importance of species in moss – phyllosilicate transformation. Soils under Pleurosium schreberi and Ptilium crista-castrensis mosses were investigated. Both formed on glaciofluvial terrace material. The C horizon soil phyllosilicates were mica, chlorite with minor kaolinite. In both profiles chlorite is lost in the A horizon under the moss. The Ptilium moss leaves an expandable mineral after glycol treatment but in a Ca – saturated air dried state the sample shows a 14.7 Å spacing instead of the

Fig. 2.5 A more recent event, 20th century in the church yard of Saint Nonna, is shown on a grave stone where crustose lichens have begun their work. (photo DV)

normal 15.2 Å hydrated ion state. Thus one can conclude that the soil clay contains some high charge smectites. Potassium saturation leaves a clay assemblage with a range of spacings from 13.8 to roughly 10 Å spacings with a peak at 12.4 Å. It is clear that the altered chlorite structures have a varied charge and charge site, forming either independent mineral units (ex. the material giving a 12.4 Å peak) or units of random mixed layering of different proportions. Hydration states vary: reflections between 15 and 12.4 Å indicate some bi-hydrated material while peaks at 12.4 Å indicate monohydrated ions and those near 10 Å indicate anhydrous interlayer ions. Here there is clearly a range of charges or charge sites in the mineral structure in the mineral which respond differently to the presence of potassium, some not responding particularly.

The Pleurozium moss clays show a 14.1 Å spacing with Ca saturation under air dried conditions and no change with glycerol treatment. However, potassium saturation leaves the clays with a range of spacings from 12.6 to 10 Å. The 12.6 Å spacing indicated a monohydrate interlayer ion occupancy while the 10A peak indicated an anhydrous inter layer ion state. Again a high charge mineral with differing states of charge or charge site produces potassium interlayer ions. However, in the case of Ptilium moss the mineral is clearly an expandable phase while in the case of the Pleurozium moss the mineral is not expandable after saturation with a divalent cation.

The different behaviours of the clays under these two mosses indicated that each species does not produce exactly the same 2:1 mineral from the chlorite precursor. In one case the new mineral is most likely based upon the 2:1 portion of the chlorite with most of the hydroxyl – interlayer ions extracted and replaced or replaceable by hydrated divalent cations. This is a classic smectite behaviour. In the other case, the mineral does not accept hydrated divalent cations but monohydrate and anhydrous potassium replaces various portions of an initial probable exchangeable hydroxy-interlayer ion layer.

These results indicate the complexity of bio-silicate interactions. However, in general the changes in the bio-horizons are more rapid than those in the "mineral" horizons (saprock and saprolite). The minerals formed are not quite the same either. Climate affects this bio-activity as one could suspect, as shown in a montane series reported by Castaldini et al. (2002). Climate changes the bio-agents (plant biomes) and this changes the chemistry of the soil pore solutions which affects the clay mineral and silicate mineral assemblages present.

2.2.4 Summary

Biological degradation of rocks appears to be much more active, in a reaction rate sense, than that of simple water – rock interaction with silicate material. Various observations have estimated differences of several times to orders of magnitude faster rates for biologically driven reactions compared to water – rock silicate mineral change. However one can assume that such rates are more or less restrained to the biosphere of alteration, where root exudates and plant debris are sufficiently active to encourage microbial interaction or in the case of lichen, such action is restricted to the rock surface. In alteration profiles, the absence or roots at depth should signal the slower rate of reaction transforming high temperature minerals into clays, essentially one of water – rock interaction. Since the root zone is to a large extent restricted to the upper 20 or 30 cm of an alteration profile for most of the root mass (Schenk and Jackson, 2002), one should find that the types of clays in this zone will be dictated by microbial interaction and interstitial solution chemistry which is guided by the type of plants present. Clays in the deeper portions of the profile will be the result of local mineral – water interactions. As we have seen above, microbial action favors the conservation of clay minerals, 2:1 minerals being little altered. However other silicate minerals can be dissolved leaving oxides or kaolinite as an insoluble residue (Khademi and Arocena, 2008).

One point that is important to retain is that the clay minerals found at the surface under weathering conditions are of moderate alkali content to relatively high content. However, the alkali present is almost uniquely potassium, fixed in the clays or calcium as an exchangeable ion. Both are present only in 2:1 minerals, those containing the most silica. Sodium is lost to the altering solutions and ends up in the sea. This is why the term salt, refers to a sodium compound. Magnesium is lost to the sea also. Potassium is largely retained in clays. Plants found along the saline bodies of water, oceans and other evaporitic lakes, are sodium bearing. They are called halophyllic. Halophyllic plants, found near sources of saline water, can adapt to less sodic conditions having a mixture of Na and K ion concentrations in their biomass (Barkaudah and Henderson, 2006). The fact that potassium is retained in clays and not sodium was crucial to the development of plant physiology. It appears that an alkali ion is necessary to plant metabolism, either Na or K which are the most abundant of these elements on the surface of the earth. Hence terrestrial plants use potassium whereas plants living near a saline body of water use sodium.

Terrestrial plants then need potassium, and it is one of the three major fertilizer components, in the NKP triad. This need for alkali ions is extremely important in

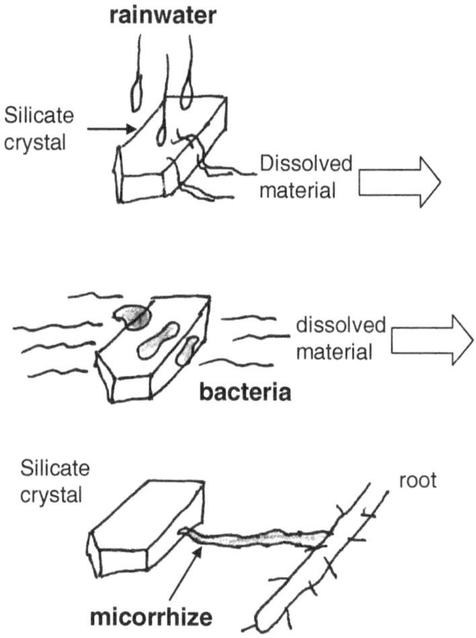

Fig. 2.6 Indication of the different types of alteration processes in a soil between (1) rain water and silicate mineral with the dissolved material going into the aqueous solution, (2) bacterial action where the dissolved material goes into pore water solutions and (3) micorrhize action where the largest part of the dissolved material is channelled to the plant through the root – micorrhize interface

the chemical impact of plants on surface chemistry and mineralogy as we will see later.

Figure 2.6 shows in a schematic way the different types of interactions that a silicate mineral might encounter in the saprolite to soil horizons of an alteration profile. The water – rock interaction is indicated b the action of rainwater with a crystal of a silicate mineral. Bacteria can colonize the surface of the crystals, and introduce themselves into fractures in the crystals with the effect of dissolution and release of dissolved material, being nourished by root exudates substances and finally micorrhize fungal action linking the plant roots direct to the silicate mineral under dissolution can be seen.

What needs to be done

More complete investigation of the interactions of bacteria and silicate minerals
More complete investigation of micorrhize – silicate interactions
Development of rates and impact of bio modifiers on soil solutions and silicate chemistry

2.3 The Alteration – Soil Profile: Transformation of Rocks into Clay, Sand and Organic Matter Complexes

The title here is at first rather ambiguous. A number of pedologists prefer to call the upper portion of an alteration sequence the soil, that part which is in fact intimately involved with plant action. In this part of an alteration profile one finds organic matter which indicates the influence of plants through decay of leaf matter and plant roots. However the largest part of most alteration profiles goes much more deeply below the surface where a gradually decreasing amount of new surface minerals is found until a zone of "bed rock" or un-altered material is apparent. In principle one finds a more or less zoned structure of horizontal "horizons" each showing specific characteristics related to alteration processes which transform rock into clay – rich assemblages. These "horizons" are outlined in Fig. 2.7. The soil portion (A horizon) is the contact of inorganic materials, rocks and clay minerals, with the living plant environment. Below this one finds very clay – rich segments which indicate a high rate of rock transformation. Depending on the intensity of alteration and the alterability of the bed rock there is a mixture of clays and rock debris in this zone. However there is no or little trace of the old rock structure, i.e. the relicts of the disposition of the old minerals in the original rock. This is the saprolite zone. The contact zone between clay – rich and un-transformed rock is the saprock zone where one finds the initial stages of rack alteration into clay minerals and mineral debris. This idealized zoned structure is the basis for soil classifications. However often one finds that the rock and saprock zones are very irregularly inter-fingered depending upon pathways for water flow which hasten alteration in a non-horizontal manner. Cracks, fault planes, and other geologically initiated structural irregularities are expressed in this un-even alteration process.

If we consider the different types of alteration processes, physical degradation, plant – rock interaction and water rock alteration, it is evident that the first two will dominate at the surface while the last will be less evident at greater depth in the alteration profile. Intuitively the interaction of rainwater and silicate or rock matter should be most intense in the intial contact zone, the soil – air interface where unsaturated rain water first comes into contact with soild matter. However, the materials present at the surface in the soil zone are already, for the most part in equilibrium with the new, low temperature, hydrating conditions of the earth's surface. Thus

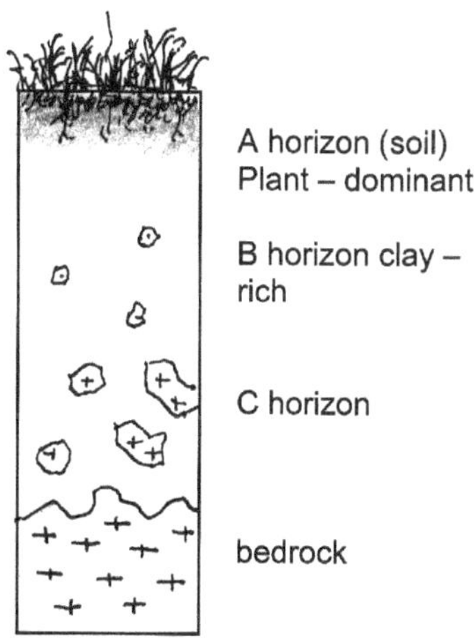

Fig. 2.7 Simplified schematic representation of an alteration profile where the plant – silicate interface is shown as the A horizon, rich in humic matter and roots, the B horizon of clay accumulation, the C horizon of alteration minerals (clays and oxides) and rock debris, and the bedrock which is initially altered at the interface with the alteration zone

the impact of percolating rainwater will be great at the rock – alteration interface, perhaps more so than in the initial contact zone. Even if the rainwater is saturated and in equilibrium with soil clays, it will be decidedly out of equilibrium with high temperature minerals constituting the hard rocks at the bottom of the alteration profile. Therefore, the rainwater percolated to the rock interface will interact to a large extent and become charged with such elements as alkalis, alkaline earths and silica. Dethier (1986) has indicated that the largest part dissolved material in water making its way to streams and rivers reflects the alteration equilibria found at the rock – water interface. In fact the measure of dissolved material in stream water is a measure of the activity at this interface and not the chemical reactions in the upper portions of the alteration and soil profile. The importance of pathways, residence time, and frequency of rainfall in this interaction zone is detailed in Chap. 5 of Velde and Meunier (2008).

The question can be posed then: Are the chemical and mineral interactions the same through out the soil alteration profiles or are they different? Is the alteration process one continuum imposed by rain water abundance and contact or are there differences in the profile?

2.3.1 The Transformation of Bedrock into Soil Clays

It is striking to note in treatises on geochemistry the change in average composition of rocks in the upper crust of the earth and the result of surface alteration processes which produce soils. The material which results from these chemcial processes is clay rich and when deposited in the oceans becomes shales. This is in fact a concentrate of the average processing of silicate materials. In shales one finds roughly the same concentration of potassium as in the upper crust average (see Li and Schoonmaker, 2004) as well as magnesium but much less Ca (1.50 compared to 3.15%) and far less sodium (0.59 compared to 2.54%). The major elements silica and alumina are roughly unchanged also, with some loss of Si and a gain of Al. Thus the transition from high temperature rock mineralogy to one of clay minerals and residual sand (quartz) is one of conservation of most elements excepting Ca and Na. Sodium makes the sea salty, as we know, and hence is almost totally lost during the process of alteration. Calcium is also lost to a great extend making up the carbonate part of sedimentary rocks through the action of animals which make their shells of it combined with CO_2 form the atmosphere. This leaves us with K and Mg being retained in the clays of the soils. How is this effected and why?

One method which can be used to answer this question is to observe the clay – forming processes and their kinetics in the different parts of the profiles. We will start with the water – rock interaction zone of the bedrock and saprock zone.

2.3.1.1 Saprock Clay Formation and Transformation Kinetics

The rate of interaction in this zone is governed largely by the inherent thermodynamic instability of the minerals in the bedrock. In principle minerals of higher temperature of formation will be more likely to alter. However the overall effect on weathering kinetics is difficult to assess in natural situations given differences in rainfall and vegetation. The most direct measurement is the total depth of alteration, i.e. the distance to the rock interface. One measurement is the thickness of weathering rids on volcanic boulders or clasts. This situation, a "fresh" rock in an essentially unweathered environment indicates the pure rock – water interaction is a very unsaturated regime concerning the mineral charge of the aqueous fluid Coleman (1986). Rind formation rates, very partial mineral alteration, is measured in millimeters per thousands of years with only 2 mm after 160,000 years alteration in Western United States gravel terraces. This is very little interaction. However, in Costa Rica, the rind alteration zone is measured in centimetres per 240,000 years (Sak et al., 2004). The starting materials are highly out of equilibrium, being basic volcanic rocks, hence on can expect maximum reaction rates for eh physical conditions. It is clear that climate, rainfall, can greatly affect the alteration by water – rock interaction, roughly by a factor of ten in the cases cited.

Another measurement of weathering rate at the rock alterite interface can be the total developed thickness of an alteration profile of granite. Evans and Cameron (1979) indicate that the development of weathering progresses about 1 cm/1,000 years. Here we are in a very inactive climate, the cold region, Baffin Island. The relations of age and depth of solum (alteration zone) are rather linear for this time range. Samples from Rhodesia on granite also (Owens and Watson, 1979), indicate an alteration formation of 11 and 4 mm/1,000 years. The differences between Baffin Island tundra, Rhodesian savanna and tropical Cameroon climates are near one order of magnitude in the differences of their effect of rock alteration as defined by the development of the saprock portion of the alteration profile.

Still another measure of reaction progress in the rock to clay transformation can be made using estimations of clay content for different aged alterites near the rock interface. Data for tephra alteration (Lowe, 1986) in New Zealand indicate similar rates of transformation even though the New Zealand material is much more fine grained and would be expected to be more reactive (Fig. 2.8). The tephra data record to a time span of near one million years and near 80% development. In Fig. 2.8 one sees that the distributions reasonably close to an exponential curve, which is to be expected for first order reactions where the relation $a = a_0e^{-kt}$ should pertain, a being the amount of reacted material. The function of temperature (t) seems not to be all that important for the temperature ranges considered here in that the Baffin Island data are in accord with those for New Zealand.

Given the above estimations of alteration rate, either in the development of the alteration front, basalt alteration rinds or saprock front progression, or that amount of reaction accomplished in the development of clay from high temperature rock (in the examples cited all of the bedrocks were magmatic in origin) it appears that the times of interaction are relatively long, centimetres depth per 1,000 years alteration in progress of the saprock interface, and near a million years for the development of a totally altered rock. Such lengths of time are on the magnitude of climatic changes induced by glacial epochs which we experience now. These can be considered to be on the order of magnitude of geological events.

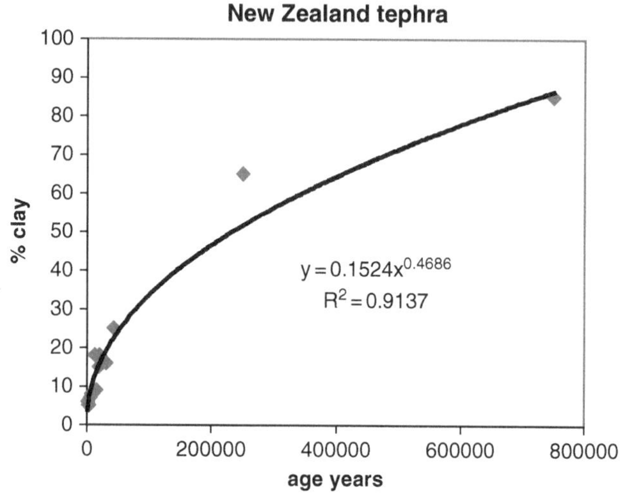

Fig. 2.8 Indication of he time scale necessary for the transformation of a silicate rock into clay minerals under conditions of water – rock interaction at the surface. The alteration of a volcanic tephra to a very high clay content occurs only after 800,000 years (Lowe, 1986)

2.3.1.2 The A Horizon of Plant – Soil Interaction

The interaction between living matter and mineral matter in the alteration zone is of several orders of magnitude greater than that for simple water – rock interaction. Some mineral transformation reactions can be effected much more rapidly than others. However, we must be very careful here to distinguish between clay – forming reactions, i.e. transformation of a non-phyllosilicate into a phyllosilicate mineral and clay transformation reactions where the nature of the clays is changed and not the existence of the clays themselves. Clay change such as that where interlayer ions are displaced, for example the change from smectite to illite (introduction of anhydrous potassium in a 2:1 structure) or the transformation of chlorite to smectites (extraction of interlayer Mg(OH) ions from the 2:1 structure) are probably not in the same range of energetic values as those where feldspar is transformed to kaolinite with loss of alkalis and silica.

In the laboratory the effects of micro-organisms can be very strong and rapidly accomplished. Barker et al. (1997 and 1998) Ullman and Welch (2002) indicate that the dissolution processes induced by bacteria can enhance reaction rate by two to three orders of magnitude compared to water – mineral reactions in the laboratory. Barker et al. (1997, p. 407) give data showing a typical S curve for reaction progress which is typical of biological reaction rates as described by Odum (1971, p. 186) which including those animal (sheep, Odum, 1971, p. 188) and eventually, according to sociologists of human activity! These types of reaction progress are well known in studies of forest growth and change (Bormann and Likens, 1994). Tree growth and forest recovery follow the same patterns as those of feldspar dissolution by bacteria. However interaction of organic molecules with feldspars in laboratory experiments

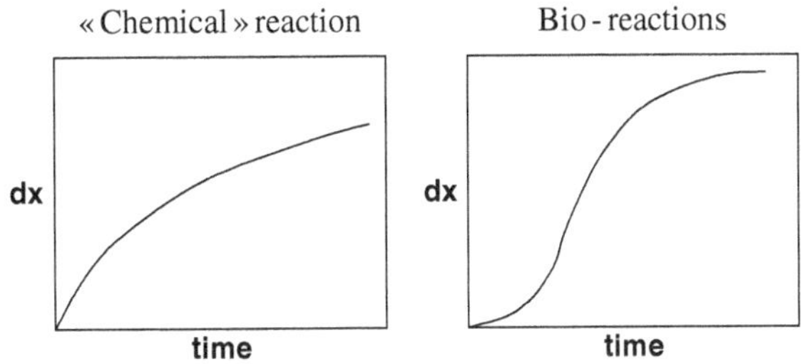

Fig. 2.9 Schematic diagrams of reaction rates observed for chemical reactions (rock – water interactions for example) and biologically driven reactions. The first is a "C" shaped curve while the bio reactions most often show an "S" shaped curve

(Ullman and Welch, 2002) show exponential curves common in normal first order reaction kinetics (Fig. 2.9).

The reaction rates (measured dissolution) for the bacteria experiments are about 1,000 greater than the organic compounds alone in contact with feldspars. Thus even though the chemical agents are similar, organic acids of various sorts, the bacterial function is much stronger. The S curve reaction progress is apparently a signature for bio-controlled processes in soil environments. An example of bio and mineral reaction rates in the same soil profile context is given in Fig. 2.10. Here the development of the thickness of the A horizon, measured by the organic – rich layer thickness is

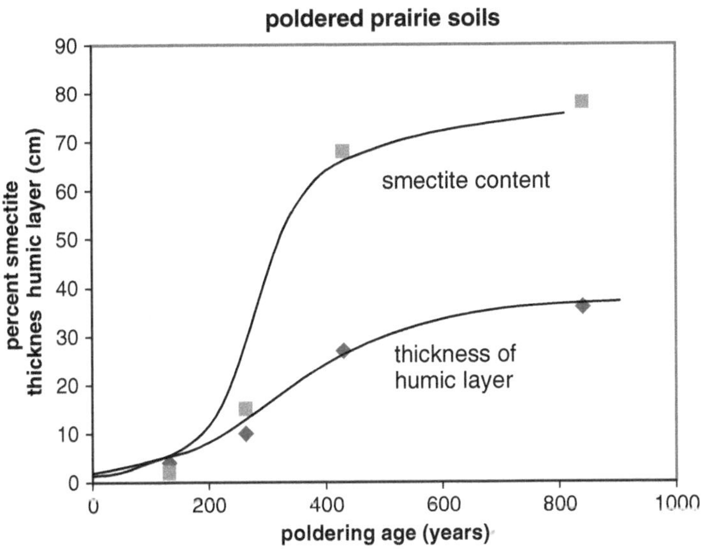

Fig. 2.10 Illustration of the relations of strictly organic reactions, thickness of the humic layer in developing poldered soils based upon newly sedimented clay – sand sediments, and the development of the smectie mineral, formed from more illitic clay minerals types (illite, smectie – illite mixed layer clays) under the humic zone in the poldered soils; The similarity of the shapes of organic depositon and clay mineral change suggest that the changes in the clays are driven by the biological processes. Both have "S"shaped reaction curves as a function of time. (data from Velde et al., 2003)

compared to the development of smectites from mixed layer minerals. The setting is a series of poldered zones along the Authie Bay near the Baie de Somme in northern France along the Channel coast, where soil profiles were observed as a function of the age of polderization. In each polder the sediment deposition was interrupted and a grazing (natural prairie) system installed. As time progresses (older polder age) the thickness of the humic layer increases indicating organic activity with the installation of the prairie ecotome. Below this humic zone one finds that the clay minerals change, with an increase in the smectites – rich mixed layer mineral at the expense of more illitic (potassic) clays. In that the clays in the humic zone remain potassic in nature, one can suspect that the effect of water – rock interaction is one of potassium loss. However, in that the humic horizon clays maintain potassium content, we deduce that the resources for this maintenance are those of clays below the humic zone exploited by the roots of the prairie plants. More about this later (5.1). In any event it is clear that the two rates of reaction progress are of the same sort, S – shaped in nature. This indicates that not only the organic activity is of this dynamic but it has an effect on the mineral parts of the system. Walker and del Moral (2003, p. 107) give data concerning the dynamics of forest growth after a landslide event. The biomass was measured as well as the potassium in runoff water from the watershed of the slide area. One sees an S shaped curve for the biomass established over a 40 year period and a reciprocal decrease in potassium in runoff water (Fig. 2.11). Here one sees the impact of the growth of plants on the mineral content of the soil and alterite substrate. It is clear that most of the decrease in loss is due to the forest growth as it re-establishes a normal stable state of production.

A similar change in clay mineralogy, S – shaped dynamics, can be seen in the data presented by Mahaney (1986) where the formation of smectites from Quaternary

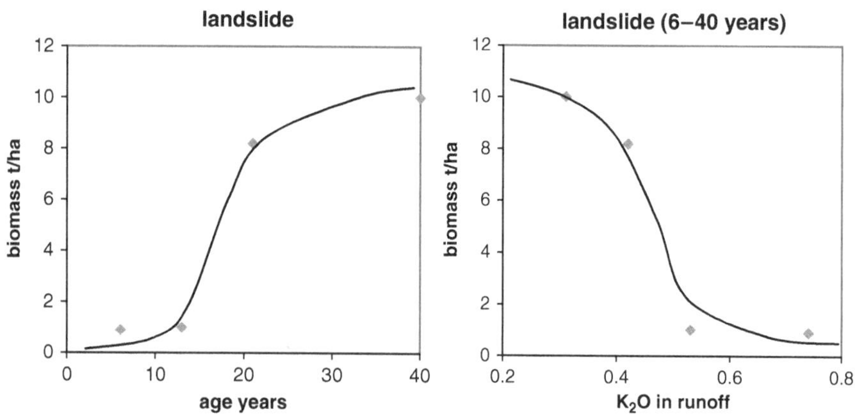

Fig. 2.11 Control of weathering by plant growth in the case of a landslide in a forested area. Biomass increases according to the "S" curve after the landslide while the potassium present in the runoff (streams) shows a decrease in the opposite sense. This indicates that chemical loss due to the water – rock weathering process is rapidly controlled by plant uptake and sequestration of this element in the biomass (Walker and del Moral, 2003)

soils developed on till in the Wind River mountains of Wyoming show a typical S – shaped change in abundance, formed at the expense of an illite – chlorite mineral assemblage. Here one sees that the smectite is replaced in its turn by soil vermiculite after 3,000 years of smectite increase in the A horizons. In fact one can find traces of this type of reaction dynamics in the generalisations given by Birkeland (1999, p. 225) concerning the development of soil charactistics which attain a "steady state" according to the author. The generalisations of the author are based upon a vast amount of knowledge and observations over a long period of time. Although the data is presented in a very generalized manner, it certainly can be used with confidence. In the development of soil properties the S curve dynamic is quite evident. The author indicates that the time constants for this development vary with the type of soil developed which is essentially one of climatic influence and hence plant – soil interaction. The attainment of steady state according to the author can vary from 1,000 years for organic – rich, wet soils to 1,000,000 years for soils in more arid climates where oxidation is dominant. In looking at the specific parts of a soil profile, A–C horizons, the author indicates that there is a difference in the time necessary to attain the steady state depending upon the horizon, and hence the degree of plant – mineral interaction. The A horizon is the first to come to equilibrium in the example given. The time necessary to attain steady state for the A horizon, that where organic activity is greatest is on the order of hundreds of years whereas the lowest horizon takes near a million years The factor of 10^4 years for differences in development of the A horizon (bio driven) and the C horizon (water – rock contact) is similar to those found in other studies quoted above.

2.3.1.3 The Effects of Bio-agents

Different bio-agents appear to effect different change in silicate chemistry and dissolution of silicate materials. In some cases there is dissolution and apparent loss of all material to the solutions, in others the minerals are only partly altered maintaining most of the silicate network intact. The relations of the bio-agents to the roots of plants are of great importance. If the minerals from the bedrock are dissolved or modified releasing mineral elements to the soil pore solution this will affect the stability of other clay minerals present. If the bio-agent is directly "connected" to the plant through the roots as are mycorrhizal fungi, the materials dissolved and released will be directly transmitted to the plant without coming into contact with other silicates in the soils and without affecting mineral stabilities. Such a system is suggested by Balogh-Brunstad et al. (2008). In their experiments pore waters in a soil where pine seedlings were introduced we analyzed and observations made on the soil solution compositions as a function of time. The developing plants used more and more mineral nutrients but less and less were present in solutions. The authors attribute this effect to direct up take through mycorrhizal action in conjuncture with the roots of the pine seedlings. Bacteria by contrast would directly affect the silicate minerals and release the dissolved elements into the soil solutions. Hence the relations of bio-agent and plant are important to the overall mineralogy and chemistry of soils.

Feldspar minerals are often cited as being transformed or affected by bio-agent actions in soil zones (see Barker et al., 1997 for a review). Some experiments have been made extracting and culturing bacteria in potassium feldspar and injecting these strains in various crop seeds (Sheng, 2005). The plants of course benefited from such increased biological action close to their roots. It has been observed that some species of bacteria will attack micas for potassium whereas feldspars are much less affected (Liu et al., 2006). Thus, as one would expect, there are different species of bio-agents for different types of work. It appears that bacteria can be influenced by the minerals which they alter, which affect the diversity of the microbial communities (Hutchens et al., 2008) This seems to be another instance of "feed back" or inter influence between the living and mineral worlds.

Escudero et al. (2007) found that lichens in soils can have species – specific effects on plants. This suggests a specialization in bio-agent link ups with plants. From another view point, Augusto et al. (2000) found that different species of trees in a forest had different impacts of feldspar weathering rates. Different plants have different needs and extract the elements necessary for their development at different rates which affects the chemistry and mineralogy of a soil. This can explain, partially, why some soils maintain different minerals in the soil zone whereas others have lost them. Of course there is an interaction between the mineral resources in the source rock of a soil and their use. To a certain extent the type of rock will dictate the plant community which will be present under given climatic conditions. Hence the inter-relations of plant species, bedrock composition and bio-agents in the soils will determine the overall chemical reactions and define the clay mineral stabilities in the A horizon of alteration sequences. In a very general manner, eruptive rocks, either acidic or basic, tend to influence the plant regime to a great extent. Most farmers know that it is not a good idea to plant wheat on granite soils nor on those of basic magmatic rocks. It is better to stick to the sedimentary rock – based soils, where in fact the high temperature minerals have gone through the geochemical transformations to produce sediments, via plant – soil interactions and provide a fertile basis for plant growth of crops used for human sustenance. More on this in Chap. 6.

2.3.2 Summary

The analysis of available data concerning differences in reaction kinetics in the bio horizons (A) and water – rock interaction horizon (saprock – bedrock) suggests two different types of reaction dynamics or kinetics; this means that the processes are quite different, one obeying normal mineral or chemical rate laws and the other with an as yet poorly explained dynamics in three parts. The S curve in initiated by a stage of slow change initially, followed by a strong growth or change dynamic which eventually reaches a stage of slow and more steady change. In fact the S curve is probably in reality a superposition of two reaction rates, one slow which is eventually overtaken by a more long lasting and stronger one.

Taking into account the above analysis, one can designate two active zones in an alteration profile:

1. The A horizon is where plant – silicate interaction is primordial. Reaction rates are relatively rapid, on the order of hundreds of years. The micro-biological elements are largely responsible for these reactions and reaction rates. This being the case, one would expect such systems to be highly affected by temperature changes. When biological activity is possible, i.e. under conditions of adequate rainfall and humidity, the temperature effect drives biological activity which determines the amount of chemical substances available to affect silicate minerals. We will call this horizon or reaction zone the soil portion of an alteration profile.
2. The saprock – bedrock horizon is where water rock interactions are dominant and normal mineral chemical processes dominate. Here the time scale of reaction kinetics is on the order of hundreds of thousands to a million years. The transformation of silicates to clays is long and produces multiphase assemblages in the first instances of reaction.

In general, the thickness of the A, soil, horizon will probably stabilize whereas the depth of the saprock – rock interface will continue to move downward with time (Fig. 2.12).

The bio-agents, bacteria and fungi inter-react with the silicates and plants to alter and transform the minerals present in the alterite zone. Different agents and different plants will change or modify the silicate minerals in the soil zone to different extents depending upon the needs of the plants and the constraints imposed by the bedrock materials.

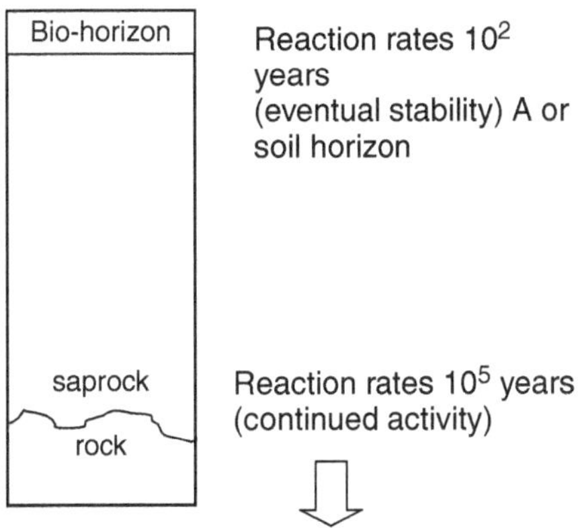

Fig. 2.12 Indication of the relative changes in mineralogy in an alteration profile. The reaction rates for clays in an organic dominated environment (A horizon) is on the order of hundreds of years, where the interlayer ins in 2:1 structures change composition, and sometimes new minerals form, i.e. kaolinite or chlorite. Reactionrates involving high temperature silicate minerals take much longer periods of time to come to equilibrium with surface chemical conditions. Here the time scale is on the order of 10^5 years

2.3.3 The Alteration Profile

In between the A, soil, horizon of plant – silicate interaction and saprock – rock interface one finds various mixtures of clays (new minerals in equilibrium with their surface environment) and relicts of the old minerals or fragments of rock which have not been altered into more stable mineral assemblages.

It is evident that the zone adjacent to the rock – saprock interface is that of saprock. The name of this alteration zone indicates its transitional nature with the rock below the alteration zone. In fact here the old rock structure persists, i.e. one can imagine the forms of old minerals which have been transformed into new clay mineral assemblages. This is the "conserved structure" zone. Gradually as more and more minerals are altered, the structure becomes less dense due to loss of material by dissolution in the transformation reaction. This change is dealt with in length in Velde and Meunier (2008, Chap. 4). The loss of material due to alteration can represent 30% of the initial rock. Gradually the structure of the old minerals is lost, with a small increase in density, to form what is called the saprolite zone. Here the old rock structure, localisation of the alteration minerals with respect to their parent phases, is lost and the new clays are mixed with other assemblages. This is the saprolite zone, essentially the same mineralogically as that below but one of gradual change where the segregation of the initial mineral assemblages is effaced to form new ones of a general chemical equilibrium. In the saprock zone one speaks of primary plasma, that space occupied by the initial mineral assemblages. In the saprolite zone one speaks of secondary plasma which calls chemical potentials of a more general nature. Often these secondary plasma mineral assemblages reflect an increased loss of mineral constituents, such as alkalis or silica. This change is seen in Fig. 2.2 where we have followed the change in feldspar alteration products to those of the secondary plasma, number 3 in the diagram. The secondary plasma is composed of new minerals, illite, smectites and kaolinite which contain less alkali (potassium) and silica than those of the initial feldspar and the first reaction products, illite in the presence of feldspar (number 1 in the figure) and the primary plasma which contains smectites, illite and the feldspar in the form of fragmented grains (number 2 in the figure). At each step of transformation, some elements are lost, K and Si in this case, which leaves behind a less dense material which is more open to water flow. The replacement minerals are less dense, containing water or OH units, and this also changes the density of the alteration material.

As alteration proceeds the saprock – saprolite zone is extended and the zone of partial alteration is extended. Partial alteration, because the change of old minerals to new one is gradual, leaves islands of old minerals or assemblages (rock fragments) which are often found in these parts of the alteration profile. In fact it is not uncommon to find rock or mineral grain fragments present in the A horizon also. In zones of intense alteration the old mineral grains and rock fragments are quite rare, and one finds at best quartz grains (SiO_2) which are very resistant to alteration. The presence of these old mineral grains is very important to the dynamics of plant growth in that deep roots can obtain needed elements in this zone without going to the rock – saprock interface. Figure 2.13 illustrates this disposition of rock and clay mineral material.

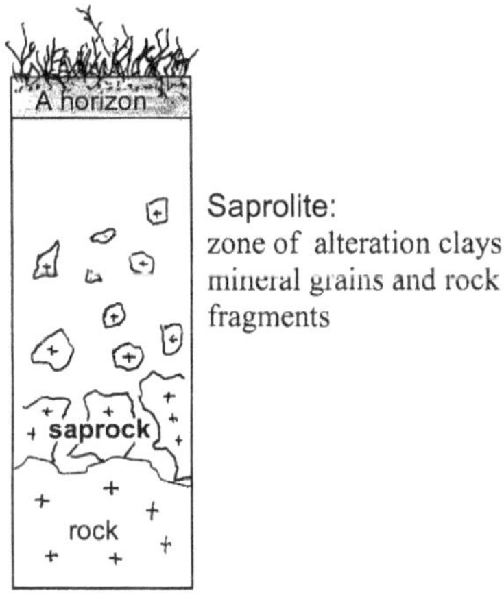

Fig. 2.13 Illustration of the relations between rock, saprock and saprolite zones of an alteration profile

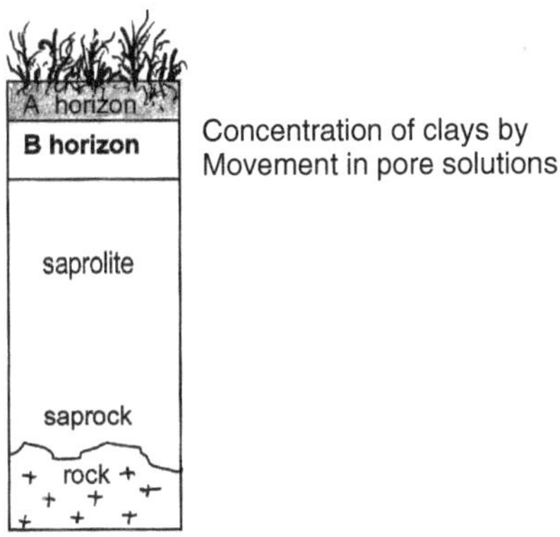

Fig. 2.14 Situation of the B horizon where clay minerals are concentrated by whysical movement in soil pore solutions

One major feature of many alteration profiles is the concentration of clays in a horizon below the A – organic zone. We will call this zone or horizon the soil, indicating the specific situation of a mixture of organic material and silicates which is the result of organic influences on the silicate and oxide materials. The movement of clays is due to percolation in pore fluids (rainwater) as the surface water moves downward to find its way to the water table. Clay concentration gives a new name to this zone, called the B horizon (Fig. 2.14).

The formation of such clay concentrations is relatively slow. This is especially true for considering the recent (geologically speaking) where glacial epochs have changed the surface geomorphology and plant regimes due to climate change to a great extent.

With a strong chemical attack, due to acid soil conditions engendered by plant activity (usually conifer forest) clays can be further displaced and the remaining material dissolved forming what is called the E horizon. In this zone one finds sand, in the form of the chemically resistant quartz and iron or other metal oxides (Fig. 2.15).

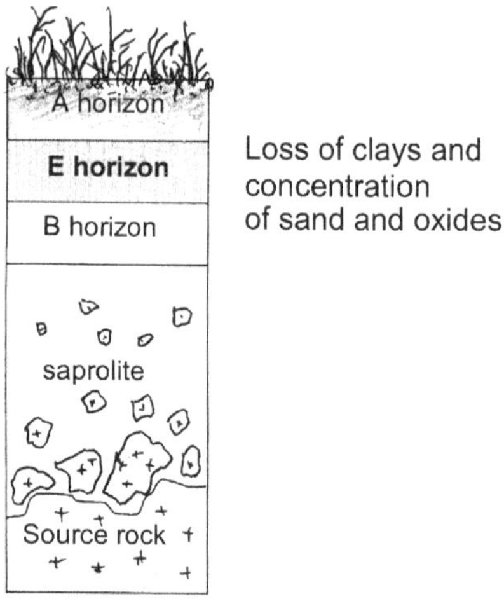

Fig. 2.15 Illustration of the formation of the E horizon, a zone essentially devoid of clay particles and organic matter. The fine grained material is transported to the B horizon, below, leaving an essentially sandy horizon below the organic, plant silicate contact zone. The E horizon is particular to forest soils or those of high acidity

Soils with a distinct E horizon are most often found in forest biomes except where alteration is very intense, tropical forests for example, where even the refractory quartz is rare. In these simplified examples one can suspect that the structure of soil profile, the different parts designated by the relation of alteration minerals and their place in the profile, can be a function of the climate (alteration intensity) or the type of vegetation present.

Generally then one finds more clays, products of alteration, at the surface of an alteration profile. The consensus is that the clay content reflects alteration intensity, which is greater at the surface of a profile than in the lower parts where pore water in soils and alterites would be more saturated with dissolved elements. However, the analysis found in Velde and Meunier (2008, Chap. 4) indicates that the pathways of major fluid flow are filled with aqueous solutions still un-saturated with respect to 2:1 phyllosilicates. The kaolinite minerals and oxides are formed along the walls of major pore pathways in alterites (saprock and saprolite) indicating that the solutions are highly under-saturated with mineral elements. This suggests that the total reaction of rainwater and silicates in soil – alterites is not accomplished just at the surface of the profiles. The information in Fig. 2.11 where the forest regime as it becomes established strongly controls the total loss of elements to stream water indicates also that the plant regime strongly dominates chemical transfer and reaction in the surface portion of a profile. Yet it has been seen in a multitude of studies that the clay content of the materials in alterite profiles is concentrated in the soil zone (organic interactions). Hence clays can be associated with the A horizon biozone of activity, more so than in the lower parts of the profiles, at least where some of the material present in the alterites is not completely transformed into hydrous silicates or oxy-hydroxides such as large portions of tropical soils.

Jenny (1994, p. 154) correlates data for soil clay content and temperature for soils on different substrates, granites and basic rocks in the Eastern United States Clay content increases from 10 to 60% with annual average temperature increases of 8°C. Such a great change in alteration activity can only be attributed to plant activity which is governed by climate. Here one can conclude that a large portion of the clays in the soil horizon are due to the bio-activity which transforms most of the alterable high temperature minerals into new phases stable in the surface environment.

2.3.4 *Summary*

The alteration profile is then divisible into different sectors where different chemical parameters are present which determine the physical and mineral properties of the altered materials. The base is the rock from which the alterites and soils are formed. This rock is then transformed by mineral dissolution and crystallization of new, hydrous phases, clays for the most part, as rain water percolates to the bottom of the alteration profile. This is the saprock zone where the initial stages of mineral reaction occur. The reactions are determined by the proximity of available water and its state of under-saturation with respect to the mineral elements in the old, high temperature minerals. The reaction rates for such interaction are very slow. The total dissolution of a rock takes nearly one million years.

By contrast the soil horizon (A) shows much greater rates of reaction. For example, data given in Ruhe (1984) suggest that the rate of transformation of loessic materials across the Midwestern United States into new clays is on the order of 20% in the space of 10,000 years. Thus, if undisturbed, one could expect a complete transformation of the silicate material susceptible to change in the space of less than 50,000 years or a rate greater than 2 10^3 times that of water – rock interaction. The zone intermediate between the A soil horizon can be divided into two horizons, one where the movement of clays by pore water transport forms a clay – rich zone, the B horizon. Below the B horizon where no trace of old rock structure is found, the material is composed of varying proportions of mineral and rock debris along with clays formed by water – rock interaction. Pedologists term this zone the C horizon including the saprock zone of conserved rock structure and the saprolite zone where the rock structure has disappeared. The important point is that throughout the C horizon the reaction rate of mineral conversion is essentially governed by mineral water – rock reaction kinetics. In the soil horizon (A) the reaction rate is much greater due to biological interactions. Plants and microbes work 24 h a day, rain or shine and they maintain an aqueous environment which is the vehicle of chemical reaction for silicate minerals.

2.3.5 *The Plant – Mineral Continuum, the A Horizon*

The A horizon is what we can call the soil, i.e. the horizon where organic matter is present and silicate minerals are present. This is the zone of plant – mineral interaction. Plants use this portion of the alteration profile to form and extract their necessary elemental mineral needs. Of course water is a part of their interchange as well as portions, i.e. ionic constituents, of minerals. This portion of the plant – mineral interface is very busy, with plant remains, and living matter, bio-agents such as fungi, bacteria and more developed living forms such as earthworms, beetles, moles, termites and so forth. Each is seeking its sustenance in this rich zone of interaction. Each modifies the material it encounters and leaves some different material behind. The most important part of the A horizon, concerning growing plant activity is that concerning roots. Plant roots are the interface between the living, organic world and the mineral world. Concerning plant bio-mass, in forest soils 15–25% of the plant is found in the root mass while up to 95 can be found in the root systems of prairie plants (Gobat et al., 2003, Sect. 4.1). One normally thinks of dead leaves, stems and twigs as representing the organic inheritance of the plant in the A horizon, but this is not the case for prairies whereas it is for forests. Thus a variable portion of the organic matter in the A horizon comes from aerial parts of plants. Roots not only leave behind an organic inheritance but throughout their active lives they produce exudates, organic compounds of varying complexity which are rapidly recuperated by bacteria and micorrhize for sustenance.

Within the surface layers bacteria, fungi and other active bio-agents work on the material or root or aerial plant origin to extract sustenance which is not available due to lack of an ability to capture sunlight and transform it into usable organic compounds. This is the beginning of symbiosis. Given the variable origins of the organic material in soils one would expect it to give rather different intermediate products of degradation depending upon the source material. In Fig. 2.6 we give a very general idea of the cycles of activity. One can consider that the bacteria are a major factor in the transformation of detrital organic material, along with fungi of various sorts and more developed or higher order organisms such as worms, termites and so forth. The most important action is bacterial, at least as that considering the clay – organic interface. The bacteria form two sorts of organic material from the detritus; one is a complex, less oxygen – rich organic material often called humus. The second is a series of smaller molecules, often rich in polysaccharides, which is easily attached to clays. The evolution of the composition of organic matter in bulk is such that with depth the ratio of carbon to nitrogen changes. This is a measure of the maturation of the humic matter. The mineralization of organic matter includes the production of CO_2 and nitrogen compounds. The ratio of change (C/N ratio) is a measure of this humification process, where the nitrogen is left behind and the carbon released to the atmosphere through bacterial action. The ratio of C/N is a function of the starting material and maturation processes which are driven by bacterial action and hence temperature. In a sequence of soils under different types of vegetation from the low Alps in France (elevations of 1,100–1,800 m) data from Mariotti (1982, p. 236) one sees the effects mainly of vegetation type, where the most carbon – rich (C/N = 30) is from a larch forest and other soil profiles dominated by grasses of different types have lower initial C/N rations (Fig. 2.16). The acidity varies with vegetation type and average temperature, ranging from near 7 under grass to 4.3 under larch. Although the initial C/N values are quite different, the end product of bacterial action is to give a rather constant value at depth, below the active A horizon.

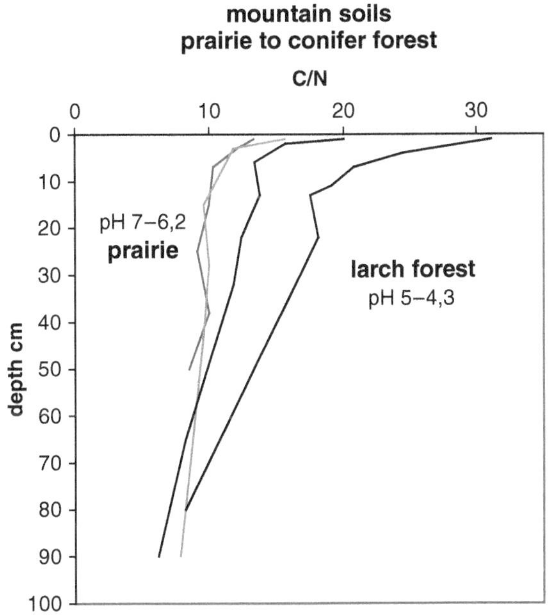

Fig. 2.16 Relations of soil C/N rations and pH and depth for different biomes in plateau mountain profiles (data from Mariotti, 1982)

This concordance of elemental values with maturation (bacteria action) suggests a driving force in the same direction that enriches the humic matter in nitrogen relative to carbon. Since the C-O bond is easier to break than the different C-C and nitrogen bonds, bacteria tend to use the most efficient and energy saving path to obtain their necessary energy levels.

On the other hand the low molecular weight materials exuded by bacterial do not contain much if any nitrogen. This material is highly reactive in the soil zone. However current debate suggests that the lower molecular weight material could be a major source of clay linkages along with or instead of the humic matter as previously proposed although most past studies indicate a strong relation between humic and clay particles (Cheshire et al., 2000; Cornejo and Hermosin, 1996 for example). Some information suggests that smaller molecule root exudates can affect clay organic interaction greatly (Oades and Waters, 1991). Puget et al. (2000) indicate that young (i.e. less humic) organic matter is responsible for aggregate formation and hence clay – organic interaction. Since these smaller molecules are more carbon – oxygen rich, and hence more easily reduced by bacteria, their longevity should be shorter but is a subject of much conjecture and research effort.

The activity of roots, by the products of exudation is extremely important to the functioning of the A horizon. Root exudates are the fundamental mechanism of transfer of mineral nutrients from mineral to plants. Roots exude complex compounds, such as polysaccharides as well as simple hydrogen ions (see Gobat et al., 2003 for example). Hydrogen for potassium ion exchange would be the most simple means of transfer of mineral to plant. The end result is the acidification of the soil itself. However, many soils are near neutral in pH, especially grasses and prairie soils. Thus although the intense root activity of such plants emits much acid, the overall balance of resident and evolving organic matter and other substances maintains a near neutral soil. However conifer forest soils are most often in the decidedly acid range, ranging from pH of 5 to 4. Soil acidity is than a multifunction phenomenon. In general acid soils are associated with higher dissolved alumina content which is toxic to grasses and other prairie plants. One finds that organic rich soil are often associated with drainage problems or more frequently with cold and wet climates. These soils contain much alumina in solution. The acidity and organic content can be attributed at the same time to climate and the plants which grow in the soils. The lower the average temperature, the slower the development of the soil and hence the evolution of the organic matter present.

2.3.5.1 Development of the A Horizon

Organic Matter

The evolution of the deposed organic matter and internal root material in the A horizon accumulates and evolves to a steady state rather rapidly under conditions of temperate climate. The data from poldered prairie soils (Fig. 2.10) shows a typical "S" shaped curve due to biological action. The thickness of the organic horizon increases until it reaches a more or less stable thickness after about 800 years. Here the initial materials are soil type clays deposited in salt marshes which are grazed as prairies in polder episodes from the year 1,058 until the late 19th century. The deposition and humification process create the humic or organic – rich layer typical of grasslands. According to the schema presented by Birkeland (1999, p. 225) such a development is relatively rapid compared to other types of soils developed under different climate regimes and hence biomes. The time necessary to develop a "steady state" in an alteration profile would be on the order of magnitude of 10^2–10^4 times to reach alteration equilibrium of the silicates compared to the development of the organic A horizon. This is to be expected in that the biological processes concerned are much more reactive than those typical of silicate minerals.

Organic matter production by plants and its residence in soils are not necessarily related. A major factor is the biologic activity which tends to mineralize this material and release it in the form of gases to the atmosphere; CO_2, nitrogen compounds and water. Jenny (1994, p. 149) indicates an exponential relation as a function of temperature for the amount of organic matter in soils in the A horizons across the Central United States where the mineral substrate is very similar, loess materials and where primary production of organic matter by the plant regimes is similar. Cooler climates promote the conservation of organic matter in both prairie and deciduous forest soils. He gives an empirical rule that for each decrease in 10°C in average temperature the conserved organic matter increases by 2–3 times. The conservation of organic matter is more critical in the tropics where very little if any humic material is present despite a very high primary production by the plants.

Typically, the most organic – rich soils are found in cold climates, tundra and others, while the most organic poor soils are found in the tropics. Bacterial action, driven by temperature is the dominant factor in the conservation of organic matter in the A horizon of alteration profiles.

The A horizon is the zone where organic matter is decomposed into different organic chemical components but it is also the zone where mineral elements are released to the mineral – organic interface. Notable are phosphorous and potassium. Potassium is only lightly held in the plant material, being rapidly released upon deposition. It is essentially re-cycled from plant accumulation into the soil solutions. Phosphorous is more difficult to follow in its specific interactions but it is clear that it too is re-cycled in the life cycle of the surface. Ammonium or nitrogen compounds are also released in a soluble form into soil solutions by bacterial action in the soil zone, and hence are also re-cycled. However the contribution of bacterial action in the soil zone associated with plant roots is of very great importance to its availability for plant growth.

Clays

Jenny (1994) gives data showing the relationship between clay content soils (horizons to 100 cm depth) as related to temperature. The relationship is the reverse to that of organic content. The higher the temperature, the more clay present in the alteration profiles formed on basic igneous rocks. In the samples studied, coming from temperate climate areas, the rainfall seems to be less important or directly applicable to a determination of the alteration of rock to clay. The soils studied are from non-glaciated areas and hence the profiles were probably forming actively for hundreds of thousands of years. Nevertheless the relations of temperature seem to be very important for the conservation of organic matter and production of clays, in opposite relationships. However Jenny (1994) reports the relationship between clay content and rainfall for A horizon samples from iso-thermal (average temperatures) situations in loess soils. The clay content is less clearly related to rainfall. Therefore one can say that both temperature and rainfall will be active functions of clay formation in alteration sequences but the temperature factor, fundamental to plant activity, is more dominant than the water – rock interaction factor (Fig. 2.17).

Clay formation is often difficult to follow in that the fine grained material is moved from the site of its initial formation to lower levels in the case of the upper horizons. Depending upon the type of plant regime, the local geomorphology (slope of the area where the profile is formed) soil clays can be moved easily. The clay content of the A horizon is variable depending upon the source material upon which the profile is formed. For example loess tends to have a strong component of clay removal for example due to its strong fine grained sand content. These effects give a high variability of the clay content depending upon local conditions. Typically the clay content is developed at a much slower rate than that of the organic matter, several orders of magnitude in most cases.

However, generally, clay content is lower in the A horizon than the B horizon due to the transport of the fine particles by aqueous solutions. The A horizon, that of plant – mineral exchange, is dominated by plant roots and accumulated organic matter. Clay content is variable but usually lower in the A horizon than that just

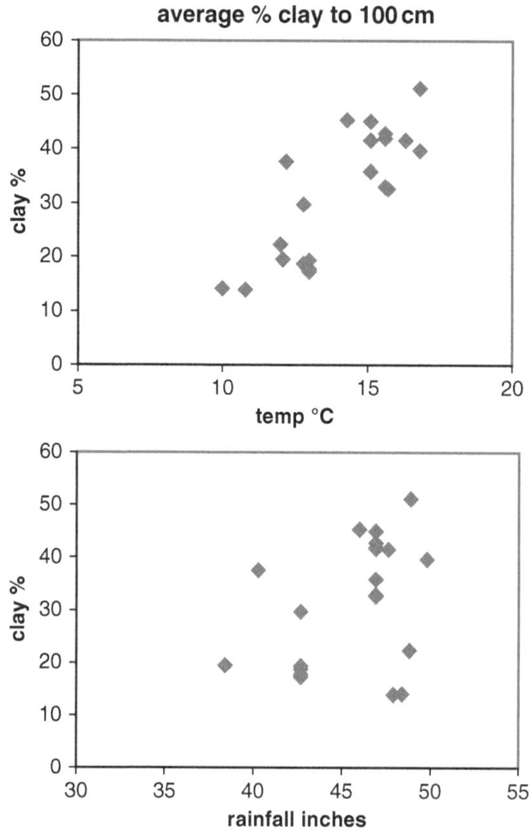

Fig. 2.17 Data from Jenny (1999) concerning clay content to 100 cm depth in soils from central United states as a function of climate variables. It appears that there is a better correlation between temperature than rainfall suggesting that plant activity, which is temperature driven, is more important than the rainfall or water – rock interaction parameter

below, where root activity is less. The chemistry of this surface zone is dominated by the influx of mineral elements coming from the degradation of plant materials and the acids developed by organic transformations or root exudates, either organic acids or hydrogen ions.

Since rain input tends to move elements either dissolved in solution or in suspension (clay particles) the overall tendency is to lose the chemical elements in solution which control the stability of clays and the availability of plants to recuperate such material. One means of retaining the nutrients present in the soils is to stabilize the clay content.

2.4 Soil Classifications: Descriptions and Factors Affecting Alteration Profiles

Our discussion here is centered upon an idealized soil and alteration structure or a surface, A horizon, where plant and biotic activity is dominant. Below this layer, which is generally a true layer, one can find a zone of clay accumulation, the B horizon, where soil solutions have moved newly formed clays downward. Below the B horizon is what is generally called the C horizon where rock and mineral debris are mixed with clays formed from water – rock interactions. This covers the saprock and saprolite zones. The boundaries of the B and C horizon are often irregular and the relations between altered material and the bedrock are most often irregular depending upon the passageways tht give access to altering soil solutions. The assemblage of or at times absence of one or the other zone is used in the soil classification or description system. It must be remembered that the classification system is used as a short hand for an alteration profile description. This saves a lot of time in that soil and alteration profiles are often complex and quite varied from one site to another. It is necessary to identify rapidly the salient features of these natural objects in order to deal efficiently with them.

2.4.1 Classification

The most widely referred to classification system is that of the United States Department of Agriculture Soil Survey Staff (1975). There are others, notably Russian, Chinese and European but in general most often one tries to refer to the categories determined in the USDA system. The concept is based upon essentially a three tier system of soil Order, Suborders and Great Groups. Further details observed in the field can lead to qualifying terms which further modify the nomenclature. In usage in the United States, one often has a type soil defined for an area, with a specific name which embodies the descriptive nomenclature in a single word. The discussion below is based upon the work in Birkland (1999), Jenny (1994), Akin (1991) and Duchaufour (1997).

In the USDA system there are 11 Orders, 48 Subgroups and 230 Great Groups. Usually a description name is based upon a combination of subgroup and Great Group terminology. It is clear that such a complex system of nomenclature reflects the extreme complexity of the objects described. Soil formation and the resulting alteration profiles depends upon many factors. The major factors are

Climate
Vegetation
Bedrock
Duration of the alteration process
Superposition of materials: wind blown, landslide or slump movements, stream carried, etc.

Objects in a five dimensional system are difficult to describe succinctly.

The Orders can be classified in general as being dependent upon:

Climate	*Aridisol*	desert climate
Vegetation	*Histosol*	organic rich accumulations (bogs, swamps, wet and cold climate vegetation)
	Mollisol	prairie
	Alfisol	deciduous forest
	Spodosol	conifer forest
Mineralogy	*Vertisol*	smectite – rich soil (expanding and contraction properties very important)
	Andosol	amorphous minerals dominant wher the bedrock is highly vitreous (volcanic rocks)
Duration of alteration	Entisol	little alteration visible
	Inceptisol	rapidly changing soil, hence young
	Ultisol	old, evolved soil
	Oxisol	strongly oxidized and hence old

The Suborder classification of each Order describe various general properties of the alteration profiles and the Great Group designations further define certain properties often more detailed such as color, humic content, and so forth. Terms concerning source rock type can be found such as sammitic (sandy), rendzina (calcareous rock) or andic (based upon glassy rocks). Rather rarely one finds specific reference to vegetation type although a certain number are generally assumed to represent plant regimes, usually under temperate climatic conditions as seen in the Order classifications above.

The general conditions of alteration are determined by climate which is rainfall and temperature. The frequency of rainfall can determine very different climates which combined with temperature conditions the type of vegetation growing on the alteration profile. One can find strikingly different climates within short geographical distances by climbing a mountain range. These changes in climate are roughly the same as one finds in going from the lowest part of the area under consideration and the change to tundra or polar climates as outlined in Sect. 1.2. However there are also significant changes in the structure of a soil and alteration profile as a function of time. Birkeland (1999, p. 47) outlines the changes in nomenclature and hence physical – chemical properties of soils as a function of time. One moves down the table below as a function of time:

Entisol
Inceptisol
vegetation – dominated soils types (Millisol, Spodosol, Alfisol)
Ultisol
Oxisol

In general the initial stages show a small but increasing soil (A) horizon as plants establish themselves but in many cases the organic horizon is confounded with the other parts of the profile (C). The formation of an A, B, C structure where soil is distinct from the underlying material indicates a certain maturity (Inceptisol). Overall, the thickness of the A horizon reaches a stable thickness with the B horizon increasing and then the C horizon can dominate is conditions of tropical or Ultisol weathering. Generally speaking the overall alteration profile loses alkali then alkaline earth elements during the formation of clay minerals, especially of the 2:1 type. Movement of the new minerals is the dominant factor in re-distribution of elements in the profiles. As alteration becomes more intense, as a function of time or climate (increased rainfall) there is a loss of silica to produce clays of lower silica content such as kaolinite, and in the extreme oxy-hydroxides are present such as gibbsite, aluminium hydroxide, and iron oxide minerals in different hydration states. The overall tendency is a convergence on the elemental concentration of Al and Fe. These elements hold no interest for plant activity.

This is the classical view of chemical alteration and clay mineral formation during alteration and soil formation.

For each stage of development in the alteration sequence, when there are differences in starting material (substrate), climate and vegetal cover one finds differences in color, humic layer content and grain size distribution in the different parts of a profile. These differences have a terminology used to describe the visible changes (color or texture) or the differences in grain size distribution. Thus the terminology is a mixture of genetic factors, climatic factors and physical appearance. All is necessary to describe and classify the great variation o soils which the product of many variables in natural settings.

2.4.2 *Results of Soil Forming Processes on the Structure of an Alteration Profile*

The major processes of soil formation, plant – silicate interaction and water – rock interaction have a strong tendency to stratify the profile into zones where one or another of the formation processes is dominant. This gives the alteration profiles their "horizon" aspect. Such displacements and concentrations are due to water movement in the profiles.

2.4.2.1 Downward Movement of Clays and Dissolved Material

The interaction of water and rock or silicate materials is essentially one of dissolution in the first stages of contact where rainwater is very under saturated with possible dissolved elements found in silicate minerals. One of the dominant facts of the initial weathering in saprock and eventually saprolite zones is a loss of density (Velde and Meunier, 2008, Chap. 4) which indicates loss of solids from the rock as they are dissolved in the percolating solutions. This is evident from the observations at the base of the alteration profile but one can expect some of the same action to occur at the surface where the initial contacts are made, in the A horizon.

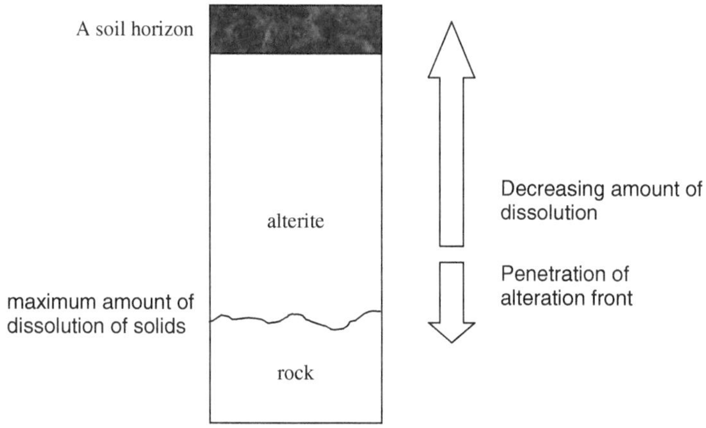

Fig. 2.18 Representation of an alteration profile indicating the relative amounts of material dissolved from the silicates as a function of position (*vertical*) in the profile. The alterite zone, interface with alterite (clay) rich and un-altered rock is the source of most of the dissolved material due to mineral transformations. The alteration front (saprock zone) descends in the profile as a function of time

Thus the overall amount of material will decrease throughout the alteration profile. However, since the clay minerals developed by the alteration process in the saprock zone should be more stable in the surface environment, the amount of dissolution can be expected to be less than at the rock alterite interface. Therefore one can say that the effect of dissolution will be less important as one goes up in an alteration profile (Fig. 2.18).

However, it is very well known that the new clays are liable to be transported in the pore fluids as they descend in the profile. The deposition of these clays forms the B horizon. Thus although the amount of dissolved materials decreases towards the surface of the alteration profile, a significant amount of solids can be transported downward (Fig. 2.19). Soils developed in the Great valley of California show and increase in clay content below the surface horizon at depths of 50–75 cm as a function of the age of the profile (Foth, 1990, p. 16) shown in Fig. 2.20. This is the developing B horizon. The presence of a developed B horizon enters into the definition and nomenclature of soils.

Under certain chemical regimes in the A horizon induced by the chemistry of the decayed organic matter in the form of litter, a strong chemical dissolution can occur just below this horizon. This is the E horizon developed under acid humic conditions in the A horizon. Essentially every thing is dissolved except the chemically inert quartz grains, leaving a sandy layer in the soil alteration profile. Curiously the clay content of the A horizon is typically greater than that of the E horizon (see Birkeland, 1984, Fig. 10–4) for typical clay content profiles). Since the dissolution of clays is responsible for the E horizon's lack of this material, one wonders why the clays persist in the zone where the acid chemistry is produced, the A horizon. We will treat this subject later in Sect. 5.2. The overall result is the production of another

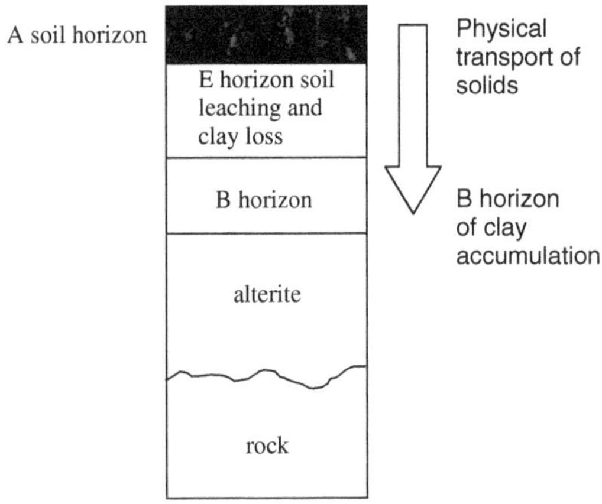

Fig. 2.19 Schematic diagram of the physical movement of clays from surface A horizons to the new, clay – rich B horizon. Under acid conditions one often finds a clay depleted zone, where sand is predominant; This is the E horizon due to chemical leaching and clay transport

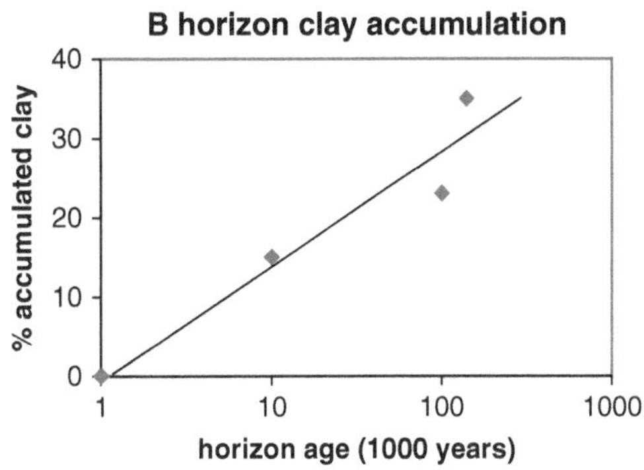

Fig. 2.20 Illustration of the concentration of clay minerals through the transportation due to pore water circulation in the soil profile. Data from Foth (1990) on clay accumulation in soils of the Great Valley CA USA.at depths near 50 cm

specific soil horizon, the E horizon which is due to chemical dissolution (Fig. 2.15) Again the presence of a sandy layer below the B horizon leads to descriptive terms and classification in soil terminology.

The movement of dissolved material or solids carried by the vector of downward moving solutions in the alteration of rocks gives rise to different zones or horizons which have specific chemical (notably pH) and mineralogical – textural features. The initial movement of clays enriches the B horizons with fine materials while the dissolution of clays and oxides from the E horizon gives rise to a more coarse grained horizon.

Data from Birkeland et al. (2003) indicate that the development of the B horizon is gradual being barely noticeable in Colorado piedmont soils near Denver CO after 10,000 years of formation and quite apparent after 140,000 years of soil – plant activity.

Some variations can be found, according to the type of plants present where a soil is forming. Graham and Wood (1991) indicate that the activity of earthworms can significant enrich the A horzon in clay sized particles compared to those under pine. This is a counteraction to the normal downward movement of the clay fraction.

2.4.2.2 Upward Movement in Alteration Profiles

Carbonates

If water in pore solutions brings clays downward to follow fluid flow, the upward movement of dissolved matter during soil moisture evaporation and capillary movement upward can influence the mineralogy of the profiles to a large extent in arid climates. Two instances are considered here, those of semi-arid steppe or grasslands and more arid evaporitic soils. In the compilation given by Jenny (1994, p. 122) one sees that there is a zone of carbonate accumulation, probably due to capillary transport upon soil drying, that is closely related to rainfall (climate) for loess soil in the Midwestern United States. Under conditions of low rainfall (25 cm/year) carbonates can be found in the soil portion of the profiles, i.e. at 25 cm or so. As annual rainfall increases to 100 cm the depth of carbonate accumulation increases to around 175 cm. The mechanism of carbonate accumulation is not all that well detailed but it can be described in a general way as shown in Fig. 2.21. In general the soil zone, through the activity of plant root exudates and microbial action on plant debris, create a slightly to decidedly acidic environment. The pH at the surface is typically lower than at depth in soil profiles. Jenny (1994, p. 134) demonstrates that the

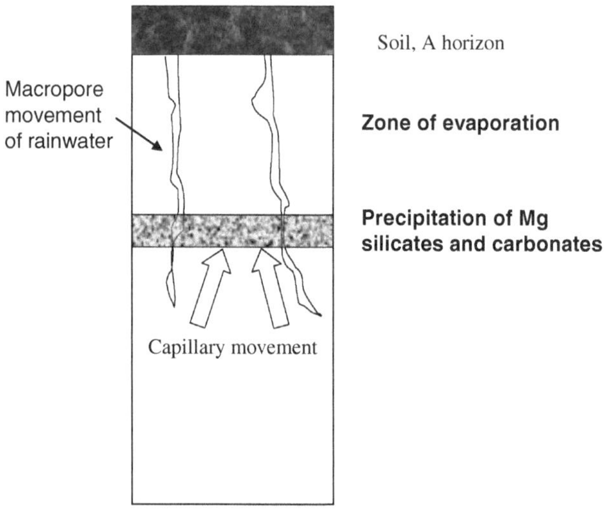

Fig. 2.21 Schematic representation of the effect of strong evaporation from a soil surface. Macropore movement of rain water moves dissolved material deep in the profile. Subsequent surface evaporation brings capillary water towards the surface but at a certain, variable depth in the profile the capillary water and minor pore water becomes strongly concentrated in dissolved material which precipitates to form new phase in the C horizon, usually of high Mg and Ca concentrations forming carbonates and magnesian silicate clay minerals, sepiolite and more often palygorskite

surface pH of soils on loess in the Midwest region of the United States follows rainfall, pH is highest at lowest rainfall and lowest as rainfall increases. Then one can deduce that the relation of the depth of carbonate concentration is related to rainfall, but more directly to the plant activity which is directly dependent on rainfall. The more plant material, the more acidic the soil, and the thicker the soil (organic zone) will become. Here it is evident that the relations of movement of dissolved mineral material (Ca ions) is largely governed by plant activity.

Magnesium Silicates

A more striking case of elemental transport is that found in soils of desert or arid conditions. These soils form magnesium – rich silicates at a horizon well below the surface. These are often called caleche or, when significant carbonate is present, calcrete soils. A summary of their occurrence is made in Velde (1985). In these soils the organic horizon is small and thus not associated with the movement which controls the chemical factors associated with the deposition of the new minerals. The new clays occur below the B horizon of clay transport, and are well within the C horizon (Vanden Heuvel, 1966). The materials which enter into the active system are those of dissolved elements in aqueous solution and the clay minerals in the alterite horizons. Most often the clay sized material in these horizons is smectite and kaolinite – rich. Therefore the clays present contain a significant amount of alumina as well as silica. The presence of a strongly expressed carbonate horizon indicates that there is a strong effect of evaporation, concentrating the dissolved elements in the soil solutions. Along with the carbonates one finds magnesium silicates, which are not stable in the soil, organic, horizon (Millot, 1964). The dominant mineral is palygorskite, the most aluminous of the magnesian silicate minerals. Here on can suspect that there is a dissolution – precipitation process which is necessary to convert the phyllosilicates into the chain silicate palygorskite structure (see Sect. 4.1.6 for further details). The important point here is that the movement of dissolved materials can be stopped, by capillary action of soil solutions which move upward and are concentrated in the upper portion of the humid zone of the soils which is well below the soil zone, in the C horizon (Fig. 2.21). Such a zonation is again produced by movement of dissolved materials in the alteration process.

2.4.3 Summary

The development of soils and alteration profiles is one of displacement of material. The salient characteristics of soil – forming and alteration processes is the re-organisation of the materials present at the surface which have or are coming into chemical equilibrium with the chemistry of the atmosphere, i.e. slightly acid, unsaturated rainwater and the products of destruction and mineralization of organic matter. The two agents of change are those of rain water and the by-products of organic activity. Whether or not plants and microbial agents alter directly the silicates present or whether the rain water comes into contact with the unaltered and unstable rock minerals, the transformations are toward hydrated phases, either silicates or hydroxyl-oxides. The action of this alteration is one of homogenization of the mineralogy in zones roughly parallel to the surface of the system, the air – solid interface. Of course irregularities in rock competence (fractures, changes in mineralogy, etc) will create non-parallel alteration zones but overall the changes are in zones which can be described and used to identify the processes and origin of the different materials and the altering agents. The pedologic nomenclature is in fact a method used to describe the chemical forces acting on a given type of material over different periods of time. The natural system has several variables and hence a relatively large number of descriptive terms are necessary to adequately describe a given result of these factors.

References

Adamo P and Violante P (2000) Weathering of rocks and neogenesis of minerals associated with lichen activity. Appl Clay Sci 16: 229–256.

Akin W (1991) Global Patterns Climate, Vegetation and Soils, Univ Oklahoma Press, 370 pp.

April R and Keller D (1990) Mineralogy of the rhizosphere in forest soils of the eastern United States. Biogeochemistry 9: 1–18.

Arocena J, Glowa KR, Massicotte HB and Lavkulich L (1999) Chemical and mineral compositions of ectomycorrhizosphere soils of subalpine fir (Abies lasocarpa (Hook.) Nutt.) in the Ae horizon of a Luvosol. Can J Soil Sci 79: 25–35.

Arocena J, Zhu L and Hall K (2003) Mineral accumulation induced by biological activity on granitic rocks in Qinghai Plateau, China. Earth Surf Process Landforms 28: 1429–1437.

Augusto L, Turpault M-P and Ranger J (2000) Impact of forest tree species on feldspar weathering rates. Geoderma 96: 215–237.

Badr MA, Shafei AM and Sharaf El-Deen SH (2006) The dissolution of K and P – bearing minerals by silicate dissolving bacteria and their effect on sorghum growth. J Agr Bio Sci 2: 5–11.

Balogh-Brunstad Z, Keller C, Gill R, Bormann B and Li C (2008) The effect of bacteria and fungi on chemical weathering and chemical denudation fluxes in pine growth experiments. Biogeochemistry 88: 153–167.

Banfield JF and Nealson KH (1997) Geomicrobiology: Interaction Between Microbes and Minerals Reviews in Mineralogy, Vol. 35, Min Soc Ame, p. 448.

Barkaudah Y and Henderson J (2006) Plant ashes from Syria and the manufacture of ancient glass: Ethnographic and scientific aspects. Jour Glass Studies 48: 297–321.

Barker W, Welch S and Banfield J (1997) Biogeochemical weathering of silicate minerals, 392–428. In JF Banfield and KH Nealson (eds.), Geomicrobiology: Interactions Between Microbes and Minerals Reviews in Mineralogy, Vol. 35, Miner Soc Ame, p. 448.

Barker WW, Welch SA, Chu S and Banfield J (1998) Experimental observations on the effects of bacteria on aluminosilicate weathering. Ame Mineral 83: 1551–1563.

Berthelin J (1983) Microbial weathering processes, 223–262. In WE Krumbein (ed.) Microbial Geochemistry Blackwell Scientific p. 330.

Birkeland P, Shroba R, Burns S, Price P and Tonkin P (2003) Integrating soils and geomorphology in mountains – an example from the Front Range of Colorado. Geoporphology 55: 329–344.

Birkeland P (1999) Soils and Geomorphology, Oxford Univ Press, 429 pp.

Blake RC, Shute EA and Howard GT (1994) Solubilisation of minerals by bacteria: Electrophoretic mobility of *thibacillus ferooxidans* in the presence of iron, pyrite and sulphur. Appl Environ Microbiol 60: 3349–3357.

Bormann FH and Likens GE (1994) Pattern and Process in a Forested Ecosystem, Springer, p. 253.

Calvaruso C, Turpault M-P and Frey-Klett P (2006) Root-associated bacteria contribute to mineral weathering and to nutrition in trees: A budgeting analysis. Appl Environ Microbiol 72: 1258–1266.

Carter D and Arocena J (2000) Soil formation under two moss species in sandy materials of central British Columbia (Canada). Geoderma 98: 157–176.

Castaldini M, Mirabella A, Sartori G, Fabiani A, Santomassimo F and Miclaus N (2002) Soil development and microbial community along an altitude transect in Trentino mountains, 217–228. In A Violante, PM Huang, JM Bolling and L Gianfreda (eds.), Develoments in Soil Science Vol. 28B, Elsevier.

Chen J, Blume HP and Beyer L (2000) Weathering of rocks induced by lichen colonization – a review. Catena 39: 124–146.

Cheshire M, Dumat C, Fraser A, Hillier S and Staunton S (2000) The interaction between soil organic matter and soil clay minerals by selective removal and controlled addition of organic matter. Eur Jour Soil Sci 51: 497–509.

Clark, S (ed.) (1966) Handbook of Physical Constants, Vol. 97, Geol Soc Ame Memoir, 587pp.

Cochran M and Berner R (1996) Promotion of chemical weathering by higher plants: Field observations on Hawaiian basalts. Chem Geol 132: 71–77.

Coleman SM (1986) Levels of time information in weathering measurements with examples from weathering rinds on volcanic clasts in the Western United States, 379–393. In SM Coleman and DP Dethier (eds.), Rates of Chemical Weathering of Rocks and Minerals, Academic Press.

Cornejo J and Hermosin MC (1996) Interaction of humic substances and soil clays, 595–619. In A Piccolo (ed.), Humci Substances in Terrestrial Ecosystems, Elsevier 720 pp.

Denbigh K (1987) The Principles of Chemical Equilibrium, Cambridge Univ Press, p. 494.

Dethier D (1986) Weathering rates and the chemical flux from catchments in the Pacific Northwest, USA, 503–529. In SM Coleman and DP Dethier (eds.), Rates of Chemical Weathering of Rocks and Minerals, Academic Press.

Duchaufour Ph (1997) Abrégé de Pédologie, Masson, p. 291.

Escudero A, Martinez I, de la Cruz A, Otalora M and Maestre F (2007) Soil lichens have species – specific effects on the seeding emergence of three gypso-phile plant species. J Arid Environ 70: 18–28.

Evans L and Cameron B (1979) A chronosequence of soils developed from granitic morainal material, Baffin Island. Can J Soil Sci 59: 203–210.

Foth H (1990) Fundamentals of Soil Science, Wiley, p. 360.

Garrels R and Christ G (1965) Solutions, Minerals and Equilibria, Harper and Row, 450 pp.

Glowa K, Arocena J and Massicotte H (2003) Extractionof potassium and/or magnesium from selected soil minerals by *Piloderma*. Geomicrobiol J 20: 1–13.

Glowa K, Arocena L and Massicotte H (2004) Properties of soils influenced by ectomycorrhizal fungi in hydrid spruce [Picea glauca x engelmannii (Moench.)]. Can J Soil Sci 84: 91–102.

Gobat J-M, Aragno M and Matthey X (2003) Le Sol Vivant, Presses Polytechniques et Universitaires Romandes, 568 pp.

Graham R and Wood H (1991) Morphological development and clay redistribution in lysimeter soils under chaparral and pine. Soil Sci Soc Ame J 55: 1638–1646.

Hutchens E, Clipson N and McDermott F (2008) Mineralogical influence on the structure and diversity of bacterial communities associated with silicate minerals. Geophys Res Abstracts EGU2008 10: A 3930.

Jenny H (1994) Factors of Soil Formation, Dover, 280 pp.

Johnson N and Gehring C (2007) Mycorrhizas: Symbiotic mediators of rhizosphere and ecosystem processes, 73–98. In Z Cardon and J Whitbeck (eds.), The Rhizosphere an Ecological Perspective, Elsevier, 203 pp.

Kawano M and Tomita K (2002) Microbiotic formation of silicate minerals in the weathering environment of a pyroclastic deposit. Clays Clay Miner 50: 99–110.

Khademi H and Arocena L (2008) Kaolinite formation from palygorskite and sepiolite in rhizosphere soils. Clays Clay Miner 56: 429–436.
Kostka JE, Haefele E, Viehweger R and Stucki J (1999) Respiration and dissolution of iron (III)-containing clay minerals by bacteria. Environ Sci Technol 33: 3127–3133.
Li Y-H and Schoonmaker J (2004) Chemical composition and mineralogy of marine sediments, Chap. 7.01, 1–35. In F Mackenzie (ed.), Treatise of Geochemistry, Vol. 7, Elsevier 567 pp.
Liu W, Xu X, Wu X, Yang Q, Luo Y and Christie P (2006) Decomposition of silicate minerals by *Bacillus mucilaginosus* I liquid culture. Environ Geochem Health 28: 133–140.
Lowe D (1986) Controls on the rates of weathering and clay mineral genesis in airfall tephras: A review and New Zealand case study, 265–319. In SM Coleman DP Dethier (eds.), Rates of Checmial Weathering of Rocks and Minerals, Academic Press.
Mahaney W (1986) Rates of mineral weathering in the Wind River Mountains, western Wyoming, 147–167. In SM Coleman and DP Dethier (eds.), Rates of Chemical Weathering of Rocks and Minerals, Academic Press.
Mariotti A (1982) Apports de la Géochimie Isotopique a la Connaissance du Cycle de l'Azote, Univ P et M Curie Paris, p. 488.
Millot G (1964) Géologie des argiles, Masson and Cie, Paris, p. 400.
Newman DK (2001) How bacteria respire minerals. Science 292: 1312–1313.
Oades J and Waters A (1991) Aggregate heirachry in soils. Aust J Soil Sci 29: 815–828.
Odum E (1971) Fundamentals of Ecology, Saunders, 574 pp.
Owens L and Watson J (1979) Rates of weathering and soil formation on granite in Rhodesia. Soil Sci Soc Ame J 43: 160–167.
Pedro G (1966) Essai sur la caractérisation géochimique des différents processus zonaux résultant de l'altération des roches superficielles. Compt Rendus Acad Sci Paris Ser D 262: 1828–1831.
Proust D and Velde B (1978) Beidellite crystallisation from plagioclase and amphibole precursors: Local and long range equilibrium during weathering. Clay Miner 13: 199–209.
Puget P, Chenu C, Balesdent J (2000) Dynamics of soil organic matter associated with particle size fractions of water – stale aggregates. Eur Jour Soil Sci 51: 595–605.
Raulund-Rasmussen K, Borggaard O and Hansen H (1998) Effect of natural organic soil solutes on weathering rates of soil minerals. Euro J Soil Sci 49: 397–406.
Ruhe R (1984) Soil – climate system across the prairies in Midwestern USA. Geoderma 34: 201–219.
Sak P, Fischer D, Gardner T, Murphy K and Brantley S (2004) Rates of weathering rind formation on Costa Rican basalt. Geochim Cosmochim Acta 68: 1453–1472.
Scheinost AC and Schwertmann U (1999) Color identification of iron oxides and hydroxysulfates. Soil Sci Soc Ame J 63: 1463–1471.
Schenk H and Jackson R (2002) The global biogeography of roots. Ecol Monographs 72: 311–328.
Sheng X (2005) Growth promotion and increased potassium uptake of cotton and rape by a potassium releasing strain of *Bacillus edaphicus*. Soil Biol Biogeochem 37: 1918–1922.
Smits M, Hoffland E, Jongmans A and van Breeman N (2005) Contribution of mineral tunnelling to total feldspar weathering. Geoderma 125: 59–69.
Soil Survey Staff (1975) Soil Taxonomy US Dept Agriculture Handbook no 436, 754 pp.
Taylor L, Leake J, Quirk J, Hardy K, Banwart S and Beerling D (2009) Biological weathering and long-term carbon cycle: Integrating mycorrhizal evolution and function into current paradigm. Geobiology 7: 171–191.
Ullman W, Kirchman D, Welch S and Vandevivere P (1997) Laboratory evidence for microbially mediated silicate mineral dissolution. Chem Geol 132:11–17.
Ullman WJ and Welch SA (2002) Organic ligands and feldspar dissolution, 3–35. In R Hellman and SA Wood (eds.), Water – Rock Interaction, Ore Deposits and Environmental Geochemistry, Vol. 7, Geochemical Society Special Publication, 672pp.
Valeton I (1972) Bauxites: Developments in Soil Science, Vol. 1, Elsevier, 226 pp.
Vanden Heuvel RC (1966) The occurrence of sepiolite and attapulgite in eh calcareous zone of a soil near Las Cruces N Mex. Clays Clay Miner 13: 193–207.

Velde B (1985) Clay Minerals: A Physico-Chemical Explanation of Their Occurrence, Elsevier, 427 pp.

Velde B, Goffé B and Hoellard A (2003) Evolution of clay minerals in a chronosequence of poldered sediments under the influence of a natural pasture development. Clays Clay Miner 51: 206–218.

Velde B and Meunier A (2008) The Origin of Clay Minerals in soils and Weathered Rocks, Springer, 406 pp.

Walker L and del Moral R (2003) Primary Succession and Ecosystem Rehabilitation, Cambridge Univ Press, 456 pp.

Welch SA, Barker WW and Banfield J (1999) Microbial extracellular polysacchrides and plagioclase dissolution. Geochim Cosmochim Acta 63: 1405–1419.

Welch SA and Ullman WJ (1999) The effect of microbial glucose metabolism on bytownite feldspar dissolution rtes between 5°C. and 35°C. Ame Miner 63: 3247–3259.

Wellman C (2003) Dating the origin of land plants Chapter seven. In P Donoghue and M Smith (eds.), Telling the Evolutionary Time: Molecular Clocks and the Fossil Record, Taylor and Francis, London, 288 pp.

Wierzchos J and Ascaso C (1996) Moprhological and chemical features of bioweathered granitic biotite induced by lichen activity. Clays Clay Miner 44: 652–657.

Wu L, Jacobson AD and Hausner M (2008) Characterisation of elemental release during microbe – granite interactions at T = 28°C. Geochim Cosmochim Acta 72: 1076–1095.

Chapter 3
Physical Factors Affecting Soil Profiles: The Three Dimensional Aspect of Plant – Mineral Interaction: Displacements

Contents

There is an age old maxim: what goes up must come down. It is no less true for mountains and high land masses. As tectonics attempt to shorten the crustal distances making mountains arise, the action of rain, wind and gravitation are designed to put the mountains back in their place. The natural atmospheric forces tend to level the surface of the earth which results in a filling of the oceans, at least at the borders of continents. These actions pass by a destruction of the structure of the rocks making up mountains, either by physical means or mechanical ones so that the smaller pieces can be transported to a situation of lower gravitational potential.

Soils, plant – silicate interaction, are developed upon altered rock material. However, if plants are not present, the altered rock material does not stay in place but is swept to lower level aqueous deposits by streams and rivers into lakes and seas. Thus plants stabilize the material produced by water – rock interaction. Without plants, there is no soil development if one considers that the definition of a soil is the presence of an A horizon. Nevertheless, despite the fixing power of plant regimes and the formation of soils, superficial alteration materials find their way

B. Velde, P. Barré, *Soils, Plants and Clay Minerals*,
DOI 10.1007/978-3-642-03499-2_3, © Springer-Verlag Berlin Heidelberg 2010

eventually into aqueous driven deposition areas. Given this, it is useful to look at the different aspects of the displacement of alteration materials and their relationship to the forces which move them as well as the ways that plant regimes fix the surface materials. It is very easy to understand that a tree cannot develop well without a certain root – friendly soil zone where some moisture is kept and nutrients can be transferred outside of rainy periods. Most plants develop in soil (silicate alteration products and organic matter). This chapter is designed to help one understand the important points in these relations.

3.1 Movement of Clay and Soil Material

The soil surface is the source of much of the material attributed to erosion. One initially thinks of the physical displacement of clays and finer materials by surface run off. However there is a variety of displacements which can be classed as erosive. Landslides, slumping of soils on slopes and clay displacements within the soil profiles all contribute to the movement and re-making of surface soils. All of these processes contribute to a construction of a new material at the surface which plants colonize and re-model to their needs. Each time a landslide occurs, a new soil and alteration sequence is developed on the old surface which has lost its mantle of fine grained material. Also, new profiles and plant regimes are developed on the displaced material where it has settled for different periods of time. The surface of the earth is in continual motion, and plants are motors or brakes on this development.

In fact for a geologist, there is nothing more normal than erosion. Sedimentary rocks are largely the result of the physical transport of fine grained or other material from mountains or upland to deposits in large bodies of water. This is the way the largest part of sedimentary rocks form. Further, it is clear to all geologists that without erosion one would not be able to study the interior of the earth, at least the upper several kilometres. Mountains are created by compressive forces which push rock strata upward to be dismantled at various rates and by different means. This presents the depths of the earth to the inspection of geologists. Volcanoes are a slightly special case, being composed of molten rock which rises rapidly to the surface, but they are affected by the same levelling forces as are tectonically formed mountain chains. Inevitably, these high edifices are all brought to or near the level of the sea. Hence there is nothing more normal, in the long term, than erosion. Further, the transport and accumulation of fine grained material along rivers and sea coasts, clays and silt – sand materials, is the matrix for much of traditional farming. Most of the initial sites of man's agricultural activity have been on secondary deposits of soils transported by either water or wind action. Stream banks, flood plains, marsh areas and loess plains were sought by the first agricultural groups where soils were easily cultivated and water resources plentiful.

In a very general way one can view such displacements, due to the forces of gravity and accentuated or aided by rainfall, in the context of a hill side (Fig. 3.1). The high ground is usually relatively densely covered by vegetation, while the slope is less so. Along the slope one finds displacement of material, and hence erosion. At

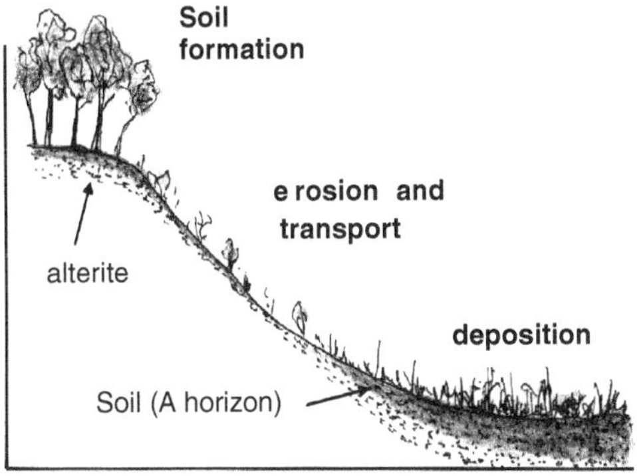

Fig. 3.1 Diagram indicating the different zones of surface alteration and accumulation on a typical semi – arid savanna type landscape. The alterite is formed on the stable uplands where plants are well situated to prevent displacement of surface, soil, material. Erosion, i.e. physical displacement of material, occurs on the slopes and accumulation of soil material occurs on the *bottom* land areas. The material present at the base of the slopes is a composite of old alterite and newly accumulated soil material coming from areas of higher altitude

the foot of the slope one finds a zone of deposition. There is in general a contrast in the type of vegetation which occurs in the different parts of the profile. The uplands show a developed zone of alterite and soil, The slope has a small extension of these zones while the foot of the slope shows a zone of accumulation of the fine grained material which is missing from the slope. The soil development is usually different, being thickest at the foot of the slope. Differences in plant regime and displacement of material lead to these differences. The end result is a strong inhomogeneity in the development of soils and plant regimes.

This being the case it is useful to understand the relations between the movement of materials at the surface and plant actions which affect them.

3.1.1 Movement of Clay Particles Within the Soil Zone: Internal Loss of Material

3.1.1.1 Translocation of Clays From A to B Horizons (Local Vertical Transfer)

The movement of clays developed in the A horizon by movement in pore water solutions has been outlined in Sect. 2.3. The physical movement of clays is effected by water movement where the small clay particles are held in suspension by Brownian motion through thermal agitation. For various reasons, the velocity of water movement lessens, often due to a greater density of the materials in lower parts of the alteration profile below the soil horizon. There are fewer pores and crack passage ways in deeper parts of alteration zones. The soil horizon is characterized by a relatively great amount of pore space through biologic action. Earthworms, moles and plant roots create open spaces in the altered silicate material which allow water to pass through and be replaced by air in drier periods which is necessary for healthy plant growth for non-aquatic species. These spaces are the preferential passages for pore water and the vectors of clay transport. As one moves downward in the alteration profile these spaces are less frequent, pore water movement slows and clays are deposited. These materials form what pedologists call cutanes, zones of oriented clay particles which fill pore space (Fig. 3.2). The accumulation of clays forms the B horizon, where clay accumulation is the primary and diagnostic characteristic. Loessic deposits fine grained deposits due to wind transport (see 3.1.3) show a typical accumulation of clays below the A horizon forming the new, clay rich B horizon as a function of time (Ruhe, 1984). The time factor for this accumulation is not negligible, tens of thousands of years. This time factor is in between the soil clay (bio-influenced) and water rock interaction time frames (Sect. 1.2). The case of clay

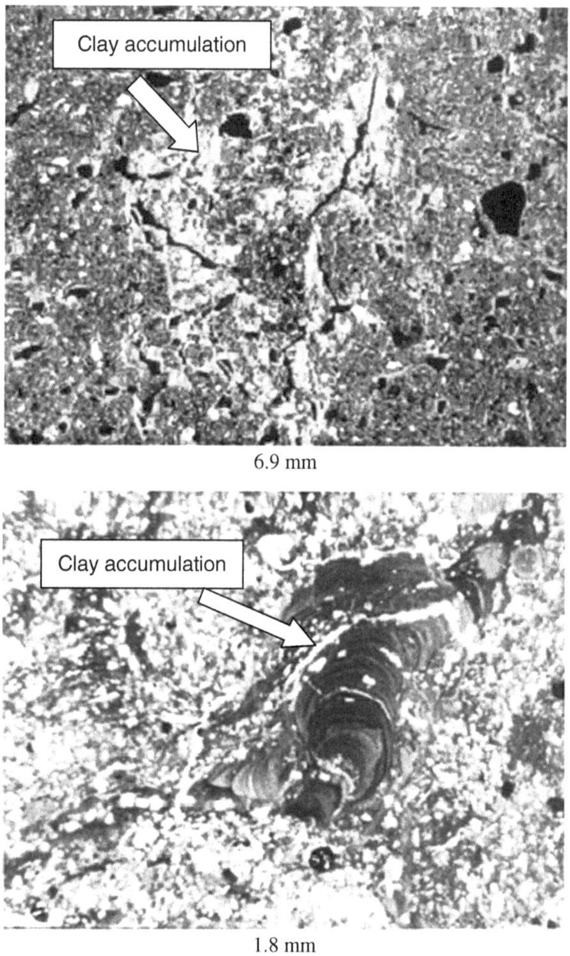

Fig. 3.2 Micro photographs of soils where clays have been transported within the alteration profile and deposited at a lower level. The accumulation of clays is called a cutane. This material follows pore passageways in the soil structure

migration in loess is probably one of the most rapid in that the basic altered material is fine grained where much of the material is quartz sand and not susceptible to chemical or physical change, hence many small passageways are immediately available for clay transport around these grains.

3.1.1.2 Translocation of Clays Within Profiles Along a Sequence of Soils: Down the Profile and Down Slope Transport of Clays Within Soils: Lateral Displacement

Movement within soil – alteration profiles is extremely important under specific circumstances. The local movement of clays down from the A organic horizon to that just below, the B horizon, is a very well recognized phenomenon. However the lateral movement of these materials within the alteration zone is less well recognized and reported. Down hill movement of clays in the subsurface can be an important factor in the redistribution and accumulation of matter in alteration zone. This concept enters into the ideas of soil Catenas (see Birkeland, 1984, p. 235). Lateral movement of clays by gravity and fluid movement within the alterite can concentrate clays in toe slope areas while the slopes themselves are more clay – poor and the uplands above the slope much more normal concerning the vertical movement of clays into the B horizon. It is clear that there is often lateral movement of the clays in the materials on the slope. The concentration of clays at the bottom of slopes gives rise to many different characteristics which affect the vegetation through differences in water retention, and texture. Basically, grasses can do much better in a clay – rich environment, more so than trees especially in arid or semi – arid climates. This is one reason why one finds trees on the uplands, shrubs on the slopes and grasses on the bottom lands in semi – arid areas. Frequently the slopes have thin soils and shrub vegetation. The accumulation of clay as a function of position in the soil catena in Mediterranean climates can be seen in the data given by Paquet (1977). Different effects occur in such landscapes, not the least of which is a change in clay mineralogy as a function of position in the topographic sequence. In general, illite and kaolinite are found in the uplands areas, kaolinite and oxides on the slopes and smectites dominate in the lower part of the topography. The change in clay mineralogy is most likely not a function of transportation but one of plant – soil interaction favoring different types of clays according to the vegetation present (see Chap. 5). One result of clay transportation within the soils in a lateral sense is that the clay content accumulated down slope tends to decrease drainage and hence to change the parameters which influence the type of vegetation present.

In more arid areas, the decrease in drainage favors the implantation of grasses while more drained soils, upland favors trees and slopes shrubs due to the unstable situation of the materials at the surface. In more temperate climates, one finds more trees in the bottoms (stream valleys) and grasses on the uplands. The vegetation sequence is reversed.

One very important, and very simple observation that can be made is that in almost any stream, small to large, during a period of constant rainfall heavy or light up to soil saturation, one finds clay sized particles in the water. Stream water tends to be brown or yellow during periods of constant rainfall, colored by clay particles. This is the case even in areas of pastures or grasslands where there is little slope to the landscape and where the vegetal cover is maximal and surface runoff is minimal. The reason one finds these clays in the streams is that they are transmitted through the soil profiles themselves into the water table and then into the streams. It appears then that a significant amount of clay material is evacuated from soils within the structure of the alteration profile itself. This clay material is essentially suspended in the aqueous medium by Brownian motion due to its small grain size and will hence be transported long distances before deposition. Often such material ends up in oceanic, salt water bodies where the electrolyte (NaCl) promotes flocculation and deposition along the shore lines or on Continental platforms.

An example of such transport of clays within the profile is given in Fig. 3.3 (data kindly supplied by B Zhang, Univ Nanjing, China). The example is found at the Red Soil Station of the Nanjing Academy of Science Soils Institute. The slope is less than 5 m from hill top, the upland forest site, to an accumulation on the edge of a rice paddy which is an old stream bed. Diagrams of clay content show an accumulation in the forest site at near 50 cm, typical of B horizon accumulation while in the slope site, there is no accumulation and in the toe slope position one finds a stronger accumulation near 1 m depth.

It is important then to consider that a non – insignificant portion of clay transportation is made within the soil and alterite profiles where the fin grained material is taken in subsurface transport into streams and thus into areas of deposition.

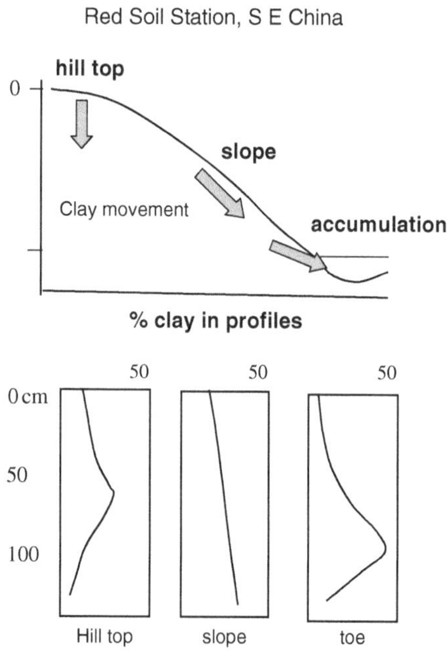

Fig. 3.3 Illustration of a situation of low topographical change where clays are seen to accumulate in the upland profile on the hill top under forest conditions, clay movement within the alteration profiles is indicated by a loss of the B horizon of accumulation on the slopes and a strong accumulation at the base of the slope. Data courtesy of B Zhang, Univ of Nanjiang Department of Agriculture and Director of the Chinese Academy of Sciences Red Soil Station in Southeastern China

3.1.2 Erosion of Fine Grained Materials From the Surface and Transportation by Water

A very large amount of work has been done on soil erosion, to say the least (see Toy et al., 2002 for example). This has been a subject of intense research for several decades and a known problem, especially in agricultural areas, for some time although classic texts on agricultural practice in the first half of last century did not seem to be preoccupied with the problem (Cox, 1925 for example). Severe climate change, relatively short term, in the 1930s brought about an awareness of the strong consequences of certain managerial practices on grasslands in the United States. A bit more recently, Xu (2005) has demonstrated the relation between forest cover (density) and rainfall on erosion of soils on the Chinese loess plateau. Even though this area is highly modified by human activity, the effect of plants, forests in this case, is primordial in the displacement and retention of soil materials due to erosion processes. Thus not only human intervention as agriculture but also more natural vegetation such as forests can have an important effect on the loss of surface material by erosion. Probably the greatest part of present day surface erosion is due to interaction of rainfall on cultivated land. However, under conditions of natural vegetation probably little surface erosion would occur in densely forested areas and humid grasslands. But in more arid areas of grasslands, shrub lands and sparse forests significantly higher amounts of erosion will occur. This was expressed by Langhein and Schumm (1958) in their observations on the non-linear relations between sediment yield (displacement of soils) and rainfall. The drier climate areas, those of sparse vegetation, the higher the sediment loads in streams compared to those areas with a greater vegetative cover.

It is difficult to say where the masses of sediments have come from which constitute the geological depositions in basins and the formation of sedimentary rocks found in materials of more recent geologic history, the Tertiary or Quaternary periods for example. Whatever the source, erosion of fine grained material certainly did occur, probably at a slower rate than in areas of intense farming today but it did take place in any event. There is no particular evidence that less deposition of sediments occurred at higher latitudes than those in more desertic parts of the globe. However climate has changed frequently and the further one goes into the geologic record the less information is available on this aspect of erosional history.

Given that the clay and silt fractions are the most generally affected by surface runoff, one would expect that they would be rapidly taken from their sources before being deposited. The smaller the particles are, the longer they stay in suspension in water and the further they will be transported. Typically near shore deposition in ocean basins is of fine grained, clay – rich material which eventually forms what are called shales by sedimentologists and geologists. The more the surface is covered by plants, especially grasses, the less surface erosion will occur. If the vegetation becomes sparce, leaving bare spaces open to rain fall influence, the greater will be the tendency to have surface run off and translocation of fine grained material. Numerous scientific journals are dedicated to this problem.

As mentioned above, surface transport is the most often studied and important aspects of erosion concerning agricultural effects. The action of raindrops falling on bare soil, which is especially important to studies concerning agricultural practices, is the initiation of the loss of soil fertility. The impact of a rain drop is such that it dislocates fine material making the way for further dislocation. As the matrix of a soil loses its fine materials, clays and silts, the larger grains are destabilized from their positions and become vulnerable to water movement (Fig. 3.4).

Once included in a moving rivulet of rain water, the clay and silt increase the density of the fluid and give more force to dislodge other material as it flows down slope. Of course erosion processes need a gradient to move material and the greater the difference in altitude, the greater the effect of the erosion. The phenomenon of surface erosion is one of increasing intensity as the system begins to increase in magnitude of the slope, rilling (surface movement) along plowed lines, becomes gullying (deeper erosion into the soils structure, often just to the bedrock) and eventually this leads to failure such as slumps and land slides.

Rain water surface erosion is greatest when the soils have no vegetative cover. In a natural state, i.e. without the intervention of agricultural man, arid areas of sparce vegetation are the most susceptible to erosion of this sort. However the normal agricultural practices leave the soil without vegetal cover for at least a part of the year in many climates and hence vulnerable to erosion. Even without such effects, plowed surfaces, the vegetal cover can be primordial in stabilizing soil material, as in the relations of forest cover and erosion on Chinese loess plains reported by Xu (2005). Basically, the more surface area covered by plants (leaves) the less erosion per surface area. This is simply due to the change in velocity of the rain drops as they encounter vegetal matter, diminishing the impact on soil grains.

Methods of countering or minimalizing soil erosion of agricultural surfaces are numerous, usually concerning questions of reducing pathways down slope (contour plowing) and stabilizing down slope movement by creating small retainers or vegetal zones along the contours (see Toy et al., 2002, Chap. 7). The most sure strategy overall is to maximize vegetal cover during periods of highest rainfall.

The distance of movement of materials originating from rainwater surface erosion is proportional to the size of the grains present. The clays tend to be carried the furthest while gravels will end up at the bottom of the local slope for the most part. Some of the material which finds its way into the rivers or streams will be moved

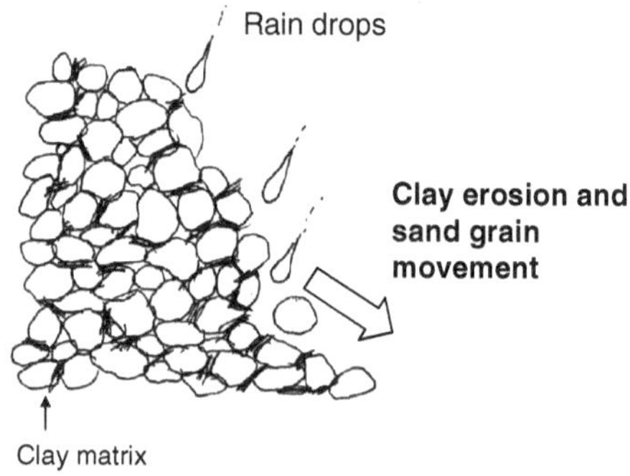

Fig. 3.4 Illustration of the effect of rain, in the form of raindrops, on a clay – sand soil without vegetal cover. The rain loosens the sand grains by displacing the clay particles, which enter into the run off water moving the material into a water transport mode

yet farther due to the increases force of the flowing water, i.e. velocity and cross section of the flowing water. Also, the higher the slope gradient, the more material will be moved and the further it will be moved from its source. Therefore one can expect more surface erosion by water in areas of low vegetal cover.

3.1.3 Surface Erosion by Wind Action and Consequent Loess Deposition

Wind action is similar to that of rainwater but it operates on much finer material. The capacity of displacement is a function of the viscosity of the moving fluids and water is then much more effective than air. However, wind erosion can be very important under certain climatic conditions. In the Continental United States one can divide the country roughly in half, east – west, the western plains being susceptible to wind erosion and soil loss while the eastern half is subject to water erosional loss (Toy et al., 2002, p. 9). Most of the arid areas north and south of the tropical rain belts are subject to wind erosion such as one finds in Australia, Africa, and China which are classic examples. These are the hot deserts (Obruchev, 1945). These are the areas of present day dust transport followed by satellite photography (see Prospero et al., 2002). In order for dust to be put into the atmosphere there must be a scarcity of plants at the present day surface and a relative abundance of fine grained material. Barren mountain tops do not produce much dust. However, the large and flat sedimentary deposits in arid areas formed by intermittent flooding and sediment transport are ideal producers of dust. Eventual sedimentation of this wind born material creates what is called loess. Displacement can be over relatively short distances, hundreds or over thousands of kilometres depending upon the energy of the events that put the fine grained material into the air. The pre-requisite for airborn transport is a lack of plant cover at the surface.

A general schema explaining the phenomenon is given in Fig. 3.5. Two pre-requisits are necessary, (1) a source of fine sediment which is eroded to given a water born deposition and (2) a low plant growth regime where wind erosion can occur because the surface deposits of clays are not anchored in a soil.

Several areas of today's land masses have strong imprints of past loess deposits. Most often one thinks of China where the process is very important, and even today 2 mm of dust is deposited annually in Xian, the ancient capital in western China. This is the equivalent of the A horizon of soils being renewed every 100 years. As far as one can tell the origin of the fine grained material is from sediments deposited in the great central desert areas. For the most part this loess is largely composed of illite and chlorite, typical low grade metamorphic materials (observations by the author for samples from west central China). These are the hot deserts (Obruchev, 1945).

A second type of loess is that of a cold origin where continental glacial outwash plains, periodically dried but free of vegetal cover, provided sources of fine grained material in abundance. These wind born materials accumulated over the last 2.4 million years or so (Catt, 1988). Similar material to that of the hot desert loess

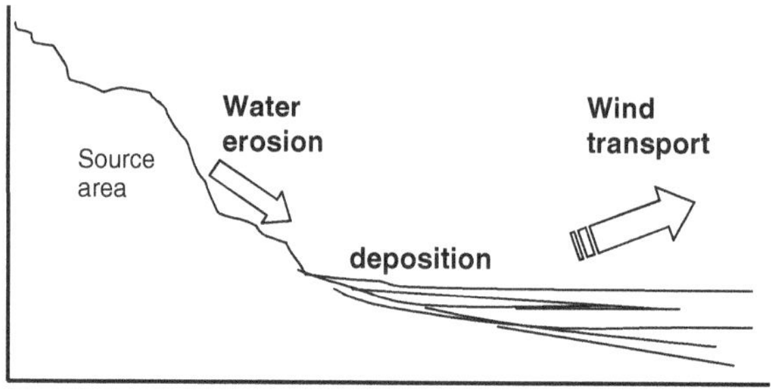

Fig. 3.5 Schematic representation of the origin of loess. Water erosion moves fine grained material onto depositional areas, usually very flat and of a significant surface area, where the sedimented material is moved again, this time by wind effects moving across the flat, dry sediments. The principle factor of less displacement is the depositon on a flat surface where plants cannot establishes foothold either because of climatic conditions, dryness, or because the arrival of material is too rapid and each season produces a new sediment before plants can anchor the fine grained material

of China is found in sediments of the Po river delta and glacial sediment found on the shelf of North Eastern United States (Velde and Church, 1999). These are sediments originating from the most recent glacial epochs where outwash of fine grained material coming from sedimentary and metamorphic rocks was eroded by wind and deposited further from the glacial front. The same minerals, illite and chlorite, are often dominant in till and other glacial deposits in Northern Europe (Gillot et al., 1999, Teveldal et al., 1990).

However the clays in loess deposits in the United States and Europe (Burras et al., 1996; Ransom et al., 1988; Kuzila and Lewis, 1993; Jamagne, 1973, for example) originating from continental glaciation (cold) phenomena indicate that such minerals can in fact be rare in the loess itself. The clays in European loess and those loess and tills of the continental United States are illite/smectite assemblages which are common in soils. Given that the loss of chlorite and the formation of illite/smectite mixed layer minerals can be rapid, on the order of tens of years under mild climates in salt marshes to thousands of years under climates of northern latitudes (see p. 309 in Velde and Meunier, 2008), one can suspect that the soil clay mineralogy can change rapidly under the influence of plants as they fix and transform the clays in the A horizon. The loess deposits in the humid areas of northern Europe and Continental United States could well have been modified before transport as fine grained wind born material or during the periods of deposition. In fact the tills found in Illinois, near the loess deposits (Willman et al., 1963) can contain little chlorite and large amounts of smectites mixed layered minerals. This suggests that these tills, materials moved by the glaciers themselves and deposited in zones which become resources for wind transported materials, were in fact based upon soil materials scraped from the landscapes to the north. When this material is eroded on flood plains in front of the tills and subsequently moved by the action of wind, it already has the imprint of soil processes. However other tills, from nearby areas can show illite – chlorite mineralogies which are altered in soil development sequences (Drooste et al., 1966).

In Europe and the Central Plains area of the United States, the loess accumulation, wind born material, is not great, less than a meter over flat areas and tens of meters along rivers and streams where secondary deposition is important. However in China the loess mantle can be very important, tens to hundreds of meters. Whatever the thickness, the impact on plant life is very great. Loess provides a ready source of potassium, through the alteration of illite or muscovite, and magnesium and iron through the loss of interlayer ions in chlorites. Thus loess deposition can be important to the type and vigour of plants installed upon it but more importantly, it provided an ideal substrate for agricultural experimentation in the Neolithic period.

If the observations confirm the following, one can surmise that the "cold" loess of the glacial periods was of a complex nature, much having passed through a soil stage or having experienced a soil forming process after deposition and cover by other material. This is particularly the case for the North American loess. In Europe some loess is of the illite – chlorite mineralogy, in the Po valley accumulations. However this loess was deposited on the Alps where plant action was reduced due to the altitude and the reigning cold climate. However the loess found in France and Germany shows evolved illite/smectie mineralogies with little illite – chlorite mineralogy. This suggests that a soil forming process has affected the material. In the arid areas, the loess comes more from the direct disintegration of low grade metamorphic rocks which are deposited in such abundance that the plant regimes in the rather arid climate which prevailed did not change the mineralogy greatly. This is still the case.

The "hot" loess developed from desert alteration of rocks is more illite – chlorite rich, and has thus experienced little plant driven soil – silicate interaction before its wind born deposition. In principle the desert type of loessic materials should be more liable to change and alter under plant interaction giving a high rate of potassium and magnesium, iron flux to the soil solutions.

Loess deposition is probably one of the more underestimated modes of displacement concerning the continental surfaces. It is very difficult to discern the moderate amounts of deposition when it falls on a growing and active plant community. Loess is fine grained and a large portion of the materials can be composed of clays similar to those found in soils, specifically 2:1 minerals. Also if the sedimented minerals are unstable, being of high temperature origin, they can be converted into surface or soil clay minerals in a short period of time. If the deposition rate is relatively low and plant activity high, much of the soil clays could be of loessic origin but would go un-noticed in routine analysis of clays.

3.1.4 Summary

The renewal or displacement of clay materials within a soil profile either vertically or laterally can be a strong factor in soil and alteration profile development. It is probable that much of the clay concentrations in valley bottoms or zones at the base of long slopes can be due to lateral transport. This enrichment of clays will change the structure of the soil which can favorize a specific type of plant regime, grasses for example. Increased clay content tends to keep the soils more humid, through somewhat poor drainage, and thus change the hydric regime. Development of the clay poor E horizon under forest growth indicates that grass plants will not be favored by clay transport out of their soil environment. Thus the dynamics of particle movement within the soils can influence the type of plants which will be present.

Movement of fine grained material through wind transport deposits the clay sized fraction evenly on the surface, on hills or in valleys, but it will also have an influence of the plant regimes. Most likely the addition of clay materials will increase the fertility of the surface materials replacing 1:1 clays (kaolinite) by 2:1 clays which are more prevalent in loess materials.

The formation of fine clay mineral materials by alteration or plant action is not static. Once the materials have formed their displacement can be an important factor in further plant – soil interactions.

3.1.5 Transportation and Displacement by Mass Mechanical Failure: Landslides and Slumps

3.1.5.1 Rock and Soil Slides

Massive transport of soil and alteration materials, as well as bedrocks, is the second aspect of displacements of material at the surface. Rugged mountainsor hilly terrains are more likely to be eroded by massive movement events than water transport of clays in flowing solutions. Mountain streams are notoriously clear, indicating that they transport little clay sized material. However landslides, rock slides and earth slides are common alongside of these clear streams.

In a very general view one can consider the mass movement of materials in two ways. One is due to high relief, or slope, where the material becomes unstable by having the base of the slope undercut, with the result of a large slide movement forming a C shaped displacement zone (Fig. 3.6). This is the sliding mass with rotational failure schema of Gray and Sotir (1996). The instability created by a discontinuity at the base of land mass determines the change in resistance of the slope material at a given, punctual site, which allows gravitational forces to cause a large scale displacement based upon the mass of the material on the slope.

This action, rotational, is treated in most texts on landslides and the physics of rock mechanics (Cornforth, 2005, Chap. 1; Duncan and Wright, 2005, p. 57). Much interest for this type of phenomenon has been given to denuded slopes (man made events caused by the removal of plants) which become highly unstable. In this type of displacement the surface (soil zone) and subsurface material (alterite and bedrock) is affected in the same manner. The physics of the process slope and mass, determiners the depth of material affected. Here the major factor is the disturbance of the shape of the lowest part of the slope. The undercut factor decides the movement of the material to the largest extent.

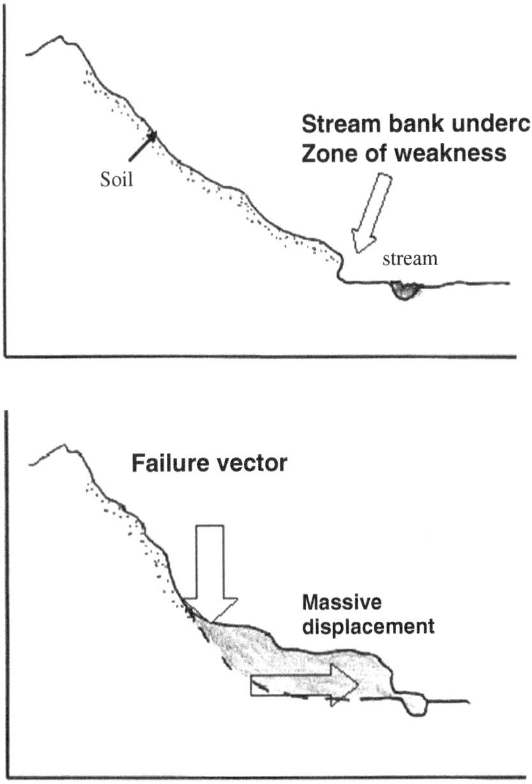

Fig. 3.6 Scenario of massive slope failure due to a destabilisation of the slope material by an undercut at the base of the slope. The failure movement is directed by a vertical movement, which moves to the surface below the slide area on the lower slope. This is a rotational movement, or "C" shaped failure profile

A second cause of movement on a slope is the development or augmented contrast in the physical properties of different layers of material. The contrast in physical properties between two layers of material produces potential instability at the interface between the two layers. For example, if a bedded sediment is uplifted on a mountain side, the contrast of physical competence between the different layers can cause planes of weakness in the material at the interface of the layers and encourage dislocation along these planes. The effect is much more likely to be prominent when the materials on a hill side or mountain slope are bedded roughly parallel to the surface. In such an example the failure in competence is guided by a contrast in the materials of the bedrock.

In both types of slope failure described above one will find a re-organization of rock materials and some soil or alterites materials. However the large part of the material deposited down slope is that of un-altered rock. Thus the large part of the displaced material is much the same, chemically, as that of the mountain or hill side.

3.1.5.2 Soil Slides

Another type of failure is formed along planes of weakness in which the difference in physical properties is engendered by surface alteration and soil forming processes which contrasts the properties of the soil, alterite and bedrock. This is the sliding mass with planar failure schema of Gray and Sotir (1996) called infinite slope and plane slip surface failure (Duncan and Wright, 2005, p. 57). On a hill slope, one often finds the development of a soil layer, often under grass. Here the alteration process works to produce a less dense material than that of the bedrock. If the alterites is of an important thickness, there can be another interface of contrast between the soil and the alterite. Such an instance is indicated by Morgan (2005, p. 35) where surface soil has a significantly greater shear strength (4–9 kPa) compared to the underlying alterite with 2–3 kPa shear strength resistance. The fact of the topography of a slope indicates that the bedrock is rather competent otherwise there would be no hill. Alteration creates a surface layer of very different physical properties: one of lower density, one which can change its mass when rainfalls and enters the porous soil and one where plants put down roots into their substratum.

However the fact of its development, changing the density and porosity of the surface material into a more or less homogeneous layer of ever increasing thickness creates with time, a more and more unstable situation. The interface between rock and alterite is highly unstable. The increased mass at the very surface of this layer by plants and rain filled soil periodically heightens the contrast between the two layers as rainfall events occur. Hence it is almost inevitable that the soil layer will fail and become unstable, sliding to a lower region of potential energy. Such a situation is outlined in Fig. 3.7. The "soil slide" is a very common feature of mountain landscapes.

It might seem paradoxical that plants would tend to create a situation of instability where they will be displaced and destroyed by their own development, yet it is a common case in land movement on slopes. Here the soil developed and the alterite zones are thickened downward creating a new landscape and habitat for plant life, one built on altered material for the most part and amenable to rapid colonization by other or the same species.

In the cases of slump and landslide one finds a mixture of altered materials and those of the bedrock. In such situations the alteration processes again take over, water – rock interaction and plant modification of alterite. However in the case of surface soil failures (soil slump or slide) there is a mixture of materials which strongly represents a situation of the C portion of an altering rock sequence where new clays are present with old parts of the disaggregated and altered rock present but with the new mineralogy of soils, the plant – silicate interaction zone. Here the plants can take hold rapidly in that they have a favourable base for root growth and the access to nutrients furnished by the clays (2:1) present.

If what goes up comes down, it occurs with or without the help of plant – rock interaction. The different forms of slope failure take place depending upon the importance of the slope, and the relative competences of the materials present or the geometric development of the slope itself.

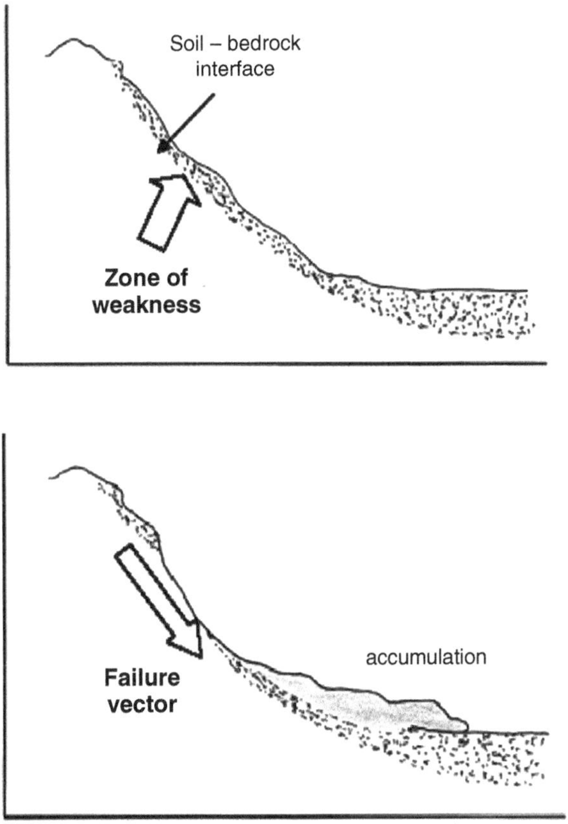

Fig. 3.7 Massive failure due to a parallel vector to the slope. The major cause is the contrast of alterite and soil structure compared to the bedrock or more consolidated substrate material. The origin of he weakness in the development of an alterite of low competence and the more stable root zone created b plants and their roots

Hence one has two basic types of failure, one rotational, with deeper movements initiated at the surface down slope and the other planar where the instability occurs at a mor superficial level.

3.2 Retention of Soils and Clay Material by Plants

3.2.1 Surface Displacement and Plants

The principle function of plants is the preservation of an environment that is amenable to their survival and growth. They need a source of water, lasting beyond the periods of rainfall and abundance. They need certain mineral nutrients, such as potassium, calcium and phosphorous for example which are found in the mineral environment that they grow and prosper from. Most plants need air around their roots a significant portion of the time, in the soil, in order to maintain a balance of chemical processes of degradation of their own dead material and an extraction of other elements through biological action. Some plants of course have adapted to anoxic environments but since the largest part of the surface of the continents is drained, the normal procedure is to keep this situation functioning. Soil, a mixture of clay minerals, debris of other minerals and organic matter is the habitat of the plant substrate. It is to the advantage of plants to develop and preserve the alterite zone of the surface where these interactions can occur, the A soil horizon. Hence the roots in the soil not only are the vehicles of chemical exchange but they can also provide a sustaining structure to preserve this environment. Roots must anchor the soil and try to foster it. However different plants have different root types and they do a varied job of anchoring the soil to the bedrock from which it is developed and develops. Such differences are a part of the history of the plants (genetic inheritance) and adaptation to different climatic and bedrock – alterite situations. As we know plants on a calcareous plateau are not the same as those on a granite massif. The types of plants, their adaptation to the local chemistry and the local soil substratum are reflected in their shape, root structure and plant chemistry. As climate changes, the substrate has perhaps less effect on the physical and chemical characteristics of the alterites zones and the thickness of the soil substratum. Plants must adapt and have over the several million years of their existences. Thus the relations of plant species, soil type and climate are all variables in the statement of how plants behave. Further plants of a single species can adapt to their environment, slopes, hydrodynamics and so forth. This also gives a specificity to each situation in which one finds a plant in its environment.

Given the rather large variety of variables in soil – plant relations (species, climate, slope, soil chemical substrate) one must be aware that any generalization is likely to have as many exceptions as cases confirming principles. However, we would like to trace several principles here to act as a guideline for further thought.

The largest body of information on the effects of roots in the stabilization process of soils involves trees. This is normal in that the disciplines which are most interested in retention of surface materials and the stability of the plants present on a long term basis are those of forestry. Forests grow, now, for the most part on hills and slopes not to mention mountains. These are areas where agriculture is difficult. In the past much agricultural land was present on slopes and hill sides (up to the late 19th and early 20th century). As tractors replaced animal traction, the hill side cultures were gradually abandoned with the exception of vineyards. Thus the areas of land slides and massive failure were those where trees were left to grow. By contrast surface erosion, the displacement of clays and sand to gravel sized material in moving water, is the domain of agriculture where flat land is more often laid bare to the effects of rain without the protection of plant cover. The lack of vegetation leads to the displacement of fine grained material.

How do plants intervene in the stabilisation of surface materials? In the first place it is in their interest to do so. Plants need a substrate of fine-grained material which is

and silicate grain size) failure can occur and the work of the plant biomass has to be re-initiated. In a sense the strong development of plants, especially those that have rather shallow root systems, can be a danger for their immediate survival on hill slopes.

Coming back the classification of failure as being or a rotational or planar type (Gray and Sotir, 1996), in the case of plant occupancy of the surface, one could expect a planar failure in the case of grass cover and a rotational failure in the case of tree cover. In the case of grasses, the structure of the surface is highly occupied by a dense and rigidifying mass of roots near the surface. Here the zone of instability is nearly planar and near the surface. In the case of shrubs and trees the root structure is less dense and presents a greater depth for some roots. The zone of root intensity is more dispersed. In this case the resistance to movement is more evenly distributed in a way and will lead to resistance at the surface and to a certain depth. In this case the failure zone will have a more curved aspect, rotational. The mass of material displaced will come from greater depth than in the case of grass cover. Hence we see the reasons for the differences in the amount and frequency of landslide displacement shown in Fig. 3.9. The tendency for grasses to create a rigid and resistant zone a the surface (A horizon) leaves them un-coupled from their substrate which is more plastic and liable to failure, given the importance of the discontinuity of physical properties between the root mat and the underlying clay – rich alterite zone (Fig. 3.14). The case of woody plants with a more vertically oriented root mass gives a more gradual change in resistance properties between the A horizon and the alterite substrate. Here when failure occurs it is more deep seated, and gives a "C" shaped structure of rotational failure. However a more deep seated root mass leaves the surface layer in contact with rain fall more vulnerable to surface soil erosion.

In order to overcome the inconveniences or each type of root structure resistance it would be advised to intersperse woody plants (trees) with grasses so that the surface rainfall erosion will be reduced and the deeper rooted trees will protect against planar, surface sliding effects. Mono-species on a hill side is not a good thing. This is another way of calling for bio-diversity, but in a very simplified manner.

3.3 Fixing and Accumulating Clays in the Surface Horizons

3.3.1 Plants as Accumulators of Soil Materials

3.3.1.1 Salt Marsh Grasses

The role of plants in maintaining soil material in place is quite well know and well studied. The role of plants in accumulating soil materials is perhaps less well known. If one looks at the structure of salt marshes, one can see that the limit between reworked muds, deposited and eroded by tides, and the stable, plant anchored marsh is rather well defined. The interface is well marked by grasses and sedges which are rooted in the muds forming an approximate ten centimetre escarpment. The advance of this vegetal material into the realm of the sea is encouraged by the accumulation

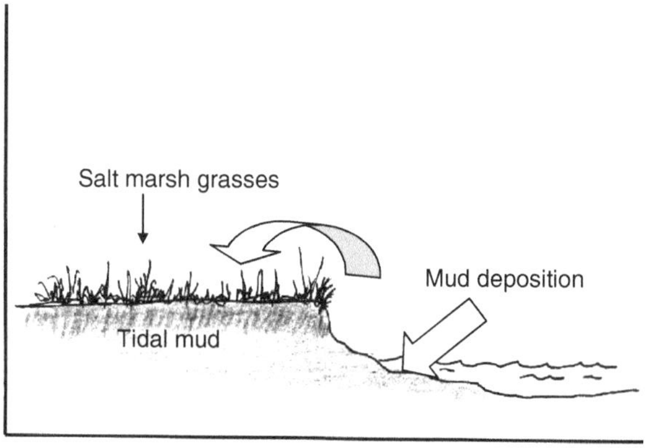

Fig. 3.15 Diagram of deposition of coastal clay – rich sediments in tidal mud flat zones. The establishment of plants on the shore fixes sediments which accumulate with high tide movements. The plants allow the sediment mass to increase and grow seaward as the plants fix more and more of the sediment. Deposition seaward of the plants is subject to tidal erosion and displacement

of sediments, deposited during high tides, into the grass marsh areas. Figure 3.15 indicates these actions. The continual movement of suspended materials is stopped by the grasses which create a barrier to erosion and which capture the sediments deposited on the grasses at high tide and stabilize them making a plateau of material which advances into the tidal zone as clay sediments are deposited on the grasses which establish root zones further into the mud flats with time.

Such actions can occur only when deposited materials exceed those taken into the sea, i.e. when sedimentation exceeds erosion along the coast of sea – land contact. If plants are not present, the mud banks move and are moved continually by tidal movement. However, if plants anchor the suspended and deposited materials, the land front can advance into the sea. Such actions are probably responsible for much of the shallow water deposition of muds in the past which form organic – rich sedimentary units which in turn develop hydrocarbons through the maturation of the organic material deposited in the clay – rich sediments of the marshes. Since organic materials are responsible for the black color of sediments, to the largest extent, the black shales known for their organic content can probably be associated in many cases with the plant – sediment interface common in salt marshes.

3.3.1.2 Grasses and Loessic Materials

If the accumulation of sediment in salt marsh grass due to entrapment of clay material in water born vectors is clearly established, and visible in many instances, the capture and accumulation of wind born material in grassed areas is less well established. However, the dense foliage and high root mass density at the surface typical of grass vegetation is ideal for trapping fine – grained materials. The problem is that wind born clays are usually not abundant, on a yearly massic basis, and tend to be difficult to detect within an abundant vegetal structure. However, one often finds that grasses have clay rich root zones, especially near the surface. In a garden, when one trims the grass along a cement or stone walk way, there is a very important amount of clay material associated with the roots growing over the stone or cement substrate. This trapping effect can probably be extrapolated to an accumulation of wind born material at the surface of grass vegetation. In forests, the vegetal cover is less dense, and a larger portion of the wind born material will be transported downward in the soils, to be evacuated in the ground water or deposited in the B horizon, often below the clay deficient E horizon.

3.3.1.3 River Sediments

The accumulation of clays and clay – rich sediments in valley bottoms is not an accident in that when it occurs, the valley is normally covered in grasses and/or trees. These elements tend to slow the movement of flowing water and promote deposition and enhancement of the soil materials. The frequent inundation and deposition typical of such geomorphologic settings is conducive to growth of grasses which can regenerate their root structures and move "upward" to follow sedimentation, as is the case in salt marsh environments. One must remember that the fine grained sediment which is entrapped by grasses is rich in organic mater (nitrogen – bearing in many instances) and potassium – rich clays developed in soils. In fact the entrapment process is a system of re-cycling of the surface resources in soils that have been lost through erosion of different sorts. What is lost in one place is often re-used and becomes an important resource in another. The clay (soil) rich valley bottoms have always been prized land from the time of the Neolithic farmer to the present. Displacement of soil resources is frequently not lost to the bio-geological cycle.

Figure 3.16 indicates the relative emplacements of the different types of clay and soil movement in a landscape. This is valid for temperate climates where rainfall is more or less regular throughout the year. Slump and landslide occurs on the strong slopes, movement of clays on or in the soils occurs on the less inclined slopes, and deposition of clays occurs through the action of flowing water which leaves a deposit when the flow exceeds the capacity of the stream bed to convey the water present. The effect of flooding, extension of the water flow path, leads to a slowing of the velocity of the water and a lowering of its capacity to carry matter in suspension. The slowly moving water cannot maintain the material in suspension and it is duly deposited on the surface of the stream banks and flood plain. These lead to terrace deposits and stratified layers of sediment accumulated over reasonably long periods of time. In the example given, one can deduce that the material once deposited can be re-mobilized as soil profiles are developed and clays move down slope within the soil profile or on its surface. Significant erosion of clay material from vegetated soils in temperate climates is rare. The erosion that one sees these days is largely due to the effect of denudation of soils materials by agricultural practices in temperate climates.

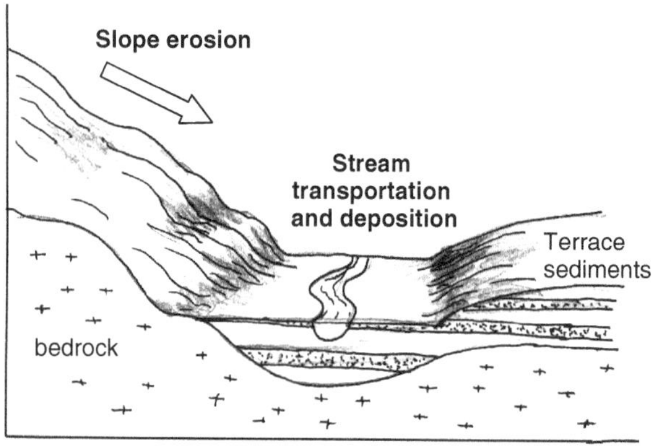

Fig. 3.16 Diagram of the general situation of surface movement and accumulation of soil materials with erosion. Slopes lead to displacement of fine grained material, stream beds and flood plains lead to the accumulation of this material. The movement of materials by streams and rivers is of course in general towards the sea but a temporary zone of accumulation often occurs along the stream beds in the forma of terrace sediments which can eventually be eroded and transported seaward

Toy et al. (2002, p. 9) indicate that significant erosion (tolerable according to the authors) occurs by water on cultivated lands east of the Mississippi River in the United States, while significant erosion west of this approximate boundary is by wind action. The boundary is essentially that of old forest growth to the east and prairie growth to the west. This leads one to believe that the old structure and soil conditioning effected by plants is capable of out living the intensive mechanical and chemical conditioning of modern agriculture, to a certain extent.

3.3.2 Plants and Clays in Profiles: Stabilization and Regeneration

3.3.2.1 Control of the Physical Environment by Plant Cover

Several studies have been conducted on the effects of denudation of a terrain and recovery of plant cover (see especially Bormann and Likens, 1994 among others). The recovery process which in effect controls the physical impact of rainfall, gravitation al forces and wind effects is essential to the well being of plant regimes. Walker and del Moral (2003, p. 106) give data which indicates the type of change effected by plant re-growth after landslide in the Himalayan mountains (Fig. 3.12). It is evident that the growth of plants follows the S curve dynamics and the loss of soil material by erosion and loss of potassium from the remaining soils diminishes according to the strong slope dynamics of the S curve. In this instance the alterite material is largely devoid of soil (the plant – silicate interaction zone) and hence plants must re-conquer and change the alteration substrate in order to come to a state of sustenance for forest ecosystems. Bormann and Likens (1994) show similar data for forest recovery after clear cutting in a New Hampshire forest area. Rapid changes occur after the clear cutting which start from a soil based system, i.e. a layer of clay – organic conditioned materials developed by a plant regime. All that is necessary is that plants re-conquer the soil materials and develop to protect them from erosion. Here loss of potassium in stream water is reduced 50% after just 4 years of plant growth. Biomass storage of K_2O increases 14 fold over the time span. A portion of the potassium comes from organic litter in the soil, but also a significant portion comes from the mineral resources in the soil necessary to increase the biomass rapidly. This material is used by the plants instead of being lost by dissolution and stream transport.

The dynamics of plant regime recovery are of great importance in that not only landslide or clear cutting of forests can occur but also forest fires can burn over different types of forest, leaving them open to a re-establishment of a protective shield over the soil. Bormann and Likens establish a three part dynamic system where the initial stages of recovery are those of recovery and re-organization, the initial low slope of the S curve. Once the system gets established, root growth and extraction of mineral resources among other things, the dynamics of growth increase sharply forming the aggradation part of the growth curve. This is the steep slope of change. The maximum of the curve represents the transition stage where maximum possible change has been reached and a tendency to establish a steady state of slow change takes over. The steady state, one of very gradual change, shows a lower amount of biomass production than that attained previously and changes in other chemical properties of the ecosystem are decreased somewhat. Bormann and Likens are careful not to propose a true steady state for the ecosystem, as constant conditions do not really exist for long periods in nature, but the rate of change is minimal during the mature state of an ecosystem.

One of the principal overall observations based on such studies are that plant regimes re-install themselves on perturbed sites minimizing loss of minerals and mineral resources. The plant system attempts to use extracted resources to a maximum in re-cycling them through the decay of plant litter. Hence the surface soil zone tends to be one of small change with time while the water – rock interface continues to change at roughly the same rate with time. Plants establish a more or less closed chemical system at the surface and attempt to maintain the soil zone intact physically through the stabilization due to root growth and colonization. Chemical loss is minimized although some material is lost through mineral dissolution and release of matter from plant and mineral decay as rain water percolates through the system to the inevitable evacuation by stream flow. Also, on slopes, the inevitable development of the soil zone and the alterites produces an inherent unstable situation where the underlying rock has strongly different properties from the solum leading to physical instability.

Plant regimes also minimize the effects of degradation of their soil substrate. Leaves disperse the energy of rain drops which can erode soil (clay) particles. Roots reinforce the inherent coherence of soil materials at the surface thus preventing slump displacements and eventually landslides. Plants cover the surface and retain the water that arrives via rains. They maintain water in place in the soils which enables them to survive during periods of dryness. These are all truisms known to each of us but the impact is often overlooked. In fact plants attempt to control the physical forces which would otherwise degrade the soil substrate that they need to thrive. One must look at this aspect, control of external forces, in order to understand the structure of the bio – geo communities.

Egli et al. (2004) present data for a series of soil profiles along a mountain valley where slopes vary from 35 to 60%. The vegetation is forest of different dominating species depending upon the altitude (ranging from 100 to 200 m). In all but one profile the AE or E horizon (in the upper 20 cm of the profile) show a clay content superior to that of the parts of the profile immediately below. Despite the strong slopes, the plants (trees and under growth) were able to maintain more clay than that below this surface layer. Thus the importance of plants on soil clay stability cannot be overlooked.

3.3.2.2 Grasses

Plants found in prairies and savannas tend to have a dense root system at the surface. A part of the reason is that *graminae* can reproduce from root growth extension, which they colonize a surface area even though they are cut systematically in lawns and prairies. The dense root mat is a strong characteristic of these plant regimes. Casual inspection of the root mass, frequently called turf, is its high clay content. In fact the root mass captures and holds clays which enter the system from air transport of lateral displacement, erosive processes. Wider spaced and more coarse roots present in trees do not fix as much clay at the surface. In this manner grasses are better conservers of soil clays. Perhaps this is due to the lack of general leaf cover that they provide. In forests the strong erosive effect of falling rain drops is mitigated by leaf cover and as such rain fall is much less erosive than it is in prairie landscapes. Hence it seems that there is a compensation for soil loss depending on the morphology of the plants in a given biome.

3.3.2.3 Clay Content and Possible Re-generation of Clays

Clay minerals, the <2 μm fraction of soils and alterites, are generally found in greater abundance in the upper parts of an alteration profile. Barshad (1959) indicates that grass vegetation is more effective in clay formation than tree or forests. This suggests that in fact that the clay content of the A horizons of grass soils are more clay rich than those commonly found in forests. Generally one attributes the difference in clay content to the accentuated movement of clays by percolation and dissolution in the A horizons of forest soils, assumed to be to high acidity. However this is not always the case, even in sandy soils, one can find an enhanced clay concentration in the upper most layers of forest soils (Teveldal et al., 1990) and Birkeland (1984) suggests that such might be a common case, using his Fig. 10-4. Thus the loss of clays within the soil profile is not necessarily inevitable, but highly probable especially in forest soils. However, an intriguing data set is given by Tice et al. (1996). In an experiment with forest growth, a conifer forest landscape was planted in pine in one area and oak in another in mountain California. After 40 years of forest growth, an analysis of the soil – alteration profile indicates that the pine forest has, as would be expected, less clay in the A, surface, horizon. However, the oak forest shows a higher percent of clay (Fig. 3.17). Thus the pine forest shows clay loss, but the oak forest shows an increase in clay in the surface A horizon. One can pose the question of whether or not the clay content at the surface was not a product of oak tree – soil silicate interaction. Very similar results were obtained by Graham and Wood (1991) for soils in southern California (Mediterranean climate). Deciduous tree cover showed an increase in clay content at the surface whereas chaparral and pine cover showed a lower clay content at the surface. However these authors indicated that the increase in clay content under oak compared to pine was due to a transportation to the surface of clays by earthworms. Thus in these cases the clay content can be a factor of biological activity outside of the plant – soil interaction itself.

In general, as indicated by Barshad (1959), grasses tend to have a higher clay content of the A horizon in their alteration profiles than that for other plant types.

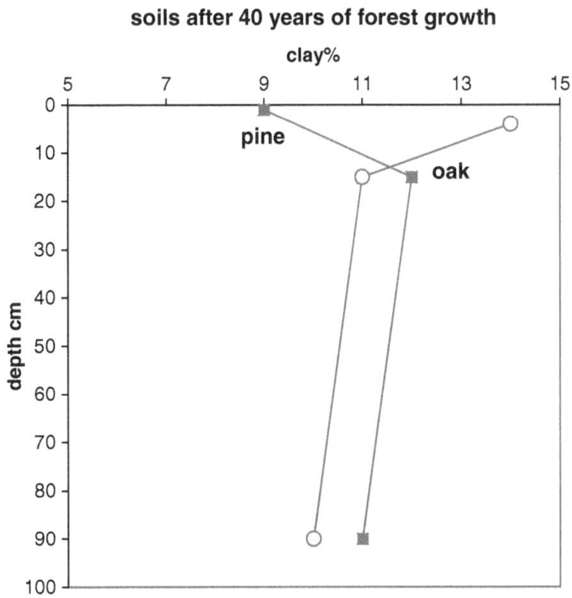

Fig. 3.17 Instance of an increase in soil clay content engendered by change of plant regime (Tice et al., 1996). Soils under pine trees planted in a former pine grove are compared to oak tress planted a the same time, after 40 years of growth. The clay content of the oak forest soils is significantly greater than that of the opine plantations. This effect could be due th the formation of new clays (subject treated in Sect. 5.5) or the effect of earthworms or other fauna brining clays to the surface due to a change in nutritional advantage by the oak tree litter

Forests tend to loose clays by either fluid suspension and movement downward in an alteration profile or perhaps by chemical destruction and leaching. The root structure of grasses is more dense than that of trees, in the surface horizon, and hence one can propose that grasses tend to fix clays more than do trees. This indicates that the type of plant will engender a lower or higher clay content at the surface. Then the movement of clays can be retarded by plant type or perhaps augmented by the plants themselves due to the density of the root structures. The possibility of soil clay creation, i.e. crystallization is an intriguing prospect and will be further pursued in Sect. 5.5.

It is probably fair to say that without plants there would be almost no clay accumulation at the surface of continental masses. The retention of clays is the key to the system. Plants get several nutrients from clay masses: mineral nutrients and water which is absorbed or held in soil aggregates in micropore sites. The root masses are a barrier to water and erosion. Different climates have fashioned different plants, by adaptation, and each situation is one of constant change. Plant succession sequences change the populations and proportions of different types; fire, heavy rains and floods change the proportions of soil – rich material available and new plants form on sites where the environment has been ruptured or changed. As with most situations in nature there is a reciprocity in each situation, one element gets something from another and eventually a sort of steady state is attained for a certain time. One must always remember that the nature of nature is to efface the continents and reduce them to mud at the bottom of the ocean. In doing so, the structures of the continental surfaces changes and plants adapt to these changes. Plants attempt to maintain an optimum for their survival, any other reaction is suicide. However some things cannot be overcome and changes occur. Happily geological forces occur which renew old structures that have been levelled. Mountain chains are erected and resulting local climates change also. Conifers can be found in temperate climates due to the high altitudes of mountains. As erosion occurs the high altitudes are lessened and the climatic situations conducive to high altitude tundra growth are lost. However, wait awhile (several millions of years) and another mountain chain will be formed. Glaciation changes (in the last several millions of years) the climatic zones favoring different types of vegetation. Conifers grow in areas where deciduous trees were formerly present, grasses invade the edges of former tropical zones and so forth. In fact the surface of the earth is in a constant state of change and plants adapt to these forces. Along with these adaptations one finds mutation and change within the different groups or types of plants. Thus the state of plant and soil clays is in a constant state of flux with few constants from the scale of tens to hundreds of years to millions of years. Perturbations solicit adaptation and this changes the chemistry of the soil and alterite zones. New plant successions can have other new needs and promote a new chemistry that will affect the clays. We will look into this in the following chapters.

What needs to be done

Establish the physical characteristics of roots

Resistance to rupture

Effect of directional growth

Relations of root properties (density, resistance to rupture, etc) to deformation modes

Relations of rooting to resistance to rainfall erosion

References

Abe K and Ziemer R (2001) Effect of tree roots on shallow-seated landslides. USDA Forest Gen Tech Report PSW-GTR–130.

Amato M, Belfiore G, Mazzoleni S and Urciuoli G (2005) Analisi contributo meccanico della vegetazione alla resistenza dell coltri ei terreno superficiale, 1–12, Convengo Nationale La Mitigazione del Rischioda Colate et Fango, Sarno.

Barshad I (1959) Factors affecting clay formation, 110–132. In E Ingerson (ed.), Clay and Clay Minerals, Vol. 2, Pergamon Press, 411 pp.

Birkeland P (1984) Soils and Geomorphology, Oxford Univ Press, 351 pp.

Bormann F and Likens C (1994) Pattern and Process in a Forested Ecosystem, Springer Verlag, 253 pp.

Burras L, Smeck N and Bigham J (1996) Origin and properties of smectites in loess-derived soils of Western Ohio. Soil Sci Soc Ame J 60: 1961–1996.

Catt J (1988) Economic significance, 113–142. In A Lerman and M Meybeck (eds.), Physical and Chemcial Weathering in Geochemical Cycles, Kluwer Academic, 375 pp.

Cornforth D (2005) Landslides in Practice, Wiley, 596 pp.

Cox J (1925) Crop Production and Soil Management, Wiley, 516 pp.

De Baetz S, Poesen J, Knapen A, Barbera G and Navarro J (2007) Root characteristics of representative Mediterranean plant species and their erosion-reducing potential during concentrated runoff. Plant Soil 294: 169–183.

Devkota B, Omura H, Kubota T, Paudel P and Inoue S (2006) Revegetation conditions and morphological characteristics of grass species observed in landslide scars, Shintategawa watershed, Fukuoka, Japan. J Appl Sci 6: 2238–2244.

Docker B and Hubble T (2008) Quantifying root-reinforcement of river bank soils by four Australian tree species. Geomorphology 100: 401–418.

Drooste J, Bhattachanya N and Sundermann J (1966) Clay mineral alteration in some Indiana soils, 329–342. In E Ingerson (ed.), Clays and Clay Minerals, Vol. 11, Pergamon press, 614 pp.

Duncan J and Wright S (2005) Soil Strength and Slope Stability, Wiley, 297 pp.

Egli M, Mirabella A, Mancabelli A and Sartori G (2004) Weathering of soils in Alpine areas as influenced by climate and parent material. Clays Clay Miner 52: 287–303.

Encyclopaedia Britannica (1969) William Benton, Chicago.

Gabet E and Dunne T (2002) Landslides on coastal sage-scrub and grassland hillslopes in a sevier El Nino winter: The effects of vegetation cover on a sediment delivery. Bull Geol Soc Ame 114: 983–990.

Gillot F, Righi D and Räisänen M (1999) Formation of smectites and their alteration in two chronosequences of Podzols in Finland. In H Kodama and AR Mermut and JK Torrance (eds.), Clays to Our Future, Proc.11th Int. Clay Conf., pp. 725–731.

Graham R and Wood H (1991) Morphologic development and clay distribution in lysimeter soils under chaparral and pine. Soil Sci Soc Ame J 55: 1638–1646.

Gray D and Sotir R (1996) Biotechnical and Slope Bioengineering Slope Stabilization, Wiley, 323 pp.

Greenway D (1987) Vegetation and slope stability, 187–230. In M Anderson and K Richards (eds.), Slope Stability, Wiley, 648 pp.

Gyssels G, Poessen H, Bochet E and Li A (2005) Impact of plant roots on the resistance of soils to erosion by water: A review. Prog Phys Geogr 29: 189–217.

Jamagne M (1973) Contribution à l'étude pédologique des formations loessiques du Nord de la France (Thesis), Faculté des Sciences Agronomique de l'Etat, Gembloux, Belgium, 445 pp.
Kuzila M and Lewis D (1993) Properties and genesis of loessial soils across a South-central Nebraska basin. Soil Sci Soc Ame J 57: 155–161.
Langhein L and Schumm S (1958) Yield of sediment relation to mean annual precipitation. Trans Ame Geophys Union 39: 1076–1084.
Morgan R (2005) Soil Erosion and Conservation, Blackwell, 304 pp.
Obruchev W (1945) Loess types and their origin. Ame J Sci 243: 256–262.
Paquet H (1977) Evolution géochimique des minéraux argileux dans les altérations et les sols des climats méditerranéens tropicaux à saisons contrastées. Mém Bull Carte Géol Alsace Lorraine 30: 1–212.
Pollen N (2007) Temporal and spatial variability in root reinforcement of stream banks: Accounting for soil shear strength and moisture. Catena 69: 197–205.
Prospero J, Ginoux P, Torres O, Nicholson S and Gill T (2002) Environmental characterization of global sources of atmospheric soil dust identified with NIMBUS 7 total ozone mapping spectrometer (TOMS) absorbing aerosol product. Rev Geophys 40: 1–29.
Ransom M, Bigham J, Smeck N and Jaynes W (1988) Transitional vermiculite – smectites phases in Aqualfs in Southwestern Ohio. Soil Sci Soc Ame J 52: 873–880.
Roering J, Schmidt K, Stock J, Dietrich W and Montgomery D (2003) Shallow landsliding, root re-enforcement, and the spatial distribution of trees in the Oregon Coast Range. Can Geotech J 40: 237–253.
Ruhe R (1984) Soil – climate system across the prairies in Midwestern USA. Geoderma 34: 201–219.
Schenk H and Jackson R (2002) The global biogeography of roots. Ecol Monographs 72: 311–328.
Teveldal S, Jorgensen P and Stuanes A (1990) Long-term weathering of silicates in a sandy soil at Nordmoen, Southern Norway. Clay Miner 25: 117–465.
Tice KR, Graham RC and Wood HB (1996) Transformations of 2:1 phyllosilicates in 41-year old soils under oak and pine. Geoderma 70: 49–62.
Tosi M (2007) Root tensile strength relationships and their slope stability implications of three shrub species in Northern Apennines (Italy). Geomorphology 87: 2268–2283.
Toy T, Foster, G and Renard K (2002) Soil Erosion: Processes, Prediction; Measurement and Control, Wiley, 337 pp.
Velde B and Church T (1999) Rapid clay transformations in Delaware Salt marshes. Appl Geochem 14: 559–568.
Velde B and Meunier A (2008) The Origin of Clay Minerals in Soils and Weathered Rocks, Springer, 406 pp.
Waldron L and Dakessian S (1981) Soil reinforcement by roots: Calculations of increased soil shear risistence from root propeties. Soil Sci 132: 427–435.
Walker L and del Moral R (2003) Primary Succession and Ecosystem Rehabilitation, Cambridge Univ Press, 456 pp.
Willman H, Glass H and Frye J (1963) Mineralogy of glacial tills and their weathering profiles in Illinois circular, 347, Illinois state Geological Survey, 55 pp.
Xu J (2005) Precipitation – vegetation coupling and its influence on erosion on the Loess Plateau, China. Catena 64: 103–116.

Chapter 4
Clay Minerals in Soils: The Interface Between Plants and the Mineral World

Contents

4.1 The Silicate Clay Mineral Types Present in Soils

The presence of different clay minerals in soils, the plant – silicate interaction zone, is most often due to an incorporation of clays formed in the water – rock zone into the bio-zone and the initial transformation of rocks by living organisms such as lichens, and bacteria. Thus there is some inheritance from underlying mineral reactions and those engendered by the successive stages of bio-activity.

The first stages of surface interaction which form clays, appear to be under the influence of bio-agents, such as lichens and mosses. Eventually higher order plants can establish themselves in the substrate of humic and clay minerals formed. Without plants and bio activity, the products of physical breakdown of rocks and those products of water – rock interaction which are of low enough intensity to produce clays are likely to have been washed away by rainwater. One must be reminded

B. Velde, P. Barré, *Soils, Plants and Clay Minerals*,
DOI 10.1007/978-3-642-03499-2_4, © Springer-Verlag Berlin Heidelberg 2010

that all chemical reactions forming new minerals from rocks involve water. Deserts and ice packs are poor reaction vessels.

Among the silicate clay minerals one normally finds the aluminium hydroxide phase gibbsite but the iron hydroxides and hydroxyl phases are rarely mentioned.

4.1.1 Kaolinite and Halloysite, 1:1 Minerals

1:1 clay minerals are composed, for their silicate part, by a two layer structure where one layer is occupied exclusively by silicon cations and the other in large majority by aluminium ions. The aluminium is in octahedral coordination with oxygen atoms, shared with Si cations, and hydroxyl cations. This mineral is highly hydrated (hydrogen content) and therefore less thermally stable than the 2:1 minerals. The ratio of Si to Al is very near one. The kaolinite structure is not subject to cation substitutions for the most part and its composition is thus near Al, Si, O and H. The absence of alkali ions suggests that its conditions of formation and stability have low alkali ion activites. The presence of Si in equal quantities compared to Al indicates that the activity of silica in solution is still reasonably important. Kaolinite and halloysite can be considered to represent an intermediate step in water – rock interaction and alteration of silicate materials to oxides and oxy-hydroxides.

However, kaolinite is often found in the initial stages of rock- water interaction. It is formed locally, in a given mineral type due to very local (scale of the crystal) chemical potentials (see Velde and Meunier, 2008, Chap. 4). Usually, the kaolinite formed in the early stages of rock alteration is a minor part of the newly formed clays. Kaolinite persists up the alteration profile under very many climates and biomes. Its presence as a majority of the clays in alteration clay mineral assemblages indicates strong weathering, or strong chemical action which reduces the silica content in the clay assemblage.

Halloysite (see Joussein et al., 2005 for a review of halloysite occurrence and crystallochemistry) is a hydrated form of kaolinite, in that a layer of water molecules is "interlayered" between the 1:1 structural layers. Further, halloysite can contain some iron is substitution for aluminium ions, up to 7 molecular percent substitution The interlayer spacing is 10Å. However, the presence of water makes a curved surface of the sheet forming a tubular morphology. The interlayer water is not stable, being expelled rather easily upon heating. There is a certain affinity of halloysite for potassium, intercalated with the water between the 1:1 layers. This explains the somewhat anomalous high cation exchange capacities when these minerals are present where there are apparently few or no smectite minerals present in soils.

Halloysite seems to be present in soils when the alteration zone is frequently saturated with water. Some tendency to lose the water forming kaolinite is seen in upper parts of soils where wetting and drying occur compared to deeper zones below the water table or in more permanently wet conditions. It seems that halloysite is an initial stage in the formation of kaolininte under conditions of strong weathering, i.e. surface soil and alterite zones.

4.1.2 The 2:1 Clay Minerals in Soils

4.1.2.1 The End Members

The « backbone » of soil clay mineralogy in areas of fertile soils are the 2:1 minerals. These minerals can fix potassium and other base cations (Ca; Mg), major nutrient and probably also ammonium, another major nutrient. They are the only materials in soils to have a significant amount of cation fixation capacity which can retard the inevitable loss of these elements to the sea. Weathering is an inexorable loss of alkalis and alkaline earth elements. Local conditions of chemical potential or chemical activity of different ions in solution can determine the mineralogical expression of these minerals. In fact the 2:1 minerals are similar in nature and crystallographic expression, but the chemistry of the interlayer site determines the chemical and physical response of these minerals to plant solicitation and the treatments in the laboratory used to identify the different characteristics of the 2:1 mineral base in soils.

The most reasonable method of designation of these minerals is their initial nature in the soils themselves. This means a first investigation without chemical treatment of the soil material. Basically one would like to understand the interaction of the clays with plants, their chemical action, either by release of ions or fixation of ions. Hence the "un-treated" state can give insight into the function and chemical action of the clays in their environment. After such an initial investigation has been completed and recorded, some chemical treatments can elucidate the reactions of the clays under different conditions of chemical solicitation engendered either by plants and the bio-sphere of the natural water – rock interactions. These steps can be used to describe a function- reaction situation which will be used to designate the behaviour in terms of a mineralogical classification. Of course the use of such a classification does not obey the rigours of official mineralogy as currently practiced, but it can be useful to understand the function and interaction of these clays with their plant – dominated environments.

Our classification of 2:1 soil clay minerals, as outlined briefly in Sect. 1.3, is based upon the occupation of the interlayer ion sites between the 2:1 units and the stability of the ions found in these sites in natural specimens. Three basic types of interlayer ion occupation can be defined:

1. Anhydrous potassium ions forming a stable, non-variable cell size. These minerals are classically called illite. The basic cell dimension perpendicular to the 2:1 layers is 10 Å.
2. Hydrous cations of various types forming either one or most often, two water layer complexes which can be changed by changes if temperature or relative humidity.These minerals are classically called smectites. The double water layer hydration state is 15.2 Å and the mono – hydrate gives a spacing of 12.5 Å. Relatively easy removal of interlayer ions and replacement by other ions indicates a 2:1 structure with a low charge.

3. Hydroxy coordinated interlayer ions, usually Al but also probably Mg form a structure with a 14.2 Å spacing. These minerals are usually called HI or soil vermiculite when the interlayer ion is dominated by Al ions in an incomplete layer (see Meunier, 2005), and chlorite when the Al interlayer complex is more complete and in some cases magnesian. If the hydroxy layer ions are displaced by potassium these minerals are high charge smectites.

These three poles of interlayer ion occupation form the basis for the soil clay minerals with a 2:1 phyllosilicate structure. The interlayer site occupation can be either inherited from higher temperature phyllosilicates, such as mica or metamorphic chlorite, or possibly be formed under conditions of soil formation as suggested by the laboratory studies of Bauer et al. (2006). The behaviour of the minerals in fixing different cations in soils is dependent upon the chemistry of the soil solutions and the affinity of the different ions to enter into the interlayer ion sites. The attraction of different ions is most likely a function of the total charge on the 2:1 structure (charge imbalance usually below a positive charge of 22 for a unit cell). The excess negative charge, non-compensation of the cations present, is compared to the structural anionic charge generated by the ten oxygen and two OH ion units in the structure. The tendency to hold a cation or attract a specific ion into the interlayer ion site can probably be attributed to the amount of charge and/or the site where the charge imbalance originates, i.e. in the surface silica dominated layers or in the central octahedrally coordinated sites. Competition between Ca and K ions for sites in synthetic smectites shows a variability depending upon the site and intensity of the charge generating the electronic imbalance in the structure (Velde, 1985).

Given the nature of the likely precursors of the soil 2:1 mineral structures, it is highly likely that the material present is not homogeneous. If the high temperature bedrock contains phyllosilicates, as does a very large portion of continental rock, the individual grains will be of different chemical compositions to a varying degree. Thus, initial chlorites of diagenetic or metamorphic origin can reflect the orogenic history of the rock maintaining crystals of different composition formed under different physical conditions. The differences in composition will most likely give rise to slight differences in charge and charge site in the crystal from one grain to another. Thus individual layers or crystallites will behave slightly differently under the same chemical solicitations. In this way a type of mineral behaviour, say potassium incorporation, will be of slightly differing intensities giving a range of responses to ambient chemistry and when investigated by X-ray diffraction (the only viable method of routine clay mineral analysis). Instead of having a single crystallographic response one will see a slight range of responses which widens the diffraction maximum. One must consider that the clay "phases" in soils are in fact a group of crystallites which have similar but not identical physio-chemcial properties. This leads to what is called mixed layering in 2:1 clay minerals, common in soil environments. This being the case, inspection of many X-ray diffractograms allows one to be rather optimistic concerning the formation of groups of mineral grains having the same or very similar characteristics. One can in fact use the term minerals, in the sense of a type of identifiable behaviour for a given range of chemical composition.

4.1.2.2 Mineral Identification and Mixed Layering: Case Studies

Accepting these premises, one can proceed to establish a working definition of soil clays with a 2:1 structure as proposed above. Of the three types of structures one can propose that the smectites have a relatively low deficit in layer charge, around 0.3 charges per unit cell (22 negative charges). This allows cations to move in and out of the interlayer site depending upon local chemical gradients of bi-valent cations in solution. These cations are hydrated. The selectivity between bivalent ions and monovlent ions is strongly in favour of the bivalent ions. Thus in smectites Ca is favored over K ion occupation of the interlayer ion site. Treatment of the clays with concentrated potassic solutions still leaves portions of the material in a hydrated cation state even though the ion is monovalent potassium.

In high charge 2:1 structures either Al or K is selectively fixed under natural conditions. The high pH and relatively high Al content of pore solutions favors the formation of Al (OH) cations and their fixation to form HI minerals. However a strong presence of potassium will favour its incorporation as an anhydrous cation forming what is called illite. Laboratory experiments by the authors indicate that although much cation occupation can be changed by using concentrated solutions of potassium for example, the Al hydroxyl (or Mg) interlayer sites will not be completely exchanged. In the case of the HI minerals it is probable that the fixed material is well structured and polycationic which gives the mineral a more stable chemical character. The same is true for potassium minerals, so called illites. These minerals can in part be transformed into apparent smectites when submitted to strong bivalent cationic solicitation. The potassium is replaced by hydrous bivalent cations. However, not all of the material becomes smectitic in behaviour and some remains in the anhydrous potassium state.

The situation of some layers having a given chemical response to cation selection and others another response gives rise to what is called mixed layering. In this state some of the 2:1 units in a crystallite have a given cation interlayer occupancy while others have another. Since most crystallites appear to have a basic two component behaviour, series of mixed layer minerals can be identified between the three poles of behaviour; illiite, smectites and HI minerals. Thus illite – HI minerals can be found in soils, illite – smectites also, and HI – smectites as well. Figure 4.1 indicates these relations. The zones around the poles can possibly give rise to some three component behaviour, as suggested in the figure but at present it is not know how much, if any, there is to be found commonly in soil clays.

The main interest in these minerals is their capacity to fix with varying degrees of chemical attraction potassium or eventually ammonium. These are mono-valent ions, or ionic groups of similar diameter which find a place in the 2:1 minerals and can sooner or later be extracted by bio-activity. The fixed, anhydrous potassium and eventual ammonium is a vital reservoir for plant nutrition.

One characteristic of 2:1 soil clay minerals is that they most often occur as a mixture of layer types; high charge, low charge or Al, MgOH, K, Ca, Mg H_2O interlayer occupancies. This is not the case for clay minerals formed in the water – rock interaction zone where non-phyllosilicates are altered to new phyllosilicate clay minerals. The process of formation of the clays, 2:1 structures, is surely

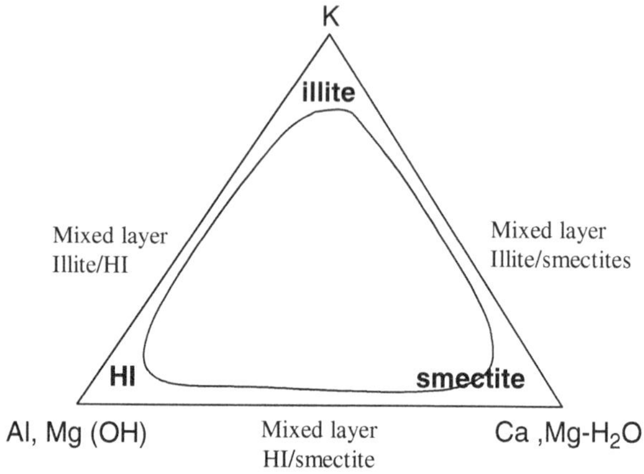

Fig. 4.1 Representation of major sol clay types as a function of interlayer ion type. K (potassium) = illite: Al, Mg(OH) = HI hydroxy interlayer ion minerals: Ca;Mg-H_2O = smectites minerals with hydrated exchange ions, dominantly divalent. Mixed layer type minerals are formed between the HI-K (hydroxyl-interlayer/illite), K-Ca, Mg-H_2O (illite/smectites) or Ca,Mg-H_2O-HI (smectite/hydroxy interlayer) mineral types

complicated but certainly not that of normal water – rock interaction. We will look briefly at the major groups of mixed layer minerals, those phases with characteristics between the poles of the diagram shown in Fig. 4.1.

Illite – Smectite

In general, the water – rock interactions in the initial stages of weathering do not produce mixed layer clay minerals (see Velde and Meunier 2008, Chap. 4). The structures formed during the initial destruction of high temperature minerals form more or less mono phase clay minerals, either smectites or kaolinite or illite, in various proportions in a given alteration site. As the minerals are brought to the surface, their characteristics change. Old, high temperature phyllosilicates such as muscovite, chlorite or biotite are likewise transformed in the soil zone (see Righi et al., 1997)

Several examples are given to indicate the changes which can be expected using moderate chemical treatment of these soil clays. The first example is the effect on several illite/smectite minerals. Initially the clays are saturated with a weak solution (0.01 M $SrCl_2$). Potassium saturation was made with a strong solution of KCl, 0.5 molar. In the sample proposed in Fig. 4.2, two samples of high smectite content mixed layer illite-smectites are shown, one with a significant amount of illite present and the other almost without an illite component in the clay assemblage. Potassium saturation affects the mixed layer clays in similar manners, shifting the major peaks to slightly lower *d* spacings or higher two theta values (less than the double water layer hydration state indicated b a peak at 15.2 Å) indicating the increase in potassium in interlayer sites either as a mono – hydrated ion (giving a one layer water structure and a 12.5 Å component) or by creating an anhydrous illite – like potassium inter layer population with a 10 Å peak. One suspects that the one water

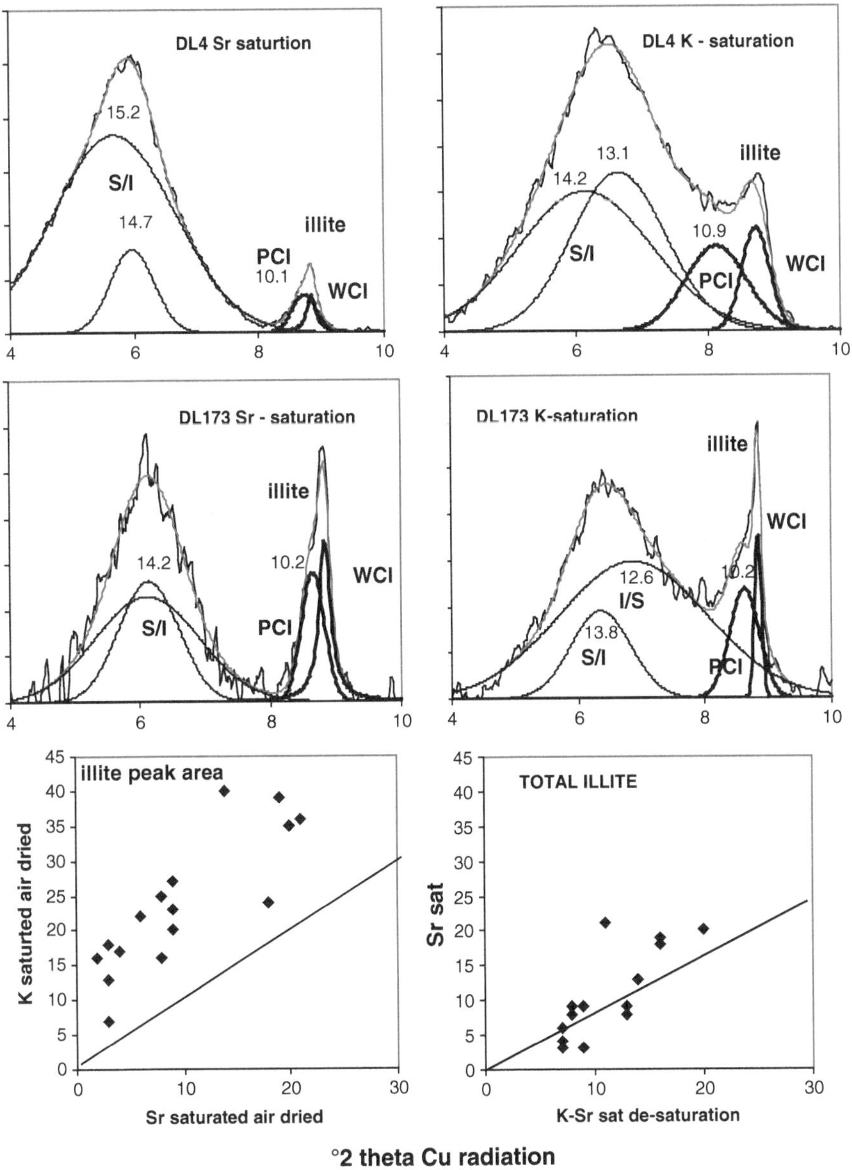

Fig. 4.2 Iowa prairie soils in K (O.5 M KCl) saturated state and comparison of the low concentration Sr-saturated illte peak areas with K saturated areas with the Sr – desaturated KCl clays (K-Sr de-saturation) Potassium saturation increases the illite peak ares relative to the Sr – saturated samples whereas the de-saturation of potassium brings the clays back to an initial weakly Sr – saturated illite peak area. This indicates that the potassium saturation producing illite peaks is reversible. One would expect that such a process would occur in soils which are treated with potassium fertilizer in the Winter to be de-saturated by plant growth needs during the Spring and Summer. Background – subtracted X-ray diffractogram treated using decomposition methods

layer potassium structure is rare in that the major peaks remain in the range of an interlayered two water layer structure (towards 15.2 Å).

The most important effect in the first example (DL4) is the formation of a stronger illite component, especially the 10 Å well ordered illite of rather narrow peak width (WCI). This effect is important in that it indicates that portions of the mixed layer, illite – smectites material, has large zones of high charge layers which all become anhydrous upon potassium saturation. This effect is visible in cases where illite is initially absent (DL 4), or almost so, and where it is strongly present (DL 173). In the first example the poorly crystallized illite peak (PCI) is stronger also. We see in both examples shifts in the mixed layer mineral (S/I where smectites is the major component) and the formation of a more illite – rich mineral (I/S) with a peak at 12.6 Å. Thus potassium saturation produces illite, anhydrous potassium interlayered mineral and illite layers in the mixed layer structures.

Potassium saturation in strong concentrations of KCl was made on 16 samples from a geographic area of several kilometres squared in central Iowa. The results show that there is a systematic change in relative peak area for the total illite peaks (PCI and WCI, poorly and well crystallized material). A systematic increase of about 10% of the peak areas is seen for a range of initial illite content, 2–20% (Fig. 4.2). It is apparent from these data that high charge smectites are present in all of the samples. When the same samples are de – saturated using a 0.5 M SrC12 solution, the initial illite content is found, indicating that the formation of these anhydrous potassium layers in the high charge smectites is exchangeable if a strong exchanger is used. This action is known in agricultural soil testing where "non-exchangeable potassium" is identified and extracted when a strong exchange solution is used, such as ammonium acetate or others. Lower concentration solutions of divalent ions do not affect the anhydrous potassium layers it appears.

Another data set, taken from samples presented by Velde (2001), for samples from various top soils across the central prairie United States (Fig. 4.3) shows very similar clay reactions but to weak potassium concentration saturation (0.01 M KCl). Shift in illite – smectites peak positions toward the 10A illite position and increase in the surface areas of the illite peaks. These are all old prairie soils under cultivation. It is clear that illite layers in a strongly random mixed layer structure can form as well as groups of illite layers in a crystallite which form the "new" illite minerals. In fact the changes are nothing but cation selection by layers with different charges and affinities for cations in solution. The reversibility of the change in cation abundance to a smectites (divalent cation with two water layers present in the interlayer site) shows that the clays react to their local chemistry, changing their X-ray diffraction characteristics but not their basic nature.

Fig. 4.3 Illustration of the effect of potassium on mixed layer illite/smectites mineral assemblages of diffeent proportions of illite layers in the different phases. Highly smectitic clays show little effect while those with more illite layer present in the initial state (Sr-saturation) show a greater increase in illite (10 Å) peak area. Not only is there more of the illite mineral present ut also the mixed layer peaks shift to a position nearer to 10 Å indicating more illite layers in the structure

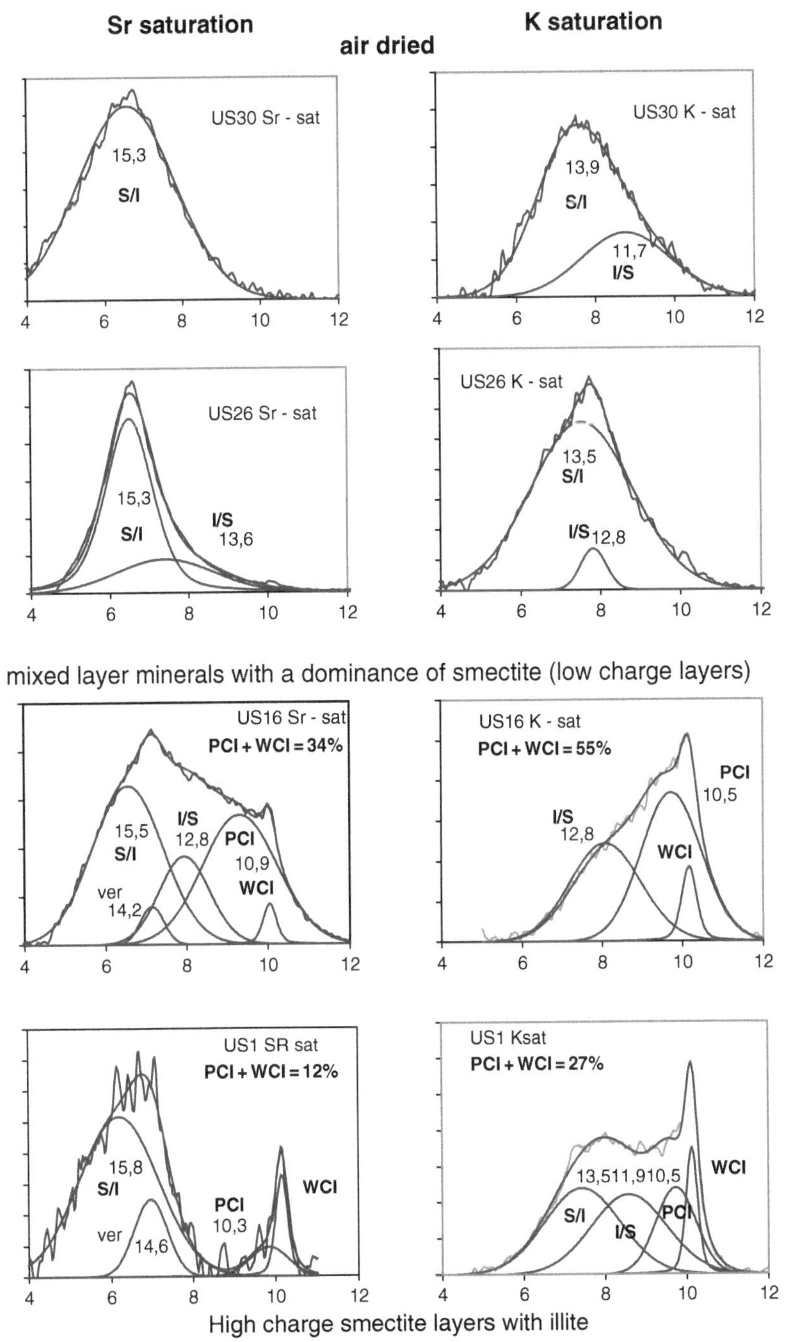

Fig. 4.3 (continued)

The reverse, potassium de-saturation reaction, can be effected by plants, of course. Here the extraction of potassium produces a smectitic layer. In Fig. 4.4 which presents data given by Velde and Peck (2002) for clays from the experimental farm at the University of Illinois (again old prairie soils). The extraction of potassium reduces the peak surface area of the well crystallized illite peak, (WCI), increases the surface area of the poorly crystallized illite peak (PCI) and does not significantly shift the illite content of the illite – smectites (S/I) phase evidenced by the same peak position. This occurs over a time period of 40 years where only corn was grown without fertilizer addition to the soil. In a second step, over 40 years again under the same conditions, the poorly crystallized illite (PCI) loses illite layers to form a mineral at 11.2 Å, more smectites – rich. The mixed layer mineral loses illite content seen in a shift in the peak position to higher values. The increased change in clay mineralogy is due to the introduction of new hybrid corn varieties which significantly increase the grain production and hence the amount of potassium extracted per year. Hence in this experiment it appears that the illite is the first mineral to lose potassium, and this is followed by the illite layers in the illite – smectite structures. Similar results are reported by Barré et al. (2007) for laboratory experiments using rye grass growing from a clay substrate of illite and mixed layered minerals. Here the demonstration includes potassium saturation and plant extraction from a complex clay assemblage.

HI Vermiculites, Chlorites, Smectites and Illites

HI minerals are identified by the presence of a 14.2 Å 001 peak position. This spacing indicates the presence of an organized hydroxyl ion interlayer mineral. High temperature minerals are called chlorite. These minerals have both the divalent ions Mg, Fe present in the hydroxyl complexes and Al ion complexes. Soil clay minerals can have Al ion complexes, most commonly admitted, but also most likely a significant amount of divalent ion complexes and this most often Mg ions (Righi et al., 1993). Potassium can displace some of the hydroxyl complex ions to form illite, the anhydrous interlayer ion mineral. In some layers H_2O can be displaced by organic molecules as in smectites. However, much of the material remains in the hydroxyl form and this is called HI, hydroxyl interlayered mineral or soil chlorite. Thus the behaviour of HI minerals under chemical treatment can form different interlayer ion occupancies and apparent different soil clay mineral types. If laboratory treatment can change the apparent mineral type by substitution of different interlayer ions soil pore solutions can certainly effect the same changes if they are sufficiently concentrated in the appropriate cations.

Often, a given soil sample will have material of these different types present at the same time. In Fig. 4.5 the Sr – saturated (0.01 M $SrCl_2$) sample shows a strong, but slightly complex 14.2 Å peak with minor amounts of mixed layer minerals (12.3 Å peak) and illite minerals dominated by the PCI type (peak at 10.4 Å). One can see little if any smectite behaviour in an initially largely HI mineral assemblage upon glycol treatment. However with low intensity potassium treatment (0.01 M KCl) illite (PCI) is much more apparent in peak surface area. A significant amount of

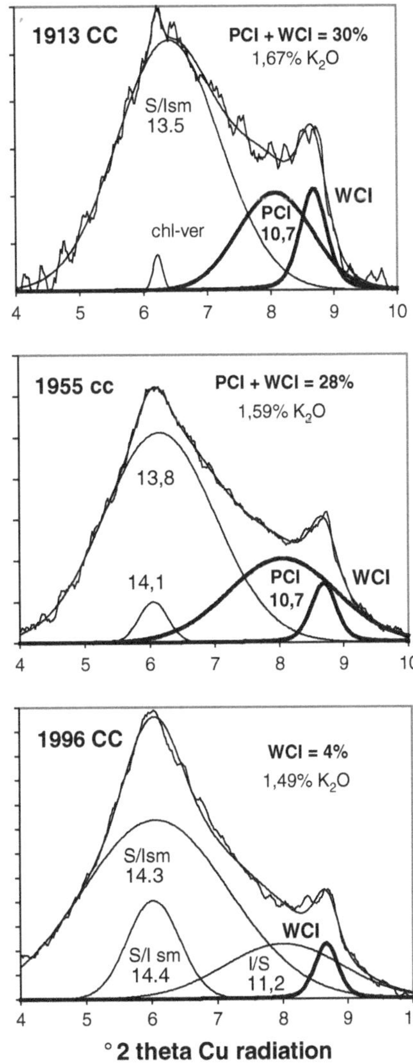

Fig. 4.4 Effect of potassium extraction on illite/smectites mineral assemblages in an experiment of unfertilized continuous corn growth for 80 years in an experimental plot at the University of Illinois (Velde and Peck, 2006). The extraction of potassium is shown by the decrease in illite content (WCI and PCI; well crystallized and poorly crystallized forms of the 10 Å phase). Background – subtracted X-ray diffractogram treated using decomposition methods

mixed layer mineral (shift in peaks to the illite position, 14.1 and 11.5 Å) shows that there is an important portion of high charge material present where the hydroxyl ions can be displaced in favour of anhydrous potassium ions. Hence here we have an HI mineral with an important portion of exchangeable hydroxyllated cations present, showing a high charge smectites (vermiculite) behaviour. Much material has both HI and vermiculite layers present in the same structure forming mixed layered minerals.

A similar behaviour but one showing apparently only high charge layers and HI layers can be seen in different samples from the Po delta plains near Pavia (data

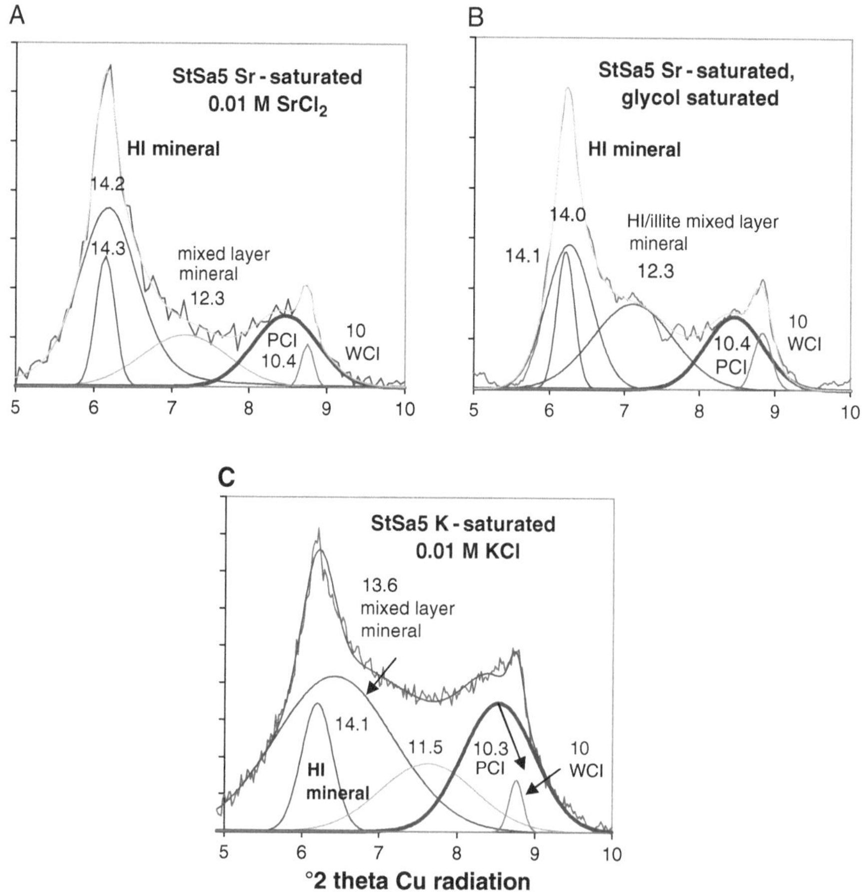

Fig. 4.5 Effect of potassium saturation on a HI – illite mineral assemblage (Saint Savignien, Charents, France). The HI mineral, peak position at 14.2 Å is not displaced by glycol treatment, but a significant portion of the HI layers are filled with potassium upon saturation in weak concentration of KCl solution. The poorly crystallized HI mineral shifts upon potassium saturation from 14.1 to 13.6 Å, becoming slightly more illite – rich, whereas the PCI (poorly crystallized illite) peak area increases significantly. This suggest a significant exchange of hydroxyl interlayer cations with potassium in many of the HI mineral crystallites. Background – subtracted X-ray diffractogram treated using decomposition methods

supplied by F Terribile, Univ Frederico II, Naples). Figure 4.6 shows Mg treated samples (0.5 M $MgCl_2$) and those treated with strong KCl solutions (0.5 M KCl).

The Mg samples show strong sharp 14 and 10 Å peaks indicating the presence of chlorite and illite. There is no indication of a smectites mineral (peak towards 15.2 Å) in the samples. Potassium saturation only slightly widens the 14 Å peak but strongly increases the intensity of the 10 Å illite peak. The potassium saturation also changes the relative intensity of the first order and second order chlorite-vermiculite peak (14.2 and 7.1 Å), showing a weaker intensity 14 Å peak which indicates a

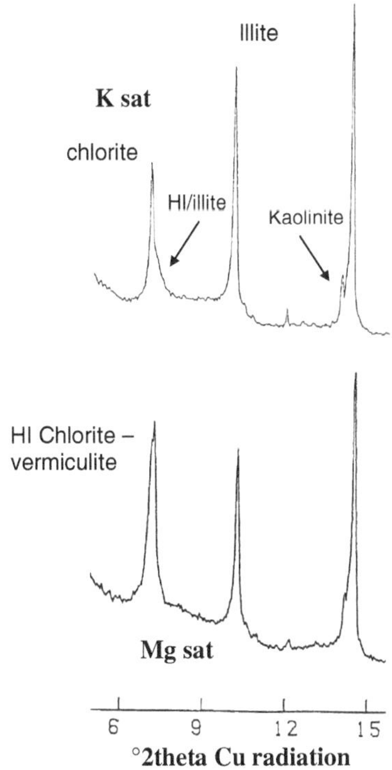

Fig. 4.6 Effect of potassium or magnesium saturation on an apparently simple illite – chlorite clay mineral assemblage from the Italian Po valley sediments. Peak widths are narrow, indicating a high degree of order in the interlayer ion position and a relatively large number of coherently layers in the crystallite structures. The relative intensities of illite change between Mg and K saturation, and the relative intensities of the 14 and 7 Å peaks of the chlorite minerals indicate the presence of a well organized HI component which is replaced by potassium. Lower intensity of the 14 Å indicates an increase in Mg content of the exchanged minerals

change in interlayer ion composition of the 14 Å minerals, most likely due to a loss of aluminium ions being replaced by potassium. An identical behaviour was noted by Ross et al. (1985) for HI clays in soils which had experienced manure fertilizer treatments for 6 years.

It is quite apparent that HI minerals, identified as such in the untreated conditions, can have a variety of behaviours under different chemical treatments. These chemical treatments can occur under conditions of plant activity giving different types of minerals due in fact to different chemical compositions of pore solutions in the soils due to the influence of plants as will be seen in Sect. 4.2.

4.1.2.3 Origin of Mixed Layer 2:1 Clays in Soils

Laird and Nater (1993) have indicated that the illite-smectite mixed layer minerals can be found preferentially in the fine fraction of prairie clays which is shown in Fig. 4.7.

This is a rather well known fact, but interesting to recall here. The grain size is smaller when the expanding minerals are present in the illite – smectite paragenesis. This must indicate some genetic paths in the formation of this material. In general one thinks of crystallization procedures when one invokes small grain size materials. However, we believe that most of the clay material in soils (in the

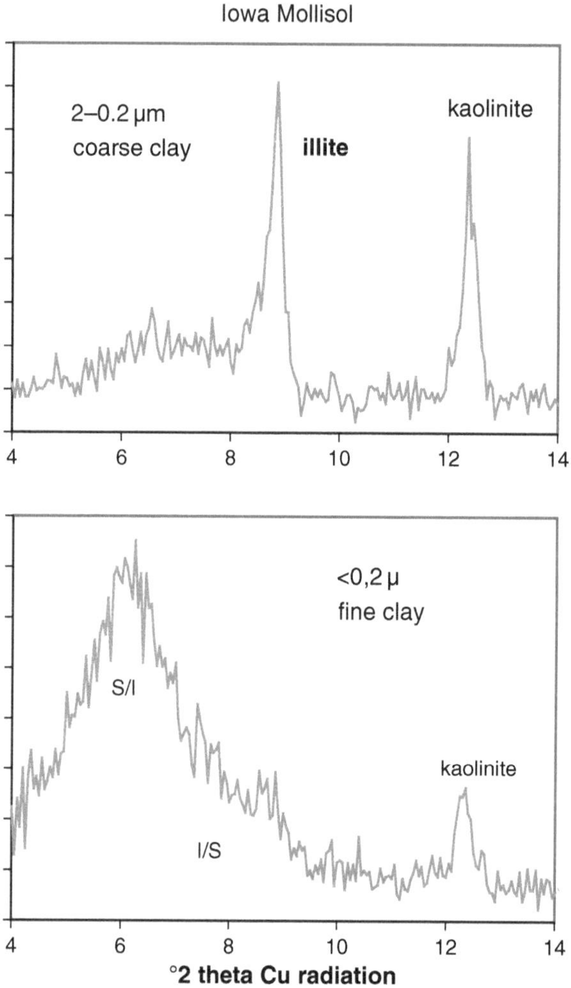

Fig. 4.7 Indicaion of a change in clay mineralogy with particle size. The example is an Iowa Mollisol where the coarse clay has a dominantly illite – kaolinite mineralogy while the fine clay fraction shows a dominant smectite – illite mineralogy

plant – silicate interaction zone) has been formed to a large extent before it is present in this zone. The transformation of various high temperature phyllosilicates into soil clays of smaller dimensions and of mixed layer character can be seen in the descriptions of Righi et al. (1999), Gillot et al. (2000), Olsson and Melkerud (1989), Moberg (1990) among numerous studies. The characteristic of these transformations is the formation of smaller and smaller particles, but the chemical properties of the basic 2:1 mineral are retained to a large extent. Thus the 2:1 clay minerals in soils appear to be the product more of transformations, larger grains to smaller grains, than dissolution and crystallization reactions. However Gillot et al. (2000)

demonstrate that the individual crystals of high temperature phyllosilicates can change their composition within the 2:1 layers, in fact observations indicate becoming dioctahedral when initially trioctahedral (Righi et al., 1993).

Using TEM morphological observations Gillot et al. (2000) on soil minerals in eluvial E horizons of Spodosols in Finland have shown that nearly all clay particles exhibit dissolution features, which are seen both on edges and basal faces of the high temperature crystals. Moreover, the development of dissolution features results in a mosaic pattern: large particles are divided into smaller platelets by a continuous network of fine fissures and finally aggregates with diffuse boundaries and fluffy or cloudy pattern are formed. Large particles slightly affected by dissolution display an electron-diffraction pattern typical of ordered or semi-ordered stacking structures whereas electron-diffraction patterns from aggregates display ring patterns similar to turbostratic disorder.

Thus it appears that the weathering of high temperature phyllosilicates, micas and chlorites, produces mosaic like structures, partially dissolved where the 2:1 layers can have a modified compositon. The overall trend of phyllosilicate mineral weathering is to form dioctahedral lower charge 2:1 minerals from micas. Change in layer charge indicates change in chemical composition of the 2:1 unit layers.

An illustration, by X-ray diffraction observations, is given in Fig. 4.8 where an initial detrital, high temperature phyllosilicate assemblage of clay sized material, glacial flour deposited on the Atlantic shelf and re-deposited by storm action, is deposited in a Delaware Bay salt marsh where *spartina* grasses cover the sediments and fix new material. The development of soil clay minerals from high temperature chlorite – mica mineral assemblages in this instance is from the top downward in that the constant influx of high temperature mineral sediments is incorporated into the soil from above. The initial fine grained mineral assemblage is illite (mica) and chlorite concerning the phyllosilicate phases. In the figure the top most sample is the least altered with transformation of the phyllosilicates increasing downward. The initial alteration reaction is to form S/I, a smectite – rich mixed layer mineral. The relative proportion of the illite (peak area) decreases, whereas chlorite appears to remain relatively un-changed. The relatively less important in peak surface area of illite and the presence of a S/I mixed layer mineral are the major features of the diagram. One can propose an illite to mixed layer mineral transformation accompanied by perhaps a portion of chlorite. The second stage (going from root zone to lower root zone) indicates a strong loss of chlorite and the formation of an illitic I/S mineral. Poorly crystallized illite (PCI) is lost in the process of forming an illitic mixed layer mineral. The S/I mineral peak broadens and increases in relative surface area. Both mixed layer clays have distinctly wide diffraction peaks indicating fine grained and poorly organized structures. The interaction of the chlorite structure with the other minerals is not certain but it appears that the presence of a tri-octahedral interlayered mineral is largely lost. From the deposition rate in the salt marshes one can estimate that the reactions or transformations observed here occur over a period on the order of less than 15 years.

Here soil clays developing from a fine grained mixture of high temperature material are present in a short period of time. This indicates the action of bio-agents. One

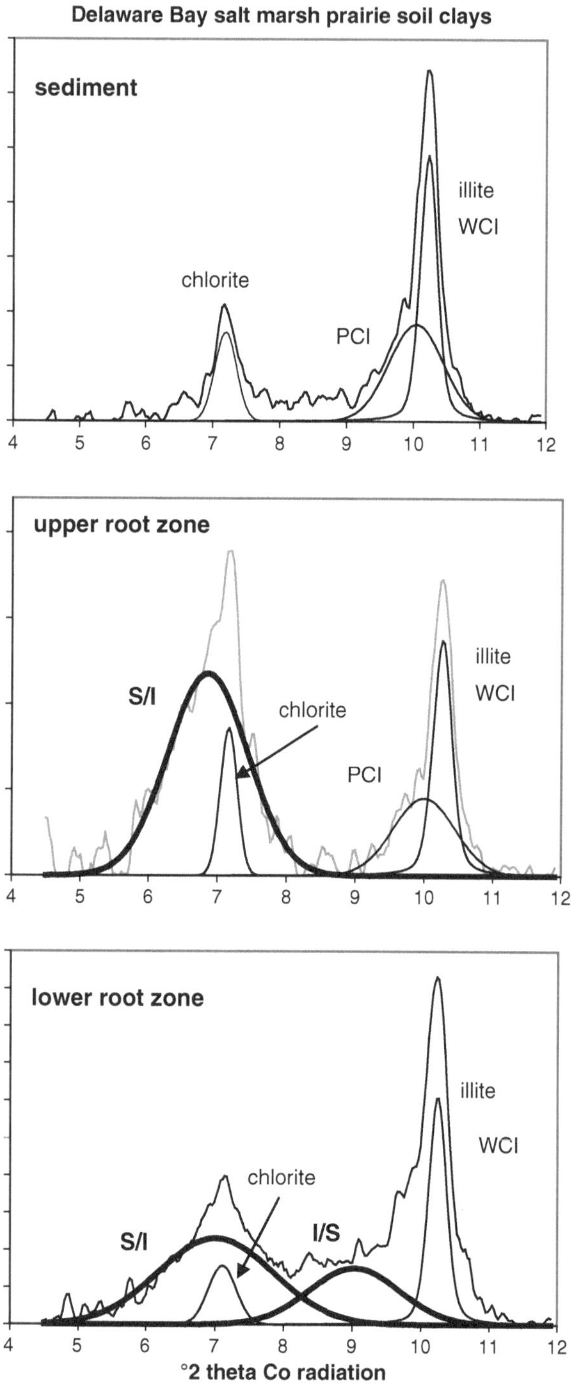

Fig. 4.8 Indication of mineral change in a sequence of soils developed in a Delaware Bay (USA) salt marsh (Velde and Church, 1999). The sequence shows alteration of an initial illite – chlorite

would not expect the existence of a dissolution – recrystallization process over such short time spans (See Sect. 2.2). The appearance of less well organized crystallites of a mixed layered nature indicates that the process or transformation is not totally homogeneous and that the different layers in the structures will probably have slightly different properties, such as charge and charge site which produce two types of mixed layer minerals, S/I and I/S.

4.1.2.4 Chlorite Alteration

Millot and Camez (1963) attributed most of the vermiculite or HI mineralogy to the alteration of chlorite in soils from mountain terrains of eastern France. According to these authors and many others since, chlorite can be a major source of clay minerals in soils.

Chlorite minerals altered by biological action (micorrihze) can be transformed into pure smectites in their laboratory behavior (Arocena et al., 1999). This is an apparent case of stripping the interlayer hydroxyl ion layer from the 2:1 substrate. Since it is assumed that most of the ionic substitution in chlorites which compensates charges in the tetrahedral site occurs in the interlayer site (Deer et al., 1962) one can expect that the a chlorite stripped of its interlayer ions will have a significant charge from one 2:1 layer to another. Several studies have indicated that the charge on the 2:1 layer changes and the ionic composition of the octahedral layer also. The alteration minerals change from a trioctahedral character to a dioctahedral one which means a loss of Mg, Fe and a gain of Al. These ionic displacements seem to occur in silicates at low temperature. Bauer et al. (2006) have shown that such changes can occur over short periods of time, however at temperatures of 80°C and under strong chemical conditions (several molar KOH solutions).

When the chlorite 2:1 layer has been significantly modified by such elemental displacements, the altered chlorite becomes indistinguishable from altered dioctahedral micas an forms part to the mixed layer "illite/smectite" mineral assemblage common to evolved soils. Thus altered chlorites (trioctahderal high temperature minerals) become to a large extent indistinguishable from mica alteration products.

From the above observations one can conclude that the reactions forming 2:1 clay assemblages is a complex process, using several materials to produce inhomogeneous phases. The different behaviors of the structures after treatment in the laboratory indicate the ability of the layers to react to chemical agents. These laboratory behaviors can probably be found to a lesser extent in the soil clay situations as they minerals are subjected to different pore solution chemistries, developed by plant action and rain water dilution.

Fig. 4.8 (continued) mineralogy deposited at the surface of the salt marsh sediments which progresses downward I the profile. This is the reverse of normal rock weathering alteration profiles. The initial chlorite component is altered as is the illite to produce two illite – smectites minerals typical of soil clays. The new minerals are the result of interlayer ion extraction of both the chlorite and the illite Background – subtracted X-ray diffractogram treated using decomposition methods

4.1.3 Complex Mixed Layer Minerals (KS): Smectite, Kaolinite and Kaolinite/Smectite Mixed Layer Minerals

The overall tendency in alteration sequences is to lose alkali then silica compared alumina in soil clay assemblages. Then alumina in its turn is lost (Pedro,1966). The loss of silica relative to alumina marks the difference between temperate climate soils and humid tropical alteration. This change in overall chemistry is applicable to the clay minerals, phyllosilicates and hydroxyl – oxides. Most commonly one finds kaolinite as the clay mineral present for several meters depth in such soils (Schultz and Ruhiyat, 1998). In general, it is apparent that soils under high rainfall tend to contain kaolinite at the expense of 2:1 minerals. Kaolinite and muscovite have the same Si/Al ratios (near one) whereas smectite has a higher relative silica content than kaolinite. In fact if one considers the phyllosilicate transformations from micas and chlorite (high temperature minerals) to soil clays under temperate climate conditions, the initial equilibrium is towards a more silica-rich phase, smectite in most cases. One can contrast then the clay assemblages in temperate climate to humic tropical soils as being more silica-rich than those formed under high rainfall. Often, kaolinite is identified with strong chemical alteration but it can be produced in the initial stages of water – rock alteration.

However, an increasing number of reports of a "mixed" interlayer mineral, kaolinite/smectite, indicate that there is a possibility that a transition between 2:1 and 1:1 minerals can be found in alteration sequences. Among the situations where one finds kaolinite/smectite mixed layered mùinerals is in alteration sequences of eruptive rocks, where silicate glass is probably a factor of alteration (Wilson and Cradwick, 1972; Schultz et al., 1971). This seems to be independent of climate conditions.

4.1.4 Oxides and Hydroxides

4.1.4.1 Al-Oxy-Hydroxides

In soils and weathered rocks, aluminum may form different hydroxides, the most abundant being gibbsite (α-alumina trihydrate: $Al(OH)_3$) while its polymorphs bayerite and nordstrandite are much more rare. *Boehmite* (alpha-AlOOH) is common in bauxite deposits but less common in soils.The aluminium oxides and hydroxides have a non distinctive grayish-white color, which is easily masked in soils except when large concentrations occur. These minerals form platelets in which the *(001)* faces are considerably more developed than the *(hk0)* ones. This is due to the fact that the Al-OH-Al bonds in the a-b plane (layer) are much stronger than the hydrogen ones between layers in the c direction. Consequently, the gibbsite morphology looks like that of clay minerals. For this reason, it was called "hydrargillite". Aluminum hydroxides may accumulate in the most weathered rock levels in tropical countries forming bauxite deposits. As do Fe-hydroxides, gibbsite may form also in the earliest steps of the weathering process. Thus, this phase is competing with clay minerals for aluminum ions. We will not consider these minerals in detail in this work. For detailed descriptions see the review papers of Schwertmann and Taylor (1989) and Hsu (1989) for Fe and Al oxy-hydroxides respectively.

4.1.4.2 Gibbsite (Aluminum Hydroxide)

The hydroxyl aluminium mineral gibbsite is generally restricted to two conditions of formation: one where highly acid conditions produce the precipitation of the mineral indicating the destruction of other alumino-silicate minerals, with loss of silica. It is the buffer to aluminium exportation. Gibbste is found in the early stages of granite alteration under temperate climates. Braga et al. (1989) have made some very interesting observations on the presence of gibbsite in the early stages of granite alteration, the arenite facies. They find a zonation from north to south in Europe, through temperature and more arid climatic zones essentially. Gibbsite is however rare in soil clay assemblages, the plant – clay interaction zone, based on these altered rocks. As one moves upward in the alteration profiles more 1:1 and then 2:1 minerals are present.

One reason that gibbsite is rare in the plant – silicate interaction zone is that the amount of soluble alumina necessary to produce this phase is often sufficiently high that it will poison many plant species (Wolt, 1994). High concentrations of alumina cannot be tolerated by many plants. However, if the plants send roots to sufficient depths, they can probably extract silica that is the product of mineral dissolution in the deeper saprolite or saprock zone and bring it into their biotic sphere deposited in the aerial parts of plants as phytoliths. This silica, in an amorphous and highly reactive form, will be combined with gibbsite to form kaolinite and probably also smectites (becoming HI mineral in most cases) in the soil zone.

It must be remembered that the water which reaches the water – rock alteration zone at the base of alteration profiles is often present due to a rapid drainage along major fracture or pore systems. It is not charged with ions from dissolution of minerals nor through the equilibration of clays with the aqueous solution. Thus the presence of gibbsite in the early stages of weathering, in the saprock zone, is due to a structural cause in the geometry of weathering. However, in the cases of tropical weathering, the great influx of unsaturated aqueous solution induces the dissolution of most minerals and only the most insoluble ionic species, Al and Fe, remain, forming gibbsite and hydrated iron oxides. Of course this analysis is valid for aluminous rock types, shales, gneiss and granite. More basic or calcareous rock types, alumina poor, will not lead to the formation of gibbsite in the early stages of alteration.

4.1.4.3 Transition Metal Oxy – Hydroxides

Fe-Oxy-Hydroxides

Different species of Fe-oxides and hydroxides are formed in weathering or soils reflecting the moisture, pH, Eh and microbial activity conditions. The Fe^{3+} cation is framed in different oxide or hydroxide crystal structures. These iron minerals

Table 4.1 Color of Fe oxides (simplified from Scheinost and Schwertmann, 1999)

Fe oxide	Most frequent color	
Hematite	Moderate reddish brown	α-Fe_2O_3
Goethite	Strong yellowish brown	α-FeOOH
Lepidocrocite	Moderate orange	γ-FeOOH
Ferrihydrite	Brownish orange	$Fe_5HO_8 \cdot 4H_2O$
Akaganeite	Strong brown	ß-FeOOH
Feroxyhite	Strong brown	δ-FeOOH
Maghermite	Dark yellowish brown	γ-Fe_2O_3

The most common coloring agents in soils are goethite and hematite. The ratio of hematite (red) to goethite (yellow brown) seems to be controlled by the pH of the soil (Schwertmann and Taylor, 1989).

become a dominating phases in the final stages of weathering because of their low in aqueous solutions. They form thick indurated horizons in the inter-tropical zone (laterite). However, they are present in all the weathering process including the initial steps where they are associated with different types of clay minerals. They have a great importance in the distribution of iron since after oxidation, Fe^{3+} ions expelled from the silicate structures spontaneously form oxides or hydroxides. Scheinost and Schwertmann (1999) show that the colour of oxides, hydroxides and hydroxy sulfates is determinative for that of soils (Table 4.1).

Mn-Oxy-Hydroxides

Manganese oxide minerals have a black color. They show even a greater tendency than iron oxides to occur in concretions. Manganese-rich micromorphological zones in peds that are black, often also contain large amounts of iron oxides. The main mineral species are *pyrolusite* (MnO_2), *birnessite* $(Na,Ca,K,Mg,Mn^{2+})Mn_6{}^{4+}O_{14}{}^{*}H_2O$, *lithiophorite* $(LiAl_2Mn^{2+}Mn_2{}^{4+}O_9{}^{*}3H_2O)$, and *hollandite* ($BaMn_8O_{16}$).

4.1.4.4 Laterites

A special case of alteration under tropical conditions is the formation of laterites. These are depositions of strongly bonded iron oxides and aluminum hydroxy oxides forming a crust. These alterations are found in the present savanna areas bounding the tropical rain forests of Africa and South America. The difference in the mineralogical expression of the two alteration types is probably due to climate; in the savannas the seasons are contrasted, with high rainfall during one season and a contrasting dry season. Tropical forest have constant rainfall all year round. The vegetation is of course different also, rain forests are tree dominated (broad leaved evergreens) and the savannas are grass dominated. However, Ségalen (1995) believes that true laterites (crusts of iron oxide above gibbsite layers) are relict soils having formed under climates no longer existing where the laterites are now found.

Intense weathering can lead to the formation of concentrated aluminum in the form of gibbsite or boehmite, hydrous and anhydrous alumina, and especially iron in the form of hydroxyl – oxides. Such soil concentrations are called laterites. Much work has been done on these soils or soil residua due to their economic importance as resources of aluminum, their potential as a resource of iorn and their qualities as construction materials. The striking feature of these phenomena is a zone of accumulation of oxides and hydroxides below the soil zone. This accumulation zone is one due to intense leaching and transport of fine grained material (Aleva, 1994) and its deposition below the surface. Laterites are complex, showing the formation of concretions and nodules which indicate local, intense migration of ions to form specific mineral concentrations (Boulangé, 1984, pp. 136–149). In certain, typical or ideal, cases the alumina is separated from the iron oxides into two layers, the iron forming a top crust or duricrust (Millot, 1964; Lelong, 1969). These cases are due to formation in tropical contrasted season climates under savanna vegetal cover (see review by Temgoua, 2002, pp. 4–6). These soils sequences were formed under conditions of intense rainfall for a part of the year and a more dry regime for the part of the year. This is the contrasted season climate of the tropics.

Since such intense concentration and segregation of elements in a soil context is likely to occur over long periods of time. One can assume that laterites are the products of alteration which must have persisted over at least tens of thousands to millions of years (Aleva, 1994, p. 73). Thus one can assume that the formation of laterites necessitates a stable climate or at least one that has not changed drastically over long periods of time contrary to those in the northern hemisphere,which are strongly affected by recent glacial cycles. In most places where they are found today the oxide crusts appear to be unstable and can be seen to be breaking up. Hence one can assume that the laterites are a fossile of sorts, corresponding to climatic conditions no longer active.

The saprock zone (recent water – rock interaction) usually contains kaolinite and smectite along with iron oxides (goethite). Gibbsite and goethite – hematite become important in the bauxitic and duricrust zones. Quartz appears in most horizons indicating its chemical inertia.

Temgoua (2002, p. 7) indicates that the laterite structures formed in contrasted season tropical conditions are not stable under tropical forest conditions, where the concentration zones tend to be ruptured by climatic change. Thus the formation of mono-elemental zone concentrations of Al and Fe depends upon wet and dry seasons, not those of a humid tropical forest.

According to Boulangé (1984) the intital stages of alteration or mineral transformation in the rock – water interaction zone produces kaolinite and gibbsite and biotite-HI phases in granites. Quartz and goethite are present as are white micas. As one moves up the alteration profile, gibbsite is more abundant at the expense of kaolinite. More basic rocks (amphibolite) produce gibbsite and goethite with less kaolinite. Hence the major change is a decrease in the silica content of the clay assemblages.

Pion (1979, p. 190) gives a general scheme for the occurrence of different clay minerals formed from basic rocks under climates of relatively low rainfall (400–1,100 mm/year) to moderate (850–1,400 mm/year) to high rainfall (1,200–2,200 mm/year). Under high rainfall one can find vermiculites type minerals at the rock – alterite interface whereas above this zaone gibbsite and kaolinite (plus iron hydroxides) are dominant. The alteration minerals in profiles formed under less rainfall show the strong presence of more siliceous 2:1 soil clay minerals.

4.1.4.5 Red Soils

One can consider red soils to be the product of contrasted season alterations. Here strong dissolution occurs during the rainy season and strong evaporative movement and dehydration of hydroxyl-oxides occurs during the dry season. This process is slow to occur and hence much of what one sees today as laterite material is frequently being degraded by forces of new climatic actions.

Ségalen (1995) indicates the special distribution of soil types or alteration facies for tropical climates. His classification shows gradations of soils by color and hence oxidation state and iron content along with silica content as it affects clay mineral stability (smectites). Basically, and schematically, one finds yellow soil types with iron oxides and kaolin present in tropical forest zones, where the iron oxide is hydrated (goethite). On the edges of the forests under contrasted season climates one finds grasslands with red soils but where silica content is higher and smectites are present along with kaolinite. These relations are above all valid for soils formed on pelitic or aluminous rocks such as granites and sedimentary or meta-sedimentary rocks.

4.1.5 Poorly Crystallized Silicates Imogolite and Allophone (Allophone Si – Rich, Imogolite Si = Al)

These minerals, found in alteration sequences, are often precursor phases to others. They are formed through the rapid precipitation of Si-Al, forming a rather nebulous phase, allophane, with various Si/Al ratios and Si-Al imogolite with Al=Si. Both are formed during the alteration of vitreous rocks or under conditions of alteration under high pH control (Farmer and Lumsdon, 2001; Farmer, 1999).

Allophane and imogolite are called minerals. But only imogolite has ionic coordinations ordered over short distances which can be identified by infrared or perhaps Raman spectral methods. There is a regularity only in small atomic molecular domains so that X-ray diffraction does not show coherent domains. Allophone by contrast appears to have much more variable chemical relations and only on the molecular scale so that there is no characteristic or much less so as would be seen by infrared means of identification. Allophane is pratically undeterminable using X-ray diffraction. On the contrary imogolite presents some broad diffraction peaks. Both minerals are currently detected in soils using a combination of infrared spectroscopy, thermal and chemical analyses. Even if still debated, the short range "crystalline" structures of these minerals have been approximated (Cradwick et al., 1992; Wada, 1989). Allophane contains a hydrated network of Si-Al-O atoms. Both

H_2O and OH units are present. The stoichometry is slightly variable but usually is between 1:1 to 1:2 Al/Si atoms. Imogolite a hydrated network of SiAlO atoms with a rather well organized tubular structure which is not detectable by XRD methods. The ratio of Si to Al is near 1.

The key to the presence of both imogolite and allophane is a continous humid condition of alteration environments. Under conditions of contrasted climate they are no longer present and are probably transformed into gibbsite for allophone and kaolinite or 2:1 minerals for imogolite. They represent hence humid alteration conditions, and the presence of highly reactive substances, for the most part. The allophone and especially imogolite minerals are most likely very important in processes of chemical transfer from surface to lower horizons in acid soils under temperate climate conditions.

4.1.5.1 Rainfall and Vitreous Rocks

Imogolite and allophane are typical of soils formed on glassy rocks in various types of climates (andisols). The nearly constant moisture content of andisols maintains the stability of these minerals. Under drying conditions these minerals become unstable and rapidly transform into halloysite. They form complex associations with organic matter in the humic horizons of these soils. The preferential complexation of Al by humic compounds may inhibit the formation of allophane. Imogolite has been identified in spodosols where its formation is related to the degradation of the Al-organic matter complexes.

Andosols represent a combination of two specific variables which combine to produce very characteristic soils. Andosols are noted for their high water content, high content of organic matter in the A horizon (15–30%), high capacity to fix phosphorous, and moderate to low cation exchange capacity (CEC) often related to the amount of amorphous or extractable alumina (Shoji et al., 1993, p. 2; Garcia-Rodeja et al., 2004). Although fresh or immature Andosols usually select Ca over K, it has been noted in several instances that slightly more mature Andosols (containing 1:1 minerals) can strongly select potassium (Ndayiragije and Delvaux, 2004). In general, Andosols are fertile. This fertility is most likely due to a continued high weathering rate of glassy materials which furnishes plant nutrients (Shoji et al., 1993, p. 210). As an example the volcanic context of central and southern Italy and the development of Andosols undoubtedly attracted populations from regions with less fertile soils, such as the limestone Aegean area, and as a result the Italian peninsula became an area of constant population movement from the times of human agricultural practice.

Andosols occur under various latitudes from tropical Africa and Oceanic Pacific regions to mid continent areas such as Italy, Germany (Kleber et al., 2004) or France (Quantin, 2004) to northern climates such as Iceland (Arnalds, 2004). Temperature is obviously not a significant variable in the formation of Andosols (see the review Catena 2004 Vol. 56). However continuous humid conditions are very important for their formation.

Several principal factors then are necessary to produce these special soils, high rainfall, equal to or less than the tropics but above what is normally considered to be necessary for temperate climates. The soils should be well drained. The second factor is a source rock which contains significant amounts of glassy material, volcanics. It appears that there is not much difference between basic, basaltic material and more acidic types as they form Andosols except that the transformation or alteration rate for basalt is significantly higher than for acidic glass up to two times greater (Shoji et al., 1993, p. 38) The essential material for the formation of Andosols is a glassy substrate.

A distinct problem in the study of the properties and the development of Andosols is that they are normally found in active volcanic areas and hence subject to renewed accumulation of fresh material on the top of the profile. Andosols profiles are frequently interrupted by paleosols, indicating renewal of material by ash flow or lava flows which buries and fixes the soil zone with a renewed starting material upon which soils are formed.

Major phases characteristic of Andosols are (see summary by Shoji et al., 1993, pp. 113–123) are:

Allophane, *Imogolite*, *Gibbsite* the hydroxyl form of aluminum, *Opaline silica*.
Halloysite, the hydrated form of kaolinite.

4.1.5.2 Weathering Process

Generally the high rainfall interacts rapidly with the highly unstable silicate glass to form an alkali and alkaline earth – depleted mass. Since there are few micro-sites (places where pre-existing minerals transform into new clays, see Sect. 2.1), there is little tendency to form new clay mineral phases, i.e. those with continuous layer structure edifices. Thus the more glass present, the more alteration. One finds that there is also a loss of silica to the soil solutions which leads to the formation of the silica – poor phases, imogolite and allophone. These are the water – rock interaction minerals found at the rock – alteration interface (Sifferman, 1973, pp. 149–151). As one moves upward in a profile, towards the surface one can find gibbsite (Al hydroxide) and some halloysite. Depending upon the drainage, age and development of the profile. The presence of halloysite appears to be dependent upon a continuously humid environment. In soils which experience periods of drying, kaolinite is present. In more crystalline eruptive rocks one can find halloysite forming from feldspars in the early stages of alteration (Bates, 1962).

The different variables of ground water flow, presence or not of primary crystals in the volcanic material, seasonality of rainfall and so forth can and do modify the mineral succession relationships in Andosol sequences. The zones of contrasted seasonal humidity tend to destabilize the allophone and favour gibbsite – halloysite and eventually kaolinite mineral assemblages.

One can say in general that Andosols and amorphous minerals are found as young soils developed from highly glassy materials. Even this is not universal in that kaolinite and smectite mineralogy was found in the surface horizons in soils formed on recent (Krakatoa eruption, 1883) ash fall material developed in a rainforest environment (Fauzi and Stoops, 2004).

4.1.5.3 Clay Mineral Origin in Andosols

In the brief description above, it appears that Andosols, soils based upon the alteration of glassy silicate materials, will first produce un-structured materials which cannot be called minerals because they lack a determined composition and mineral structure (crystallographically identifiable). Allophane is one material and the hydroxyl – iron complexes are the other. Further there is a strong association between organic matter and these elements, Al and Fe. These materials are typical of the initial stages of rock – water alteration in Andosols which are themselves due to high water input and constant humidification. Alkalies, alkaline earths elements and silica are evacuated by percolating solutions. Another important phase is imogolite, a much more stoichiometric phase of nearly equal portions of Al and Si. The diffuse organisation of this material makes it essentially non-detectable by XRD methods, infrared is the major method of identification (See Farmer et al., 1983). In this phase one finds Si along with Al.

Under different conditions, either as replacement of crystalline phases such as feldspar or in zones nearer the surface of the alteration profiles, one can find halloysite ("hydrated" kaolinite) and gibbsite. It is not clear from the data in the literature if the presence of these minerals is due to a kinetic process, taking thousands of years, or specific chemical conditions. One can imagine that gibbsite will replace allophone, which is largely of the same composition. However, the presence of halloysite indicates the conversion of imogolite, which has a similar Si/Al ratio. Thus both gibbsite and halloysite may be successors of allophone and imogolite, or perhaps form under slightly different chemical conditions that the first two phases. The minerals which give the physical and chemical characteristics of Andoslos (high water content, low density, high capacity to fix phosphorous) and perhaps not the stable end product of silicate equilibria.

One can outline the following classification based upon phases present:

Imogolite, allophone Early ANDOSOL	gibbsite, halloysite (imogolite) mature ANDOSOL	kaolinite, smectite, imogolite andic soil

In any event, the phases typical of Andosols are probably not stable under long term equilibria.

The effect of vegetation on Andosols has been reported to be one of enhancement by grasses forming imogolite and allophane and destruction of Andosol mineralogy to form more silica – rich Podsols under forest growth (Shoji, 1993, p. 130) in Hawaiian soils. Meijer and Buurman (2003) give data for soils on an acidic volcanic edifice in Costa Rica where the plant regime varies from fern – grass at altitude, through decidous forest, tropical forest to grazed land at low altitude. In the chemical data reported one frequently sees an increase in potassium in the uppermost samples of the profiles, the Ah horizons. This is especially true for the fern – grass soil at high altitude and tropical forest profiles. In the tropical forest profiles, one can note the presence of kaolinite, and vermiculite (HI) in these horizons whereas lower portions contain the less siliceous mineral assemblage gibbsite, goethite and halloysite. In these Costa Rican soils it is possible that the plants have an effect on the chemistry of the A organic horizon.

Typically, gibbsite is abundant at the rock – water interaction interface. Also one finds kaolinite or the hydrated form halloysite. Towards the plant – clay mineral interface one can find smectite in some abundance. It appears that the alteration intensity of silica loss is greatest at the base of the profile. This would seem to paradoxical in that the water which reaches the rock interface should be the most highly charged in dissolved ions and hence at least a little silica. Conversely the rainwater falling on the soil surface should be the least saturated. However, if one considers that significant silica can reach the surface by plant action (phytoliths) one can imagine a source of silica beyond that of rainwater – kaolinite interaction (Derry et al., 2005; Farmer et al., 2005). This can be stretched to the existence of podsols in tropical forest areas (Lucas, 2001, p. 138). Clearly plants can have a strong impact on soil clay mineralogy.

4.1.6 Some Minerals Formed and Found at the Surface but Not in Soils

Not all minerals formed at the surface are associated with soils, i.e. the zone of plant activity. Two are notable, as follows:

4.1.6.1 Palygorskite

Palygorskite and sepiolite are minerals which can be formed in alteration profiles due to the effects of evaporation. These magnesian minerals formed in desert soils where capillary action precipitates saturated solution material to form magnesian silicates. They are not stable at the surface, in the soil zone (Paquet, 1970; Badraouri et al., 1992). Khademi and Arocena (2008) demonstrate one possible reason for this instability. Bacteria are efficient in extracting Mg from thse non-silicate minerals leaving kaolinite in their place. Hence they are most likely to disappear in many soil formation processes, in the biosphere. These minerals are relegated to areas of very low water activity near the surface where rain water is not likely to be saturated with Mg, Al and other ions which form these minerals. In fact much of their presence is due to highly concentrated solutions, desert lake and other closed basin sedimentation in the presence of carbonates or sulfates (see Velde, 1985, Chap. 7). Hence these minerals are in fact present in the C horizon of alteration profiles but not in soils, and usually in sedimentary deposits where soil is little developed at the surface (see Sect. 2.4.2). An interesting case of palygorskite formation is documented by Hillier and Phandande (2008) where the mineral is formed in arid climate agriculture in cases of flood irrigation. It is found in below surface horizons due to salt accumulation. Surprisingly the formation of this mineral can occur over periods of 40–50 years or less.

4.1.6.2 Zeolites

Zeolites are tecto silicates classified in roughly the same mineralogical and structural family as feldspars. The contain alkali ions and have a high silica content. Such minerals would not be expected to form in soils nor in regimes or water – rock interaction. However, in regimes of intense evaporation of surface waters, zeolites are known to form. Spiers et al. (1984) describe such a situation in surface environments. Zeolites as well as palygorskite can be found in the evaporate crusts of saline lakes, features which are not related to lateration profiles. Jacob and Allen (1990) observed that in fact alkali zeolites were unstable in the A horizons of arid soils in Texas. But as with palygorskite, when soil pore waters are drawn to the surface under conditions of intense evaporation, one can find under certain circumstances the formation of zeolites as deposites in saline sediments. These are minerals which contain the elements which have been taken from alteration zones under conditions of water – rock interaction. However, if the water carrying these elements does not reach a large body of water (the ocean) then one can at times find a re-deposition of the dissolved elements in the alteration sequence. In the case of zeolites, the zone of deposition through evaporation of pore waters is in the A and B horizon unlike the palygorskite occurrences which are found in the B or C horizons.

The zeolites are high in silica and alkali content, especially potassium, forming clinoptolite, the potassic low temperature zeolite phase. They are accompanied by amorphous silica. The alumina needed to form the zeolites appears to come from smectites in the soil profile. The zeolites are accompanied by an increased clay content of illite, also a potassium mineral. Such occurrences are rare, of course in that they go counter to all of the normal alteration chemical principles, but it is logical that waters draining alteration sequences, enriched in alkalis and silica would produce such minerals if they were found in an evaporative environment.

It is unlikely that any plants will be found in such evaporitic environments, or only as sporadic islands of vegetal activity. Hence it is not appropriate to classify zeolites as soil clay minerals as is the case for palygorskite.

This overview of clays and clay minerals present in soils and alteration zones indicates a vide variety of minerals, with different structures and chemical compositions. One can classify them into the silicates where silica is linked to aluminum ions through bridging oxygens. These minerals are the the 2:1 phases (HI, illite and smectites), 1:1 phases (kaolinite and halloysite) which are crystallized and can be identified by X-ray diffraction. Minerals of similar composition but lacking long range organization can be present and usually dominant when present. Oxides and hyroxy-oxides can be important and at times dominant. Such are the results of interactions of rainwater with rocks and the modulation of this material by plants.

4.2 The Critical Clays, the 2:1 Clay Minerals in the A Horizon: Examples in Soil Profiles Showing the Effects of Plants and Their Bio-agents

In looking at the literature on clays and clay types found in natural settings, it is apparent that certain features of plant regime can affect the type of clay mineral present in the soil zone. Given the same climate conditions, differences in topography or plant succession can effect transformations in the clays. The best examples, or most clearly observed, are those where human intervention has varied the vegetal cover of essentially the same soils based upon the same materials under the same climatic conditions.

4.2.1 Trees of Different Types: Conifer and Deciduous

Tice et al. (1996) report changes is soil clays over a 41 year period where oak and pine forest were compared. Observations were made on soils developed from granitic material under a Mediterranean climate. The initial and surrounding vegetation was shrub and sparse grass. Plantation of pine and oak was made and the soil clays observed after a 41 year growth period. The soil – alterite profile clay mineralogy shows a strong soil vermiculite (HI) component which becomes illitic upon potassium saturation in the alterite zones. This is the case for both both oak and pine forest soils. However in the A horizon, the pine clay mineralogy is the same as that at depth but the oak forest shows a large loss of vermiculite and the formation of illite. Archived samples show the vermiculite mineralogy in all samples. Hence the oak forest has provided enough potassium to change the soil vermiculite into an illite by forming an anhydrous potassium interlayer mineral. Here the 2:1 structures of the minerals are most likely very little affected by the plant regime. The only change has occurred by a strong enough presence of potassium furnished by oak leaf litter to saturate the high charge sites in the HI material.

The plant regime controls the chemistry of the soil solutions sufficiently to change the interlayer ion site populations.

4.2.2 Trees and Prairies: Contrasts Brought About by Plant Regime

4.2.2.1 Fluvio – Glacial Sediment Clays

Clays in soils in Denmark, (Madsen and Norberg, 1995) based upon fluvio – glacial materials were determined for different plant regimes. The regimes have been present for 60–75 years. The different plants were spruce (*Picea abies*), soils under a large oak tree, under heather – crowberry and another under grass vegetation. Clays in the lower, B horizons of the sandy soils showed similar clay suits essentially smectite minerals and a lesser amount of kaolinite. The clays are dominantly 2:1 types with exchangeable ions present in the interlayer positions.

1. Under spruce trees the clays develop a strong HI or soil vermiculite character in the surface soil horizon. The interlayer sites are filled, most probably, with an incomplete Al hydroxyl layer. However a large portion of the clays remain as smectites.
2. Under heath cover the clays develop a HI – vermiculite mineral suite and a strong illite phase is present.
3. Grass produces a similar mineralogy to that of the heath, with illite and HI vermiculite.
4. The oak tree changed the clay suites the least, but a definite shift to an illite phase is visible.

The initial clays, those deeper in the profile which were probably present at the surface in the early stages of the experiment, were affected by the plant regime producing phases that indicate high charge layers of 2:1 minerals, expressed either as illite (anhydrous potassium interlayer) or HI – vermiculite with Al-OH interlayer ions. One can suppose that the initial clays have a lower charge, and most likely they have less access to hydroxyl-Al and K ions. In the cases of grass -heather, a significantly greater availability of potassium produced the illite mineralogy.

Augusto et al. (2001) have identified the effects of Al incorporation into high charge 2:1 mineral layers (vermiculites) as induced by different types of tree regimes, conifer and deciduous at two forest soil sites. Test vermiculites were imbedded in soils under different tree forests. The results, after a three year experiment, indicate that the type of tree present can effect fixation of hydroxyl-cations and that the soil type is also important. Thus the plants and the overall alterite chemistry combine to determine the activity of different ions in the soil solutions. None of the tree types appears to have induced an illitic (potassium) mineral in the vermiculite test minerals. However, not only Al but also Mg hydroxyl ions were detected as being fixed in the clay interlayer sites. The experimental design was such that the effect of plant regime was tested at different depths (0–40 cm) in the soil – alterite materials. There is a significant difference in reaction effects, decreasing with depth, in all cases. The greatest clay change is effected in the upper 5 cm of the profiles, essentially in the leaf litter zone. This is direct evidence that the chemistry of soil solutions affecting clay mineralogy is driven by plants and that the results are seen in a short period of time, in this case 3 years.

4.2.2.2 High Mountain Wet Prairies

Unpublished data gathered by Velde, illustrate some of the effects of plant control of clay minerals in young soils in these cases based upon granite (Bighorn mountains, Sheridan Wyoming) and rhyolite (Hebegen Lake, western Montana). Three series of samples were taken from zones near the natural prairie – forest interface in high mountain prairies. Soils below isolated trees or forming small groups of 2–5 individuals surrounded by prairie or soils from natural prairie were sampled in triplicate. Sampling was within 20 m one from the other at each of the sites. The soils were relatively young with a maximum of 30 cm depth to bedrock. No significant layered or horizon structure was observed, i.e. A, B and C horizons. Strong organic content was found throughout the soil zone. In general, the prairie soils showed strong illite content and much less interlayered illite-smectite material. Soils under both aspen (deciduous) and spruce (conifier) trees showed significantly more S/I and I/S minerals in the XRD diagrams (Figs. 4.9 and 6.10).

The spectra are very similar for the three types of forest cover. Illite is the dominant soil clay mineral present. Hence in the initial stages of weathering, the potassic 2:1 mineral dominates. In all spectra one finds a small HI – soil vermiculite peak indicating the presence of a hydroxyl interlayer ion in some of the clays. The major difference from one biome to the other is in the mixed layer mineral content. The illite/smectite mineral is most present in the spruce forest soils and least present in the prairies soils. These observations suggest that the prairie regime favors the formation of a high charge 2:1 mineral where an abundance of potassium, from the decomposition of the granite, is fixed as an anhydrous layer in the minerals.

If we assume that the conifer and aspen trees are replacing prairie vegetation in these boundary zones, it is apparent that the first soil clay mineral formed in abundance from the underlying rocks by interaction with prairie plants is illite. Since both the granite and rhyolite contain much fine – grained muscovite, this is not surprising. A small but definitely HI mineral peak is present. One should note that there is a significant amount of PCI present (poorly crystallized illite, PCI), indicating the possible presence of soil illite. Illite remains a major component of the soils under conifer and deciduous trees but very much interlayered illite – smectite is present as well as a persistent HI mineral. It appears that in these cases the initial stages of tree colonization engender the formation of smectitic minerals.

These observations are similar to those made on the four vegetation types in Danish soils where spruce develops more smectitic (low charge 2:1 structures) and HI – vermiculite soil clays (Madsen and Norberg, 1995).

4.2.2.3 Sequoia and Grass

A comparison of clay minerals developed under *Sequoia gigantea* trees and surrounding grass (prairie) (BV unpublished) gives similar information to the above observations of mountain prairie and forest clay suites but in a more limited time frame. In France the sequoia, a tree from northern California, was introduced and

Fig. 4.9. (continued) unexpected to see illite (10 Å mica – like mineral) present. However it appears that the tree soils are more active in extracting the potassium of the clays in the top soil forming smectitic minerals

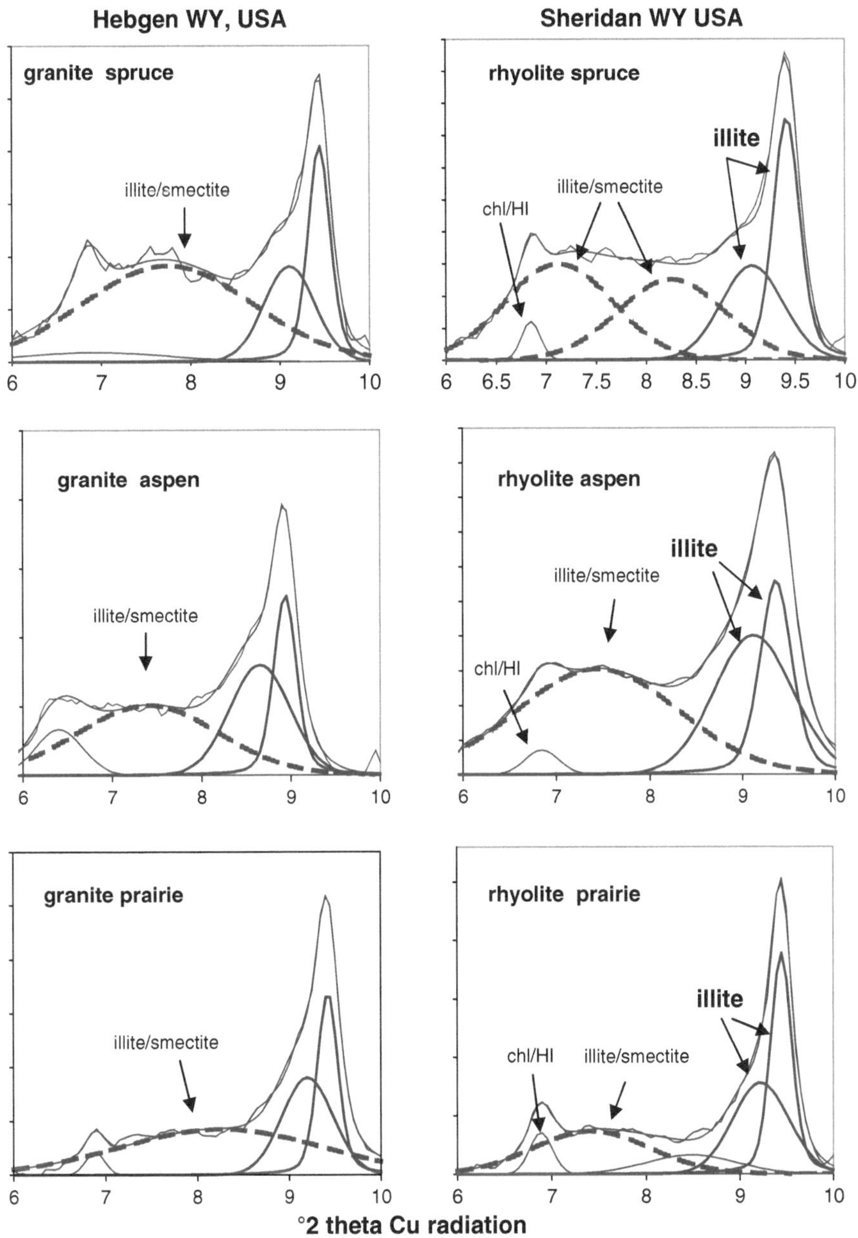

Fig. 4.9. Example of the effect of different biomes on clays during alteration of granitic rocks in the Rocky Mountains (WY) USA. Background – subtracted X-ray diffractogram treated using decomposition methods. The samples came from soils beneath trees or on prairie within tens of meters of one another. Spruce tree soils show a strong illite/smectites component, aspen (deciduous trees) show a similar mineralogy while the prairie shows a much stronger illite component. All soils have a small HI mineral content. Since the granites and rhyolite contain muscovite, it is not

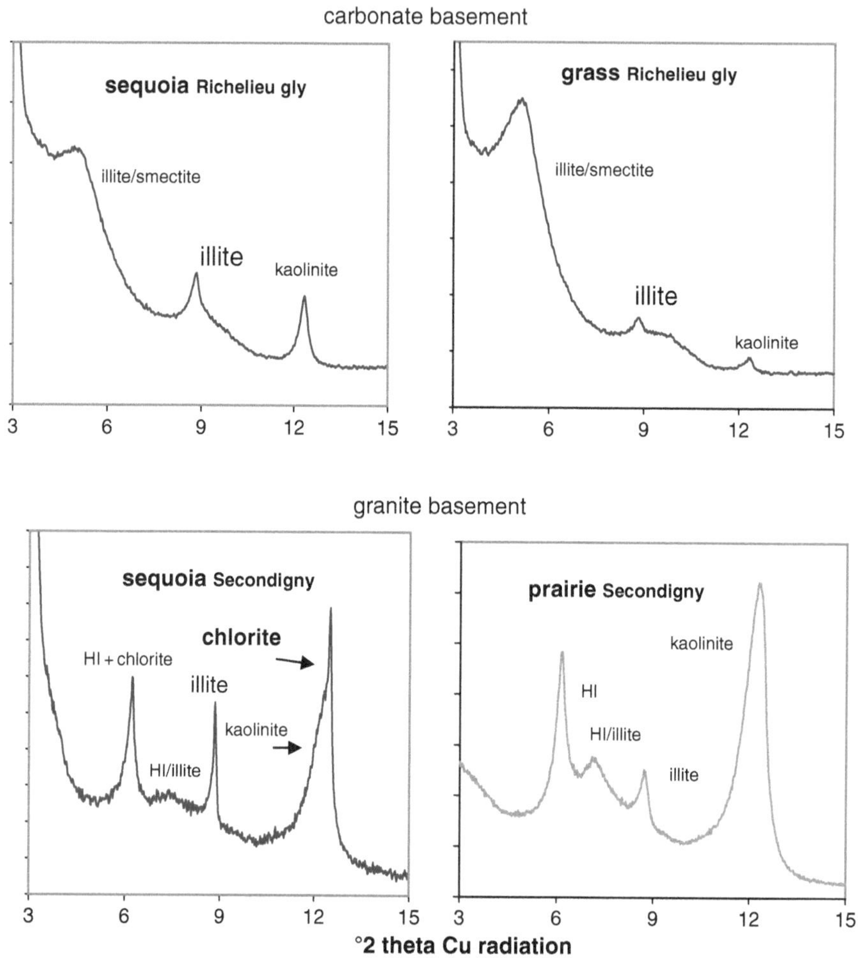

Fig. 4.10 Example of the soil clays developed under sequoia trees after 150 years growth compared to those of surrounding prairie (Richelieu, Vienne and Segondigny, Deux Sevres France). The clay mineralogy is similar in the carbonate basement soils with an decided increase in illite under sequoia whereas the granite basement soils produce Hydroxy interlayered minerals either HI (aluminous or magnesian) under prairie or chlorite (magnesian) under sequoia. This sequence indicates the importance of rock chemistry in forming new soil clays as well as the influence of the types of plants growing in them

used to form a part of the romantic style landscaping on great domains in the 1850s. The trees planted in the parks around older castles grew and developed surrounded by trimmed grass or natural prairie. The previous land use of the terrain in the parked areas is not known. However we do know that the plant regime was determined and maintained from given dates. Thus in such instances we have a means of comparing the effects of two given plant types on the soils under them.

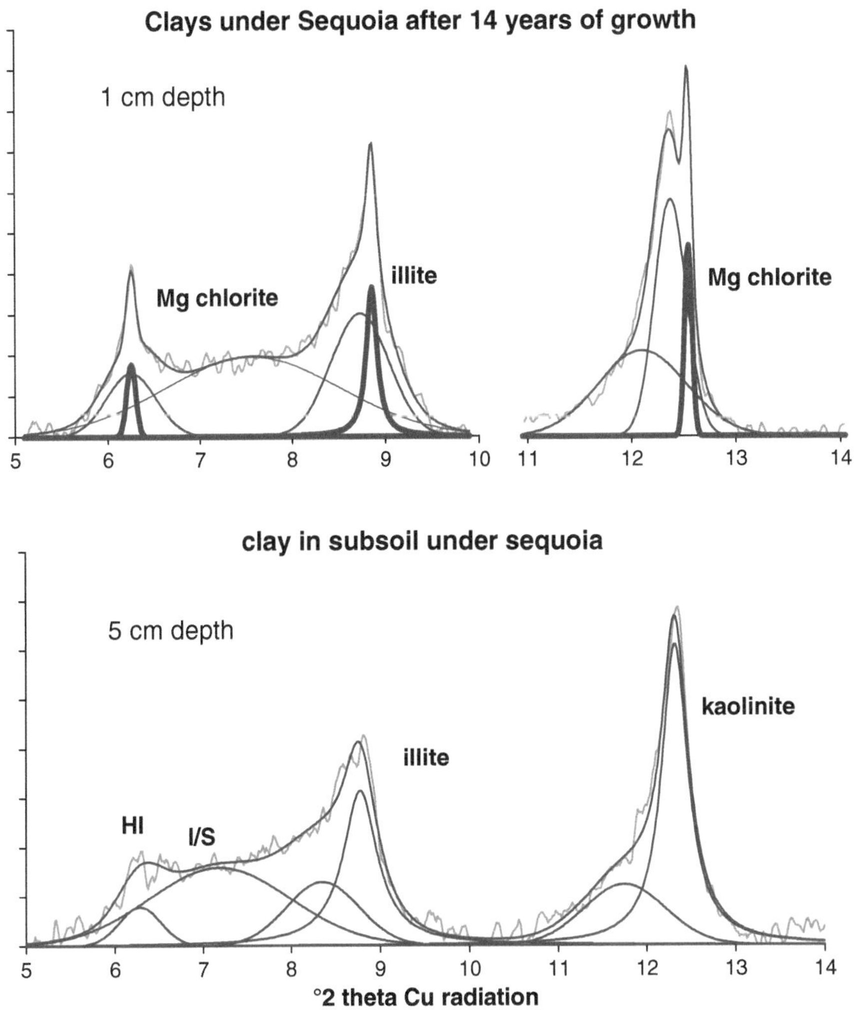

Fig. 4.11 (**a**) Soil clays under sequoia after 14 years of growth. Background – subtracted X-ray diffractogram treated using decomposition methods. The upper diagram indicates the presence of a magnesian chlorite, formed by insertion of a Mg hydroxyl interlayered layer in a smectites. The 5 cm deep sample below the top soil shows the initial clay mineralogy found at depth and around the tree in the grassed area (La Percerie, Vienne, France). This rather rapid change in interlayer ion occupancy type changes the apparent clay mineralogy, where the chlorite is chemically inert compared to the smectites minerals it replaces; However potassium saturation extracted the interlayer Mg hydroxyl material, indicating that the interlayering was not polymerized and hence rather unstable. (**b**) Photograph of the 14 year old sequoia. (photo DV)

Comparison of the soil clays under the two plants shows differences in fine clay potassium content, usually higher under sequoia, and the clay mineralogy where illite tends to be more distinguishable. Six pairs of sequoia – grass soil samples

Fig. 4.11 (continued)

gave an average K_2O content of 3.21weight percent for sequoia clays and 2.78 for those under grass. This is not enormous but still 15% more potassium under sequoia is potentially important. The samples were collected on essentially two types of subsoils, one on carbonate and the other on granite, gneiss and loessic materials. The carbonate soils, more basic pH being controlled by carbonate fragments in the top soil, showed mostly smectite/illite mineralogy while the granite – loessic soils, more acidic, showed strong HI mineralogy. All soils contained kaolinite in the clay assemblage. Thus initially it is apparent that the type of subsoil, or alteration substrate, is important in defining the clays present. The 2:1 clay minerals indicate a hydroxyl component in the interlayer sites which is lacking in the carbonate substrate soils. In all soil types the sequoia showed a tendency to have more potassium minerals and an equivalent or a higher potassium content that that found for the grass soils. X-ray diffractograms are shown in Fig. 4.10 which indicate the differences between plant regime clays on different substrates. Two minerals differentiate the two plant regime clay assemblages: illite present in all cases but more abundant in the sequoia soils, and chlorite present in three of the more acidic soils under sequoia but not present in the corresponding prairie soils. The chlorite is identified by the sharp peak at 14.2 and 7.1 Å. In the sequoia Secondigny sample the 7.1 Å peak is distinguishable but less so for the 14 Å peak where HI minerals are strongly present and mask the chlorite peak. Such a chlorite was also identified under sequoia for a tree planted for just 14 years compared to the surrounding grass (Fig. 4.11) where the both the 7 and 14.2 Å peaks can be clearly seen. Here there is also an increase in a sharp illite peak indicating an increase in the potassium component of the clays as well as the presence of the chlorite peaks. Thus the increase in the potassic (illite) and (chlorite) components of the clay assemblages can occur over short periods of time (14 years). However, as the image in Fig. 4.11b indicates, the size of the tree in question is such that one could expect significant control of the soil surface chemistry.

When identified the intensity of the 7Å (002) peak of the chlorite was more intense than that of the 14.2 Å peak (001). This indicates that this new soil clay mineral contains magnesium in a hydroxyl interlayer site (see Sect. 1.3). This comparative study indicates that the soil substrate (chemistry) can influence the phases produced under two different plant regimes in a parallel manner, a similar conclusion to that of Augusto et al. (2001) concerning the effects of soils under different types of trees at two different sites as they affect a high charge 2:1 vermiculite mineral.

Plant regimes can nevertheless control to a certain extent the type of interlayer ion present in the interlayer ion site of 2:1 minerals (illites and chlorites) by controlling soil pore fluid chemistry. However, the effects are modified by the general chemistry of the soil substrate, the bedrocks. For the moment we do not know the dynamics of such processes or the reversibility of ion exchange in interlayer sites. The presence of anhydrous potassium indicates a relatively fixed site where normal ionic concentrations in soil solutions will not significantly modify the interlayer ion population. The time necessary to build up a strong hydroxyl layer is less well understood. In the sequoia experiment, the chlorite formed after 14 years was unstable in 1 M KCl solution, whereas the chlorite formed over 150 years was not affected by the treatment. Hence the chemical stability of the hydroxyl phases is perhaps a function of greater periods of time than the formation of anhydrous potassium layers.

In certain instances the forest – prairie plant regime can concord to produce the same type of clays at the surface. In Fig. 4.12 oak forest and prairie soils (maintained for nearly 100 years) show a change in clay composition (clay type and hence soil chemistry) developed on a sandy soil on a site near Paris France. The deeper samples (45 cm) show a strong HI mineral content, whereas the oak forest and prairie both show a dominance of illite (potassic 2:1 clay mineral) in the surface (soil) A horizons. In these examples it appears that both oak forest and grassland replace a significant portion of the interlayer HI hyroxy ions with potassium, which is held in the anhydrous state giving a 10 Å spacing. In this case there is a convergence of clay mineral types by the action of the plant regime at the surface in the organic A horizon (Fig. 4.12).

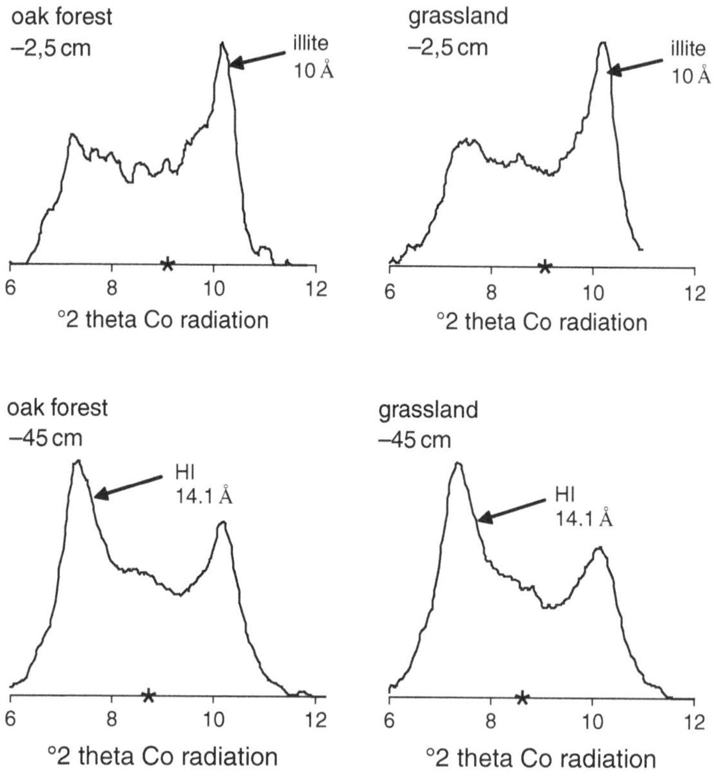

Fig. 4.12 Grassland and oak clay mineralogies at two different depths on sandy soil (ENS facility at Folljuif, Yvelines, France). Spectra are an average of four sampled sites

4.2.2.4 *Shrubs and Prairie:* Red and Black Soil Sequences

In a number of semi-arid steppe climates, it has been noted that basalts and basic rocks produce alternating red and black soils, most often depending upon topography (Bühmann and Grubb, 1991; Herbillon et al., 1981;Vingiani et al., 2004). Typically the black soils are found in small basins or shelves along a topographic sequence in rolling hill country. On the slopes one finds red soils. The red soils are composed dominantly of kaolinite and iron oxides and the black soils are more smectite – rich. However, close inspection indicates that the soil clays contain smectite/kaolinite mixed layer minerals. The mixed layer minerals in the red soils have a larger kaolinite component and the black are more smectites – rich. The proportions vary from one site to another. Typically, the soil clay composition becomes more potassic at the surface and the clays have a small illite component whereas it is almost absent in the deeper C horizons. On the slopes iron oxide and kaolinite rich mixed layer minerals dominate. Further the black soils, as one would expect, have a higher organic material component.

In general, the clay content of the black soils in the depressions is greater than that of the red soils on the slopes. Hence one finds more clay and more organic matter in black soils. It is well known that grasses thrive more in wet, clay rich soils, especially in semi-arid climates. On the slopes one finds scrub and sparse grass. The studies showing clay mineralogy were conducted on grazing or farmed land so that these vegetation types have been effaced for the largest part. Nevertheless one can propose that the depressions were grass – rich and the slopes with less vegetation. The difference in biome led to a difference in clay mineralogy, where reducing conditions fixed the iron in a divalent state in the smectites (Vingiani et al., 2004). The chemistry of the bedrock, basalt, leads to the strong effect of iron on the clay mineralogy. The plants determine the local chemical constraints, reducing environment in the black soils and higher potassium content at the surface producing some illite. This favors the presence of iron-rich smectite at the surface horizon. In soils with kaolinite/smectite mixed layer minerals developed on normal pelitic (aluminous) materials, one finds a lower smectite content in the surface horizons in the clay mineral structures (Churchman et al., 1994).

Here it is clear that plants and bedrock, guided by topography can lead to special clay assemblages under the influence of the chemistry engendered by the biome. In this case, the limit between 2:1 minerals and 1:1 minerals, the major change in silica content of the phyllosilicate clay minerals.

From the above studies one can conclude that the type of biome can influence greatly the type of 2:1 clay interlayer ion population, from hydroxyl interlayer ions (HI-vermiculite) to illite (anhydrous potassium) to smectite (hydrated exchangeable cations). Similar soil substrates, bedrocks or sediments, give similar results. However if the substrate (source rock) is strongly different, the clay suite will be different. Nevertheless the tendency of sequoia trees to form potassic clays seems to be trans-source rock in nature, even though the clay assemblage is significantly different under different chemical (bedrock) influences. The lesson here is that one should remember that the type of clays in a soil can be a function of the plants as well as the source rock. The source rock imposes a basic chemistry, acidity and relative abundance of different cations, whereas the plants impose a chemical flux which conditions the clays present. We do not have sufficient data at present which can indicate the extent to which plants can influence the charge on the 2:1 minerals which will in turn determine the type of cation fixed in the interlayer sites.

4.2.3 Climate, Plants and Soil Clays

An example of the effect of climate, essentially determined by rainfall and its seasonal variations is given by Sifferman (1973) for soil clays developed in Cameroon. One finds tropical rain forest climate and biomes and a shift, to the north, towards the savanna climate and biomes. The author follows these variations as they affect the clays in soils. Soils developed on both acidic and basic eruptive rock substrates were investigated. The major climatic variation considered was annual rainfall, under similar temperature regimes. Such climatic control inevitably varies the vegetal regime, going from tropical forest to savanna in a contrasted seasonal regime. The rainfall varies from 12 to 1.7 m per year. In the high rainfall regime (mount Cameroon) one finds, on these glassy eruptive rocks, a strong development of amorphous materials strongly associated with organic matter in the surface soil horizons. The silicate and other mineral phases are dominated by the amorphous allophone and the crystalline, low silica minerals gibbsite and halloysite (hydrous kaolinite). A small amount of smectite is present, a 2:1 more silica rich mineral. As rainfall decreases and becomes more contrasted into two seasonal regimes, dry and wet, the clay mineralogy changes becoming less rich in allophane and containing different forms of kaolinite. Gibbsite (aluminium hydroxide) decreases in proportion and is replaced by vermiculite, the aluminium hydroxyl interlayered form of smectite. The proportion of 2:1 minerals increases, initially representing only about 5% of the mineral phases. The amount of organic matter also decreases. The final stage observed is the transformation of vermiculite to an aluminous chlorite, i.e. a 2:1 mineral with non-exchangeable aluminium hyroxy units present.

The author attributes the increase in interlayer aluminium to the presence of organic matter, very abundant in the smectite forming regime and much less abundant in the chlorite forming regime (p. 156). It is possible that the type of organic matter could have a role in the availability of the aluminium. Different plant regimes tend to produce different types of organic matter as it decomposes in the soil. Andosol soils are known to fix aluminium on organic mater forming a very stable humic- mineral complex. In the grassland soils of Mount Cameroon this association has not been noted. One should then consider that the presence of free alumina in soil solutions can be controlled to a certain extent by the type of organic matter present, highly complexing or less so. The end result is that the high rainfall clays contain much less Si than those of the savanna soils formed in drier climates. A higher proportion of 2:1 clays are present in this regime also, towards 20% of the soil minerals.

In this sequence the climate differences engender a difference in plant regime, and probably a change in the chemical control of dissolved elements in the soil solutions depending on the complexing power of the organic material issued from the biologic degradation of the plant detritus. A schematic diagram of the climate – clay relations is given in Fig. 4.13.

In this climate – biome sequence, formed on the same rock substrate, the high rainfall tropical forest soils contain very much amorphous or poorly crystallized material, as well as gibbsite. As rainfall decreases the amorphous materials are replaced by 1:1 clay minerals, initially kaolinite and then mainly halloysite. The relative amounts of 2:1 minerals remain roughly constant, increasing some in relative amount with initially smectite, then soil vermiculite (HI mineral) and finally one has noted the presence of chlorite, probably a magnesian interlayer 2:1 aluminous mineral. As one sees more 1:1 clays there is less organic matter present indicating the strong link between organic matter and amorphous silicate and oxides in Andosol soils which globally become 1:1 clays. It is interesting to note that the relative amounts of 2:1 clays remains roughly constant, being present in the upper parts of the soil – alteration sequence.

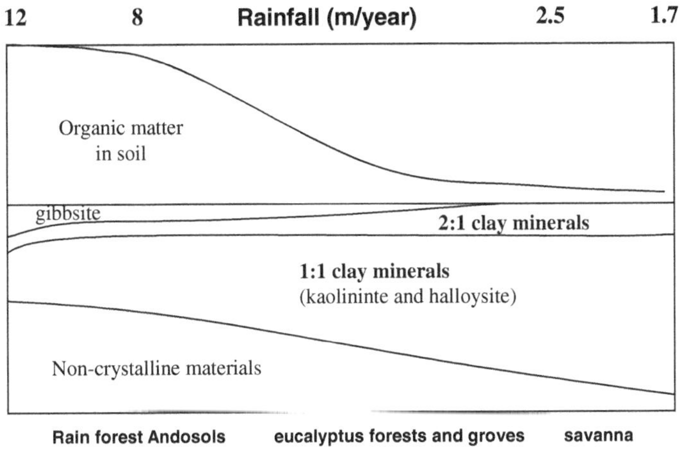

Fig. 4.13 Representation of the effects of rainfall on vegetation and clay mineralogy in the African rain forest – savanna zone Andosols developed on volcanic materials of Cameroon (after Sifferman, 1973). Organic matter follows rainfall in this sequence, as does the importance of non-crystalline materials. High rainfall produces gibbsite while lower rainfall produces more kaolinite and a certain amount of 2:1 clays

Climate, as it controls the vegetative materials and affects the through put of water in the system, then should be considered in assessing the types of clays present in a given soil. Plant regimes can vary the clays present under similar climate conditions but the second variable is certainly that of climate.

What needs to be done

Studies of rate of change of clay minerals when plant regime is changed. Studies of effect of different plant species on clays

References

Aleva G (1994) Laterites: Concepts, Geology, Morphology and Chemistry, ISRIC, Wageningen, Netherlands, 156 pp.

Arocena J, Glowa KR, Massicotte HB and Lavkulich L (1999) Chemical and mineral compositions of ectomycorrhizosphere soils of subalpine fir (Abies lasocarpa (Hook.) Nutt.) in the Ae horizon of a Luvosol. Can J Soil Sci 79: 25–35.

Arnalds O (2004) Volcanic soils of Iceland. Catena 56: 3–20.

Augusto V, Ranger J, Turpault M-P and Bonnaud P (2001) Experimental in situ transformation of vermiculites to study the weathering impact of tree species on the soil. Eur J Soil Sci 52: 81–92.

Badraouri M, Bloom P and Bouabid R (1992) Palygorskite – smectite associations in a Xerochrept of high Chaouia region of Morocco. Soil Sci Soc Ame J 56: 1640–1646.

Barré P, Velde B, Catel N and Abbadie L (2007) Soil-plant potassium transfer: Impact of plant activity on clay minerals as seen from X-ray diffraction. Plant Soil 292: 137–146.

Bates, T (1962) Halloysite and gibbsite formation in Hawaii, 315–328, 9th Conference on Clays and Clay Minerals, Pergamon Press.

Bauer A, Lanson B, Ferrage E, Emmerich K, Teubald H, Schild D and Velde B (2006) The fate of smectite in KOH solutions. Amer Miner 91: 1313–1322.
Boulangé B (1984) Les Formations Bauxitiques Latéritiques de Côte-D'ivoire, Travaux et Documents de l'ORSTOM, Paris, 341 pp.
Braga M-A, Lopes Nuenes J, Paquet H and Millot G (1989) Essai sur les arènes de l'Europe atlantique. Zonalité climatique, Compte Rendu Acad Sci 309: 1955–1962.
Bühmann C and Grubb P (1991) A kaolin-smectite interstratification sequence from a red and black complex. Clay Miner 26: 343–358.
Churchman G, Slade P, Self P and Janik L (1994) Nature of interstratified kaolin-smectites in some Australian soils. Aust J Soil Res 32: 805–822.
Cradwick PDG, Farmer VC, Russell JD, Masson CR, Wada K and Yoshinaga N (1992) Imogolite, a hydrated aluminum silicate of tubular structure. Nature (Lond) Phys Sci 240: 187–189.
Deer W, Howie R and Zusmann J (1962) Rock Forming Minerals: Phyllosilicates, Vol. 3, Longmans, 353 pp.
Derry L, Kurtz A, Zeigler K, Chadwick O (2005) Biological control of terrestrial silica cycling and expert fluxes to watersheds. Nature 433: 728–730.
Farmer VC, Adams MJ, Fraser AR and Palmieri F (1983) Synthetic imogolite: Properties, synthesis and possible applications. Clay Minerals 18: 459–472.
Farmer V (1999) The solubility characteristics of organic Al precipitates in allophonic Bs horizons could mimic those of Al hydroxides and indicate the solubility of associated allophanes. Eur J Soil Sci 50: 713–715.
Farmer V and Lumsdon F (2001) Interactions of fulvic acid with aluminium and proto-imogolite sol: The contribution of E-horizon eluates to podzolization. Eur J Soil Sci 52: 177–188.
Farmer C, Delbos E and Miller J (2005) The role of phytolith formation and dissolution in controlling concentrations of silica in soil solutions and streams. Geoderma 127: 71–79.
Fauzi A and Stoops G (2004) Reconstruction of a toposequence on volcanic material in the Honje Mountains, Ujung Kulan Peninsula, Java. Catena 56: 45–66.
Garcia-Rodeja E, Novoa J, Pontevedra X, Martinez-Cortizas A and Burman P (2004) Aluminum fractionation of European volcanic soils by selective dissolution techniques. Catena 56: 155–183.
Gillot F, Righi D and Elsass F (2000) Pedogenic smectites in podzols from central Finland: An analytical electron microscopy study. Clays Clay Miner 48: 655–664.
Herbillon A, Frankart R and Vieloye L (1981) An occurrence of interstratified kaolinite-smectite minerals ina red-black soil toposequence. Clay Miner 16: 195–201.
Hillier S and Phandande A (2008) Contemporary pedogenic formation of palygorskite in irrigation – induced, saline – sodic, shrink – swell soils of Maharashtra, India. Clays Clay Miner 56: 531–548.
Hsu P (1989) Aluminium oxides and hydroxides. In J Dixon and S Weed (eds.), Minerals in Soil Environments, Soil Science Society of America, Madison, WI Book series 1, pp. 331–387.
Jacob J and Allen B (1990) Persistence of zeolite in tuffaceous soils of the Texas Trans-Pecos. Soil Sci Sci Ame J 54: 549–554.
Joussein E, Petit S, Churchman J, Theng B, Righi D and Delvaux B (2005) Halloysite clay minerals – a review. Clay Miner 40: 383–426.
Khademi H and Arocena L (2008) Kaolinite formation from palygorskite and sepiolite in rhizosphere soils. Clays Clay Miner 56: 429–436.
Kleber M, Mikutta C and Jahn R (2004) Andosols in Germany – pedogenesis and properties. Catena 56: 67–83.
Laird D and Nater E (1993) Nature of the illitic phase associated with randomly interstratified smectite/illite in soils. Clays Clay Miner 41: 280–287.
Lelong F (1969) Nature et genèse des produits d'altération des roches cristallines sous climat tropical humide. Sciences de la Terre Mémoire, Vol. 14, Nancy, 188 pp.
Lucas Y (2001) The role of plants in controlling rates and products of weathering: Importance of biological pumping. Ann Rev Earth Planet Sci 29: 135–163.

Madsen H and Norberg P (1995) Mineralogy of four sandy soils developed under heather, oak, spruce and grass in the same fluvioglacial deposit in Denmark. Geoderma 64: 233–256.

Meijer E and Buurman P (2003) Chemcial trends in a perhumid soil catena on the Turriaba volcano (Costa Rica). Geoderma 117: 185–201.

Meunier A (2005) Clays. Springer, Heidelberg, 472 pp.

Millot G and Camez T (1963) Genesis of vermiculite and mixed layered vermiculte in the evolution of soils in France, 90–95. In A Swineford (ed.), Clays and Clay Minerals Earth Sciences Monograph, Vol. 12, Pergamon Press, 509 pp.

Millot G (1964) Géologie des Argiles, Masson and Cie, Paris, 499 pp.

Moberg, J P (1990) Composition and development of the clay fraction in Danish soils: An overview. Sci Géol Bull 43: 193–202.

Ndayiragije N and Delvaux B (2004) Selective sorption of potassium in a weathering sequence of volcanic ash soils from Guadeloupe, French West Indies. Catena 56: 185–198.

Olsson M and Melkerud P-A (1989) Chemical and mineralogical changes during genesis of a Podzol from till in Southern Sweden. Geoderma 45: 267–287.

Paquet H (1970) Evolution géochimique des minéraux argileux dans les sols des climats méditerranéens tropicaux à saisons contrastées, Mémoire Service Carte Géologique Alsace – Lorraine 30, 212 pp.

Pedro G (1966) Essai sur la caractérisation géochimique des différents processus zonaux résultant de l'altération des roches superficielles. Crompt Rendus Acad Sci Paris Ser D 262: 1828–1831.

Pion J-C (1979) Alteration des massifs cristallins basiques en zone tropical Sèche. Sci Géol Mem 57: 230.

Quantin P (2004) Volcanic soils in France. Catena 56: 95–109.

Righi D, Huber K and Keller C (1999) Clay formation and podzol development from postglacial moraines in Switzerland. Clay Miner 34: 319–332.

Righi D, Petit S and Bouchet A (1993) Characterization of hydroxy-interlayered vermiculite and illite/smectite interstratified minerals from the weathering of chlorite in a Cryorthod. Clays Clay Miner 41: 484–495.

Righi D, Räisenän ML and Gillot F (1997) Clay Mineral transformations in podzolized tills in central Finland. Clay Miner 32: 531–544.

Ross G, Phillips P and Culley J (1985) Transformation of vermiculite to pedogenenic mica by fixation of potassium and ammonium in a six-year field manure application experiment. Can J Soil Sci 65: 599–603.

Scheinost AC and Schwertmann U (1999) Color identification of iron oxides and hydroxysulfates. Soil Sci Soc Ame J 63: 1463–1471.

Schultz L, Shepard A, Blackmon P and Starkley H (1971) Mixed-layer kaolinite-montmorillonite from the Yucatan Peninsula, Mexico. Clays Clay Miner 19: 137–150.

Schultz A and Ruhiyat D (1998) Soils of Tropical Forest Ecosystems, Springer Verlag, 204 pp.

Schwertmann U and Taylor RM (1989) Iron oxides, 379–438. In JB Dixon and SB Weed (eds.), Minerals in Soil Environments, 2nd edition, Soil Science Society of America, Madison, 948 pp.

Ségalen P (1995) Les Sols Ferrallitiques et leur Répartition Géographique, Vol. 3, Editions ORSTROM, Paris Tome 189 pp.

Shoji S, Dahlgren R and Nanzyo M (1993) Volcanic Ash Soils, Elsevier, 288 pp.

Sifferman G (1973) Les Sols de Quelques Régions Volcaniques du Cameroun, Vol. 66, ORSTOM Mémoire, 163 pp.

Spiers G, Pawluck S and Dudas M (1984) Authigenic mineral formation by solodization. Can J Soil Sci 64: 515–532.

Temgoua E (2002) Cuirassement ferrugineux actuel de bas versants en zone forestière du Sud – Cameroun. Mem Géol Lausanne 38: 135.

Tice K, Graham R and Wood H (1996) Transformation of 2:1 phyllosilicates in 41-year-old soils under oak and pine. Geoderma 70: 49–62.

Velde B (1985) Clay Minerals: A Physio-Chemical Explanation of their Occurrence, Elsevier, 429 pp.

Velde B (2001) Clay Minerals in the agricultural horizon of loams and silt loams in the Central United States. Clay Miner 36: 277–294.
Velde B and Church T (1999) Rapid clay transformations in Delaware Salt marshes. Appl Geochem 14: 559–568.
Velde B and Peck T (2002) Clay mineral changes in the Morrow experimental plots, University of Illinois. Clay Clay Miner 50: 364–370.
Velde B and Meunier A (2008) The Origin of Clay Minerals in Soils and Weathered Rocks, Springer, 28 pp.
Wada K (1989) Allophane and imogolite. In JB Dixon and SB Weed (eds.), "Minerals in Soil Environments", 2nd edition, Soil Science Society of America, Madison, pp. 1051–1087.
Vingiani S, Righi D, Petit S and Terrible F (2004) Mixed layer kaolinite-smectite minerals in red-black soil sequence from basalt in Sardinia (Italy). Clays Clay Miner 52: 473–483.
Wilson M and Cradwick P (1972) Occurrence of interstratified kaolinite – montmorillonite in some Scottish soils. Clay Miner 9: 435–437.
Wolt J (1994) Soil Solution Chemistry, John Wiley, 345 pp.

Chapter 5
The Chemistry and Mineralogy of Plant and Soil Interactions: Plant as Manipulators of Their Environment

Contents

5.1 Soil Properties Pertinent to the Prosperity of Plants

5.1.1 Terrestrial Plants and Peri-Maritime Plants: Alkali Content

Terrestrial plants are tied to the earth. They are relatively immobile, at least for each generation, and they must adapt. The earth forms a base for growth in what is called soil. However this material, necessary for plant development, is

B. Velde, P. Barré, *Soils, Plants and Clay Minerals*,
DOI 10.1007/978-3-642-03499-2_5, © Springer-Verlag Berlin Heidelberg 2010

composed of a variety of components. Much of the interaction of plants and the surface of the earth is in the surface tens of centimetres where most of their roots are developed.

If one looks at the composition of plants from peri-maritime and adjacent arid areas in Syria (Barkoudah and Henderson, 2006) it is clear that there is a reciprocal relation between sodium and potassium. This stems from the relative availability of the two alkali elements, apparently interchangeable in plant metabolism chemistry. The sea is salty, a simple truism but not all that innocent. Seawater concentrates elements that are not incorporated into solids during the weathering cycle and sedimentation cycle. Elements that find their way to the sea as dissolved ions are basically Ca, Mg, and Na. These are not retained in alteration products of the water – rock interaction process. Sea animal activity efficiently decreases the concentration of Ca, which combines with Mg to make carbonate material, decreasing the overall content of these elements in sea water. Potassium is present in only low abundance mainly because it is held in clays. Sodium is not held in clays. Hence terrestrial plants are potassium – based, not sodium based. The plot of Na and K in Fig. 5.1 indicates the gradual changes in plant chemistry as the relative proportions of sodium and potassium are available in the plant environment. The data represent plant ash compositions for four *salosa* shrub species, which indicates a small range of K-Na variation for each. This indicates that the different *salosa* types adapted to more and more potassic environments but also that each plant type can adapt to a certain extent to its chemical environment. However, by and large, most terrestrial plants are potassic, and sodium is of very low abundance in plant ash. Thus the needs of potassium are of great importance for plants and the soil clay resources of this element are key to plant – soil interaction. This demonstration indicates that plants are highly adaptable to their chemical environment. Not only do they adapt, but they manage to constrain some chemical parameters so that they can prosper given certain necessities such a mineral element components.

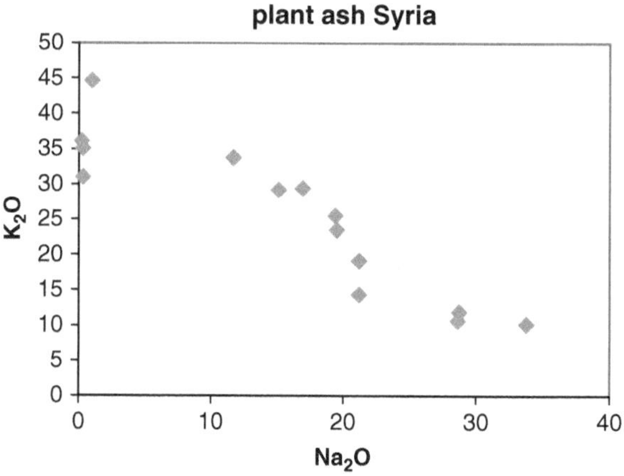

Fig. 5.1 Data for plant ash of different salosa plants in Syria (Barkoudah and Henderson, 2006). The reciprocal relations of sodium and potassium indicate that plants need alkalis for their metabolism and that depending upon the local chemical resource they can adapt to the use of one or the other

5.1.2 Elements Needed by Plants

There are basically two sources for the elements which make up plants. The atmosphere supplies carbon and nitrogen. The soil and alterite supply water and mineral elements. Sun energy is captured and transformed into the complex chemistry of plants. This is one of the most efficient methods of energy transfer that we know of. Thus the materials necessary for plant growth are found in two distinct environments air and earth. One normally thinks of rock and mineral destruction by transformation as the major source of mineral elements for plants. These elements are brought into the aerial plant by the root system. However, much of the mineral matter used in plants is in fact re-cycled. Leaf and other litter on the surface of the soil is gradually incorporated into the soil through bacterial action. During this process, and by simple rain water washing, much of the mineral elemental material is released into the soil solutions. The relatively dense root system in the soil zone captures a large part of this material which is re-incorporated into the plant. This occurs on a short time scale, yearly input and decomposition, or longer scale through plant death and decay.

Figure 5.2 indicates these cycles. The soil is the major source of water and nitrogen and a large quantity of the mineral elements such as K, Ca, P, Mg and Si. The

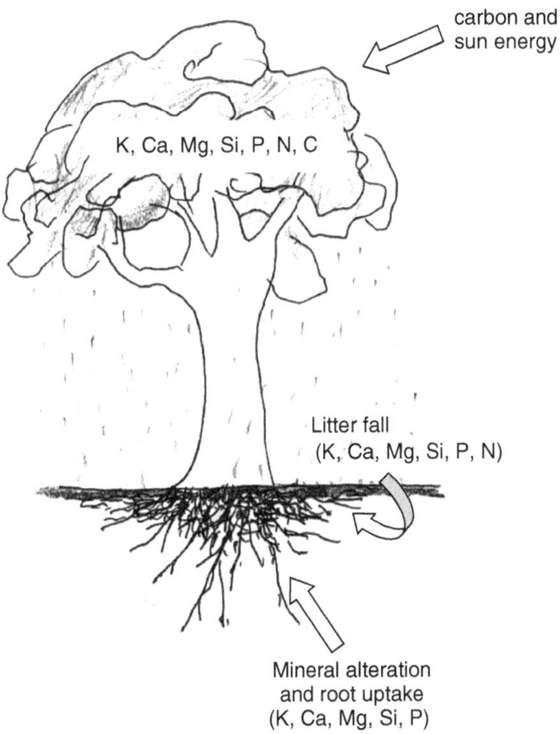

Fig. 5.2 Illustration of the mineral element cycling in the biosphere. Water – rock alteration transfers some elements into the pore solutions where plants can absorb them. Also bio-engendered alteration of minerals releases mineral elements that can be incorporated into the biomass. Litter fall and root decay release these elements which can be re-cycled into the biomass. The movement of mineral elements from the rock sources into the biomass and their re-cycling is a fundamental concept in exchanges at the earth's surface

action of rain water, dilution the soil solutions and carrying away some of the various elements represents a loss of nutrients to the system. Some of this material is recuperated by root action at depths below the soil interaction zone. Plants basically try to keep an equilibrium between deep resources and re-cycled resources. Data from numerous studies can be used to point out this chemical structure. Potassium is often a striking example as it can be concentrated in the soil horizon accumulation through root action. It is most evident when the source rock is potassium – poor, however in certain cases, as in the example given in Fig. 5.3 for a young mountain prairie soil (Righi et al., 1999) one can see the change in K_2O content and SiO_2 as a function of time in the A (plant) horizon compared to the B horizon. The plant zone maintains the potassium content whereas the lower zone more clearly affected by water – rock interaction shows loss of these elements.

Beyond the mineral elements, of course carbon and nitrogen are concentrated in the soil zone through accumulation of plant activity and re-cycling. There are other interactions necessary in order to keep a balance between the needs of plants in the different environments where they are found. Some of the vital elements, those readily soluble, must be kept out of the immediate soil chemistry cycle. These are among the ingenious uses of minerals to the advantage of the plants in the biosphere.

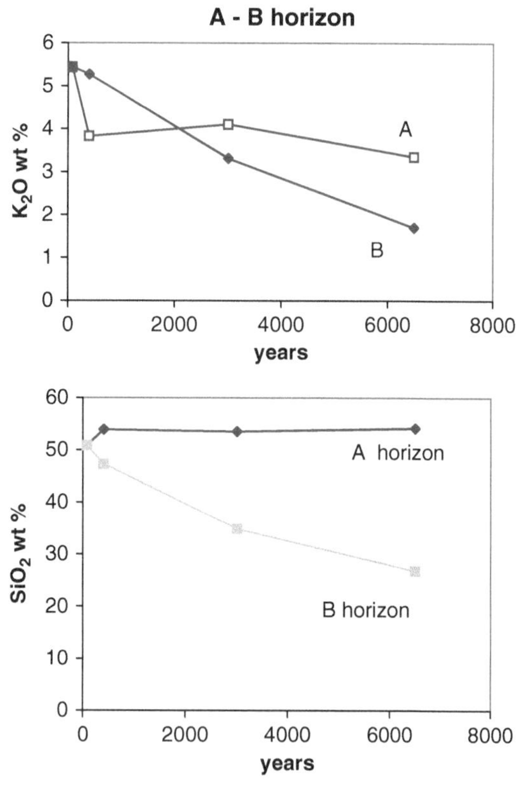

Fig. 5.3 Demonstration of element loss in the organic A horizon compared to the deeper C horizon where water – rock alteration reactions prevail as a function of time. Both potassium and silica are conserved in the A horizon whereas they are progressively lost in the deeper C horizon. Data from Righi et al. (1999)

Some of the major problems for a plant are assuring the availability of the elements necessary for growth. The transfer and retention mechanisms must be assured in order to have a ready supply throughout the growing season. For this plants use the capacities of clay minerals and organic matter.

5.1.3 Clay Minerals

5.1.3.1 CEC Cation Exchange

Exchange cations are in fact those elements necessary for plant growth such as K, Ca, Mg and perhaps ammonium. Their capture and release are primordial for plant growth.

Clay minerals have a large surface area compared to their volume due to their sheet structure which has a small development in one direction compared to the others. The ratio of 1:20 is common. This fact gives a clay particle a large surface area compared to a sphere for example. In some minerals the mineral surface is largely un-charged, i.e. the structure has all atoms with electronic compensation. The kaolinites are such minerals where the large surface of the sheets is without charge whereas the crystal edges are charged due to incomplete bonding of oxygen and Si, Al cations. Other minerals with cationic substitutions within the structure (see Sect. 1.3) leave residual charges on the crystal surfaces. These charges attract cations from aqueous solutions. Thus the surfaces of clays are "covered" with cations in proportion to the charges on the minerals surfaces. Kaolinite, (a 1:1, TO structure) has no residual surface charge due to internal substitutions for example, and has only charges on the crystal edges where un-terminated bonding leaves cations or anions with residual charges. Kaolinite has an attraction of 5–10 milli equivalents of a cation charge per 100 g of clay due to un-terminated bonds on crystal edges. In such cases the amount of charge and cations fixed to the surface varies as a function of the crystal size, the smaller crystals having more edge sites per mass. The more coarse sizes (10–20 μm) have an exchange charge of 2 milli equivalents of cation per 100 g of clay while the fraction 1–0.5 μm have a charge of 10 milli equivalents. (Grim, 1954, p. 137). These are the lower limits of cation exchange for clays.

TOT or 2:1 minerals with a low charge due to internal cation substitution of ions with different charges in a given coordination site (tetrahedral or octahedral in the TOT structure). The charge deficit can be 0.4–0.6 per unit cell $O_{10}(OH)_2$ which necessitates compensating cations to migrate between the layers of the structure (TOT or 2:1 layers) where they maintain electrostatic neutrality within the overall clay silicate structure. These ions are easily exchanged with other ions in solution following the laws of mass action (elemental concentration in the solutions). Smectites are such minerals with low layer charges that have cation attractions (exchange capacities) of near 135 milli equivalents. The amount of charged ions attracted is of course due to the amount of substitutional charge but also to the site of the charge (see Velde, 1985, p. 123).

The measurement of exchange ion capacity, ions easily attracted to the surface of or in interlayer sites in a mineral, is its cation exchange capacity or CEC. Measurements of exchangeable ions are usually made using solutions of 0.5–1 molar cationic strength. Commonly one uses ammonium ions (NH_4^+) to displace cations on natural clays and measures the ammonium present in order to estimate the number of ions and hence charge attracted to the clay particles.

Clays then have charge sites of two sorts, compensated between the crystal structural layers and on the edges, but the cations in solution have different electrostatic and chemical properties. The cation properties are multiple, one being a function of ion size and charge which is in fact concentration of charge per unit surface to the ionic spherical influence of the cations. If two ions are of similar cationic diameters and different charges the selectivity for fixation on a charged surface is in favour of the higher charged ion. Ca and K ions are a good example. The double charged Ca cation is much more strongly selected by clay surfaces than potassium. Not only do relative charges on cations affect selectivity of different species but residual charges on clay structures that attract ions can effect cation selection. The charge intensity and the site of the charge (tetrahedral or octahedral site) play a role in cation electivity. Figure 5.4 gives an idea of such selectivity. This indicates that the clay mineral itself can determine the relative propositions of different cations selected as exchange ions.

The selectivity of one cation over another can also be affected by temperature and ionic concentration which is due to the differences in dissociation constants

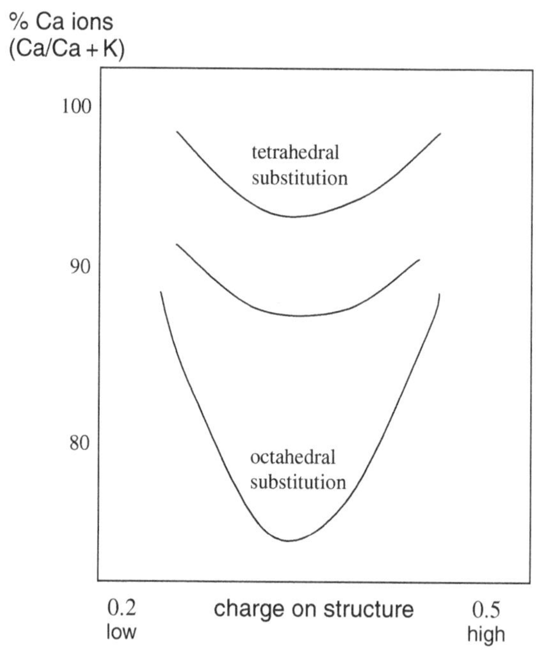

Fig. 5.4 Illustration of the importance of charge on the 2:1 structure in the caion selectivity between K and Ca cations (mono- and di-valent) and the site of the charge imbalance (data from Velde, 1985). The amount of charge changes the selectivity as does the site of the charge in a non linear manner. It is clear that the cation selectivity for mono- and di-valent cations is a complex system

of the different salt forms which release the cations in solution as individual ions. (see Velde and Meunier 2008 for a more detailed discussion). The major factor in clay induced CEC (cation exchange capacity) for a soil is the mineralogy or crystal structure and internal ionic substitutions of the crystallite.

In general, in most soils the CEC (exchange ions) is dominated by the divalent ion Ca and, to a lesser extent, Mg ions. These ions are divalent, and thus have a higher surface charge density than mono valent ions such as potassium. Usually potassium represents only about 10–15% of exchanged ions in soil samples. Thus the selectivity has done its work leaving the divalent ions as dominant cations present and readily available to plants.

There are two principles which are very important to remember when considering cation exchange and the availability of exchange ions. In fact the exchange principle, one ion being replaced by another, is just part of the equation of ionic availability. A major effect is due to dilution of the system. If a clay – organic complex contains a given number of ions on exchange sites under a given set of conditions, the total number of ions can be changed according to the concentration of ions in the aqueous solution which equilibrates with the ion exchanger. As aqueous solutions become more dilute in exchangeable ions, the amount of these ions fixed on clays and organic matter diminishes. Thus in climates of high rainfall input, one would expect that the exchange cation content would be small. In soils where periodic rainfall produces flooding and evaporation (arid climates) the exchange cation population should be high. This principle is outlined in Fig. 5.5.

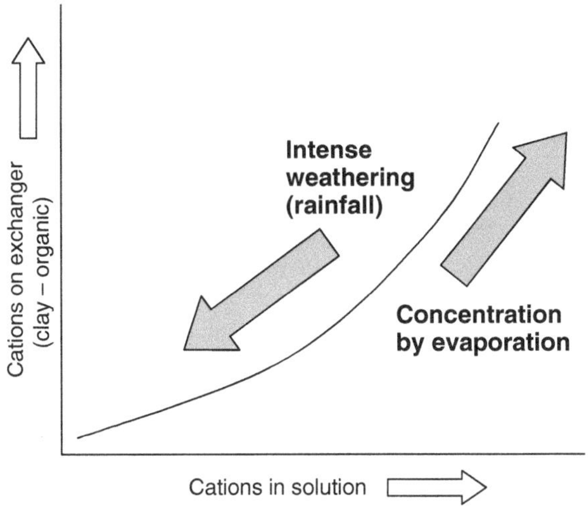

Fig. 5.5 Relations between the amount of water relative to the soil materials and the cations found on the exchangers in the soil The more it rains, the more dilute the solutions and the more cations will be desorbed into the dilute pore water solutions. If the soil solutions are concentrated by water loss due to evaporation, the amount of ions exchanged onto the soil materials will increase. Hence the amount of exchanged cations on soil materials is determined relatively, by the dilution of the pore solutions. If a soil is charged with cations due to an influx at the surface, this amount of absorbed cations can be reduced by rainfall cycles

A second principle, also very important, is that hydrogen ions (H^+) behave in fact as cations in aqueous solutions. That means that a K^+ ion adsorbed on a clay can be displaced into solution when it becomes acidic and one has the relation $H^+ = K^+$. Acid soils show a high hydrogen ion content, reducing the number of what are called the base exchangeable cations. The normal cations are usually associated with OH units. As plants need mineral elements for growth, acidic soils or those under high rainfall tend to diminish these resources by displacement or dilution.

5.1.3.2 Fixing Potassium in 2:1 Clays

Exchange on high charge minerals can induce apparent irreversible exchange, fixing the ions within the structure. The case that occurs in soils is when potassium is in sufficient quantity to form a layer of cations which is anhydrous, i.e. the cations are not surrounded by water molecules which renders the ions for all intents and purposes immobile or inaccessible to normal cation exchange in solutions of relatively dilute concentration, i.e. less than 0.2 M. However, experiments by the author (BV) on soils from central Iowa USA indicate that much of the apparently fixed, anhydrous potassium, can in fact be extracted suing strong (1 m $SrCl_2$) exchange solutions (Fig. 5.6). In the figure the amount of anhydrous potassium mineral (illite) is estimated from X-ray diffractograms using peak decomposition methods (see Lanson, 1997). In the figure the proportion of illite present in the Sr saturated state at low concentration levels (0.01 M $SrCl_2$), normal smectites with two water layer hydrated cation layers, is compared to that for the same samples in the K – saturated state, where there is a tendency to form an anhydrous layer of potassium cations. In Fig. 5.6 the same samples after the potassium treatment have been de-saturated using strong (1 M) $SrCl_2$ solutions. The peak positions are nearly coincident with an equivalence of illite content between the initial state and the K-sat to de-saturation state. Hence in these clays the collapsed (anhydrous ion after K treatment) illite layers can be exchanged with strong cationic treatment. When the solutions are less concentrated in strontium (0.01 M) the potassium remains fixed in the layers. Barré et al. (2007) have shown that the capacity to fix potassium in mixed layer illite/smectite clays can be a function of the solution concentration of potassium ions. Hence the concept of exchange ions and CEC (cation exchange capacity) is to a certain extent ambiguous when considering potassium as the exchange ion. The higher the ionic concentration, the more ions enter into the structure. Other ions do not seem to show such a behavior in that they do not tend to form anhydrous cation layers in the 2:1 structures and increasing the cation concentration in solution does not change their presence greatly.

5.1.3.3 Potassium and Ammonium

It is of course well known that potassium is an important part of plant physiology and that nitrogen (and its soil form ammonium) is important also. As it turns out, the potassium ion and the ammonium complex (NH_4^+) have roughly the same "ionic" dimensions. In soil chemistry practice, the potassium ion is displaced by strong

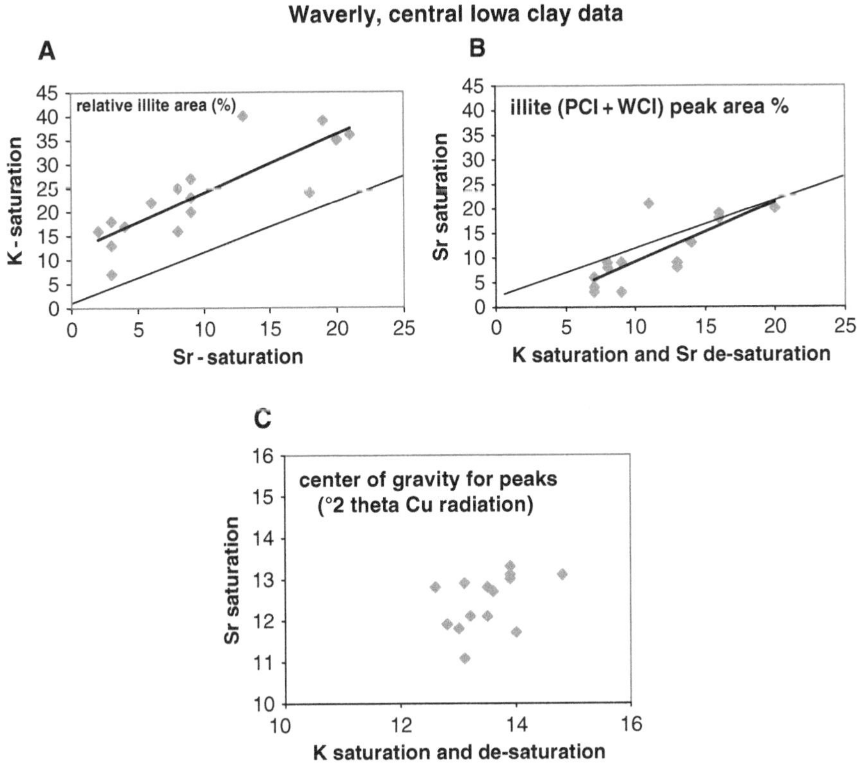

Fig. 5.6 Reversible effect of potassium saturation and de-saturation. Saturation of several Iowa soil clays shows an increase in the illite content (fully saturated interlayer sites where the potassium is anhydrous) shown in Figure (**a**). In comparing (**a**) to (**b**) one sees a subsequent return to the initial state by de-saturation with strong $SrCl_2$ solutions. In this case the formation of a "new" phase, illite, from smectiet layers in the clay aggregates is simply one of cation exchange, where the illite, is stable under conditions of low concentration of exchangeable ions in solution but can be restored to its "initial" smectite state under conditions of strong activity of Sr in solution. However some of the smectite layers in the mixed layer minerals (illite-smectie) fix potassium in a more irreversible manner, indicated by the overall weighted peak position (center of gravity, cg) measurement (**c**). This indicates that much of the potassium in the apparent illite minerals in these soils clay can in fact be of an exchangeable nature, whereas the illite layers formed by potassium saturation in the mixed layer minerals can be less exchangeable mineral forms

solution concentrations of ammonium. This is a standard practice used to measure the amount of "exchangeable potassium" and other cations in the soil materials. However in soils, it seems that there is not an easy inter-relation between the two ions. In mixed layer minerals and illites from prairie soils, ammonium seems not to be selected in preference to potassium (experiments of the authors). However soils which have been treated with continued potassium fertilizing agents show a propensity to capture ammonium under laboratory conditions (Liu et al., 2008). There is little indication of the presence of anhydrous ammonium interlayers in illites in soils even though they can be produced in the laboratory and are found in metamorphic rocks. Thus even though ammonium could be found in the clay structures it seems that it is not present under normal plant growing conditions. This situation could be due to the strong mineralizing power of bacteria in soils. When fertilizing agents are used in agricultural practices, the sequence of treatments and the concentrations of each ion in the soil solutions should be considered so that one or the other is left for plant use.

Numerous studies have been made on soil materials and soil clays in order to establish the relative fixation capacities of K and NH_4^+ in clays. It is known that fully ammonium clays of the mica type can be produced in the laboratory (Sucha et al., 1998) and they are found in natural high temperature rocks (Juster et al., 1987). Therefore the pure ammonium phase can be found in nature as well as in the laboratory. However, what is the selectivity between ammonium ions and potassium in natural soil systems? Laboratory experiments (Evangelou and Lumbanraja, 2002) using high charge vermiculite (smectites) indicate that the presence of potassium enhances ammonium fixation compared to Ca. This suggests that there is a relation between potassium anhydrous layer formation and ammonium capture. Such an effect has been seen in studies of field experiments using long term (20 years) fertilizer treatments on rice paddy soils (Liu et al., 2008). In fact potassium saturation and ammonium saturation of the same clays indicates that ammonium forms more 10A mica like layers than potassium as seen in Fig. 5.7 based upon data from Liu et al. (2008).

The soils investigated in the studies cited have a strong HI (vermiculite) content similar to the material examined in the laboratory studies noted above. More information is necessary in order to assess the K^+-NH_4^+ relations in the smectites/illite mineral series.

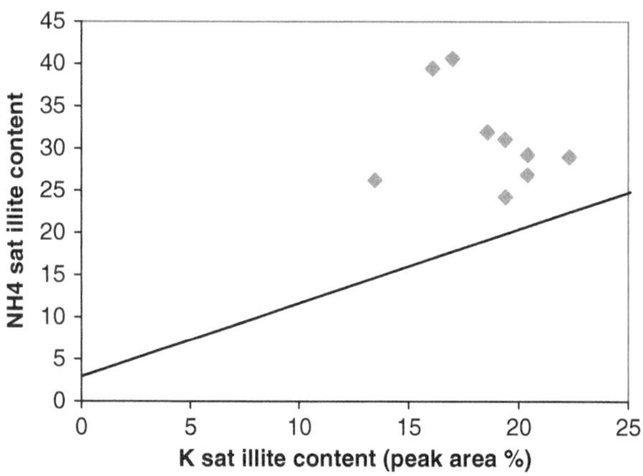

Fig. 5.7 Relations of illite peak surface area under conditions of potassium and ammonium saturation (illite %) for some Chinese red soils (loess bases) which had received potassium fertilizer treatments for 20 years. The long term of potassium treatments enhances the fixation of ammonium producing an ammonium illite mineral (anhydrous ammonium interlayer ions)

5.1.3.4 Plant Extraction of Potassium

The possible formation of anhydrous potassic 2:1 layer structures is in fact rather common in different types of soil clay assemblages. Samples from several sites throughout the United States Great Plains (Velde, 2001b) area show significant formation of illitic layers upon contact with relatively dilute KCl solutions (0.01 M). These clay assemblages in the soils are essentially smectite/illite interstratified minerals plus kaolinite. Also samples from cultivated soils in the Po river delta area near Pavia Italy (data communicated by F Terribile, Univ Frederico II, Naples) show significant illite formation upon K-saturation. The clays are basically two mineral assemblages, a 14.2 Å HI (HI chlorite – high charge vermiculite) plus illite in these agricultural soils. The spectra of Ca – saturated clays show a slightly more intense 7 Å than 14 Å peak indicating magnesium occupation in the interlayer site. The HI chlorite minerals are not affected by Ca exchange ions, retaining the OH interlayer ion structure instead of a double water layer associated with smectites as hydrated Ca cation interlayers. However they are affected by the presence of potassium, which apparently expels some Al and possibly Mg ions to form illite (anhydrous potassium layers) as seen by an increasing intensity of the 10 Å peak. Further the ratio of 14 to 7 Å peak of the HI mineral changes to a lower relative intensity for the 7 Å peak indicating a higher magnesium interlayer occupation for the 14 Å clays that remain. This is seen in Fig. 5.8. Therefore, the initial assemblage contains HI – chlorite (some Mg chlorite and some aluminous HI minerals) which is affected by potassium saturation to form a more magnesian chlorite and more illite.

The potassium saturated samples, compared to Mg saturation, generally show an increase in illite with the presence of plentiful potassium, indicating that in most samples there is exchangeable Al and Mg-OH present, but mostly the Al-OH is extracted by strong cation exchange. However, when looking at the A horizon samples only (Fig. 5.9) there is little difference if any between the two treatments. This indicates that the farming practices provide more than enough potassium to the soils to sustain the extraction needs of the plants grown in them. The results of K – saturation in deeper soils indicates that they have less available potassium than the surface agricultural zones.

Of course plants can effect the de-sorption or extraction of potassium from clays also. This has been demonstrated by Barré et al. (2007) where laboratory experiments using rye grass grown for several weeks effectively stripped out near 40% of the potassium from the clay assemblage they were grown in. From this study it appears that all of the common 2:1 clay types, smectite – rich interstratified mineral (S/I) illitic interstratified mineral (I/S) illite and soil HI phases can be affected by addition or extraction processes.

The quantities of potassium fixed and released as anhydrous interlayer ions is far greater than the potassium available as exchangeable ions by CEC mechanisms in soils. Hence the resources of potassium in clays are not in the exchange sites classically measured as base exchange ions (exchangeable basic cations, i.e. Mg, K, Ca) but those fixed under conditions of higher potassium concentration in solution. The

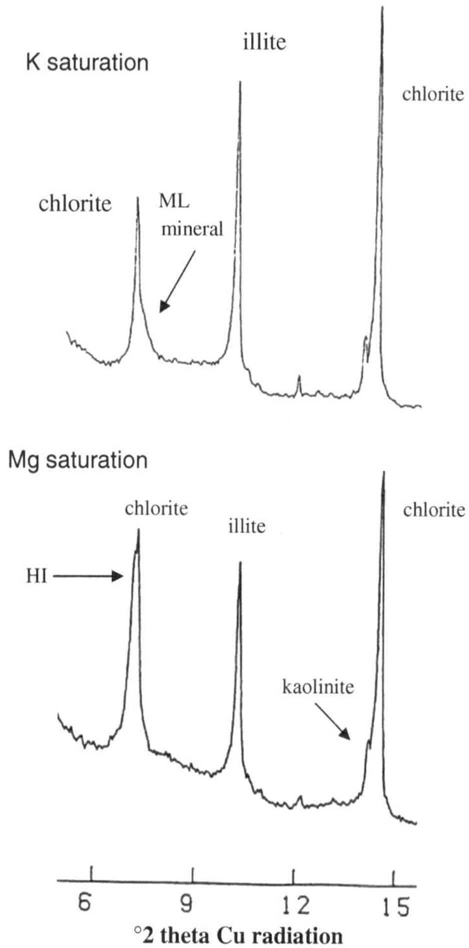

Fig. 5.8 X-ray diffractograms for some soil clays of Po valley sediments under agricultural exploitation. The differences between potassium saturated and magnesium saturated samples indicates that the increase in the relative intensity of the illite peak compared to the chlorite 7 Å diffraction peak shows an increase in minerals with an anhydrous potassium layer. However the lower intensity of the 14 Å compared to the 7 Å peak of the chlorite shows that the loss of chlorite is from an aluminous interlayered mineral (HI) which has exchangeable aluminum hydroxyl ions present, replaced by potassium ions. Here the apparent chlorite mineral, with a narrow and sharp peak at 14.2 Å has an HI component which responds to potassium treatment

example of grass and sequoia on clay types given in Sect. 4.2 indicates that the amount of fixed potassium (illite) can be greatly influenced by the plants present which determine the activity of this element in the pore solutions which eventually saturate the clays. In this way the clays function as a reservoir for potassium, filling when un-needed material is present in solution and emptying when there is a need manifested by the plants.

Potassium is not the only ion useful for plant growth, Ca and Mg are also among those found in plants and their availability through de-sorption from clays is certainly important. And of course phosphorous is of great importance. However, we do not at present know all that much about the cycle of phosphorous fixation and release on the organic material and on clay mineral materials.

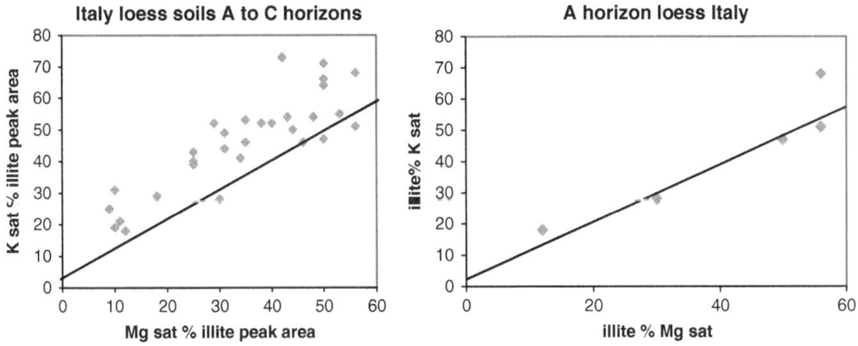

Fig. 5.9 Measurements of the clay peak relative intensities of the Po delta soils shows that the potassium saturated clays have, on average, a higher peak area than the chlorites but when one looks at A horizon clays, the difference disappears, due to the use of potassium fertilizer which saturates the HI minerals with an anhydrous potassium layer to form an illite – like mineral

5.1.4 Cation Exchange in Soils

Each of the two chemically active major components of soils, organic matter and clay minerals, have a capacity to fix in a temporary manner cations. Cation exchange is just that but the total exchange of cations is a function of the amount of ions in the exchanging solution: the more dilute a solution the fewer ions will be fixed on the exchangers. If we know the exchange capacities of each major component, how do these exchange values fit into observed CEC (cation exchange capacities) and ions exchanged in soils? Sposito (1989, p. 171) indicates values for soil type which reflect at the same time organic matter content and clay type:

Ultisol (very mature oxide, hydroxide minerals)	0.06 moles of charge/kg
Alfisol (forest type, HI minerals)	0.12
Mollisol (prairie type, smectite/illites)	0.22
Vertisol (smectite – rich)	0.37
Histosol (essentially organic matter)	1.4

Looking in more detail at data for soils with moderate amounts of organic matter and clays one can observe in several studies that the CEC of soils is within the range of clay mineral cation exchange capacities for the amount clays present in the soil. Figure 5.10 shows the relations of several studies for CEC, soil carbon and clay content. The values of clay CEC are indicated for values between the low values for kaolinite and illite, at 5–10 milli-equivalents charge for 100 g of clay to 130 milli equivalents for smectites (black lines in Fig. 5.10). CEC values are usually between several milli equivalents per 100 g and about 30 milli-equivalents. This is not a large range of values to be sure. For the most part the exchange capacity observed in soils does not exceed the theoretical maximum capacity for pure clay samples as is the case for the four soil series shown in the figure. Thus it is clear that the chemically and exchangeably active part of the organic matter does not overshadow the clays as cation exchangers. This is not the opinion of Sposito (1989) who advocates a more important role for organic matter (humic components) as cation exchangers. He uses values of 2–4 times the CEC of clays attributed to humic materials in order to calculate the CEC potential of a soil. However the values of exchange capacity vary apparently according to different specific field examples. Black (1968, p. 224) indicates a value of 6 milli equivalents per percent carbon in a sandy soil which is four times that of smectites clays. CEC for soils with several percent organic carbon rarely show such high cation exchange capacities. In the figure of examples cited here, the CEC is not related to the amount of organic present, ranging from 10 milli equivalents for 1% organic C to less than less than 0.5 milli equivalents, if one does not take into account the exchange capacity of the clays present. The problem of the role of organic matter as a cation exchanger undoubtedly resides in the assessment of the type of organic matter present. Overall, there appears to be a tendency in the series shown to decrease CEC as clay content increases. the active fraction in the soils. The estimation of organic matter is solely based upon organic carbon content. This does not take into account the type of organic matter; immature, evolved or mature. If the organic matter is immature, not much of the material will have active bonding capacities. If the organic mater is highly evolved, towards a carbon-rich type, the number of active sites will be limited. Thus the type of organic matter present will determine the role it will have in cation exchange.

Concerning clay minerals, if kaolinite is dominant, the CEC contribution will be low, as is the case in the Hainan soils in Fig. 5.10 where the clays are mostly kaolinite. However there are other mineral materials which have varying and at times very high cation exchange capacities such as vermiculites. The soils from the east Coast of the USA shown in the figure indicate variable cation exchange capacity but higher at lower clay amount, suggesting that the vermiculite has an effect as an exchanger in these soils. The samples from Sweden (Andrist Rangel, 2008) from soils with some vermiculite but significant amounts of illite-smectite mixed layer minerals and illite show which higher cation exchange capacities for lower clay contents following rather closely the theoretical exchange values for smectites (lower line in the figure).

Other materials in soils, can contribute to the cation exchange capacity but in a rather erratic manner. These are the amorphous silicate materials (imogolite and allophone) and oxides (iron) and hydroxy-hydroxides (iron and gibbsite). Such minerals or materials found in soils are highly active on their surface sites but can combine with organic matter or other elements such as phosphorous (see Dixon and Weed, 1989). Silica – alumina amorphous matter is well known in Andosols, those based upon glassy eruptive rocks. These soils are usually extremely fertile, largely due to the fixation of cations and the presence of large amounts of organic matter.

From the above, it appears that the prediction of CEC based upon the constitution of a soil is difficult to foresee. Knowledge of a number of variables is necessary in order to estimate the chemical surface activity of a soil. One major piece of information necessary is the type of exchanger present, be it organic matter in different stages of evolution or clays of different structural and hence exchange types. As we will see further on, the effect of plants on their environment largely determines the parameters of cation exchange, in forming or controlling the type of clays present or in the formation of organic components which are of course entirely due to the vegetal regime present, in natural settings.

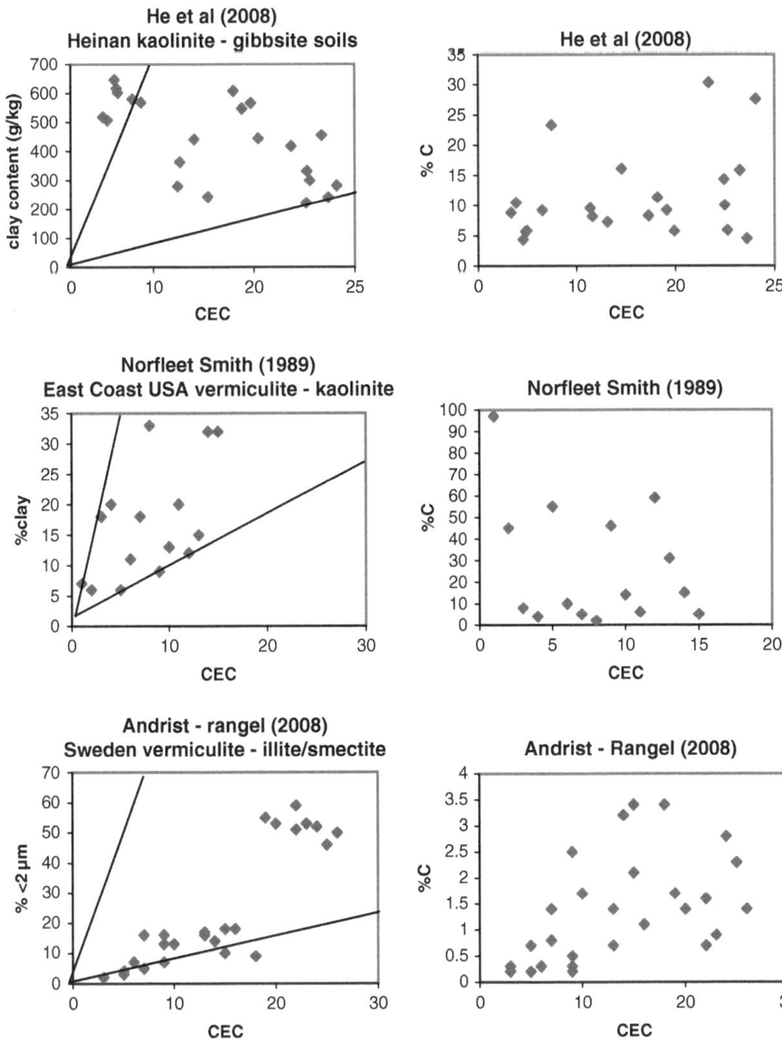

Fig. 5.10 Assemblage of clay content, organic carbon content and cation exchange capacity (CEC) for several soils (Mariotti, 1982; Andrist-Rangel, 2008; Righi et al., 1999; Norfleet and Smith, 1989; Graham and Southard, 1983; He et al., 2008). The relations are relatively clear concerning clay content and CEC, where the lines in the figures represent the estimated maximum CEC for smectites and minimum CEC for kaolinite. Most of the CEC values fall between these limits. However there is little or no correlation between organic content and CEC. One can assume that in soils with modest amounts of organic material present in a relatively evolved state (humic) the dominant cation exchanger is the clay material

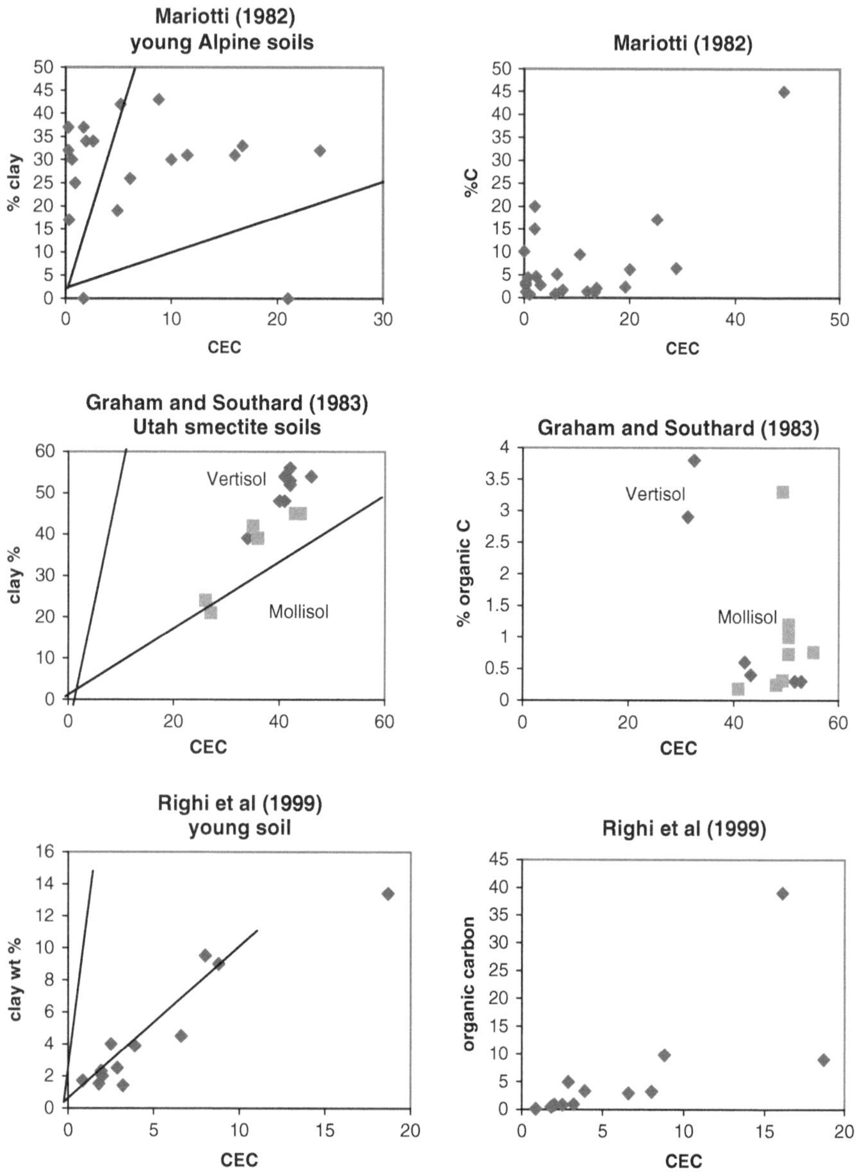

Fig. 5.10 (continued)

Cation exchange is a complex system dependent upon several factors at the same time. In the first place, it is extremely rare to find that exchange follows ionic activity in solution or more less so ionic concentration.

5.1.5 Role of Clay Minerals in Critical Nutrient Exchange: Potassium and Ammonium

As we have outlined above, clays are present and respond to the needs of plants in the A horizon where the interaction between plants and soil mineral materials is the greatest. Clay minerals are, initially, the response of interaction mechanisms between rain water and silicate rock material as it is brought in contact with the surface environment. These reactions continue as long as there is rock and rain water which comes into contact with the surface materials. A prolonged contact is necessary for such an effect to be useful in the formation of clay minerals. If water is moved through the system at too great a rate, it will be below saturation with the silica and alumina of the silicate phases and will not produce a new mineral. Thus bare rock appears to be little altered in that there are few new minerals formed. Plants slow down the water flow, by root action and accumulation of mineral debris at the surface which allows the rein water solutions to equilibrate with the silicate minerals and eventually to produce new minerals by an over-saturation with dissolved elements, notably alumina.

The alteration front in alteration profiles continues to proceed downward into the rock surface once the mechanism of plant water retention becomes important. As time goes on, the thickness or the alteration material, largely clay minerals, becomes more important. On old cratons where little vertical movement of the rock basement has occurred over several millions of years, one finds alterite zones of several to tens of meters thickness. Without plants the alterite materials would be washed away into streams and oceans leaving a largely little altered rock surface. However, the protective cover of plants which allows the formation of soils is such that the alterite zone of fine grained material, clays, is stabilized. The relatively soft and little resistant clay mantle is where roots can penetrate with ease, where water is fixed and released easily and hence a zone where plants can develop a strong root system with ease. The plants need, or at least benefit from, the alterite cover that they stabilize. Clays provide a medium where chemical exchange is enhanced and where the development of root systems necessary for this exchange and the mechanical stability of the plants can occur. As it turns out several mineral elements from the underlying rocks are necessary for plants to survive well. Of course different types of plants can do without with less of different elements as they have adapted to different chemical environments.

There are essentially two types of mineral elements which are used by plants, those with alkali or alkaline earth chemical characteristics which are ionically charged and those more co-valent in character. In the first category are the elements such as Ca, K, and Mg. and the second elements such as Si and P. The ionically charged elements are strongly associated with clays. Cations are attracted to the surfaces of clays or the interlayer sites where they are fixed in different states of ionic bonding intensity. The weakly bonded ions are fixed onto the clay surfaces or interlayer sites and can be exchanged for other ions easily depending upon the ionic concentrations of elements in solution, or their lack. If a clay has a certain number of cations attracted to its charged surfaces they can be exchanged according to the laws of mass action. The higher the concentration of another element in solutions in contact with the clay, the more of the element fixed on the clay will be released to the solution. The ion concentration tends to attain an equilibrium of concentrations according to the affinities of the elements for the charged clay surface. The lower the concentration of elements in the solution, the more of those fixed on the clays will be taken into the solution. This is the effect of dilution or leaching by rainwater which contains little mineral matter. Leaching is assumed to be the major factor in alteration processes. In the end this has to be true in that rain keeps flowing over and into soils bringing more mineral matter into solution. However, plants need a certain number of these elements for their development and growth. They cannot afford to have them all swept from the surface layers of the earth. Thus there is a subtle interchange of elements from different parts of the altering surface towards the plant horizon of interaction. If plants wish to retain some elements they need to be available throughout the growing season and extractable by biologic processes.

Clay minerals are then the vehicle of mineral element exchange for a certain portion of those elements necessary for plant growth and prosperity. The exchangeable elements as outlined above are made available to plants in the surface layers of the alteration zone. The relatively abundant elements such as magnesium and above all calcium, are fixed on surfaces of clays as exchangeable ions. However certain elements are in fact present in more tightly bound states in the interlayer sites. These are potassium and ammonium ions. Although ammonium is not a mineral element, it is exchanged from the atmosphere to bio active organs in the soils, usually driven by bacterial action. The ammonium ion, NH_4^+, is easily absorbed into plant root systems and a valuable source of nitrogen for plants. As it turns out, the ammonium ionic unit is roughly the diameter of a potassium ion and often can be found in clays or on clays as is potassium. It is known that ammonium can form anhydrous layers in 2:1 minerals producing a sort of ammonium mica (Sucha et al., 1998). Smectites accommodate ammonium also as an interlayer exchange ion (Sucha and Sirahova, 1991). Potassium has been studied more as a mineral component than has ammonium. Although ammonium is commonly assumed to be present as an exchange ion it is not commonly assumed to be present as an anhydrous interlayer ion in a mica – like mineral. However there is no reason that ammonium not be present in a 2:1 structure as an anhydrous interlayer ion.

5.1.5.1 Non Exchangeable or Extractable Cation Occupancies

We will use potassium as an example according to the concepts of soil chemistry. Huang (2005) gives a resume of the types of potassium chemical relations in soils, and hence on soil clay minerals. Usually one has exchangeable potassium, assessed using strong exchange ion solutions, ammonium acetate is frequently used, to displace the potassium form the clays. Fixed potassium, or non-exchangeable potassium is that which resists the exchange treatment. It is most often assumed to be present in interlayer sites as an anhydrous state. Estimations of fixed or non exchangeable potassium are often made by treating the clays with strong HCl solutions to dissolve the fine fraction minerals and extract potassium with hydrogen cations or a less destructive method using tetraborate solutions can be used (Moody and Bell, 2006). Coarser minerals such as micas or feldspars remain intact after these treatments and the potassium found in solids is then considered to be structural potassium, in fact material of high temperature origin. The question which is very important is: how much of the fixed potassium or even structural potassium is available to plants? The same question can be asked for ammonium, often a proxy for potassium in clays but virtually absent from silicate minerals. It is unlikely that ammonium will be present in detrital or high temperature micas or other phyllosilicates but it can be found as "non-extractable" ion material in soil clays. Thus the origins of potassium and ammonium ions in soil materials are different, potassium being a mineral element while ammonium has its source in atmospheric nitrogen. However both can be found and exchanged between clays and plants.

The non exchangeable or non-extractable potassium is in fact a type of ionic fixation that is not that of the high temperature minerals but not that of the superficial or poorly held ionic species on clays.

5.1.5.2 Nutrient Cycling

In general it is clear that potassium and ammonium are important for plant growth. Extraction of potassium from soils seems to be related with plant activity. Tripler et al. (2006) report that the stream water content of potassium and ammonium are seasonally related. During growing seasons of forest cover, the stream water concentration of these elements is greatly diminished, by a factor of two or more. Moments of leaf fall, and return of potassium to the soil in the autumn, outside the growing season, greatly increase the potassium content of stream water. Calcium and magnesium stream water content are much less sensitive to growing season driven forces. Thus it is clear that there is a strong relation between potassium retention and use, as well as ammonium, in natural forest systems among others. In more detailed experimental observations, Scherer and Ahrens (1996) show that the ammonium fixed on clays in soils in a non-exchangeable state can be extracted by plant action localized near the roots. This action, exchange of hydrogen cations for ammonium, is plant species related as well as dependent upon the mineral present, in this case either illite (anhydrous potassium interlayer ions) and vermiculite (hydroxyl ion and other cation interlayer populations). Thus the conditions and mechanisms of potassium and ammonium fixation on and in clays seems to depend upon several variables according to the solicitation of the plants present which are extracting these nutrients.

If plants extract these elements at variable rates how do they get fixed in or on the clays? Initially one can observe the effects of amendments which are renowned for their efficiency – animal manures. Ross et al. (1985) show that the use of such fertilizing agents increases the illite content drastically after 6 years of use. In the experiments vermiculite (HI) minerals are largely converted to an illitic form with both potassium and ammonium present.

The initial starting point is the CEC, or easily exchanged cation population, essentially on organic matter and on expandable clays. This material does not select potassium nor ammonium in that the ionic radii are relatively large and the charge small, plus one. Calcium is about the same diameter in ionic size but it has a charge of plus two, hence a significantly higher charge density. This gives it an advantage in the cation exchange process. Ca and Mg are selected before the large mono-valent cations. But since these potassium and ammonium ions are vital to plants some mechanism must be available to sustain their needs. As it turns out this mechanism is the formation of anhydrous ionic interlayers resulting in illite – like clays. Liu et al. (2008) have demonstrated that in soils which had a long history of potassium fetilization, with a creation of a significant amount of illite clays, ammonium is more easily fixed in the interlayer sites of the clays as "non-extractable" ions. In other words, it takes illite, a supposedly closed mineral, to fix ammonium ions, by cation exchange with the anhydrous potassium in the illite structures. It appears that the ammonium is more easily extractable than the potassium in these minerals. There is then a relation between potassium fixation and ammonium fixation in the more permanent interlayer ion sites. The more illite that is "created" by ion exchange in highly potassic soil solutions, the more ammonium can be accommodated. Plants that bring an excess of potassium to the surface which is in its turn fixed to make illite then create a mineral capable of fixing more ammonium.

The stabilization of 2:1 clays, such as high charge smectites, is a decisive step in forming means of conserving the fertilizing agents of potassium and ammonium (nitrogen). As one can expect, there is a certain amount of these elements which escapes into stream water which represents the free pore water flow through soils. This material, essentially lost to the plant cycle, must be replenished. Ammonium (nitrogen) comes from the air, and is fixed by bacterial action. Destruction of roots by other bacteria releases this nitrogen in the form of ammonium compounds to the plant. Potassium must be sought either in the unaltered minerals in the soil or in the zones below this zone where water – rock interaction occurs to produce potassium ion sin solution. Some deep roots, through the agents of micorrhize can undoubtedly extract this element directly from the un-altered silicate minerals.

The problem is to conserve these elements in the soil zone throughout a growing season, or for several seasons. Fixing potassium, then ammonium in the anhydrous interlayer sites of high charge 2:1 minerals seems to be the method used. These "non-exchangeable" ions can be extracted by plants, but not by normal ion exchange mechanism under the chemical conditions of most soil pore waters. Field experiments have shown that there is a build up of potassium, and eventually ammonium in systems where the supply exceeds the demand of the plant regimes. This is a sort of "holding" mode of clays where the vital elements are kept in a situation which allows extraction by plants but not by soil solution chemistry (dilution or cation exchange). One can envision a system whereby plants bring more elements into the clays than needed but when for diverse reasons the plans cannot furnish these elements, the stock in the clays can be used for some period of time, i.e. several growing seasons. This is a sort of a pantry system, dear to our great grandmothers, whose living and lively hood at times depended upon a supply in reserve for an unforeseeable future.

One of the roles of clays is then to provide a reserve of fertilizing elements or future use due to fluctuations in supply from one year to the next.

5.2 Organic Matter, Clay Minerals and Soil Aggregation (Water Resources for Plants)

5.2.1 Clay Minerals and Organic Molecules Interactions: Soil Structure Development

Soil profiles most commonly show an increase in organic carbon in the A horizon, due to the presence of plants. Plants more or less define the formation of soils as a matter of fact. Thus the measure of carbon content is an index of the amount of carbon that remains in the soil after plant deposition (root or aerial material) and bacteriological action. In Fig. 5.11 one sees a typical profile of carbon content, in this case in two un-plowed prairie soils from a slightly hilly situation, one on a plateau and the other in a valley bottom from Missouri (USA). Combination of organic and clays results in what is called soil structure or the capacity to form clay-organic aggregates.

The soil structure corresponds to the three dimensional organisation of solid particles and pore space. In soils, solid particles (mineral particles and organic molecules) are associated in aggregates. These aggregates are the "negative part" of the porosity. Clay minerals, due to their high reactivity, interact strongly with

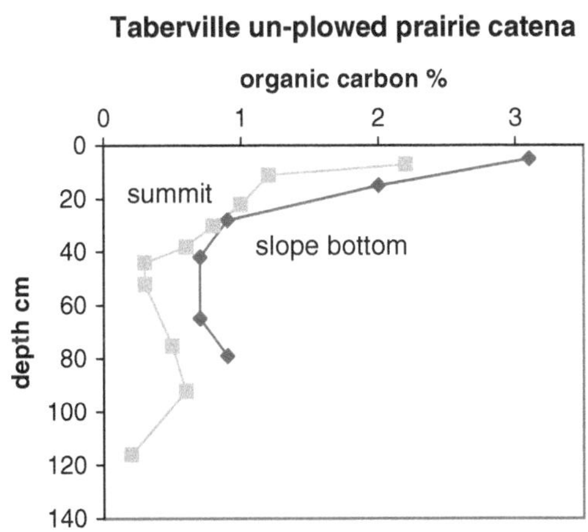

Fig. 5.11 Relations between organic carbon content and depth in a catena sequence of soils in Missouri (Fenton et al., 1999). The carbon content increases towards the surface in the upland and slope bottom prairies soils but there is more present in the slope bottom prairie soils where humidity is greater, plant growth is probably greater and possibly bacterial action is less intense

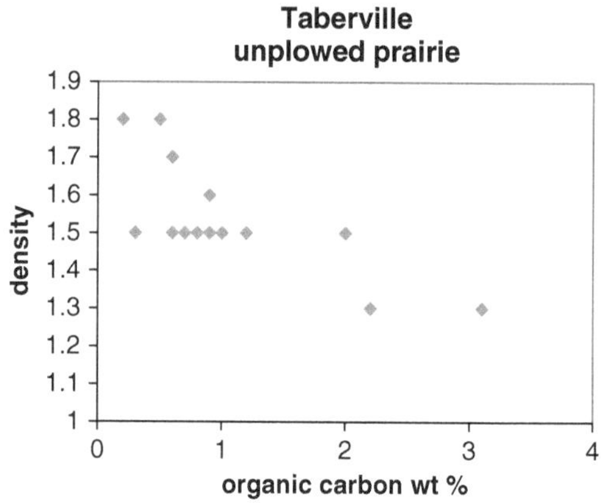

Fig. 5.12 Relationship between the density of the soils in the Catena sequence of Figure 5.9 from Missouri and the organic carbon content (Fenton et al., 1999). Higher organic carbons forms more soil aggregates which have a capillary porosity filled with water in humid periods

organic molecules and are very efficient at building aggregates. Virto et al. (2008) showed that the clay fractions (<2 μm) of silt-size aggregates are enriched in clay minerals compared to the "free" clays in the soil. Among organic molecules, polysaccharides exuded by micro-organisms or roots have a central role for aggregate formation and stability.

For healthy plant growth, a soil should possess aggregates with both pores larger than 75 μm to remain aerobic and it should have a sufficient number of pores 30–0.2 μm diameter to retain water by capillarity and large pores between aggregates to allow rapid infiltration and drainage (Tisdall and Oades, 1982). Such ideal structure would provide available water to plants and a suitable environment for micro-organisms whose activity is crucial for nutrient cycling are living on pore surfaces. A demonstration of the results of aggregate formation related to organic – clay interaction can be seen in the density of soils related to organic matter content. Fenton et al. (1999) cite data concerning several soil profiles in un-plowed prairie Missouri (USA). In Fig. 5.12 one sees a good correlation between density (measured in oven dried samples) and organic carbon content of soils. The higher the carbon content, up to near 3 wt%, one approaches a density of near one. Since clays have a density of near 2.5, this means that with a relatively small amount of organic matter one creates structures of low density, where the voids are filled with air. These voids will contain water in a natural state in most soils. This is one of the major reasons that agricultural specialists insist on determining the organic carbon content of soils. It is a rough measure of water holding capacity (capillary sites) and hence a ready receive of water for plants.

Due to its crucial role, soil structure has stimulated much research. This effort can be separated into several complementary approaches:

5.2.1.1 Soil Structure Stability and Dynamics

Soil structure is largely related to the intimate association of organic matter and clays. The organic materials tend to fix clay particles together and to form aggregates of particles. The size and stability of these aggregates is vital to the existence of plants in soils. Thus it is important to understand

- The underlying mechanisms of soil structure development and stability and the means by which the soil aggregates form
- Aggregation related to the biologically mediated structure modification in the rhizosphere

Soil structure is highly dynamic because of the dynamics of organic matter stability. It is important to be able to follow and to quantify soil structure evolution. Several tests have been designed to test the stability of soil structure. They are all performed on aggregates.

5.2.1.2 Understanding the Underlying Mechanisms of Soil Structure Development and Stability

Numerous studies have made observations on the effects of different organic molecules on the formation of aggregates. It is clear that root exudates are highly important in this action (Troare et al., 2000 for instance). Also other secondary products of bacterial action can be important as well as the residual humic material (Dinel et al., 1991; Cornejo and Hermosin, 1996). Much of the interactions appear to occur at crystal edges (Varadachari et al., 1995) but the interaction of organic material and clays can be observed on clay surfaces and these more humic organo – clay complexes are bound by lighter molecule materials (such as polysaccharides) to one another (Chenu, 1993). Despite the complexity of the systems observations allow us to understand the fundamentals of the process.

Aggregate Formation at the Molecular Scale

Aggregation begins when two mineral particles or a mineral particle and an organic molecule are bonded together. Due to their high reactivity clay minerals interact with each others and with organic molecules. They are therefore of primary importance for aggregate formation. Many authors studied more precisely the interactions between clay minerals and organic molecules. Usually, such experiments were conducted in dilute suspension with pure clay minerals and model organic molecules. These studies gave interesting insights on the mechanisms of aggregate formation. It has been observed that the adsorption of negatively charged organic molecules on clay minerals increase colloidal stability, that is, reduces interaction between particles. The authors proposed the edge site reversal mechanism. They postulated that the adsorption of negatively charged molecules on the positively charged edge-site of clay layers reverses the charge from positive to negative and prevents edge-to-face clay mineral interactions.

This mechanism may occur for 1:1 clay minerals (Kretzschmar et al., 1993, 1997) and 2:1 clay minerals (Heil and Sposito, 1993; Tarchitzky and Chen, 2002a and b). The edge-site reversal mechanism does not lead to any colloidal stabilization when polyvalent cations are added to the suspension. In this case, polyvalent cations establish bridges between negatively charged particles leading to flocculation. Majzik and Tombacz (2007) established that the aggregation of mineral particle is enhanced by the joint effect of negatively charged organic molecules and calcium ions. Concerning neutral molecules, it has been reported than small neutral molecules do not really influence colloidal stability whereas larger neutral molecules, on the contrary, dramatically decreased colloidal stability of montmorillonite (i.e. diagenetically formed pure smectites mineral) suspension (Tarchitzky and Chen, 2002b). In this case, the bridging occurs through hydrogen bonding between the planar oxygen on the clay surfaces and the hydroxyl groups in the polymer.

Polysaccharide-Induced Aggregate Stabilisation

According to soil structure models soil organic matter acts as a cementing agent between mineral particles (e.g. Tisdall and Oades, 1982). Among organic molecules, polysaccharides have been described as very efficient binding agents. This has been illustrated in the previous part for neutral polysaccharides and negatively charged polysaccharides in presence of polyvalent ions. In several studies authors attempted to establish whether or not polysaccharides also increase the stability of aggregates. Chenu and Guérif (1991) showed that scleroglucan (a fungal polysaccharide) increased tensile strength of pure clay samples. Czarnes et al. (2000) observed the same trend with polygalacturonic acid, PGA (a root mucilage analogue) on natural soil.

Importance of Clay Mineralogy in Aggregate Formation and Stability

Although the importance of clay minerals on aggregate formation is of course recognized the importance of clay mineralogy, i.e. the type of clay present, on aggregate formation and dynamics have been less studied. Denef and Six (2002, 2004, 2005) and Six et al. (2000) observed that aggregate stability in 2:1 dominated soil depends more on organic molecules than in 1:1 dominated soil. These authors proposed that organic molecules are needed to cement negatively charged 2:1 particles whereas electrostatic interaction between kaolinite particles and iron oxides can occur without a cementing agent. However, they only studied two soils (one kaolinite dominated and the other illite dominated). It is difficult therefore to generalize their results.

Barré and Hallett (2009) studied the effect of the intensity of the organic – clay bond using rheological measurements of clay – organic mixtures in different water concentration solutions. These experiments were designed to describe the intensity of the bonds through the physical coherence in a fluid system subjected to physical shearing stress. They demonstrate that the effect of bonding increases shear strength on all soils studied and variably affected it depending upon the type of organic molecule in contact with the clay and upon the type of clay mineral present, 2:1 or 1:1 mineral. Hence it appears that the intensity of bonding depends upon several factors which can vary of course in a complex natural system such as a soil.

The stability of aggregates, i.e. their presence over a growing season or more, is determined by the rate of destruction of the organic material which binds the particles together. The smaller the molecules, less humic material dominated by C-C bonds, the more rapidly they will be destroyed by bacterial action. Hence the root exudates and bacterial exudation products, of lower molecular weight and less humic in nature, are fundamental for aggregate formation but liable to a short half life due to their vulnerability to bacterial action. The aggregates and their structure in soils then form a very dynamic system, dependent upon renewal by the living materials present, roots and bacteria which degrade plant detritus. Constant renewal is the rule, and since most matter is supplied by plants in one way or another, they are the major agents of aggregate formation and maintenance.

A simplified vision of the organic – clay processes in soils can be given as follows: organic matter is transformed by bacterial into gases and high carbon content molecules (humic matter). The most efficient bonding agents are less humic molecules produced by plants by root exudation or bacterial action; However these less carbon – rich molecules, chemically active by their external functional groups, have a shorter life span due to eventual bacterial action. The cementing agents need to be renewed in the soils in order to maintain a strong clay – organic structure to form aggregates.

5.2.2 Consequences of Clay – Organic Interactions: Soil Structure and Aggregation

There is nothing as rock – like as a clay mass that has dried. This is true for clay lumps or surfaces of clay deposits such as those found in deserts or along sea coasts on mud flats. Clays dry to a very hard and coherent mass. Yet when wetted they become more supple, and one can find aggregates in them which are individual masses of clays separated from others in the clay sample. In soils one often looks at the dried surface, for example in a plowed field. Here the soil is relatively dry, otherwise one would not be walking in it unless with stout rubber boots on. One is familiar with cracked surfaces of soils as they have dried. These cracks are a manifestation of contraction of the material due to loss of free water. When wetted, clay masses usually expand about 30% or so due to adsorbed water and absorbed water in the case of smectites. Interlayer water, two layers associated with a cation, expands the mineral by 50%. In extreme cases the major cracks can be quite wide, several centimetres.

5.2.2.1 Soil Pore Structures, Aggregation and Fractures

Soil Structure Observation

Soil structure is obviously not directly visible in three dimensions in that soil materials are largely opaque. Thus its observation is challenging in itself. Some methods have been used in the past which have given useful results. More are needed in the future. The first method that provides soil structure visualization is by the thin section method. A soil core is impregnated by a resin which penetrates the pore system to different size domains depending upon its viscosity. It is then cut with and treated to produce either flat surfaces or in the form of a thin section which is on the order of 15 μm thickness. The thin sections are then observed and photographed using a petrographic microscope. The image is reduced to a black and white figure and methods of image analyses of the pictures allow one to quantify the porosity of the soil core. Usually this method is limited to pore spaces on the order of 25 μm width or diameter. Serial cuts of a block can be used to give the same effect, taking pictures at each stage of evolution (See Moreau et al., 1999 for example). However, soil structure is a 3-D object. In particular, more than porosity, the key property is the continuity of the poral network. The continuity of the poral network can only be studied using a 3-D visualization of the soil structure.

This visualization was first obtained by doing several parallel cuts in an embedded soil core. Several parallel thin sections were obtained and analyzed. The 3-D structure was interpreted from these images. This technique that is particularly time consuming has now been replaced by the direct observation of the soil core using C-T scanners (Perth et al., 2008). These scanners coupled with more and more powerful imaging programs allow looking inside the soil core and provide direct measurements on soil structure. For instance, data processing programs allow quantification of porosity and the connectivity of the poral networks.

Surface Features

Soils have holes and passageways in them as well as cracks. For a pedologist and above all a farmer, a primary criterion for assessment of a soil is its structure, i.e. the extensions and discontinuities in three dimensions. Cracks, holes and passageways due to animal activity are the characteristics of soils. In between these pores are the solid parts, or at least the parts with only micro porosity. The "reason" for the presence of macro porous structure of soils (the organic surface horizon) is that most plants cannot live in a water saturated environment. Air is a necessity for roots and plant growth. Most plants cannot survive in an anoxic soil environment. Some do but they are specialized to say the least. One of the reasons for this is the need to have a source of air to extract nitrogen from. Most of the nitrogen exchange between the air and plants occurs in the soil zone. Thus soils have pores, i.e. passages in which air is present. As it turns out these pores are also a vehicle for water flow, out of the soil system and hence they evacuate water to let air into the system. Pores are then vital to the well being of most plants. Certainly some plants have adapted to the total wetting of their roots and anoxic conditions of soils. Such are the plants that thrive in swamps and bogs. However, in surface environments if clays have no plant – soil interaction, i.e. organic content, they tend to have almost no macro – porosity or passageways for air circulation. Organic matter interacts with clays to form aggregates or self contracting units which as they release water form pore spaces between themselves. This occurs by wetting – drying cycles (Li and Velde, 2003). The structures, spatial distribution of these pores, can be quite varied.

Internal Features and Porosity

Porosity is due to the disposition of the solids in a soil. Most soils, those at the surface with some organic matter content have soil aggregate formation. This is a state where the clay particles are bound or fixed together to form a more or less dissociable mass of roughly equant dimensions. The finer the aggregate sizes, the smaller the pores. The more irregular the aggregates, the greater will be the range in size distribution of the pores, from fine to large. Aggregate formation is affected by cycles of wetting or drying (Fig. 5.13). In this figure one sees samples of roughly broken soils in the dried state loosely packed in a cylinder (0 cycles of wetting and drying) and a second stage after nine cycles of wet – dry after the methods described in Li and Velde (2003). There is a clear tendency to form large aggregates with wet – dry cycles and pore sizes become somewhat smaller as the soil particles settle downward in the cylinder. This settling effect is common to plowed and cultivated fields under agriculture use. One can see that the aggregation process

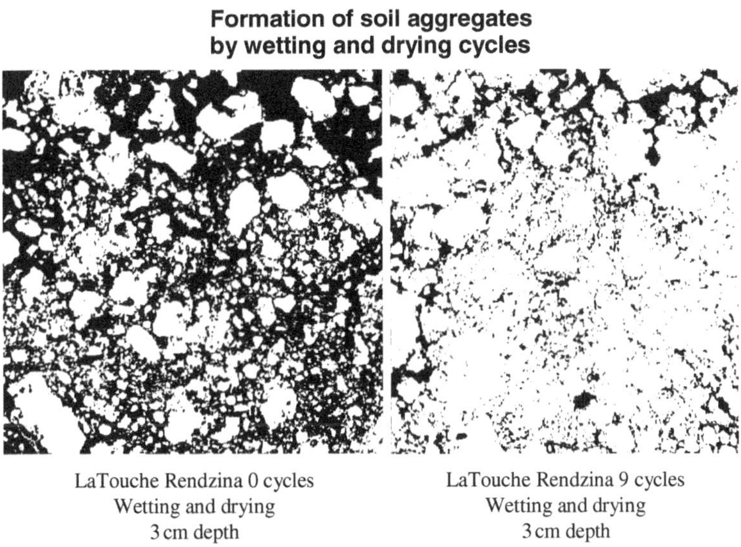

Fig. 5.13 Illustration of the effects of wetting and drying on soil aggregates. The same soil material is shown in an initial roughly broken state and after nine cycles of wetting and drying. The tendency to form larger aggregates at the expense of smaller ones is quite evident (data after Li and Velde, 2003)

tends to maintain more or less equant soil aggregates. Eventually the aggregates are sufficiently large to be unable to shrink homogeneously and they are cracked. One can see the beginning of such a process in Fig. 5.13 in the lower left hand corner.

The existence and nature of soil aggregates is normally attributed to the organic matter present and a relation between clay particles and organic matter is assumed to be largely responsible for aggregation. However these are not necessarily the only variables affecting pore sizes and aggregates.

Figure 5.14 shows the differences in pore spaces developed in the same Mollisol soils by three different regimes of plants growth. In the figure one sees the differences between pores and their distribution over a 6 cm depth in cores from an experimental farm in central Iowa, USA (samples courtesy of David Laird NSTL, Ames Iowa). The variations are not extreme but one sees the influence of the depth on soil porosity due to the type of plants grown in the soils. The natural prairie shows the deepest zone of macroporosity distributed in various pore sizes. The continuous soybean cropping shows a more or less homogeneous pore distribution with an abrupt loss of macropores near 10 cm depth while the rotation of corn and soybeans shows a more inhomogeneous distribution towards the surface with larger and smaller pores present and an abrupt of more or less evenly dsitributed macroporosity at 7 cm depth. In the lower zone one sees evidence of cracking, shown by the linear macropore. In these examples the soil type and physical characteristics of

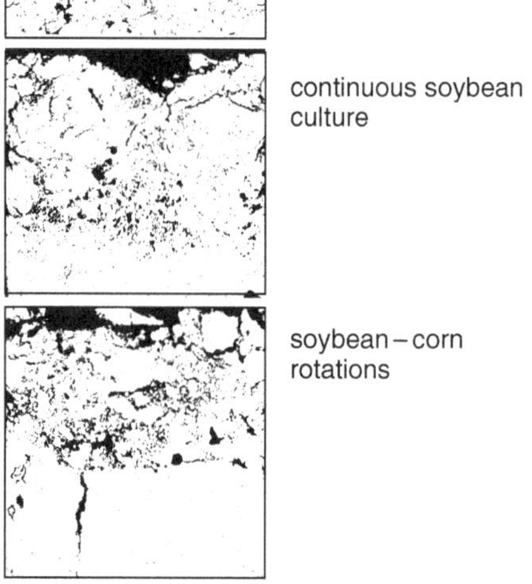

Fig. 5.14 Example of the effect of crop types on soil texture (aggregation) as a function of depth in cultivated soils after harvesting (samples courtesy of D Laird, USTL Ames IA USA). The depth of disseminated pores (in *black*), indicating aggregation is quite variable depending upon cropping management. The zone of low aggregation and sporadic cracking rises to the surface as cropping is more important and the presence of corn increases the effect. The soils with the greatest distribution of soil aggregates are those of native prairie followed by soybean cropping and the least aggregates soils are those where corn (*zea mais*) was grown. Vertical cuts are near 7 cm in depth

grain size distribution (clay and sand content) are almost identical but the influence of the type of plant present affects the surface macropore sizes and distribution and the depth to which it extends. The most irregular distribution of porosity, variety of pore sizes, is found in that of the native prairie soil. Is this an advantage? Most likely yes. The problem for roots is to find a place to live and develop in a macropore. Roots can penetrate soil masses (aggregates) but it is easier to grow around the soil aggregates where they are in contact with the solid phases, which have microporosity which is usually filled with water and where they also have access to air.

In Fig. 5.15 we can see a two dimensional distribution of porosity and aggregate type as a function of depth in a soil, in this instance a Mollisol from central Illinois taken at the end of the growing season and numerous cycles of wetting and

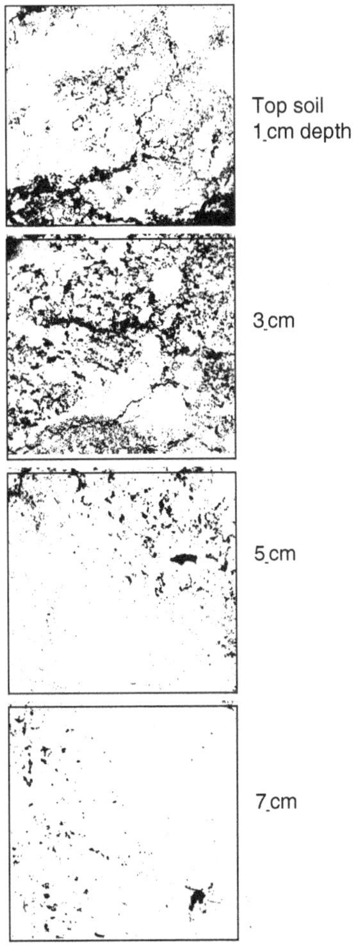

Fig. 5.15 Sequence of pore structures in sequential cuts at different depths for a corn field in Central Illinois. The aggregates are quite similar down to the depth of around four centimetres where cracking is the only form of pore structure except for worm holes (round pore in the lower figure) Dimension of the soils shown is 6 cm on a side

drying since the tilling and planting of the corn crop. In each section the porosity type changes, from an initial fine pore size distribution at 2 cm depth, to a more coarse distribution at 3 cm, and below 5 cm one finds a loss of macropore size distributions and the presence of cracks. It is interesting to note the trace of a worm burrow (round pore) in the upper right hand portion of the 5 and 7 cm depth figures. This shows that aggregate formation is not the only cause of soil porosity.

In Fig. 5.15 one can follow the change in macropore structure by horizontal sections as a function of depth in a profile of a soil which had been strongly watered, by rainfall and the formation of standing pools for several days, with subsequent formation of surface cracks. Here the fine clays were moved towards the pools by surface erosion, on a minor scale, and formed a strong clay, low organic and silt – sand, textured material. This material formed cracks with little or no aggregation. This is the type of soil cracking that is most often remarked when on walks in fields or open spaces. However the surface cracks quickly give way to the underlying aggregate structure it the soil due the clay – organic interactions (Fig. 5.16). At less than 2 cm depth one sees the trace of the surface cracks but already much smaller porosity is apparent. At 3 cm depth there is no trace of this surface crack expression.

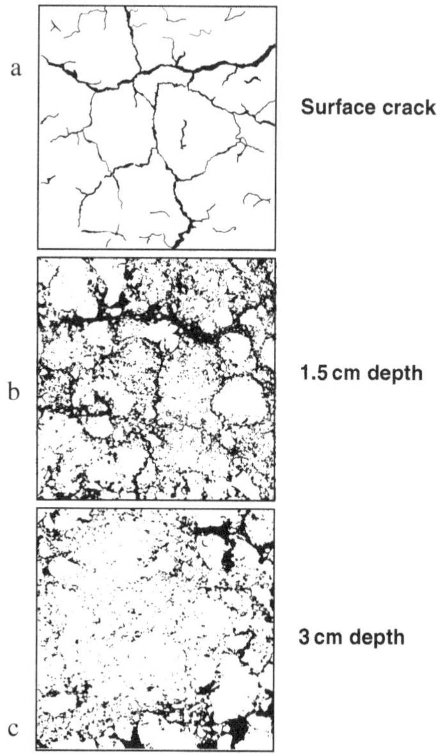

Fig. 5.16 Illustration of surface cracking on a soil and the effects it has in the organization of cracks and aggregate porosity as a function of depth. Sample is from a rendzina soil in center western France (see Velde, 2001a). The obvious cracks at the surface are quickly dissipated into the soil aggregate structure at minimal depth

This example illustrates the importance of having a soil of balanced components: clay, silt, sand and organic matter. Such material promotes "good" soil structures with a variety of irregularly distributed pores created by soil aggregates, i.e. association of organic matter and clay minerals.

Deep Fissures

Some soil types are reputed for their strong cracking and fissure formation. Often they are classified as Vertisols or when having such characteristics they are designated by a vertic adjective. Such materials are clay – rich, but often not excessively so, in the 20–35% range. However if the organic matter is of low content or poorly fixed in the clay-organic complexing system, aggregation is not sufficient to prevent the formation of deep and persistent cracks. Surface cracking is seen in the presence of wide cracks which descend to significant depths. Well below the plant – soil interaction zone one finds very wide and deep cracks extending to several meters depth. An example is seen in Fig. 5.17.

The surface of the soil is under prairie, as shown in the figure. Here cracks are apparent on a large scale (4–5 cm large). In fact they are not just the expression of

Fig. 5.17 Example of the succession of crack features in an Italian Vertisol (Moreau et al., 1999) showing the importance at eh surface, tehe formation of aggregates in the soil (plant – clay zone) and the formation of larger cracks at depth in the C horizon of the alterite material (photos BV)

drying in an organic – poor and clay – rich medium. As one can see, prairie has installed on the soil surface and on can expect a relatively significant content of organic matter. Still, the high smectite content and the significant clay content of the soil (near 30%) produce a vertic structure. Low organic matter content allows the full expression of water loss from the smectites which is moderated by clay – organic associations (personal communication, F Terrible, Univ Frederico II, Naples). In the adjacent photograph one can see the extent of the crack formation. In fact the large cracks are more apparent at depth than at the surface. In the photograph (the geologist's hammer is a scale factor at the bottom left) below a meter, one finds the development of very large cracks or fissures. In fact it appears that the soil horizon (plant – silicate interaction zone) has mitigated, again, the effect of fissure creation and such activity has resumed at depth below the A horizon.

However, the internal soil structure, i.e. crack and aggregate structure, can be important concerning the properties of a soil (Velde, 2001a; Logsdon, 1995).

5.2.2.2 The Importance of Soil Structure (Organic Matter and Clays) on Some Soil Properties

Aggregate Formation

Aggregate formation essentially divides the soil water into two compartments, that found in large passages where water can flow and air can enter when gravity has emptied the pore space and that water present in only small capillary sized pores which can contain water but where the water is much more tightly held (Coppola et al., 2008). The micro pore structure can be visualized and treated numerically using high energy scanning devises (Perth et al., 2008). Such features as average path length, pore size distribution and so forth for the micropore network are values that are now available to modern research. The micropore structure is vital to plants in that it is the major reservoir for water under normal growing conditions in most climates. The two compartments of pores in soils, intra and inter aggregate rhythm the life of the soil – organic structures. Low organic matter content or soil clays that do not aggregate well (kaolinite) limit the capacity of the soils to retain water and hence the growing capacity of the soil.

Smagin et al. (2004) show the importance of organic matter on the water retention capacity of soils from a macroscopic point of view. Using measurements on long term field studies and laboratory experiments they demonstrate the strength of retention of water depending upon the carbon content of the soils and hence the organic – clay contacts forming soil aggregates. This sustains the age old adage that organic matter content is important for soil fertility, in the sense of water availability. Thus on of the foremost functions of soil aggregates is to hold water in the micropore positions which is a reservoir for plants throughout the growing season.

Figure 5.18 indicates the structure of such soil – organic complexes with micropore space and macropore, inter-aggregate space.

The decoupling of macropore water, which moves through gravitational forces from the aggregate micropore water which is strongly retained in the soil is the

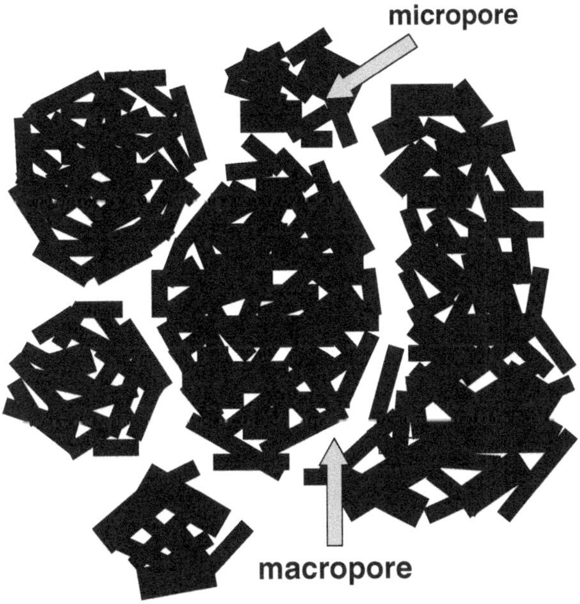

Fig. 5.18 Diagram indicating the organization of clay particles forming aggregates with micropore sites and inter – aggregate porosity

key to understanding the chemical hydrology of soils. The water in the pores of aggregates will come into chemical equilibrium with that of the soil clays and the organic matter while the water in the macropore solutions will be less influenced by the solids. Data from a forest soil where certain mineral elements were analyzed in fluids coming from macropores and micropores (Ranger, 2002) shows a decided difference depending on the site where the water is found for certain elements and for others no significant difference can be noted. Potassium and Si are notably higher in the capillary, micropore water whereas Al seems to be present in roughly the same amounts in either case. Hydrogen ions seem to be more concentrated in the capillary, aggregate water. This is just one example for one situation, a conifer forest soil, which does indicate different levels of saturation for the different types of water. Such studies would be valuable in order to understand which mineral elements are controlled by the clays present and which are more likely to be extracted from the system by water flow through the macropore structure.

The existence of aggregates, controlling water flow and retention then not only controls water availability in the soil but also the flow of mineral elements from the solids in the soil into free flowing water due to rainfall.

Importance of Aggregate Formation on Physical Properties

The aggregate structure in soils is not only of chemical importance but can also play a role in the structure of the soil in a physical sense. Aragon et al. (2000)

show the importance of organic matter, and hence aggregate structure, in the overall resistance of soil to compaction. The more organic matter in the soil, the more aggregates are formed and these aggregates resist re-organization which determines compaction. Such resistance to compaction pressure is important for agricultural practice, especially when heavy wheeled vehicles are used to treat the soil. But in a general sense it is better to have a compaction -resistant soil structure which will hold its aggregate structure under heavy rainfall for example, and maintain the pore microstructure and almost as important the inter-aggregate macropores which convey oxygen to the roots when the free – flowing water content diminishes.

Higher bulk densities of soil aggregates, i.e. lower abundance of micropores, impede root growth (Logsdon et al., 1987). This reflects the resistance to root pressures made by high density, low porosity soil material. An organic rich soil of good aggregate structure is less resistant to root penetration.

Yet another effect of the organic – clay aggregation is that of maintaining soil clays in place. The more tightly bound, in the assemblage, the more resistant to transport in runoff water by suspension of soil clays (Rhoton et al., 2002). This study is oriented to show the effects of different types of soil tillage practices, no till and conventional tillage. The no till practices leave a maximum of organic matter in the field after harvesting which is incorporated into the soil as clay organic complexes upon bacterial reduction. Normally in plant covered terrains, runoff loss of clays is not important. However in agricultural terrains, the soil is left without plant cover during the non-growing season and hence subject to runoff of soil clays. The same situation exists in arid and semi-arid climate terrains when plant cover is not complete and bare zones without plant cover can be easily attacked by rainwater carrying clay and other material in suspension. Aggregate stability under conditions of wetting is greatly increased by overall organic matter content and as such the soils are more stable. Water dispersible clay content in run off and streams decreases as a result of aggregate stability. Hence the presence of organic matter stabilizes and conserves clay content of soils.

The inter-relation of organic matter and clays is fundamental to the environment that plants grow in. The formation of soil aggregates, which are essentially sponges for the storage of water, is an extremely important of the functioning of soils (the A horizon). This structure is strongly dependent on the presence of 2:1 clay minerals which are more chemically active than are the 1:1 kaolinites. The stability of the aggregates depends upon the presence of smaller organic molecules which are more chemically active. However these smaller molecules are more easily degraded or mineralized by bacteria due to the presence of C–O and other bonds instead of the predominance of C–C bonds in larger molecule humic material which are resistant to bio degradation. Thus the continued activity of plant deposition of their aerial parts and decay of roots is necessary to microbial action which produces some low molecular weight material. Also, the roots themselves exude this low molecular weight organic material which also maintains the aggregate structures. Loss of input, from aerial deposit, decreases the content of the low molecular weight material. Hence the cropping and exportation of plant bio mass impoverishes the soil in its organic content and destabilizes the aggregate structures. This leads to more difficult conditions of plant growth and to a loss of clays due to surface rainwater transport. The organic – mineral relationship is fundamental to plant growth.

5.3 Plants and Soil Chemistry

5.3.1 Plant Compositions and Chemical Needs

Plants do not live on air and pure water. There has to be something in the water. The reason is that plants are not composed uniquely of components found in the air, though largely so. Most of plant substance is found in the air, water, carbon dioxide and to a lesser extent nitrogen. These major components are freely available to all and with sunshine as a source of energy plants do very well. However, minor mineral elements are necessary for the proper growth and well being of vegetal matter. For terrestrial plants these elements are found in the soil they develop in or on the rocks that they colonize. Agricultural science has developed around the needs of the mineral elements found in soils and nitrogen which is to a large extent transmitted in the soil zone from air to plant via bacterial action. It has been found that certain elements are especially important. They are potassium and phosphorous for the mineral parts. Traditional fertilizer contains nitrogen, phosphorus and potassium. Other cations such as Ca, Fe and Mg are usually sufficiently available in soils as to pose little problem in limiting plant growth. However some cases show a lack of these elements which needs to remedied by application of a mineral supplement.

One of the functions of plant roots is to gather the necessary mineral elements needed for proper growth and development and to prospect for water. The methods of obtaining the necessary mineral components appear to be varied and at times complex. Often plants use or encourage micro-organisms to provide the mineral resources via actions of mineral dissolution or elemental extraction certain elements from minerals. There is a strong interchange between root exudates material and microbiological activity in soils which promotes the extraction of mineral elements form the soil, rock and mineral substrate. The concentrations of different mineral elements is not constant in the plant, roots, stems or trunks, branches and leaves (Gain, 1895) indicating their involvement with different metabolic functions in the plants. However, in general mineral substances are important though present in much lower concentrations than CO_2 and H_2O which make up about 90% of plant matter. Mineral elements vary from one plant species to another. As an indication of the relative importance of each component, Bormann and Lichens (1994) give the following ratios for an active hardwood forest in Northeastern United States: Ca 10, Mg 1, Na 0.02, K 0.6, P 0.2, Fe 0.3, N 0.8. The relative importance of potassium and nitrogen as far as their influence on plant growth is much greater than their relative ponderal importance.

5.3.1.1 Silica

Among the mineral elements that the roots gather one finds silica. The role of silicon is usually not mentioned in studies of plant physiology except that it is present in most species. The presence of Si is higher in more primitive species (Hodson et al., 2005) such as horse tails, liverworts and club mosses. Non-woody plant parts show up to 1.5 wt% of Si of the total plant material (i.e. including water components or "wet" weight). The silica is present in plants as phytoliths, concentrations of amorphous silica in a pure form. These silica aggregates can persist in soils for significant periods of time. Borrelli et al. (2008) indicate that the phytolith populations can remain un-changed for at least 50 in soils subjected to different vegetal cover in Argentina. However it is generally considered that the phytoliths control the silica content in soil pore by their dissolution water and in the water that eventually drains in to rivers and the sea (Farmer et al., 2005; Derry et al., 2005). Thus the concentration of silica in soil pore solutions appears to be controlled by plant action, in the form of phytolith concentrations of pure amorphous silica. Amorphous silica is far more reactive, i.e. can dissolve much faster and at higher concentration that most silicates and especially quartz, the pure crystalline form of silica. Quartz has a very low dissolution constant and rarely controls the silica activity of pore water solutions. Hence with one form of silica, phytolith, the silica concentration is solution is very high and the other, quartz, it would be very low. The impact of the activity of silica in solution is very important to the stability of the different types of clay minerals present which we will seen below.

Data from Gain (1895) shown in Fig. 5.19 indicate that the different parts of crop plants contain different proportions of potassium and silica. The silica content of the structural parts, stems of grasses, is particularly high in that it forms a structural element within the stems. This is one of the reasons why we like to eat the fruits of plants and not the stems, most often.

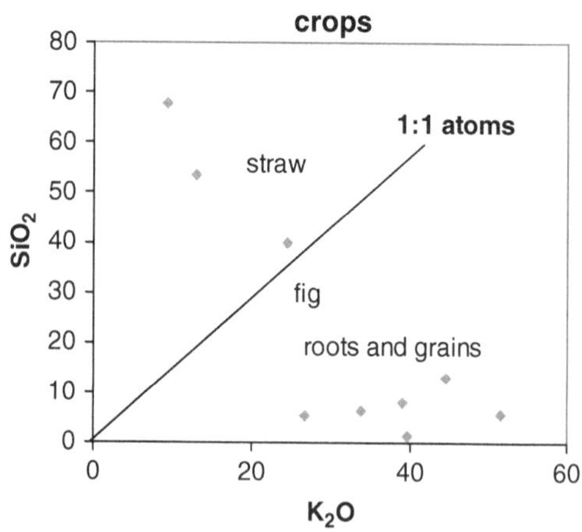

Fig. 5.19 Relative atomic proportions of potassium and silica in several examples of grain crops comparing straw and other portions of the above ground plant (data from Gain, 1895)

5.3.1.2 Calcium and Magnesium

Calcium and magnesium are present in plants, but seem rarely growth limiting. However they have been used as fertilizers (Larbaletrier, 1891) but they are more rarely used as fertilizers today but instead used to correct the soil pH. The common presence of calcium in rock silicate minerals and carbonates is such that calcium is almost ubiquitous. However Ca is largely present in plant compositions, equal to or about half that of potassium. Knecht and Göransson (2004) report similar abundances for a large variety of plant types but Broadley et al. (2004) indicate lower values for angiosperms, and especially herbaceous plant types. Magnesium is present in plants in much lower proportions, about 10–20% of Ca weight content. Calcium seems to be equally present in coniferous, deciduous and herbaceous plants.

Hence calcium being almost omni – present soils is hence rarely critical to plant development. The importance of magnesium is less well documented.

5.3.1.3 Potassium

Potassium is another story. Very much of agricultural practice has been focused on the availability of potassium for terrestrial plants. It is one of the more abundant mineral elements in these plants. Plants fixed nearer the sea, or in arid areas of salt deposition are much more dependent on sodium, because of it s relative availability. For terrestrial plants potassium is an element which changes their growth factors greatly, as seen in demonstrations of the use of fertilizers in 19th century texts. Potassium is frequently a limiting factor in plant growth. However although potassium is found in most parts of plants its metabolic and functional role is poorly known (Treshaw, 1970, p. 182). It does not appear to be fixed in plant tissue units. Since the 1860s in Europe the use of potassium chloride has been know as a major element that could be used in synthetic fertilizer agricultural use (Larbaletrier, 1891). The material could be mined in essentially the pure state. It replaced the use of wood ash as a source of potassium. In fact such ash materials were imported to Europe from Russia and the United States as a fertilizing agent before this time. The use of potassium has been much studied and its consequences on plants, not only crops but tree growth has been well documented.

From the assembled data of Knecht and Göransson (2004) herbaceous plants contain more potassium than trees on a nitrogen based comparative scale. Gain (1895) indicates that in agricultural crops, grains have less potassium in roots than do legumes and cotton. Stems and grains have two times as much potassium present than do roots but there is a strong variation from one crop to another. Usually the material harvested is the most potassium rich. However the needs and plant composition do not remain constant as a function of maturity stage. Clark et al. (1980) show that the potassium content in grasses and legumes can change by a factor of about three during different stages of growth. Potassium uptake is strongest in the early stages of maturity. Thus the exportation by agriculture of potassium is critical to the soil – plant balance. The ratio of potassium to calcium in different parts of trees can vary to a large extent, oak for example can have 25% of calcium content in the wood and 17% in the leaves of the tree. By contrast beech and confiers show a lower potassium content in leaves and wood than in calcium. These differences are important in the short term in that trees seem to stock more calcium than potassium especially in the wood whereas perennial plants appear to stock more potassium in the plants. Thus the annual return of mineral elements to the soil shows a higher return of potassium for herbaceous plants and oaks (leaves) than trees. Overall herbaceous plants take out more potassium (and return it to the surface in the soil if left to mature) than do most trees. The herbaceous plants can be seen as potassium storers and gatherers more than are trees.

Potassium is found in clay minerals and some high temperature minerals (feldspars) which are found in the soil zone. The major source of potassium in soils is from clay minerals. Potassium can be extracted from high temperature minerals in the water – rock interaction portion of alteration profiles, and deeper plant roots surely use this as a source of potassium renewal as this rather soluble element is leached out of the plant soil horizons. The potassium is transferred from root zone to plant and plant to soil by leaf and other plant decay mechanisms. Then the potassium is incorporated into the clays as illite, an anhydrous layer of potassium ions in the 2:1 structures. Release of potassium from leaf matter can be rapid, where leaf decay and transformation by bacteria is not necessary to release the K in the leaves (Lupwayi et al., 2006). Potassium is not held in the plant structural elements.

5.3.1.4 Phosphorous

The use of phosphorous as a fertilizing agent was known and practiced in Europe from the 1840's (Gain, 1895). It is relatively abundant in many countries as reasonably concentrated calcium phosphate deposits, usually found in sedimentary rocks. Since most agriculture is accomplished on sedimentary terrains, this availability was quickly exploited. Phosphorous is, as is potassium, fundamental to good plant growth and lack gives poor plant growth. However the chemical path of phosphorous in soils is not well understood. Phosphorous is fixed strongly to amorphous silicate materials (allophone) and oxy-hydroxides especially those of iron. This capacity to fix phosphorous is a definition for Andosols in soil classification (Dixon and Weed, 1989). However the chemical path of phosphorous in soils with crystallized silicate materials is less sure. It is apparent that phosphorous is little available in the organic matter. Thus one must presume that a large portion of available phosphorous is associated with clays somehow. The major source of mineral phosphorous is from apatite, a calcium phosphate. This mineral is strongly affected by low pH solutions. The source of phosphorous in rocks is often that of animal concentrations, as bones and teeth deposited in phosphate deposits, made mainly from fish remains. Some phosphorous is found in igneous rock, especially granitic type materials but is some basic volcanics it can be abundant, especially in eruptive rocks of high potassium content. Such are found in the Roman co-magmatic province and have given rise to the rather fertile material around Naples and Rome. It is not an accident that this reliable source of soil fertility and hence crop production gave rise to a strong and independent nation in the distant past.

It is proposed by many authors that the phosphorous is fixed by the Al and Fe hydroxyl bonds on the amorphous materials in such soils (Dixon and Weed, 1989). Such bonds exist, in smaller numbers per mineral mass, on clays at the edges of crystals. It would seem possible that such electronically active bonds could also fix phosphorous. The amount and intensity of bonding on clay edges, according the type, 1:1 or 2:1 structure would determine the phosphorous availability for plants. At present we do not know about the relative values of these bonds and their intensity but it is perhaps a fruitful avenue for future research.

5.3.1.5 Nitrogen

Nitrogen is the major component of air. However, as far as one can see now, most nitrogen in the bio-cycle passes through soils before entering into the plants, initially. Thus the air in soil, in the pores at the surface is the vector of nitrogen fixation in the life cycle of plants. The importance of nitrogen and the nitrogen cycle has been the subject of much research and discussion. There is a very large body of data on the subject, related to the growth of domestic plants (crops) and forest growth. Essentially, nitrogen is a large component of air, about 80% by volume. It is found in many components of plants. It has been known for a long time that nitrogen containing components can be used to facilitate plant growth. Nitrogen seems to be a sort of catalyst for plant growth being present in relatively low concentrations in many parts of a plant. The basis of farm manure use is based upon a knowledge of the management of this essential fertilizing agent. Ninteenth century texts on agricultural practice, preceded by those of other practitioners of the art (de Serres, 1600 for example) who quote ancient Greek authors, all acknowledge the importance of recuperation and proper treatment of animal excrement in the nitrogen cycle. The problem in agriculture is to avoid the "mineralization" or reduction to a gas state of the nitrogen by bacteria in the soil. Nitrate is a useful form of nitrogen for plants as is ammonia. The ammonia state is often found directly produced by bacterial action in manure accumulations and in animal urine. Hence the barn yard was an important cite of recuperation and fixation of nitrogen for fertilizer use. Today, barn yards no longer exist. Animal excrement is considered to be a noxious by-product and mostly misused by direct application to the earth where it is very largely mineralized by bacteria being lost for about 80% to plant use. The intermediate step of mixture of excrement and urine with straw or other plant materials fixes and develops a more stable form of nitrogen – bearing substances. Russell (1945, p. 237) illustrates this effect as one lasting for at least four years which increases crop productivity by a factor of two. Of course synthetic sources of nitrogen can be more powerful but are not as long lasting.

Plants most often extract nitrogen from organic matter via bacterial action on such detritus as provided by the plants themselves. Further sources of nitrogen are accumulated in soils via bacterial action associated with plant roots forming nitrogen – bearing nodules, especially prevalent on legume plants roots. The ultimate source of nitrogen for plant growth is in the air present in the pores of soils. This transfer replaces the nitrogen lost from the biomass through bacterial action producing gasses and nitrogen – bearing molecules dissolved in aqueous pore solutions which drain from the soils eventually into the water table and stream sources.

5.3.1.6 Distribution of Elements in the Plant

As might be expected, elemental distribution is different in different plants and is variable within the plant itself. These variations have been used by farmers for some time, perhaps from the beginning. Basically one can consider the roots, the stems and woody parts and the leaves of plants. Overall woody plants appear to store much K and Ca in stems, trunks and branches. Some specialists (personal communication D Righi, Univ Poitiers) attribute the acidity of forest soils in large part to the exchange of the bases, Mg, Ca, K for hydrogen ions as the mineral elements are taken up across the root membranes. Since a great mass of these elements is permanently stored in a forest, the overall effect is to permanently acidify the soil as the bases are taken from the soil and acids left behind in an electro – neutral exchange. Overall it appears that deciduous trees and forest plants take up more potassium than do coniferous trees (Knecht and Gôransson, 2004). Herbaceous plants have even more present than woody plants. Forest floor plants such as ferns have roughly twice as much potassium present in their leaves that beech trees for example (Larbaletrier, 1891). It appears that much potassium can be stored, above and beyond incorporation in cellular material in the leaves of herbaceous plants leading to "luxury consumption" by certain species (Broadley et al., 2004). This potassium is readily available to the soil solutions either by rainfall washing or rapid release from fallen leaves.

Root compositions of "old prairie" given by Larbaletrier (1891) are comparatively less potassic that shoot material of herbaceous angiosperms (Broadley et al., 2004). However compositions of root crops and grain from traditional agriculture (Fig. 5.11) indicate high potassium levels. Thus all depends upon the plant concerned and this will have an impact on the chemical balance in a given soil setting.

For the most part, modern agricultural practice is focused on the NKP fertilizing agents, while other mineral elements are used in cases of special deficiencies due to specific local soil, and substrate conditions. The fundamental necessity of these three elements and their use in encouraging plant crop growth have led to what has been called the green revolution in changing the productivity of agriculture in developing countries in the period 1960–1980. The systematic use of these chemical fertilizing agents gave rise to an unprecedented increase in agricultural productivity in less developed countries as was the case earlier in developed countries. The importance of these chemical agents was of course not unknown before the use of synthetic agents. The traditional agricultural practices used conservation and concentration practices which enhanced or preserved such element sin the soil horizons. Thus the modern agricultural practices have not discovered new truths, as far as the chemistry of plant growth is concerned, but they did use such information in a scientific way to concentrate the effects in crop science. More on this subject in Chap. 6.

5.3.2 Chemical Profiles in Soils and Alterites Engendered by Plant Activity and Resulting Clay Mineralogy

The chemical effect where plants have been seen to transfer certain elements from the alterite zone to the soil zone has been known for some time (Duvigneaud and Denoyer – DeSmet, 1970 for instance). These authors indicate that the uptake of potassium the alterite and soil zones is greater than that retained in the soil zone. The transfer and fixation are dependent upon several factors, not the least the type of plants present such as conifer versus deciduous tree for example (Homann et al., 1992; Quideau et al., 1996). Nevertheless, in many instances, one can clearly see a trend of potassium increase at the surface (soil horizons) and frequently an increase in silica where free quartz is not present (Fig. 5.20).

This is true for alteration profiles in temperate climates on basalts (Chesworth et al., 1981) serpentinites (Rabenhorst et al., 1982; Bonifaccio et al., 1997) and various eruptive rocks under humid and tropical climates (Leumbe et al., 1993; Meijer and Buurman, 2003; Pion J-C, 1979; Mohr and Van Baren, 1959, p. 147; Zarin and Johnson A, 1995; Nguetnkam et al., 2007; Jahn and Asio, 1998) highly weathered podzolic soils in Alabama (Bryant and Dixon, 1964) for contrasted season semi

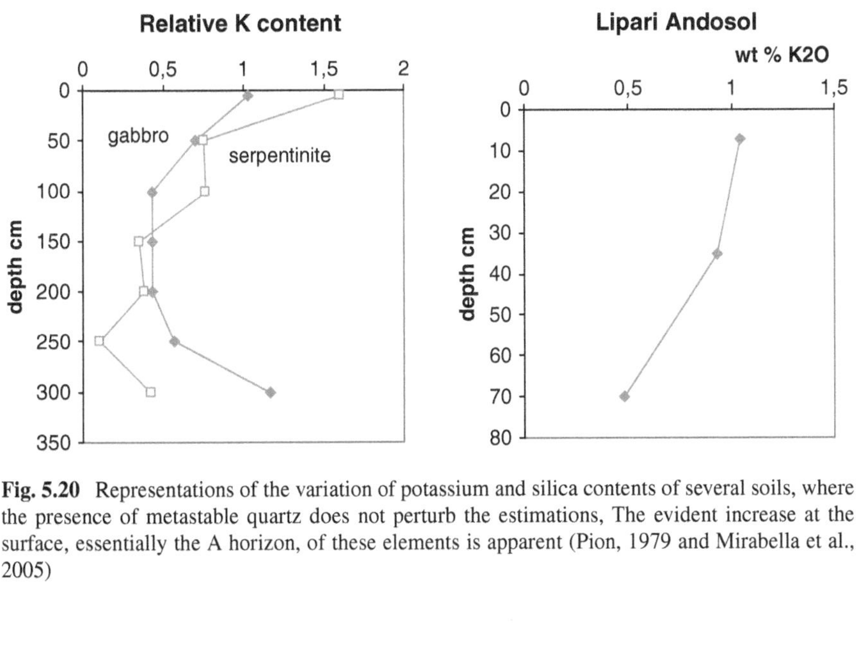

Fig. 5.20 Representations of the variation of potassium and silica contents of several soils, where the presence of metastable quartz does not perturb the estimations, The evident increase at the surface, essentially the A horizon, of these elements is apparent (Pion, 1979 and Mirabella et al., 2005)

tropical forest soils on terrace deposits (Ahmed et al., 1971) and on a basalt in New South Wales, Australia (Loughnan, 1969, p. 89).

Silica is found in plants as a pure, amorphous phase, phytoliths, which has been considered to be the major controlling phase of silica activity in soil solutions (Farmer et al., 2005). This effect has been modelled comparing the results to long term observations on soil pore solutions under conifer forest in a temperate climate (Gérard et al., 2008). Similar observations have been made by Derry et al. (2005) for soils with various plant covers on the Island of Hawaii. Potassium is found in most parts of the plant. Plant decay by leaf shedding or loss of branches and eventually the entire tree deposit these mineral elements at the surface (Fig. 5.19). De Connick et al. (1987) show data demonstrating that the transfer effect can be concentrated in the fine fraction of the soils in the a horizons. The presence of phytolith material is the controlling factor of silica activity in solutions because phytoliths are highly reactive materials susceptible to dissolve much more rapidly than other forms of crystalline silica such as quartz (Fraysse et al., 2006).

Righi et al. (1999) give data showing the rapidity with which this uplift or plant engendered chemical transfer can occur in young montane soils. One compares the A (plant influenced) horizon to the C horizon where the major activity is due to water rock interaction. In Fig. 5.21 given above one sees that potassium is lost in both horizons but much less in the A horizon with perhaps a stabilizing composition near 4% K_2O. However potassium is lost in the C horizon regularly with time. The same effect is seen for silica for the two horizons. The time span is 6,000 years, and one sees the effect of plant transfer in the very early stages of development of the soil profile, probably less than 400 years. It is clear that plants can control the silica activity, which determines mineral stability, in soils.

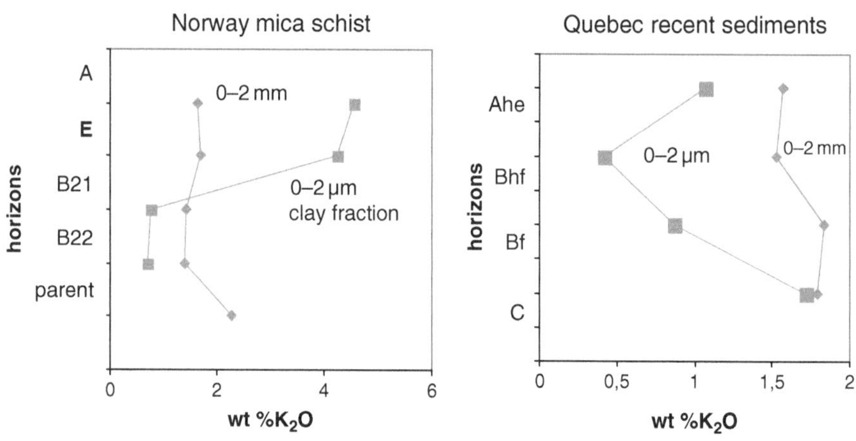

Fig. 5.21 Illustration of the relative increase of potassium as a function of size fraction in two sandy till and loess based soils. Data from De Connick et al. (1987)

5.3.2.1 Clay Minerals

Lucas et al. (1993) and Lucas (2001) have emphasized the importance of the phytolith uplift of silica to the stability of kaolinite, the alumino-silicate 1:1 clay, compared to gibbsite, the alumium hydroxide clay mineral. The maintenance of silica activity at the surface favors the presence of kaolinite in tropical soils. The normal weathering trend is to lose alkalis then silica (Pedro, 1966).

However the stabilization of other clay minerals is possible and numerous studies suggest their presence due to the silica activity engendered by plant phytoliths. This is a sort of reverse weathering.

Reverse weathering would occur as follows:

Gibbsite	Al
to	
Kaolinite	Al-Si
to	
2:1 mineral	Si-Al-Si

The 2:1 portion of the minerals can vary from muscovite composition with an Al:Si ratio of one as is the case of kaolinte, to aluminous interlayered ion hydroxyl vermiculite where the Al:Si ratio is slightly greater than one due to a lower substitution of Al for Si in the tetrahadreal sites, to a smectites where the charge is low due to more silica in the tetrahedral site. The mineral formulae are something like the following:

It is clear that the Si/Al ratio will be quite different if one takes into account the interlayer ion population of Al in the hydroxyl interlayered minerals. In the case of accounting for these Al atoms one changes the Si/Al ratio from 1.45 to 0.8, below that of either kaolinite or muscovite, placing it chemically between kaolinite and gibbsite. This seems to hardly be the case in that hydroxyl vermiculites are not usually associated with gibbsite in soils. On the contrary they are found with kaolinite. Thus it appears that the essential chemistry of the clay minerals, comparing 1:1 to 2:1 structural types, should be based upon the covalently bonded cations in tetrahedral and octahedral sites and not the interlayer ion population. Interlayer ion occupancy will be determined by the general ambient chemical conditions of the soil pore solution while the same solutions will determine the stability of either silica-poor kaolinite or silica deficient gibbsite, compared to the silica – rich 2:1 minerals such as smectites, HI minerals or illite.

The chemistry of the soil profiles shown in Fig. 5.22 indicats that for the Red Bay series, Mississippi oxysoils developed on Mesozoic sedimentary rocks (Glenn and Nash, 1964) where one finds kaolinite – gibbsite in the lower horizons and vermiculite in the soil A horizon. The presence of more silica at the surface is reflected

	Interlayer	*oct*	*tet*	*Si/Al in 2:1 unit*
Muscovite	K	Al_2	Si_3Al	1
Al hydroxyl vermiculite	$Al_{1.6}$	Al_2	$Si_{3.5}Al_{0.7}$	1.45
Smectite	Ca, Mg	Al_2	$Si_{3.7}Al_{0.3}$	1.6

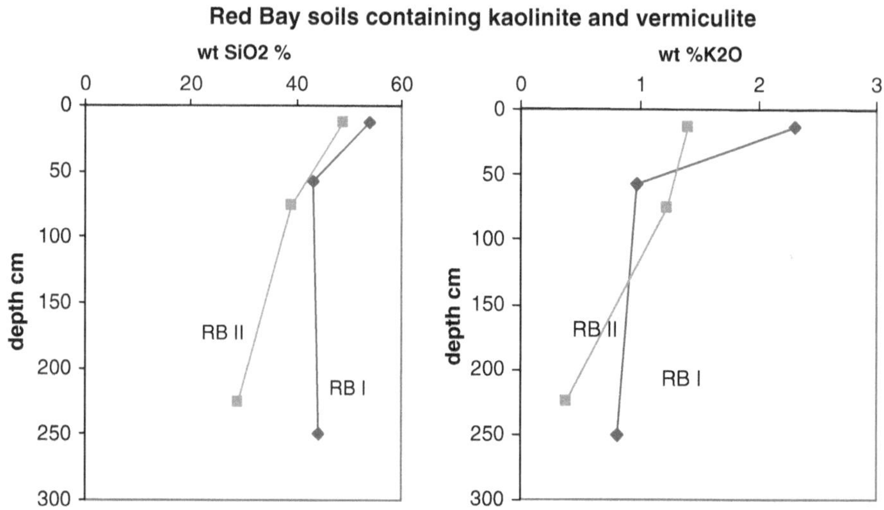

Fig. 5.22 Red Bay series of soils from the Mississippi Bay region showing the increase in potassium and silica in soils where gibbsite and kaolinite are dominant in the C horizons and HI minerals and illite in the A horizon. (Glenn and Nash, 1964)

in the change in soil clay mineralogy from a silica – poor type at depth to a more silica – rich one at the surface where plants have a greater influence on the soil clay minerals.

An example of the influence on the filling of interlayer ion sites in soil clay minerals effected by plant enrichment of the surface horizons, can be seen in the chemistry of the clays present in the various horizon, as is shown in Fig. 5.23 (data from Glenn et al., 1960). Here the clays are essentially illite-smectite interlayered minerals with some illite and kaolinite present. For different size fractions of the clay sized particles there is a significant increase in potassium content in the A (plant) horizon. Here it is clear that the clays are affected by the presence of plants and that the interlayer ion sites are impacted by the availability of potassium in the soil zone.

The higher availability of potassium at the surface fills the interlayer ion sites of the clays, increasing it by about 1 wt% or around 25–30% of the clay K_2O content. This clearly demonstrates the importance of plant uplift of potassium seen within the same family of clays, where the interlayer ion sites fix more K_2O. Ahmed et al. (1971) indicate that illite is favored and possibly formed at the surface in soils formed on terraces under conditions of strongly contrasted humid climate.

5.3.2.2 Fixation and Loss of Translocated Elements

If there is a strong tendency to increase potassium and silica in the surface A horizons of alteration profiles, one cannot assume that the potassium and silica transfered to the surface are permanently captured. Duvigneaud and DeSmet (1970) have

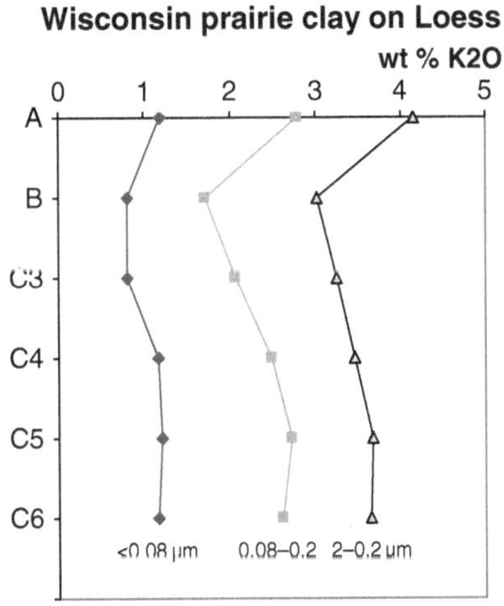

Fig. 5.23 Plot of potassium content of various clay fractions in a Wisconsin loess soil (Glenn et al., 1960) as a function of depth. The higher potassium content in the more coarse fraction of the surface indicates the illite content, often found in the coarse clays of soils

studied the net change in these elements under forest growth in very old forest land (from 1,836 to the present day). Different elements are moved upward in different proportions for different tree species. However in this study on two oak forests in Belgium, the ratio of MgO:CaO:K_2O is on the order of 1:5:4. In a larger geographic comparison, these authors find that the potassium transfered is retained in a ratio of 3:1 of uptake to retention in the Belgian forests, but 5:1 in German and Russian oak forests. Beech retains potassium in a 1:3 ratio of retention to loss, and pine in a 2:1 ratio. Depending upon the tree species more or less of the uplifted potassium is retained in the A soil horizon. But in any case more potassium is lost than retained.

Walker and del Moral (2003, p. 106) show how important vegetation is to the stabilization of alteration processes as seen in the composition of stream water. After landslide event, forest recovery significantly decreases the loss of potassium from the land area as seen in stream water compositions. Thus forest vegetation fixes potassium but there is still loss compared to the uplift or transfer of potassium from mineral to vegetal sources (Fig. 5.24).

The balance between translocation of mineral elements will depend upon the rainfall and vegetation type. Some plants bring up more of these elements than others, some have higher turn over than others and the different types of soils and water retention along with the overall rainfall will determine the amount of material fixed and re-cycled by plants at the surface.

5.3.2.3 Climate, Plants and A Horizon Clays

If we can identify a transfer effect where plants bring mineral elements to the surface and if we accept that this will affect the surface clay mineralogy, can we identify effects according to broad type of vegetation and climate? Our investigation to answer this question includes information for soils of different age and hence most likely different stages of evolution. The more humid the climate, the older the soils. This is a general rule due to happenstance to a certain extent. As it turns out,

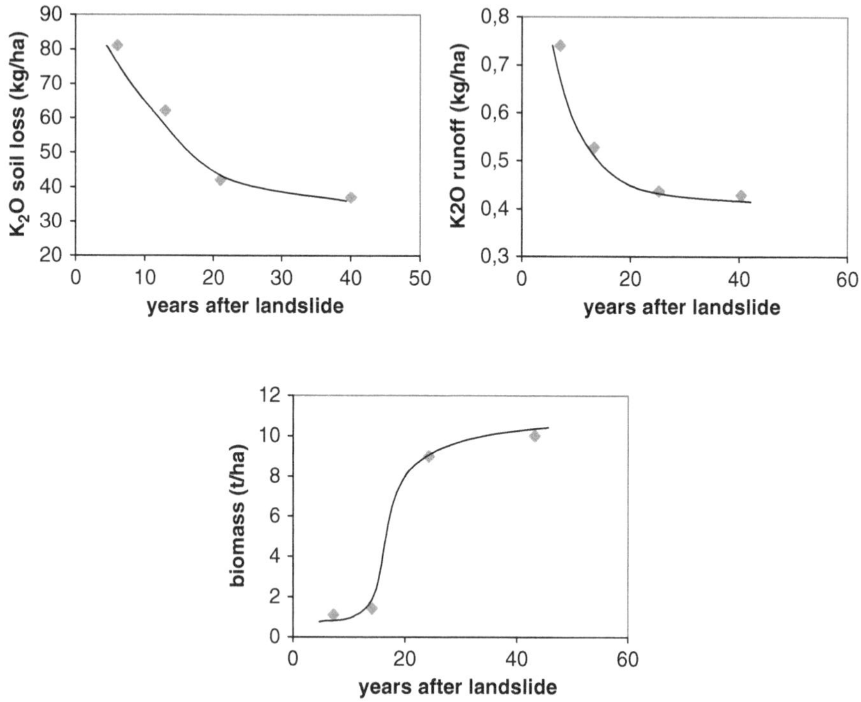

Fig. 5.24 Relations of potassium loss form water – rock interactions after a landslide in a forest area. Potassium in stream water decreases with time as does the amount of potassium present in soils. The biomass increases in a reciprocal manner showing the stabilization of potassium in the surface portions of the alterites due to plant activity. Data from Walker and del Moral (2003, p. 160)

the tropical areas and areas of high rainfall are frequently based upon old cratons where tectonic renewal of surface material is small. However in some areas, the south Pacific for example, significant new deposits of volcanic material are present and thus one can find younger soils and older soils (see Mohr and Van Baren, 1959). These authors indicate the different stages of chemical alteration from juvenile to laterite stage induced by heavy rainfall. Chemical alteration is attested by the color of the earth, going from brown to red and also by the amount of organic matter in the surface layer of the alteration profile. This organic content is also a testimony of the fertility of the soils and the plant growth activity. There is then an evolution of plant action and capacity to modify the surface chemistry as a function of time. He et al. (2008) indicate this for clays formed on basalt substrates under sub tropical conditions. In the initial stages of alteration (60,000 years) the surface clays contain 2:1 minerals with the presence of illite, in small quantity. With time the2:1 clays change in character and diminish in relative quantity. The longer periods of alteration indicate that the translocation of certain elements is not sufficient to modify the clay mineralogy: illite is lost and replaced by and HI mineral.

Huang and Uehara (1968) demonstrate thet climate (rainfall) can have a significant impact on the transfer of elements to the surface. This effect is of course indirect in that the agent of uplift is the vegetation present and with increased rainfall the intensity of plant activity increases. The study observes the change in clay mineralogy in A horizon soils on Hawaii where the rock substrate is similar, a basalt of rather low alkali content ($K_2O = 0.23$ wt%). The formation of illite in the plant – silicate interaction zone is a direct function of the rainfall parameter, increasing from near zero percent with 800 mm rainfall to being the dominant clay present at rainfall values of 6,000 mm per year. No information is given for the vegetation regime but one can consider that it changes from steppe – like prairie to heavy tropical forests in the range studied. The presence of illite, when no phyllosilicates are present in the basalt is striking. Here the uplift effect, transport of potassium necessary to form illite, creates a new mineral in the soil clay assemblage which co-exists with a diminishing amount of kaolinite. The change with depth is from illite kaolinite in the A horizon to kaolinite – gibbsite in the deeper horizons.

Thus the plant – clay mineral alteration system is a function of time to a certain extent, possible governed by the relative biomass produced which reflects the total available resources. In the case of basalt, there is little potassium available in the rock and gradually some is lost to rain water percolation even though plants attempt to bring it to the surface.

In the brief review of surface 2:1 clay mineralogy given here, we attempt to give data for the younger soils in a given situation.

In Fig. 5.25 we indicate the topsoil (A horizon) mineralogy for the 2:1 minerals when present. Data are from Netttleton et al. (1973), Ruhe (1984), Barré et al. (2007), Graham and Southard (1983), Boettinger and Southard (1995), Righi et al. (1999), Mirabella and Egli (2003), April and Keller (1990), April et al. (2004), Gillot et al. (2001), Fichter et al. (1998), Pevear et al. (1984), Teveldal et al. (1990), Tice et al. (1996), Ahmed et al. (1971), Bryant and Dixon (1964) and Graham et al. (1990). Birkeland et al., (2003), Huang and Uehara (1968)

Some points should be reiterated: cold climates are associated with glacial events which are recent, on a continental scale or in mountain ranges. Hence soils associated with cold climates today are by nature young and may not represent a stable clay assemblage. Tropical climates are biased toward old, stable craton geological features and have been active for a long time, hundreds of thousands of years or more. This is the case for African craton soils or those in certain areas of the Amazon regions. Here the time factor would lead one to think that most of the material has come to an equilibrium with the climatic elements. However, if one allows a cold climate to come to ultimate equilibrium it might well converge in its mineralogy with that of a warm and humid climate. The overall trend in weathering is to lose alkalies and silica, no matter how much plants work to overcome this inevitable fact. The soils will become impoverished in the long run. The end assemblage of a very old soil would be iron oxide plus gibbsite and perhaps some kaolinite. This means that what we see in soils today is part of a dynamic system, one that will change with time. This is no surprise to a geologist but perhaps unsettling to a biologist or ecologist.

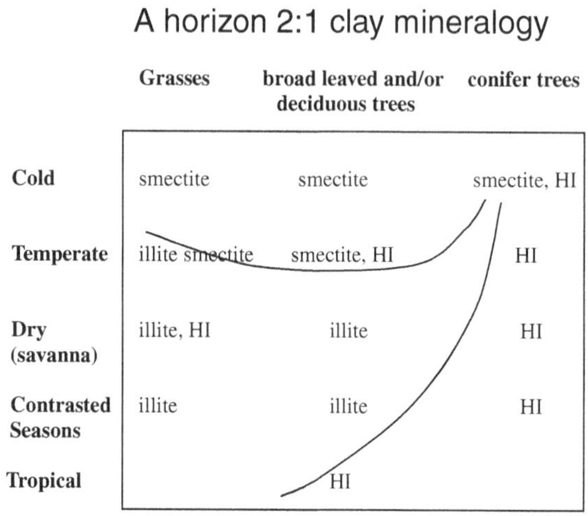

Fig. 5.25 Relations between climate and vegetation as they determine the dominant 2:1 clay minerals types in the A horizons of the soils. Different vegetation types under similar climatic conditions affect the soil solutions differently producing different interlayer ion site occupations. Conifer trees tend to produce HI (Al hydroxyl complex) clays, while grasses and deciduous forests favour potassium as the interlayer ions. Under conditions of lower temperatures smectites are favored. Data are from Nettleton et al. (1973), Ruhe (1984), Barré et al. (2008), Graham and Southard (1983), Boettinger and Southard (1995), Righi et al. (1999), Mirabella and Egli (2003), April and Keller (1990), April et al. (2004), Gillot et al. (2001), Fichter et al. (1998), Pevear et al. (1984), Tedeval et al. (1990), Tice et al. (1996), Ahmed et al. (1971), Bryant and Dixon (1964), and Graham et al. (1990). Birkeland et al. (2003), Huang and Uehara (1968)

In looking at the assembled observations in Fig. 5.25 one sees a tendency to have HI minerals present as 2:1 minerals under conifer trees for various climates. The HI mineral listed under tropical climate is perhaps misleading in that most broad leaved trees in tropical forests are in fact evergreen and not true deciduous types. Smectites seem to be more or less restricted to cold climate regimes. Here one must be careful in that the young soils may evolve over slightly longer periods of time. Data from Mahaney and Halverson (1986) for soils of Holocene and Pleistocene soils in montane regions of the Rocky Mountains (USA) indicate that the older soils would tend to have HI minerals present after 8,000 years instead of smectites for shorter periods of 3–4,000 years, similar to those reported by Righi et al. (1999) and Mirabella and Egli (2003). It is not sure then whether or not smectites would be the stable 2:1 mineral phase in older A horizons of soils. These authors do describe the change in octahedral site occupancy. In the stage of smectites formation one finds a gradual change from trioctahedral occupancy to diocthahedral occupancy changing the clays from ferro – magnesian to aluminous in character. The aluminisation of the clays is a general characteristic, well documented by these authors.

The presence of illite in grass or prairie soils as well as deciduous forests under temperate to drier climates seems to be rather universal. Here the tendency to stock potassium in the clays is quite apparent.

Overall then one finds that the 2:1 clays in the A horizons of soils are largely of two types, both of high charge. HI and illite fix cations in the interlayer site which can most often be extracted by plants. Some illite is of a very high charge where it will be difficult to remove the potassium as is the case for the strongly polymerized Al-OH in HI minerals. In a general way, conifer forests produce the HI minerals while prairies and deciduous forest produce illites. Younger soils will show important amounts of smectites forming in the surface horizons especially in cold climates.

These are that clay minerals "useful" to plants in that they can store potassium and eventually ammonium in the interlayer site. There are however other clays present, such as kaolinite, which are less useful but with time become omni-present. In a number of cases of advanced weathering the 2:1 clays are present only in the A horizons.

Overall, the presence of kaolinite, and eventually gibbsite is a function of rainfall, age and to a certain extent temperature. The kaolinite is found in the initial stages of mineral weathering under water – rock interactions. It is more and more important as climate provides more rain and more dilute pore water solutions. With time the kaolinite content increases further. Old soils show high kaolinite and often gibbsite in the C horizons. In soils of the Coastal Southeastern United States one finds a loss of A horizon 2:1 minerals, essentially HI minerals, after 1 My, where the soils of older age show, rather abruptly, a loss of significant amounts of 2:1 minerals, a change which was accompanied by a change in bulk composition of the soils (Howard et al., 1993). The 2:1 minerals are replaced by kaolinite and gibbsite. In soils of semi tropical Hainan Island (China) clays formed from basalts have a very low, but persistent 2:1 mineral content which is reduced with time changing from an initial illite facies, to a smectites – illite and eventually HI mineral facies (He et al., 2008). The ratio of kaolinite to gibbsite changes in favor of gibbsite (aluminium hydroxide) with time affecting the interlayer ion of the 2:1 minerals changing from a potassic form to an Al-OH form. In these cases the plant regime cannot maintain the necessary elements in the A horizon for stabilization of potassic minerals due to the gradual loss of this element in the total alteration profile.

5.4 Impact of Plants on Clay Minerals and Soil Formation According to Biome Type

Numerous studies have given results showing the specific change in clay mineralogy from the source rock to the surface (A) horizons (Drever, 1994 for example). If one correlates these different results with the type of plant regime, and hence climate, a general pattern can be initially observed. There are many questions to be asked and specific situations not yet studied, but nevertheless a certain pattern emerges. In Fig. 5.26 we represent the general characteristics usually attributed to prairie and forest soils in manuals of pedology. It is important to note tht the point of view of most pedologists is of their local surroundings, and hence the inhabited Northern

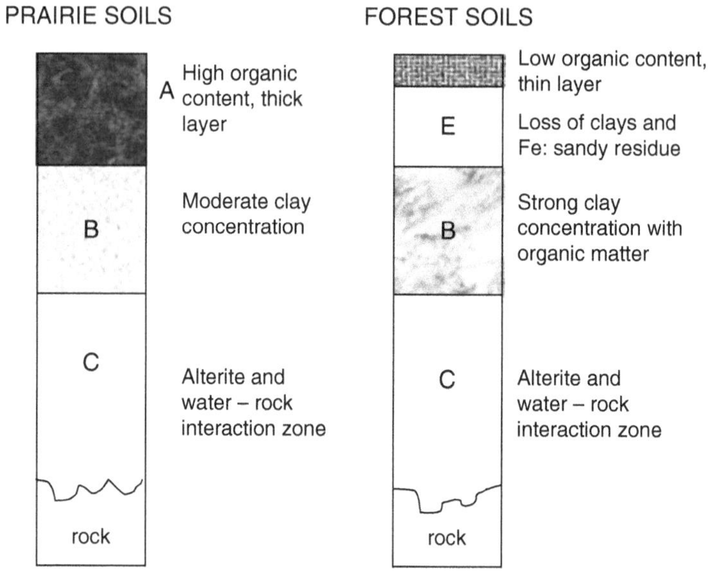

Fig. 5.26 Comparison of some of the major characteristics of forest (especially conifer) and prairie alteration profiles. Notable are the differences in the thickness of the layer of organic matter and the presence of the low clay E horizon of the forest soils

Hemisphere in middle latitudes. These generalisations are valid for temperate climate biomes to a large extent (see Duchaufour, 1977). In fact one can find certain characteristics of these two type profiles in other biomes and climate alteration situations. Further, the rock substrate can alter the profile characteristics by influencing the pH and clay – organic matter stabilities.

5.4.1 Prairie and Forest Type Profiles

We will use the temperate climate vegetation as a model or starting point for the description of the two major and often competing types of plant regimes. Much work has been done on these types of plant – soil interaction regimes because they concern a large part of the environments in which human activity has evolved and manipulated the two plant regimes.

5.4.1.1 Prairie

In prairie soil profiles one finds the epitome of a rather continuous change from a highly transformed mineral matrix, rich in clays and organic matter with a gradual decrease in clay content, to an increase in mineral debris and rock fragments until one finds the bedrock. Usually, as in most more mature profiles, one finds clay accumulation below the organic horizon (see Sect. 2.4.2). There is transport of clays and organic matter in a humic state well below the dense root mat characteristic of grasses. Normally prairie soils are of neutral to basic pH characteristics. However, in mountain or tundra prairies the pH is most often acidic. The difference is due to the relative activity of bacteria. In colder climates there is a stronger accumulation of organic matter in a less evolved state which leads to acidic conditions. This is true for swamps or lowland situations also where anoxic conditions lead to incomplete bacterial action. The organic mater in most prairie soils lends a black color to the soil and alterite. (Fig. 5.26).

Jenny (1994) has assembled much data concerning soil properties across the United States. If one considers the data correlated to rainfall, one of his major variables, values up to about 35 in. of rain (about 85 cm) per year on average covers the dry land and prairie range where grass dominates or dominated before agricultural practices became universal in prairie lands, between 180 and 100 years ago in the region of study. Clay content, organic carbon content, nitrogen content and the depth of concentration of these components in the A horizon, presence of large soil aggregates found in the alteration profile are all correlated with rainfall. There is a strong correlation of pH and rainfall, changing two units from 5.5 under conditions of more abundant rainfall to 7.5 under conditions of arid conditions. In general, the more rain the greater the biological activity in the A horizon. Thus under the same general biome conditions there can be strong changes in the physical and chemical characteristics of prairie soils.

Ruhe (1984) demonstrates the importance of time as a factor in controlling the different effects of plant – soil interaction. In looking at soils based on the same starting material, loessic depositions, as a function of age, <9,000 and 14,000 years. In particular the clay content of the B horizons of the prairie soils in a transect from western Iowa to eastern Nebraska, shows that the B horizon, clay accumulation zone is on the order of 80 cm in older soils, whereas in the younger soils the thickness of the B horizon varies from 80 cm to less than 50 cm. Jenny (1994, p. 138) shows in data from soil profiles from different points in these sequences that in all cases the Si/Al ratio increases in the A horizon, indicating the influence of grasses in extracting silica from depth and depositing it on the surface.

Prairie soils in temperate climates then are rich in organic matter and clays in the surface horizons. Soil color changes from grey in more arid soils to very dark grey or black in full prairie regimes bordering on forest biome areas. Acidity in soil A horizons tends to be slightly acid to basic in well to moderately drained situations of prairie growth.

5.4.1.2 Forest

In forest soils the organic layer is less developed, brown in color and it often has a low amount of clay material present. The layer just below the organic zone is typically sandy, i.e. where few clays are present, and devoid of organic matter. Below this zone (called the E horizon) the accumulation of clays and some organic matter is apparent. The stark contrast between the surface A horizon and the sandy E hori-

	mmole/litre Al
Pine, oak, birch	0.8
Spruce, fir	0.3
Poplar	0.1

Wheat	0.02
Barley	0.015
Cotton	0.0015
Graminae	<0.0009

zon is the characteristic which is used to describe these soils as Podzols. These soils are typically acid, with pH in the 5–4 range or less.

Miles (1985) gives a detailed review of the effects of various types of tree regimes on soil properties. He indicates that confiers are the most acidifying and deciduous trees less so and further these trees have different effects depending upon the soil substrate. Acidity is identified with the formation of a clay – poor E horizon just below the humic zone of organic activity. The intensity of this effect (podsolization) is variable depending upon the climate, nature of the substrate and type of tree present. Overall, deciduous trees have a less strong podzolization effect, with higher pH values than those of confier forests. Some shrubs, such as heather, show even stronger effects of acidification and podzolization. Quideau et al. (1996) conclude that depending upon the vegetation type, deciduous or conifer trees, the weathering reactions are different in that they produce different concentrations of and relations between elements in pore solutions associated with the two types of plants. Thus not all forests are the same.

Low pH (acidity) is assumed to be the cause of the loss of clays in the E horizon by dissolution of silicates due to organic acids. This description best fits the forest soils developed under coniferous forests but can be seen under other types of plant regimes. Alumina in soil solutions is relatively high in acidic soils. This is reflected in the aluminium toxicity threshold levels for several types of trees (Wolt, 1994, p. 235).

One can compare this to graminae and similar food crops where the toxicity tolerance is much lower:

There is an obvious antagonism between the chemistry of forest species and prairie species tolerance. The soil solutions typical of forests, are of course tolerated by trees but not by prairie plants. Robert (1996, p. 85) indicates that the alumina content of soil solutions is related to acidity and that the pH or alumina content of soil solutions affects crop plant growth such that of corn (*zea mais*) and soybean plants.

It is obvious that a forest tree, especially conifer, is subject to destruction by fire and relatively slow re-growth. Prairies burn but are green again in the following spring. The time lag in re-establishing vegetation is greater for forests than prairies.

If there is no retarding mechanism, the forests will be covered in prairie after a fire. However this does not happen. Taking the example of Yellowstone National park where forests were allowed to burn in the 1980s, the re-growth of the pine forests fifteen years later can be traced to within several meters of the old stumps without invasion by the adjacent prairies in the high country meadows inter fingering the forests (personal observation of BV). The adjacent grass biome could not invade the old conifer forest soils. The answer can lie in the inherent pH and more importantly free alumina content of the soils which impedes the installation of grasses and non-woody plants. A deficit in clays in the E horizon will also hamper the installation of the dense grass root system usually found at these depths in grass soils. Thus the structure of a forest soil (often classified as a Podsol) is such that it is antagonistic to the installation of grasses.

5.4.2 Comparing Forest and Prairie Soils

5.4.2.1 Temperate and Semi Arid Climates

Comparisons of different properties in forest and prairie soils in proximity to one another indicate consistent differences. Jenny (1994, p. 223) compares acidity in forest soil profiles and prairie profiles showing a strong difference in the E horizon (Fig. 5.27) for soils in the temperate climate area of Illinois. This data can be compared for prairie and forest (oak) soil in semi-arid Utah (Graham and Southard, 1983). The differences in pH are found in the E horizon, below the A organic-rich horizon in both examples. The A horizon pH is similar in the two sets of soils. However the pH values in the semi-arid soils is higher than those of the temperate, more organic – rich soils of Illinois. In general one can correlate the pH of the soil A horizon with the amount of organic matter present which is a function of climate, in these cases in biomass production, greater under conditions of higher rainfall.

Jenny (p. 148) also shows systematic differences in nitrogen content, a measure of humic content in A horizons as a function of climate. In general the prairie soils have two to three times the nitrogen content and hence humic matter. Both diminish as a function of latitude from Wisconsin to Mississippi (USA) showing the influence of humification and lower bacterial activity in colder climates.

The organic – rich layer in forest soils is less thick than that in prairies, the color is not grey to black but different colors of brown. Brown acid soil is a typical description of forest soils. However, it is necessary to distinguish between the two major types of trees present, confier and deciduous, in forests under conditions of temperate climate growth. Confiers develop more acid soils while deciduous forests are closer in their soil development characteristics to prairies. Both forest plant regimes do however develop the E horizon structure where clays are absent and sand (chemically resistant quartz) predominates. Acidity is usually greater than in prairie soils, especially below the humic A horizon. The difference in soil chemistry and the resulting stability of silicate minerals is due to the difference in the organic mater and the mummification of this material in the soil horizon by bacterial

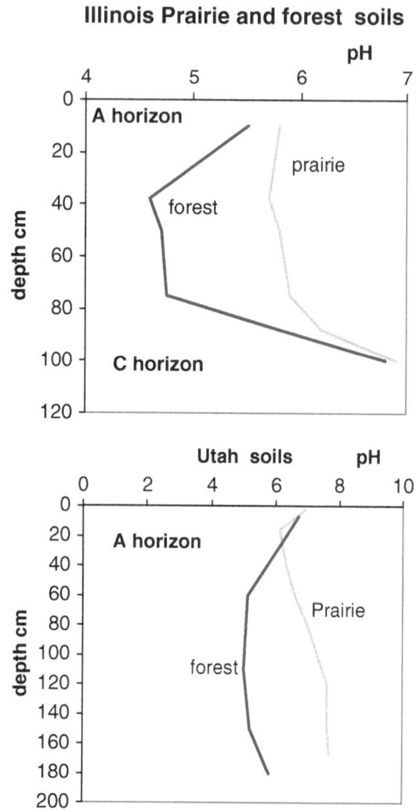

Fig. 5.27 Comparison of forest and prairie soil pH values in two different climate areas, one in the Midwest prairies of Illinois (USA) and the other in the semi arid climate of Utah (USA)

action. The fixation of alumina by organic matter is known, but poorly understood in a specific manner. However the free alumina is lower in prairie soils than in forest soils, especially confier soils. This difference allows the change in clay interlayer ion occupation where fixed alumina allows potassium to be fixed by 2:1 clays whereas free alumina produces HI minerals typical of forest (conifer) soils.

Cold Climate Prairie and Forest Soils

Cold climates usually lead to the accumulation of organic matter in the A horizon. Biologic activity is reduced and the humification process is slower as well as the mineralization of the organic litter. This effect is important for the pH of the soils. In general, cold climate soils have a lower pH. The relationship between pH and organic content of cold climate soils is indicated in Fig. 5.28.

In general the forests have a lower soil pH as in other climates. Cold climate soils are usually poorly developed because of recent glaciation and hence their young age and the slower chemical activity due to temperature when water is ice, little chemical activity can occur. Biological activity is equally slowed by lower temperatures. Conifer forest soils are more acidic, and have more organic matter present. In

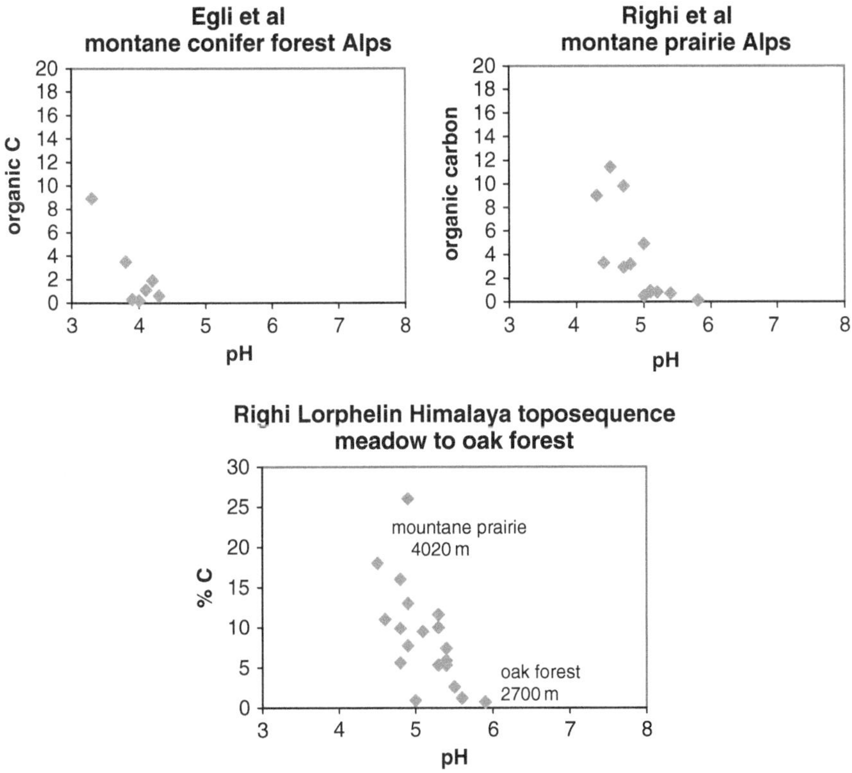

Fig. 5.28 Relations between organic carbon content % of soils and pH in forest – prairie situations in colder climates. Righi et al. (1999), Egli et al. (2002), Righi and Lorphelin (1986). Notice the tendency to higher pH values corresponding to the type of vegetation and altitude

Fig. 5.28 one also sees data for a toposequence in the Himalayas from an oak forest soils at lower altitudes to high altitude a prairie soil. Here the prairie is more acidic and the carbon content is higher than that in the oak forest.

The major difference in properties of the humic zone, organic content and resulting pH are then controlled by the type of vegetation and the climate, the colder and wetter the more the organic matter tends to be retained in the soil giving it specific chemical characteristics. In these regimes the soil is not well developed, usually less than a meter deep due to the relatively short periods of time that have elapsed since the rock substrate was colonized by plants.

5.4.2.2 *Hot Wet Climates:* Laterites and Plant Regimes

These climates are conducive to laterite formation, i.e. strong oxidation of the iron in soil materials. The term oxidation intimates a change of state of the iron from incorporation silicate minerals to free oxide phases and hence a dismantling of silicates either the initial rock minerals or at times clay minerals formed deeper in the alteration profile by water rock interaction processes. There is often a strong association with plant regime. Laterites are not restricted to hot wet climate areas today because former climates changed during the glacial periods of the recent past, have left traces of former climatic and plant regimes in areas today less humid. The expression of iron oxides as a reddish color is not restricted to the tropics of today but covers a rather larger band extending from the equator (See p. 26, Segalen, 1971). Govinda Rajan and Murthy (1971) demonstrate the relationship between intensity of laterite formation and vegetation, of course guided by climate, in southern India. The strong laterite soils are found under broad leaved evergreen forests.

The problem of iron accumulation and the formation of laterite crusts in alteration profiles is very important in areas of the world where the tropical rain forests border savanna climates. Due to recent climatic fluctuations, the stronger accumulations of iron observed today are frequently not active but relict. (Bardossy and Aleva, 1990, p. 81; Temgoua, 2002, p. 5; Mohr and Van Baren, 1959, Chap. 13). However, in many regions laterite accumulation is still active. The movement of laterite iron oxide crusts towards the surface is of great importance in many areas. The formation of a hard pan of oxide renders the soils very difficult to work. Laterite formation indicates the last stages of chemical alteration (water – rock interaction) where silica content is very low and there is a strong accumulation of iron and aluminium hydroxy – oxides.

Numerous authors have indicated a relation between present day plant regime (biome) and the accumulation of iron oxide in particular. Essentially authors coming to these conclusions have looked at forest and savanna situations under conditions of warm and humid climates. The importance of such biomes in many regions of lateritic accumulation is that the relations of forest to grasslands are constantly perturbed by the action of agricultural or pastoral man. This changes the equilibrium which affects the accumulation and dispersion of iron oxides. This is a problem concerning natural equilibrium and those attained by perturbation due to human action.

Forests are considered to be conducive to the formation of lateritic soils (iron – rich) and the accumulation of a lateritic crust (duricrust) at depth, below most of the tree root masses (Maignien,1958, p. 65). The soils are friable near the surface with an increase in iron oxide nodules at depth in many cases. The entire alteration zone is of course oxidized to a large extent but the accumulation of iron is found at depth relative to the active soil zone. By contrast grass biomes (tropical savannas) near the forest areas are known to have iron crusts closer to the surface than those found in forests. This can be illustrated by a toposequence (Maignien, 1968; Bowden, 1987) where forests on slopes or hill tops have an iron accumulation at depth and prairies landscapes down slope show iron accumulations nearer the surface (Fig. 5.29). Under forests the deeper crusts have both Fe and Al accumulations whereas the prairie accumulations are richer in iron. Bardossy and Aleva (1990) have observed that iron is more likely to migrate, down slope or down an alteration profile, than alumina.

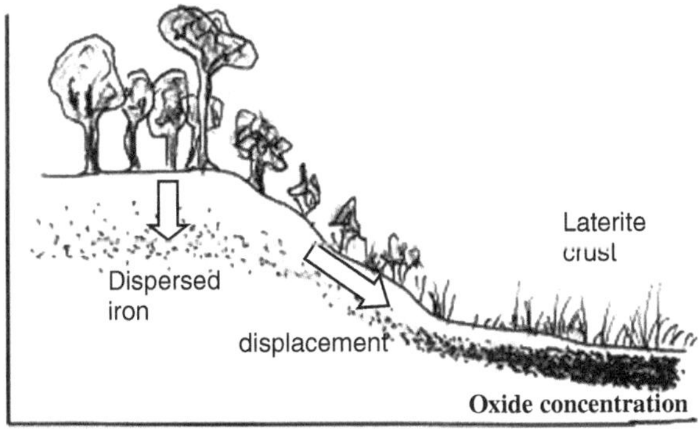

Fig. 5.29 Illustration of the movement of iron oxide in soils with a tendency to produce laterite crusts. The oxidation is greatest on the uplands under forests where the iron oxide accumulation is below the A horizon. Lateral movement down slope moves iron oxides to accumulate in thicker crusts nearer the surface in grassland soils

Figure 5.29 indicates the extent of development of the different soil horizons of temperate to tropical soils and the dominant clay minerals present over the depth of the profile. In general, the clays in the A horizon are more silica – rich than those in the C horizons.

These observations have led authors to imagine a migration of iron down slope to form the duricrusts under prairie biomes. The formation of iron duricrusts is then often considered to be a function of topography. However, along with the topography factor one finds a change in plant occupation where forests are on hill tops and the upper slopes whereas grasses are found along the slope base and flat down slope bottoms, nearer rivers and other water resources.

McFarlane (1978, p. 50) has made further observations on the relations between vetegal cover and duricrust (laterite) formation. He has observed the relations of forest to duricrust under forests formed on old laterite surfaces or stable biome situations. Here the iron is concentrated well below the tree root zone. It has been observed that in many instances the inrtroduction of trees on duricrust (laterite crust) has softened them and contributed to a downward movement fo the iron oxides. Trees break up duricrusts. However when trees are replaced by grasses with change in landscape hydrodynamics, the iron oxides have been observed to "move" upward or at least be more present towards the surface. Temgoua (2002, p. 101) proposes that the initial formation of laterites in Cameroon was due to a general savanna biome during the Cretaceous which was gradually replaced by trees. The iron oxide masses were broken up under the forest regime and moved down slope to be found frequently under prairie.

It seems that in the case of iron accumulation trees are more efficient in moving iron oxide downward from the root zone whereas grasses do not do this to such an extent. McFarlane (1978) suggests that forest organic litter and root exudates are more efficient at moving the iron to lower levels where it accumulates, one imagines due to the loss of biotic activity and stability of iron complexing agents associated with the litter of the trees. The difference in depth in accumulation of iron oxides associated with the toposequences or catena of hill and valleys show the strong inter-relations between plant and mineral regimes which are often controlled by local climatic or microclimatic variations such as water availability and depth of the water table (Fig. 5.29). The needs of one plant regime are not those of another and the manipulation of the local chemistry by plant litter and root exudation change according to the local circumstances.

5.4.3 Successions and Soil Properties Influenced by Change in Plant Regime

The action of change in plant regimes has been observed in several cases. For the most part the observations have concentrated upon the chemistry and soil properties such as the formation of aggregates and presence of organic matter. An initial consideration is to follow the changes in soil chemistry when plants invade clay rich substrate of water – rock alterite. These conditions occur after landslides. Zarin and Johnson (1995) give data concerning the chemistry of the soils re-colonized by plants over a 60 year period in a region of Puerto Rico normally covered by a montane tropical forest. Potassium is particularly interesting in that it follows a rather regular sequence of change in above ground biomass, roots and soil. Figure 5.30 indicates the changes in concentration of this element in biomass and surface soil (upper 60 cm). The accumulation of potassium in the biomass follows the classical S shaped curve, with a possible decrease after an optimum is reached. The loss of potassium from the soil appears to decrease gradually and becomes, probably more or less stabilized to a lower steady state after 100 years or so. Potassium uptake in plants is of the same order of magnitude as the exchangeable K in the soil and follows the same dynamics. There is an obvious inter-relation between K in the solutions that is fixed on the clays and the change in biomass and hence activity of the plants which bring potassium to the surface and re-cycle it into the upper part of the alteration zone.

From these data it is clear that the plant regime in the montane forests not only stabilizes potassium loss but brings it to the surface from depth below the plant – soil interaction zone.

Bormann and Likens (1994) give information concerning a de-forestation experiment in northern New Hampshire. Loss of mineral elements from the soil was measured as concentration of elements in stream water. The initial stages of forest recovery over 4 years showed a strong increase in potassium and calcium concentrations from 10 to 85 kg/ha equivalent which decreased to a near initial value 6 years after the de-forestation event. This indicates that the tree cutting event is much less dramatic to chemical balances in that roots are still present and most likely plant re-growth much more rapid than in the case of a landslide event where the soil cover and chemical structure is totally taken away. Nevertheless the loss of tree cover and growth leads to significant leaching of mineral elements for a period of time. The comparison of the two data sets indicates that the loss of vegetation and re-growth

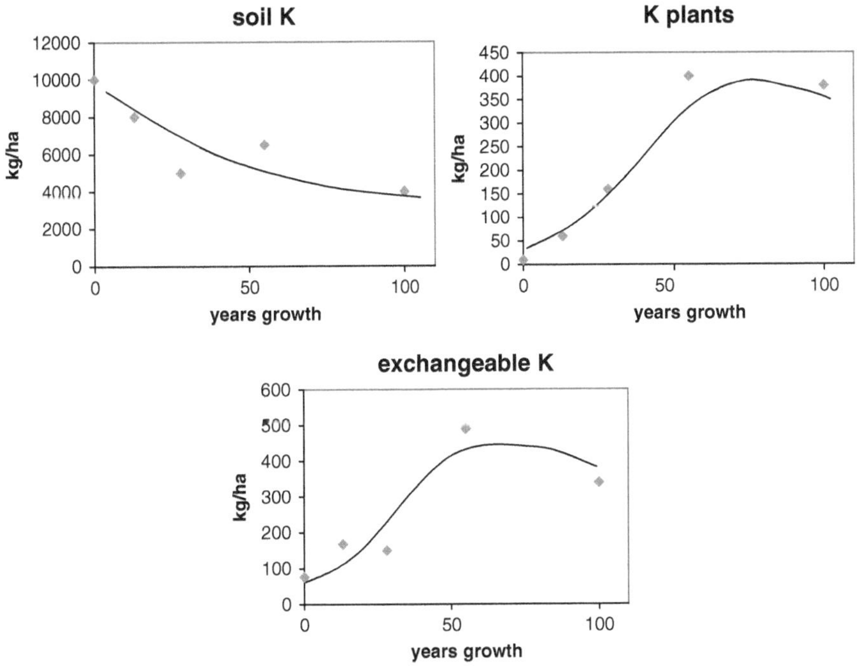

Fig. 5.30 Effects of the re-establishment of forest vegetation on soil potassium balances after 100 years in Puerto Rico (Zarin and Johnson, 1995). Loss of potassium from soil is found to be resident in the plants as growth proceeds. The result of plant litter fall can be seen in the amount of exchangeable potassium in the soils (CEC exchangeable) which can be re-used by the plants

is less important for mineral element transport than the loss of topsoil and alterite and subsequent slower re-establishment of a vegetal cover. The root structures and soils structure play a role of absorbing the shock of plant loss.

5.4.3.1 Effect on Soils of Changing Vegetal Cover

The change in vegetal cover can be of importance to the chemical and physical characteristics of a soil. Such effects are important when considering human intervention is land use. Miles (1985) discusses the changes effected by major change in plant species under conditions of a temperate climate. The examples given by this author concern the change in podzolic soils, developed under confier, heather and juniper stands which are acid, have low clay content below the top organic – rich A horizon and are generally of poor agricultural quality. One step to change is accomplished by deciduous trees which change pH by a unit or so and favor the decomposition of organic litter on the surface, which is at the origin of the low pH values of coniferous tree soils. Of course the reverse is true. When heather is introduced it will decrease pH by several units over periods as short as 10 years. Nielsen et al. (1987) show the effects of an oak invasion of a previous heather soil. The oak invasion of the over story cover is accompanied by a cover of grass in the understory. The type of organic matter changes from black to brown, indicating more transformation by bacteria, and pH increases.

Herbaceous species (grass and prairie vegetation) raise the pH by a unit over several decades of influence. The studies cited are instances of the introduction of grazing animals which destroy shrubs (heather) and favour grasses. This is an important concept which will be discussed in some more detail in Chap. 6. The higher the pH the easier it is for other graminae species to prosper such as grain crops.

It is evident that the change in plants affects the humic layer first and this determines the pH of the soil. This in turn influences the types of clays present, especially

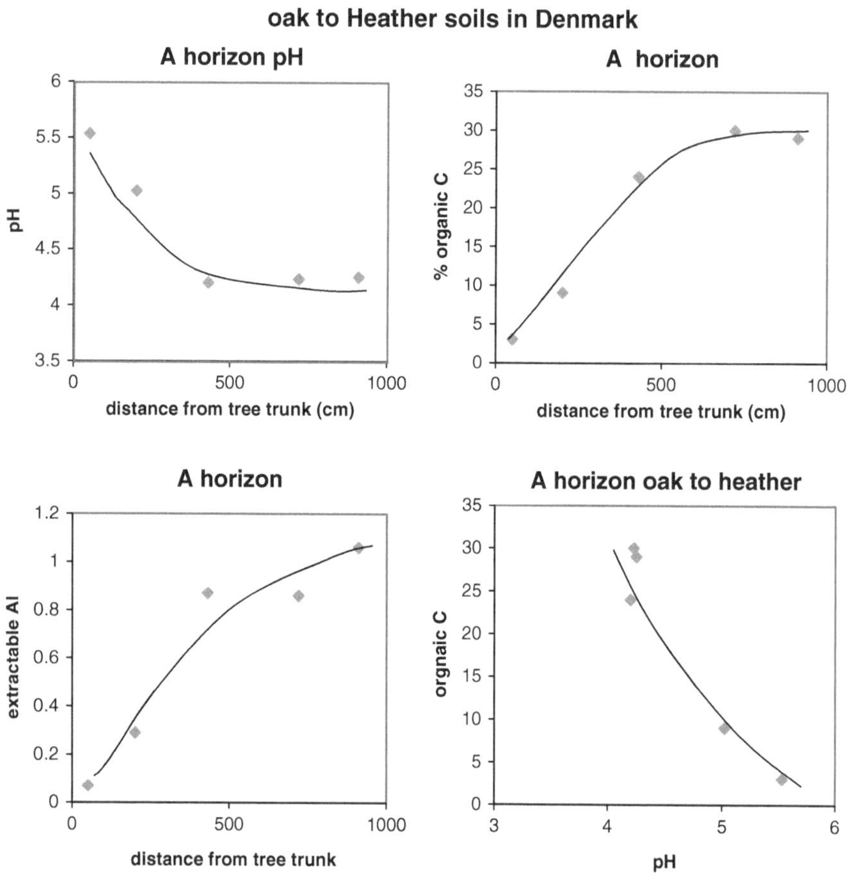

Fig. 5.31 Effect of plant types on soils. An example used here is of soil characteristics under a large oak and moving to the surrounding heather growth seen in Denmark. Changes in pH, organic carbon % and Al content % of soil solutions are compared showing the effect of vegetation on the same soil substrate (data from Nielsen et al., 1987). The oak tree promotes higher pH in the soil, lower extractable Al in the soil solutions but a lower organic content of the soils. Heather is an acidifier, resulting in increased available aluminium, often a limiting factor for plant survival

the occupation of the interlayer ion sites. Such changes can occur in relatively short periods and are thus susceptible to manipulation by human activity.

Graham and Wood (1991) suggest that the effect of the introduction of oak on a chaparral and pine developed soil is to increase the clay content at the surface. The mechanism of this change, visible over 41 year period, is attributed to earth work activity. The increase in faunal activity is due to the change in leaf litter, and subsequent pH change. Organic matter is lighter and more brown indicating a higher bacterial activity and with a change in these parameters it is assumed that earth worms move upward and with them clay particles in their soil bio-perturbation activities.

The increase in faunal activity is due to the change in leaf litter, and subsequent pH change. Nielsen et al. (1987) present data to document such a change where samples were taken under a large oak tree and at distances from the tree trunk into a heath cover in Denmark. Figure 5.31 shows the changes in organic content, pH and extractable aluminium from the A horizons. As one might expect organic matter decreases under the oak due to better microbial action, and thus pH increases. The change in extractable alumina is very interesting in that it shows the inter-relations between organic matter and pH as they affect the critical content of alumina which determines the limits of chemical tolerance for many plants. Also the availability of alumina will determine the interlayer ion occupation in many cases for charged 2:1 layer minerals. Under conditions of heather one would expect to find HI (soil vermiculite) minerals while under oak, illite or smectite as demonstrated by the mineralogy of an oak on pine soil experiment by Tice et al. (1996).

5.5 Mineral Element Transfer in Soils: The Translocation Effect

5.5.1 Chemical Variability as a Function of Alteration Profiles

The transfer of mineral elements from depth to the surface was understood long ago, and considered as a fundamental part of plant growth and soil fertility (see for example Larbaletrier, 1891, p. 104). However the overall effects and importance in a general sense have been ignored for much of the recent past in soil science circles. For some time the observation has been made that the chemical composition of the upper parts of soil profiles does not follow the accepted alteration scheme which would be produced by simple water – rock alteration. The interaction of silicates and dilute water (rain) will initially hydrate the minerals present, exchanging hydrogen cations for alkali and alkaline earth cations:

$$nH^+ = Ca^{2+},\ K^+, \text{etc.}$$

The second step or stage of alteration is the systematic loss of silica. Most clay minerals are less silica – rich than their precursor tectosilicates (feldspars) and hence the formation of clays such as illite, smectite or kaolinite from these minerals releases silica into solution. Micas of course are near the clay compositions in terms of silica balance, to those of clays. Other silicate minerals in rocks tend to be as siliceous as clays that form from them. However, the most common clays are dioctahedral, and hence aluminous. Thus the silicate minerals with low alumina content, such as olivine, tend to form non-clay minerals, or tend to combine dissolved elements in solution with released alumina from other altering minerals (see Velde and Meunier, 2008, Chap. 5).

Continued alteration favors the low silica clay kaolinite. Eventually this mineral loses its silica content to be transformed into the aluminous hydroxide gibbsite.

1. $K_{0.8}(Al_2)(Si_{3.2}Al_{0.8})O_{10}(OH)_2 \Rightarrow 1.4Al_2Si_2O_5(OH)_4 + 0.4SiO_{2(aq)}$
   Illite kaolinite
2. $Al_2Si_2O_5(OH)_4 \Rightarrow 2Al(OH)_3 + 2SiO_{2(aq)}$
   kaolinite gibbsite

These are the classical steps of de – silicification. The clay assemblage of kaolinite and gibbsite is typical of soils in regions of very high rainfall (Keller, 1967 for example). The normal trends of chemical alteration in surface environments would affect clay minerals as shown diagrammatically in Fig. 5.32 where aluminium content is constant.

In principle the part of the alteration profile which has been subjected to the most water – rock interaction, the A horizon (surface) one should find the most altered and hence the most potassium poor materials.

Some examples of chemical data reported in the literature are quite persuasive in indicating that the strict chemical alteration of soils is not always obeyed. Several cases are given in Fig. 5.33 for soils developed on rocks containing little potassium and being low in silica content. In these examples one sees a decided increase in potassium content in the A horizons of the profiles. In the example based upon andesitic volcanic rock where the profiles were observed at different altitudes, the

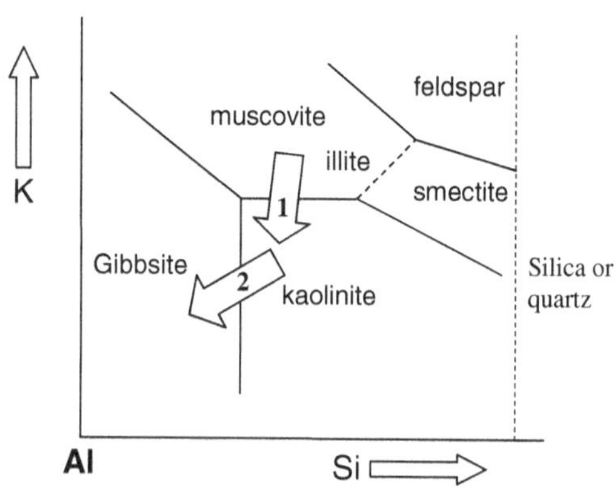

Fig. 5.32 Diagram indicating the chemical relations of clays and magmatic minerals (muscovite and feldspar) as a function of the activity of Si and K in a soil or alteration zone. Alumina is considered to be of constant content in the system. *Arrows* indicate the path taken by the chemical system in a soil under normal conditions of water – rock interaction. The first step is loss of potassium and the second is loss of silica

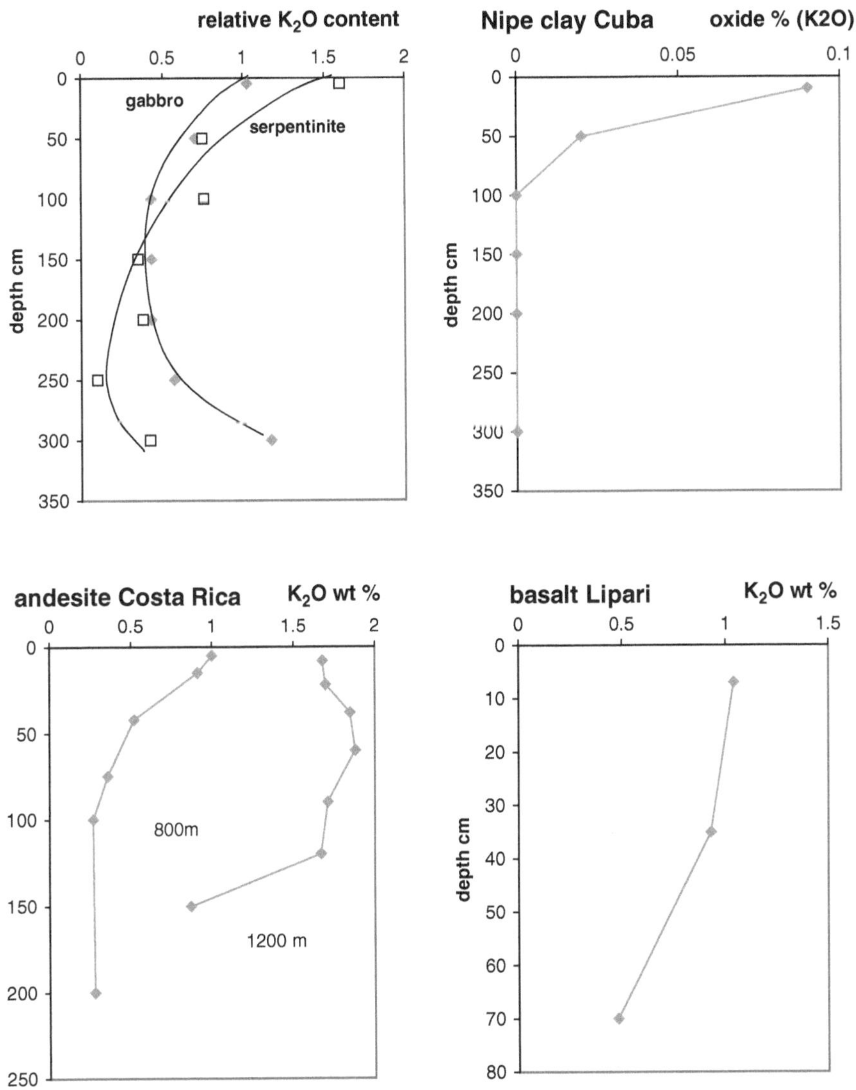

Fig. 5.33 Illustration of concentrations of potassium in various profiles based upon parent material of low potassium content; gabbro and serpentinite (Pion, 1979), Nipe clay from Cuba (Mohr and Van Baren, 1959) volcanic andesite deposit forming a mountain in Costa Rica (Meijer and Buurmann, 2003) and basic volcanic rocks on the island of Lipari Italy (Mirabella et al., 2005). The andesite example shows soil compositions under forest at two altitudes, the 800 m soil shows a greater effect of potassium loss and relative gain at the surface

effect is accentuated at lower altitudes where plants are more abundant and alteration profiles are deeper and chemical weathering more advanced. The same observations can be made concerning silica content (Fig. 5.34).

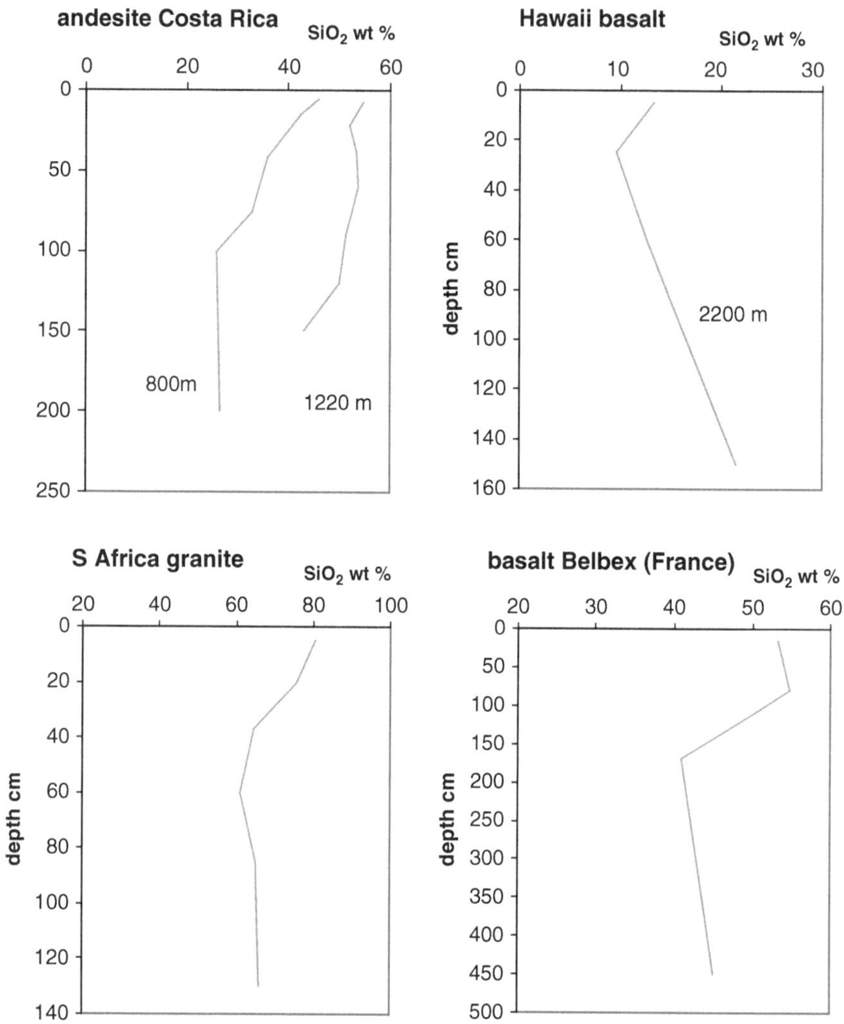

Fig. 5.34 Changes in silica content of profiles on low silica rocks, most without initial quartz mineralogy. Andesite in Costa Rica shows values for two altitudes of forest growth where the lower altitude has a stronger effect of silica loss and gain at the surface due othogher rainfall and greater biomass production. Data from Meijer and Burmann (2003), Mohr and Van Baren (1959). This indicates that under various conditions of climate and plant growth, it is clear that the transfer of silica to the surface is an important factor in the overall chemistry of alteration profiles

These chemical surface anomalies can be attributed to the effects of plants which transport potassium and silica to their living parts, especially the aerial parts. Leaves, stems and woody parts of plants contain potassium and silica. Schactmann and Schroeder (1994), Zairn and Johnson (1995), Nettleton et al. (1973), Juang and

Uehara (1968), Ahmed et al. (1971) among others observed the formation of potassium – bearing minerals in the surface horizons (A) of soils under different climatic conditions, ranging from dryland to semi tropical. These authors attributed the presence of this potassic mineral to plant action where the concentration of this element occurs at the surface. However, these authors did not find a general audience for their conclusions, and most clay mineralogists and soil chemists ignore in fact that the minerals found in the soil in the A horizon can be influenced by plant action. Most authors and generalizations presented in text books indicate or state that plants tend to destroy clay minerals by interaction with organic acids especially when the presence of potassium minerals is concerned.

As mentioned before, it has been recognized for some time that there can be a translocation of elements from depth to the surface, alkalis and other cations (Larbaletrier, 1891 to Jobaggy and Jackson, 2004). The translocation of silica has also been recognized and considered to be a fundamental factor in forming or stabilizing more silica – rich minerals (Lucas et al., 1993; Lucas, 2001; Kleber et al., 2007 for example). The cycle of silica in soils has been considered from the standpoint of a geochemical model based upon the deposition of this element in the form of amorphous phase, the phytolith (Derry et al., 2005; Farmer et al., 2005; Gerard et al., 2008).

Interestingly, the silica content (phytolith) in plants can reflect the silica availability in the soil in which the plant grows (Henriet et al., 2008).

Plant action then affects the A horizon by a translocation of key elements which are fundamental for the stability and perhaps formation of certain clay minerals. The most important element is silica, which determines the present of the type of clay mineral in the soil: either the silica – rich 2:1 minerals (illite, smectite, HI) 1:1 minerals (kaolinite, halloysite) and eventually the silica – free gibbsite. One can consider this situation as a function of the presence of either K or Si which is the reverse of the water – rock weathering sequence of Fig. 5.32. This transfer or translocation effect is shown in Fig. 5.35 where again Al content is considered to be constant.

In this representation of the impact of chemical variables, engendered by plant activity, the final clay mineral can be smectite or illite, depending upon the potassium content of the solutions. However, in reality, the representation does not take into account fully the variability of the 2:1 soil clay mineral. We do not know what the chemical parameters are that determine the charge on the 2:1 mineral, smectite (low charge) or illite – HI mineral (high charge). It could very well be the amount of available silica in solution which would form the more silica – rich and low charge smectite mineral. Perhaps the presence of smectite and 2:1 minerals is initially determined by the silica activity in the soil solution and then the interlayer ion occupancy is determined by potassium availability and alumina availability. In any event, there must be a certain amount of potassium present in soil solutions to saturate the interlayer sites with the anhydrous potassium layer characteristic of illite. Further it seems that he presence of HI minerals depends upon a relatively high charge on the 2:1 mineral. This is normally due to a higher alumina content compared to silica for the dioctahedral, aluminous minerals.

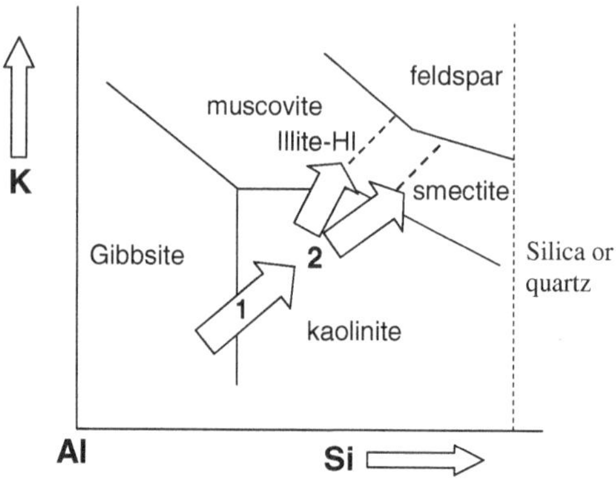

Fig. 5.35 Diagram of mineral relations as a function of chemical parameter of their stability. The arrows indicate the general change in A horizon chemistry as a function of plant (biomass) deposition. Deepest level clay assemblages in highly altered rock profiles often show gibbsite (aluminium hydroxide) which is replaced by kaolinite and eventually 2:1 minerals HI or illite of high charge and smectite of low charge

Translocation then has the major effect of stabilizing, anr/or creating 2:1 minerals in the A horizon of alteration sequences. One can pose the question of its durability.

5.5.1.1 Dynamics of the Translocation Effect

Initially plant regimes must stabilize the soil chemistry in order to create a translocated mass of nutrients in the soil (A horizon). In order to do this they need to create an A horizon, i.e. a layer with organic matter renewal, a stable clay fraction and so forth. How long does this take? One can give a tentative response using data from two sources, one for a landslide area in New Hampshire USA (Walker and del Moral, 2003) and another analogous situation in Puerto Rico (Zairn and Johnson, 1995). Both studies show that the potassium content appears to stabilize after a hundred years or so (Fig. 5.36). This indicates that the dynamics of plant action is rather rapid in establishing a steady return of soluble elements to the surface of an alteration sequence.

Studies in areas of semi-tropical to tropical climates (Howard et al., 1993: He 2008) show that there is an initial maintenance of higher Si content in soils with plant growth but after a certain period of time the transfer of silica from depth cannot compensate the water – rock interaction and dissolution of silicate minerals. On the Atlantic Coast of the South eastern United States the soils have a strong 2:1 mineral (HI) content in the A horizon for periods of up to about one million years, and after this period the presence of these silica – rich minerals decreases notably. In the Hainan basalt sequence (southern China) the period where the presence of 2:1 minerals can be detected is on the order of 600,000 years. Given that the climate in Hainan is now, and perhaps was, one of more rainfall than that of the south Atlantic Coast of the United States, one can assume that rainfall and hence climate

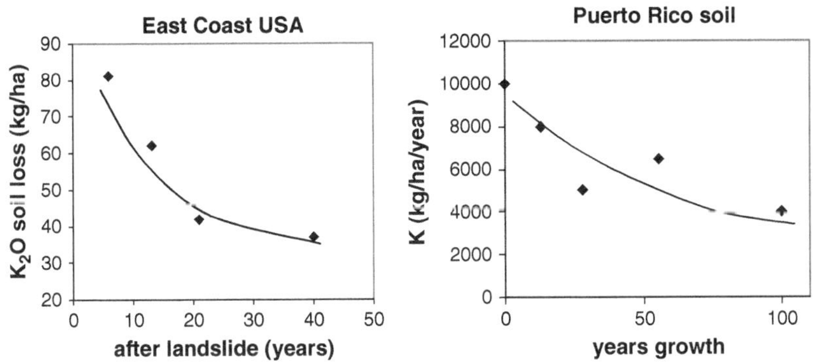

Fig. 5.36 Illustration of the effects of plant growth on landslide areas in stabilizing the soil chemical balance. Potassium loss from the soils is slowed significantly after a period of about 100 years in two different areas after the re-establishment of forest cover

determine the time period of 2:1 mineral stability in soils. The relative proportions of 2:1 minerals are greatly different in the two cases, largely due to the initial lack of alumina in the basalts compared to the mica schist substrate for the American soils. However, even though the chemistry is quite different (seen in the difference in Al./Si ratios for the two horizons, A and C shown in Fig. 5.37). It appears that the effect is the same. Also the potassium content of the A horizon is shown by the presence of illite in the youngest soil sampled (Fig. 5.38). One can conclude that the effect of chemical transfer by plants can overcome the forces of high rainfall for about a million years after which the chemical context of the soils changes to the disfavour of the more siliceous 2:1 clays. The 2:1 clays respond to the changes

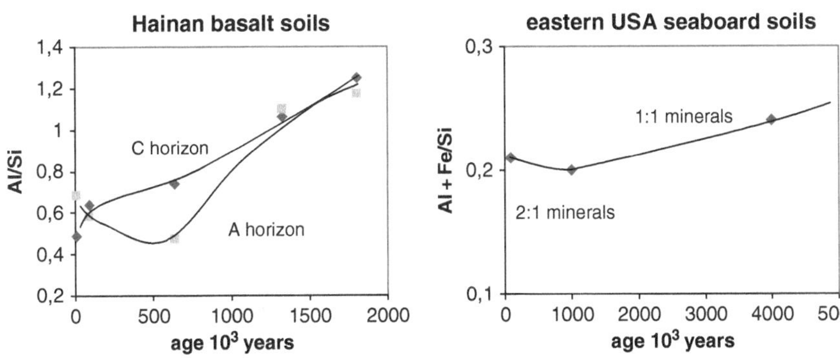

Fig. 5.37 Ratios of Al and Si in two soils series where weathering and plant growth have occurred over periods of more than a million years. In both series there is a lowering of the Al/Si ratio in the initial stages of plant implantation indicating the transfer of silica to the surface but after 500,000 years or more this effect is lost and silica loss dominates in the soils. In both instances the climate is of rather high to high rainfall and the plant regime is of strong growth. Data from He et al. (2008) and Howard et al. (1990)

in general chemistry as shown in the Hainan series (Fig. 5.37) where the first mineral formed is a mica, then a smectite – illite mixed layer mineral, and eventually an HI mineral, even though it is of very, very low abundance. The prevailing pore water chemistry determines the interlayer ion content of the 2:1 minerals which are present. The high intensity of the gibbsite peak in Fig. 5.38 for the older A horizon clay assemblage shows that the overall chemical equilibrium will be of high alumina activity in solution, where the aluminium hydroxyl mineral becomes a very important part of the mineralogy.

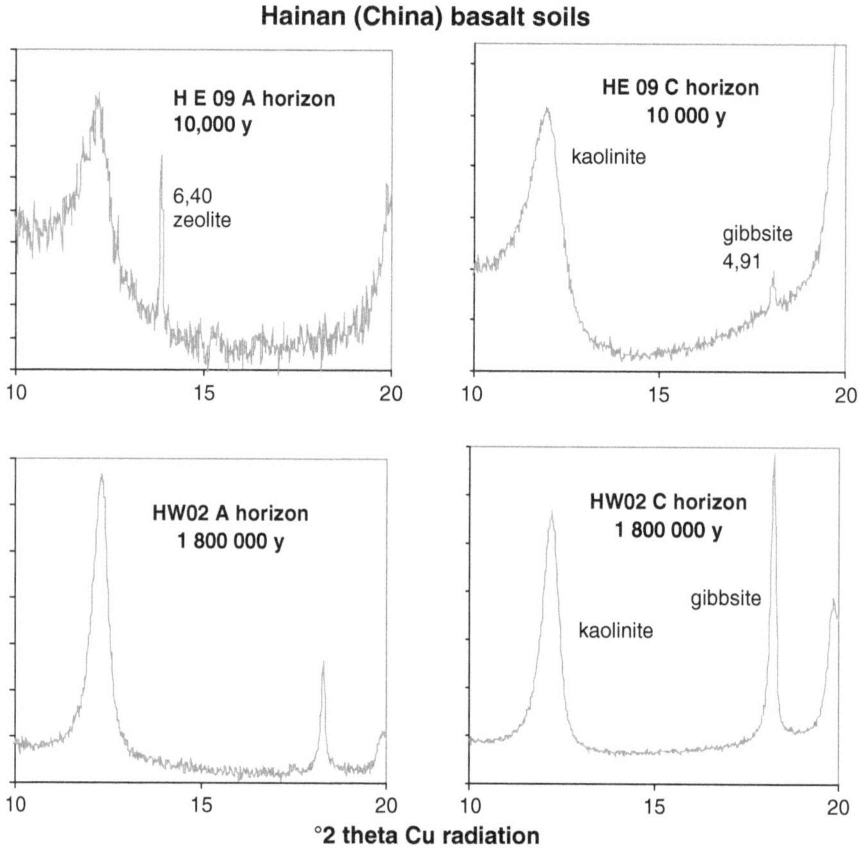

Fig. 5.38 (**a**) Hainan (China) soils based upon basalt substrates showing the differences between the A and C horizons. In the younger to older samples one sees a strong increase in the gibbsite clay mineral component in the C horizon, especially marked in the deeper part of the alteration sequence. This indicates that the silica transfer mechanism functions even though the ratio of Al to Si has begun to change significantly in the A horizons (see Fig. 5.37). Data from He et al. (2008). (**b**) Diffractograms of the A horizon soils in the Hainan basalt soils showing the changes in the 2:1 mineral fraction of the clays. Initially the phase is a mica, changing to a smectite mixed layer mineral and eventually an HI mineral, with aluminium hydroxide interlayer ions

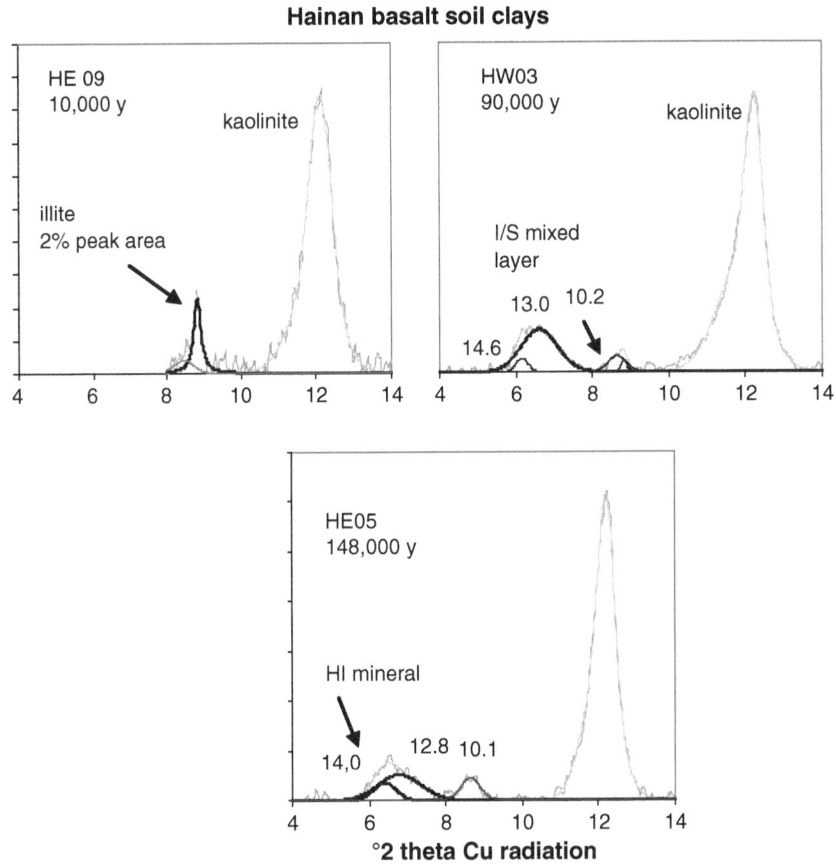

Fig. 5.38 (continued)

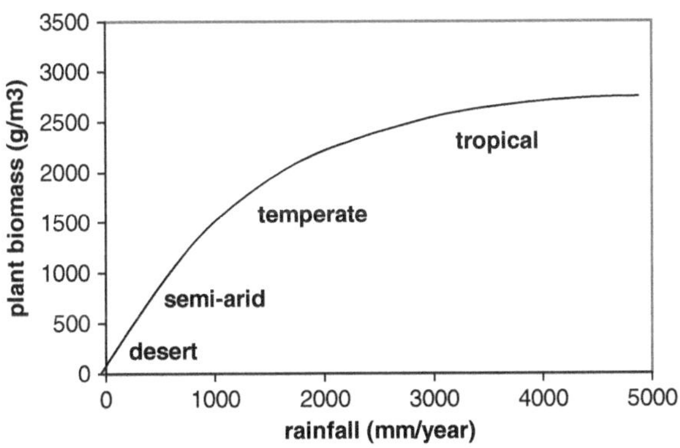

Fig. 5.39 Data from Lieth (1975) on biomass production as a function of average annual rainfall

The limits of the efficiency of the transfer mechanism operated by plants seem to be within the range of several tens to a hundred years to establish the system and stabilize an effective 2:1 clay mineral A horizon and near a million years under high rainfall conditions when the uplift effect can no longer compensate for the loss of alkalis and silica.

5.5.2 Parameters of the Translocation Effect

5.5.2.1 Primary Production of Biomass

The main parameters of the translocation effect are those of climate and plant species. Obviously both climate and plant species are inter-related. In the same way, climate and plant growth are inter-related. These inter-relations determine the effect or impact of plants on the chemistry of the surface, A horizon of the alteration zone of the earth's surface. If we postulate that the translocation of mineral elements to the surface is due to plant accumulations, the amount of plant material produced is an important part of the equation. The more plant growth, the more vegetal matter is deposited on the surface and as a result the more mineral elements are present. Thus the factor of climate can increase the translocation of elements from depth to the surface. However, the more plant growth, the greater the need for rainfall. Rainfall, precipitation, is the motor of dissolution and dispersion of the elements formerly in the bedrocks in pore solutions. The more rain, the more material is dissolved from the overall profile and the more is evacuated to rivers and eventually the sea. Thus at the same time rainfall promotes the translocation of material to the surface, and it engenders its dispersion and extraction from the soil zone. This is a sort of dilemma for plants as they try to keep the needed mineral resources present for the formation and stability of 2:1 clays.

On the one hand rainfall promotes translocation and enrichment of the surface layer in mineral elements and on the other it extracts these same elements from the soils. Lieth (1975) has given a compendium of plant productivity (primary biomass production) as a function of rainfall for various regions of the earth. In Fig. 5.39 one can see that the average line deduced from these data is not at all a linear function of rainfall. Initially primary production increases very rapidly with rainfall increase but then finds a sort of plateau value at high rainfall values. In fact in the original data presented by Leith there is significant scatter in the initial portion of the rainfall – productivity curve. This is precisely due to climate in the full sense of its meaning. The numbers most easy to deal with are those of average rainfall, which are used as a general measure of climate. However the frequency and duration of the rainfall affects the bio – productivity very greatly. For example, the average rainfall in much of France is similar to that of northern Morocco. The vegetation is quite different. Infrequent but abundant rainfall does not produce the same vegetation. This is the syndrome of the savanna alternating season climates near the tropical rainforest areas. There is water but it is not distributed evenly enough throughout the year to give consistent plant growth. Thus the measure of average rainfall is

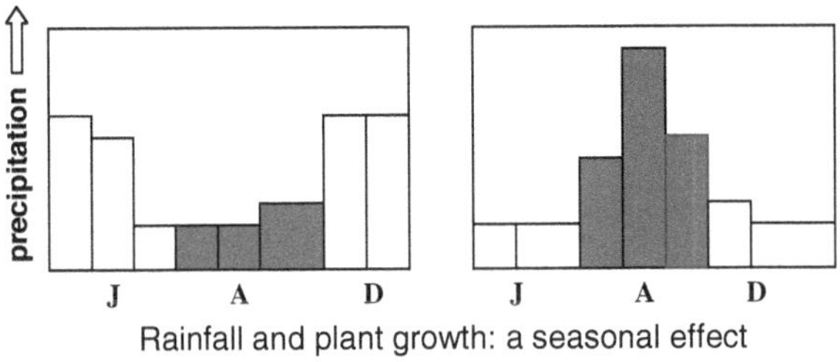

Fig. 5.40 Illustration of the possible effects of rainfall distribution as a function of season where rainfall in the summer months produces much more biomass than that falling in the winter, low temperature, months

useful but does not tell the whole story. Nevertheless average rainfall is the most efficient tool to measure climatic effects, and relate this to plant growth and primary biomass production. Again if it rains very much during a short period, the overall effect on mineral dissolution will be less than a constant rainfall over the entire year. Since mineral dissolution is a slow process, it will occur during the periods of soil pore saturation where the solutions can eventually be drained from the soil.

5.5.2.2 Mineral Elements and Biomass

Of course a second and very important variable in the transfer equation is that of the mineral content of different plant species. Silica and potassium can vary very much from one plant species to another, as well within different parts of the plant itself. Hodson et al. (2005) and Broadley et al. (2004) indicate the variations in silica content for plants of different orders. The variations can be very great. The same is true for potassium (Gain, 1895). Then as could be expected the ratio of potassium to silicon can vary from one plant or plant assemblage to another. The mass of mineral elements and the ratio of potassium to silicon is a very important part of the translocation effect.

5.5.2.3 Rainfall and Biomass

Biomass responds to water input and the temperature when the water is available. Precipitation can be variable throughout the year (Oliver, 2007). For instance over much of the central and northeastern United States rainfall is rather constant throughout the year varying by about 15% from one month to another In other areas, Los Angeles or Oklahoma, the change is near 80%. In Singapore the change is near 80% month to month but the total rainfall is twice that of the American sites. The period of rainfall concentration is important in higher or lower latitude areas in that during colder periods vegetal activity is low, plant production is low and little transfer of mineral elements is accomplished. In Fig. 5.40 we suggest the potentials of rainfall frequency during a yearly cycle according to the temperature, warm in summer (A) and cool in winter (J,D). The greyed areas show monthly rainfall which is roughly proportional to plant growth and plant production of mineral element bearing substances.

Thus the biomass present under different climates will be controlled by total rainfall (precipitation) and the periods in which it occurs. It would be necessary to use a function of biomass linked to the climate; rainfall, frequency and temperature. This would give a better insight into the mineral input through biomass input.

5.5.2.4 Rainfall and Pore Water

Rainfall is fundamental for the production of vegetal material and hence mineral elements transported to the surface. However rainfall has a second effect of first dissolving silicate matter and then transporting it out of the chemical system of the A horizon. The more rain, the more dissolution will occur due to the contact of under – saturated aqueous solutions with the silicates. All minerals come into equilibrium with their aqueous environment by dissolving into the solution. When a certain limit is reached dissolution no longer occurs and if one adds more of the mineral elements to solutions the minerals will increase in mass through mineral growth. Rain water is rarely saturated with mineral elements.

The variation in the amounts of rain per period in the year is a basic part of climate. If rainfall occurs in short periods, and in abundance, throughout the year the soil is essentially dry when it rains the effect will be dilution of any water present and low amounts of dissolution of silicates. The contact time of the aqueous solution would be short and little dissolution of minerals would occur. If rainfall is regularly spaced over the year, pore water will be more likely to come into equilibrium with the silicates in the soil. Upon release of mineral elements from vegetal input, the silicates could be stabilized or perhaps the input from plant matter would create new silicate minerals. High rainfall throughout the year would promote dissolution of silicate materials in the soils.

The above assumes that the pore space is equally distributed in the soil. However this is far from the case. Many studies have been performed to determine the water holding capacity of soils as it is related to the availability of water for plants. Much of the water held in soils is in the smaller pores and capillary pore spaces. This is the ultimately available water for plants. However, Coppola et al. (2008) demonstrate the importance of the two types of pore sites, in soil aggregates and in the inter-aggregate pores.

One can visualize this function using images of pore space in different soils and depths in soil profiles (Fig. 5.41).

The example is an agricultural soil, where horizontal cuts in the profile have been used to show porosity. The top zone has a range of aggregate sizes and pore sizes which are both more or less regularly distributed in the sample. At the base of the agricultural zone, one sees the development of more large scale pores and cracks. Below the agricultural horizon only crack and biologically caused (worm

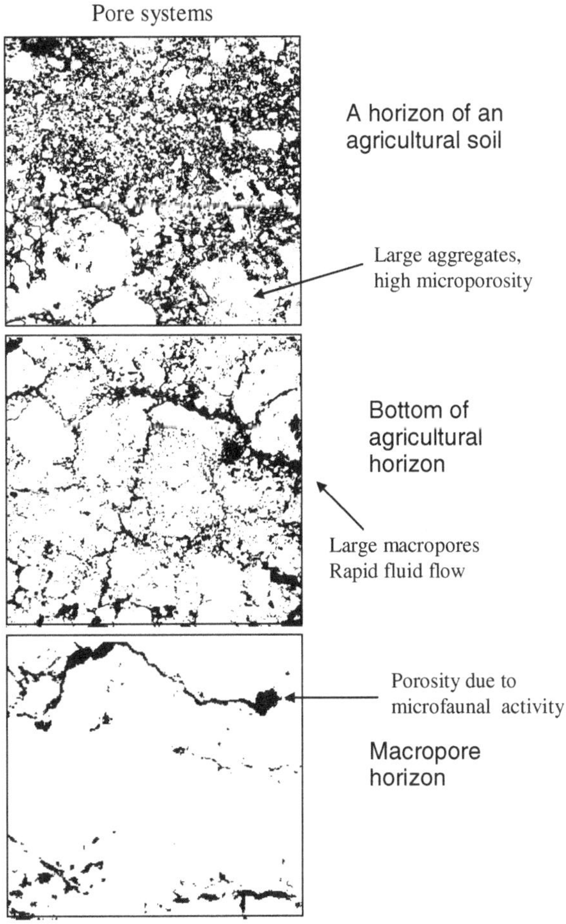

Fig. 5.41 Illustration of soil porosity types as a function of depth in a profile of an agricultural soil. Horizontal soil cuts show that at the surface one finds large aggregates with capillary porosity and micropores with significant macroporosity developed between the aggregates. At the *bottom* of the agricultural horizon cracking macropore porosity begins to occur forming passageways of rapid fluid flow and the lower macropore zone where only fractures and biotic porosity occurs leaving little micropore spaces and allowing water to flow out of the system rapidly (images are 6 cm squares)

galleries in this case) pores. These large scale events are interconnected and allow rapid passage of rainwater downward into the soil to reach the water table. In the large aggregates one finds small pores and capillary porosity. Figure 5.42 shows a vertical cut of a native prairie soil where the aggregates are present throughout the soil and interconnected crack like pores are found also.

In both examples there are two sets of pore types, one where the aqueous solution is largely in equilibrium with the soil clay minerals, and the large inter-aggregate pores and cracks which become filled with air after rainfall events. The larger pores are zones where plant roots grow and develop.

Depending upon the frequency of rainfall, there will be a different average occupation level of the macropores, high in climatic zones of regular rainfall and low in zones of irregular rainfall events. The more fine pores and aggregates of smaller size, the more small pores and the more water will remain in the soil zone. The

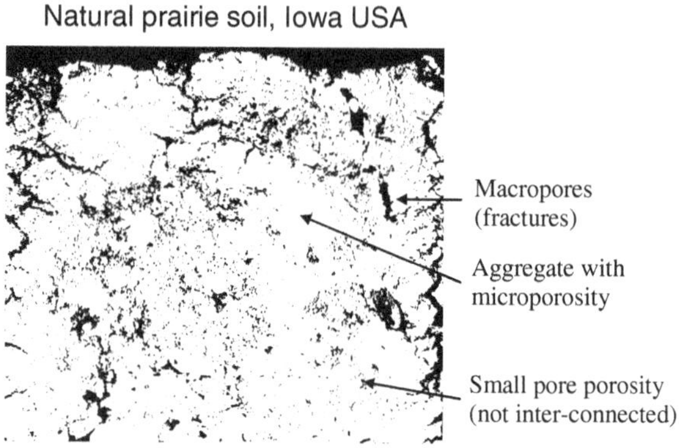

Fig. 5.42 Example of soil porosity types under native prairie where large aggregates hold capillary water and micropore structures with less frequent fracture porosity. The system is relatively homogeneous through the vertical 6 cm depth shown

fewer aggregates and a higher proportion of macropores and cracks will leave the soil with less residual water.

Consequences of such pore size distributions can be seen in the chemical data of Ranger (2002) for residual macropore and capillary water in a forest soil (Fig. 5.43). The potassium is essentially in dis-equilibrium between the capillary pores in soil aggregates and the remaining macropore water and silica is significantly more concentrated in the capillary water. Since potassium is present as a mono – ionic species in solution, its mobility would be expected to be relatively important and one would expect it to diffuse into the macropore solutions rapidly. However, silica is a more strongly in dis – equilibrium in the two types of water indicating that the complex ion and has lower mobility remaining associated with the major source of silica in solution, the clays of the soil aggregates. These data indicate that in this case there

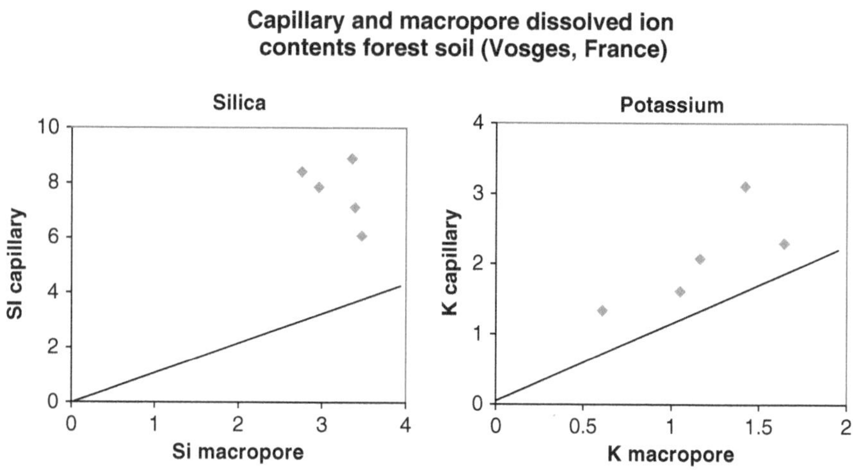

Fig. 5.43 Dissolved ion contents (weight %) of micropore, capillary water solutions and macropore solutions. The line indicates chemical equilibrium between the two types of solutions in the same forest soil (data from Ranger, 2002) which is the more the case for potassium content than for silica

is not a balance between micro and macroporosity in soils for these elements but that the flow of water through the soil will control the amount of mineral material in solution and its residence time in the soil, potassium coming into closer equilibrium than silica. Higher rainfall, at constant amounts will tend to move out less dissolved material. Less frequent rain events will allow mineral elements which remain in the soil for longer periods of time to approach equilibrium to a greater extent, but there is a differential diffusion and certain elements will be selectively evacuated.

The presence of organic matter in soils enhances the stability of siliceous clays under the same porosity conditions (Karathanasis and Wells, 1989)

It is evident that the transfer effect operated by plants to move mineral elements to the surface A horizon will depend upon several factors:

1. Rainfall and frequency which influence primary production of *biomass* and hence the capture of mineral elements by plants,
2. Rainfall and frequency which effects the differential *movement of fluids* will determine the evacuation of elements from the aggregate environment or the inter-aggregate pore space through capillary or macropore space.

5.5.3 Modelling Mineral Stability

Given the understanding that plants move key elements to the surface that stabilize certain clay minerals, one can attempt to model the effects of climate on biomass and mineral dissolution or crystallization (Barré et al., 2009). In a first approximation one can consider the relations of rainfall and biomass, one a dilution and destabilizing factor and the other a mineral forming factor. Taking simple assumptions of biomass as a function of annual rainfall and annual rainfall as an average dilution factor one can propose the mineral stability relations as a function of these two variables (Fig. 5.44).

There are several key points which confirm, in a general way, the model. One is at the low rainfall end of the curves. Nettleton et al. (1973) indicate that mica (illite) forms in semi-arid soils, near the low precipitation end of the diagram. Mahizhnan et al. (2006) give evidence for the formation of kaolinite in a very a rid desert in Australia. These two points are very closely related at the low rainfall end of the diagram, the semi-arid climate could be expected to produce some illite while the very arid climate should form kaolinite. In very low rainfall areas the stability of potassic, siliceous and 1:1 minerals (kaolinite) are not very fair apart in their regions of chemical stability.

The second area of interest is the high rainfall portion of the diagram where the stability of 2:1 minerals cedes to that of 1:1 and eventually silica – free minerals, gibbsite. In the initial stages of rock – plant interaction, 2:1 minerals can form but they are lost over period of high rainfall. This indicates that despite the high biomass productivity the effects of dissolution cannot be counteracted. The biomass productivity reaches a limit where increase in rainfall does not produce that much more

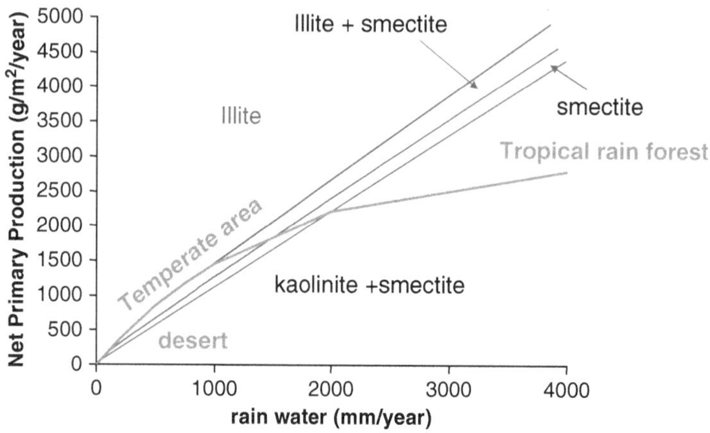

Fig. 5.44 Results of equilibrium mineral calculations of aqueous solutions with different amounts of mineral elements from biomass input (net primary production) and annual average rainwater quantities (Barré et al., 2009). Net primary production averages (heavy line) taken from Lieth (1975)

plant material and the translocation effect of mineral elements cannot compensate for loss to percolating rainwater through the dissolution of the clays in the A horizon of the soil. This state exists well below the A horizon in older soils with high rainfall inputs. This is indicated by the data or Glenn and Nash (1964) for soils from the Mississippi (USA) costal plain (Fig. 5.45). Surface (A horizons) show the presence

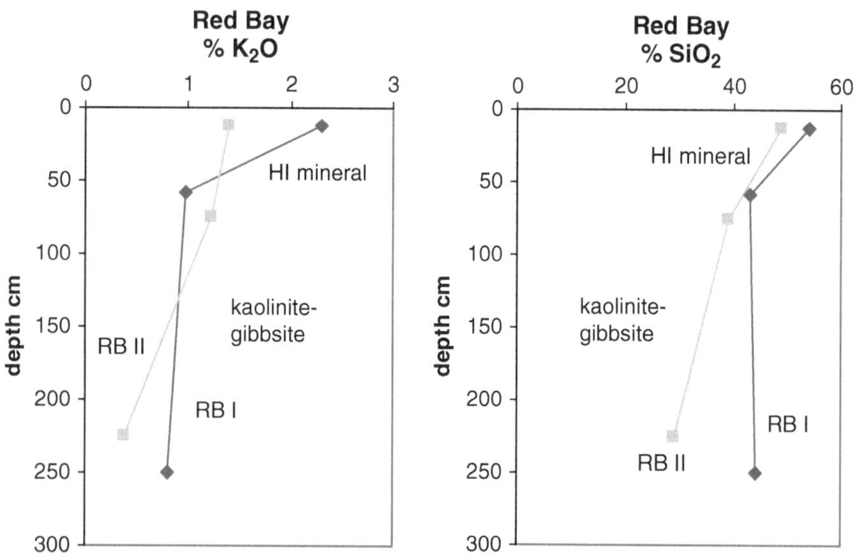

Fig. 5.45 Comparison of clay mineralogy and potassium and silica contents of soils in a south east coastal plain soil sequence (Glenn and Nash, 1964)

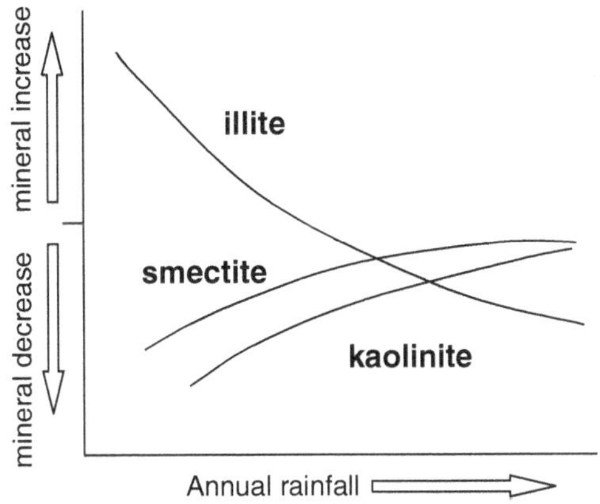

Fig. 5.46 Comparison of relative stabilities of the minerals illite, smectite and kaolinte as a function of annual rainfall (adapted from Barré et al., 2009). Illite is strongly affected while smectite and kaolinite increase less rapidly in relative amount

of 2:1 minerals (HI phases) while less siliceous minerals (kaolinite and gibbsite) are present at depth. When the biomass cannot keep up the needed supply of Si at the surface, under conditions of high rainfall, or in older soils where resources have been exhausted, the 2:1 minerals in the A horizon are lost.

Given the parameters of biomass and rainfall, what are the components of these factors that affect mineral stability? Barré et al. (2009) give examples of the individual effects of the rainfall – biomass variables. For example, all other factors being constant, what is the effect of dilution or total water through put to a clay mineral assemblage? In Fig. 5.46 we indicate the relative changes in mineral abundance with change in average annual yearly rainfall. This does not take into account the frequency nor flow through the system. It is just an estimation of the dilution effect or the relative rates of mineral dissolution. The illite tends to decrease in content and smectite then kaolinite increase. This reflects the pore water compositions reported by Ranger (2002) for capillary and macropore fluids. The equilibration of K between the solutions indicates that it will be lost to the flowing solutions and that the Si and Al remain in the solid phases, kaolinite and smectites in the example given.

On the other hand net primary production of biomass, taken independently from rainfall shows a trend of favoring illite as more biomass, and more element transfer is effected to the surface. Figure 5.47 indicates these relations.

At a constant ratio of Si and K in the plant mass, 5% K_2O and 0.75% SiO_2, the increase in plant biomass favors illite compared to smectite and kaolinite If one changes the relative amounts of Si and K in the plant biomass brought to the surface A horizon, the effects are such that increase in K_2O favors illite, as would be expected, but if Si is increased relative to K then smectite is favored at the expense of both kaolinite and illite whose Al/Si ratios are similar, being near one (Fig. 5.48).

The inter-relations of elemental content of the critical elements potassium and silica are fundamental to the stability of illite, smectite and kaolinite. Of course a deficit of silica will in the end favor the aluminous mineral gibbsite, with the loss of silicate clay minerals.

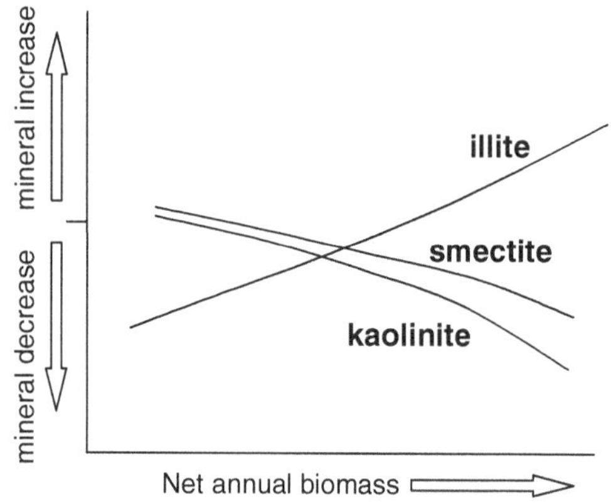

Fig. 5.47 Effect of annual biomass on relative mineral stabilities at constant amounts of Si and K in pore solutions (adapted from Barré et al., 2009)

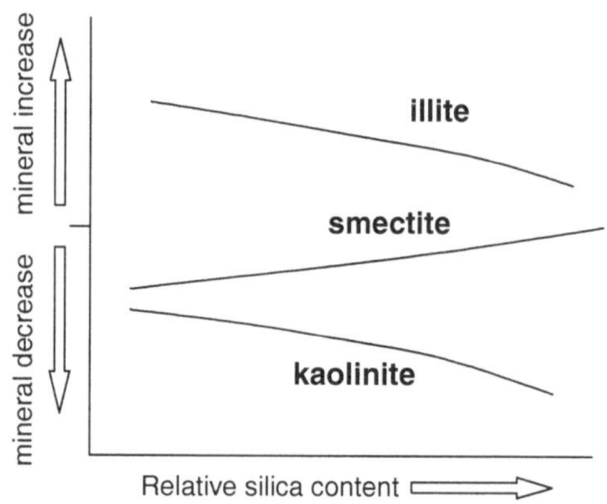

Fig. 5.48 Effect of different silica to potassium ratios in the biomass affecting a soil system. Increase in silica increases smectite while it decreases relative to illite and kaolinite content

These simulations indicate that the simplified diagram of Fig. 5.44 should be considered as a general concept but not applicable to all specific situations. One factor is of course the relative amount of rain that moves through the soil in macropores compared to caplillary and micropores. The factor of dilution, rainfall in Fig. 5.46, is moderated by the soil structure, i.e. the rate of flow of rainwater for a given rainfall event. The plant productivity affects the clay mineralogy in the total of the matter brought to the surface but also, and perhaps more importantly, in the proportions of the mineral elements contained in it. Bacterial action and root exudate production can affect the acidity of the soil solutions in the A horizon so that some minerals are favored over others. The higher the acidity, the more hydrogen ions will be incorporated into the clays, (OH units) and hence kaolinite will be produced compared to illite or smectite. The ratio of OH ions in kaolinite compared to silica – alumina in illite is 1–3. Hence acid solutions favour kaolinite greatly. Thus the system is multidimensional but not impossible to understand.

The idea of translocation of mineral elements to the surface horizons by plants has been in the minds of agronomists and soil scientists for some time, even in the 19th century (Larbaletrier, 1891, p. 104 for example) but up until very recently no systematic attempt has been made to use numerical data to explain and eventually predict the minerals present in the A horizon of an alteration profile according to the variables of plant recycling and climate. The formulation of transfer as a function of mineral stabilities in equilibrium with aqueous solutions is certainly a progress in understanding the nature of the mineralogy of soils and the importance of plant interactions concerning their stabilities according to the parameters of climate and plant growth.

What we need to know

This chapter has presented much information which pertains to the possibility of building a chemical model which can be used to describe and eventually predict the effects of plants on clay mineralogy in the A surface horizons. Several points are inadequately known:

1. *What are the kinetics of mineral dissolution compared to mineral formation in the A horizon*? Studies must be done to establish realistic reaction rates for dissolution and crystallization of 2:1 minerals under surface conditions.
2. *What is the transit time of rainwater in different soils? What is the time constant of equilibration of the different elements dissolved within the micropore and macropore compartments*? This information can be used to estimate the effect of soil texture on the clay mineral stabilities.

References

Ahmed I, Hussain M and Khan M (1971) Distribution of mica in the soils of the Madhupur Tract; East Pakistan 61–64, Proceedings UNESCO Bandung Symposium 149 pp.

Andrist-Rangel Y (2008) Quantifying mineral sources of potassium in agricultural soils Doctoral Thesis Faculty Nat Res Agric Sci, Univ Uppsala, Sweden, 426 pp.

April R and Keller D (1990) Mineralogy of the rizosphere in forest soils of the eastern United States. Biogeochemistry 9: 1–18.

April R, Keller D and Driscoll C (2004) Smectite in Spodosols from the Adirondack Mountains of New York. Clay Miner 39: 99–113.

Aragon A, Garcia M, Filgueria R and Pachepsky Y (2000) Maximum compactibility of Argentine soils from the Proctor test: The relationship with organic carbon and water content. Soil Tillage Res 56: 197–204.

Bardossy G and Aleva G (1990) Lateritic Bauxites, Elsevier, 624 pp.

Barkoudah Y and Henderson J (2006) Plant ashes from Syria and the manufacture of ancient glass:Ethnographic and scientific aspects. J Glass Stud 48: 297–321.

Barré P and Hallett P (in press) Rheological stabilization of wet soils by model root and fungal exudates depends on clay mineralogy. Eur J Soil Sci 60: 525–538.

Barré P, Berger G and Velde B (2009) How element translocation by plants may stabilize illite clays in the surface of temperate soils. Geoderma 151: 1–2.

Barré P, Velde B, Catel N and Abbadie L (2007) Soil-plant potassium transfer: Impact of plant activity on clay minerals as seen from X-ray diffraction. Plant Soil 292: 137–146.

Birkeland P, Shroba R, Burns S, Price A and Tonkin P (2003) Integrating soils and geomorphology in mountains – an example from Front Range of Colorado. Geomorphology 55: 329–344.

Black C (1968) Soil Plant Relationships, Wiley, 792 pp.

Boettinger J and Southard R (1995) Phyllosilicate distribution and origin in Aridisols on a granitic pediment, Western Mojave Desert. Soil Sci Soc Ame J 59: 1189–1198.

Bonifaccio E, Zanini E, Boero V and Franchini-Angela M (1997) Pedogenesis in a soil catena on serpentin in north-wesern Italy. Geoderma 75: 33–51.

Bormann F and Likens G (1994) Pattern and Process in a Forested Ecosystem, Springer Verlag, 253 pp.

Borrelli N, Osterrieth M and Marcovecchio J (2008) Interrelations of vegetal cover, silicophytolith content and pedogenesis of Typical Argiudolls of the Pampean Plain, Argentina. Catena 75: 146–153.

Bowden D (1987) On the compositions and fabric of the footslope laterites (duricrust) of Sierra Leone, West Africa. Zeit Geomorph N F 64: 39–53.

Broadley M, Bowen H, Cotterill H, Hammond J, Meacham M, Mead A and White P (2004) Phylogenetic variation in the shoot mineral concentration of angiosperms. J Exp Bot 55: 321–336.

Bryant J and Dixon J (1964) Clay mineralogy and weathering of the red-yellow poszolic soil from quartz mica schist in the Alabama Piedmont. Clays Clay Miner 16: 509–521. E Ingerson editor 691 pp.

Chenu C (1993) Clay – or sand-polysaccharide associations as models for the interface between miro-organisms and soil: Water related properties and microstructure. Geoderma 56: 143–156.

Chenu C and Guérif J 1991. Mechanical strength of clay minerals as influenced by an adsorbed polysaccharide. Soil Sci Soc Ame J 55: 1076–1080.

Chesworth W, Dejou J and Larroque P (1981) The weathering of basalt and relative mobilities of major elements at Belbex, France. Geochim Cosmochim Acta 45: 1235–1241.

Clark F, Cole C and Bowman R (1980) Nutrient cycling, 659–706. In A Breymeyer and G Van Dyne (eds.), Grasslands, System Analysis and Man, Cambridge, 851 pp.

Coppola A, Comega V, Basile A, Lamaddino N and Severino G (2008) Darcian preferential water flow and solute transport through bimodal porous systems: Experiments and modelling. J Contam Hydrol 104: 74–83.

Cornejo J and Hermosin MC (1996) Interaction of humic substances and soil clays, pp. 509–625. In A Piccolo (ed.), Humic Substances in Terrestrial Ecosystems, Elsevier, 753 pp.

Czarnes S, Hallett PD, Bengough AG and Young IM (2000) Root- and microbial-derived mucilages affect soil structure and water transport. Eur J of Soil Sci 51: 435–443.

De Connick F, Jensen W and De Kimpe C (1987) Evolution of Silicates, Especially Phyllosilicates, during Podzolisation, In Righi D and Chauvel A (eds.), Podzols et Podzolisation, Association Française pour l'Etude des Sols, 231 pp.

Denef K and Six J (2005) Clay mineralogy determines the importance of biological versus abiotic processes for macroaggregate formation and stabilization. Eur J Soil Sci 56: 469–479.

Denef K, Six J, Merckx R et al. (2002) Short-term effects of biological and physical forces on aggregate formation in soils with different clay mineralogy. Plant Soil 246: 185–200.

Denef K, Six J, Merckx R et al. (2004) Carbon sequestration in microaggregates of no-tillage soils with different clay mineralogy. Soil Soc Soc Ame J 68: 1935–1944.

Derry L, Kurtz A, Zeigler K, and Chadwick O (2005) Biological control of terrestrial silica cycling and expert fluxed to watersheds. Nature 433: 728–731.

de Serres O (1600) Le Théâtre d'Agriculture et Mesnage des Champs, Reproduction by Actes Sud 2001, 1545 pp.

Dinel H, Mehuys G and Lévesque M (1991) Influence of humic and fibric materials on the aggregation and aggregate stability of a lacustrine silty clay. Soil Sci 151: 146–158.

Dixon J and Weed S (eds.) (1989) Minerals in Soil Environments, Soil Sci Soc Ame Book Number 1, Madison, WI, 1245 pp.

Drever J (1994) The effects of land plants on weathering rates of silicate minerals. Geochim Cosmochim Acta 58: 2325–2332.

Duchaufour P (1977) Pédologie, Vol. 1, Masson, 475 pp.

Duvigneaud P and Denoyer-DeSmet S (1970) Deciduous forests, 197–225. In D Reichle (ed.), Analysis of Temperate Forest Ecosystems, Springer, 304 pp.

Egli M, Zanelli R, Kahr G, Mirabella A and Fitze P (2002) Soil evoloution and development of the clay mineral assemblages of a Podzol and a Cambisol in "Meggerwald", Switzerland. Clay Miner 37: 351–366.

Evangelou V and Lumbanraja J (2002) Ammonium – potassium – calcium exchange on vermiculite and hydroxyl – aluminium vermiculite. Soil Sci Soc Ame J 66: 445–455.

Farmer VC, Delbos E and Miller JD (2005) The role of phytolith formation and dissolution in controlling concentrations of silica in soil solutions and streams. Geoderma 127: 71–79.

Fenton T, Brown J, and Mausbach M (1999) Effects of long – term cropping on organic matter content of soils: Implications for soil quality, 89–123. In R Lal (ed.), Soil Qualtiy and Soil Erosion, CRC Press, 329 pp.

Fichter J, Turpault M-P, Dambrine E and Ranger J (1998) Minerral evolution of acid forest soils in the Strengbach catchment (Vosges mountains N-E France). Geoderma 82: 315–340.

Frayssc F, Pokrovsky O, Schott J and Meunier J D (2006) Surface properties, solubility and dissolution kinetics of bamboo phytoliths. Geochimica Cosmochim Acta 70: 1939–1951.

Gain E (1895) Précis de Chimie Agricole, J-B Baillière et Fils, Paris, 436 pp.

Gérard F, Mayer K, Hodson M and Ranger J (2008) Modelling the biogeochemical cycle of silicon in soils: Application to a temperate forest ecosystem. Geochim Cosmochim Acta 72: 741–758.

Gillot F, Righi D and Räisänen M (2001) Layer – charge evaluation of expandable glacialmarine drift, Northwestern Washington. Soil Sci Soc Ame 48: 208–215.

Glenn R, Jackson M, Hole F and Lee G (1960) Chemical weathering of layer silicate clays on loess-derived Tama silt loam of southwestern Wisconsin. Clays and Clay Minerals, 8th Conference, pp. 63–83 (Pergamon press, New-York).

Glenn R and Nash V (1964) Soils from the lower Mississippi coastal plain. Clays Clay Miner 19.

Govinda Rajan S and Murtrhy S (1971) Trends in rock weathering in the southern part of peninsular India – its expression in morphological genesis 65–75, UNESCO Proceeding of the Bandung Symposium, 149 pp.

Graham R and Southard A (1983) Genesis of Vertisol and an associated Mollisol in Northern Utah. Soil Sci Soc Ame J 47: 552–559.

Graham R, Diallo M and Lund L (1990) Soils and mineral weathering on phyllite colluvium and serpentinite in Noathwest California. Soil Sci soc Ame J 54: 1682–1690.

Graham R and Wood H (1991) Morphologic development and clay distribution in lysimeter soils under chaparral and pine. Soil Sci Soc Ame J 55: 1638–1646.

Grim R (1954) Clay Mineralogy, McGraw Hill, 384 pp.

He Y, Li DC, Velde B, Yang YF, Huang CM, Gong ZT and Zhang GL (2008) Clay minerals in a soil chronosequence derived from basalt on Hainan Island China. Geoderma 148: 206–212.

Heil D and Sposito G (1993) Organic matter role in illitic soil colloids flocculation. 1. Counter ions and pH. Soil Sci Soc Ame J 57: 1241–1246.

Henriet C, Bodarwé L, Dorel M and Draye X (2008) Leaf silicon content in banana (*Musa spp.*) reveals the weathering stage of volcanic ash soils in Guadeloupe. Plant Soil 313: 71–82.

Hodson J, White P, Mead A and Broadley M (2005) Phylogenetic variation in the silicon composition of plants. Ann Bot 96: 1027–1046.

Homann P, Van Miegroet, Cole D and Wolfe G (1992) Cation distribution, cycling and removal from mineral soil in Douglas-fir and red alder forests. Biogeochemistry 16: 121–150.

Howard J, Amos D and Daniels W (1993) Alluvial soil chronosequence in the Inner Coastal Plain, Central Virginia. Quart Res 39: 201–213.

Huang P (2005) Chemistry of potassium in soils, 227–282. In M Tabatabai and D Sparks (eds.), Chemical Processes in Soils, Soil Sci Soc Ame Book Series number 8, 723 pp.

Huang T and Uehara G (1968) Mica genesis in Hawaiian soils. Soil Sci Soc Amer J 32: 31–35.

Jahn R and Asio V (1998) Soils of the tropical forests of Layete Philipines, 27–36. In A Schultz and D Ruhiyat (eds.), Soils of Tropical Forest Systems, Springer, 323 pp.

Jenny H (1994) Factors of Soil Formation, Dover, 281 pp.

Jobbagy EG and Jackson RB (2004) The uplift of soil nutrients by plants: Biogeochemical consequences across scales. Ecology 85: 2380–2389.

Juang T and Uehara G (1968) Mica genesis in Hawaiian soils. Soil Sci Soc Ame Proceed 32: 31–35.

Juster T, Brown P and Bailey S (1987) NH_4 – bearing illite in vey low grade metamorphic rocks associated with coal, northeastern Pennsylvania. Ame Miner 72: 555–565.

Karathanasis A and Wells K (1989) A comparison of mineral weathering trends between two management systems on a Catena of loess – derived soils. Soil Sci Soc Ame J 3: 582–588.

Keller W (1967) Processes of origin and alteration of clay minerals, 3–76. In C Rich and G Kunze (ed.), Soil Mineralogy, Univ North Carolina Press, 330 pp.

Kleber M, Schwendenmann L, Veldkamp E, Rossner J and Jahn R (2007) Halloysite versus gibbsite: Silicon cycling as a pedogenetic process in two lowland neotropical rain forest of La Selva, Costa Rica. Geoderma 138: 1–11.

Knecht M and Göransson A (2004) Terrestrial plants require nutrients in similar proportions. Tree Physiol 24: 447–460.

Kretzschmar R, Hesterberg D and Sticher H (1997) Effects of adsorbed humic acid on surface charge and flocculation of kaolinite. Soil Sci Soc Ame J 61: 101–108.

Kretzschmar R, Robarge WP and Weed SB (1993) Flocculation of kaolinitic soil clays – Effects of humic substances and iron oxides. Soil Sci Soc Ame J 57: 1277–1283.

Lanson B (1997). Decomposition of experimental X-ray diffraction patterns (profile fitting): A convenient way to study clay minerals. Clays Clay Miner 40: 40–52.

Larbaletrier A (1891) Les Engrais, Baillière et Fils, Paris, 351 pp.

Leumbe Leumbe O, Bitom D, Tematio P and Lucas Y (1993) Etude des sols ferralitique a caractères andiques sur trachytes en zone de montagne humic tropical. Etude et Gestion des Sols 12: 313–326.

Li D and Velde B (2003) Aggregation in some clay-rich agricultural soils as seen in 2D image analysis. Geoderma 118: 191–207.

Lieth H (1975) Modeling the primary productivity of the world. In H Lieth and RH Whittaker (eds.), Primary Productivity of the Biosphere, Springer-Verlag, Berlin, Heidelberg, New York (144 pp).

Liu Y-L, Zhang B, Li C-L, He F and Velde B (2008) Long term fertilization influences on clay mineral compositions and ammonium absorption in rice paddy soil. Soil Sci Soc Ame J 72: 1580–1590.

Logsdon S (1995) Analysis of aggregate fractal dimensions and aggregate densites back calculated from hydraulic conductivity soil. Sci Soc Ame J 59: 1216–1221.

Logsdon S, Renean R and Parker J (1987) Corn seedling root growth as influenced by soil physical properties. Agron J 79: 221–224.

Loughnan F (1969) Chemical Weathering of the Silicate Minerals, Elsevier

Lucas Y (2001) The role of plants in controlling rates and products of weathering: Importance of biological pumping. Ann Rev Earth Planet Sci 29: 135–163.

Lucas Y, Luizao FJ, Chauvel A, Rouiller J and Nahon D (1993) The relation between biological activity of the rain-forest and mineral composition of soils. Science 260: 521–523.

Lupwayi N, Clayton G, O'Donovan J, Harper K, Turkington T and Soon Y (2006) Potassium release during decomposition of crop residues under conventional and zero tillage. Can Jour Soil Sci 86: 473–481.

Mahaney W and Halverson D (1986) Rates of mineral weathering in the Wind River Mountains, Western Wyoming, 147–167. In SM Colman and DP Dethier (eds.), Rates of Chemcial Weathering of Rocks and Minerals, Academic press, 603 pp.

Mahizhnan A, Anand RR and Wilde SA (2006) Nanoscale investigation of red-brown hardpan in Western Australia. Abstract Bridging Clays, joint meeting of the Clay Minerals society and the French clay group, June 3–7, 2006.

Maignien R (1958) Le cuirassement des sols en Guinée, Afrique Ocidental, Mem Carte Géol Als Lorr 16 239 pp. Compte Rendu de Recherches sur les Latérites UNESCO.

Majzik A and Tombacz E (2007) Interaction between humic acid and montmorillonite in the presence of calcium ions. 2. Colloidal interactions: Charge state, dispersing and/or aggregation of particles in suspension. Org Geochem 38: 1330–1340.
Mariotti A (1982) Apports de la Géochimie Isotopique a la Connaissance du Cycle de l'Azote, Univ P et M Curie, Paris, 488 pp.
McFarlane M (1978) Laterite and Landscape, Academic Press, 151 pp.
Meijer E and Buurman P (2003) Chemcial trends in a perhumid soil catena on the Turriaba volcano (Costa Rica). Geoderma 117: 185–201.
Miles J (1985) The pedogenic effects of different species and vegetation types and the implications of succession. J Soil Sci 36: 571–584.
Mirabella A and Egli M (2003) Structural transformations of clay minerals in soils of a climosequence in an Italian Alpine environment. Clays Clay Miner 51: 264–278.
Mirabella A, Egli M, Raimondi S and Giaccai D (2005) Origin of clay minerals in soils on pyroclstic deposits in the island of Lipari. Clay Clay Miner 53: 409–421.
Mohr E and Van Baren F (1959) Tropical Soils N V Uitgeverji Van Hoeve Hague 498 pp.
Moody PW and Bell MJ (2006) Availability of soil potassium and diagnostic soil tests. Aus J Soil Res 44: 265–275.
Moreau E, Velde B and Terribile F (1999) Comparison of 2D and 3D images of fractures in a Vertisol. Geoderma 92: 55–72.
Nettleton W, Nelson R and Flach K (1973) Formation of mica in surface horizons of dryland soils. Soil Sci Soc Ame Proceed 37: 473–478.
Nguetnkam J-P, Kama R, Villièras F, Ekodeck G-E and Yvon J (2007) Altération du granite en zones tropicals: Exemple de deux séquences étudiées au Cameroun (Afrique Centrale). Etude et Gestion des Sols 14: 31–41.
Nielsen K, Dalsgaard K and Nornberg P (1987) Effects on soils of an oak invasion of a Calluna heath, Denmark, I Morphology and chemistry. Geoderma 41: 79–95.
Norfleet M and Smith B (1989) Weathering and mineralogical classification of selected soils in the Blue Ridge Mountains of South Carolina. Soil Sci Soc Ame J 53: 1771–1778.
Oliver J (2007) Encyclopedia of World Climatology, Springer, 854 pp.
Pedro G (1966) Essai sur la caractérisation géochimique des différents processus zonaux résultant de l'altération des roches superficielles. Compt Rendus Acad Sci Paris Ser D 262: 1828–1831
Perth S, Horn R, Beckman F, Donath T, Fischer J and Schmucker A (2008) Three-dimensional quantification of intra-aggregate pore-space features using synchrotron-radiation-based microtomography. Soil Sci Soc Ame J 72: 879–907.
Pevear D, Goldin A and Sprague J (1984) Mineral transformations I soils formed in glacialmarine drift, Northwestern Washington. Soil Sci Soc Ame 48: 208–215.
Pion J-C (1979) Alteration des massifs cristallins basiques en zone tropical Sèche. Sci Géol Mem 57: 230.
Quideau S, Chadwick O, Graham R and Wood H (1996) Base cation exchange and weathering under oak and pine: A controlled long-term experiment. Biogeochemistry 35: 377–398.
Rabenhorst M, Foss J and Fanning D (1982) Genesis of Maryland soils formed on serpentinite. Soil Sci Soc Ame J 46: 607–616.
Ranger J (2002) Le Cycle Biogeochimique des Éléments Nutritifs dans les Écosystemes Forestières, INRA Nancy, 168 pp.
Rhoton F, Shipitalo M and Lindbo D (2002) Runoff and soil loss from Midwestern and southeastern US silt loam soils as affected by tillage practices and soil organic matter content. Soil Tillage Res 66: 1–11.
Righi D and Lorphelin L (1986) Weathering of silt and clay in soils of a toposequence in the Himalayas, Nepal. Geoderma 39: 141–155.
Righi D, Huber K and Keller C (1999) Clay formation and podzol development from postglacial moraines in Switzerland. Clay Miner 34: 319–332.
Robert M (1996) Le Sol, Masson, 241 pp.

Ross G, Phillips P and Culley J (1985) Transformation of vermiculite to pedogeneic mica by fixation of potassium and ammonium in a six-year field maure application experiment. Can J Soil Sci 65: 599–603.
Ruhe R (1984) Soil-climate system across prairies in Midwestern USA. Geoderma 34: 201–219.
Russell E (1945) Soils and Manures, Cambridge Press, 296 pp.
Schactmann D and Schroeder J (1994) Structure and transport of a high – affinity potassium transporter from higher plants. Nature 370: 655–658.
Scherer H and Ahrens G (1996) Depletion of non-exchangeable NH-N in the soil root interface in relation to clay mineral composition and plant species. Eur J Agron 5: 1–7.
Segalen P (1971) Metallic oxides and hydroxides in soils of the warm and humic areas of the world: Formation, identification and evolution, 25–38. UNESCO Proceeding of the Bandung Symposium. 149 pp.
Six J, Elliot E and Pausitan K (2000) soil structure and organic matter: II A Normalized stailityindex and the effect of mineralogy. Soil Sci Soc Ame J 64: 1042–1049.
Smagin A, Sadovnikova N, Nasrova T, Kiryushova A, Mashika A and Eremina A (2004) The effect of organic matter on the water retention capacity of soils. Eurasian Soil Sci 17: 267–275.
Sposito G (1989) The Chemistry of Soils, Oxford Univ Press, 278 pp.
Sucha V and Sirahova V (1991) Potassium and ammonium fixation in smectites by wetting and drying clays. Clay Miner 39: 556–559.
Sucha V, Elsass F, Eberl D, Kuchta L, Madejova J, Gates W and Komadel P (1998) Hydrothermal synthesis of ammonium illite. Ame Miner 83: 58–67.
Tarchitzky J and Chen Y (2002a). Polysaccharides and pH effects on sodium-montmorillonite: Flocculation, dispersion and rheological properties. Soil Sci 167: 791–801.
Tarchitzky J and Chen Y (2002b) Rheology of Sodium-montmorillonite suspensions: Effects of humic substances and pH. Soil Sci Soc Ame J 66: 406–412.
Temgoua E (2002) Cuirassement ferrugineux de bas de versants en zone forestière du Sud-Cameroun Mémoires Géologie (Lausanne) no 38, 135 pp.
Teveldal S, Jorgensen P and Stuanes A (1990) Long-term weathering of silicates in a sandy soil at Nordmoen, Southern Norway. Clay Miner 25: 447–465.
Tice K, Graham R and Wood H (1996) Transformations of 2:1 phyllosilicates in 41-year-old soils under oak and pine. Geoderma 70: 49–62.
Tisdall J and Oades J (1982) Organic matter and water – stable soil aggregates. J Soil Sci 33: 141–183.
Treshaw M (1970) Environment and Plant Response, Mc Graw Hill, 423 pp.
Tripler C, Kuashal S, Likens G and Walter T (2006) Patterns in potassium dynamics in forest ecosystems. Ecol Lett 9: 451–466.
Troare O, Groleau-Renard V, Plantureux S, Tubeileh A and Boef-Tremblay V (2000) Effect of root mucilage and modelled root exudates on soil structure. Eur J Soil Sci 51: 575–581.
Varadachari C, Mondal A and Ghose K (1995) The influence of crstal edges in clay-humus complexation. Soil Sci 159: 185–190.
Velde B (1985) Clay Minerals; A Physico-Chemical Explanation of their Occurrence, Elsevier, 423 pp.
Velde B (2001a) Surface cracking and aggregate formation in Rendzina soil, LaTouche (Vienne) France. Geoderma 99: 261–277.
Velde B (2001b) Clay Minerals in the agricultural horizon of loams and silt loams in the Central United States. Clay Miner 36: 277–294.
Velde B and Meunier A (2008) The Origin of Clay Minerals in Soils and Weathered Rocks, Springer, 406 pp.
Virto I, Barré P and Chenu C (2008) Microaggregation and organic matter storage at the silt-size scale. Geoderma 146: 326–335.
Walker L and del Moral R (2003) Primary Succession and Ecosystem Rehabilitation, Cambridge Univ Press, 456 pp.
Wolt J (1994) Soil Solution Chemistry, John Wiley, 345 pp.
Zairn D and Johnson A (1995) Nutrient accumulation during primary succession in a montane triopical forest, Puerto Rico. Soil Sci Soc Ame J 59: 1444–1452.

Chapter 6
Man and Agriculture: Manipulating Soil – Plant Interactions

Contents

6.1 The Adventure of Agriculture: the Manipulation of Nature by Man

Plants manipulate soils and rocks debris to stabilize and produce clays that form good properties to their benefit. Without plant there would be no soil i.e. association of organic matter and clay minerals. This stabilizing influence and the ecological interactions between rocks, silicate minerals, clays, organic matter and micro-organisms is the basis of plant life on the surface of the earth. Animals live from plant material, either first hand (herbivores for example) or second hand (lions, jackals and other hunters). The trick is to use the ecology of plant actions to produce vegetal material for direct consumption (wheat for example) or for the production of meat (cattle and other meat animals). Man needed to understand how to manipulate

B. Velde, P. Barré, *Soils, Plants and Clay Minerals*,
DOI 10.1007/978-3-642-03499-2_6, © Springer-Verlag Berlin Heidelberg 2010

the original manipulators, how to use the natural tendencies and inter-relations of plant and soil materials in order to become less dependent on the vagaries of nature. This chapter gives a brief resume of the steps of plant manipulation related to soils constraints and the eventual impact of agriculture on the soil materials, especially the clay minerals.

6.1.1 Introduction

There are many books in most languages concerning the evolution of mankind. The use of natural materials and concentration of resources mark the evolution of mankind in general in the different regions of the inhabited earth. Most histories of this evolution are centred on the material and intellectual side of the evolution. Housing, graves, the written word, use of stone and metals as tools and so forth dominate books on the pre-history of mankind. These aspects are of course most important to an understanding of today's societies but they do not indicate much of the crucial developments that were accomplished and which shaped the life of mankind. One obvious reason for study of non-perishable artefact is that they are non-perishable and relatively abundant. Stone is well preserved in all archaeological sites as are ceramic materials, wood much less so and the remains of crops and grain even less so. It is not incuriosity on the part of archaeologist that leads to the concentration on durable goods. The biggest problem is to have representative populations of objects and artefacts which can be used to describe the life a given community in the distant past. One of the major if not the major transformations of human societies was the development of agriculture. This step in changing the local environment to the advantage of human sustenance represents a step from being submitted to the whims of nature to controlling the natural system necessary for human welfare. Here we would like to try to get an understanding of the problems faced and the uses of nature that were effected by early communities as they began to use the vegetal resources available in their environment.

One does not often think that agriculture is a manipulation of nature. However this was the case up to the mid – 1950 revolution of agricultural practice in the west. Before this period man was using nature to produce things in nature. The genius of farmers was to use the natural tendencies of plants and animals in order to control and regulate the productivity of vegetal and animal species. The basis for animal sustenance is plant matter, in one way or another. Cultivating plants is not only useful to produce food for humans but also for the animals he uses for traction and eventually food also. The idea of a factory product which could be used to increase the productivity of the earth was for the most part a dream for much of the agricultural world. However, the use of the tractor, factory produced ammonium and herbicide, insecticide products made in factories from petroleum changed the concept of the earth and the relations of man and the earth. The progress of this idea was not immediate nor was it done with conscience. The idea of progress due to Science is to a large extent a part of post – War culture; change is good, change is Science applied to everyday endeavors, change is progress (for the better of course). Yet the means of using nature to produce the material needs of food and clothing was practiced from time immemorial. This initial step we think of as the Neolithic Revolution, one where man became independent of the whims of nature and natural availability which permitted him to multiply enormously and become a more than major actor in the world of the living. Here in the distant past, some 11,000 years ago depending on where in man's past habitat one looks, there was a revolution. Instead of submitting to the laws and vagaries of nature, man became to a certain extent independent of yearly changes in climate and could ameliorate his future supplies of basic substances.

Man did not become a farmer first. The development of the human species passed through various stages of using natural resources for sustenance and comfort. Hunters became hunter – gatherers and eventually herdsmen. The gatherers became more specialized and developed into farmers. The process was long, covering many hundreds of thousands of years.

Earlier crucial steps in human development, pre-farming, included the invention of the shaped stone or flint which produced a cutting edge. The process of shaping flint to make the variety of tools we have found today was quite long, much more than a million years, By forming a strong cutting edge of a flint stone man made not only a useful penetrating weapon to kill game with thick hides but this sharp edge also allowed him (or her) to cut meat from a carcass and take off the skin with efficiency. One could transport and store the kill and further use part of it (the skin) for decent clothing. Heaven knows how one skinned a mammoth, bison or musk ox before that. In any event great material progress was made using simple tools and a lot of human ingenuity. Try to make a blade from a piece of flint.

However the true manipulation of natural resources seems to have come together in what is called the Neolithic revolution where not only selection of seeds and amelioration of the species for food sustenance was practiced, but there was a selection of heretofore wild animals and a development of specific characteristics of the animals. Roughly at the same time heating clays to the point of becoming artificial rocks (ceramics) came into use. Instead of chipping stones to make sharp edges the concept of stone polishing developed which allowed the production of a much more robust tool, useful to cut tree trunks or effectively dig in the earth. These various things came together in a surprising way and with relative rapidity in several centers of human experience, so that in cities or communal concentrations humans could live longer and more plentiful lives. In a sense man was re-born when he could manipulate his natural surrounding with success. His everyday life changed radically. For one thing specialization occurred, where one family or person could do what it did best and not have the necessity of doing all things necessary for sustenance. Potters, axe makers and other specializations could occur albeit we know little of just how nor by whom. In any event it is clear that life in a large community allows portions of it to specialize in their activities to the benefit of all. Providing food for a large stable population fixed in one location was a key to development of humankind.

The processes was not universal nor did it come at the same time nor in the same succession of events or mutations over all of the human (*homo sapiens sapiens*) inhabited earth. However the major changes occurred in a more or less parallel manner over a similar time period following the climate – controlled plant regimes which were presented to man as glaciers retreated. The control of climate over plant regime shaped the agricultural future for some peoples, others chose different paths to arrive at about the same stage of self sufficiency in the major areas of the world. Often times one dealt with things as they were. The radical changes and agricultural progress of the latter half of the 20th century have effaced these differences between developed cultures largely, putting different human cultures and climates on a more or less similar agricultural footing. One can make an argument for a better or poorer choice and future, assuming changes in human needs and conscience, but for better or worse we are in the post industrial agricultural world.

How did we get there or here? How can we use past experience to benefit from the fantastic evolution of human kind based upon the manipulation of natural processes? We will not answer these questions but propose a short, simplified history of the development of agriculture in order to understand the relations of plants and soils as they were understood and used by man over the millennia. We have used several texts as references to describe the general sequence of events in the agricultural adventure but many more texts can be found (Bell and Walker, 1992; Guilaine, 1989; Masurié de Keroualin, 2003; Clark, 1955; Shane and Cushing, 1991; Stamp, 1961; Chambers, 1993; Dalfes et al., 1994; Ruddiman, 2005; Duby and Wallon, 1975; Ferdiére et al., 2006).

6.1.1.1 Multiple Centers and Multiple Problems

The spread of human communities was completed some millions of years before the time we wish to describe here. Most habitable zones we know today were colonized by the time of 10,000 years before our era. Arctic and sub arctic areas, tropical rain forests and the great steppes below the tundra stretching across climatic and geographic zones were known to the hunters and hunter – gatherers. If one were to use the natural vegetal sources of sustenance to develop a reliable food supply, the obstacles of natural vegetation had to be overcome. Gathering is fine but one rapidly runs out of resources at a given site and migration is necessary. The shift from migration to sedentary communal structures necessitates a greater concentration of food sources over a small area. One can only go so far to bring back food to a village each day or fortnight. Concentration of resources necessitated an intensification of exploitation of vegetal sources and an amelioration of their productivity. This is the story of agriculture.

Roughly four “centers” of agriculture can be identified; the Indo- Near East, Central China, the Meso and South American zone and the Southeast Asian and Polynesian areas. Concerning the climatic structures and hence vegetation biomes, the Chinese and Near Eastern areas were similar in these aspects at the end of the last major glacial period, around 11000 BC. Steppes supplied abundant food for foraging animals, and a certain sustenance for food gathering nomads who eventually became sedentarized in a general way. The landscape was dominated by grasses. This biome developed a specific type of soil and alteration structure with much organic matter preserved in the soil zone having a significant pH (usually above 6). Here the obvious plant types which could be used for agriculture were the grasses and associated legumes so abundantly present. Further climate change, increase in temperature and changes in rainfall in some areas, necessitated further adaptation of agricultural practice, irrigation in some instances and after several thousands of years eventual displacement of the major population centers occurred. The great development of agriculture took place during the climatic optimum, around 7000–2000 BC. However the exact time span varies depending on the location (Bell and Walker, 1992, p. 70)

The situation in Mezo and Andean South America is less well known but certain generalizations can be made. The introduction of agriculture as such was later than in the Chinese and Near Eastern areas and certainly more diverse. Significantly different types of crops were produced coming from both forest and prairie resources where rapid climate change occurred due to topography. Avocado along with squash can be found in the Andean zones while more grass – like plants were used in the Meso American area in the much more arid and steppe like areas. Valley bottoms were often preferred in initial stages of development. Overall the areas of agriculture seemed to be temperate and somewhat dry, dominated by grasses and prairie type vegetation.

A very interesting experience occurred in the South East Asian and Island areas of the Southwest Pacific. Here the climate was and is dominated by tropical or at least very dense forests. This is not where grasses such as the precursors of millet, rice and wheat would grow. However the forests did have abundant root resources, or at least they were discovered and used by the inhabitants as a source of food. Taro and Manioc became the stables of the vegetal part of diets here. Root culture is quite different from that of grasses and as such agricultural development was not the same. In such areas drainage of water is often more important than irrigation, totally unnecessary there. The forests provided fruit trees which eventually became domesticated. There is a strong contrast between the problems faced by rain forest farmers and those on the steppes. Nevertheless some of the oldest traces of agricultural manipulation of plant species are found in the rainforest areas. Essentially the agriculture here remained in the same biome throughout time whereas that in the old steppes moved to new areas by necessity or choice. The agricultural practice did not necessitate a complete change in biome as it did in other areas.

However in the European experience the needs of agriculture led to a major change in the vegetal cover. This very interesting manipulation of nature is the subject of the next chapter.

6.1.2 History of Traditional Agriculture in Europe: An Example of Soil – Plant Manipulation: Changing the Landscape Through Agriculture

Due to a reasonably complete body of information, the history of agriculture is exemplified here in using the European experience. The Neolithisation process occurred in different areas under different conditions and the response of mankind was adapted to these local conditions. One of the more striking examples of plant manipulation is seen in this conquest of the European continent where the relations of plants to climate and the development of soils had to be strongly modified in order to accommodate the ever increasing numbers of humans. It is not the only case to be sure but one of interest in that the manipulation of nature is rather striking.

6.1.2.1 The Context of the First Steps in Agriculture: Steppe Plants and the Hunter-Gatherer Cultures

Homo sapiens sapiens emerged at the end of the last cold stage of the last ice age. Also the initial steps towards agriculture in the European area occurred on the Near East in what is often called the "fertile crescent". The Palestinian – now Israeli – coast, Lebanon and the Tigris – Euphrates valleys descending to the Red Sea were the early centers of agriculture from what we know today. This is of course not what we consider to be Europe, but things started there. The Neolithic era marked by domestication of grazing animals, the development of new species or strains of wild grain plants, and the making of pottery all of which were practiced at different centers in the area at roughly the same times. Uncertainty is on the order of a thousand years. Different authors give slightly different dates and different sequences of use of the different elements of the Neolithic "package". However a strong and recurrent use of these methods of food productions seems to have begun by the time of 9000 BC and firmly fixed in most areas of the Near East by about 8000 BC. The northern point of the crescent, in present – day Turkey seems to have been the zone of diffusion of the Neolithic methods toward more northern European areas.

Different types of wheat and oats were domesticated as were lentils and peas. Thus not only grasses were used and developed but also legumes. This is a very important step. As we know, legumes enrich the soil by symbiotic transfer of nitrogen to the soil via bacteria – root actions. Surely the initial farmers had observed the beneficial results of crop rotation as they had the two crops at hand. Animal husbandry occurred by the domestication of wild sheep, goats, pigs and cattle. These animals could be herded especially sheep and cattle or be kept in enclosures such as often was the case for pigs or goats. We do not have a very good idea of the spatial distribution of animals and farming practices relative to dwellings and other more distant areas. Perhaps the herders, who preceded the farmers stayed out in the larger spaces beyond the concentrations of the towns or other population concentrations bringing their produce home at different intervals. The information we have on animal use is of course from population centers, villages or towns, as is that for the use of plants, although palynology could be practiced in non-inhabited areas but there would be great problems in dating the materials outside of inhabited sectors. Hence we know about life in areas of human concentration but less in areas of temporary habitation or the "open spaces". However the general landscape seems to have been one composed of grasses upon which the sheep and goats could graze.

The climate in this period of discovery was Pre boreal in Europe, with temperatures similar to those today but with slightly less rainfall (Calwe et al., 2007; Watson, 1991; Bell and Walker, 1992) following the end of the last glacial effects 134000–9000 BC. The boreal climate favored forest growth in Europe with cooler and wetter periods followed by a period of Atlantic climate type (warmer) in the period 7000–3800 BC. These climates reigned in Europe for the period which concerns us. The climate in the Near East was such that savanna with scattered oak patches dominated by grasses, becoming more or less humid as climatic patterns shifted. Thus there was a contrast between the Near Eastern area of Neolithic agricultural discovery and the European area. Somewhat minor climatic fluctuations appear to have affected the Near East slightly modifying the proportion of grass to scattered forest. Some authors feel that such fluctuations caused the migration of Neolithic practices (Maellart, 1975, p. 69). The area of interest to us is that of present day Greece and Turkey which is the obvious passage from the traditional Near East and the Fertile Crescent (Palestinian coast and Tigris – Euphrates river systems) culture to the forests of Europe.

In the Near East just where and how the initial settlements of hunter gatherers and herders settled down and why is open to question. Surely sedentarisation occurred during a time of abundance in food supply (Bell and Walker, 1992) both for animals and man. The economy was based upon a strong supply of prairie plant materials. Numerous archaeologists pose the question as to why man decided to settle down to farm, a seemingly more difficult task than roaming around to hunt and have the women gather. If one considers it, hunting is lots of fun, running around after game and killing for a living. Farming means staying at home, i.e. one place, and working each day in the same place. However, if one looks at it from the stand point of women, hunting means picking up all of your belongings almost daily, moving a lot and breaking things along the way. Sedentary living means that one is able to use and keep those nice ceramic vessels that one just recently learned how to make that do so well to cook the grains harvested and transformed into more nourishing food stuff. Perhaps the forces of civilisation were driven by feminine good sense.

In any event it happened, gradually over a period of several thousand years that Near Eastern communities formed and prospered based upon agriculture, vegetal and animal food production. Possibly as important, among the early crops were fiber bearing plants, hemp, cotton and flax, which could be used to make clothing and utilitarian articles such ropes, string and cords. The use of pottery was un-even in chronology compared to the agricultural process. Nevertheless it became a staple of the Neolithic civilisation centers by the time of its spread to other regions. In particular the fertile crescent area, Palestine-Syria and the Tigris Euphrates regions, the core areas of neolithic development, began to expand or diffuse the new technologies to the north and west into the area of present day Turkey. This in the period around 8000 BC and by 6500 BC or so had reached the eastern parts of Greece (Ruddiman, 2005; Masurié de Keroualin, 2003). At this point climate changed and forest began to encroach on prairie – steppe vegetation moving eastward, the boreal climatic period (Bell and Walker, 1992). Now things became more difficult in that expansion of the "method" faced a new biotope, the forest. Grasses do not like acid soils in that they most often allow a build up of aluminium ions in the soil pore solutions. Most of the cultivated species have a low toxicity limit (see Sect. 5.4). Of course not all trees are alike, nor are all soil substrates the same. Conifers tend to produce the lowest pH values in soils and show the presence of aluminium hydroxyl interlayer ion clays (HI minerals) which indicate high alumina content of soil solutions. Oaks tend to be less acidifying and so forth. Nevertheless, forest soils are typically of lower pH, lower organic content and less rich in clay minerals. The characteristics of the A horizon of forest and prairie soils are quite different. If Neolithic man wished to move further onto the European continent he had to find ways to change the basic soil characteristics of the forests that had recently established themselves there. This task was accomplished but only over significant periods of time. Progression was steady to the northwest (Fig. 6.1) where the conquest of the forested areas was probably guided by topographic – geomorphologic situations. Agriculture is most easily effected in areas where the soil contains few rocks and is relatively loose. Such is the case for loess soils, based upon fine grained dust deposits and sediments accumulated along stream or river valleys as terraces and along the shores of lakes. These were the zones or pathways of penetration for the Danubian civilisation moving northward around the alpine massif moving roughly at about 600 km/1,000 years. This rate of progress suggests cultural change more than conquest. Just why the daunting task of changing forest into prairie was adopted remains somewhat of a mystery. Archaeologists still debate the subject. In any event it happened and one should ask the question of how.

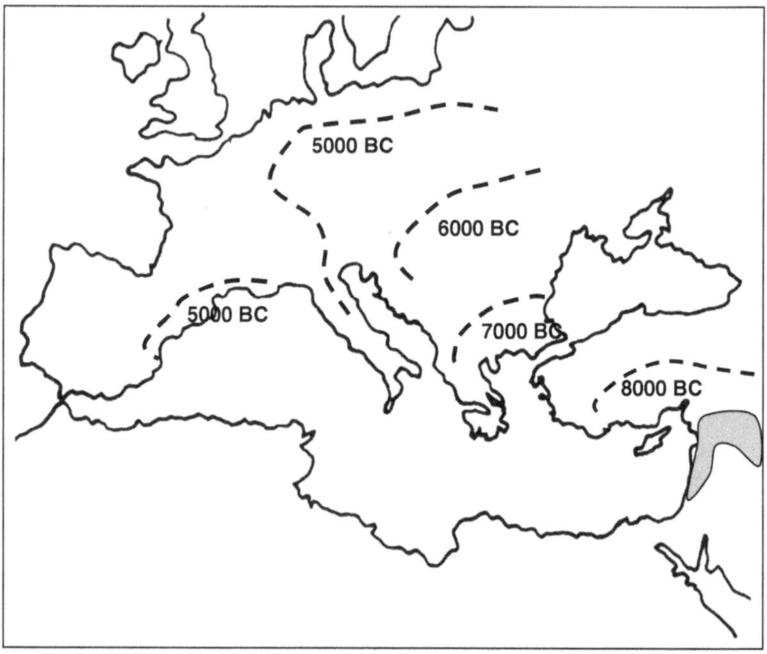

Fig. 6.1 Spread of Neolithic culture across Europe (adapted from data and representations in Ruddiman, 2005; Masurié de Keroualin, 2003; Calwe et al., 2007; Watson, 1991; Bell and Walker, 1992; Maellart, 1975). Note the rather slow extension of the Neolithic culture, which was based upon agriculture. The movement to the north and west from the Near East was toward forested zones originating in the initial steppes of the Fertile crescent area of the Near East

6.1.2.2 The European Experience: Changing Forest Soils into Prairie Soils and the Neolithic Revolution

In the list of first plants domesticated from the prairie areas one finds grasses or *graminae* such as wheat and oats as well as lentils and peas. Thus at the same time Neolithic man domesticated grain producing grasses as well as the legumes which have the property of accumulating nitrogen in quantity in the soil zone. Thus a simple crop rotation will enhance the fertility of the soil. Undoubtedly this was not lost on these first farmers. The modern skills of eliminating trees and inducing prairie – based farming were acquired very early. The technological advance which made such forest clearing possible was the polished stone used as an axe. Thus the first major tool for the neolithisation of Europe was probably this instrument shown in Fig. 6.2. Simple enough but not all that easy to conceive. The idea of shaping stone was very old at this stage. Cutting instruments and spear or arrow

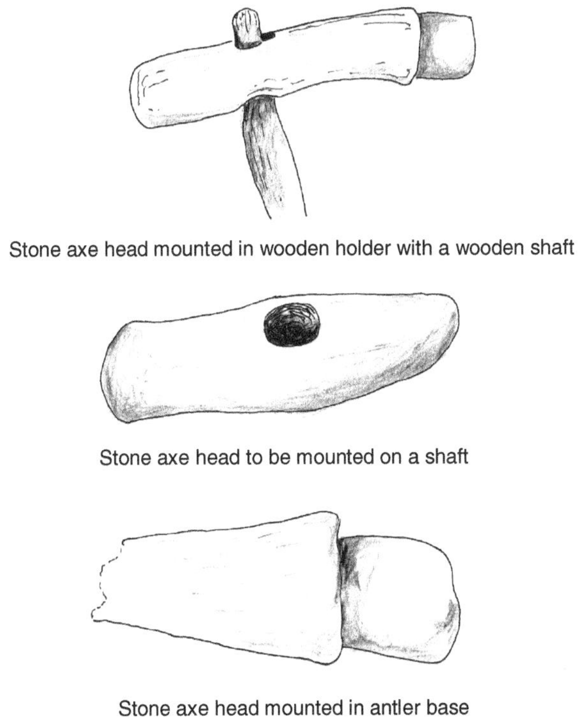

Fig. 6.2 Sketches of the different types of Neolithic axes one has found in different areas of Europe. The polished stone is the technical advance that allows one to chop down trees and to use the fallen wood to better advantage than by simple burning of the trees. Cutting and stacking allows a concentration of wood ash which can then be proe efficiently distributed on the land. Also cutting the branches and trunks allow one to build houses, cut firewood among other things. The stone axe was the key to de-forestation in a systematic manner and the eventual exploitation of the land for agriculture (sketch BV)

points were quite common. But the smoothed surface on a more robust piece of stone was much better for the work of percussion. Such stones could not be shattered or chipped but needed to be smoothed and ground down into the shape desired. Thus the cutting edge was less sharp but much more robust against the impact of a tree trunk. Accessorily, the same instrument could be fixed on a longer handle to break and cultivate the earth. It is still used in some parts of the world. Cutting trees and breaking the ground was possible using the same type of instrument. The importance of cutting a tree down is that one can more efficiently burn it to a complete state of ash. If burned standing, most trees leave a charred trunk, which is not easy to farm around and which losses a great amount of potential ash material for fertilizing.

But how does one change the fundamental chemical parameters of forest soils, their chemistry, which hinder prairie plant growth? This was the fundamental problem of progression of farming into the European forest areas. The initial step is to eliminate the trees, by cutting them down. This act then gives light and eliminates competition for soil resources to low growing plants. Edwards (1993) and Clark (1955) indicate the steps used to eliminate the overgrowth and in certain circumstances use of the felled trees. When a forest is cleared, it is necessary to dispose of the fallen wood. The easiest method is to burn it. This produces alkali – rich wood ash which when spread over the cleared area increases the pH. Soil pH values were changed from 4.8 to 7.8 in an experiment by Firman (1986) when wood ash was spread on the forest soil from which they came. Also this method increases the potassium content of the soil zone. In the experiment exchangeable potassium (by cation exchange) changed from 38 to 1,200 milli equivalents of charge but dropped to 370 milli equivalents after the first year's harvest. This is a source of increased fertility for crops coming from prairie biomes, but it appears that it will be rapidly exhausted, usually after about 7 years of culture according to Clark (1955). This leads to the slash and burn type of agriculture where cleared areas are cultivated for relatively short periods and abandoned for new land, creating another type of nomadism. If the land is put into pasture, animals browsing on the vegetation, the plants which remain and grow are of the prairie type, grasses and various other low growing non-woody plants, which provide sustenance for domestic animals, sheep, goats and eventually cattle. Browsing maintains a higher pH level, decreases the high aluminium content of pore waters due to the forest type humic matter and builds up the soil type that fosters prairie type domestic plants. The key factor is the elimination of forest plants and the maintenance of prairie plants which will eventually build soil characteristics compatible with crops. Thus it would seem that the cutting and burning of trees is a first step to change in soil chemistry but the second fundamental and more long – lasting step is that of pasturage with a continued strengthening of the chemical properties of prairie soils necessary for cropping. Hence the initial stages of clearing the land were based upon the use of domestic animals to maintain the new biome. Clark (1955) shows that the system of cutting, burning and grazing to engender farmable land was successful for Neolithic farmers in deciduous forest situations but not strong enough to transform the conifer forests of the north in Norway and Sweden.

Miles (1985) gives examples of the effect of burning and browsing of forest soils, first creating a less acid Podsol and then forming a prairie – type soil. This is accomplished by the preservation and encouragement of grass and fern growth on the grazed forest soils. The pH increases and the type of organic matter changes being more like that of the prairie vegetation.

The initial, long term conquest of the forest to produce stable farmable land was not made just with the polished stone axe (see Fig. 6.2) but a judicious use of the wood ash and a succession of grazing which fostered the development of a prairie soil type. The advantage of such a system was that it was applied in areas of greater rainfall than the steppes, so that the planted crops would benefit more from their managed culture. A second advantage of farming in former forest land was that there was always the possibility of developing a wood lot. If one wishes to cook the food one produces and to make pots of ceramic materials to cook it in, one needs fuel, in the form of wood. In a natural wood land this is easy to do whereas in a steppe land it is more difficult. Thus it made sense to go to the forests to obtain a long lived sustenance for permanent communities. Not only did one have a supply of wood, forage for animals and eventually a good land for farming but the land was relatively un-exploited with a low intensity of habitation: the "open west" to a certain extent.

A very important technological innovation occurred in the late Neolithic, not very well documented in most cases but on some cave walls one finds a representation of the use of cattle to pull plows (Fig. 6.3).

The details of agricultural practice in the Neolithic (pre-historic) period are not well documented either, only the types of plants cultivated, the loss of forest trees (pollen records) and development of animals, mainly for food (carpology) can be stated with certainty. However the use of animals for traction can be determined by bone deformation. One can assume that use of oxen (cattle) could well have been for pulling a plow over the ground to replace human energy. The first evidence of such agrarian systems occurs some 3,000 years BC in Alpine cave paintings (Guilaine, 1989, p. 126). Such a step most likely created the necessity of keeping these animals close to the habitations, and hence created a ready source of manure. Clark (1955, p. 196) indicates this necessity and use for communities at a slightly later date (perhaps 1000 BC and later further north) for areas to the north of the Alps in the deeper forests. The earlier models seem to be present in the early bronze age in the Mediterranean area and iron age in northern Europe. It appears that there is an association of metal working with the appearance of the ox – drawn plow. This is reasonable enough in the problem of fixing a sufficiently resistant stone on a plow point was surely not a trivial problem.

The presence of work animals is of double importance, one for a multiplied capacity to work the earth, digging and conditioning it. A second is the presence of stabled animals and a source of manure. Manure as we know (or will see in Sect. 6.2.3) is a very powerful fertilizing agent and one that keeps fertility of the prairie soils at an optimum despite the extraction of soil resources in the proceeds of the activity. The use of animal manure to maintain or increase fertility is of course a major advance in agricultural practice. It is apparently of quite ancient use then,

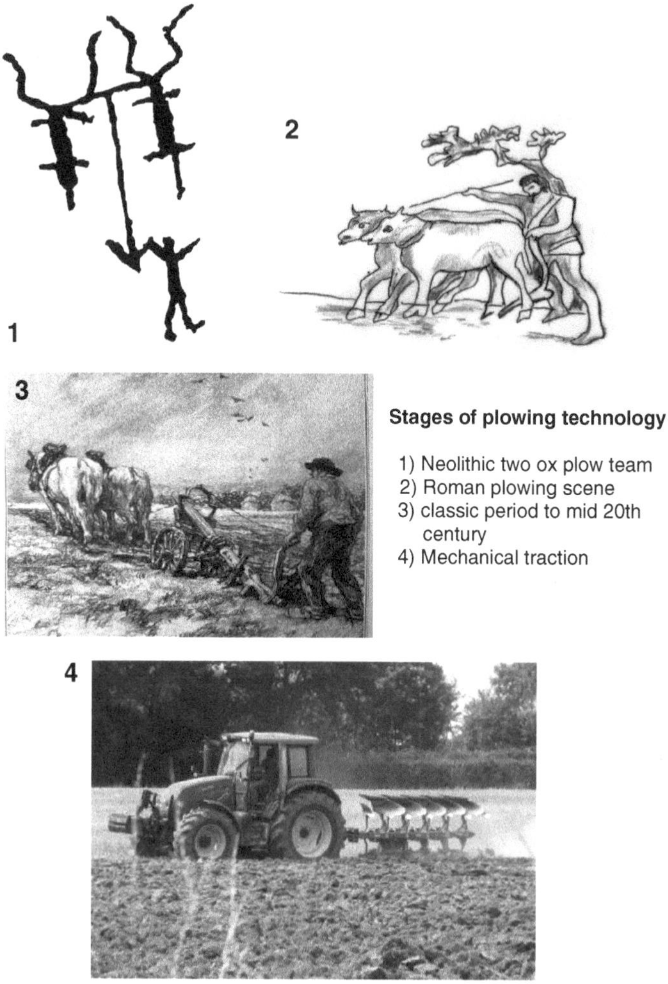

Fig. 6.3 Stages in the methods of plowing the earth with a different types of traction. (1) Initial stages shown in a sketch of a late – Neolithic cave painting. Note the two oxen team used to pull the plow, adapted from Duby and Wallon (1975) (2) Sketch of a mosaic representation of Roman plowing techniques (sketch, BV, of a floor mosaic, National Museum of Antiquities Saint Germain-en – Laye, France (3) Plowing scene from the late 19th century in France, Leon Lhermite (4) advent of motorized agriculture, late 20th century (photo DV)

being practiced for 5,000 years or more. In this way the foundation of the renewal and maintenance of crop land fertility was established in Europe with the change in soil chemistry by forest cutting and the introduction of prairie growths using animals grazing. The use of animal traction and use of manure finished the gains made by potash fertilizer use, and crop rotation. The fundamentals of agriculture as we knew it up to the mid 20th century had been acquired.

A second vector of neolitithisation occurred. Instead of going north and turning around the alpine and associated massifs, a vector of Mediterranean coastal change occurred. along the Italian coast and along the French eventually reaching Spain. Phillips (1982) indicates the steps in this process along the French coast. Here the initial sites of settlement agriculture and domestication or use of domestic animals was first practiced along the coast and in river delta situations, and massif edges but not in the deeper forests. Initial lodgings were in caves and rock shelters. After some time dwellings expanded into open air sites. And around 3500 BC the Neolithic system was installed, farming, pottery, herding and the use of domestic animals. The process was less expansive but followed basically the same pattern as in north eastern Europe, except perhaps that the problem of forest soils was less evident, agriculture being implanted in less vegetated areas along stream bottoms and river deltas. Progress along the coast was clearly accomplished but penetration into the land mass was much slower that the corresponding movement to the north of the Alps.

Kleber et al. (2003) found that the best way to identify the agricultural activity of Neolithic farmers in loessic soils of Germany was to analyze the soils for potassium content and remains of charcoal. The mark of the new agriculture in forest soils was one of wood transformation to potash bearing ashes which was maintained in the surface horizons by the development of prairie soils through the use of grazing animals and then crop rottion.

6.1.2.3 Agriculture in the Classical Age

The stage is now set for classical agriculture, that lasted in Europe roughly from a thousand years BC up until 1950. We have some written accounts of these systems in ancient Greek texts, quoted by de Serres (1600) and Toutain (1927) where Pline, Virgil, Columelle and Plutarch among others writing in the period of 300 BC 100 BC indicate that crop rotation and fallowing were approved practices, manure and green fertilizing was also important. Using lime is also mentioned. Burning in the fields in autumn was used as a control of weeds and pests. However by this time climate change, warmer and drier, and probable intensive agriculture made it necessary for the Greeks to resort to importation of grain from other Mediterranean regions and the Scythian plains.

North of the Alps and to the west, iron age agriculture was practiced in much the same way. It appears that fallow – cultivation cycles were practiced (Ferdiére et al., 2006). It also appears that planting legume type crops was not practiced as much as it was to the south. From archaeological evidence the use of an ox pulled plow was common, using a metal pointed plow shear. Also different types of hand tools probably used for hoeing and weed control have been unearthed. The initial heavy work done by animals and the more detailed and specific work being done by man. Animals became an intimate part of the daily farming practice. The use of fibre and animal skin was certainly of great importance. In the early Iron age (700–400 BC) it appears that a rather wide diversity of grains was used, perhaps to counter climatic influences on the success of a single type of crop. Later there is a tendency to have only one type of grain present. Here one finds evidence of cutting the prairie to make hay, which indicates the presence of animals in enclosures over the winter period. This reflects the writings of classical Greece which indicate the use of manure in the fields. Also the use of lime or carbonate rock on acid fields was practiced. The abundance of nitro – philic plants, or weeds (nettles for example) indicates the use of manure on agricultural fields.

Here we have a mastery of the concept of crop rotation and fallow, where the extraction of mineral resources from the soil is compensated in part by additions from below the soil zone (deeper root penetration and uplift of nutrients) and a period of time where plants and bio-agents could extract elements from mineral sources available but of slow release rate in the soil zone itself. Legumes were used to enhance the nitrogen content of the soils, and corrections for natural acidity were made using lime and calcareous rock. A portion of the plants taken from the soil were re-employed in the form of fertilizer, rich in potassium and also in ammonium which aided the bacterial increase in the soils. It appears that in the Classical age the means of plant manipulation were clearly mastered, using the resources of animal and other transformations of natural and plant material from the cultivated land.

Diversification and specialization occurred, trees and fruits were present and vineyards became the source of a large amount of commerce between different parts of the European world. Agriculture became a longer range commercial affair. Better transportation, using wheeled vehicles on paved roads (Roman) allowed and encouraged exchange and specialization of agriculture. Wine amphora of various regions found virtually all over the Empire attest to this. Instead of being tied to local conditions, use of the land in one area could influence the use of its products in another.

It seems that the mastery of plant manipulation and the use of animal by – products had become mastered techniques before our era. The transformation of forest soils into those useful for prairie plants was accomplished and maintained by grazing, prairie and fallow cycles, hay cutting and manure use. Wood ash was employed as a source of potassium and a means of increasing pH and lime was used to this effect also. These measures were adopted to change the chemistry of the soils so that plants of prairie origin could prosper. The forest lands were those of higher rainfall than that of the natural prairie – steppes, and thus the output of agricultural land was increased by installation in more northern areas than those of the Near East. The climate change of drier and warmer conditions experienced in the period of near 2000 BC in the Near East (1994) creating the collapse of the great bronze age civilization centers was a final push for development of these wetter northern sources which had already succumbed to the Neolithic revolution. The centers of power shifted as the Mediterranean was subjected to a more Mediterranean drier climate.

The agricultural cycles would then be gradually adapted and improved using different methods of manipulation of plant soil relations.

6.1.2.4 Agriculture in the Post Middle Ages in Europe

The continuation of agriculture was practised after some interruption during what are called the dark ages, the period of about 700–1000 AD. Significant re-organization and better practises were accomplished during the times of Charlemagne, but progress was slow. Fragmentation of land resources and markets did not encourage efficient production. For a long period land owners essentially owned the working class on their lands, the feudal system, which did not encourage production. However gradual sales of land to peasants and the use of money encouraged a surplus economy of agricultural production (Duby, 1962). In the 12–13th centuries things got better, land was reclaimed from forests developed on abandoned areas after economic stagnation and as a result agriculture progressed. Crops were more diversified, sale of animals and products in local markets gave a bit of relief to the limitations of isolated production and reliance on individual resources. Markets meant that a good crop would be stored for same at a later time, and not only at the profit of the land owner be it the local lord or the church, which owned and had much land farmed.

The structure of agriculture in the 16th century was one of diversity, relative, and a more market oriented economy. The following is based upon the account of agricultural practice given by Olivier de Serres in his study commissioned by King of France and published near the end of the 16th century as well as details of 17th century agricultural practice in Center Western France (Merle, 1958).

Movement of agricultural products was important in certain regions where an excess over consumption was possible as was generally the case after the periods of plague and ensuing wars of the 14th and 15th centuries. Wine was produced for export outside of the local consuming area, grain moved over some distances and useful items of manufacture found their ways into farm households. However, the farmer, and people writing in his stead, proposed numerous ways of managing natural forces and of manipulating nature. The old methods of crop rotation with fallow and prairie use were institutionalized. Cycles were established which often took into account the local vegetation and geological (soil chemistry) constraints. Here we have descriptions and reasons given for specific agricultural practices which have become lost to us to a large extent today.

Judicious Use of Fire

The use of fire was very important to farming in the period considered (Sigault, 1975). It had been practices in various ways before, clearing land, providing wood ash as fertilizer and a means of changing pH, and one of controlling insect, bacterial and other pests. Thus burning un-need straw and plant refuse in the fields was an advised practice. At times and when available understory plants such as ferns were gathered, placed over the fields and burned. Here the use of fire for prophylactic purposes was evident but also the understory growth in forests is well known as a source of potassium. In fact during these periods ferns were used as a source of potash in glass making in France. Burning these materials on the soil then served two ends addition of fertilizer and, equally important, pest control. Weed seeds were destroyed and pests were largely eliminated from the soil surface.

The practice of burning was also used in a slightly modified manner to another end. The use of prairies was common, one where land was but into prairie for 6 years or more in order to restore fertility for grain crops. The land was of course not left un-tended, animals were grazed which was an important source of income for the polyvalent farmer as most were. In these times specialization was rare except in mountain areas where grains could not be successfully grown. The problem with a prairie is that it leaves a very dense root mass and a lot of weed seeds behind. If this is plowed and seeded, most of the crop is unwanted material. The solution to this problem was fire. However one needed to reduce the root mass and disaggregate it. The method is called *ecoubouage* in French, practised in most European countries north of the Alps.

The system consists of stripping the sod from the prairie and reducing it to ash and charcoal by slow burning. The sod is piled into mounds and set fire slowly from the center. This is the same method used to produce charcoal. Here the root mass is reduced greatly and totally de-structured. The clays are heated to temperatures high enough to de-hydrate them and rupture the crystal structures thus releasing the potassium held in mica flakes. The roots that become charcoal present a highly active surface that can fix cations and release them increasing the cation exchange capacity of the soil. In recent "discoveries" based upon observation sin Brazilian steppe – forest areas, this system is called "biochar", where there is a distillation of organic matter to produce charcoal and recuperate the organic gasses to be used as fuel. This latter aspect of course escaped the farmers of the middle ages to modern times in that they had no vehicles that needed to be propelled using paleo – biofuels. Since petroleum and the internal combustion engine were to be found in the future, the farmers let the organic gasses escape, but as slowly as possible in order to gain a maximum of carbon in the residual matter.

Once burned, the residue is spread over the field and it is ready to be plowed for a grain crop. Numerous tools were developed for this process, starting with hand tools similar to flat hoes and eventually plow like wheeled tools were developed to slice under the root structures and loosen them for stripping.

Thus not only did one produce ash fertilizer, de-structured clays and charcoal but also the insects in the soil and bacterial and other diseases harboured there were destroyed. This soil could then be planted with significant hindrance due to the development of unwanted materials. One must remember that weeds were the greatest enemy of the classical farmer. This is why fallow fields were not left to the installation of new plants but were plowed during the year and weeded by successive cultivation.

Use of fire was then an integral part of agriculture for a long period, mentioned by classical Greek authors and continually practised until late modern times.

Crop Rotations and Green Manuring

Again, the ancient Greeks (those of the classical period at least, around 300 BC) found that the use of leguminous plants to enrich the soil was a good idea. In fact the methods persisted into the 17th century where such use is clearly described. Flat beans (feves), alfalpha and peas were turned into the earth before maturity in order to enrich the soil. This of course increased the nitrogen content since these plants are known for their symbiotic use of bacterial to fix nitrogen in great quantities, but also the potassium content of the soil. Legumes and especially alfalfa are "luxury consumers" of potassium, bringing up more to the surface than they need for their proper growth. This practice was also used to increase fertility in vineyards where crop rotation is of course difficult. The legumes were sown between the rows of vines and plowed under in mid – season (de Serres, 1600, p. 209). The concept of green manuring is essentially that of using mineral resources brought to the surface by plants in order to increase the abundance of these elements when less efficient extractors are planted such as grain crops. The green manuring is simply the direct transfer of mineral and nitrogen into the A horizon of the soils. The elements concentrated in the plants are returned directly to the soil. Since much of the material in the plants has been taken from different parts of the soil but also alterite horizons, this represents an increase in the soil zone of nitrogen, potassium, calcium and so forth.

These methods were of course combined with various methods of crop rotation. The specifics of changes in plant cycles depended upon the type of terrain at hand and hence no one single crop rotation scheme was proposed in these times. However, it was proposed to plant a grain crop and alternate it with fallow (no plants growing in the fields) and gain several times and then a use of prairie established for more than seven years to increase the nutrients in the soil. Usually the cropping sequence was similar to or less than the time in prairie culture.

Fertilizers

In this category the agricultural writers put mineral as well and vegetal materials. The use of wood ash was still practised, and a certain amount of importation occurred. This changed the pH and the potassium content of soils. Mineral components were mostly designed to change pH, lime and clay – rich carbonate rocks were used as modifying agents. The pH value was of course important for the well being of prairie plants, and the preservation of the organic matter which captured and fixed alumina.

The most important fertilizer was that of barnyard manure. This was the golden resource of a farm. Since the working animals and others were stabled in the winter, and some for other parts of the year such as milk cows, this material was abundant and extremely efficient in transferring ammonia and potassium to plants. Since animals do not retain potassium and not much nitrogen, the residue going through their systems is deposited on the stable floor. The easiest way to clean it up is to put a layer of straw down before the animals come in the stable. This mixture of straw and animal excretions is a very, very powerful agent when applied to soils with a reasonable cation fixing capacity. Its use of course dates from the post Neolithic times, roughly the beginning of the Bronze Age when oxen were used for pulling plows and farming had settled down for long periods of time in the same place. Returning much of the mineral resources to the soil and extracting only the grain was an important procedure. Russell (1945) shows that the effects of farmyard manures are strong in the first year after application diminishing some over the following four or more years. Fallowing increases nitrogen availability for the first year but the effect is negligible after that. In grain cropping one has an average yield 3.5 times greater using manure than not. Taking away all of the plant materials is of course not possible for a sustained agriculture system. One must remember that the farmland of Europe was used for agricultural purposes for about 6,000 years or more before modern methods managed to "wear it out" and exhaust its naturally maintained fertility.

The use of lime, calcium oxide, known from Greek times or earlier, can be associated with problems of pH in changing or modifying forest soil acidity to be more compatible with prairie soils. However, another aspect is or was known under traditional farming methods. Liming was used in conjunction with manuring, after the organic matter was deposed on the fields (Richeyre, 1884, p. 80; Cox, 1925). This action was reputed to make the manure more efficient, releasing ammonium from organics and probably potassium by cation exchange.

Use of wood ash as fertilizer was practiced from Greek times (Sigault, 1975, p. 102) which compensated the extraction of mineral elements taken away by crops. It was practiced until the 19th century.

6.1.3 Summary of Classical Farming Practice

From the beginning much of European farming was concerned with maintaining a prairie type of soil where forests where converted into farming land. The process found its limits in the northern areas where dense conifer forests proved too much for the process uses (Clark, 1955). Crop rotation was used, using legumes (lentils or alfalfa) alternating with grain crops or by leaving the land fallow (plowing and cultivating but not planting it). Nevertheless long periods were needed to re-generate or maintain the soil characteristics of pH, and type of organic matter by putting the land into prairie which was grazed. As animals were used in greater numbers for traction, stabling increased and the resulting manure was spread, used as a re-generative method on all fields especially those that were cropped. Thus the manipulation of plants and use of animals and mineral materials changed the character of the soil from forest to prairie type where graminae plants could grow.

Further manipulation of plants was used in order to maintain a fertility of the soil which could then bear the extraction of natural soil resources, especially mineral ones. Fallow allowed clays to be slightly dissolved by rain water interaction. Prairie, grazed but not cropped, was used to bring mineral resources to the surface from depths by plant root extractions. Crop rotation allowed enrichment of different mineral resources or at least differences in extraction rates from one year to the next.

Manuring was a method of recuperation of a part of the mineral resources extracted and transported from the land. The return in a more assimilable material aided greatly to the renewal of soil fertility year in year out.

Control of pests was accomplished in part by the judicious use of fire to get rid of weed seeds and various pests in the soil. Liming had this effect to a certain extent also (Richeyre, 1884).

The practice of farming changed but little, in concept, from late Neolithic times until the 17th century in Europe. Tools were improved, had hoeing and ground breaking was replaced by ox drawn plows, different metallic tools for hoeing and cutting were developed in the iron age but the essential concept changed but little. Animal – plant relations remained much the same. Crops were much the same and crop rotation was similar depending upon the soil substrate and climate regime. Once the basic pattern was found during the Neolithic expansion the structure of farming and its concept were very similar over the ages. Agriculture was waiting for the scientific revolution.

6.1.4 Science and Agriculture

6.1.4.1 Eighteenth Century Experimentation and Improvements in the Use of Plants

The big problem, for many agronomists in this period, was that the prairie cycle of crop rotation schemes was quite long. About half of the time a parcel of land was under grass and the rest in grain and fallow cycles. If the major product of meat was easily exported and sold, this was fine. But if one needed more durable goods, transportable over longer distances and capable of distribution throughout a year or several years, animals and meat were not practical. However wheat was. An effort was thus made to reduce the prairie regeneration cycle and deliver the agriculture of the period from the cycle of prairie and culture. The remedy was proposed by Henri Pattullo to Madame de Pompadour, a good friend of the French king, in 1763. Undoubtedly he was not alone in finding out the secret but he seems to have wished to have been. However his proposition was a concrete one, more tax money for the crown due to an increase in productivity. This was by no means a small thing, if it could be realized.

Two things needed to be reduced in the cycle of alternating, one the long period of prairie (6–10 years) roughly half of the time and the other the fallow periods alternating with grain crops. Often one had viable crops but only one quarter of the time. Pattullo proposed to reduce the prairie portion of the cycles by introducing new plants: clover, alfalfa or sainfoin. Three successive grain crops and fallow were followed by a crop of rye grass mixed with the above plants. The productivity of the non-grain artificial prairie crop was such that it could nourish more animals than the ordinary prairie. Here then the grain crops were present half of the time and artificial prairie and fallow half. This is an increase which doubles the periods of grain production on the same land. The use of clover, alfalfa or sainfoin depended upon the type of soil, clay – rich, intermediate or sandy. Thus the artificial prairie was adaptable to different types of soils and climates. This was very useful in France where climates change and the geology dictates different types of soil chemistry.

Henri Pattullo gives due credit to his predecessor, Olivier de Serres, in 1600, in that the use of clover, alfalfa and sainfoin was already prescribed but not in such a concentrated way and in such a short succession after grain crops.

This new method or revolution of agricultural practice, doubling the productive cycle for grain crops was one of using the natural resources in a prairie but in a concentrated manner. In the natural prairie one has grasses, legumes, and other plants of less nutritive value. Garola (1918) indicates the extreme variability of natural prairies which depend upon climate and soil substrate. Grasses dominate, but not always, with different proportions of deep rooted plants and legumes. In a natural prairie, and the artificial one of course, when animals are pastured, they leave behind them all of the mineral resources, taking away only nitrogen and of course carbon – hydrogen compounds. Thus grazing of the artificial prairie was recommended. One can compare the relative amounts of nitrogen and potassium in clover, alfalfa and grains (Richeyre, 1884, p. 23):

	N%	K_2O%
Clover	21	19
Alfalfa	22	15
Grains	4	5–6

In fact the new agriculture proposed by Pattullo (1758) was the selection of the deep – rooted, leguminous plants which were present in minor abundance in natural prairies. The selection of the nitrogen – rich and potassium – rich plants, those that concentrate these elements by different methods, was the innovation. Of course animals could well eat grass but the more deep – rooted plants brought more mineral elements to the surface soil zone. Here again one sees manipulation of natural elements to the overall goals of agriculture. Reduction of the long prairie cycle was effected by concentrating the useful plants in the prairie so that their benefit was concentrated and as such could be used to reduce the long period of non grain farming. Along with other uses of methods already known, such as liming, claimed by the English as a remedy to many problems of agricultural production, the system of agriculture was much the same up to the 19th century. However, at that time significant changes in agricultural practice occurred, if not on a massive scale, but on a philosophical scale.

6.1.4.2 Nineteenth Century Innovation Through Chemical Investigation

In the 19th century modern science entered into the agricultural picture in Europe. Chemistry and plant husbandry joined forces in an attempt to ameliorate the economy in a rational and modern way. Agriculture was an early field of application of modern science. The method of investigation was by chemical analysis. First the composition of plants was determined, thus establishing their potential needs. When the different proportions of mineral elements were determined, an analysis of the soil components was made. Then depending upon the potential needs of plants, different fertilizer agents were dissolved and poured over soils. The amount of the different components retained by the soil was measured. Using these determinations the scientists determined what elements could be retained by soils and which could not. The consensus was that potassium, and ammonium (nitrogen) were retained as well as phosphorous. Thus soils could be reservoirs of the vital elements needed for plant growth. If traditional methods of fertilization did not fulfil the needs of a given soil and crop situation, one needed to remedy this by the use of concentrates of the needed elements.

This knowledge initiated the use of artificial fertilizers. Larbaletrier (1891, p. 55) notes that the elements which nourish plants (N, K, P) are fixed in the soil (A horizon or cultivated zone thereof). The logic is to replace those that lack by adding them from external sources. One needs only to analyze the soil chemically and add the needed materials to assure the growth of plants which extract these elements from the soil. This was of course understood implicitly by previous farmers who used plant alternation mechanisms to bring up nutrients to the surface and returned a maximum of nutrients after they had been cycled by animals. Use of ashes increases the potassium content of the soils also.

This being the case scientists began to search for sources of these elements found in nature and methods of rendering them more active chemically in soils (Larbaletrier, 1891). Two dates are important, the early 1840s when the source of phosphorous was associated with geological deposits rich in the mineral apatite, which was rendered chemically active by the Liebig process to make what is called superphosphate. One advantage of such sources is that such natural phosphate deposits (essentially concentrations of animal bones of different sorts often fish remains) are reasonably abundant and widespread in regions with sedimentary rocks of Mesozoic and Tertiary age. Chemical treatment using sulphuric acid in factories allowed widespread use of this fertilizing agent.

The second date is 1861 when the mineral sylvite or potassium chloride was discovered at Halle in Germany. Here no significant chemical treatment was necessary for use and the mineral could be shipped directly to the user. Other deposits were discovered in Europe, in Alsace notably, after this. Thus two of the major mineral fertilizers were made available in a large supply. Previously in the century wood ash from the United States and Russia was imported as fertilizer.

Earlier in the century sodium nitrate, "caliche" evaporate deposits were imported from South America but their price prohibited widespread use. Potassium nitrate, saltpetre, was also known for some time, the 17th century and before, but its relative scarcity made its use more restricted to industrial processes such as glass making.

The use of cheaper, more concentrated agricultural mineral chemical promised much for the development of agriculture. However, the availability of these products did not much change the average productivity curve, at least in France (Fig. 6.4). The trends of average production in one of Europe's greater agricultural countries were not much changed by this new knowledge.

One of the major problems was the relative cost of the new chemicals. Most farmers were too traditional to invest systematically in chemical fertilizer when climatic differences from year to year could wipe out such investment. The gains were not seen as being worth the risk. Thus although the technology said a new way was

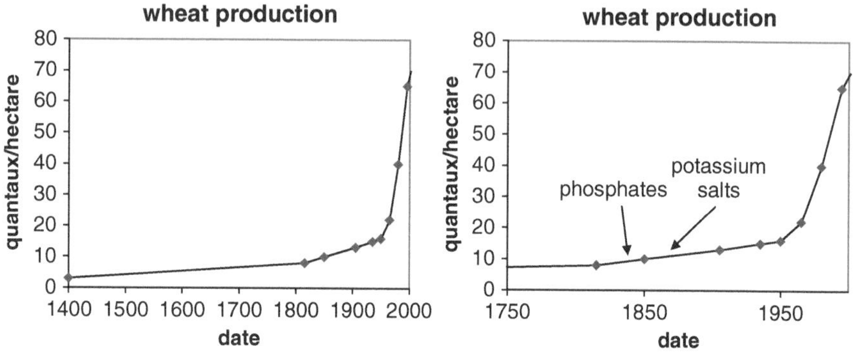

Fig. 6.4 Wheat production in France (data from Duby and Wallon, 1975, 1976a, b). Overall the productivity per surface area changed gradually over the centuries from 1400 to 1900. The discovery and introduction of natural chemical fertilizers (potassium and phosphorous) did not the average productivity significantly in the 19th century. Major changes in productivity occurred in the latter half of the 20th century

possible, major change did not occur. The productivity increased by general better management over the period 1750–1950 by about 2.5 times all the same, in a gradual trend. There was no great incentive to invest more in the production of a crop if the value was not a significant compensation. Population pressures were not great yet.

6.1.4.3 The Truly Revolutionary Agricultural Revolution

From the earliest times agriculture in Europe rested upon the same sets of knowledge, managing and manipulating plants and soils. The crop rotation system, the use of a fallow period and the re-generation of the basic prairie soil chemistry by alternating crops with prairie natural or artificial, were essentially natural practices. Recuperation of chemical elements by the use of manure was also in this tradition. In the United States mechanization in the form of replacing animals with motor driven vehicles began before the Second World War but in Europe the change took place massively in the 1950s. Perhaps the Marshall Plan had an influence in that money was available to buy American products, such as tractors. The effect was stunning for many reasons. In Fig. 6.5 one sees the change in the number of tractors in France.

The tractor population increased seven fold in 20 years while the cultivated surface area decreased by only 10%. Thus one sees an intensification of agricultural effort. The results are rather evident in looking at the average productivity for wheat in France over the period (Fig. 6.4).

Wheat productivity increases 4.5 times. Since the surface area did not increase but slightly decreased, gain in productivity is due to the use of machines but more so the use of artificial fertilizers. In Fig. 6.6 one sees that the use of ammonium was greatly increased also.

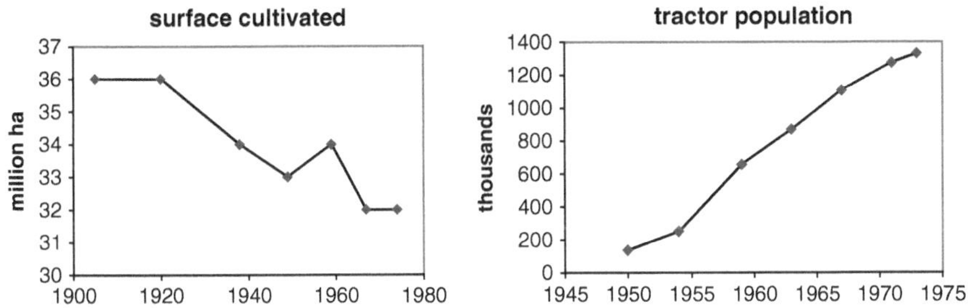

Fig. 6.5 Graphs of tractor population and cultivated area in France (data from Duby and Wallon, 1975, 1976a, b)

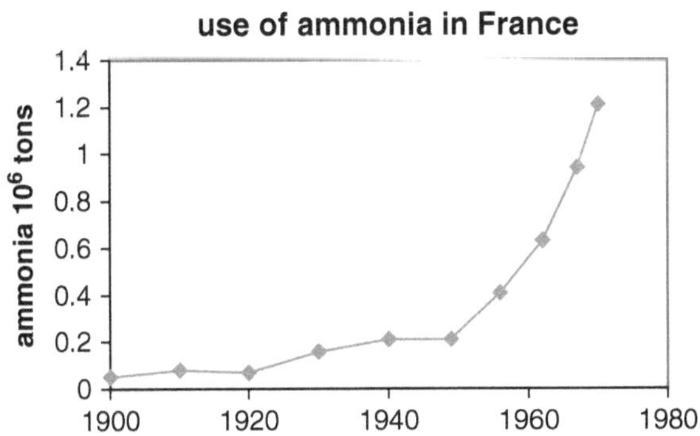

Fig. 6.6 Use of ammonia in French agricultural practice

Artificially captured nitrogen was a necessary additive in that the traditional source, manure, essentially no longer existed in that traction animals, horses and oxen, were replaced by non-manure producing machines. Since draught animals used about 25% of the productivity of a farm before the mechanical revolution, this effectively increased the cultivated land areas but the fertilizer that they provided had to be replaced, and in this case it was by factory transformation of nitrogen from the atmosphere into a stable chemical form, ammonia or ammonium salts. All of these processes, mechanical traction and artificial fertilizers, were obtained by the use of fossil energy, or is some cases water driven electricity production such as in the Tennessee Valley Authority plants in the United States. Here the use of external materials and chemicals radically changed the relations of plants and soils. In a sense the modern world was born.

Looking again at data for agricultural France, one can compare the agricultural productivity with population growth. Traditionally since the 13th century of so the population of France held near 20 million people. However the 18th century saw a steady and eventually dramatic change in population (Fig. 6.7). The curve increases very dramatically only to be slightly interrupted by the period of 20th century wars but has resumed since that time, especially in the period 1950–1970. If one considers the productivity and populations in this period it is clear that the rather gradual

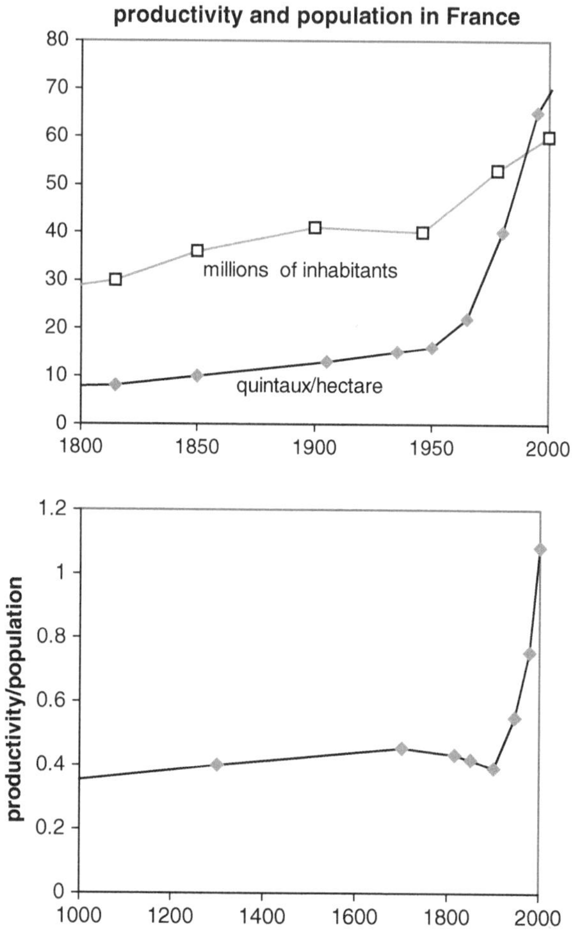

Fig. 6.7 Comparison of the productivity of wheat and the population of France in the 20th century. (Duby and Wallon, 1976a, b)

increase in productivity before 1950 more or less kept up with population increase. The graph shows crop production divided by population. The explosive growth in production made it possible to eat differently. As is generally the case today, much of the crop production went to the production of meat. Thus diet changed as more was produced from the land.

The last part of the agricultural saga in Europe shifts out of the manipulation and fostering cycle of plant soil inter – relationships to one of a type of industrial use of the soil. Since the replacement of fertility usually practised using animal by – products, manure, was no longer possible and was replaced by powerful fertilizers, the increase in plant growth seemed boundless. Traditional grain production in the Middle Ages was on the order of 6 grains harvested for one planted. Today corn production in the United States in on the order of 280 grains harvested to one planted.

There seemed to be no reason to alter this type of development. Nature was and is essentially ignored in agricultural practise.

6.1.5 The American Experience: A Counter Example of Soil Use in the Classical Period

The New World of the 17–18th centuries on the North American continent provided apparently unlimited fertile soil for farming. The major reason for immigration was for farming land which was in apparently short supply in the Old World. The methods of farming were initially probably relatively close to those practised in Europe but apparently rather rapidly they were modified. It seems that the several millennia of farming and apprentiship on European soil had been forgotten in the land of plenty. An English agronomist, Arthur Young wrote to George Washington concerning the economics of the American experience. He wanted to know what enticed or motivated the Americans to use so much land so quickly. The letters of George Washington (1801) are a fascinating response to the questions of Mr Young. Not only did Washington respond with accounts of his estate at Mount Vernon but he asked several of his colleagues, Thomas Jefferson, Alexander Hamilton, among others, about the economics of farming and we have these letters also. Young was interested to see if slavery, the cost of labor, had a role in the methods used, but Washington was convinced that the abundance of new and very cheap land was at fault.

From these letters we have a good description of farming methods on the East Coast of the United States. One must remember that the area farmed which Washington discusses was based upon deciduous forest soils. It was not until the 19th century that American farming got onto the very rich prairie land which responded differently to the farming practices of the period. On the East Coast, deciduous forest was dominant and gave rise to reasonable fertile soils. They were comparable to the types found in many places in England. Therefore the comparison sought by Arthur Young was an apt one.

Essentially Washington described what was going around him and had so for several generations. He describes the use of land from the time it was bought and then cleared for farming. This indicates essentially virgin land in many areas of his surroundings. Typically farmers used the land in three parts: cleared and farmed, parts being de-forested and land to be cleared. Initial farming of wheat gave roughly 20 bushels/acre (some 300 kg/ha). The land was cleared and the wood seems to have been used as such or burned but no mention is made of using wood ash as a fertilizing agent in the discussion of Washington. However he does mention slash and burn practices which indicate at least some return of wood ash materials to the land in the first stages of culture. This initial productivity of 20 bushels could be increased to 30 or 40 bushels with vigorous manuring and crop management.

The crops of most new farmers, the more recent immigrants of whom Washington did not especially approve, seemed to have been wheat and "indian" corn or in some places tobacco. All three were reputed to extract native fertility rapidly. Frequently the tobacco land was planted for two consecutive years followed by wheat or corn. After several cycles the productivity was reduced to 10 bushels an acre. The newly cleared land was then put into production. This wasteful method of farming seems to have been especially practised in the southern States (tobacco country) where as in Pennsylvania for example, more conservative farming practises were employed. The use of legume crops and artificial prairies was not practised on the tobacco lands. As Washington states, after this initial cycle of "scratch" farming, one has two choices: either resort to crop alternation and manuring or to move on, over the mountains to the west where more cheap and plentiful land awaited. Washington himself, in writing to Arthur Young asked for the best grades of clover and alternative prairie seed to be sent to him, and thanked his colleague when such was received. Animal manure was a vital part of his agricultural practices, green mauring also and grazing cattle on either natural prairie or planted prairie as part of the rotation scheme.

The practice of rapid clearing of virgin forest and apparent small use of the wood ash resource is mentioned by Hall (1829, p. 71) when he describes the spectre of girdled trees, dead and uncut in the farmlands of north western New York State. Land was cheap to new farmers accustomed to much higher prices in Europe, and as such its use was much neglected. A century later Cox (1925) devotes a Chap. to growing crops in well-planned rotations. Chap. VII tells how to Maintain the Fertility of the Soil by crop rotation, growing legumes (alfalfa, clover, etc.) use manure, lime and phosphate, potash and ammonium fertilizers. All but the last are of classical Neolithic practice, and rock or refined mineral resources from the 19th century.

However here in the early 19th century one is more worried about the fertility of the plains areas or woodland – prairie soils more than the deep oak forests of the East Coast. In looking at the maps of corn and wheat acreage planted in the United States in the 1920s it is clear that much of the agricultural land of the time of George Washington had been abandoned. Some land was put into production on the West Coast in California and especially Washington State and Oregon. Even though Washington demonstrated how to re-gain soil fertility, much of the land given to agriculture was abandoned on the East Coast and is now in forest. Curiously, the forests of the United States supplied wood ash to farmers in the early 19th century before mineral potash sources were used in the 1850s. This material did not appear to be used to re-generate fertility on American East Coast farms of low productivity, combined with manure of course for ammonium.

It seems clear that it is quite possible to reduce the fertility of land by using crops that extract much of the mineral nutrients. It is not that such land is no longer useful. Washington insists that by careful planning and use of manure, one can restore worn out tobacco land to a high fertility, three times that of the depleted state. In Table 6.1 we indicate the relative fertility in kg/ha in the time of George Washington.

It is interesting to note that virgin land can have a still higher productivity when using crop rotation and manure as agricultural practice. This suggests that the method developed by the Neolithic farmers in converting forest to grain agriculture

Table 6.1 Production in kg/ha using different farming practices

New land recently Cleared	Manured and crop rotations old or new land	High use without return to the land
300	450	150
Wheat 1 in 3 years	Wheat 1 in 2 years manure	Wheat 1 in 2 years
300	220	150

was more efficient and productive than using just the steppe lands for grain farming. The necessity of moving into forests can possibly be explained by an increase in productivity over the initial farming on the steppes, without an intensive use of stabled animals which produce the much treasured manure. The important point is that poorly used land can be put back into normal productivity by traditional practices. Washington was adamant on this point. No need to move over the mountains to make a living. One just needed to do some hard work, or at least spend the resources necessary to keep up a decent rate of production.

6.1.6 Change in Agricultural Land Use

Modern agriculture changes the landscape and the soil upon which it is based. Tractors become larger and larger with time and as such they are less manoeuvrable on the land. They cannot move up hill nor can they move along a slope with ease. More and more land is changed into pasture or wood lot. An example is given in land use in France between the mid 19th century and the 20th century (Figs. 6.8 and 6.9). On these sketch maps one finds cultivated, prairie and woods or other non-farm lands. Of course at the scale of the maps, many exceptions are excluded from the general trends, but overall the information is instructive. In fact the large areas of farmed lands indicate small farms of multi-crop traditions, where the crop rotation and prairie schemes prevailed. Draft animals were used on hill sides and slopes and thus the factor of topography was minimized. However the use of the tractor, especially important after 1950, had he effect of suppressing the smaller cultivated areas favoring the creation of large fields which were more efficiently farmed using motor vehicles.

Hence a new era was born and land geographic areas created based upon topography and geology (soil substrate type). If on looks at the map, the area of Normandy (central northern peninsula) changed from a multi – use agriculture area to one dominated by grazed prairies hence the importance of the camembert cheese today. The Brittany peninsula (western France) was converted from the same type of farming to an area of enclosed farming of chickens and pigs. Thus changes in overall patterns in farming practice have affected France but of course other countries. For example, the traditional beef cattle cycle between the Midwest and Western states in the United States is largely ended, that where cows were born and bred on the high planes grasslands of the west, shipped to the corn belt of the Midwest after several

Fig. 6.8 Land use in the first half of the 19th century. U-C indicates basically un-cultivated areas.(Ministère Agriculture, France 2001)

Fig. 6.9 Land use in the last half of the 20th century in France. (Ministère Agriculture, France 2001)

years to be fattened and rounded out and then slaughtered in specialized units near or in the big cites of the Midwest such as Chicago.

In fact, the presence of farm animals has not decreased but probably greatly increased in the last half of the 20th century. Roughly 70% of the crops grown in Europe and the United States are destined as feed for animals. People in these areas surely eat more meat than people did in the preceding periods. Thus there is not a lack of potential manure for agriculture. Tractors have replaced horses, decreasing the numbers of farm animals used for traction but the numbers of pigs, cows and chickens has certainly increased. The problem is that the places of animal culture have been transferred from small units and dispersed areas to become more central and concentrated. Farm animal or meat production is now closer to the markets or performed on land less productive for grain and other food stuffs. In fact there is no reason that the manure of animal production cannot be used on the grain producing fields, except for its cost and difficulty in handling.

A particular disadvantage in intensive animal production is that the efficiency of such systems excludes the use of straw or other by products of grain farming. Moving animal dejection – laden straw (traditional animal barn yard manure) is very labor intensive, or at least enough so as to exclude it at present from common farming practice. When "raw" manure is spread on a field, it is mineralized by bacteria at a very rapid rate. However, traditional straw and dejection manure has a fertilizing effect which is useful for more than four years depending upon soil type (Larbaletrier, 1891). Of course George Washington was adamant concerning the use of manure fertilizer.

As a result of concentration and specialization in farming, the disposal of nitrogen – rich animal dejections causes water pollution problems where different forms nitrogen compounds find their way into drinking water resources. This is a paradox in that the intensive use of artificial nitrogen compounds as fertilizer also pollutes drinking water in grain growing area while the effluents from intensive animal production do the same thing. There is an over abundance of nitrogen on both sides of the food production structure in modern agriculture. It would seem reasonable to invest in some research to adjust these two agricultural systems so that a reasonable transfer could be obtained and that municipalities would not have to pay for expensive de-nitrogenating measures in the processing of water for human consumption.

The use of straw – based manure compost would also enrich the soil in a form of organic matter which appears to be more resistant to bacterial action than raw manure or green manure (Larbaletrier, 1891). The action of manure formation takes place over a period of several months where during this time there is most likely, a very high rate of bacterial action. One can attest this by looking at the steaming manure pile in the dead of winter. Manure is always over heated, due to bacterial action. This is one reason that it is applied on fragile plants in the ground during the coldest seasons. This bacterial action most likely produces a pre-humic material which is important to eventual clay – organic interactions when applied on the soil.

The major result of the new farming methods has been to increase productivity of a given land surface. The increase is so great that it more than compensates for the loss of some less productive types of land that have been abandoned to crop farming. Intensification of land use has led to dislocations of the two parts of agriculture, crop and animal husbandry which co-existed in many farm areas. Lack of natural fertilizing agents has led to the intense use of synthetic or mineral extracted fertilizing agents. Overall the system is immensely more productive than the traditional type of farming. However, there are several problems with such farming practices such as fertilizing agent pollution of water resources and pollution of the same water sources with chemical agents designed to protect the fast – growing plants from insect, mould and other attacks. These problems are beginning to be addressed by research at our major Agricultural institutes.

6.2 Effects of Modern Agriculture on Soil Properties: Organic and Clay Mineral Changes

As a reference point, we should compare classical agriculture effects, i.e. pre – 1950 in Europe and slightly before this in the United States with the changes that can be effected by intensive agriculture. There is not much data but one can refer to the results given by Kleber et al. (2003) who searched for methods to detect the presence of pre-historic agricultural activity. They found that in the A horizons of chernozemic (loess plain) soils in Germany, there was a distinct increase in illite in the A horizon compared to the C horizon. Illite peak height compared to smectite were clearly dominant in the A horizon and smectite appeared at 40–50 cm depth in the reference profiles whereas it was around at near 100 cm in the pre-historic farming sites. Other signs of old farming practice were found as traces of charred organic matter and a wider C/N ration in the surface horizons. These indicators can be predicted from the outline of the progress and practice in farming in Europe presented in Sect. 6.1. The efficient use of crop rotation, use of plant ash to increase potassium content and lower pH, lead to the efficient production of plants which bring potassium, and silica to the surface so that 2:1 clay minerals can be stabilized and so that potassium can be stored in these minerals when sources are sufficient to more than replenish the potassium is extracted in the form of crops.

Here it is clear that the traditional (from about 5,000 years ago) were efficient in stabilizing the potassium filled 2:1 clays. Thus one can use this as a reference for pre – 1950 farming.

On the other hand it is important to understand what will happen under the other extreme: water – rock interaction with a soil (i.e. formed under plant silicate interactions) an essentially fallow state. An interesting experiment in this sense was and is being carried out at the Agricultural Experimental Station of Versailles (France) reported by Pernes-Debuyser et al. (2003). Here experimental plots were kept plant free for 70 years and treated with different fertilizing agents. In the reference plot (no treatment and no plant interaction), there is a loss of organic carbon of 45% in the whole soil, and a similar loss in the clay fraction. Clay content decreases by 5%. Cation exchange capacity decreases by 43%. Exchangeable potassium is small in relative amount (1.5 % of CEC) but constant under the water – rock weathering conditions of the experiment. The initial clay assemblage, without chemical treatment, is dominated by illite/smectite interstratified minerals with some minor kaolinite and illite present in roughly equal proportions and a small amount of HI mineral. The water – rock interaction essentially decreases the illlite peak by about half, and shifts the illite/smectite peak to a higher (more smectie – rich and potassium poor) peak position. Hence the water – rock inter action over a 70 year period is essentially one of potassium extraction.

Fertilizer treatment with potassium salts and farm manure increase the illite content, decrease the HI mineral content and shift the illite/smectite position only slightly. Addition of potassium (and ammonium in the case of the manure treatment) then affect the illite content of the illite peaks more than the illite content of the illite/smectites mixed layer mineral.

Water rock interaction then decreases organic matter content, potassium content and decreases the clay content, most likely through the loss of the binding agent of the organic matter which fixes the clay fraction.

Given these data, one can attempt to understand the effects of modern agricultural practice on soil properties.

6.2.1 Organic Matter and Clay Content

Darmody and Norton (1994) indicate the relative effects of crop rotation or non-rotation on the organic content in mid western loess soils of the University of Illinois experimental plots. The study plots had all been farmed for some 50–80 years prior to the beginning of the observations. In all instances of cropping, 3 year, 2 year or no rotation, the soil organic content decreased. Analysis of the grass (sod) border of the plots showed an increase in organic content of about 30%. Thus all modern farming practices decrease the organic carbon content of the soil, non-rotation (*zea mais* cropping) showed the highest loss. Fenton and Lauterbach (1999) indicate that this effect is stronger on former forest soils than on prairie soils. In studies of various soil types where paired prairie and forest soils were observed, prairies were seen to lose roughly 30% of organic carbon under cultivation practiced for 100 years of less and forest soils lose some 60%.

One can conclude that modern agriculture, i.e. motorized, strongly influences the organic content of the soils, but not all that much more than the Versailles experiment of no cropping where similar amount of loss were observed. However, alternating crops with un-planted treatments (fallow) tended to increase the effect of loss of organic matter (Fenton and Lauterbach 1999).

6.2.2 Clay Content of Soils

Numerous studies report, though often make no comment, on the change in clay content of soils under cultivation. Darmody and Norton (1994) summarize published results on the importance of water stable aggregates in soils under different cropping practices. Compared to grassed soils, the stable aggregates are of course much less present, following the loss in organic matter. We give two examples here concerning two rather different situations, one in semi-arid alfalfa farming and the other in a rice paddy environment.

6.2.2.1 Lanzhou Alfalfa

In this semi-arid steppe province of northern China (Jiang et al., 2007) experimental plots have been observed for the changes in soil quality depending upon alfalfa cropping. The soils are based upon loess with a basic chlorite – illite mineralogy. The soil textural data indicates that, in general in all horizons, the clay content is related to organic carbon content (Figure 6.10). The range of organic matter is up to 3 wt%.

The amount of clay in the A horizon (0–20 cm) decreases with time. However the change is relatively small, 10%. This occurs when the organic content changes by nearly 10 but increases in the A horizon. Here the practice of agriculture does not impoverish the soil of organic matter but it decreases the clay content slightly.

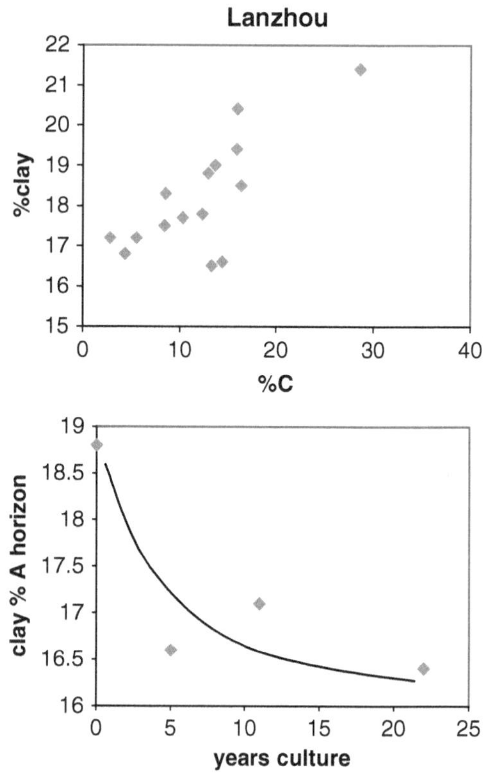

Fig. 6.10 Relations between clay content and organic carbon in the Lanzhou (China) experimental plots, semi-arid climate where alfalfa is produced. The data is for the agricultural horizon of these soils (unpublished data of Yang Yanfang, Li Decheng, He Yue, Zhang Ganlin, and Hu Feng in Nanjing Agricultural Institutions, China)

6.2.2.2 Rice Culture

Rice is grown in the Jiangxi province as in many parts of China, but here one has a red soil sequence based upon oxidized (red) Tertiary sediments. The clays are dominantly illite and chlorite. Iron content ranges from 5 to 3% varying inversely, after cultivation, with organic carbon content (Li et al., 2003). In the experimental plots of the Red Soil Station in this province, the system of flood irrigation has been used for periods ranging from 5 to 80 years. The clay content decreases rapidly with time, in the first 20 years or so, rising to above 1.5%. Here the system is basically anoxic where iron is reduced and organic matter is probably retained in a state of low maturation (bacterial action). The clay content decreases rapidly, when organic matter content is low, but stabilizes after 20 years of culture. Here the organic matter content stabilizes at near 2% and seems to fix the clay also. In the flood irrigation system it appears that high influxes of water can wash out much of the clay if it is not stabilized in organo – clay complexes (Fig. 6.11).

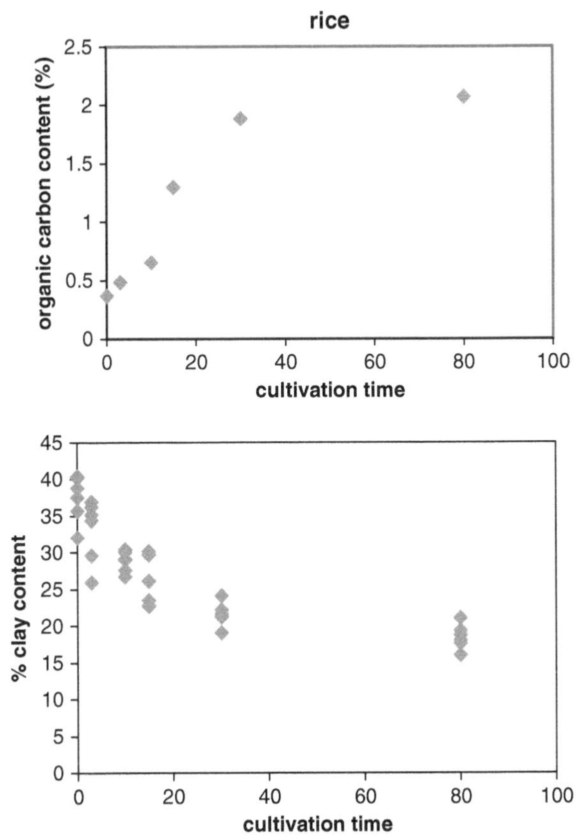

Fig. 6.11 Relations between clay content and organic carbon as a function of time on an experimental Station in southern China (Li et al., 2003). The increase in organic matter in the flood irrigated soil plots appears to stabilize the clay content after a period of increase in organic content

The clays were stable in the initial soil, probably in part because of the cementing capacity of the iron oxides, but also because of the lower flux of water in the system.

6.2.3 Tillage Practice

A study which sheds some light on the relations of organic matter and clay content under different tilling practices is rather instructive. Karathanasis and Wells (1989) give data on the effects of tilling practices, one of which, no till (NT), is designed to retain organic matter in the soil. Comparison is made of various soil properties using the two methods for 18 years in loess soils of Kentucky and Tennessee USA. The relations of soil clay content and organic matter content are interesting in that it is apparent that the tillage practice is not the only factor in determining retention of organic matter. In the soils with less clay, near 10%; the tillage practice is very important, where no till roughly doubles the organic content of the agricultural soil. However in soils with more clay the values seem to converge. Thus the organic matter will help bind the clays in clay poor soils but it is not as effective in the clay rich soils. Here one could predict that organic matter is important in forming clay complexes and aggregates when it is less abundant. The same study indicates that in soil pairs, NT and CT or no – till and conventional till, the illite content is slightly higher but other clays are very much the same. Hence there seems to be no strong selection of clays to form the aggregates or retain the organics and this does not significantly change the clay mineralogy from one tilling system to the other. However the authors do note a significant change in the chemical activity of mineral soil components in the pore solutions. Higher organic matter in the soils tends to increase the activity of Si and potassium in solutions. We can see the importance of this observation from the analysis of mineral stability and biomass (organic matter) given in Sect. 5.5.

These several examples of time studies under different conditions of agricultural practice indicate that the interaction of clays with organic matter is important in maintaining clays in the A horizon. However, the exact effects seem to be conditioned by the amount of clay and most likely also by the amount and type of organic matter (humic vs. low molecular weight substances). However, since the total of organic matter is not the only indicator of clay aggregation it seems that the system is somewhat multivariable in character. High influx of water tends to remove clays unless they are bound by the presence of several percent of soil organic matter. Higher clay content appears to decrease the importance of organo – complexion somewhat. It is clear that more work is needed to understand the relative importance of the variables.

6.2.4 Agricultural Practice and Its Results

6.2.4.1 Prairie Soils

If the analysis in Sect. 5.5 is correct in that plants stabilize clays favourable to their needs (potassium reservoirs, formation of aggregates as water resources) one can ask what is the effect of farming practices and particularly modern methods of agriculture on clays. By modern methods we mean systems where motor traction has replaced animals and hence barnyard manure and animal dejections in an artificial prairie are lacking. As shown by Kleber et al. (2003) classical agriculture practiced in former forest soil areas in neolithic times sufficiently changed the chemistry of the soils as to allow them to be used as markers of man's conquest of loessic soils in Germany. These methods of farming were more or less maintained with some ameliorations, such as replacing natural prairie with artificial prairie which increases the potassium uplift effect. The general tendency was to use green manure, prairie roots and other means to maintain or increase the organic content of the A soil horizons and to replace potassium and other elements taken off through cropping. Manure was very important not only for the potassium and organic matter but also, and perhaps more so for the ammonium source of nitrogen for the plants. The liquid parts of a manure pile were carefully kept and dispersed with the solids (Larbaletrier, 1891). Loss of animals, through the introduction of motorized machinery, lost t his natural fertilizer source. Since the objective of modern farming, in the 20th century, was to reduce labor costs, machines were employed greatly as seen above in Fig. 6.5. in the case of French agriculture.

We can investigate some of these effects and changes in soil chemistry by observing the effects of potassium extraction on clay mineralogy, i.e., 2:1 clay loss and potential gain through the presence of more organic matter, or less loss of organics.

We will look at several examples of agriculture on prairie soils in a prairie plant regime where there is a natural tendency to form or conserve illite under conventional prairie grazing and the effects of intensive hay cropping. A time series study indicates the effects of alfalfa, the crop used in the 18th century to replace the natural prairie cycle as the alfalfa effect.

Natural Prairie Development on Recent Sediments, Using Traditional Grazing Methods, Baie d'Authie (Somme, France)

The samples investigated represent the effects of natural prairie on sediments brought in from the coastal region of France near the Somme River. The zone investigated is just north of this large estuary where systematic poldering occurred for nearly 1,000 years (Velde et al., 2003). The poldering effects a well established time sequence which essentially isolates estuarine sedimentation which is put into prairie. In this region the oldest polder is dated at 1158. The sediments are essentially surface, soil based materials which are re-deposited on the surface and subjected to alteration processes (water – rock) and the interaction of plants as they establish themselves to form soils. Pasturage by cows and sheep was the agricultural practice in this region. No chemical products were added to the soils nor were they cultivated. In this instance we have a time series describing the installation of prairie interactions with as very clay-rich substrate, which contains quartz and calcite (shell material). The initial sediments are relatively illite – rich in that the illite mineral component is evident and the mixed layer minerals (illite/smectite) are proportionately smectite (S/I) and illite- rich (I/S). This is shown in Fig. 6.12. The

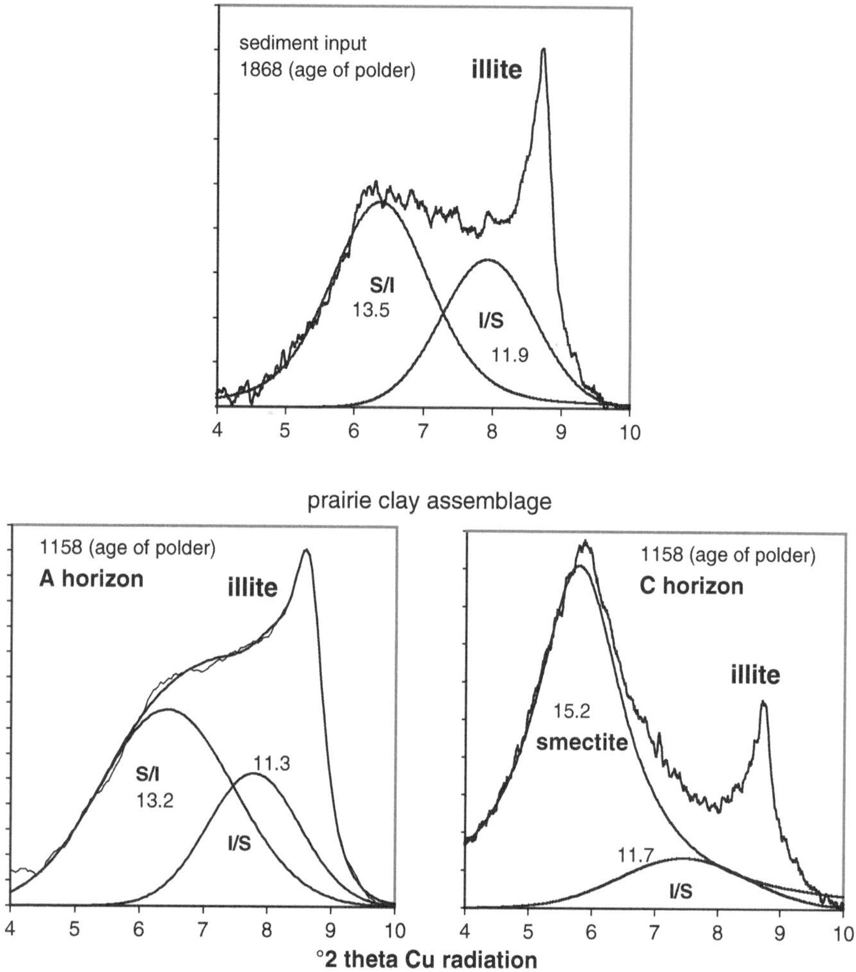

Fig. 6.12 Baie de Authie grazing on poldered land since 1158 compared to sediments grazed since 1868. Background – subtracted X-ray diffractogram treated using decomposition methods. Illite content and mixed layered minerals are similar in new sediment and old poldered A horizon soil. However in the C horizon of the old poldered material significant illite has been lost to smectites. The prairie under traditional grazing maintains illite (potassic clay) content (Velde et al., 2003)

establishment of the prairie soil changes the assemblage into an illite – rich A horizon with a shift in illite/smectite peak positions toward illite (10Å) and the presence of a strong illite component. In the C horizon, below the organic activity one finds a strong, dominant smecite – rich mixed layer mineral and a much smaller illite peak intensity. The illite (potassium) component has been greatly reduced in the C horizon whereas the illite content (potassium) has been mostly preserved in the A horizon under prairie development. We can take this to be the natural development of a grass – dominated soil system.

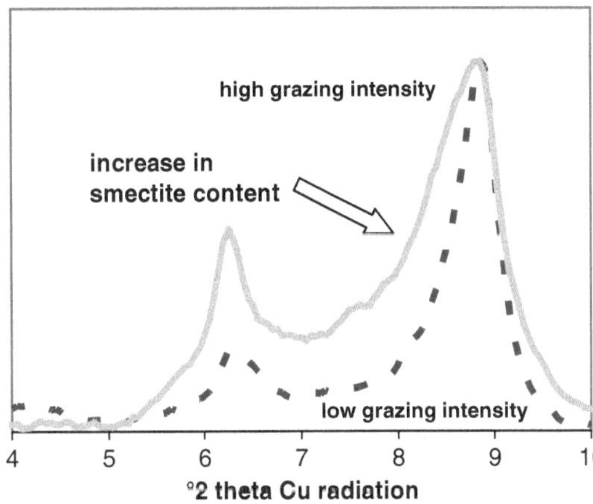

Fig. 6.13 Data for two series of experimental plots on the Clermont Ferrand INRA station (France) station. X-ray diagrams are compilations of spectra for four different plots managed under the same conditions. Background – subtracted X-ray diffractogram is an average of clay spectra for four different experimental plots. Grazing, and hay cutting is the major difference in the two series, intensive five cutting per year and low intensity, one cutting per year with animals grazing throughout the period

Intense Grazing on Prairie

An example of prairie clay systems under intensive, extractive culture is reported here by Barré (unpublished). This example is from an experimental area near INRA Clermont Ferrand Station in France. The experiment was to use the normal natural prairie system and to harvest it five times a year or one time with constant grazing. This was performed for 17 years. The impact on the clay assemblage is seen in Fig. 6.13. Essentially the illite peak is strongly affected by the intensive culture and exportation of the organic biomass. Mixed layer smectite/illite and illite/smectite mixed layer minerals are more important in the intensively farmed soils. The change is expected in that the major element which can come from the clays is potassium. This element is extracted and replaced by a divalent, bi-hydrated cation giving a peak position nearer to 15.2 Å. The importance of this experiment is to show that the extraction of potassium from the prairie soils is practically the reverse of the change which occurs in the establishment of the prairie on sediments in the baie d'Authie example given above. The process of potassium accumulation appears to be reversible in the extraction process.

Alfalfa Cropping

Agronomists in the 18th century developed an efficient method of increasing soil fertility in using alfalfa plants. It is generally assumed that this fertility is due to the bacterial development of nitrogen-bearing nodules on the roots of the plants. This is of course a strong fertilizing agent. However it is also known that alfalfa is a "luxury consumer" of potassium, i.e. it accumulates more than it needs. How does this work? Data (Velde et al., unpublished) from a Chinese farm experiment in the northern province of Gansu at Lanzhou indicates what happens as alfalfa plants determine the chemistry of a loess soil in a semi-arid climate. The initial and final state of the clay mineralogy is not extremely different, being dominated by chlortie and illite minerals. Figure 6.14 Lanzhou indicates the initial soil clay mineralogy as a result of curve decomposition analysis. Without this method of observation it is virtually impossible to determine any changes in the X-ray diffraction spectra. Throughout the 22 year experiment the chlorite peak areas are almost constant compared to the illite components and the relative peak intensities of the chlorites are similar if not identical (7/14 Å peak ratios). Potassium saturation does not change the peak ratios significantly but the width of the 14 Å peak does change and hence although there is little of an HI component present it can be noticed. The potassium saturated sample shows a stronger PCI peak and a shift of the low intensity components of the

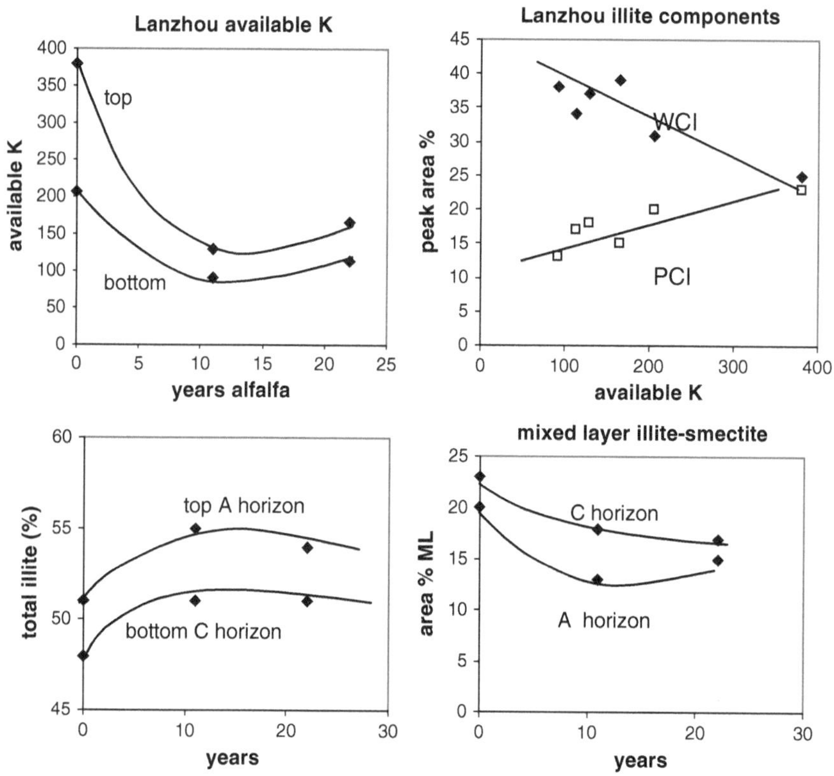

Fig. 6.14 Comparisons are made or the A horizon (*top*) and C horizon of the profiles where potassium availability (extractable K, by ammonium acetate) is compared to the different mineralogical changes observed with time of alfalfa cropping. The amount of available K decreases, the amount of mixed layer illite-smectite decreases and the poorly crystallized illite (PCI) decreases with time and illite (WCI) increases. The change in clay content is small, and the relative amount of illite and smectite-illite change only slightly with time. Overall it appears tht the easily extracted potassium is exhausted y the cropping but the more firmly held potassium in illite increases. Alfalfa appears to build up reserves of long term potassium in the soil clays. Relations of X-ray diffraction decomposition determinations concerning the peak positions, and relative surface areas of the major phases in the Lanzhou alfalfa experimental plots (China, unpublished data of Yang Yanfang Li Decheng He Yue Zhang Ganlin Hu Feng in Nanjing and Lanzhou, China)

14 Å peak, those of an HI component. It appears that the increase of PCI illite peak intensity indicates a shift of this part of the mineral components to a form of HI or interstratified HI/illite mineral.

The major change with cultivation time of alfalfa is the relative intensities of the mixed layer mineral (I/S), the well crystallized illite (WCI) and the poorly crystallized illite (PCI). In the experiments the amounts of soluble potassium and available potassium (ammonium acetate treated and water treated) decreases as a function of experimental farming in both the upper A horizon and the C horizon of these loessic soils. This can be followed by the decrease in intensity of PCI and mixed layer mineral. Alfalfa growth, cropped and exportation of the aerial biomass, increases the relative illite content of the clay fraction in the A horizon. This means that even though the potassium available to plant growth decreases, the relative amount of illite (eventual potassium resource for plants) increases, by more than 50%. Potassium saturation of the samples increases the PCI, width was lost to plant growth.

This behaviour can explain a part of the increase in fertility in alfalfa planted soils. It would appear that the alfalfa exhausts the available potassium in the soils from the more easily extractable parts of the soil assemblage but leaves the illite component (WCI) essentially intact or increases it. In other plant experiments, using rye grass the WCI illite can be a source of potassium for the biomass (Barré et al., 2008).

In this series of experiments one sees that the prairie growth tends to reinforce the potassium – bearing mineralogy, forming or maintaining illite layers in the 2:1 structures, either the illite components or in the mixed layer minerals. Intensive exploitation of the prairie plant assemblages decrease the illite (WCI and PCI) content shifting the remaining 2:1 layers into the illite/smectite minerals. However, the alfalfa plants take out less of the potassium from illite (WCI) and extract potassium from the PCI and mixed layer minerals. This exchange of potassium in different mineral types (compartments) could very well explain the use of alfalfa as a plant which increases soil fertility, not only storage of nitrogen by bacterial nodules on the roots but also by an enrichment of potassium in the 2:1 mineral clay assemblage in an illite type mineral form. The soil – enriching effect of alfalfa is probably of two sorts, one for K and the other for N.

In general, the prairie soils tend to enrich the clays in potassium. This effect is a fundamental part of prairie plant growth and prosperity.

6.2.4.2 Agriculture on Prairie Soils

Traditional Agriculture: Marais Poitevin (Center Western France)

In this region poldering was effected on aggrading sediments along the Atlantic coast. Here the newly conquered lands were put into agricultural use instead of grazing land as was the case of the Baie d'Authie example given above. The agricultural methods were undoubtedly traditional, crop rotation, prairie and manuring of the cultivated area. Sampling of the soils at the plow zone depth, near 20 cm, gave a

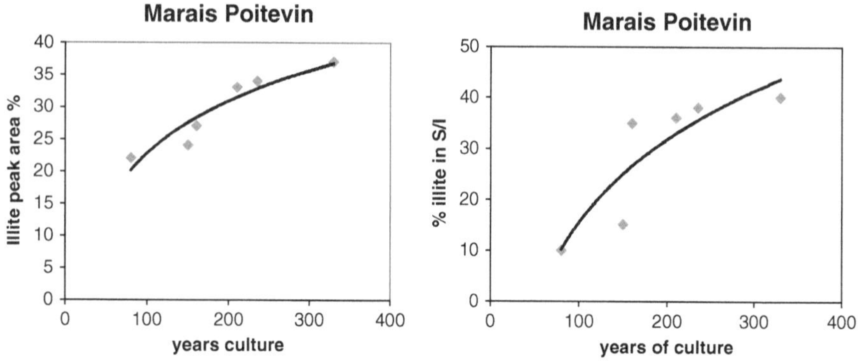

Fig. 6.15 Clay mineralogical data for a sequence of poldered soils from center western France (Marais Poitevin, Righi et al. 1995). The overall evolution of the soil clays is one of increasing illite content in the mixed layer minerals as well as in the relative amount of the potassic illite mineral itself

sequence of clay mineralogies as a function of the time the new clays were affected by agriculture (Righi et al., 1995). The time period of agriculture is 350 years. The clay minerals are dominated by the typical mixed layer illite/smectite types with an illite and a smectite rich type. Illite is also present and kaolinite to a lesser extent. In this time series one finds that illite content increases as an individual phase and the illite content of the mixed layer minerals increases also. In both cases the illite (potassium mineral) doubles its presence relatively over the 350 year period of culture (Fig. 6.15). This result is similar to the observation of Neolithic farmers made by archaeologists to determine the presence of ancient agricultural activity (Kleber et al., 2003).

This information indicates that the traditional methods of farming on soil material developed fro eroded surface materials of illite – smectite mineralogies develop illite (potassium) – rich minerals as does the prairie soil developed on similar material.

Intensive Agriculture on Prairie Soils: Illinois Loess

Morrow plots University of Illinois (USA)

Velde and Peck (2002) have reported upon the changes in clay mineralogy engendered by the cropping of continuous corn for a period of 83 years on a Illinois Mollisol. The clay mineral suite is similar to most of the prairie soils clays in the Central part of the United states (see Velde, 2001). The observations are centered upon the effect of corn crop removal of mineral elements from clays, therefore potassium as related to illite content either as a distinct phase or as an interlayered part of the mixed layered minerals present. In these experimental plots over the course of the experiment, corn hybrids were introduced in the 1950s which dramatically increased corn productivity. Hence a plot of clay composition against time is not the whole picture. Peak positions for the illite, illite/smectite and smectite/illite minerals shift with time of the experiments where continuous corn production (CC) was employed, without additional fertilizer. There is a distinct overall trend in the CC soil clays to first show a decrease in peak position, increase in illite content, and with the introduction of high productivity hybrids one sees an impoverishment in illite (potassium) as the peak position increases. In the CC plots it is interesting to note that there is a clear tendency for all clay phases to become more illite – rich (peaks shifting towards 10 Å) with time as productivity decreases in the initial stages of the experiment due to lower availability of potassium and other resources. The use of hybrid corn varieties in the early 1950s increased production and increased the amount of potassium taken from the soil with the result that the clays show less illite (potassium) content. In Fig. 6.16. We plot clay mineral peak position against the time of the experiment for continuous corn and orn – oats – hay rotations. In this plot there is a general trend of change to a postion of higher spacing, one with more smectite present for the continuous corn plots and a slight change to higher illite content for the corn – oats – hay rotation. However, in the continuous corn plots clays initially there is a change to lower spacings, higher illite content. In Fig. 6.17 the corn production is plotted against the illite content of the clay assemblage. Here it is clear that there is a direct relation between the amount of material, potassium in corn, taken off of the land and clay mineral potassium content.

In the experiment the crop rotation program (COH, corn oats hay) appeared to have been little affected the clay minerals as far as illite (potassium) content is concerned, as indicated in Fig. 6.16 where the peak position change little with time whereas continuous corn decreases the illite content of the clay assemblages as a function of the amount of material taken from the soil zone.

The X-ray diffraction diagrams for three stages of the experiment, 1913, 1955 and 1996 show that the extraction of potassium occurs in the illite and illite/smectites mixed layered minerals which are shown in Fig. 6.18. A minor amount of kaolinite is present but it does not participate in the elemental exchange

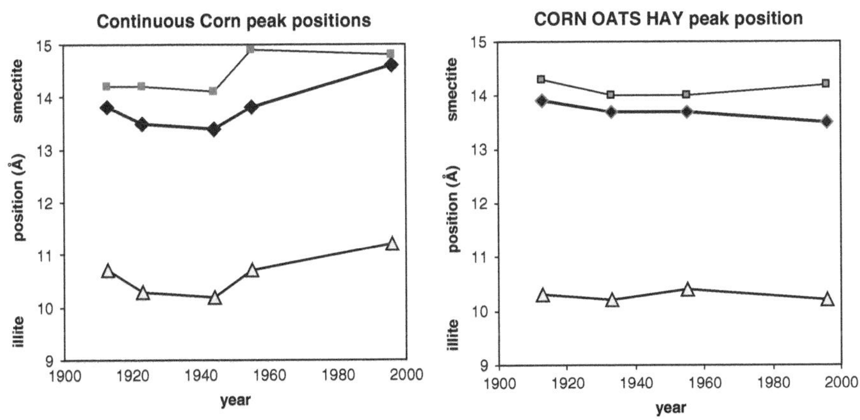

Fig. 6.16 Peak positions of the various clays in the Morrow plot experiments as a function of time of culture without fertilizer amendements in a continuous corn production scheme

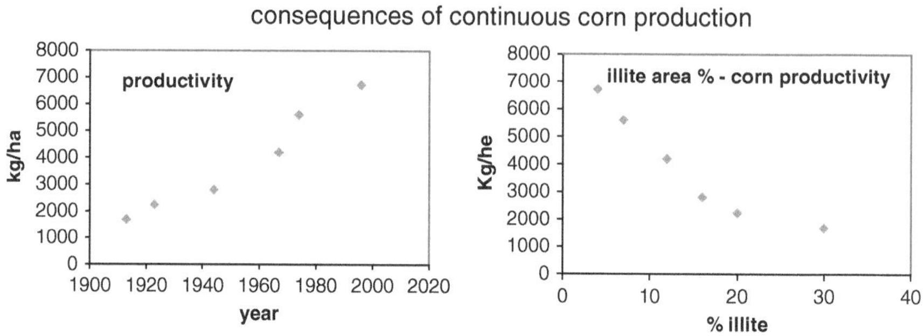

Fig. 6.17 Relations between corn productivity and clay mineral potassium content (illite) in the Illinois Morrow plot experiments (Velde and Peck, 2002)

considered here. One sees the gradual decrease in the intensity of the illite peaks (WCI and PCI) where the well crystallized illite component (WCI) is decreased before the poorly crystallized component. This means that the decease in potassium potential in the soil solutions is first compensated by the larger illite crystallites. The PCI component is then decreased leaving a small WCI peak in the oldest experimental plots. The corn was grown without chemical fertilizer nor other amendments. The plot of corn productivity against mineral change (Fig. 6.17) indicates that the overall illite content is affected by corn production, i.e. the amount of potassium extracted from the soil. The presence of smectites/illite, lower in potassium content is evident, and hence it is also a source of potassium. The peak position of this mineral shifts, indicating a change in illite content of about 25% illite content. In fact the total potassium extracted from the soil is just 10% of the amount present in the soil itself, thus the clay fraction which is strongly affected is only a part of the total potassium reserves present. Since the corn productivity increased with time, one must feel that the corn plants were able to extract potassium from the more coarse, non-clay fraction minerals such as detrital mica or feldspars as well as from the clay fraction. They did not totally exhaust the potassium reserves of the clay fraction, in that the illite/smectite minerals maintain some potassium all the same, about 8% illite content. The loss in potassium from the clay fraction is only about 11% of the amount present.

In these experiments one sees the effects of extraction of potassium from the ready reserves of the soil. However these reserves seem not to be the only ones in the soil concerned in that corn production continues to increase with the introduction of new high productivity hybrid species in the 1950s. One can suspect that a certain amount of the potassium comes from greater depths than the sampling zone of the experiment, which was at the top 10 cm of soil. Nevertheless, the shift in interlayer cation type is seen to be directly related to the extraction (corn yield) of the potassium from the soil.

The Morrow plot experiment indicates that the direct impact of plant extraction can be seen in the illite minerals, those most rich in potassium. The total of potassium extracted is small compared to the total reserves in this particular soil.

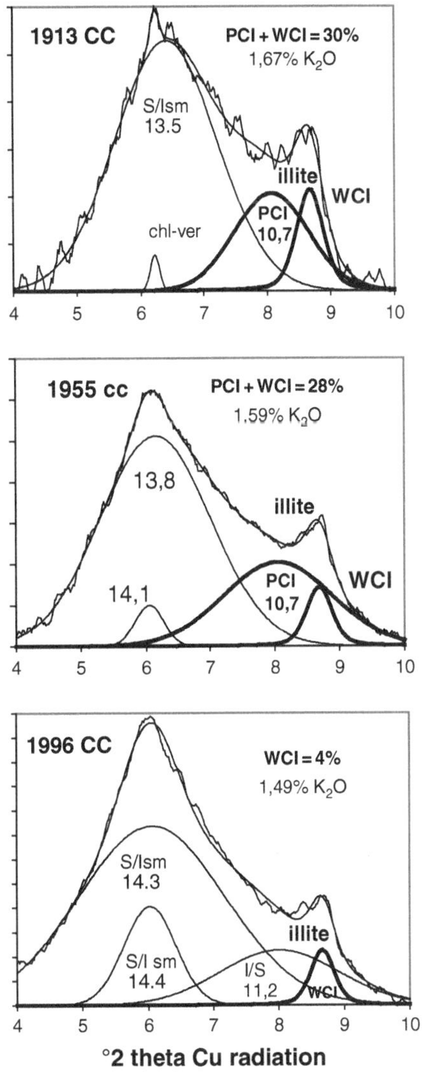

Fig. 6.18 Decomposed X-ray diffraction diagrams of three stages in the Morrow plot experiment showing the change in illite peak intensity as a function of potassium content in the clay assemblage

6.2.4.3 Former Forest Soils

A perhaps exotic and special case of forest soil cultivation is reported by Li et al. (2003). Here a red, oxidized, forest soil from a near tropical region in China was used in a rice growing experiment where the soil was flood irrigated and cropped three times per year. In this experiment NKP fertilizer treatment was used with KCl at a rate of 1,500 kg/ha. The intensive agriculture for a period of 80 years changed the clay mineralogy greatly. Illite was especially affected. In Fig. 6.19 the average illite content is plotted against the period of rice cultivation. One sees a typical "S"

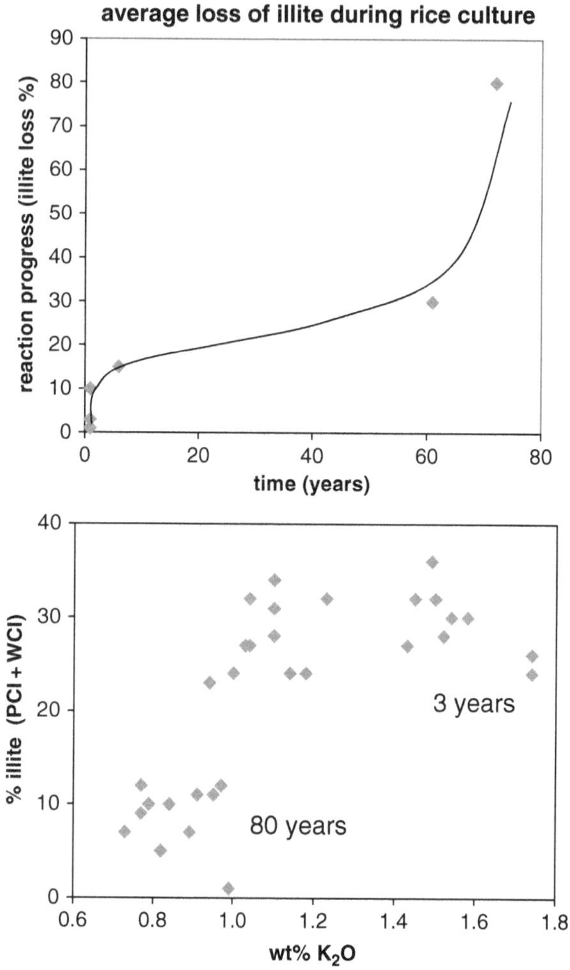

Fig. 6.19 Irrigated rice plots in China (Li et al. 2003) where loss of illite (potassium bearing phase in the clays) and potassium content of the clays is compared to the time of the experiment. The typical "S" shaped curve can bee seen for the loss of illite

shaped curve of illite loss (reaction progress) due to bio-action on soils. There is a corresponding loss of illite from the soil clay fraction. Also the relative intensity of the mixed layer mineral chlorite/illite decreases. Thus one sees a direct relation between illite and potassium content in the soils, either as an independent phase or as layers in a mixed layered mineral. However there is a corresponding loss in clay content from the soil which may affect differentially the presence of illite to a certain extent. It is unlikely though that the illite was the only clay lost in that the other minerals were chlorite and an illite/chlorite mixed layer mineral one of which lost potassium content also.

The rice culture experiment indicates that even though potassium fertilizer was used, the overall exportation of potassium as greater than the ability of the clays to capture and store some potassium.

6.2.4.4 Fertilizers

Studies by Singh and Goulding (1997) and Tributh et al. (1987) indicate the effects of potassium fertilizer use on clays on old forest soils in England. The case of rice cultivation on red soil sited above indicates that one needs a correlative amount of fertilizer to compensate the exportation of potassium from soils by crops. Barré et al. (2008) indicate the effect of different potassium treatments on soils where crops of different types were grown and harvested on the Versailles INRA experimental farm (France). The 50 year experiment using different amounts of potassium and other fertilizing agents was conducted to indicate the amount of potassium taken off of the plots as well as the amount supplied. In the case of un-fertilized plots, some 3,800 kg/ha was taken off. On a heavily fertilized series of plots an excess of 3,400 kg/ha K_2O occurred.

Analysis of the average weighted peak position of the smectite-illite mixed layer mineral and illite assemblage indicates a difference between the net gain and net loss soils. Measured potassium content of the soil clay fraction indicates that a difference of 20% of potassium content (near 0.7 wt% K_2O) occurred between the extremes of treatment. The average weighted peak position changed by about 10% of mica content assuming a simple mixed layer illite/smectite mineralogy. Since the normal potassium content of an illite is about 7% the calculations and X-ray diffraction estimates seem quite reasonable. However, in the cases of high potassium fertilization, roughly seven times more potassium was added than fixed in the clays above the starting composition. Then it is clear that some fertilizer is necessary but it is quite possible to add too much, at least as far as the fixing capacity of the clays is concerned.

Here one can see that the excess of K_2O can be stored in part in the clays, and that the systematic exportation of potassium impoverishes the clays. In the example presented the illite content is changed relative to the other phases; I/S, S/I, and HI. The excess ofpotassium and the extreaction by crops is basically found in the amount of illite present (Fig. 6.20).

It is thus apparent that the modern practices of agriculture diminish the basic properties of surface horizon (A) soils. The clay content decreases, the organic carbon content decreases, the soil structure is less aggregated and hence holds less water in reserve for plants. The clay minerals reflect the change in overall potassium availability showing loss of illite and illite layers in mixed layered minerals. This is a contrast compared to the effects of traditional farming, from the Neolithic period on to the 20th century where the organic carbon content was increased, where the potassium content was increased and where most likely the soil aggregation was reinforced. The potassium balance is one of the most studied aspects of soil change, although other aspects such as phosphorus content and capillary retention of water are also very important aspects of soil management. The factor of the loss of organic

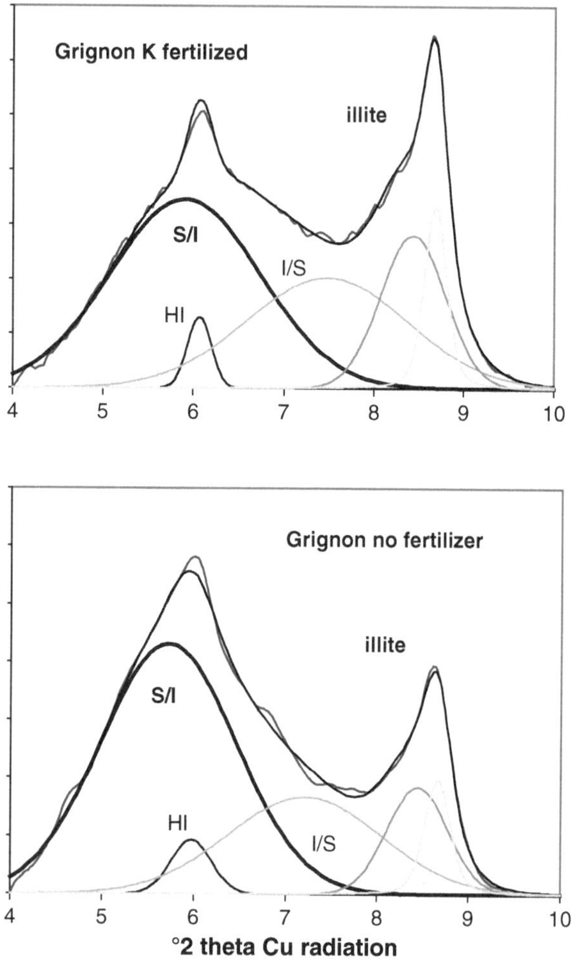

Fig. 6.20 Decomposed X-ray diffraction data for the experimental plots at the Grignon INRA station (Yvelines, France). Spectra are averages of four plots with different treatments of high potassium application and non potassium fertilizer (Barré et al., 2008)

carbon from the soil horizon is reflected in the lower aggregation ability but also in the loss of clays from this horizon. The loss of clays diminishes the ability of a soil to retain potassium, water and provide a generally beneficial context for the growth of plants from the seeding stage onward.

The major change giving rise to these changes is the loss of organic matter in the form of manure applied regularly to the agricultural surface. This loss is due to the use of mechanical power which replaces animal traction. The use of machines is of course much more efficient in the production of plants, along with and largely because of the use of factory origin fertilizers. Without machines one could not have diminished the agricultural population of the western counties from near 70% to less than 5% today. Many changes have occurred during this process, mostly as by-products of the change in traction and fertilization methods. However, the apparently inevitable change to factory methods has its draw backs, notably in the loss of soil fertility which needs to be remedied by input from factories. Hydrophonic culture is of course possible, but perhaps it is not the best solution to the problem of providing food for humankind. The problems of natural resistance to the other parts of the natural system, bacterial action, moulds, insects and so forth, are intensified with the use of artificial support systems. It is perhaps time to look back to the past and re-invent the processes which have served to well for thousands of years to keep mankind alive and reasonably well.

6.2.5 Summary

It seems that there are various results of agricultural practice depending upon the type of soil that pre-existed man's intervention of manipulating the regime of plants and soil clays. Depending upon the intensity of the extraction of plant material the soils can be more or less impoverished in their potassic components in the clay mineral assemblage. The classical regime of alternating of crops with grass – based plants (prairie) maintains or enhances the potassium resources of the soil clay minerals. Prairie grazing, on grasses, can be either an enriching mechanism or a loss of potassium depending upon the extraction rate. The same is true of grain cropping. All resides in the rate at which the mineral elements are extracted and the time or plant regime which is present to re-new the mineral resources. This concept is the key to renewable agricultural practice.

What is needed

It is clear that a better understanding of the natural extraction of mineral elements by different plants is needed.

We have at present an idea of the kinetics of such changes but more work is necessary in order to predict the time frame of reaction rates so that one can use such information to guide agricultural practice in a realistic manner.

A better comprehension of the effectiveness and mechanisms of potassium – ammonium fixation in clays by manure treatment (animal excrement and vegetal matter) is of primordial importance. Such information can lead to an effective use of animal excrement on agricultural soils with a decrease in pollution and an effective addition to crop dynamic processes.

References

Barré P, Montagnier C, Chenu C, Abbadie L and Velde B (2008) Clay minerals: A potassium reservoir in temperate terrestrial ecosystems. Plant Soil 312: 213–220

Bell M and Walker J (1992) Late Quaternary Environmental Change, Longman Scientific Technical, 273 pp.

Calwe N, Dolukhanov P, Kozlowski J and Van Berg P-L (2007) Le Néolithique en Europe, Armand Colin, 381 pp.

Chambers F (ed.) (1993) Climate Change and Human Impact on the Landscape, Chapman Hall, 303 pp.

Clark J (1955) L'Europe Préhistorique, Payot, 490 pp.

Cox J (1925) Crop Production and Soil Management, Wiley, 507 pp.

Dalfes H, Kukla G and Weiss H (eds.) (1994) Third Millennium BC Climate Change and Old World Collapse, Springer, 727 pp.

Darmody R and Norton L (1994) Structural degradation of a prairie soil from long term management, 641–649. In A Ringrose and G Humphries (eds.), Soil Micromorphology: Studies in Management and Genesis Developments in Soil Science, Elsevier, 354 pp.

de Serres O (1600) Le Théâtre d'Agriculture et Mesnage des Champs, re-edition by Actes Sud 1996, 1546 pp.

Duby G (1962) L'économie rurale et la vie des campagnes dans l'occident médiéval, Aubier, 432 pp.

Duby G and Wallon A (1975) Histoire de la France Rurale, Vol. 1, Seuil, 620 pp.

Duby G and Wallon A (1976a) Histoire de la France Rurale, Vol. 3, Seuil, 505 pp.

Duby G and Wallon A (1976b) Histoire de la France Rurale, Vol. 4, Seuil, 665 pp.

Edwards K (1993) Models of mid-Holocene forest farming for north – west Europe, 35–147. In F Chambers (ed.), Climate Change and Human Impact on the Landscape, Chapman Hall, 303 pp.

Fenton T and Lauterbach M (1999) Soil map unit composition and scale of mapping related to interpretations for precision soil and crop management in Iowa. Proceed Fourth Int Conf Precision Agriculture 239–251.

Ferdiére A, Malrain F, Matterne V, Ménil P and Jaubert A (2006) Histoire de l'agriculture en Gaule, Errance, 229 pp.

Firman G (1986) Archéobotanique et archéologie expérimentale qu Néolithique, 53–70. In J-P Demoule and J Guilaine (eds.), Le Néolithique de la France, Picard, 406 pp.

Garola C-V (1918) Prairies et Plantes Fourragèrers, J-B Baillière Fils, Paris, 572 pp.

Guilaine J (1989) Premiers paysans de l'ancien monde, 113–136. In J Guilaine (ed.), La Préhistoire d'un Continent a l'Autre, Larousse, 288 pp.

Hall B (1829) Travels in North America, Vol. 1, Lea and Carey, Philadelphia, 319 pp.

Jiang J-P, Xiong Y-C, Jia Y, Li F-M, Xu J-X and Jiang H-M (2007) Soil quality dynamics under sucessional alfalfa field in the semi-arid loess plateau of Northwestern China. Arid Land Res Manag 21: 287–303.

Karathanasis A and Wells K (1989) A comparison of mineral weathering trends between two management systems on a Catena of loess – derived soils. Soil Sci Soc Ame J 3: 582–588.

Kleber M, Rôssner J, Chenu C, Glazer B, Knicker H and Jahn R (2003) Prehistoric alteration of soil properties in central German Chernozemic soil: In search of pedologic indications of pre-historic activity. Soil Sci 168: 292–306.

Larbaletrier A (1891) Les Engrais et la Fertilisation du Sol, J-B Baillière Fils, Paris, 351 pp.

Li Z, Velde B and Li D (2003) Loss of K-bearing clay minerals in flood – irrigated, rice – growing soils in Jiangxi Province, China. Clays Clay Miner 51: 75–82.

Maellart J (1975) The Neolithic of the Near East, Thames and Hudson, 300 pp.

Masurié de Keroualin K (2003) Genèse et Diffusion de l'Agriculture en Europe, Errance, 184 pp.

Merle L (1958) La Métairie et l'Evolution Agraire de la Gâtine Poitevine, SEVPEN, Paris, 230 pp.

Miles J (1985) The pedogenic effects of different species and vegetation types and the implications of succession. J Soil Sci 36: 571–584.

Ministère Agriculture Pêche (2001) L'agriculture, la fôret et les industries agroalimentaire, Service Central des Enquêtes et Etudes Statistiques, 159 pp.

Pattullo H (1758) Essai sur l'Amélioration des Terres, Durand, Paris, 283 pp.

Pernes-Debuyser A, Pernes M, Velde B and Gessier D (2003) Soil mineralogy evolution in the INRA 42 plots experiment (Versailles, France). Clays Clay Miner 51: 577–584.

Phillips P (1982) The Middle Neolithic in Southern France. BAR Interntional Series 142, 204 pp.

Richeyre A (1884) Traité d'Agriculture, Ch. Delgrave, Paris, 322 pp.

Righi D, Velde B and Meunier A (1995) Clay stability in clay-dominated soil systems. Clay Miner 30: 45–54.

Ruddiman W (2005) Plows, Plagues and Petroleum, Princeton Univ Press, 202 pp.
Russell E (1945) Soils and Manure, Cambridge, 296 pp.
Shane LCK and Cushing EJ (eds.) (1991) Quaternary Landscapes, Belhaven Press, 229 pp.
Sigault F (1975) L'Agriculture et le Feu: Rôle et place du feu dans les Techniques de Preparation du Champ de L'ancienne Agriculture Européenne, Mouton Cie, Paris, 320 pp.
Singh B and Goulding K (1997) Changes with time in the potassium content and phyllosilicates in the soil of the Broadwalk continuous wheat experiment at Rothamsted. Eur J Soil Sci 48: 651–659.
Stamp L (ed.) (1961) A History of Land Use in Arid Regions, UNESCO, 388 pp.
Toutain J (1927) L'Economie Antique, Renaissance du Livre, Paris, 438 pp.
Tributh H, Boguslawski EV, Lieres AV, Steffens D and Mengel K (1987) Effect of potassium removal by crops on transformation of illitic clay minerals. Soil Sci 143: 404–409.
Velde B (2001) Clay minerals in the agricultural horizon of loams and silt loams I the Central United States. Clay Miner 36: 277–294.
Velde B and Peck T (2002) Clay mineral changes in the Morrow experimental plots University of Illinois. Clays Clay Miner 50: 364–379
Velde B, Goffé B and Hoellard A (2003) Evolution of clay minerals in a chronosequence of poldered sediments under the influence of a natural pasture development. Clays Clay Miner 51: 206–218.
Washington G (1801) Letters from General Washington to Arthur Young, McMillan, 171 pp.
Watson P (1991) Origins of food production in Western Asia and Eastern North America, 1–29. In L Shane and E Cushing (eds.), Quaternary Landscapes, Belhaven Press, 229 pp.

Chapter 7
Plants and Soils: The Ecological Paradigm of Plants and Clays

Contents

The pages of text and figures which precede these notes are intended to show what is the reality (perceived as such today) of plant – mineral interaction. This effort is at the same time new and a re-working of old ideas and knowledge. We have not invented a new concept to say the least. The interactions of plants and soil minerals have been suspected since people could write it down. Perhaps the Neolithic agro-pioneers would have done so if they had had the tools of such expression. However, modern Science has obscured, gradually, these basic tenets of knowledge and now they might seem new to some experts and practitioners of agricultural practice, soil science and clay mineralogy. Ours is not to decry the vices of the past, nor to re-invent a "new" science. We wish to put things in perspective in order to allow a more fundamental approach to the investigation of plant growth in symbiosis with the mineral substrate of their existence. Such an effort has allowed us to investigate the past, through the tools of archaeology, the ancient texts of Greek and Roman authors and less distant books on agricultural practice from the more recent past, 19th–20th centuries. Basically, these sources converge on the evident conclusion of a strong interaction between plants and soil mineral resources. The key to understanding the inter-relations is to understand the chemical forces exerted by the different actors on the material found at the surface of the earth.

Basically the structure of the plant – earth situation is one of loss. The forces of rainfall and subsequent dissolution of minerals at the interface of atmosphere and rock are such that some of the major elements needed for the biological cycle

B. Velde, P. Barré, *Soils, Plants and Clay Minerals*,
DOI 10.1007/978-3-642-03499-2_7, © Springer-Verlag Berlin Heidelberg 2010

of plant life are inevitably lost. The initial interaction of rainwater, aqueous solutions essentially without mineral ionic charge, is one of dissolution of the existing solids found at the interface of atmosphere and rock. The fundamental chemical force is one of an exchange of hydrogen ions with ions in the rock minerals. This exchange is one of a type of "hydration" where H^+ is substituted for ionically bound cations. The result is a transformation of old, high temperature minerals, into new, hydrated forms, stable under conditions of earth surface chemistry. Exchange means that some ions, components of the pre-existing mineral, will be dissolved in the aqueous solutions whose inexorable mission is to find their way to the sea, the final repository of the earth, or at least that part which sticks up out of the sea. Chemical and gravitational forces move matter to the depths of the ocean.

The ions lost are those that stabilize some minerals containing elements vital for plant growth. Our story for the most part is one of potassium, the major element for terrestrial plant growth dynamics. The second part of the story is one of water retention. Rain wants to find its way to the sea or to be evaporated to become rain again. However, plants are largely composed of water. They need a ready and continual source of this material. Such resources need to be kept in the surface of the plant – earth interface. Here the problem is one of physical capture as much as chemical interaction. Water should be available during the growing and vegetation periods so that plants can develop.

The obvious repository for these two, among other, resources is that of the soil. The soil is a name given by scientists, according to the real practitioners, farmers, to the surface zone of plant earth interaction where organic matter and fine grained alteration products are present. The new minerals, formed by water – rock interaction, are called clay minerals, usually of small physical dimensions, < 2 μm. The organic matter in soils is the result of vegetal life, either its exudate extension into the soil via roots, or the eventual decomposition of old debris which has been discarded from the life cycle of the plants themselves. Nothing is totally lost, and much is re-used. In the interactions of plant and soil materials one finds a great range of micro biotic materials such as bacteria, micorrhiza and other organisms. Thus one can easily call this situation one of ecological interaction. Such is the great variety of interactions between plants, micro-organisms and silicate materials from the rock substrate. The variety of actors is a response to changing conditions of climate and different chemical constraints imposed by rock substrates, temperature and rainfall variables that control the initial chemical configuration of water – rock interaction.

Plants have adapted to these constraints in, of course, a multitude of ways creating species and subspecies, communities of different plant types, symbiotic relationships with diverse microbiological communities and so forth. The complexity is staggering when one looks into it. However even though things are complicated they are not impossible to understand, to paraphrase the great 20th century scientific sage, "God is obscure, but not unkind". We have much information at hand, often more than is currently used to solve the ecological problems, but more is needed.

7.1 Stabilizing 2:1 Clays

One of the most striking examples of interaction in the soil zone is that between plants and clay minerals. Traditionally one thought of clays as a transient feature of soil zones, a stage between the surface alteration due to water – rock interaction. The soil clays are assumed to be inherited from underlying rocks either by the alteration of some high temperature minerals or by the transformation of the high temperature phyllosilicate minerals biotite; chlorite and muscovite. In all cases it was assumed that the clays would be destroyed by the chemical dissolution action of organic acids. Thus the plants make do with what is left to them.

However, in assembling existing published data and conclusions from certain studies, it is clear that plants can have an impact on the clay mineralogy of the surface, soil zone. Not only does this effect occur in tropical zones of high weathering intensity, but it also quite evident that the clay assemblage is conserved and perhaps renewed by crystallization of new minerals at the surface through the transfer of certain mineral elements, especially silica and potassium. The overall objective of the plant communities is to keep 2:1 (TOT) clay minerals present in the soil biozone. The type of clay mineral is determined by the ratios of silica, potassium, and perhaps calcium. The acidity (H^+ content) can also play a role in the stability of clay minerals. Natural water – rock interaction produces the following reaction series

Mica, feldspar =>	illite, smectites =>	kaolinite =>	gibbsite
	2:1	1:1	hydroxy

In a sequence of lower silica content clays, logically, the lower silica minerals, such as gibbsite should be found in the zone of highest contact rate with un-saturated water, the surface horizon. Thus the sequence as presented is one from bottom to top of an alteration sequence.

In a number of instances in weathering sequences from temperate climates one finds somewhat the reverse where kaolinite and at times gibbsite are found at depth, near the water – rock interaction zone and 2:1 minerals are present at the surface. This is a clear reversal of the normal weathering trend, which can only be effected by the addition of silica to the surface clay materials. Such an effect is known through the deposition of phytollitic material from plans in the surface soil zones.

Mica, feldspar =>	illite, smectites =>	kaolinite =>	gibbsite =>	smectite, illite
	2:1	1:1	hydroxy	2:1 potassic

This is a sequence from bottom to top of an alteration and plant intervention profile. The reversal can be called the translocation or transfer effect due to plant intervention. Figure 7.1 indicates these relations in a schematic manner.

Curiously enough, it appears that some bio-agents active in the root zone tend to preserve the phyllosilicate structure when present in the bedrock material (Arocena and Velde, in press). These bio-agents (micorrhize and bacteria) seem to preserve the phyllosilicates while destroying other silicate structural forms. Such an observation

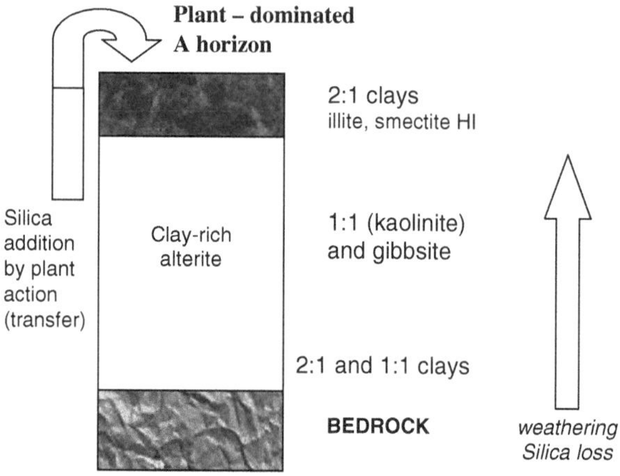

Fig. 7.1 Schematic diagram of the effect of plant transfer of mineral elements from the alterite to soil horizons. Normal weathering trends are shown on the *right hand side* of the diagram and plant action on the *left side*

indicates that plants attempt to maintain the clay minerals in soils, either by sparing them under conditions of chemical attack or by forming them in the soil surface through the addition of silica.

7.2 Potassium Retention

This is all well and good, 2:1 clays are nice objects to study but of what use are they to plants?

As it turns out, these clays are the only ones which can fix and preserve potassium in significant concentrations. This is the only alkali ion present in terrestrial soils to any significant extent, sodium being only very loosely held in clays where Ca and K ions are preferred by the clay structures. Since potassium is primordial to plant growth dynamics at many stages of development, its presence is very important. The role of 2:1 clays is then to fix and store potassium as illustrated in Fig. 7.2. One can describe such actions as nutrient exchange, where potassium is absorbed and desorbed depending upon the chemical activity of mineral and bio-agents in the soil.

The 2:1 clays, smectites but especially illite and HI minerals are capable of fixing K ions from potassium – rich soil solutions within the structure (interlayer sites) and keeping them from being re-exchanged with dilute rainwater. This is the reservoir function. When plants or bio-agents exchange organic acid or other materials with these potassium ions in the interlayer mineral sites this extraction is like taking goods from a store room. The clay suite is one of a holding and sequestering action keeping a ready supply of potassium for plant use. It is probable that some ammonium can be fixed in this same manner in 2:1 mineral interlayer sites.

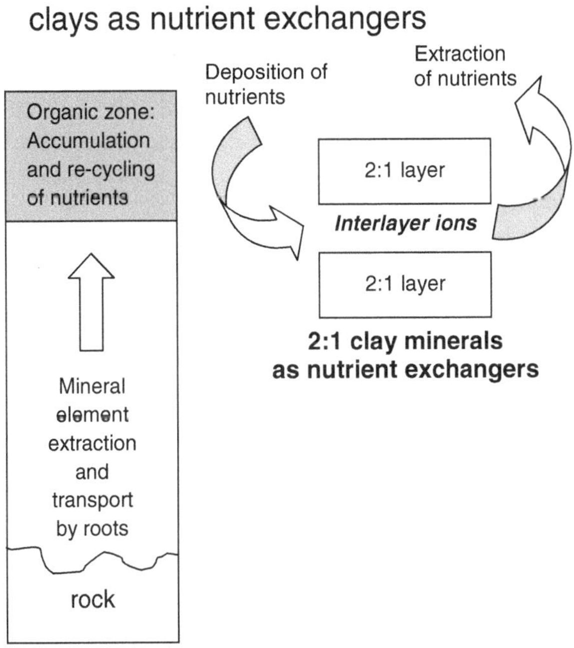

Fig. 7.2 Clays as nutrient exchangers placed in the alteration schema. The 2:1 clay minerals are found or enriched in the soil zone where they play the role of retention and release of nutrient elements for plant growth, notably potassium

Much farming practice in the past has been designed to find ways of supplying these potassium ions, and possibly ammonium ions, to the clays. First crop rotation was practiced using legumes as potassium and nitrogen providers. Then animal manure was used to provide a more potent stock of potassium and ammonium. An example was given by G Washington, Chap. 6, where newly cleared deciduous forest land could produce 300 kg/ha with wheat grown one in three years. Using no crop rotation system, no manure and growing wheat one in two years dropped the productivity to 220 kg/ha. However using crop rotation and manuring the land gave results of 450 kg/ha when wheat was grown. It is clear that the availability of potassium and nitrogen give a high growth rate to crop cultures increasing productivity by 50% compared to cropping from the normal fertility of virgin land. Substitution of chemically derived potassium gives the same effect, and as shown in several studies, the potassium content of clays can be greatly enhanced by such practices. If the right clays are not present to hold the potassium throughout the growing season most is lost to rain water dilution and a very high amount must be applied to maintain plant growth. Kaolinite is a poor potassium fixing agent while high charge smectites (illite) is a good fixing agent.

One can conceive of using silica – rich plants, and returning much of the material to the soil as a means of maintaining or increasing the amount of 2:1 minerals in a soil.

7.3 Humic Matter, Soil Aggregates and Water Potential

A second and perhaps equally important role of 2:1 clays is their ability to combine with various forms of organic matter in soils, either the evolved humic matter or lighter molecule and more ephemeral root exudates, bacterial exudates or other material to form stable soil aggregates. The soil aggregate is in fact an assemblage of clay particles held together by organic matter which is largely micro-porous. Capillary pores are filed with water for much of the year and this water is then a reservoir for plant growth. The dark color of soil, due to humic substances, indicates the presence of soil aggregates. However, 2:1 clays are much more efficient at forming these structures and hence when present give an added water resource for plants. Use of appropriate manuring practices retains much of the needed organic matter and in fact builds it up in soils. Crop management can help to retain some of the un-used parts of plants such as straw and stalks in the eventual form of humic matter after bacteria have reduced it to the appropriate form. These are means of using the natural cycle of plant growth and reduction by bacteria to maintain a more natural soil structure which is beneficial to plant growth.

Very much effort has gone into an understanding of the water retention potential of soils by laboratory measurements but as yet much needs to be done to better understand the relations between soil clay mineral types, organic matter humification, bacterial action and plant root exudates in their role in creating micro-porosity.

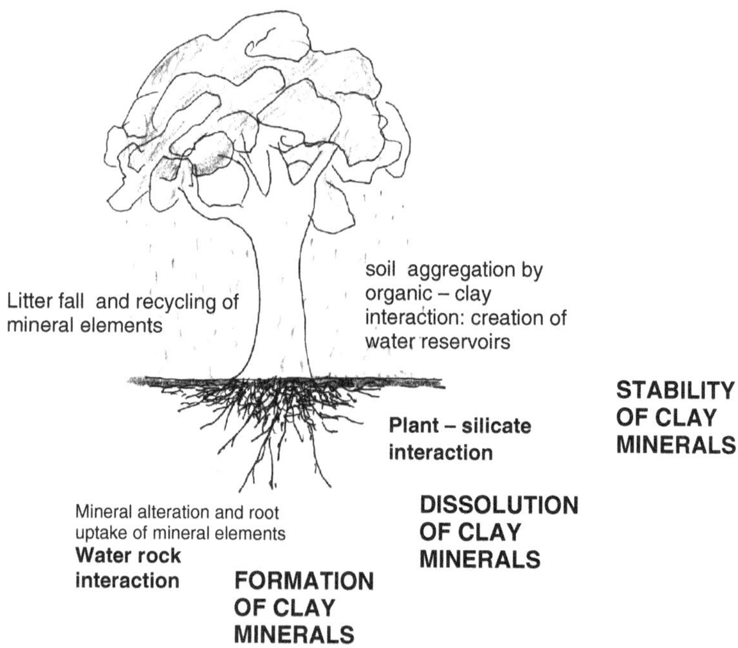

Fig. 7.3 Overall schema of plant action and the major chemical actions of alteration and the effect of plants

The overall picture is presented in Fig. 7.3. The elements of clay formation in the water – rock interaction zone form clays, more contact with un-saturated rain-water dissolve clays or changes them to less siliceous forms, and at the interface of plant and mineral clays are stabilized by plant transfer mechanisms. The interaction of evolved plant debris (humic matter and other forms) with clays is another very important aspect of the surface zone.

In sum, the mineral sphere is used and manipulated by plant activity in order to maintain a suitable type of clay mineral structure which has various interactions providing needed resources for plants. In order to achieve this state of affairs the natural tendency of alkali and silica loss engendered by aqueous interaction with silicate materials crystallized under high temperature conditions must be countered. Such is the transfer effect or translocation of needed mineral elements from the lower parts of an alteration profile to the surface where a specific mineral assemblage is stabilized.

7.4 Modelling Chemical Stability

We have seen in Sect. 5.5 that it is possible to model the chemical interaction between plant transported mineral elements and silicate minerals, clays. The variables of the model are for the most part solution chemical components incorporated in the clay silicate minerals. The variables of K-Si proportions in the plant biomass, the pH developed by bacterial action or plant exudates, the amount of rain per year and its frequency can be used as input to establish the equilibrium clay assemblage under a series of conditions imposed by plant regimes. Such a model can be used to determine and explain the effects of certain plant regimes on a given type of alerite material. More important, there is a potential of prediction of the effects of plants introduced into a given soil situation under given climactic conditions. Should one introduce silica – rich plants into a given site? What will be the consequences of introducing a conifer forest on a grassland under dry conditions? Such are the types of questions one could hope to answer using the model of chemical equilibrium for clay assemblages. However, there are a great number of questions which need attention. Some of them concern kinetics of clay mineral formation. There are indications in several studies that the change in clays can be important over periods of years if one can take the liberty to extrapolate laboratory conditions of chemical variables and temperature that are outside of the normal range known for terrestrial soils. Producing new silicate minerals is the key to the predictive power of the model. Models must take into account the change in solution concentrations due to wetting and drying during a year. Drying increases the ionic potential of the pore solutions and hastens reaction rates. Wetting dilutes the solution and slows or stops reactions. The periods of chemical activity per year must be determined according to the fluctuations of water content in a soil. Thus there is a significant amount of work to be done before one can start a serious rate of change prediction. Nevertheless it is possible to use long term experimental sites to monitor change in clay compositions due to change in vegetation. If the parameters of climate can be assessed reasonably or at least be compared to other sites, such information can be used and fit to the model in order to establish rate constants. The potential of the model is such that further investigation is sure to be useful in the long run.

7.5 Roots and Landscapes

An aspect that has been only mentioned briefly in the text is that of root action concerning the retention of soils, the stability of landscapes and the overall importance on a long term basis for landscape stability. Engineers have been concerned with embankment stability and the consequences of changing the morphology of a landscape. Some work has been done to try to understand the importance and functioning of root systems as they relate to surface land mass stability. This work is extremely important and would deserve a serious effort in developing models based upon different types of root structures and distributions in soils formed under different climates. It is not just a problem for road engineers or housing development construction but also for the change in land use for agricultural purposes. In semi-arid regions the change in plant regime engenders a change is root structures and hence potential stability in land surfaces. Slopes, clay types, alterite depth, and other factors should probably be considered as factors of choice in using plants as stabilizing agents in a given situation. Perhaps more work could be done on this subject considering its potential importance for developing countries.

In agriculture, root densities are probably an important aspect in choosing a plant type in a given context, especially in areas of sporadic and intense rainfall. Stabilizing the surface to avoid clay surface erosion is one thing but prevention of mass movement by landslides, even on a small scale, can be an important part of agricultural consideration. Loss and disturbance of a landscape are important factors to consider.

In the same direction, one can consider the impact of introducing different types of forest plantations as a function of terrain and the risk of mass movement. This is of course a concern for foresters in mountainous areas and should always be kept in mind even in less steeply sloped areas. Roots are the conduit of mineral extraction from the earth but also the mainstay of stability for an individual plant and a group of plants under given conditions. The stability of a soil zone depends upon the plants present and the stage or advancement of alteration (thickness of the clay zone).

One should take into account plants, the alterite configuration and of course climate, when and how much rainfall there will be in a given environment. Factors of soil porosity (amount of water which can be incorporated) permeability (rapidity of evacuation of water) and change in physical properties with saturation (plasticity) are all part of the equation that links plants to their substrate.

7.6 New Pathways

Given this situation, and the all encompassing interest of mankind in the plant zone of continents (agriculture and forestry) it seems only natural that mankind has manipulated the manipulators (plants) to his benefit. Classical farming practice was designed to use the different effects of plant species in order to ameliorate the productivity of the surface for man's needs. New farming methods have largely ignored the lessons of the past preferring to substitute industrial know how for natural, time tested methods. Perhaps today with the new outlook on a foreseeable future similar to that known today, one can propose a new vision of the sciences concerning the interface between plants and the earth. This biosphere is in great need of being better understood. The means are available, modern methods of interpretation of X-ray diffraction spectra of clay minerals allows one to quantify changes observed natural settings. DNA sequencing methods can be used to identify more precisely the bio-world in a soil; bacteria; micorrhize and so forth. This can change our knowledge greatly by giving much more precision in determinations of the species present and allow one to un-ravel the extremely complicated inter-relations between the living and mineral world in soils. It is time to begin a new scientific endeavor, that of the ecology of soils concerning the mineral plant and microbiological world. It must be done together, as a whole system, or else it will be in vain.

Reference

Arocena J and Velde B (2009) Transformation of chlorites by primary biological agents: A synthesis of X-ray diffraction studies. Geomicrobiology J 26: 382–388.

Annex A
Simplified Methods for the Interpretation of X-Ray Diffraction Diagrams of Soil Clay Assemblages

Contents

Annex I X-ray Diffraction Diagrams of Soil Clay Assemblages and a Simplified Method of their Interpretation

Decomposition Methods

X-ray diffraction spectra of clay assemblages are the major means of identifying the mineral types present, and more importantly in measuring, quantitatively, the changes from one environment to another of the different components of the clay assemblages. This tool, now available routinely, via internet and by other means, is the key to any modern interpretation and assessment of clay mineral change and relative abundance. Identification is one thing, reasonably important in itself, but the change effected by different environments in clay abundance is the most important factor and tool which can be used to assess the use of different practices in using plants to effect changes in soils. Soil clays respond, rapidly, to changes in their local chemistry. This local chemistry is often, or can be, determined by the plants and bio – agents in the soils. The changes in clays are an indictor of their potential fertility, or infertility, for the eventual growth of other plants. Some plants need certain elements that are found stocked in clays, other plants other elements or perhaps none at all. The effects of plants on clays can be seen only through a quantitative and comparative analysis of X-ray diffraction spectra of the minerals present.

B. Velde, P. Barré, *Soils, Plants and Clay Minerals*,
DOI 10.1007/978-3-642-03499-2, © Springer-Verlag Berlin Heidelberg 2010

This has been demonstrated throughout the preceding chapters of this book, at least we hope so. Here we would like to present a short explanation of the principles of the interpretation of spectra using decomposition methods in order to obtain quantitative estimates of mineral type abundance. One must be careful in using such estimates in that the intensities of different minerals depends upon their composition, on their grain size and on the structure, inter-stratification of different layers, all of which change the peak surface for a given mass of material. Hence absolute abundance or mass is a difficult objective to attain. However, if one looks at a series of samples with approximately the same mineralogy, one can observe the changes of abundance in a relative manner and thus interpret the importance of the different chemical parameters affecting the clay assemblages. Relative change in mineral abundance is a very important estimate, especially in time series experiments, which are very common in agricultural practice. Comparing one sample against another is often more important than an absolute estimation of mineral abundance in a single sample.

The decomposition of complex spectra of electromagnetic signals is an old art. Problems arise when electro magnetic radiations interact with complex (often multiphase) materials. Here the individual bands of intensity, having each a specific intensity distribution, often overlap and present a vague distribution of intensities. Spectra of infrared interaction, Raman interaction, and different X-ray energy interactions have been the subject of efforts at extracting the individual components from the complex, overlapping series of bands in a spectrum. The use of relatively proven methods for an interpretation of X-ray diffraction data is not new (Lanson, 1997). However it is little employed at present in studies of clay minerals in soils. Nevertheless a certain number of studies have used such methods to interpret the spectra of soil mineral samples.

The problems of interpretation can be seen very clearly in the example of Fig. A.1.

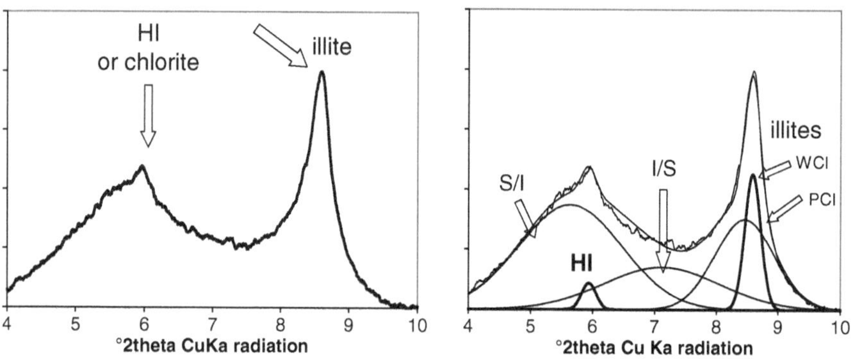

Fig. A.1 Typical background subtracted X-ray diffraction spectrum where the obvious peaks are illite and HI – chlorite types. However decomposition shows that these peaks are of minor intensity compared to the others, especially the HI peak. Most of the diffraction intensity is found in the mixed layer smectite – illite (S/I), illite – smectite (I/S) and poorly crystallized illite (PCI) peaks

In this background subtracted spectrum of a soil clay mineral assemblage one sees two sharp peaks which can be interpreted as diffraction peaks of specific minerals in the soil, illite and HI – chlorite. The illite peak is in fact a major feature of the spectrum but the HI mineral is in reality a minor component. However the large majority of diffraction intensity is contributed to by other minerals, two types of illite – smectite mixed layer minerals and the poorly crystallized form of illite. An identification based solely upon identifiable sharp peaks is very misleading in interpreting X-ray diffraction spectra of soil clays. One would say illite plus HI or chlorite when most of the diffraction intensity is contributed by mixed layer minerals of different sorts. This has been the problem with the interpretation of X-ray diffractograms for the last 50 years. It is almost impossible to estimate the relative abundance of the less well characterisable minerals present. Clearly it is necessary to go into the problem on a different scale.

Some Fundamental Concepts of Decomposition Techniques

Background

X-ray diffraction diagrams of clays usually show a highly asymmetric background intensity which increases with decrease in diffraction angle (° 2 theta). This background must be subtracted in order to observe the diffraction maxima for the different components of the clay assemblage. Such subtraction methods are routine in most analysis programs. In Fig. A.2 one can see the effect of background and its subtraction in going from A to B.

In the example of a meta bentonite mono-mineral sample, background subtraction gives a peak that is almost symmetrical, at near 7° 2 theta Cu radiation which can be decomposed into one major Gaussian shaped peak and another minor one to the low angle side. The peak is slightly assymmetrical.

Given the need for decomposing the complex X-ray spectra of clay minerals, how can one go about the process without making too many errors? If one considers again the Fig. A.1 there are obviously two sharp maxima but other peaks are present which overlap and give difficult interpretations upon visual inspection. In Fig. A.1 in the lower figure it is clear that the large part of the diffraction intensity of the spectrum is made up of large peaks that often overlap, with the result of an apparent continuum of slightly varying diffraction intensity. How then can one go about identifying the different components of this diffraction intensity?

Initially there are several principals which must be recognized:

Principles for Decomposition

Principle number one: in the region of interest for clay minerals, 4–14° 2 theta for copper radiation, or 16–7 Å, *some of the peaks appear to be slightly assymmetrical*, with a slight broadening to the low angle, high d value (Å) side. This can be an effect of peak broadening due to physical effects of X-ray diffraction as is the

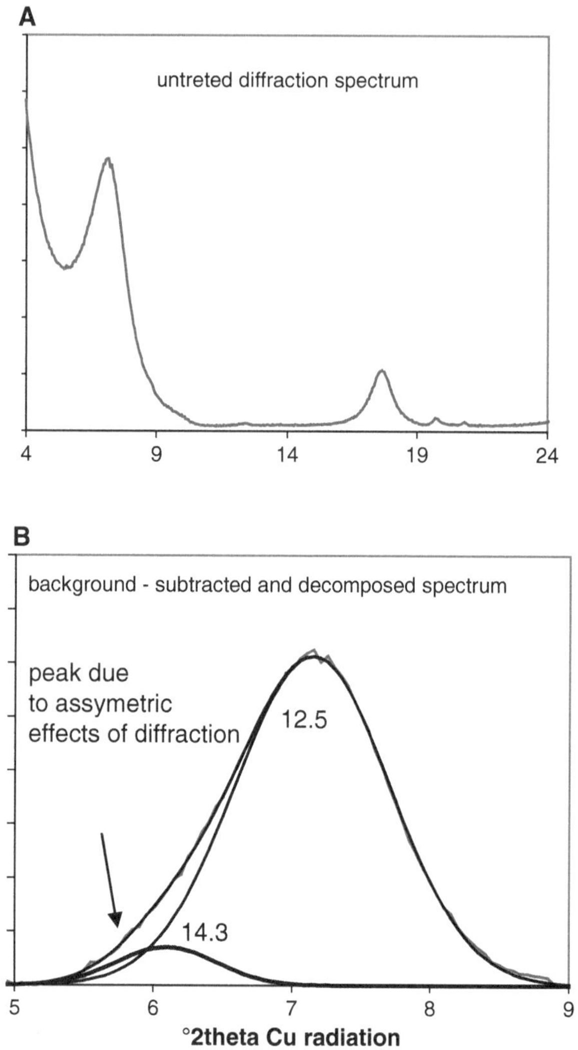

Fig. A.2 Illustration of the effect of peak assymmetry on decomposition results. The un-treated spectrum of a bentonite clay is shown and the decomposition result treating the peak near 12.5 Å. The assymmetry gives a decomposable peak near 14.3 Å which is only of minor intensity, about 7% of the intensity of the 12.5 Å peak

case in Fig. A.2. The assymmetry can be defined by a second peak of minor peak area, usually below15% of the total peak surface area of the major reflection. Figure A.2 indicates this effect for a bentonite clay (smectite) from Sweden (Kinekulle deposits). The assymmetric part of the peak can be simulated using a small peak to the low angle, high d value side of the major peak. In the example, the peak is found at 12.5 Å and the simulation of the assymmetric part of peak can be simulated with a small band at 14.3 Å. The relative intensity of the small band, simulating the assymmetry of the large peak is 7% of the area of the large peak. Thus, the problem of assymmetry due to physical effect of diffraction is minor. However assymmetry of peak shape is usually far more important in most clay spectra and thus is due to other effects. *Thus when there appears to be a small peak widening at the base of a sharp peak, it is most likely due to the presence of another mineral phase.*

Principle number two: It is possible to "fit" or simulate any complex curve with a suitable number of individual band components. This is the principle of Fourier synthesis analysis. However, there is no use for such results in the identification of specific mineral in a complex assemblage without any basis in mineralogical reality. We are interested in identifying each species present. There are not an infinite number of species present, in fact only a limited number in most cases. We in fact wish to simulate a complex spectrum by using the minimum number of components. Hence principle number two is: *use the minimum number of bands possible to simulate a complex spectrum.*

Principle number three is: *use only bands that could represent a known mineral.* Initially this is difficult to follow in that one does not know, a priori, what is possible and what is not. Experience is the best teacher, and above all use of published, established diffraction components representing different soil clay minerals is recommended.

Principle four : *peak widths range from 0.2 to 2°2 theta at half height.* Wider peaks are un-realistic. Usually wide peaks are the result of an inter-stratification of different types of mineral layers in the same crystallite or small crystallite size. If one needs a very large peak to decompose a spectrum, it usually means that there in fact two peaks present. Peak widening from grain size effects is much less than 2° 2 theta (Fig. A.3).

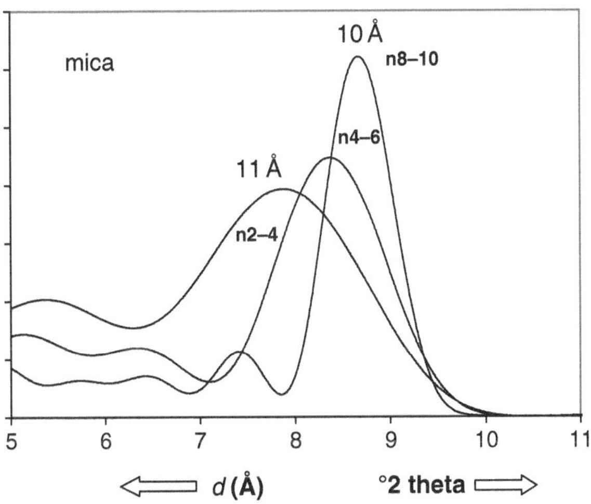

Fig. A.3 Diagram showing calculations using the Reynolds NEWMOD program where a mica peak is modelled. The peak near 10 Å occurs for well crystallized illite, i.e. where crystallites have near 10 coherently diffracting crystallographic units present. Decreasing the number of layers in the crystallites shifts the peak to higher *d* spacings, from 10 to 11 Å, and widens the peak

In Fig. A.3 the effect of coherent diffracting domain, an approximate estimation of the grain size or order in a crystallite, is shown for illite. This widening effect is especially true of illites in soils. The reason is that the grain size is variable, and some material shows a narrow peak at low d spacings (higher ° 2 theta) and a larger peak at higher d and lower ° 2 theta. This effect is usually interpreted (see Meunier and Velde, 2003) as being due to a peak shift and peak broadening due to the dispersion of grain sizes, and perhaps composition of the illite particles. The effect of peak shift due to grain size (number of coherent layers diffracting in a crystallite) is shown in Fig. A.3.

The peak shift is not negligible and peak broadening contributes to the well known apparent asymmetricity of the illite peaks, called in the past illite crystallinity. To our knowledge, peak asymmetry, with a fitted band of more than 15% of the surface area of the major peak, is due to a population of different particle sizes of the same clay species. This is typical of illite and kaolinite but we have not identified it in cases of other minerals (Fig. A.4).

In the example shown for a diagenetically formed illite, there is a decided asymmetry of the peak at 10 Å with a width at half height of near 0.7° 2 theta caused by a peak centered near 10.7 Å which is significantly wider, 1.4° 2 theta. Hence it is common to need two peaks to fit the illite peaks at 10 Å and often the kaolinite peaks at 7.21 Å. In the case of illite the more narrow peak is called WCI, or well crystallized illite and the wider peak is PCI or poorly crystallized illite (PCI).

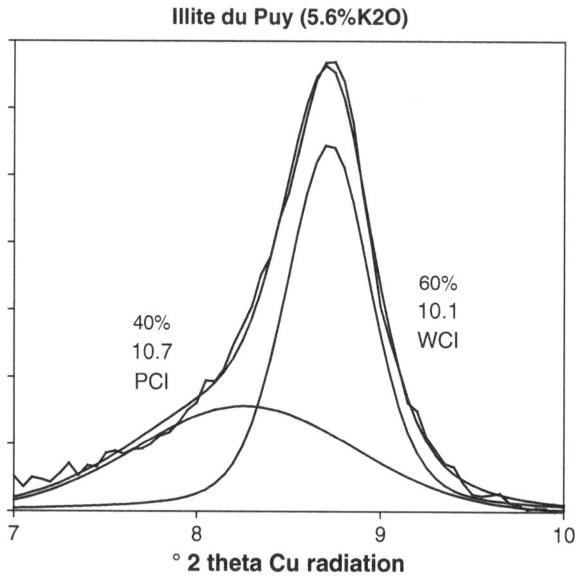

Fig. A.4 Typical peak assemblage for illite with a relatively narrow peak near 10 Å (called WCI or well crystallized innite) and a wider peak near 10.7 Å (called poorly crystallized illite, PCI). The shift and widening is due to a smaller grain size (coherent diffracting domains in the crystallites)

Decomposition Examples

All spectra should be background subtracted before an attempt is made to decompose the complex spectrum into its components. In the examples below we show the un-treated spectrum before treatment, and begin the decomposition with the background – subtracted spectrum.

The easiest procedure for decomposing a complex spectrum is to work from the edges to the center, like eating a sandwich. Once the outermost peaks are defined, one can make an attempt to determine peaks in the center. This usually means going from the 10 Å illite peak to the 14–15 Å peaks representing the HI or smectites phases. We advise using the divalent ion air dried state for the initial determination of mineral species present. It is important to use a specific protocol for curve fitting (decomposition) in that the proposition one uses to establish the best fit of a series of peaks to a complex curve depends upon a knowledge of the phases (peaks) likely to be present. In most soils one finds kaolinite present, but the position of kaolinite is such that it is detached (little or no overlap with other peaks) and it can be treated as an isolated example, usually with two peaks as indicated above. We consider the 10–15 Å region (4–12° 2 theta using Cu radiation) as the most complicated and, for plants, the most critical part of an X-ray diffractogram necessary for the identification of the soil clay assemblages. It is here that one finds the 2:1 minerals, which can contain potassium Ca or Mg hydrated ions or Al-OH ions.

In Fig. A.5 we show a case of complex mineralogy where two peaks are evident in the spectrum, one at 10 Å and the other at 14.2 Å, indicated by arrows D and B in the figure. In between one sees a distinct but wide hump, arrow C. However there are one other point to be remarked, indicated by arrow A, which is where the slope to the low angle, high d value side of the 14.2 Å peak shows an apparent asymmetry, or spreading at the base. The 14.2 Å peak is sharp and well defined but its base is wider that a normal peak shape would allow. We use a proposition to fit the wide base of peak, B, with curve A. Then one can fit the sharp peak (B), to the 14.2 Å peak. The combined peaks A + B account for much of the diffraction in the region of the first peak.

Now we can look at the high angle, lower d spacing side of the spectrum. Here the illite peak is evident and can be fitted with a band at 10 Å (peak D). However, there is a slight widening at the high angles side of the peak. Also there is some asymmetry to the low angle side of the 10 Å. This can be fitted using the peak E. As we have indicated above, illite is usually described by two peaks, one at 10 Å (called WCI or well crystallized illite) and the other slightly to lower angles 2 theta and higher *d* spacings (PCI or poorly crystallized illite). Thus a two peak proposition is most often necessary to describe the potassic illlite component in soils.

Having taken care of the exterior parts of the spectrum one can then look at the obvious peak at near 11.9 Å in the center of the spectrum (C). This can be simulated by a peak at 11.9 Å. However, there is still some diffraction intensity found between this peak C and the illite peaks (D and E). A peak at 11 Å is needed to complete the simulation. Here we have found two peaks at each end of the spectrum associated with a diffraction maximum, one in the center associated with another obvious

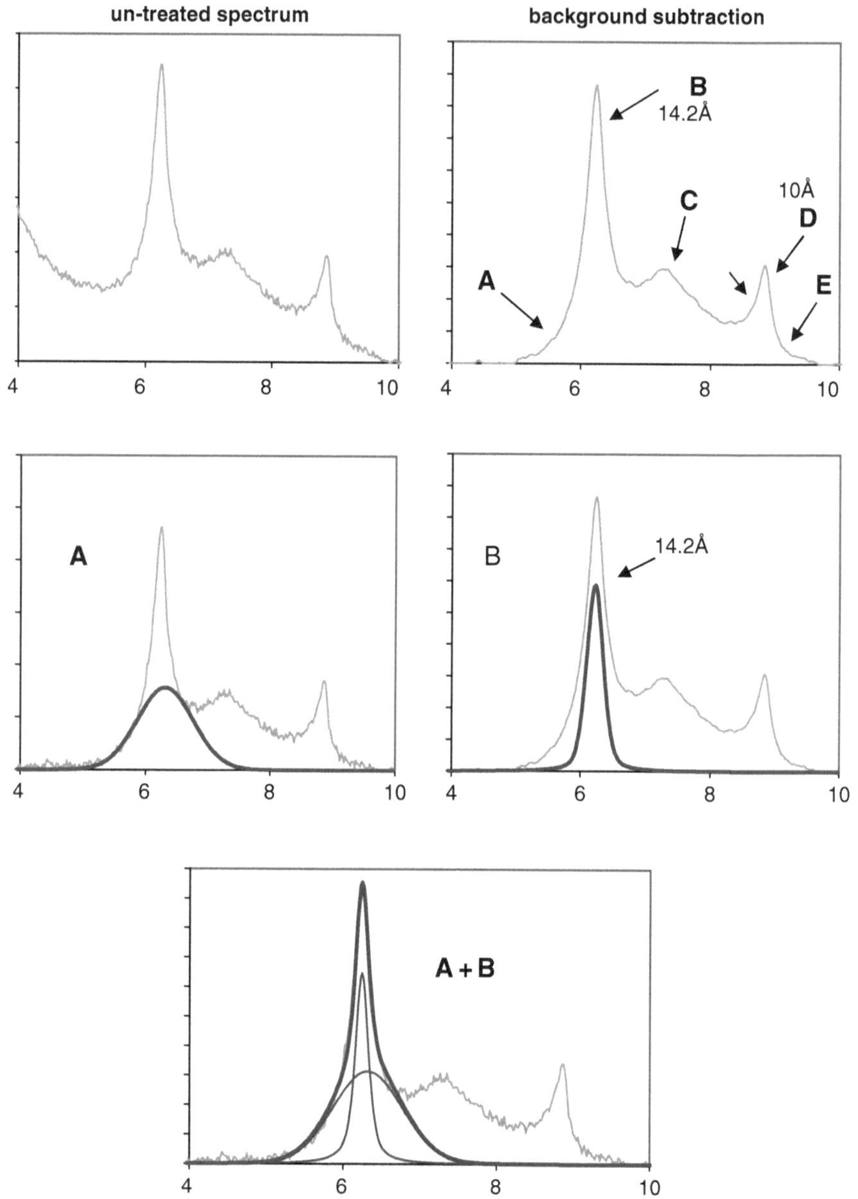

Fig. A.5 Example of decomposition of a complex spectrum. The initial spectrum is untreated and the following spectra are background subtracted. Background subtraction is the first step in decomposing spectra. One can readily identify peaks A through E visually. Peaks A and E are subtle but necessary. Peak A is fitted in order to take into account of the spread at the edge of the spectrum. Peak B is used to take into account the intense and sharp peak intensity near 6° 2 theta. On the right hand side of the spectrum, the same process is used but here the wide peak (E) which expands the spectrum base is on the inner side of the intense peak near 9° 2 theta (D). The peak near 11.9 Å is then fit into the spectrum and final decomposition shows a need for a peak at near 11 Å

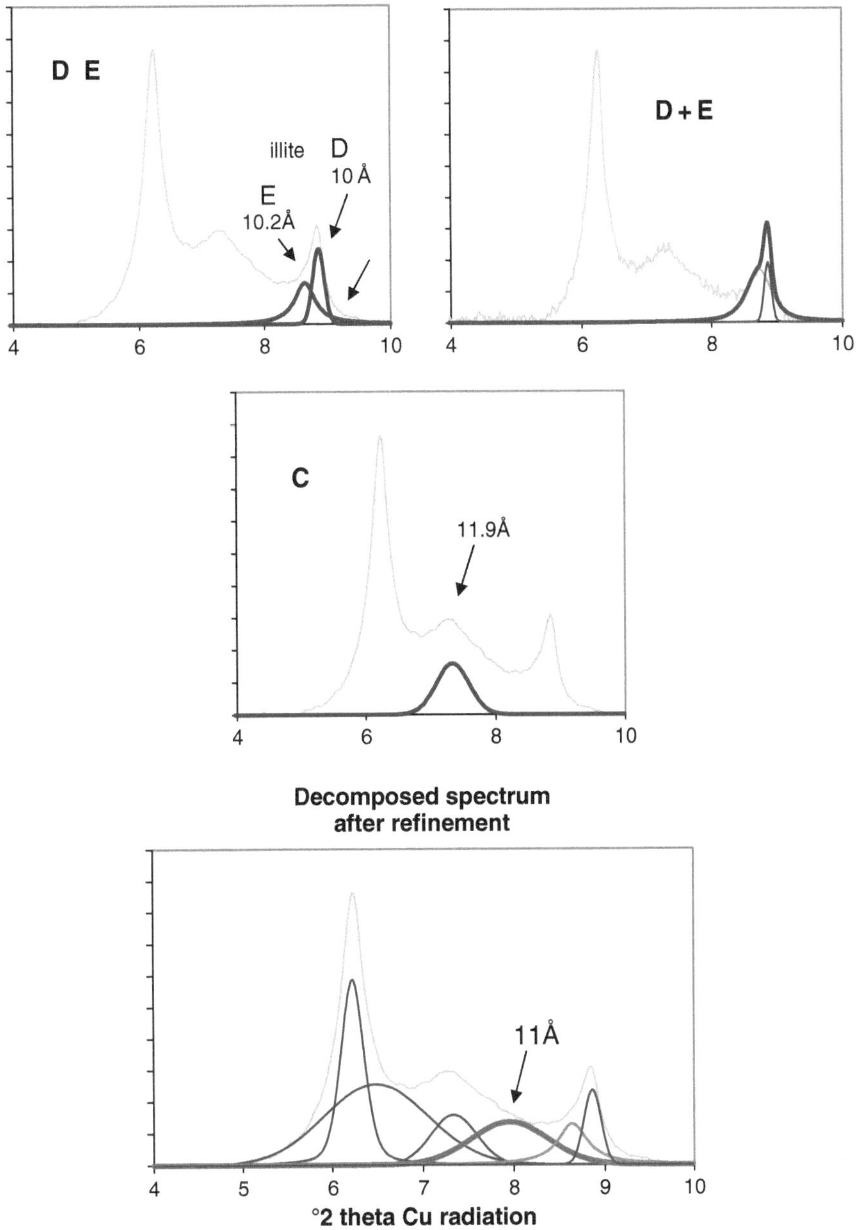

Fig. A.5 (continued)

diffraction maximum and another which is found due to the incomplete description of the initial proposition. The last peak at 11 Å is not to be ignored in that it represents 15% of the total peak diffraction area. The sharp, clearly visible peaks represent only 25% of the surface in the example given. Therefore it is extremely important to use decomposition in order to have an idea of the total diffraction surface and phases present. The final least squares refinement of the spectrum, based upon the propositions we have proposed shifts some peaks slightly, for example peak A which is widened and moved to a higher position in ° 2 theta. This indicates that the whole spectrum must be considered when all elements are fitted to find the best solution to the problem.

In Fig. A.6 we indicate another example of similar mineralogy, but with different problems of decomposition. Here the peak at 14.2 Å is well expressed and dominant (B), while the peak at 10 Å (illite, peak D) is very small, in fact a shoulder on a larger, broader peak, indicated by arrow C.

Starting at the low angle (high *d* values) we can propose two peaks as before to fit the widening at the base of the sharp 14 Å peak. This is peak A which when completed by a sharper peak at 14.2 Å (B) completes this end of the spectrum. The necessity of peak A is seen at both sides of the sharper peak B.

The low d spacing, high angle side, of the spectrum is simulated by the double peaks of illite (D and C). In this case the illite peak of low crystallinity, (PCI, smaller number of diffracting layers), is significantly larger (greater peak area) than the 10 Å, WCI or well crystallized illite peak. However there is significant diffraction intensity between the 14 and 10 Å ends of the spectrum. This can be accounted for by a peak at 12.1 Å which is relatively wide (1.3° 2 theta).

Again we see that the least squares refinement for a best fit changes the peak positions to a certain extent, again the peak A is shifted to higher ° 2 theta and widened slightly.

Peak Shape and Intensity Change: The Comparative Method

The major use of decomposition is to identify the peaks present and to compare their relative areas in order to identify changes in mineral abundance and mineral type. This can be done for similar materials which have been subjected to different chemical changes.

An example of the use of decomposition to follow changes in spectra is given in Fig. A.7. A sample of agricultural prairie soil from Illinois is shown in the strontium – saturated state and in the potassium – saturated state. In the Sr-sat diagram, one finds two mixed layer illite – smectite peaks. The first is the smectite – rich phase (S/I) which is very close to smectite in composition, given its spacing at 15.5 Å. Another illite – smectite mixed layer mineral is present, whose peak is closer to 10 Å (12.8 Å). The two illite peaks (PCI and WCI) are present, with the poorly crystallized form (PCI) of larger peak area than the well crystallized form (WCI). The two illite peaks represent 34% of total peak areas.

Potassium saturation shifts the peaks to higher peak positions and lower *d* values, indicating an increase in potassium content as the peaks approach 10 Å. Potassium in the anhydrous form in the interlayer position, produces a 10 Å spacing. In the global experimental spectrum, irregular line above the component peaks, one sees an increase in the presence of the illite peak, indicated by the arrow. The combined area of the illite peaks now represents 55% of total peak area. Further, the smectite –

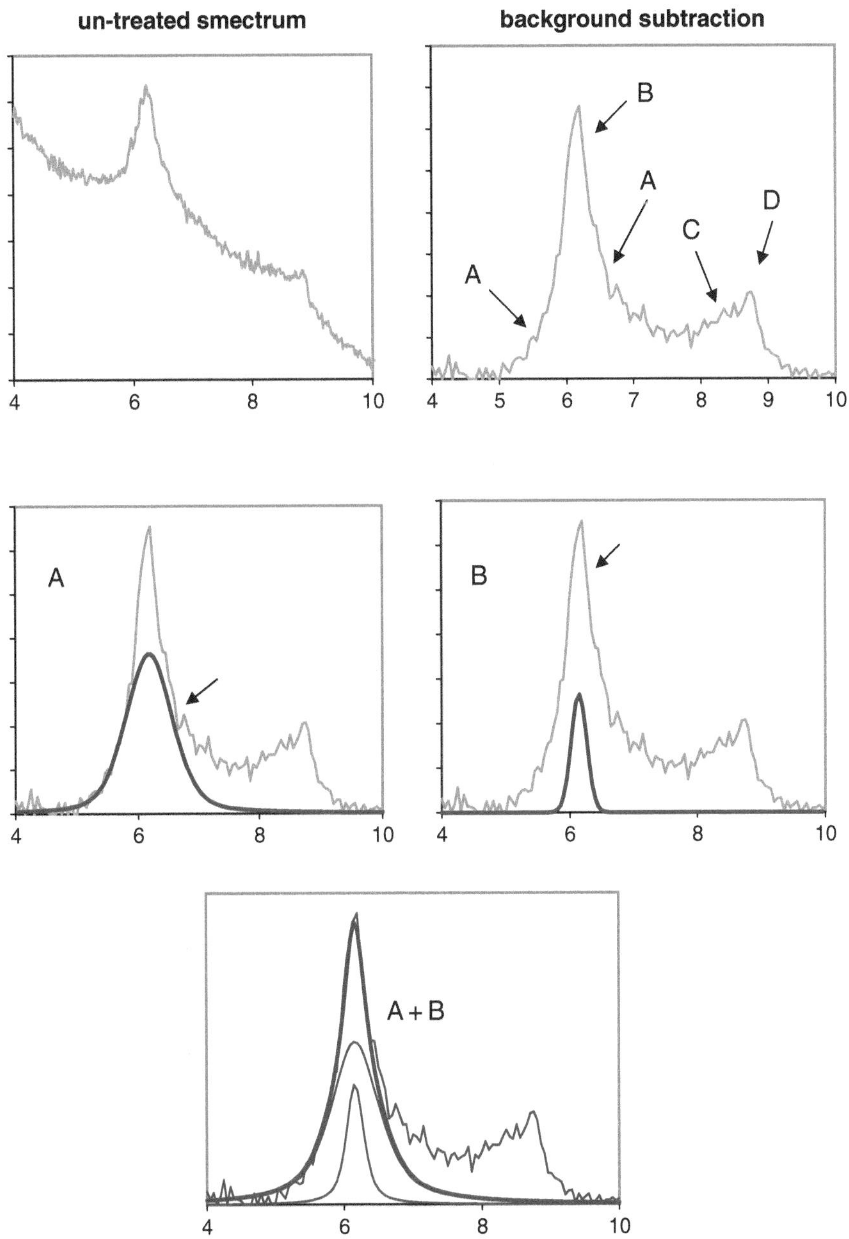

Fig. A.6 Example of the decomposition of an HI – bearing assemblage with a major peak near 6° 2 theta. The first peak on the left has a spread at its base on either side of the major peak B. this is fit with peak A. The second portion of the spectrum to be fit is on tD.he right hand side where two peaks are used to describe the illite material, one wide peak C (PCI) and a smaller illite peak (WCI). Final decomposition needs a peak to take into account of the intensity between the HI and illite peaks

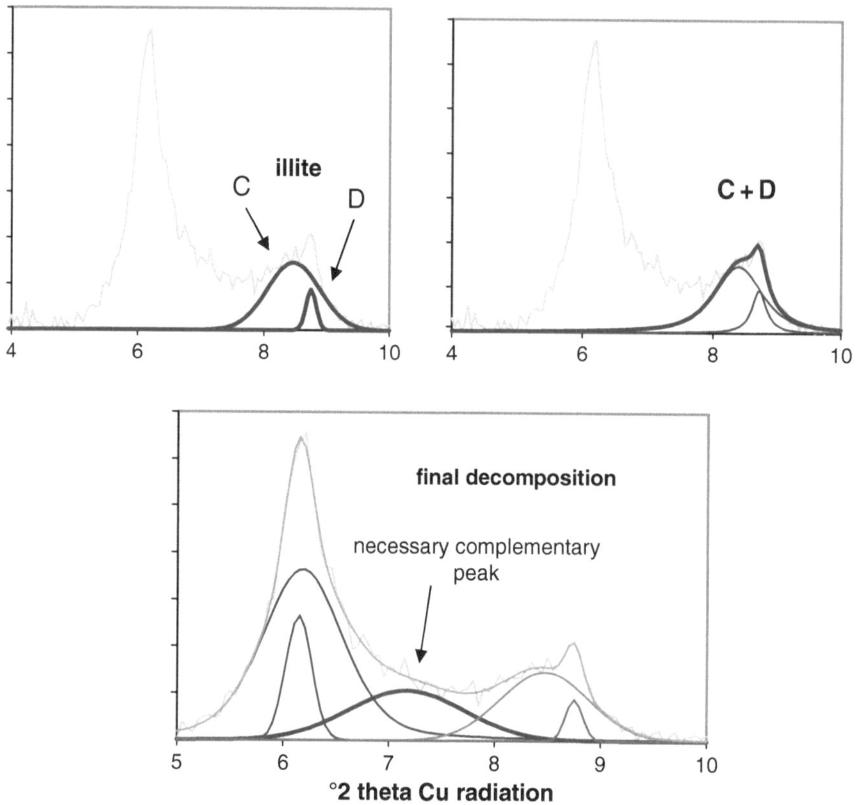

Fig. A.6 (continued)

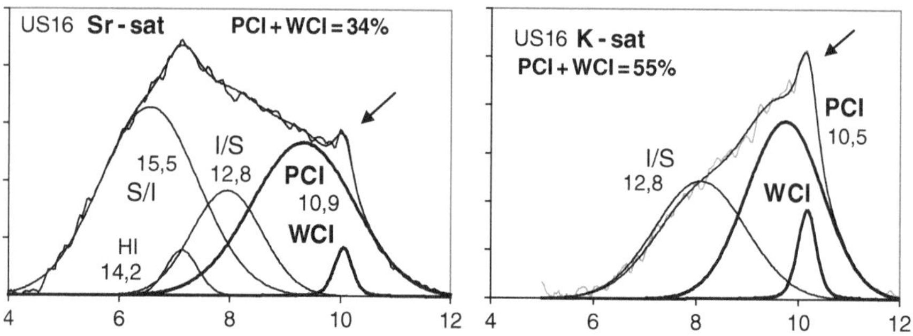

Fig. A.7 Illustration of the effect of saturation saturation a clay assemblage of primarily illite – smectite clays. Potassium shifts the peaks toward an illite component of both greater PCI and WCI peaks. The peak areas can be calculated using the width at half height (WHH) in ° 2 theta multiplied by the peak intensity determined by the decomposition routines. All peak intensities (width times intensity) are added together and each is divided by the total of determined intensity in order to estimate a relative peak area rich mixed layer mineral peak has disappeared, with an important I/S phase present as in the initial spectrum. Note that the S/I and I/S peaks define the peak envelope in Sr and K – saturated states and that the PCI peaks define the other side of the peak envelope. By comparing the decomposition peaks in the two spectra one can numerically determine changes in the mineralogy (interlayer ion occupancy) of the samples under the two different treatments

Surface Area

In this example we indicate the relative peak area of the different phases identified. The method of estimating the surface area for Gaussian peaks is simple, one multiplies the width at half height times the peak maximum value which defines a good approximation of the area. Summing the peaks and dividing the area of each by the total gives an estimation of the relative area for each phase.

Another example of the use of decomposition methods can be seen in a series of three soil samples (top of A horizon) under a sequoia tree near the trunk, at 2 m from the trunk and in the open prairie (lawn) around the 150 year old tree (park of the château Saint Fargeau, Vienne, France). The effect of the sequoia, grown on the same initial soil as that of the grass in the park, is to increase the potassium interlayer content of certain of the clays. One sees in Fig. A.8 that the smectite – rich phase (S/I in the figure) maintains its peak position and relative intensity. This peak, defining the left hand side of the peak envelope, is stable. The smectite – rich inter-layered mineral (S/I) appears to maintain its peak position also either in the grass or sequoia soils. However, one finds that there is a great increase in the intensity of a peak near 11 Å, an illite – rich inter-layered mineral (I/S) and in the intensity of the illite peak. Here it seems that the phases present initially remain essentially of the same composition (ratio of illite and smectite present in the crystallites) while a portion of the initial phases becomes either an illite – rich inter-layered mineral or illite.

Center of gravity

If one multiplies the peak area of each phase (width at half height times intensity) and compares it to the total peak areas for all peaks, relative peak area, one can then multiply by the area by the peak position (in Å). This gives the contribution for each peak. By adding the contributions one makes a weighted average of the peaks in a spectrum. This is the center of gravity approach (Barré et al, 2009) to numerical comparisons. In the Fig. A.8 the center of gravity measurements (cg) are indicated for each spectrum. One sees that the center of gravity reflects directly the shift in peak intensities and peak positions described above. The cg shifts from a position of 14.2 Å under grass showing the predominance of smectite (two water layer hydration and a contribution at 15.2 Å) which is decreased under the influence of the sequoia tree where the cg value is 12.8 Å mid way between smectite and illite positions. This indicates the change in illite content of the mixed layered minerals and the amount of illite created by potassium saturation. This measure is very useful

Fig. A.8 Illustration of the effects of plants (sequoia vs. grass) on clay mineral type. Estimation of the effect can be made by estimating the center of gravity for the spectrum. This is obtained from an estimate of the individual relative peak intensity for each peak (width at half height times intensity, divided by the sum of peak intensities) multiplied by the peak position (in Å). The results are summed to give the center of gravity or average position considering the individual peak areas

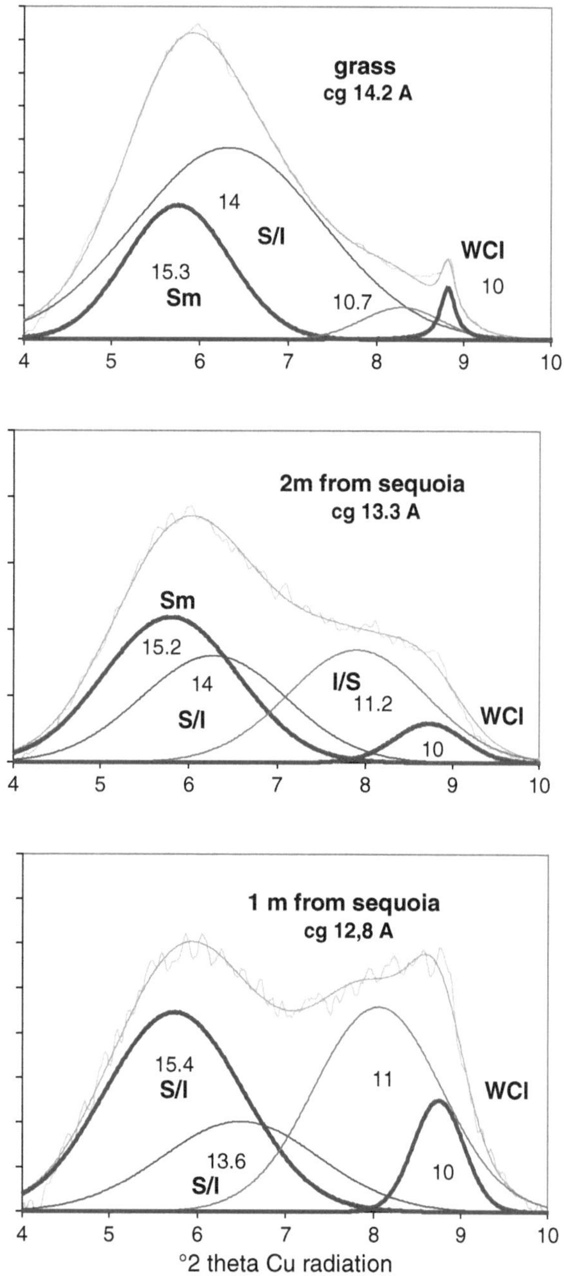

as a comparative tool for soil assemblages which are similar but have experienced different histories.

Curve decomposition is in fact rather simple when one uses a minimum number of peaks to describe a complex spectrum. The most powerful use is to follow changes from one state to another of a clay assemblage which has been subjected to chemical changes either in the laboratory or those induced by plants in the soils themselves.

Annex II Identification of the Major Groups of Clay Minerals

Major Mineral Types

As described in the initial chapter (Chap. 1.3) clay minerals in soils can be classified into two groups;

1:1 minerals essentially kaolinite or hydrated kaolinite (halloysite)

2:1 minerals which include the groups of

Illite (anhydrous potassium interlayer ions)
Smectites with hydrous interlayer cations, essentially divalent forms (Ca, Mg)
HI minerals with hydroxy interlayer ions or complexes based essentially on Al forms.

Hydroxyl – oxides such as gibbsite, Al $(OH)_3$, iron oxides and hydroxyl, hydrous oxides. These area not silicate minerals and represent advanced stages of weathering even though they may occur in the initial stages of the water – rock interaction process.

X-ray diffraction sees essentially four phyllosilicate mineral types, based upon interlayer spacings of the sheet – like structures, the first near 7 Å, the kaolinite 1:1 structure. Another spacing, based upon a 2:1 structure, is near 10 Å which indicates a potassium interlayer without water molecules present. The third is a spacing at 14.2 Å, still a 2:1 type mineral, where the interlayer site is filled with hydroxyl ion (OH) units either as hydroxyl cations or fully polymerized hydroxy ion interlayer structures. These are HI minerals. The fourth and final type of spacing is near 15.2 Å, which indicates the presence of bi-hydrated divalent exchange cations. These are smectites.

Thus one has a relatively simple nomenclature to follow based upon X-ray diffraction characteristics as a function of laboratory treatments:

1:1 mineral	*Hydrated*	*glycol treated*	*heated 200°C*
Kaolinite	7 Å	7 Å	7 Å
2:1 minerals			
Illite (anhydrous K ions)	10 Å	10 Å	10 Å
HI (hydroxy ions, OH + cation)	14.2 Å	14.2 Å	14.2 Å
Smectite ($2H_2O$ + cation)	15.2 Å	17 Å	<15.2 Å

From the above one can see that much of current laboratory treatment of conventional use is designed to identify the smectites type interlayer ions structure. Other more complicated treatments can be used to distinguish HI minerals, but essentially the peak position of HI structures is sufficient.

Inter-Layering of Structural Types

There seem to be only rare cases of inter-layering between 1:1 kaolinite and smectite, although it is not unknown to say the least. We will designate this as K/S inter-layering. The principal characteristics are a peak to the high *d* spacing side of the 7.2 Å kaolinite peak and a peak at greater than 15.2 Å in the air dried state. Normal 2:1 mixed layered minerals and smectites give peaks at 15.2 Å

Interlayering between illite and smectite (ill/sm) is rather well know and has been thoroughly studied. The peak positions are between 10 and 15.2 Å in the hydrated, air-dried state.

Interlayering between illite and HI (HI/ill) minerals gives peaks between 14.2 and 10 Å. One can call this HI/illite.

Interlayering between HI and smectites (HI/Sm) will give a spacing between 15.2 and 14.2 Å in the hydrated state.

Thus there are three types of inter-layering between the three interlayer ion poles of 2:1 minerals, and one type of inter-layering between kaolinite and smectites, which gives us four inter-layered structures to identify. Using the possibility of identifying smectite components by glycol treatment, one can simplify the identification procedure as follows:

Peak position air dried	*peak position glycol treated*	*inter-stratification type*
10–14.2 Å	unchanged	HI/ill
10–15.2 Å	increase in *d* spacing	ill/sm
14.2–15.2 Å	increase in *d* spacing	HI/sm

Peaks near the 10 Å illite position (between 10.5 and 10.0 Å) which do not change in position with glycol treatment are considered to be poorly crystallized (small diffraction domain) material called here PCI. Peaks that are un-affected by glycol and at the 10 Å position are WCI, illite. Thus for the illite pole there two peaks representing essentially the same mineral composition but of different crystallographic dimensions, PCI and WCI. The same is true for kaolinite. However we have not yet given the name of PCK and WCK to these forms.

Illite/Smectites

Very much work has been done on the mixed layered minerals illite/smectite given their regular transformation with changes of temperature over different periods of time in diagenetically affected (burial and increase in temperature and pressure constraints) sediments. This information is useful to interpret the different X-ray

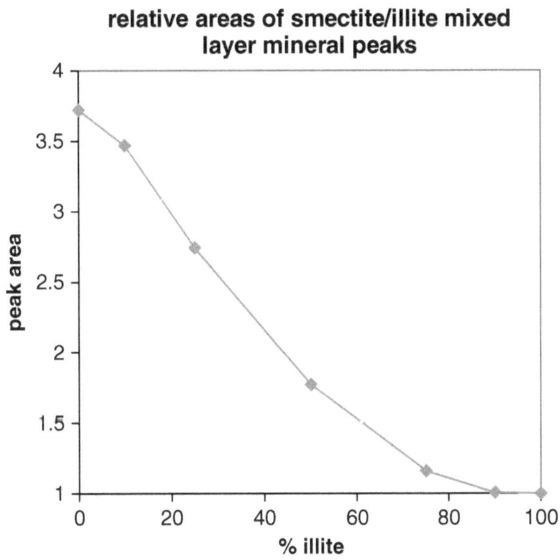

Fig. A.9 Illustration of the relative intensities of mixed layer clay mineral species in the smectite – illite series. Coherent diffracting domain is constant and hence the only variable is the change in diffraction intensity which results from changing the interlayer cations from K to hydrated Ca ions in the case of smectite. Calculations were made using the Reynolds NEWMOD program (Reynolds, 1985)

diffraction characteristics of soil clay smectite/illite interstratifications and perhaps other types.

Several relations are of great importance for interpretation of the peak positions and the relative peak areas of the different forms. Initially, when a mineral becomes more hydrated in the interlayer site, the relative intensity of the peaks increases compared to a constant mass in the sample. In Fig. A.9 we show the changes in peak areas of smectites/illite mixed layer minerals as a function of their smectite – illite composition.

The smectite form with two layers of water in the interlayer ion position changes the surface area of the peaks by a factor 3.5 while the increase in cell dimension changes by 1.5. Thus one must be very careful to refer to relative change in peak area between series of samples before any estimation of absolute peak area can be made. Other differences in structural organisation can change the peak areas even further.

Another point which is extremely important is the fact that most smectite/illite inter-layer minerals have a disordered layer stacking structure. This means that the regular succession of smectite followed regularly by illite throughout the crystal is rare. This type of structure is known, especially in minerals found in hydrothermal alterations or in the transformation by water rock interaction of the trioctahedral minerals biotite and chlorite. Most soil clay minerals have a disordered, or non-ordered, succession of layers of the different components present. As a result one can consider that the majority of cases peaks will have a position between 10 and 15.2 Å that will change position upon glycol treatment will be of a disordered type, called R0. In Fig. A.10 we indicate the change in peak position and shape for changes in illite content relative to smectite content.

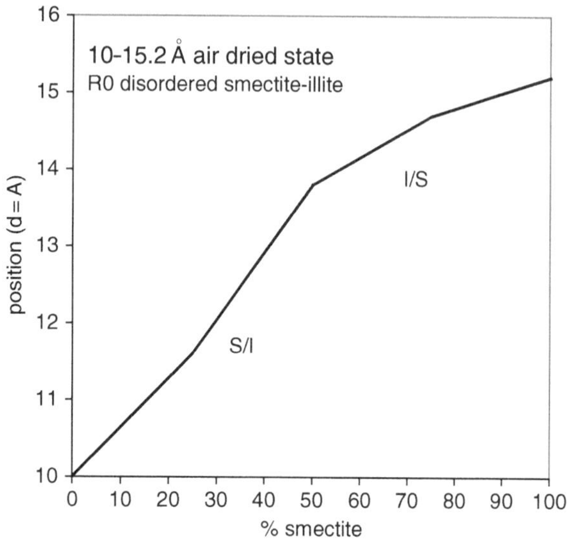

Fig. A.10 Relations of peak position in Å and smectite content in mixed layer illite-smectites. This curve is valid for random (RO) type interstratification of the clay layer types which is typical of soil clay minerals except in rare instances. Calculations were made using the Reynolds NEWMOD program

The relation of position and smectite content is not exactly linear but nearly so. One must remember that this is valid for samples with the same number of coherently diffracting layers (crystal size) in the crystallites. On the diagram sm/ill is used to designate the smectite – rich minerals and ill/sm the illite – rich minerals. Thus using a normal hydration state, divalent ion saturated and observed in the air – dried state, one can determine the proportion of smectites and illite present in the average crystallites.

The spectra for mixed layer minerals (illite-smectite) in the air – dried state where hydrous cations are surrounded by two water layers, is shown in Fig. A.11a. There is a general shift in peak position, seen in Fig. A.10, with changes in peak width.

However this is not true for the minerals when they are in the glycol treated state. In Fig. A.11b one sees that the glycol treatment tends to shift all peaks to a position near 17 Å with varying peak widths.

Hence the glycol treatment loses the potential compositional information for a portion of the smectite-illite inter-layered minerals. It essentially effaces the difference between the S/I and I/S minerals. This is a very important point to remember. The use of a moderate concentration (0.5 M) divalent ion saturation (Sr ions are preferred) to determine the composition of a mixed layer illite-smectie mineral is essential. Since we are often interested in the potassium resources in clays (essentially the illite content) the peak position of the illite-smectite minerals is a very significant index to changes in chemical environment in soils.

The two aspects one should keep in mind in the assessment of illite-smectite mixed layer minerals is that the more smectite present the greater the peak area of the same mass of material. The second aspect is that the proper assessment and identification of illite-smectites as well as smectite – illites should be done in the air – dried state for a divalent cation exchanged sample.

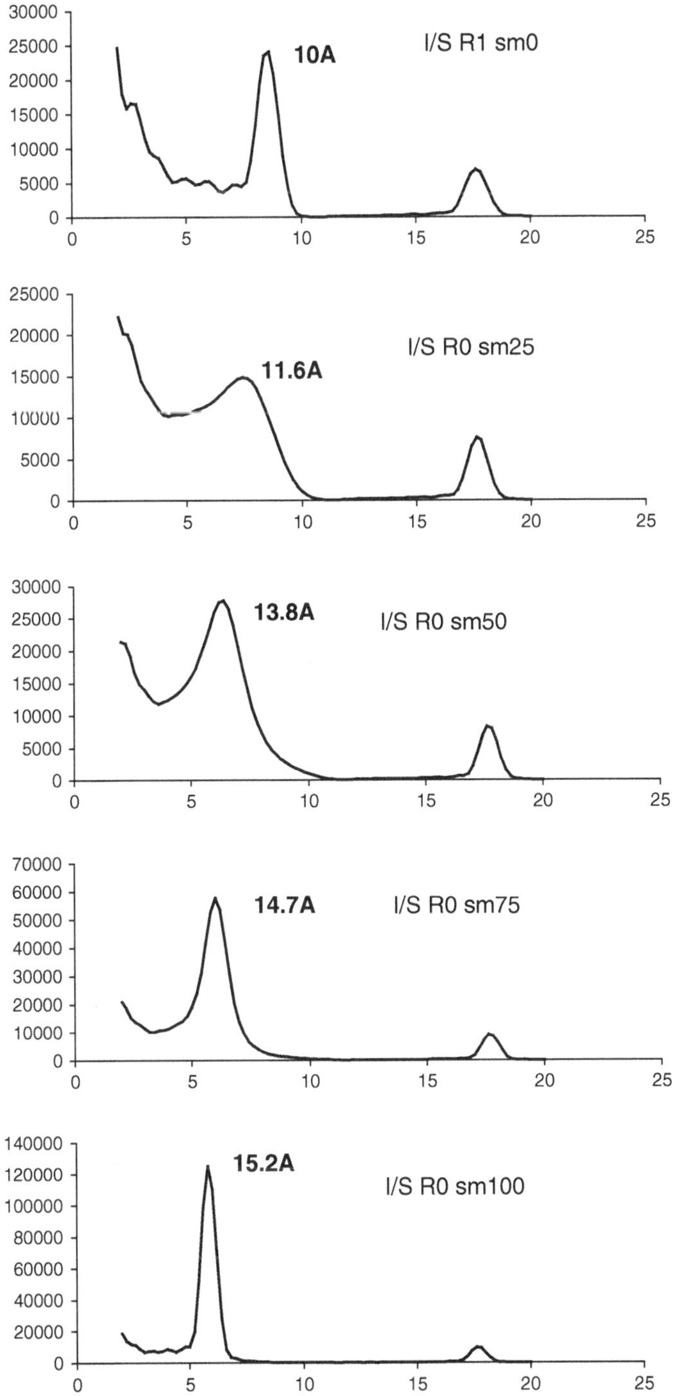

Fig. A.11 Illustration of X-ray diffraction spectra calculated using the NEWMOD program for two types of interstratified illite-smectites, two layer ordered, R1 shown in (**a**) and without ordering, type R0 shown in (**b**)

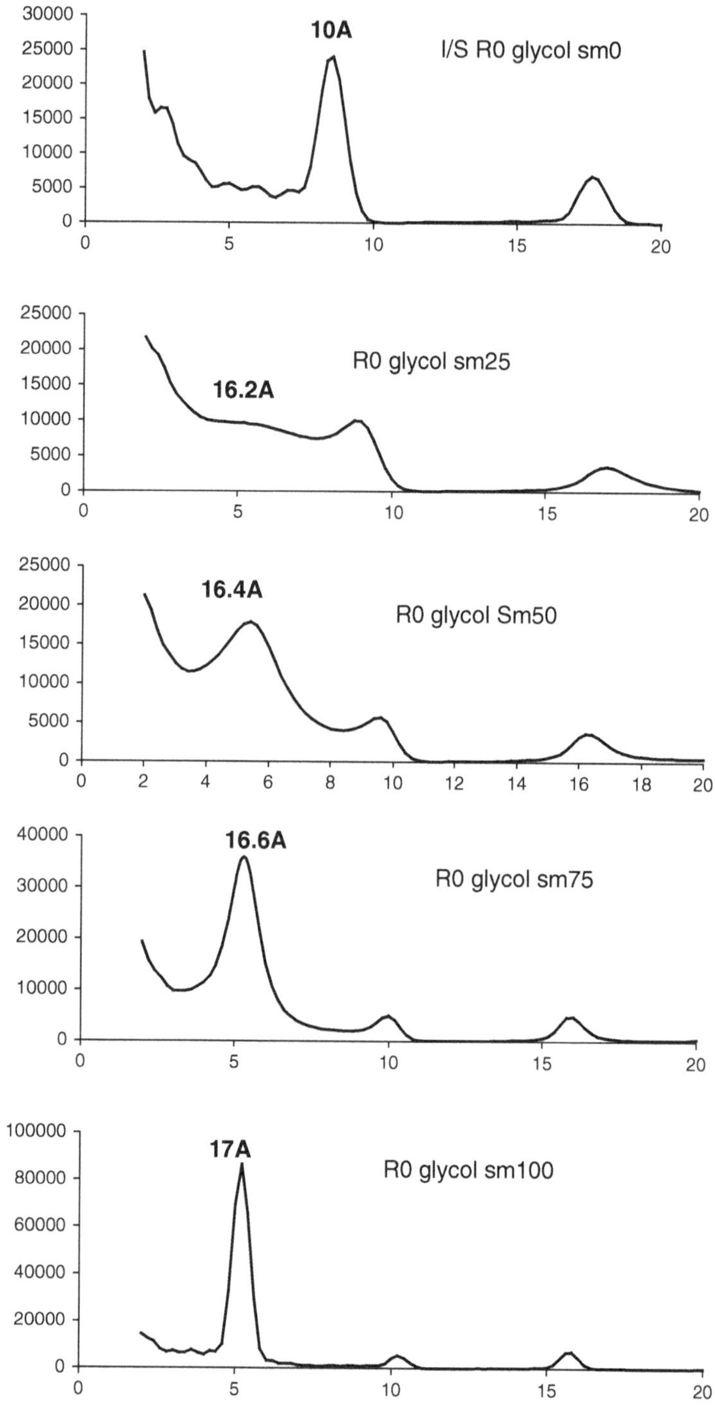

Fig. A.11 (continued)

HI – Illite Inter-Stratified Minerals

These minerals have been described many years ago (Millot and Camez, 1963). However their importance has perhaps been overlooked. Most authors consider that the HI mineral is the stable phase which can be interstratified with smectites. However, the exchangeability of Al-OH and perhaps Mg-OH interlayer ions with potassium is an important part of the mineralogy and chemistry of these soil clay minerals.

HI minerals have a basal spacing of 14.2 Å which does not change with glycol treatments or changes in air humidity. Heating changes the interlayer spacing in an irregular manner, tending to show a smaller basal spacing, moving towards illite. However, treatment with K ion solutions can, and does in most cases for soil clay minerals, change the spacing towards 10 Å and at the same time some now illite (increased peak intensity) appears. Thus there is a frequent, variable, tendency for an exchange of hydroxyl interlayer ions with potassium. These minerals can be identified by a wide peak between 14 and 10 Å of course, which does not change position with glycol treatment. Mixed layer illite – smectite minerals have a peak in this region, but it is strongly shifted to near 17 Å with glycol treatment.

This is a very brief introduction to the modern methods of treating X-ray diffraction spectra. Most new diffraction machines come equipped with decomposition programs. However, one must use them with care, so that a minimum number of peaks are defined and are well placed. Refinement by least squares methods can give good results or totally aberrant results depending upon the initial propositions that the operator defines. If one considers that in the air dried state, peaks will occur around 15.2, 14.2 and 10 Å and others at positions in between, one can eliminate many errors. The phases must correspond to true mineral possibilities as outlined above. Usually in soil clay assemblages one can have two illite-smectite minerals (S/I and I/S). The presence of an HI/illite mineral will occur with a strong HI phase. The three poles of illite, HI and smectites are the key to interpreting soil clay assemblages.

Calculations using the Reynolds NEWMOD program shown in Fig. A.12 indicate the position of this mixed layer peak and their width at half height, both values are affected by the composition (proportion of the interlayered components). Further, the number of coherent diffracting layers (*n* in the diagram) can shift the peak position and width making identification of these minerals rather difficult.

The above pages are designed to help the un-initiated to find their way in the difficult process of clay mineral identification based upon X-ray diffraction data. The only realistic method for soil clays must be based upon decomposition methods. Here one can identify peaks that are partly masked by others and in many cases one can estimate the relative proportion of these different components in the soil clay assemblage. Such estimates can help to understand changes in clay mineralogy which can be effected by plant regimes. This tool will allow one to see the effects of crop rotation or plantation of different trees on the clay mineral resources, especially the potassium present and available.

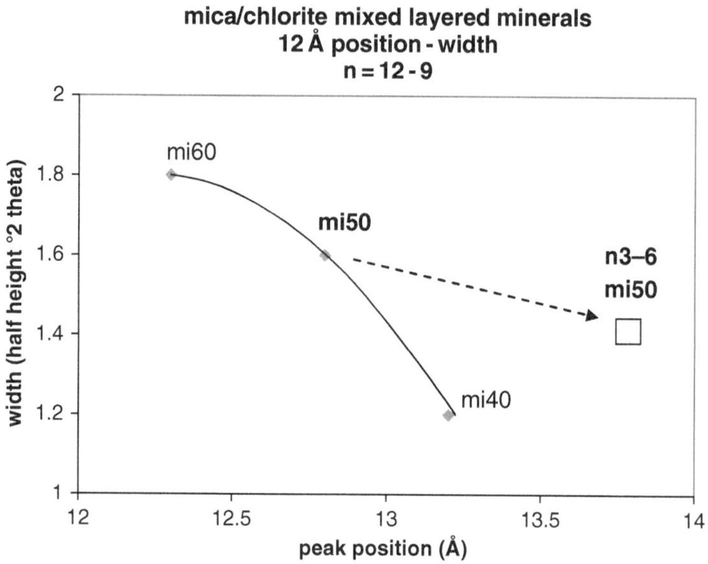

Fig. A.12 Results of calculations using NEWMOD for HI (chlorite) illite randomly interlayered minerals in the range of 40–60% illite component. Coherent diffracting domain is near 10 layers for the diamond points and 4 layers for the square. The shift in peak position and width is significant due to coherent diffracting domain size as is the case for smectite – illite interlayer minerals

References

Meunier A and Velde B (2003) Illite: Origins, Evolution and Metamorphism, Springer, 286 pp.

Millot G and Camez T (1963) Genesis or vermiculite and mixed layered vermiculite in the evolution of the soils in France, 90–95. In A Swineford (ed.), Clays and Clay Minerals, Pergamon Press Monograph 12.

Reynolds R (1985) NEWMOD, a computer program for the calculations of one-dimension diffraction patterns of mixed layer clays, RC Reynolds, 8 Brook Rd Hanover NH USA.

Index

Q

R

T

Kolibri 360 GmbH
Ferdinand-Jühlke-Straße 7
99095 Erfurt, Deutschland
produktsicherheit@kolibri360.de

On the Limits of Constitutional Adjudication

Juliano Zaiden Benvindo

On the Limits of Constitutional Adjudication

Deconstructing Balancing and Judicial Activism

Springer

Professor Juliano Zaiden Benvindo
Faculty of Law
University of Brasília
Campus Darcy Ribeiro
Asa Norte
Faculdade de Direito
Brasília - DF
Brazil 70.919-970
juliano@unb.br

ISBN 978-3-642-11433-5 e-ISBN 978-3-642-11434-2
DOI 10.1007/978-3-642-11434-2
Springer Heidelberg Dordrecht London New York

Library of Congress Control Number: 2010931461

Cover design: WMXDesign GmbH, Heidelberg, Germany

Printed on acid-free paper

Springer is part of Springer Science+Business Media (www.springer.com)

For Dani

Acknowledgements

This book is the result of the doctoral research originally entitled *Towards a Conception of Limited Rationality in Constitutional Adjudication: a Critical Response to Balancing in German and Brazilian Constitutional Cultures*, which was carried out at the Humboldt University of Berlin (*Humboldt-Universität zu Berlin*), Germany, and University of Brasília (*Universidade de Brasília*), Brazil.

I am especially grateful to Professor Dr. Bernhard Schlink, of the Humboldt University of Berlin, for supervising this research. Certainly, his precious remarks on its contents and systematization and the way he encouraged me to explore the relevant development of German constitutional culture were essential to this result. His concern for the need to develop the effective critique of the way constitutional courts make decisions was an inspiring message that accompanied me throughout the investigation. I would also like to express my gratitude to Professor Dr. Miroslav Milovic, of the University of Brasilia, who also supervised this work, for his relevant suggestions, especially the dialogue between Jürgen Habermas's *intersubjectivity* and Jacques Derrida's *différance*.

This research would not have been possible without the support of Professor Dr. Menelick de Carvalho Netto, of the University of Brasília, especially because of his suggestion to investigate this connection between Brazil and Germany in decision-making, and the problems arising from the belief in the rationality of balancing. My special gratitude also extends to Professor Dr. Vera Karan de Chueri, of the Federal University of Paraná (*Universidade Federal do Paraná*), for her careful observations on the contents of this research; Professor Dr. Cristiano Paixão, of the University of Brasilia, for encouraging me to enter into the fascinating world of constitutional history; Professor Dr. Dieter Grimm, of the Humboldt University of Berlin, for his precious lessons on the worldwide development of the principle of proportionality in his seminar; Professor Dr. George Galindo, of the University of Brasilia, for all debates on the internationalization of legal studies; Professor Dr. Márcio Iorio Aranha, Professor Dr. Marcus Faro de Castro, Professor Dr. Ana Frazão, and Professor Dr. Loussia Musse Felix, of the University of Brasília, for their support in the construction of the partnership between Humboldt University of Berlin and the University of Brasilia regarding this research.

My gratitude extends to Professor Dr. Alexandre Araújo Costa, Professor Dr. Sven Peterke, Professor Dr. Gustavo Rabay Guerra, Dr. Leonardo Barbosa, Maria Cristina Peduzzi, Paulo Maia, Paulo Blair, Fábio Portela, Guilherme Scotti, Renato Bigliazzi, Lucas Aganetti, Juliana Martins, Tatiana Baena, Thiago Jabor Pinheiro (especially for the suggestion of the title of this book), Douglas Pinheiro, Gamela, Guilherme Cintra, João Telésforo Medeiro Filho, Ricardo Lourenço Filho, and all members of the research group *Sociedade, Tempo e Direito,* of the University of Brasilia, whose debates were a fundamental source for many of the themes herein examined.

I would also like to thank Dorothea Münchberg, for her gentle support in all administrative issues at the Humboldt University of Berlin; Dr. Jakob Nolte, for his lessons on the principle of proportionality and for inviting me to participate in the invaluable debates with Professor Dr. Bernhard Schlink's and Professor Dr. Volker Neumann's staff at the Humboldt University of Berlin.

My friends Renata Camilo de Oliveira, Professor Dr. João Paulo Carvalho Lustosa da Costa, Brand Arenari, Carlos Cardonha, Fábio César Scherer, Roberto Muterle, Flaviano Isolan, Gevson Silva Andrade, Nike Bernhardt, Alexandre Hahn, Davi Tangerino, Daniela Sequeira, Ana Guimarães, Napoleon Molina, Miguel Montero-Baker, Felipe Muanis, Raphaella Luppi, thank you for all friendship and support during my stay in Berlin.

This research was supported by DAAD (German Academic Exchange Service), the Programme Alßan (the European Union Programme of High Level Scholarships for Latin America), and CAPES (*Coordenação de Aperfeiçoamento de Pessoal de Nível Superior*), to which I am very grateful. Particularly, my special thanks goes to Maria Salgado and Dr. Gabriele Althoff, of DAAD, for all kindness and helpfulness. My gratitude also extends to Konrad Redeker Foundation, for its support for the publication costs.

The dedicated and passionate support of my beloved wife, Daniela Azoubel, and the infinite presence of my family, Ângela, Francisco, Aldo, Érica, Nenzica, Regina, Zefinha, were vital to the result of this investigation, to whom I am eternally grateful.

Juliano Zaiden Benvindo

Contents

Introduction

In the book *Comunidade da Diferença,* Miroslav Milovic suggested a dialogue between Jürgen Habermas's *intersubjectivity,* in the idea of a self-reflexive community, and Jacques Derrida's *différance,* as a sign of a "sensibility towards the 'different'."[1] Without achieving any final word,[2] though, this dialogue could point towards what he called a "*self-reflexive community of différance.*"[3] These words, inherited from two complex and rather untranslatable philosophical thinkings, came out as a motivation for this research. The proposal was how to think of this idea of a "self-reflexive community of *différance*" in a particular relevant theme from which constitutional democracies have been challenged in their very basis. On the other hand, Bernhard Schlink, in his text *German Constitutional Culture in Transition,* after having criticized German constitutional scholarship and its worship of the German Federal Constitutional Court (*Bundesverfassungsgericht*), put forward the need to establish a "significant critical potential,"[4] one that could offer a critical investigation of the transformations in the interpretation and application of basic rights in German reality.

There were, therefore, two central ideas that flourished from these two suggestions: a theoretical and philosophical approach founded on this perspective of a "self-reflexive community of *différance,*" and the direct interest in the transformations German legal dogmatics has been suffering. The investigation of the German historical context of an emerging constitutional court with a movement towards activism, one that, more and more, transformed this court into a "forum for the treatment of social and political problems,"[5] and the consequent attempt to provide

[1]Miroslav Milovic, *Comunidade da Diferença* (Ijuí, RS; Rio de Janeiro: Unijuí; Relume Dumará, 2004), 131, translation mine.

[2]Ibid., 132.

[3]Ibid., 132, translation mine.

[4]Bernhard Schlink, "German Constitutional Culture in Transition," *Cardozo Law Review* 14 (1993): 735.

[5]Ibid., 729.

a rationalization of the way constitutional courts decide cases, made, finally, the link between the philosophical and the dogmatic suggestions. The historical background of German constitutional culture, the dualism between law and politics in the realm of the *Bundesverfassungsgericht,* the attempt to rationalize decision-making with these characteristics through the emphasis on balancing, as if it were "not an alternative to argumentation but an indispensable form of rational practical discourse,"[6] all seemed very interesting and relevant themes for this research. Indeed, by examining the empirical context of German constitutionalism and the recent scholarly developments, it was possible to verify that one relevant discussion that should be carried out, within the characteristics of a constitutional court assuming the authority to resolve the present and future problems of German society, in a typical political fashion, was the question of the rationality of decision-making. After all, by studying the question of rationality, in this dualism between law and politics in constitutional adjudication, the debates on rightness and legitimacy of constitutional decisions appear, showing thereby the real challenge of this movement for the comprehension of the principle of separation of powers, the quest for keeping consistent the system of rights, and, lastly, the concern for otherness, all of them premises of a constitutional court committed to constitutional democracy.

Yet, these suggestions became even more interesting and challenging when we extended the analysis to other constitutional culture, in a comparative study in which many associations, empirically and methodologically, could be established. The examination of Brazilian constitutionalism and the recent developments of the Brazilian Federal Supreme Court (*Supremo Tribunal Federal*) allowed concluding that possible interconnections exist between Germany and Brazil and, chiefly, in the way the *Supremo Tribunal Federal* decides cases, both in the comprehension of basic rights, as if they were objective principles of a total legal order,[7] and in the methodologies deployed to account for this political character it has gradually assumed. In this respect, the question of the rationality of balancing, as well as its reverberations through the themes of rightness and legitimacy of decision-making, also raises significant issues for critical investigation. Moreover, especially when the Chief Justice of this court said that "the constitutional court exists to make the most rational decisions,"[8] it seemed that the question of rationality in decision-making, and particularly the rationality in the middle of the growing deployment of balancing as a justificatory methodology for this new Brazilian constitutionalism, was not only an important matter for this research but also a necessary and actual discussion.

---

[6]Robert Alexy, "Constitutional Rights, Balancing, and Rationality," *Ratio Juris* 16, no. 2 (June 2003): 131.

[7]See Schlink, "German Constitutional Culture in Transition," 711–736.

[8]Gilmar Mendes, interview by Izabela Torres, "Entrevista - Gilmar Mendes," *Correio Braziliense*, Brasília (August 17, 2008), translation mine.

With these premises in mind, the research could then establish its main problems: (1) how was it possible that from German and Brazilian historical backgrounds constitutional courts emerged with an evident propensity for activism, assuming thereby, as their role, the discussion of the present and future problems of society in a way that challenges the principle of separation of powers?; (2) how could a concept of rationality that aims to justify methodologically this new constitutionalism, one that has a more political characteristic than indeed the concern for keeping consistent the system of rights, stem from these constitutional realities?; (3) how are both German and Brazilian constitutional cultures empirically and methodologically connected in the activity of decision-making? In sequence, this empirical analysis should lead to the debate on rationality itself, and how concepts of rationality relate to the practice of those constitutional cultures, stressing thereby the possible outcomes for constitutional democracy. For this purpose, some problems appeared: (1) which is the prevailing concept of rationality that is behind this movement towards activism in German and Brazilian constitutional courts?; (2) how does this concept of rationality deal with the tensions and complexities of constitutional adjudication, and how can it grasp the existing dualism between constitutionalism and democracy, or between law and justice?; (3) is this concept of rationality adequate to the dilemmas of constitutional adjudication stemming from the context of post-conventional societies where the indeterminacy of law reigns? Finally, this research should establish a possible reconstruction of the idea of rationality and apply it directly to the constitutional cultures that were previously investigated. The problems in this matter were: (1) how could we see another rationality in the practice of decision-making?; (2) how could this other rationality connect somehow to the idea of "*self-reflexive community of différance*,"[9] and then reveal its adequacy for the dilemmas of constitutional democracy and constitutional adjudication; and (3) how could this rationality be immediately grasped in the effective practice of decision-making and applied to the constitutional realities of Germany and Brazil? The central ideas of the research were therefore established.

We could then indicate three central hypotheses: (1) the constitutional courts' shift to activism, in Germany and Brazil, relates to the particular circumstance of a need to create a strong institution that could undertake the role of protecting constitutional democracy and the social values after a period of authoritarianism, when the government and the parliament were discredited, and the population was in need of receiving goods and benefits also through decision-making; (2) the consequent development of the idea of basic rights as objective principles of a total legal order, with an optimization nature,[10] and the progressive deployment of balancing, now seemingly rationalized within the framework of the principle of proportionality, shape a concept of rationality that may be inadequate to the dilemmas of constitutional democracy; (3) from the dialogue between *différance*

[9]Milovic, *Comunidade da Diferença*, 132, translation mine.

[10]See Schlink, "German Constitutional Culture in Transition," 711–736.

and *intersubjectivity*, in the idea of a "self-reflexive community of *différance*," it is possible to delineate a concept of rationality that, by acknowledging the boundaries of constitutional adjudication, is more adequate to constitutional democracy, because it has an explicit concern for keeping consistent the system of rights and is rooted in the quest for otherness, as the sign of justice. The central thesis, accordingly, relates to this disclosure of this concept of rationality (a limited rationality), which originates from the empirical investigation of German and Brazilian realities, passes through the perception of the troublesome consequences of the prevailing concept of rationality embedded in the practices of their respective constitutional courts, and ends in the dialogue between *différance* and *intersubjectivity*, as robust premises to account for a reconstruction of the rationality that should orient constitutional adjudication in the realm of indeterminacy of law.

Hence, this research will be carried out in three parts. The first part is concerned with the empirical problems related to German and Brazilian constitutionalisms; the second part refers to the debate on the rationality of decision-making itself, and particularly the rationality of balancing, using, for this purpose, the prevailing opinion that justifies this new constitutionalism; and the third part has a reconstructive endeavor through the stress on a concept of rationality that may be more adequate, within the contexts of indeterminacy of law, than the one critically examined in the second part. With this itinerary, it may be possible to materialize Schlink's message of a "critical potential"[11] and, likewise, connect this critical potential to Milovic's suggestion of a "self-reflexive community of *différance*"[12].

In the first part, this research will be carried out in three chapters, all of them focusing on the empirical reality of German or Brazilian constitutionalisms as well as on some constitutional cases arising from these realities. The first chapter will introduce three constitutional cases, two from Germany (the *Crucifix*[13] and the *Cannabis* cases[14]) and one from Brazil (the *Ellwanger* case[15]), which will serve as a first contact with the characteristics of this new constitutionalism and, more particularly, with the deployment of balancing as a methodological instrument that could best operationalize the discourse of this new constitutional culture. Yet, if the first chapter is centered on case analysis, the second and the third ones connect those cases to history. The second chapter will have as its main purpose the investigation of the German historical context that led to the erection of a constitutional court with a strong activism and the dilemmas arising from this movement. Similarly, it will examine the scholarship's attempt to systematize the main instrument originating from this constitutionalism, in which balancing appears as a fundamental element: the principle of proportionality. However, since balancing is not necessarily deployed within the framework German scholarship defines for

[11]Ibid., 735.

[12]Milovic, *Comunidade da Diferença*, 132, translation mine.

[13]BVerfGE 93, 1 – *Kruzifix*.

[14]BVerfGE 90, 145 – *Cannabis*.

[15]STF – HC 82.424-2/RS.

the principle of proportionality, the discussion of this principle will only appear to show how, contemporarily, balancing has seemed to acquire, according to the the German prevailing scholarly opinion,[16] a rational character insofar as it is inserted into the structural framework of the principle of proportionality. This chapter will also discuss other constitutional cases – although more briefly than in the first chapter – and introduce some German constitutional scholar's reactions against this *Bundesverfassungsgericht*'s movement towards activism and, specifically, towards the deployment of balancing. Finally, the third chapter will extend the analysis to Brazilian constitutional culture, revealing thereby the strong influence of German constitutionalism in the recent Brazilian constitutional life and in the way the *Supremo Tribunal Federal* decides cases. By the same token, it will examine some cases that expose this movement, and show the possible problems arising from this Brazilian Supreme Court's political character.

The second part, in turn, will investigate the defense of the rationality of balancing, as an attempt to justify the way the *Bundesverfassungsgericht* decides cases according to the premise that basic rights are objective principles of a total legal order[17] with an optimization nature. Briefly, it will reflexively gather the characteristics of German and, in a sense, Brazilian constitutionalisms examined in the first part through the eyes of a relevant interpretation and justification of this movement, one that aims to provide a rational comprehension of the way the *Bundesverfassungsgericht* decides cases. In this respect, it will begin, in the fourth chapter, by examining one of the most well-known and influential interpretations of this German constitutional culture in transition to activism and casuism. Robert Alexy's *Special Case Thesis* (*Sonderfallthese*) and his *Theory of Constitutional Rights* (*Theorie der Grundrechte*) will be used as the central sources to grasp how constitutional scholarship has attempted to set forth a methodology that could justify, through formulas and criteria, a rational response to the main complexities and difficulties stemming from this *Bundesverfassungsgericht*'s shift to activism, premises that could be extended to the Brazilian *Supremo Tribunal Federal*. The examination of the main premises of his thinking will raise the fundamental doubts about the belief in the rationality of balancing that he, by directly drawing attention to *Bundesverfassungsgericht*'s decisions, so strongly defends. It will also encourage the discussion of the limits of reason that might not be thoroughly verified in Alexy's approach. The fifth and the sixth chapters will, therefore, confront Alexy's premises with other viewpoints as a means to reveal that another concept of rationality within the realm of indeterminacy of law in constitutional democracies might be necessary. In fact, by disclosing the metaphysics that may exist behind the concept of rationality Alexy holds, as a direct reflex of a scholarship that interprets and justifies the *Bundesverfassungsgericht*'s activity as a rational one, it will be

[16]The most well-known representative of this opinion, as we shall stress in this research, is Robert Alexy.

[17]See Schlink, "German Constitutional Culture in Transition," 711–736.

possible to unfold another rationality, one that strongly acknowledges the boundaries constitutional democracy brings to constitutional adjudication.

The fifth chapter will introduce the intriguing and fascinating philosophy of Jacques Derrida, which will present a robust approach that can be applied to the dogmatic problem at issue. From Derrida's deconstruction and his perception that reason refers to the "reasoned and considered wager of a transaction between these two apparently irreconcilable exigencies of reason, between calculation and the incalculable,"[18] a crucial message will appear as a response to the defense of the rationality of balancing as Alexy justifies and sustains it. With the purpose of disclosing and undercutting metaphysics, Derrida's philosophy will open up a new perspective that will show the logocentrisms – or the metaphysics – that are embedded in Alexy's premises, which have some relevant implications in the German and Brazilian constitutional courts. After the introduction of the main concepts of his philosophy, as well as its extension to constitutional democracy and to legal reasoning, we will explore how Derrida's deconstruction can show that the claims to rationality, correctness and legitimacy that are at the core of Alexy's theory could, in reality, be a *logos* of rationality, correctness and legitimacy. Besides, we shall investigate which are the troublesome outcomes of this conclusion when we face the dilemmas of constitutional democracy, especially the commitment that constitutional courts must have to the principle of separation of powers. This is also where *différance* will unveil itself as a necessary message for the comprehension of the double bind of constitutionalism and democracy, and of law and justice.

The sixth chapter will continue to stress this purpose of disclosing and undercutting metaphysics, but it will work with the dogmatic problem through the eyes of theories that inherit some relevant Kantian influences, now discursively remodeled. More than the fifth chapter, now the purpose will be to investigate theories that directly examine the problem of legal adjudication in the realm of indeterminacy of law, particularly Klaus Günther's differentiation between discourses of justification and discourses of application, Ronald Dworkin's *integrity,* and, more emphatically, Jürgen Habermas's proceduralism. Even though their thinkings, when connected to the analysis carried out in the fifth chapter through Derrida's deconstruction, raise relevant insurmountable divergences, they also complement one another: while entering more directly into the institutional grounds of legal adjudication and the problems originated therefrom, they also provide powerful premises to expose how the idea of balancing, as Alexy rationally justifies it as a reflex of his interpretation of the *Bundesverfassungsgericht*'s activities, is metaphysical and can lead to problematic consequences in constitutional democracy. Furthermore, they will provide a robust response to legal adjudication within the context of indeterminacy of law, which does not result in balancing. With a distinct view, in the tension between facts and norms, they will project the question of

[18]Jacques Derrida, *Rogues: Two Essays on Reason* (Stanford, CA: Stanford University Press, 2005), 151.

intersubjectivity, which will then shape, together with Derrida's *différance,* the concept of limited rationality.

The third part has, for this reason, a reconstructive endeavor. It will gather the debates that took place in the second and first parts and, from them, unfold the *concept of limited rationality,* one that aims to express, in the practice of decision-making, the idea of a "self-reflexive community of *différance*." Now the dialogue between *intersubjectivity* and *différance* will appear as a more adequate response to the challenges of constitutional adjudication when confronted with the characteristics of a pluralistic society where the indeterminacy of law reigns. While the second part, for this reason, deconstructed the main premises of the prevailing concept of rationality emerging from this reality of constitutional courts' shift to activism, the third part aims to prove that another rationality is possible. By acknowledging its boundaries, reason releases itself from the beliefs in abstract formulas and criteria as the condition of its expression, and concentrates rather on the limits of history and the limits of justice. This is where the connection between *intersubjectivity* and *différance* will unfold its potentiality for the exercise of critique, the "critical potential" of the way constitutional courts should act, by revealing its direct application to those realities examined in the first part.

In this regard, the seventh chapter will focus on providing this disclosure of the *concept of limited rationality* by exposing, in the dialogue between *intersubjectivity* and *différance,* that reason has its limits in its incapacity to thoroughly recollect and gather the complexities of the reality and in its incapacity to fully do justice, which creates an incessant and interminable activity towards the other as a commitment to constitutional democracy. Finally, the eighth and last chapter will debate how this *concept of limited rationality,* previously theoretically discussed within the context of constitutional democracy, can be used in the practice of legal adjudication. For this intent, after having extended the conclusions of the last chapter to legal reasoning, thereby assuming the premises of the proceduralist response to the indeterminacy of law, but now radicalized by the emphasis on *différance,* it will recall the developments of the first part. This will be done through a reconstruction of German and Brazilian recent constitutional histories, showing how their constitutional courts might be forgetting the boundaries of reason when deciding cases, and by reexamining those three cases of the first chapter, in order to reveal that, when the judge acknowledges the limits of reason, the result will be strongly concerned with the consistency of the system of rights and with the quest for the other.

The final thesis of this book is that rationality in adjudication within constitutional democracies, more than the result of suitable techniques methodologically systematizing arguments through some abstract criteria and formulas, ought to be the result of the judge's posture, one that knows that it is by assuming the limits of reason that decision-making grasps the fundamental tensions of constitutional democracies. It develops then a practice that presupposes that there is no justice, when adjudication releases itself from the constraints institutional history raises, and, on the other hand, there is no justice, if this history detaches itself from the concern for otherness. Accordingly, the final thesis, while aiming to reveal a "critical potential" towards the effective activities of constitutional courts, showing how metaphysical their discourses might be, also sees, as a non-resolvable potentiality, the dialogue between *intersubjectivity* and *différance,* where the "self-reflexive community of *différance*" may unveil itself.

Part I
German and Brazilian Constitutional Cultures: Constitutional Adjudication and Activism

Chapter 1
An Approach to Decision-Making

Abstract As a practical presentation of the problems arising from the deployment of balancing in constitutional cases, this chapter aims to introduce the analysis of three relevant constitutional decisions, the *Crucifix* and *Cannabis* cases, of the German Federal Constitutional Court (*Bundesverfassungsgericht*), and the *Ellwanger* case, of the Brazilian Federal Supreme Court (*Supremo Tribunal Federal*). By exploring the opinions therein contained, different relevant discussions appear, such as the dualism between axiology and deontology in decision-making, the politicization of judicial review, the flexibility and capacity of balancing to support distinct types of arguments, opening thereby the possibilities to further critically review, directly from the practice of decision-making, how balancing can cause serious outcomes to the principle of separation of powers, and thus to constitutional democracy.

1.1 Introduction

When the theme of rationality of balancing comes into sight, a first necessary intuition is to grasp how the practice of decision-making balances different arguments in a concrete case, and how it intends to provide, with this mechanism, a rational solution that best fits the controversial reality the judge faces. Through the examination of the way adjudication deals with these arguments, it is possible to envisage the practical dilemmas this activity involves, while concomitantly revealing the starting point of a complex debate that will end in the question of which rationality should exist in the practice of decision-making. Consistent with this premise, this chapter aims to introduce, by means of case study, some practical dilemmas verified in the reality of adjudication that will serve as empirical examples for the theoretical debate that will be subject of consideration in the second and third parts of this book. It, accordingly, aims to present the initial discussion of the empirical reality of balancing in constitutional cases, normally discussed within the framework of the principle of proportionality, whose characteristics will be, in the

J.Z. Benvindo, *On the Limits of Constitutional Adjudication*,
DOI 10.1007/978-3-642-11434-2_1, © Springer-Verlag Berlin Heidelberg 2010

next two chapters, reinforced by way of a study of the German *Bundesverfassungsgericht*'s (BVG) and the Brazilian *Supremo Tribunal Federal*'s (STF) shift to a more activist approach towards the different themes of social life, and then theoretically challenged in its seemingly rational comprehension in the second part.

The purpose now is to examine the way constitutional courts sustain their arguments in some complex cases. From different words and criteria used in decision-making, many nuances of the debate on rationality in the realm of constitutional adjudication gain form. In this respect, we will analytically explore three cases: two of the BVG – the *Crucifix case* (BVerfGE 93, 1) and the *Cannabis case* (BVerfGE 90, 145), and one of the STF – the *Ellwanger* case (HC 82.424/RS). In the following chapters, when we will study the reality of the German BVG and the Brazilian STF, other important cases, although less analytically discussed, will also be subject of consideration, complementing thereby the empirical reference to the discussion carried in the second and third parts.

This chapter targets the core of some problematic issues of decision-making, as a practical introduction to a deeper debate on rationality, its connections with constitutional democracy, and with the complex question of justice. It has, at any rate, more the purpose of describing the cases than, indeed, to present a critical investigation of their contents, which will be more directly examined in the third part,[1] after we will have unfolded the *concept of limited rationality*. It is, therefore, the link with reality, which is, in this research, as crucial as the obvious sentence that we cannot examine methods only abstractly. If this chapter introduces the complexity of this discussion, this is, however, only the tip of the iceberg. It is, in any case, a fundamental aspect to understanding the role the constitution undertakes and how institutions have to work presupposing its authoritative character, not only theoretically, but mostly by reinforcing it in their practices.

1.2 The Crucifix Case[2]

"Even a state which broadly guarantees the freedom of faith and, therefore, commits itself to religion and ideological neutrality cannot ignore the opinions and the rooted historical and cultured transmitted axiological convictions, on which the social cohesion is based and on which the accomplishment of its own tasks also depends."[3] These BVG's words, stated in an important decision in 1995, denote the struggle in the interpretation of the conflict between freedom of faith and the predominant values in Bavaria, a known catholic state in Germany. The famous

[1] See the eighth chapter.

[2] BVerfGE 93, 1 – *Kruzifix*. Translation: Institute for Transnational Law. The University of Texas at Austin. http://www.utexas.edu/law/academics/centers/transnational/work_new/german/case.php?id=615 (accessed July 19, 2009).

[3] BVerfGE 93, 1 – *Kruzifix*. Translation: Institute for Transnational Law.

Crucifix case went public with a line of reasoning that indicates how complicated it is to sustain an argument that goes against a prevalent local tradition. The case, originated by a constitutional complaint (*Verfassungsbeschwerde*)[4] against the §13 (I) 3 of the School Law for Fundamental School in Bavaria (*Schulordnung für die Volksschulen in Bayern* – VSO), is still the subject of ample discussion, especially because it poses relevant questions on the relationship among democracy, constitutionalism and the principle of equality in a complex and plural society.

The case is self-evident. It is concerned with religion and tradition, on the one hand, and freedom of faith, on the other. The German Basic Law (*Grundgesetz*), in its article 4 (I), establishes that "freedom of faith and of conscience, and freedom of creed religious or ideological, are inviolable." Its article 7 in turn defines that "the entire education system is under the supervision of the state" and "the persons entitled to bring up a child have the right to decide whether they shall receive religious instruction." Afterwards, nonetheless, which demonstrates that there are still some vestiges of religious incursion into the German secular democratic constitutionalism, it establishes that "religious instruction forms part of the ordinary curriculum in state and municipal schools, excepting secular schools. Without prejudice to the state's right of supervision, religious instruction is given according to the tenets of the religious communities. No teacher may be obliged against his will to give religious instructions." Considering this norm, the School Law for Fundamental School in Bavaria (VSO) determined in its §13 (I) 3 that "the school shall support those having parental power in the religious upbringing of children. School prayer, school services and school worship are possibilities for such support. In every classroom a cross shall be affixed. Teachers and pupils are obliged to respect the religious feelings of all."[5]

The conflict was imminent. Indeed, parents, after having unsuccessfully attempted to prevent the school from hanging a crucifix in their children's classroom,[6] filed a lawsuit in the Bavarian Administrative Court questioning the

---

[4]According to article 93, 4a of the German Basic Law, the *Verfassungsbeschwerde* is a constitutional complaint "which may be filed by any person alleging that one of his basic rights or one of his rights under paragraph (4) of Article 20 or under Article 33, 38, 101, 103, or 104 has been infringed by public authority." It is the main instrument of judicial review in Germany. For a detailed analysis of its characteristics, see Bodo Pieroth and Bernhard Schlink, *Grundrechte: Staatsrecht II* (Heidelberg: C. F. Müller, 2006), 289–304; Christian Hillgruber and Christophh Goos, *Verfassungsprozesrecht* (Heidelberg: C. F. Müller, 2004); Bodo Pieroth, *Verfassungsbeschwerde: Einführung, Verfahren, Grundrechte* (Münster: ZAP Verlag, 2008).

[5]BVerfGE 93, 1 – *Kruzifix*. Translation: Institute for Transnational Law.

[6]This is the description of the conflict according to the BVG's report:

"Complainants 3)–5) are the school-age minor children of complainants 1) and 2). The latter are followers of the anthroposophical philosophy of life as taught by Rudolf Steiner, and bring up their children accordingly. Since their eldest daughter, complainant 3), went to school they have been objecting to the fact that in the schoolrooms attended by their children first of all crucifixes and later in part crosses without a body have been affixed. They assert that through this symbol, in particular through the portrayal of a "dying male body," their children are being influenced in a Christian direction; which runs counter to their educational notions, in particular their philosophy of life.

constitutionality of this norm. Preliminarily, the Administrative Court rejected the suit and argued that crucifixes in classrooms did not violate parents' right concerning upbringing nor offended the children's basic rights. In addition, it stated that the crucifix was only a way to assist parents with the religious instruction of their children. According to the BVG's report, these were the main arguments:

> The administrative court refused the urgent request. The affixation of crosses in schoolrooms infringed neither the parents' rights regarding upbringing nor the children's fundamental rights. § 13(1), third sentence, VSO did not provide that the cross be used as a means of education and made into an object of the overall school teaching. It served merely for constitutionally unobjectionable support to parents in the religious upbringing of their children. The constitutionally admissible bounds of religious or philosophical references in schooling were not overstepped. The principle of non-identification could not claim the same respect in schooling – by contrast with the purely secular sphere – because in the educational sphere religious and philosophical conceptions had always been of importance. The tension between positive and negative religious freedom had to be resolved having regard to the precept of tolerance in accordance with the principle of concordance. That meant the complainants could not demand absolute primacy for their negative confessional freedom over the positive confessional freedom of those pupils who were brought up in a religious confession and wished to manifest that. Instead, tolerance and respect were to be expected of the complainants for the religious convictions of others when encountering their exercise of religion at school.[7]

The Bavarian Higher Administrative Court also rejected the appeal against this decision, founded on two main principles: there were no irreparable disadvantages for the complainants from waiting, especially when the school somehow demonstrated "a willingness to compromise," and, also, "the sight of a cross or crucifix

---

When complainant 3) entered school in late summer 1986, in her classroom there was a crucifix with a total height of 80 cm and a 60 cm high representation of the body affixed, directly in the field of view of the blackboard. Complainants 1) and 2) asked for removal of this crucifix and declined to send complainant 3) to school as long as she was exposed to that sight. The conflict was initially settled by exchanging the crucifix for a smaller cross without body, affixed over the door. The disputes between complainants 1) and 2) and the school administration however flared up again when their other children went to school and when complainant 3) changed class and finally school, because crucifixes were again affixed in the schoolrooms. By not sending their children to school, sometimes for fairly long periods, complainants 1) and 2) repeatedly secured the compromise solution again (small cross with no body, at the side above the door) for the classrooms, but not for the other schoolrooms. The school administration, moreover, gave complainants 1) and 2) no assurance that the compromise would be kept to at every change of class.

For a time the three children attended a Waldorf school; however, for lack of the necessary funds, this remained only a transitory attempt to resolve the conflict.

In February 1991 complainants 1) and 2) brought an action against the Free State of Bavaria before the administrative court, in their own behalf and that of their children, with the aim of having the crosses removed from all rooms frequented or yet to be frequented in public schools by their children in connection with attending school. At the same time they applied for the issuing of a temporary order pending conclusion of the action for removal of crucifixes." BVerfGE 93, 1 – *Kruzifix*. Translation: Institute for Transnational Law.

[7]BVerfGE 93, 1 – *Kruzifix*. Translation: Institute for Transnational Law.

was a comparatively slight burden; the children would be confronted with this depiction elsewhere too."[8]

This was the background to open the discussion of the constitutionality of the VSO §13 (I) 3, which took place in virtue of a constitutional complaint against both decisions. Based mainly on article 4(I) of the German Basic Law, the BVG stated that both decisions offended the complainants' basic rights, and, as a consequence, the §13 (I) 3 VSO was null and void. Nevertheless, some of the BVG's Justices somehow held, with new contours, the lower courts' arguments. Indeed, they were seemingly so plausible that three Justices (among the eight that participated in the decision) did not consider the existence, in this case, of any offense to constitutional norms. For them, "This is not a problem of the relation between majority and minority, but one of how in the area of state compulsory schools the positive and negative religious freedom of pupils and their parents can in general be brought into harmony."[9] By following similar premises, those Justices stated: "The right of religious freedom is not a right to prevent religion. The necessary adjustment between the two manifestations of religious freedom must be brought about through tolerance."[10] The so-desired harmony between the positive and negative dimensions of religious freedom would be reached, according to this perspective, through the affected complainants' tolerance.

For the majority opinion, "The equipping of schoolrooms with crosses and crucifixes is said to infringe the state's duty of religious and philosophical neutrality."[11] Founded on article 4 (I) of the German Basic Law, the BVG clarified the need for minorities' protection, and thus stressed the equality principle. Its words – "The decisions challenged, by deducing from article 4 of the Basic Law a claim of the majority against the minority whereby the minority had to tolerate and respect pro-majority official acts and religious tokens in state premises as positive exercise of religion by the majority, converted the protection of article 4 Basic Law into its opposite"[12] – are a strong example of this comprehension of the role fundamental rights play in a democracy. Naturally, this was not a denial of the relevance of some communitarian traditions. By attempting to provide a methodological criterion to gather this cultural heritage in legal argumentation, the BVG understood neutrality through the application of the principle of practical concordance (*praktische Konkordanz*),[13] thereby assuming that traditions and cultural heritage are essential to social cohesion: "The Land legislature is not utterly barred from introducing Christian references in designing the public elementary schools, even if those with

[8]BVerfGE 93, 1 – *Kruzifix*. Translation: Institute for Transnational Law.

[9]BVerfGE 93, 1 – *Kruzifix*. Translation: Institute for Transnational Law.

[10]BVerfGE 93, 1 – *Kruzifix*. Translation: Institute for Transnational Law.

[11]BVerfGE 93, 1 – *Kruzifix*. Translation: Institute for Transnational Law.

[12]BVerfGE 93, 1 – *Kruzifix* Translation: Institute for Transnational Law.

[13]This principle, according to the BVG through Konrad Hesse´s *Grundzüge des Verfassungsrechts der Bundesrepublik Deutschland* (Heidelberg: Müller, 1999), "requires that no one of the conflicting legal positions be preferred and maximally asserted, but all given as protective as possible an arrangement." BVerfGE 93, 1 – *Kruzifix*. Institute for Transnational Law.

parental power who cannot avoid these schools in their children's education may not desire any religious upbringing."[14] By balancing this cultural legacy with the principle of neutrality, the court sustained that the introduction of these Christian references must be made in a reasonable way, or, as the court defined, "there is a requirement, however, that this be associated with only the indispensable minimum of elements of compulsion,"[15] which meant, in the case, "the school cannot treat its task in the religious and philosophical area in missionary fashion, nor claim any binding validity for contents of Christian beliefs."[16] From this perspective, "the affixing of crosses in classrooms goes beyond the boundary thereby drawn to the religious and philosophical orientation of schools."[17] The factual aspects of the case went beyond the limits imposed by the principle of practical concordance.

For the dissenting opinion, in turn, article 135 of the Bavarian State Constitution enforced the teaching of Christian values, whose text says that "the public elementary schools shall be joint schools for all children of elementary-school age. In them pupils shall be taught and brought up in accordance with the principles of the Christian confessions."[18] Thus, the teaching of Christian values was a principle to be followed, concerning the traditional Western heritage, whose contents were broader than its religious reference: "The affirmation of Christianity relates not to the content of belief but to recognition of the decisive cultural and educational factor, and is therefore justified in relation to non-Christians too, by the history of the Western cultural area."[19] This was also under *Lande*r's discretionary power:

> (. . .) The affixation of a cross in the classroom would, because of its symbolic nature for the supra-confessional Christian, Western values and ethical standards, also be welcomed or at least respected by a large proportion of the persons not in a church. This notion is supported not least by the fact that the provisions of the Bavarian Constitution on the Christian nondenominational school received the assent of the majority of the population (cf. BVerfGE 41, 65 [67]).[20]

Apart from this axiological argument, founded upon the Bavarian prevalent values, the dissenting opinion introduced a political one: "The state, which through compulsory schooling is deeply involved in the upbringing of children by the parental household, is largely dependent on acceptance by parents of the school system it organizes."[21] The court, for this reason, must act for the sake of finding a response that best fits the general interests of the population, a task reached by upholding majoritarian traditional values through decision-making. Similarly to the BVG's majority opinion, the principle of neutrality is also subject to questioning

[14]BVerfGE 93, 1 – *Kruzifix*. Translation: Institute for Transnational Law.
[15]BVerfGE 93, 1 – *Kruzifix*. Translation: Institute for Transnational Law.
[16]BVerfGE 93, 1 – *Kruzifix*. Translation: Institute for Transnational Law.
[17]BVerfGE 93, 1 – *Kruzifix* Translation: Institute for Transnational Law.
[18]BVerfGE 93, 1 – *Kruzifix*. Translation: Institute for Transnational Law.
[19]BVerfGE 93, 1 – *Kruzifix*. Translation: Institute for Transnational Law.
[20]BVerfGE 93, 1 – *Kruzifix*. Translation: Institute for Transnational Law.
[21]BVerfGE 93, 1 – *Kruzifix*. Translation: Institute for Transnational Law.

but is balanced with those traditions in the complete opposite way. According to the dissenting opinion, "the precept of philosophical and religious neutrality ought not to be understood as an obligation on the state to indifference or secularism."[22] Insofar as this argument may be in accordance with the principle of neutrality in this other balanced solution, the dissenting opinion concluded that this question would be resolved by the harmonization between positive and negative religious freedom: "This is not a problem of the relation between majority and minority, but one of how in the area of state compulsory schools the positive and negative religious freedom of pupils and their parents can in general be brought into harmony."[23] Again, the solution – the harmony – is provided by tolerance, which, in this case, goes in the contrary direction of the BVG's prevailing point of view. For the dissenting opinion, after having examined the particularities of the case and balanced positive and negative religious freedom, the positive freedom prevails, provided that "the psychic impairment and mental burden that non-Christian pupils have to endure from the enforced perception of the cross in class is of only relatively slight weight,"[24] especially when children do not suffer the risk of being discriminated thereby.

If we make an effort to disclose the rational motivation of both positions, we will observe some interesting controversial uses of dogmatic criteria for decision-making. The BVG's majority opinion oriented its discourse towards the protection of minorities – and hence the principle of equality – through the affirmation of state's neutrality. The dissenting opinion, in turn, sustained the prevalence of positive religious freedom through tolerance. Both attempted to balance the two types of religious freedom (negative and positive) to defend their conclusions and, for this purpose, took different dogmatic concepts into account. The first employed the dogmatic concept of practical concordance (*praktische Konkordanz*), resulting in balancing the positive and negative religious freedom. This balancing took place by recognizing the importance of traditional values to social cohesion, on the one hand, but also by arguing that those values must be imposed in a reasonable manner ("the indispensable minimum of elements of compulsion"), on the other. The unconstitutionality of VSO §13 (I) 3 as well as the incorrectness of lower decisions derived from overstepping, in the case, the limits prescribed by the principle of practical concordance. The dissenting opinion, in turn, used the precept of tolerance[25] in order to defend the thesis that the "minimum of elements of compulsion which in this respect is to be accepted by pupils and their parents is not exceeded,"[26]

---

[22]BVerfGE 93, 1 – *Kruzifix*. Translation: Institute for Transnational Law.

[23]BVerfGE 93, 1 – *Kruzifix*. Translation: Institute for Transnational Law.

[24]BVerfGE 93, 1 – *Kruzifix*. Translation: Institute for Transnational Law.

[25]The BVG's majority opinion also took the precept of tolerance into account, when it sustained that "resolving the unavoidable tension between negative and positive religious freedom while taking account of the precept of tolerance is a matter for the Land legislature, which must through the public decision making process seek a compromise acceptable to all." BVerfGE 93, 1 – *Kruzifix*. Translation: Institute for Transnational Law.

[26]BVerfGE 93, 1 – *Kruzifix* .Translation: Institute for Transnational Law.

since "the danger of their being discriminated against accordingly does not exist from the outset,"[27] especially when "pupils are there (in Bavaria) confronted, even outside the narrower church sphere, with the sight of the crosses in many other areas of life."[28] The dissenting opinion, therefore, emphasized the traditional values as the main justification for the precept of tolerance and as a justification for deciding in favor of the community's general interests instead of preserving the complainants' basic rights, members of minority not sharing the Christian beliefs. Unlike the BVG's final decision, it submitted the constitutional principle of religious neutrality to a predominant belief of a certain community, regardless of the equality principle, now relativized thanks to the precept of tolerance.

In the *Crucifix case*, the BVG assumed the cross as the primary symbol of Christianity and, particularly, as an ethical value shared by the Bavarian community: "The cross [had] thereby in the school the function of a culture symbol."[29] The question was thus how the BVG should interpret this cultural symbol. This traditional consideration took a crucial role in both majority and dissenting opinions, serving then as a source for working with dogmatic concepts. Through the principle of practical concordance, the majority opinion understood that the affixation of a crucifix overstepped the boundaries of an "indispensable minimum of elements of compulsion,"[30] inasmuch as "the cross cannot be divested of its specific reference to the beliefs of Christianity and reduced to a general token of the Western cultural tradition."[31] Although it presented a very interesting comprehension of the role legal rights assume in complex and plural societies,[32] it is undeniable that one of the main points here was the interpretation of the crucifix as a missionary spread of Christianity, which could thereby lead to a mental influence particularly on young people and, accordingly, be an offense to state religious neutrality.

The dissenting opinion, on the other hand, worked with this traditional perspective otherwise. In its opinion, this possible influence was so minimum that it could be ignored and, thus, was in harmony with the "indispensable minimum of elements of compulsion."[33] By stressing the communitarian values, the decision was

---

[27]BVerfGE 93, 1 – *Kruzifix*. Translation: Institute for Transnational Law.

[28]BVerfGE 93, 1 – *Kruzifix*. Translation: Institute for Transnational Law.

[29]Sonja M. Esser, *Das Kreuz – ein Symbol Kultureller Identität? Der Diskurs über das 'Kruzifix-Urteil (1995) aus kulturwissenschaftlicher Perspektive* (Münster, New York, München, Berlin: Waxmann, 2000), 33, translation mine.

[30]BVerfGE 93, 1 – *Kruzifix*. Translation: Institute for Transnational Law.

[31]BVerfGE 93, 1 – *Kruzifix*. Translation: Institute for Transnational Law.

[32]We can see this in the following argumentation: "The affixation of the cross cannot be justified from the positive religious freedom of parents and pupils of the Christian faith either. Positive religious freedom is due to all parents and pupils equally, not just the Christian ones. The conflict arising cannot be resolved according to the majority principle, for the fundamental right to religious freedom specifically is aimed in a special degree at protecting minorities. Moreover, Art. 4(1) Basic Law does not confer on the bearers of the fundamental right an unrestricted entitlement to activate their religious convictions in the context of State institutions." BVerfGE 93, 1 – *Kruzifix* Translation: Institute for Transnational Law.

[33]BVerfGE 93, 1 – *Kruzifix* Translation: Institute for Transnational Law.

conditioned to the "assent of the majority of the population"[34] or to the premise that "the state, which through compulsory schooling is deeply involved in the upbringing of children by the parental household, is largely dependent on acceptance by parents of the school system it organizes."[35] The precept of tolerance here was deployed as a standard for justifying a political decision through balancing.

1.3 The Cannabis Case[36]

The famous *Cannabis case* provides one of the most interesting debates on the application of balancing, embedded in the structural framework of the principle of proportionality, to support a political argument in the BVG's history. It relates to the discussion of the free development of personality combined with the right to freedom (articles 2 (I) and 2 (II) 2), the right to physical integrity (article 2 (II) 1), as well as the equality principle (article 3 (I)), all from the Basic Law, within the context of the judicial review of some German Narcotic Act provisions (§ 29 (I) BtMG). The BVG upheld the argument that the Narcotic Act did not infringe the above constitutional principles in the hypothesis of someone acquiring or consuming products derived from the plant *cannabis sativa*, using thereby relevant dogmatic concepts and methods, particularly the principle of proportionality, as a means to demonstrate how the Act provisions at issue were in conformity with the Basic Law. As usually occurs in the BVG's decisions, the principle of proportionality (*Verhältnismäßigkeitgrundsatz*) was deployed as a "general constitutional parameter, according to which the freedom of action can be restrained;"[37] thus, as a meta-principle that guides decision-making when there is a collision of principles. It was within the context of this meta-principle, as normally it is, that balancing took place.

Consistent with a longstanding tradition in German constitutionalism,[38] whenever there is a possible encroachment on some basic right – in this case, especially, the free development of personality combined with the right to freedom – the BVG proceeds to the analysis of the proportion of this intervention through the deployment of the principle of proportionality, with its three elements (suitability, necessity and proportionality in its narrow sense or balancing). This dogmatic methodology, whose features will be further historically[39] and structurally[40] examined, is nowadays one of the main mechanisms in constitutional adjudication and

[34]BVerfGE 93, 1 – *Kruzifix* Translation: Institute for Transnational Law.
[35]BVerfGE 93, 1 – *Kruzifix* Translation: Institute for Transnational Law.
[36]BVerfGE 90, 145 – Cannabis.
[37]BVerfGE 90, 145 – *Cannabis*, translation mine.
[38]See the next chapter.
[39]See the second and third chapters.
[40]See the fourth chapter.

likewise one of the highest expressions of the intent to rationally systematize decision-making,[41] reaching thereby a possible rationalization of balancing, certainly its most controversial element. In the *Cannabis case*, the argument mostly oriented towards judging the Act to be constitutional, given its obedience to those elements. This is why the *Cannabis case* is an interesting source to understand the deployment of balancing within the context of the principle of proportionality, presented as a dogmatic methodology that could strengthen the justification because of its seemingly rational nature.

The *Cannabis case* is the result of some judicial submissions (*Richtervorlage*[42]) from lower courts to the BVG as well as of a constitutional complaint (*Verfassungsbeschwerde*) questioning the validity of some of the German Narcotic Act provisions (§ 29, I, BtMG). Judged at the beginning of 1994, the case had as its main feature the judicial review of the norm incriminating illegal transactions with *cannabis* products. The investigation centered mostly on the compatibility of those provisions with article 2 (I) (free development of personality), article 2 (II) 2 (right to freedom), and article 3(I) (equality principle) of the Basic Law. Notwithstanding some dissenting opinions,[43] the final decision held the argument that the Narcotic Act provisions at issue were constitutional. These were the BVG's main arguments:

1. Article 2 (I) of the Basic Law ("every person shall have the right to free development of his personality insofar as he does not violate the rights of others or offend against the constitutional order or the moral law"[44]) does not embrace the "right to get high" (*Recht auf Rausch*), and even if considered otherwise, the intervention in this sphere was proportionate in its narrow sense.
2. In accordance with the principle of proportionality (with its elements of suitability and necessity), as well as an "evaluation and prognosis of the dangers that threaten the individual or the community," the legislator has a

[41]See the fourth chapter.

[42]According to article 100, 1, of the German Basic Law, "if a court concludes that a law on whose validity its decision depends is unconstitutional, the proceedings shall be stayed, and a decision shall be obtained from the Land court with jurisdiction over constitutional disputes where the constitution of a Land is held to be violated, or from the Federal Constitutional Court where this Basic Law is held to be violated. This provision shall also apply where the Basic Law is held to be violated by Land law and where a Land law is held to be incompatible with a federal law." Accordingly, the *Richtervorlage* is a submission of a legal matter from an ordinary court to the constitutional court whenever the possible unconstitutionality of a legal norm is at issue.

[43]Justice Graßhof agreed with the final decision; however, he defended, after a long theoretical explanation on the principle of proportionality and the social consequences caused by drugs, a more severe point of view of the general harmful effects of *cannabis* products for society. On the other hand, Justice Sommer understood differently from the majority. According to him, in this particular case, the Narcotic Act, even though establishing a mechanism for refraining from prosecution and condemnation, offended article 2 (I), (II) of the Basic Law, based, mainly, on the principle of proportionality in its narrow sense.

[44]BVerfGE 90, 145 – *Cannabis*, translation mine.

discretionary margin of evaluation, "which the BVG can only review in a limited extension".[45]

3. "By a general balancing between the severity of an intervention and the weight, as well as the urgency of their justifiable reasons, the limit of what can be reasonably demanded of the person to whom the prohibition is addressed (prohibition of excess or proportionality in its narrow sense) must be respected".[46]
4. When the consumption of *cannabis* is only occasional, in low amount, and does not cause any risk to the other, in conformity with the prohibition of excess requirement (*Übermaßverbot*), the criminal prosecution must not proceed.[47]
5. "The equality principle does not require an indistinct prohibition or permission of every potentially equally harmful drug,"[48] and, as a consequence, "the legislator [can], without infringing the Constitution, regulate, in a different manner, the dealings with products from *cannabis*, on the one hand, and with alcohol and nicotine, on the other."[49]

Some of these reasons demand a deeper analysis. The first refers to the BVG's conclusion that the collision of principles at issue revealed that there is no absolute protection to the development of personality, in conformity with article 2 (I) of the Basic Law, for there is only "an inner core of conformation of private life"[50] that is free from interference. This "inner core," nonetheless, had its contents established by an emphasis on the social consequences of the act. According to the court, "dealing with drugs, and especially the act of voluntary becoming intoxicated, cannot be reckoned [as part of this inner core], because of its numerous social effects and interactions."[51] Apart from the consideration of this "inner core," the court, grounded in article 2 (I) of the Basic Law, stated that the free development of personality cannot "violate the rights of others or offend against the constitutional order or the moral code."[52] Through the examination of the social effects and corresponding interactions, the court argued that the "right to get high" was not embedded in the "inner core" of the free development of personality nor was in accordance with the final part of article 2 (I) of the Basic Law.

The BVG reinforced these arguments by deploying the principle of proportionality. In this example, the court used this principle to "express a judgment of negative ethical-social value over a particular act of a citizen."[53] With this instrument, the court could then shape those constitutional principles according to the

---

[45]BVerfGE 90, 145 – *Cannabis*, translation mine.

[46]BVerfGE 90, 145 – *Cannabis*, translation mine.

[47]BVerfGE 90, 145 – *Cannabis*, translation mine.

[48]BVerfGE 90, 145 – *Cannabis*, translation mine.

[49]BVerfGE 90, 145 – *Cannabis*, translation mine.

[50]BVerfGE 90, 145 – *Cannabis*, translation mine.

[51]BVerfGE 90, 145 – *Cannabis*, translation mine.

[52]BVerfGE 90, 145 – *Cannabis*, translation mine.

[53]BVerfGE 90, 145 – *Cannabis*, translation mine.

ethical-social values it interpreted as relevant to that special circumstance. Balancing, in particular, was the ideal instrument for this purpose.

Although apparently respecting the legislator's duty,[54] the BVG attempted to investigate thereby how the dealings with *cannabis* products affect the society as a whole, interpreted as a condition to conclude whether the Narcotic Act provisions were in conformity with the Basic Law or not. Still, before using balancing in particular, the court considered the suitability and the necessity of the measure. For this purpose, by looking into the risks of *cannabis* products for the individual and social health, the BVG concluded that "the content [of] the penal provisions of the Narcotic Act [were] suitable to limit the distribution of the drug in society and thus limit the dangers flowing from it as a whole."[55] In this regard, the court understood that the provisions complied with the first element, suitability (*Geeignetheit*),[56] inasmuch as "the penal provisions [were] then generally suitable to promote the aim of the Act."[57] Moreover, "on the basis of the current state of scientific knowledge," the BVG acknowledged the necessity (*Erforderlichkeit*[58]) of the measure by asserting that "the legislator's conception, according to which there is no other means than criminal penalties that would be equally effective and less intrusive to attain the Act's aim, [was] defensible."[59] Indeed, as the BVG mentioned, the provisions reached until that moment the desirable goal, making then impossible to hold the thesis according to which "the unbanning of *cannabis* would be a milder means to more easily achieve this purpose."[60] This conclusion was followed by an analysis of international treaties and scientific opinions that raised some doubts whether the adoption of other different and apparently milder instruments could reach the same goal.

This connection with the social consequences becomes more transparent in the following arguments. Subsequent to this factual analysis that converged upon a teleological consideration of what was the best solution to the community, the third element, the proportionality in its narrow sense or balancing (*Verhältnismäßigkeit in engeren Sinn* or *Abwägung*), completed this procedure by concentrating upon the optimization of legal possibilities. In the case, the BVG understood that "against these important social interests [healthy risks emanating from drugs, psychological dependency and criminal organizations], there [were]

---

[54]The BVG exposed clearly that it "cannot consider whether the legislature's decision was the most suitable, reasonable or just way to solve the problem at issue. The court's role is merely to check whether the substance of the penal provision is compatible with the constitutional provisions and accords with the Basic Law's fundamental values and the unwritten principles underlying the Constitution." BVerfGE 90, 145 – *Cannabis*, translation mine.

[55]BVerfGE 90, 145 – *Cannabis*, translation mine.

[56]According to this element, the means must prove itself suitable to achieve the desirable goal.

[57]BVerfGE 90, 145 – *Cannabis*, translation mine.

[58]According to this element, the means must prove itself less harmful to the private's sphere than any other suitable means.

[59]BVerfGE 90, 145 – *Cannabis*, translation mine.

[60]BVerfGE 90, 145 – *Cannabis*, translation mine.

no interests of equal importance in unbanning dealings with the drug."[61] In this regard, the BVG determined that it should consider whether the employed means were, from the point of view of the individual, proportionate to the protection of legal interests. Hence, despite the fulfillment of the elements of suitability and necessity, there was still the inquiry of whether the "the resulting restriction on fundamental rights of the affected person significantly [outweighed] the increased protection of legal interests thereby achieved."[62] The BVG employed balancing after a long explanation of the social effects of drugs in society, leading to the conclusion that the consequences of unbanning the dealings with *cannabis* products were more serious or weightier than the entire protection of the free development of personality in combination with the right to freedom in these circumstances. The interference with the private sphere and the penal sanctions were, as preventive measures, proportionate and reasonable.[63] This conclusion applied not only to the commerce and supply of these products without the purpose of profit but also to their private consumption,[64] inasmuch as they could affect third parties and instigate an illegal market. However, in the specific case of occasional private possession and consumption in small amount of the drug, the court decided that the interference was disproportionate.

From the standpoint that the threat to the legal interests, in this last hypothesis, was petty,[65] and by affirming that they were by far the most common and discontinued prosecutions,[66] the BVG stated that the danger they caused was limited, even if they motivated somehow the illegal drug market.[67] As a result, from the argument that "the concrete danger that the drug will be passed on to a third person, in general, is not very considerable"[68] in these circumstances, the BVG deemed that "the public interest in the imposition of a penalty is correspondingly limited."[69] The prosecution of these cases and a possible condemnation based on the general provision of penalties could be considered disproportionate, for the effects on the

[61] BVerfGE 90, 145 – *Cannabis*, translation mine.

[62] BVerfGE 90, 145 – *Cannabis*, translation mine.

[63] BVerfGE 90, 145 – *Cannabis*, translation mine.

[64] BVerfGE 90, 145 – *Cannabis*, translation mine.

[65] Unlike the BVG's opinion, Justice Sommer clearly affirmed that, after a long examination of the facts and legislation, the state's intrusion was not petty, but of high intensity. According to him, the Narcotic Act provisions, as such characterized, could no longer be considered proportionate in its narrow sense in the case of own consumption of small amount of the drug, and, therefore, the criminalization was not justifiable. Moreover, even if the legislator created a mechanism, according to §§ 29 V and 31a BtMG, for refraining from condemnation and the criminal prosecution, the principle of proportionality in its narrow sense kept violated. He also criticized the perspective of a possible use of the criminal law in a "more symbolically" way and the possible interpretations of this exception by each different *Länder*. Therefore, the criminal law had to specify what was punishable or not. BVerfGE 90, 145 – *Cannabis*, translation mine.

[66] BVerfGE 90, 145 – *Cannabis*, translation mine.

[67] BVerfGE 90, 145 – *Cannabis*, translation mine.

[68] BVerfGE 90, 145 – *Cannabis*, translation mine.

[69] BVerfGE 90, 145 – *Cannabis*, translation mine.

individual offender could be inadequate. Besides, from the point of view of criminal special prevention, it was also disadvantageous.[70] In spite of that – and this was crucial in the court's analysis – the Narcotic Act's general provision of penalties, in this situation, was not deemed disproportionate in its narrow sense, provided that it established the possibility that authorities, by taking into account these facts (wrongfulness, culpability or potential offense, for instance), refrain from prosecution or imposition of a penalty. Indeed, the BVG understood that this was a mechanism that could be considered a feasible way to obey the principle of proportionality in its narrow sense.[71]

In the investigation of the principle of proportionality in its narrow sense or balancing, therefore, the main focus was on the possible effects the dealings with *cannabis* products could cause in the individual sphere and society. We could examine this discourse through the application of the formula "the more intensive the interference in one principle, the more important the realization of the other principle."[72] For this purpose, on the one hand, there is the principle of free development of personality combined with the right to freedom; on the other, there are the possible social consequences. Accordingly, since the legislator created an evaluative system of the particularities of the case to be carried out by the authorities, which could result in refraining from prosecution and imposing a penalty, and also established a general provision of penalties that could accomplish the goal of protecting the society from the disturbance caused by drugs, then the intrusion into the private sphere was constitutionally allowed, considering that this burden was justified by the principle of proportionality. Furthermore, this proportionality was also observed when article 1 (I) 2 of the Basic Law was brought into question, for: first, it is a state's duty, as it is determined by article 1 (I) 2 of the Basic Law ("to respect and protect it [the human dignity] is the duty of all state authority"[73]), to prevent the individuals from illegal interventions of third parties[74]; and, second, there is no connection between the prohibition and the potential growing consumption of other intoxicating substances (such as alcohol) that are not subject to the Narcotic Act (this argument, besides, would lead, according to the court, to the opposite meaning of the state's duty of protection[75]). In both aspects, the encroachment on the private sphere could be relativized (proportionally balanced), provided that the value – refrain society from the dangers of drugs – could be realized and was, in the hypothesis, proportionally weightier than the former.

[70]The court mentioned the possibility of pushing the individual into the drugs world and causing him to develop a sense of solidarity with it. BVerfGE 90, 145 – *Cannabis*, translation mine.

[71]BVerfGE 90, 145 – *Cannabis*, translation mine.

[72]Robert Alexy, "On the Structure of Legal Principles," *Ratio Juris* 12, no. 4 (September 2000): 298.

[73]BVerfGE 90, 145 – *Cannabis*, translation mine.

[74]BVerfGE 90, 145 – *Cannabis*, translation mine.

[75]BVerfGE 90, 145 – *Cannabis*, translation mine.

The foregoing investigation shows that collective values, interpreted in terms of social consequences, were the basis of the BVG's line of reasoning to define the proportionate extension, in this particular situation, of the constitutional principles at issue. In this case, principles were interpreted as optimization requirements, even though the court itself did not mention it,[76] and, as a result, they were balanced in accordance with the legal and actual possibilities examined through the dimension of what was the best solution to the community. The communal interests in the way the Justices interpreted them shaped somehow the free development of personality and the right to freedom.[77] Throughout the BVG's opinion, we can verify a discourse based, above all, upon goals and efficiency (what can promote better social results) by means of the deployment of a whole methodology. This seemingly rational methodology was used to sustain the argument that the Narcotic Act provisions were in accordance with the Basic Law, for it was possible to delineate the proportion to which the state intrusion could occur as well as the boundaries of the principles at stake.

The social effects were not the only reference the BVG used to properly conduce the balancing. The traditional values shaping the equality principle were also brought into question. Now, the problem was to verify a possible unequal consideration of the *cannabis* products in comparison with other also intoxicating substances, such as alcohol and nicotine. In this matter, the court, first, stated that any investigation founded on the equality principle must focus on the singularities of the case, or in its words, "the peculiarities of the concrete field that is to be regulated;"[78] second, the application of this principle is concerned with the legislator's discretionary power in his consideration of what, from the legal point of view, should be regarded as similar (and thus assigned the same consequence).[79] As a result, the court understood that "the principle of equality does not order an indistinct prohibition or permission of all drugs that are, potentially, equally harmful."[80] Besides, in order to preserve legal certainty, there was a positive list indicating the forbidden

[76]Indeed, as Bernhard Schlink describes:

"(...) The Bundesverfassungsgericht does not speak expressly of principles as rules of optimization. However, constitutional scholarship correctly observes that a relative conception of fundamental rights as rules of optimization harmonizes well with the conception of them as objective principles. The concept of principle, or rule of optimization, was coined especially to categorize the content of fundamental rights resulting from the Bundesverfassungsgericht's development from conceiving of fundamental rights as subjective rights to seeing them as objective principles" (Bernhard Schlink, "German Constitutional Culture in Transition," *Cardozo Law Review* 14 (1993): 718).

[77]The problem is not to understand that the principle of free development of personality combined with the right to freedom was not applied to the case, but its deontological weakening by the emphasis on arguments of efficiency (what is good for society and what can bring about better social results).

[78]BVerfGE 90, 145 – *Cannabis*, translation mine.

[79]BVerfGE 90, 145 – *Cannabis*, translation mine.

[80]BVerfGE 90, 145 – *Cannabis*, translation mine.

substances, whose inclusion or exclusion were based on some criteria that do not necessarily restrict themselves to a potential "risk posed to health" but could also include other factors.[81]

As these arguments indicate, the equality principle did not mean an equal treatment of all intoxicating substances in compliance with the law. In order to contradict the objection to the Narcotic Act's provisions from the perspective of the equality principle, the BVG proceeded subsequently to a comparison of the *cannabis* products with other intoxicating substances. For this purpose, it took two strategies: the exclusion of nicotine from the concept of narcotics,[82] given that it could not lead someone to "get high" (*Rausch*), and the confirmation of alcohol's cultural and historical qualities as the main attribute to account for its different legal treatment. The court avowed that, albeit the serious social damages alcohol brings about – which could be even greater than those of *cannabis* products – its permission was founded on: first, its various possible uses (comparably higher than those of *cannabis* products); second, its employment as a source of nourishing and pleasure, as we can observe in religious rituals; third, the social control that avoids mostly its consumption as a means to "get high" (in contrast to *cannabis*, whose consumption has this goal)[83]; fourth, the fact that "the legislature finds itself in the situation that it cannot effectively prevent the consumption of alcohol because of traditional patterns of consumption in Germany and in the European cultural sphere."[84]

This construction of a cultural significance of a substance, founded upon a survey of its historical patterns and uses, can become the necessary basis to complement the BVG's formerly examined teleological argument. Whereas the first established the grounds on which the principles and values at issue were considered based on their capacity to be proportionally balanced with goals (social consequences), the second defined the premises to place the principles in a practical concordance with cultural goods. Both, at any rate, converged upon a solution based on an evaluation of the collectivity in order to defend the seemingly rational argument orienting, in the hypothesis, a balanced application of legal rights.[85]

---

[81]According to the BVG, these other factors could be: "(1) the different possibilities of their utilization (...); (2) the significance of these various uses to social life; (3) the legal and factual possibilities to face the abuse with success; (4) the possibilities and requirements of an international cooperation in controlling and combating narcotics and the criminal organization dealing with them". BVerfGE 90, 145 – *Cannabis*, translation mine.

[82]BVerfGE 90, 145 – *Cannabis*, translation mine.

[83]BVerfGE 90, 145 – *Cannabis*, translation mine.

[84]BVerfGE 90, 145 – *Cannabis*, translation mine.

[85]Both Justices Graßhof and Sommer attacked the BVG's majority opinion by also introducing axiological arguments (the first emphasized the drug social effects, and the second the disproportionate state's interference with the private sphere through the principle of proportionality in its narrow sense). The debate, for this reason, centered on a conflict of personal interpretations of social values, more than a serious debate on legal rights.

1.4 The Ellwanger Case[86]

Notwithstanding its particularities, the Brazilian Constitutional Court, the *Supremo Tribunal Federal* (STF), has also remarkable decisions, some of them bringing into discussion similar criteria to those we observe in the German BVG. It is true that legal dogmatics and the deployment of consolidated criteria in constitutional adjudication have historically assumed a singular configuration, especially because of the mixed system of concrete and abstract judicial review,[87] but it is also true that they are also tied to a similar attempt to rationally systematize the STF's decisions. Brazilian constitutional reality likewise promotes a rich debate on legal principles through the STF's decisions and suffers from a gradual broadening of its influence in the main themes of Brazilian social life.[88] The problems of constitutional adjudication, despite the historical and legal singularities, are rather analogous and experience comparable outcomes.

The *Ellwanger case*, for this reason, in spite of its strong reverberant effects in Brazilian constitutionalism, is not what we could call a novelty. It is the classic discussion of the conflict between rights of personality – in particular, its offense because of racist utterances –and freedom of speech. The case relates to the publication of books by the Brazilian author Siegfried Ellwanger, whose contents were considered full of racist and discriminatory words against the Jewish community. After the *Superior Tribunal de Justiça*'s[89] decision denying the *habeas corpus* and sustaining the occurrence of the crime of racism, in accordance with article 20 of the Law 7.716/89, his lawyers filed another petition for *habeas corpus* in the STF,[90] whose judgment had wide repercussions in Brazil. As one of the richest examples of conflicting arguments among the STF's Justices, the final decision denied the *habeas corpus* by stating that those books[91] were "an apology for

[86]HC 82.424-2/RS.

[87]See the third chapter.

[88]See the third chapter.

[89]The *Superior Tribunal de Justiça* is the Brazilian highest court for infra-constitutional matters, whereas the *Supremo Tribunal Federal* is the Brazilian highest court for constitutional matters.

[90]The judgment took place in different sessions, and the final decision was made on 09.17.2003 (published on: 03.19.2004).

[91]The contents of these books could, according to the Brazilian legislation, be considered a practice of racism, and thus, a crime. Particularly in this case, although we will not enter into this debate, the title of his book was suggestive – *Jewish or German Holocaust – Behind the Lie of the Century* –, and the others he published were in this direction: *The International Jew*, by Henry Ford; *Hitler – Guilty or Innocent*, by Sérgio Oliveira; *The Conquers of the World – The Real War Criminals*, by Louis Marschalko, and the famous anti-Semitic *The Protocol of the Elders of Zion*, translated by Gustavo Barroso (In Portuguese: *Holocausto Judeu ou Alemão – Nos Bastidores da Mentira do Século; O Judeu Internacional; Hitler – Culpado ou Inocente?; Os Conquistadores do Mundo – Os Verdadeiros Criminosos de Guerra; Os Protocolos dos Sábios de Sião*). Besides, by reading Justice Maurício Corrêa's opinion, we can verify many references to the contents of these books, which clearly demonstrate their discriminatory purpose.

prejudged and discriminatory ideas against the Jewish community,"[92] which could "incite and induce race discrimination."[93]

As many other constitutional democracies, the Brazilian Constitution protects the freedom of speech (article 5, IV). Unlike other realities, however, there is a categorical constitutional norm defining the crime of racism, which is considered imprescriptible: "The practice of racism is a non-bailable crime, with no limitation, subject to the penalty of confinement, under the terms of the law" (article 5, XLII). This particularity transformed the decision into a serious discussion of the limits of freedom of speech, now in collision with the constitutional norm – and thus with equivalent hierarchy to that principle – incriminating racism. The specification of this crime was then defined by article 20 of the Law 7.716/89, later modified by the Law 8.061/90, whose contents are:

> Practice, induce or incite, by means of social communication or publication of any nature, the discrimination or prejudice of race, religion, ethnics, or national precedence. Penalty of confinement from 2 to 5 years.[94]

The debate on balancing, accordingly, appears within the context of a collision between freedom of speech and the equality principle, which, in Brazilian reality, is radicalized by the constitutional incrimination of any racist practice. Among the different Justices' opinions in this case, there were manifestations of a semantic and seriously controversial interpretation of historical facts and the extension of the protection against racism (as if racism applied merely to "black people"),[95] others

[92]HC 82.424-2/RS, translation mine.

[93]HC 82.424-2/RS – Report, translation mine.

[94]HC 82.424-2/RS – Report, translation mine.

[95]Justice Moreira Alves's opinion, normally known for his conservatism, revealed this understanding. According to him, "the question posed in this 'habeas corpus' is to determine the meaning and the reach of the expression 'racism'." The central question, therefore, involved the semantic definition of the extension of the term *racism* as a means to conclude whether it comprised, in its contents, the discrimination against Jews. Through the emphasis on the Brazilian historical tradition, Justice Alves stated that the crime of racism ought to apply merely to the black race. For him, "since the Constitution did not define racism, it seems that it ought to be restricted to the idea of race as usually understood – that is, the white, the black, the yellow, the red (. . .)." Moreover, "in Brazil there is no persecution of Jews nor, evidently, any vestige of holocaust to inspire the Brazilian Constitution's framers to include, in the Constitution, the imprescriptibility of the crime of racism." Apart from this controversial interpretation of history, Justice Alves added an originalist justification to his statement: "When the action was received, in 1991, there was not the scientific notion of genome yet, which transpired in 2000, and thus the Constitution of 1988 could not have taken it into consideration, when it refers to prejudice of race." By the same token, he, by limiting the concept of racism to "black people," brought forward a teleological apprehension about the possibility of extending it excessively, creating thereby a norm of open content: "(. . .) If we give to the constitutional term 'racism' the amplitude we now intend to give, with the meaning reaching any human groups with their own cultural characteristics, we will have the crime of racism as a norm of open content, for human groups with cultural characteristics are numerous, and not only, besides Jews, Curds, Basques, Galicians, Gypsies, these last groups we could not talk about the holocaust in order to justify its imprescriptibility" (HC 82.424-2/RS – Justice Moreira Alves's opinion, translation mine).

already pointing out the collision itself of the principle of freedom of speech and the equality principle, even though mixed up with many other arguments,[96] or entering into the discussion of international treaties and balancing,[97] the protection of human dignity and minorities,[98] the dynamic meaning of legal principles and the extension

[96]Justice Maurício Corrêa's opinion worked with a variety of usual arguments in decision making: (1) a terminological interpretation of the legal term (in the case, racism); (2) a historical investigation by stressing many facts of the Jewish past; (3) an originalist approach by disclosing the *mens legislatoris*; (4) a legal argumentation focused on the limits of freedom of speech. However, the extensive opinion – approximately with thirty-nine pages – despite its well-structured development and persuasive strength, still attempted to manage the terminological discussion of race and developed a very traditionalist perspective of history. Although interesting in this scenario, it transferred the problem of collision of legal principles only to the final part, when the limits of freedom of speech came finally to the scene. Notwithstanding that legal principles were put into perspective at the end of his opinion, the need to define a "rational criterion" for this purpose appeared. It was there again the *principle of practical concordance,* as a mechanism to "proceed to a constitutional balancing" of legal principles, with the presumption that the committed discrimination could not be erased from people's memory. Although the first intention could be seen as an argument of policy – to erase the discrimination from people's memory – Justice Maurício Corrêa's opinion, at least, defended the force of legal principles when he, by criticizing Justice Alves' opinion, remarked that "limiting racism to a simple discrimination of races by considering only the lexical or common meaning of the term implies the denial itself of the equality principle, which opens up the possibility of discussion of the limits of legal rights by a determined part of society, something that puts into checkmate the very nature and prevalence of human beings. Conditioning the discrimination as an imprescriptible crime only to black people and not to Jews is to accept as unequals those who are, in essence, equals before this guarantee. It seems to me, *data venia,* an unacceptable conclusion." (HC 82.424-2/RS – Justice Mauricio Corrêa's opinion, translation mine).

[97]Justice Celso de Mello concentrated his argumentation on the importance of international treaties, the principle of dignity and the equality principle. By emphasizing similar arguments as the ones adopted by Justice Corrêa, he supported the argument that the books had a discriminatory purpose and, therefore, configured a crime. He also mentioned that the freedom of speech could not extinguish the crime of racism in this particular case, to the extent that it "does not constitute a means that can legitimate the externalization of criminal purposes, especially when expressions of racial hate – propagated by overstepping the limits of political critique or historical opinion – transgress, in an unacceptable way, the values protected by the constitutional order." Furthermore, he made a fast reference to rational criteria to resolve conflicts of rights, for "the use of a method of balancing goods and interests does not result in the emptiness of the essential content of fundamental rights." In the confirmation of his opinion, he emphasized that, although the freedom of speech must be preserved in a democratic society, it is not an absolute principle, insofar as abuses and crimes can be committed. He also stressed the deontological force of legal principles and sustained the need to proceed, as before, to a concrete balancing of legal principles (HC 82.424-2/RS – Justice Celso de Mello's opinion, translation mine).

[98]Justice Carlos Velloso defended, from the discussion of the actual importance of human rights through the analysis of international treaties, doctrine and Brazilian legislation, the argument that minorities must be protected in constitutional democracies. He did not add any new analysis. Instead, he repeated the terminological emphasis on the concept of race to ascertain that any type of discrimination is embraced by the concept of race. Additionally, he affirmed that the freedom of speech is not absolute, and hence cannot be used as an argument when there are intolerance and incitation to violence against human dignity.

of the concept of race to any type of discrimination.[99] There was likewise the defense of the freedom of speech, in this case, grounded in the premise that those books were a manifestation of the ideology embraced by the freedom of speech and in the intellectual and scientific production the Brazilian Constitution safeguards.[100] In any case, we will concentrate on two opinions, especially because they enter directly into the core of balancing, now embedded in the structure of the principle of proportionality, leading, nevertheless, to complete opposite solutions. They are: Justice Gilmar Mendes's and Justice Marco Aurélio de Mello's opinions.

Justice Gilmar Mendes's opinion[101] is particularly remarkable due to the systematization he presented when dealing with the collision of legal principles at

---

[99]Justice Nelson Jobim also attempted to defend a broad meaning of the concept of race through the investigation of its contents throughout history. He affirmed that, notwithstanding its original purpose related to the black people, it gained nowadays a much broader meaning, and, thus, the imprescriptibility of the crime had to be applied to this particular case. In the confirmation of his opinion, he developed an analysis of the democratic procedure, in which he sustained the equality principle as a condition for democracy and for the freedom of speech. He also mentioned the dialectical evolution of legal principles in opposition to an emphasis on the *mens legislatoris*.

[100]Justice Carlos Ayres Britto, after a long explanation of the inherently conflictive character of constitutional principles, concentrated part of his opinion on discussing criminal and procedural issues. Then, a semantic and grammatical investigation took place in order to identify the extension of each term of the constitutional norm, the difference between use and abuse of freedom of speech, and the usual comprehension of the concept of racism and discrimination. However, he mentioned that, in conformity with the Brazilian legislation (article 5, VIII, of the Constitution), there are three exceptions of abuse: religious belief, philosophical conviction and political conviction. After having analyzed those books, Justice Carlos Ayres Britto understood that they were "a work of historical research" and had a revisionist content. Thus, we could regard them as manifestations of the freedom of speech and the intellectual and scientific production sphere. He remarked that expressing an ideology is not a crime, based on the premise of a plural society and Brazilian legislation, which protects political-ideological convictions. It is interesting to mention that, in the confirmation of his opinion, his interpretation of the contents of those books was severely contradicted by the other Justices.

Similarly, Justice Ellen Gracie focused almost merely on the investigation of the actual debates on the concept of race and argued that racism embraces all types of discrimination and prejudice.

In his turn, Justice Cezar Peluso sustained the point of view that the Constitution did not adopt a scientific concept of race, but rather a normative concept, which we must comprehend through a teleological interpretation in decision-making. Considering the systematic publication of books of discriminatory contents, which attack the Constitution and overstep the limits of the freedom of speech, his opinion oriented towards denying the *habeas corpus*.

Justice Sepúlveda Pertence, as well, adopted a social-cultural concept of race to defend the position that racism embraces discrimination against Jews. Besides, he argued that a book is able to incite racism, and, in this particular case, the contents of the works investigated led to the conclusion that they could not be considered a revision of the traditional history.

[101]It is important to mention that Justice Gilmar Mendes has a solid constitutional-dogmatic knowledge, particularly originated from his doctoral studies in Germany, where he worked hardly with judicial review and comparative studies between German and Brazilian reality. He has written important books in this field, such as *Die abstrakte Normenkontrolle von dem Bundesverfassungsgericht und vor dem brasilianischen Supremo Tribunal Federal* (Berlin: Dunker Humblot, 1991); *Jurisdição Constitucional: Controle Abstrato de Normas no Brasil e na Alemanha* (São Paulo: Saraiva, 2004); *Controle de Constitucionalidade: Aspectos Jurídicos e Políticos* (São

stake. Here, more than the others, he materialized the intent to provide a rational justification for the decision. From the view that anti-Semitism is a form of racism, an argument he introduced from different relevant sources[102] and other opinions, Justice Mendes entered into the intriguing and complex field of the collision between freedom of speech and racism. Unlike the previous opinions, he attempted to ascertain, through methodological criteria, whether the freedom of speech, within this context, could be used as a means to contest the crime of racism. Two legal principles should be the main focus in this case: the freedom of speech and the right to nondiscrimination, as a consequence of the equality principle. In this regard, he concluded that it was indispensable to proceed to balancing, or, in his words, the "criminalization of discriminatory manifestations such as racism must be done by means of a judgment of proportionality."[103]

The principle of proportionality appeared as a crucial criterion to achieve the reasonable answer to this problem. According to Justice Mendes, "the open – I would say inevitably open – character of the definition of the legal norm, in this case, and the dialectical tension placed in front of the freedom of speech impose the application of the principle of proportionality."[104] The main issue was to define the extension of the allowed space for the exercise of freedom of speech within the context of a plural and complex society, or, in other words, its boundaries as a way to avoid any practice of intolerance or discrimination. In this matter, the principle of proportionality "[constituted] a positive and material exigency related to the content of restrictive acts of basic rights, in order to establish a "limit of limit" or a "prohibition of excess" when restricting those legal rights."[105]

By referring to Robert Alexy's *Theory of Constitutional Rights* (*Theorie der Grundrechte*[106]), he argued that every basic right has an "essential core," whose "last limit of possible legitimate restriction"[107] is determined by the principle of proportionality. Indeed, the application of this proportional criterion was, in his opinion, evidently embodied in the nature of any basic principle, as Robert Alexy ascertains when he defends that "the principle character implies the principle of proportionality, and this implies that one."[108] This logical consequence of the nature of legal principles, which is likewise deductible, for Robert Alexy, from

Paulo: Saraiva, 1990); *Direitos Fundamentais e Controle de Constitucionalidade* (São Paulo: Saraiva, 2004).

[102]Texts of Norberto Bobbio, Kevin Boyle, Pierre-André Taguieff, as well decisions of the American Supreme Court and the British Chamber of Lords.

[103]HC 82.424-2/RS. Justice Mendes' opinion, translation mine.

[104]HC 82.424-2/RS. Justice Mendes' opinion, translation mine.

[105]HC 82.424-2/RS. Justice Mendes' opinion, translation mine.

[106]See Robert Alexy, *Theorie der Grundrechte* (Frankfurt a.M: Suhrkamp, 1994).

[107]HC 82.424-2/RS. Justice Mendes' opinion, translation mine.

[108]Alexy, *Theorie der Grundrechte*, 100, translation mine.

its application by the constitutional court,[109] represents, in Justice Mendes's words, a "general method to solve conflicts between principles, that is, a conflict between legal norms, which, rather than the conflict between rules, is solved not by the invalidation or the teleological reduction of the conflicting legal norms, nor by making explicit the distinct application field among the legal norms, but before and only by balancing the relative weight of each one of the legal norms that are, in principle, applicable and able to justify decisions in opposite senses."[110]

As a means to solve the problems of collision between legal principles, Justice Mendes deployed the principle of proportionality, similarly to the *Cannabis* case, as a method to determine how the restriction on a basic right can be verified in a concrete situation, inasmuch as it offers some criteria to define the relative weight of each principle at issue. After having explained the three elements (suitability, necessity and proportionality in its narrow sense or balancing) in abstract, Justice Mendes began working on its application to the particularities of the case. By examining some German and Brazilian constitutional decisions and scholarly analyses, he sketched a very interesting argument, remarkably when the contents of the books were directly used for this aim. Along with the presentation of different passages of their contents, Justice Mendes concluded that they did not present a historiographic or revisionist goal. Rather, they had a clear purpose of blaming Jews for all humanity's misfortunes.

Considering this premise, he could then deploy the principle of proportionality, now contextually examined, as a rational method for adjudication. First, the act was suitable, considering that the condemnation fulfilled the function of safeguarding "a pluralistic society, where the tolerance reigns."[111] Additionally, the state's position was assured, provided that the intent to "defend the fundamentals of the dignity of human beings (article 1, III, FC[112]), the political pluralism (article 1, V, FC), the principle of repudiation of terrorism and racism, which rules Brazil in its international relations (article 4, VIII), and the constitutional norm that establishes racism as an imprescriptible crime (article 5, XLII)"[113] was attained. Second, the act was necessary, inasmuch as there was no other means less harmful and equally efficient, and, since the condemnation was implemented in a reasonable manner, the element of necessity was also respected. Ultimately, the decision was in accordance with the principle of proportionality in its narrow sense, for the proportion between the pursued goal (the preservation of the inherent values of a pluralistic society, the human dignity) and the burden imposed on the defendant's freedom of speech[114]

[109]According to Robert Alexy, "The German Federal Court has said, in a somehow vague expression, that the principle of proportionality already results fundamentally from the very nature of basic rights" (Ibid., 100, translation mine).

[110]HC 82.424-2/RS. Justice Mendes' opinion, translation mine.

[111]HC 82.424-2/RS. Justice Mendes' opinion, translation mine.

[112]FC: Brazilian Federal Constitution of 1988.

[113]HC 82.424-2/RS. Justice Mendes' opinion, translation mine.

[114]HC 82.424-2/RS. Justice Mendes' opinion, translation mine.

led to the conclusion that this freedom does not embrace racial intolerance and the incitation to violence,[115] which means that it has boundaries in democratic societies.[116] Accordingly, all principles have a latent limitation according to the features of the case. In this singular one, as the investigation of the contents of the books revealed, there was an abuse in the practice of freedom of speech.

As we can observe, in order to achieve this decision, Justice Mendes took three steps: (1) a comprehension of the problem by defending, since the beginning, that the main question was not terminological, but originated from the collision between two principles, that is, the freedom of speech and the equality principle, more precisely the right to nondiscrimination; (2) the abstract analysis of the principle of proportionality by establishing the main concepts it encompasses, such as the elements of suitability, necessity and proportionality in its narrow sense; (3) the concrete application of the principle of proportionality by explaining how we could examine each particularity of the case through those concepts introduced before. The conclusion displayed, therefore, a seemingly rational argument, provided that there was an adequate fit between the concrete aspects of the case and the abstract concepts derived from the principle of proportionality. Thus, instead of justifying decisions with an unorganized structure of reasoning, this criterion afforded a whole logical way for decision-making.

On the other hand, Justice Marco Aurélio deployed the principle of proportionality to defend the opposite argument by examining the freedom of speech and the equality principle (right to nondiscrimination) and balancing them with a naturalistic premise of the characteristics of Brazilian society. Through categorical arguments directed to defending the freedom of speech, he interpreted Brazilian history and traditions to dissent from the STF's majority opinion. Again, the discussion of balancing appeared in this realm: "Balancing [is] a criterion that allows a medium term between the binding and flexibility of legal rights,"[117] resulting then in their restriction or sacrifice.[118] It was necessary to establish, from the elements of the case, a *practical concordance* of the values at issue. Justice Marco Aurélio, hence, refined the naturalistic argument with some methodology.

In this regard, unlike Justice Mendes's words, this balancing should take into consideration a crucial variable: the freedom of speech, as long as it is a primary aspect of democracy, could only be justifiably restricted in the way someone exercises it, that is, how someone diffuses the idea.[119] Hence, the simple publication of books containing a political conviction or the author's or editor's intellectual

[115] HC 82.424-2/RS. Justice Mendes' opinion, translation mine.

[116] In the confirmation of his opinion, Justice Gilmar Mendes introduced many other interesting examples of important American and European decisions that worked with this conflict between the freedom of speech and the right to nondiscrimination (equality principle).

[117] HC 82.424-2/RS. Justice Marco Aurélio's opinion, translation mine.

[118] HC 82.424-2/RS. Justice Marco Aurélio's opinion, translation mine.

[119] "The only possible restriction on the freedom of speech, in a justifiable manner, is in its way of expression, that is, how this thought is diffused." HC 82.424-2/RS. Justice Marco Aurélio's opinion, translation mine.

expression[120] could not be deemed an abuse of the exercise of such legal right. Rather, only the distribution of "pamphlets in the streets of Porto Alegre with words like 'death to the Jews,' 'let's expel these Jews from our country,' 'take the guns and let's exterminate them,'"[121] which was not the case, would have the potential to reach this qualification. By assuming that only publishing books does not correspond to an effective and aggressive form of expressing opinion nor a physical threat exposing someone to an imminent risk,[122] which was the first parameter, he introduced the second one. A satisfactory balancing of values, when freedom of speech is at stake, must be detached from an "opinion based only in abstract expectations or personal fear that are dissociated from an examination that does not take into account the social and cultural elements or traces already present in our bibliographic history."[123] Therefore, apart from the argument of effective aggression, he put forward the one of social and historical basis. He used both as the justifications for balancing.

From the standpoint that the contents of the books were not racist, nor instigated "hateful prejudice,"[124] nor even "caused a national revolution,"[125] especially in a country where people are not used to reading,[126] but a simple manifest of political and ideological convictions, Justice Marco Aurélio set forth the naturalistic argument. Arguments such as the one based on the impossibility of those books "become an imminent danger of exterminating the Jewish people, especially in a country that never cultivated any repulsive feelings against these people,"[127] the one sustaining that those books could only be considered dangerous "when a determined political community have these 'pre-requirements' and has the aforesaid ambient,"[128] or, finally, the one questioning whether the "Brazilian society is predisposed to practice discrimination against the Jewish people"[129] were continuously mentioned in his opinion.

In addition, he made reference to the *mens legislatoris* to defend the argument that a restriction on legal principles demands an "almost literal" interpretation of them, and thus a comprehension of how the Framers debated this question.[130]

[120]HC 82.424-2/RS. Justice Marco Aurélio's opinion, translation mine.
[121]HC 82.424-2/RS. Justice Marco Aurélio's opinion, translation mine.
[122]HC 82.424-2/RS. Justice Marco Aurélio's opinion, translation mine.
[123]HC 82.424-2/RS. Justice Marco Aurélio's opinion, translation mine.
[124]HC 82.424-2/RS. Justice Marco Aurélio's opinion, translation mine.
[125]HC 82.424-2/RS. Justice Marco Aurélio's opinion, translation mine.
[126]HC 82.424-2/RS. Justice Marco Aurélio's opinion, translation mine.
[127]HC 82.424-2/RS. Justice Marco Aurélio's opinion, translation mine.
[128]HC 82.424-2/RS. Justice Marco Aurélio's opinion, translation mine.
[129]HC 82.424-2/RS. Justice Marco Aurélio's opinion, translation mine.
[130]Justice Marco Aurélio expressed this understanding in the quotation below, whose naturalistic character was clearly reinforced:

"I did not find, by analyzing the Constitutional Assembly's proceedings, any mention, even one, to the Jewish people when racism was discussed. The explanation, for me, is evident. The Constitution of 1988 is a Constitution of the Brazilian people, to be applied to the Brazilian people

Especially in these circumstances, Justice Marco Aurélio argued that the court must "restrict itself to an almost literal interpretation in the hypothesis of limiting these rights,"[131] because "this court and any interpreter of the Constitution is not allowed to interpret precepts that imply a reduction of the efficacy of basic rights in an open and broad manner."[132] In this respect, the appeal to the Framers becomes the solution. The statement arguing that no mention to the Jewish people in the Constitutional Assembly was a result of the fact that "the Constitution of 1988 is a Constitution of the Brazilian People, to be applied to the Brazilian people and tended to solve our own problems," as if discrimination against Jews were not a Brazilian problem, as well as the assertion that "the Constitution of 1988 is not a Constitution for the German, French, Italian, Polish, Austrians or European people in general,"[133] are patent examples of this understanding.

The argument of *mens legislatoris* reinforced the naturalistic argument that the crime of racism, in the case, only applies to certain groups, recalling thus Justice Moreira Alves's opinion, which restricted it to "black people."[134] As Justice Marco Aurélio mentioned, "the imprescriptibility can only apply to the case of practice of racist discrimination against black people, under penalty of creating an imprescriptible open criminal constitutional norm."[135] Indeed, he sustained this reasoning by creating a gradation of discriminatory acts in conformity with the affected group: "The racism against black people, this one, definitely, established by the constitution, is only one of the forms of discrimination, and, since it is the most serious of them, because it is rooted in the Brazilians' life, it emerges as imprescriptible."[136] This argumentation would then be inserted into the framework of the principle of proportionality.

Formerly, when we examined Justice Mendes's opinion, the three elements of the principle of proportionality resulted in considering the defendant's practice a crime of racism. By using the same three elements, Justice Marco Aurélio defended the nonexistence of this crime. First, the condemnation was not suitable, that is, able to cause the desired goal, insofar as the prohibition of publishing books or their seizure or destruction were not the "suitable means to stop the discrimination against the Jewish people."[137] After all, the simple transmission of "his version of history does not mean that the readers will agree with it, and, even if they will, it

---

and tended to solve our own problems. There is not any statement of racism against Jews and, on the other hand, there are pages and more pages of manifestations to stop the racism against black people, because the Constitution of 1988 is not a Constitution of the German, French, Italian, Polish, Austrian or European people in general." HC 82.424-2/RS. Justice Marco Aurélio's opinion, translation mine.

[131] HC 82.424-2/RS. Justice Marco Aurélio's opinion, translation mine.

[132] HC 82.424-2/RS. Justice Marco Aurélio's opinion, translation mine.

[133] HC 82.424-2/RS. Justice Marco Aurélio's opinion, translation mine.

[134] See note 95 *supra*.

[135] HC 82.424-2/RS. Justice Marco Aurélio's opinion, translation mine.

[136] HC 82.424-2/RS. Justice Marco Aurélio's opinion, translation mine.

[137] HC 82.424-2/RS. Justice Marco Aurélio's opinion, translation mine.

does not mean that they will begin to discriminate Jews."[138] Second, the condemnation was not necessary, that is, the comparative less harmful measure in the circumstance. In this analysis, Justice Marco Aurélio simply stated: "The observance of this sub-principle leaves to the Court only a possible solution: to grant the *writ*, in order to guarantee the right of freedom of speech, preserving then the books, for the restriction on this legal right will not even guarantee the protection of the Jewish people's dignity."[139] Third, Justice Marco Aurélio argued that the condemnation was not proportionate in its narrow sense, as long as the contents of those books could not give rise to a "revolution in Brazilian society,"[140] and there were many other books with discriminatory and racist contents[141] still available to the public. Accordingly, considering these activities a crime of racism was against a proportion between the adopted means and the encroachment of this measure on the freedom of speech.[142]

In order to solve the problem of how to manage his premises and balance the legal principles at issue (equality principle and freedom of speech), the solution he implemented was the application, through balancing, of a naturalistic justification tied to a semantic approach. There was the affirmation of the need to interpret restrictively or "almost literally" the restrictions on basic rights, as if it were possible to proceed in this way. Afterwards, he balanced those legal principles with a peculiar interpretation of the facts by placing the naturalistic and originalist argument in a prevailing position,[143] thereby relativizing those legal principles according to a choice of what he regarded as teleologically best[144] (people are

[138]HC 82.424-2/RS. Justice Marco Aurélio's opinion, translation mine.

[139]HC 82.424-2/RS. Justice Marco Aurélio's opinion, translation mine.

[140]HC 82.424-2/RS. Justice Marco Aurélio's opinion, translation mine.

[141]HC 82.424-2/RS. Justice Marco Aurélio's opinion, translation mine.

[142]In his opinion, besides, many other worldwide well-known judgments were used as examples for the prevalence of freedom of speech in similar contexts, such as the *Lüth Case* (BverfGE 7, 198, 01.15.1958), *Book of War Case* (BverfGE, 90, 1–22, 01.11.1994), *Murderers Soldiers Case* (BverfGE 93, 266–312, 10.10.1995), *Pornography Romance Case* (BverfGE 83,130, 11.27.1990), of the German *Bundesverfassugsgericht*; *Terminiello v. Chicago Case* (337 U.S. 1 – 1949), *R.A.V. v. City of St. Paul* (505 U.S. 337 – 1992), *Texas v. Johnson* (491 U.S. 397 – 1989), of the American Supreme Court; and the case of a comic publication against the Jewish people, judged by the Spanish Constitutional Court – *Sentencia 176/1995*, 12.11.1995.

[143]After all, based on what could he defend that a group, and not another, was discriminated in a society and why is a general predisposition to discrimination the main argument to qualify an act as racist? Following the words of Nicklas Luhmann in a brilliant text that demystifies naturalistic considerations, after all, "it cannot be assumed anymore that the relations between cause and effects are objective facts of the world, based on which it would be therefore possible to proceed to true and false judgments" (Luhmann, N. (1995). Kausaliät im Süden. *Soziale Systeme 1, 1*, 7–28.), http://www.soziale-systeme.ch/leseproben/luhmann.htm (accessed June 2nd, 2009), translation mine.

[144]Justice Marco Aurélio, moreover, attacked the opposite argument by emphasizing that relativizing freedom of speech, in this case, would represent a symbolical function, for it would project a political correct image before the society. Thus, instead of examining the particularities of the

not historically predisposed to practice discrimination against Jews and, therefore, could not suffer any real damage)[145] for the whole society.

1.5 Final Words

This chapter introduced three relevant cases in constitutional adjudication with a more descriptive and instigating purpose, for they will be further subject to critical review in the third part.[146] Yet, they could already indicate relevant aspects for this research. The *Crucifix* case demonstrated how a traditional value, the Christian belief of the Bavarian community, is assimilated into decision-making and how balancing appears as an interesting instrument to account for the discussion of what is the "indispensable minimum of elements of compulsion" the state can inflict on individuals or how much tolerance these individuals, members of a minority, should have within this context. Insofar as balancing appeared in both opinions, the majority and the dissenting ones, it is also an interesting example to show how this instrument can be easily used to sustain the enforceable character of constitutional principles, such as the freedom of faith and the state's duty of religious and philosophical neutrality (the majority opinion), or, on the contrary, to assume an axiological point of view. We can observe this last characteristic in the dissenting opinion through its stress on Christianity as a heritage of "the history of the Western cultural area"[147] or the political argument of state's dependency "on acceptance by parents of the school system [the state] organizes."[148] The *Cannabis* case in turn revealed how complex the dualism between law and politics in constitutional adjudication is. Indeed, it showed how far the BVG is involved in political issues either by means of the equality principle (article 3 (I) of the Basic Law), when it carried out a comparative evaluation of the *cannabis* products with other intoxicating substances, or through the discussion of how to balance the principle of free development of personality (article 2 (I) of the Basic Law) combined with the right

---

case, the court should, in reality, act in order to please a social clamor. See HC 82.424-2/RS. Justice Marco Aurélio's opinion.

[145]As we can observe in his opinion, the discrimination in Brazil can only historically and traditionally be deemed against other groups. In his words, "it would be easier to defend the idea of restricting the freedom of speech, if the issue in this *habeas* directed to the crucial problems challenged in Brazil, such as the theme of integrating the black people, the Indians or the people from the Northeast into the society." Racism, thus, could only be effectively practiced if a particular group had been traditionally victim of discrimination, which was not the case of Jews. Were other group the victim in this case, the decision would point to other direction, inasmuch as a prejudged book against black people would have much more chance of representing a real threat to the dignity of those people, because in Brazil it would not be difficult to find adepts at those thoughts. (HC 82.424-2/RS. Justice Marco Aurélio's opinion, translation mine).

[146]See the eighth chapter.

[147]BVerfGE 93, 1 – *Kruzifix*. Translation: Institute for Transnational Law.

[148]BVerfGE 93, 1 – *Kruzifix*. Translation: Institute for Transnational Law.

to freedom (article 2 (II) of the Basic Law) with the possible social consequences of *cannabis* products. This case is particularly relevant, as we will further examine,[149] because it illustrates the way for a constitutional principle to achieve an objective content that can be directly applied to solve most problems of social life, raising thereby the doubts whether constitutional adjudication is not possibly intervening in the constitutional functions of parliament. Finally, the *Ellwanger* case extended the debate to Brazilian reality, showing how another constitutional culture deploys methods and criteria, such as the principle of proportionality and, particularly, balancing, with the same structural framework we can observe in Germany, and how the same doubts about this instrument arise. By showing Justices Gilmar Mendes's and Marco Aurélio's arguments, one deploying balancing to reinforce the equality principle and the other to relativize it in favor of a naturalistic interpretation of Brazilian history and an originalist approach to the Constitution, this chapter ended by suggesting how balancing increases the risk of discretionary rulings.

Having examined the main arguments of these three cases, the next step is to situate them in their respective constitutional realities. The intent is to verify how balancing, as we could remark in their contents, is associated with the constitutional court's shift to activism. In this regard, while understanding the context where those decisions were made, it is possible to verify the birth of a rational approach to adjudication that follows a historical constitutional development that, as interesting as it is, contributes fundamentally to the comprehension of the second and third parts. The next two chapters of this first part, for this reason, will concentrate on the recent constitutional history of Germany and Brazil. They will complement this empirical research and uncover many singular characteristics that make these two realities crucial examples of a movement whereby constitutional adjudication, while seeking rational solutions as the condition of its legitimacy, shifts to activism, resulting then in significant questions related to the principle of separation of powers. The next two chapters, accordingly, will connect history, the evolution of legal institutions, to decision-making, expressing thereby the perception that there is no possibility to investigate the rationality of adjudication without understanding that every constitutional problem, even though reaching the philosophical discussion as we shall develop in the second and third parts, is also a debate on concrete causes and consequences.

[149] See the second chapter.

Chapter 2
Balancing Within the Context of German Constitutionalism: The *Bundesverfassungsgericht*'s Shift to Activism

Abstract The historical development of the German Federal Constitutional Court (*Bundesverfassungsgericht*) and its progressive assumption of the role of "Guardian of the Constitution" through the interpretation of basic rights as objective principles of the total legal order represent a crucial movement in the contemporary constitutionalism. Particularly because of its inclination towards the definition and discussion of the main themes of social life, as if they were constitutional problems to be decided by the court, and the construction of instruments and interpretations, such as the principle of proportionality and the shift from subjective principles to objective principles, the *Bundesverfassungsgericht* is an important representative of the current worldwide judicial activism. The historical context of an emerging constitutional court after the Second World War and the consequent process of democratization where there was a vacuum of political legitimacy led to the transference of the discussion of many social themes to this court, raising thereby serious questions about a possible encroachment on the other institutional powers. In this respect, the transformations in German constitutional culture, the reactions of relevant part of constitutional scholarship, and the perception of the problems originating from this movement expose the connections between balancing and judicial activism, and demonstrate how constitutional democracy deals with the dilemmas of a process of juridification of politics.

2.1 Introduction

When the discussion of balancing appears in the center of a quest for rationalizing constitutional adjudication, the immediate connection, as the previous cases could already suggest,[150] is its comprehension as an element of the principle of proportionality, identified as the principle of proportionality in its narrow sense, whose characteristics are intimately related to the German recent historical development

[150]See the previous chapter.

J.Z. Benvindo, *On the Limits of Constitutional Adjudication*,
DOI 10.1007/978-3-642-11434-2_2, © Springer-Verlag Berlin Heidelberg 2010

and a particular interpretation of this principle that has been widely welcomed.[151] Indeed, the quest for providing a "rational" justification for decision-making through the principle of proportionality, embracing thereby the purpose of "rationalizing" balancing, temporally coincides with the BVG's shift to activism. It is therefore not unreasonable to affirm that the history of the recent German constitutionalism closely relates to the history of the principle of proportionality, or, more specifically, that the development of the principle of proportionality connects to the BVG's history. Bernhard Schlink, for instance, remarks that, not only thanks this principle its career to the BVG's decisions, but also its decisions are overall decisions based on proportionality.[152] Jürgen Habermas, similarly, critically suggests how this principle, especially in Germany, appears as a key concept that serves to provide the norms in collision with a view of "unity and consistency of the constitution,"[153] and Robert Alexy associates it with the very nature of principles.[154] In the last fifty years, we could observe a clear expansion of its deployment by constitutional courts, from Germany to Europe,[155] Israel,[156] Canada,[157]

[151]In this respect, it is notorious the interpretation of this principle with the triadic structure (suitability, necessity and proportionality in its narrow sense or balancing). An influential theory in this matter, which will be the main source for the analysis here of the rationality of balancing, is Robert Alexy's *Theory of Constitutional Rights*. See the fourth chapter.

[152]Bernhard Schlink, "Der Grundsatz der Verhältnismäßigkeit," in *Festschrift – 50 Jahre Bundesverfassungsgericht*, ed. Peter Badura and Horst Dreier (Tübingen: Mohr Siebeck, 2001).

[153]Jürgen Habermas, *Between Facts and Norms* (Cambridge: Polity Press, 1996), 248.

[154]Robert Alexy, *Theorie der Grundrechte* (Frankfurt a.M: Suhrkamp, 1994), 100.

[155]See Evelyn Ellis, *The Principle of Proportionality in the Laws of Europe* (Oxford: Hart, 1999); Søren Schønberg, "The Principle of Proportionality's Many Faces: a Comparative Study of Judicial Review in English, French, and EU Law," in *Justitia*, ed. Søren Schønberg (København: Jurist-og Økonomforbundets Forl, 2000); Nicholas Emiliou, *The Principle of Proportionality in European Law: A Comparative Study* (London: Kluwer Law Internat, 1996); Oliver Koch, *Der Grundsatz der Verhältnismäßigkeit in der Rechtsprechung des Gerichtshofs der Europäischen Gemeinschaften* (Berlin: Duncker & Humblot, 2003); Enzo Cannizaro, *Il Principio della Proporzionalità nell' Ordinamento Internazionale* (Milano: Giuffrè, 2000); Sadursky Wojciech, *Rights Before Courts: a Study of Constitutional Courts in the Post-Communist States of Central and Eastern Europe* (Dordrecht: Springer, 2005).

[156]The principle of proportionality is nowadays the basis for constitutional adjudication in Israel, and is deployed continuously through the adoption of the three-step proportionality test. See, for this purpose, *Hamdi v. Commander of Judea and Samaria* (1982), *United Mizrahi Bank Ltd. v. Migdal Village* (1995), *Ben-Atiyah v. Minister of Education, Culture & Sports* (1995).

[157]In Canada, the expansion of the principle of proportionality could be seen especially after the enactment of the Canada's Charter of Rights and Freedoms, in 1982, whose §1 establishes that the Charter "guarantees the rights and freedoms set out in it subject only to such reasonable limits prescribed by law as can be demonstrably justified in a free and democratic society." The Charter's extensive catalogue of rights, as Sweet remarks, is structured in a way that invites the deployment of the principle of proportionality (See Alec Stone Sweet, *Constitutionality, Balancing and Global Constitutionalism*, http://www.law.columbia.edu/null/Stone-Sweet+-+Proportionality+Balancing?exclusive=filemgr.download&file _id=101159& showthumb=0 (accessed July 14, 2009)). Nowadays, Canada adopts, similarly to Germany, the three-step proportionally test (suitability, necessity, and proportionality in its narrow sense), but, unlike Germany, the stress is usually on the

South Africa,[158] Central[159] and South America, particularly in Colombia[160] and Brazil,[161] as well as in Australia.[162] It could even be possible to argue for its existence in the United States of America,[163] although some objections exist in

examination of necessity, instead of the proportionality in its narrow sense. See, for this purpose, *R. v. Oakes*, Supreme Court of Canada, [1986], S. C. J. No. 7. An interesting analysis of the principle of proportionality in comparison with the United States can be seen in Vicki C. Jackson, "Ambivalent Resistance and Comparative Constitutionalism: Opening Up the Conversation on 'Proportionality', Rights and Federalism," *University of Pennsylvania Journal of Constitutional Law* 1 (1999): 583 ff.

[158]In South Africa, after the end of the apartheid regime and the introduction of an interim Constitution in 1993, which established the judicial review by the South Africa's Constitutional Court, the principle of proportionality gained a very strong diffusion. In the permanent Constitution of 1996, the principle of proportionality received the constitutional status, as a "standard operating procedure for adjudicating limits on rights" (Sweet, *Constitutionality, Balancing and Global Constitutionalism*, 29), although not applied with the same systematization and analytical basis as in Germany. See, for this purpose, State v. Makwayane (1995).

[159]See Ruben Sánchez Gil, *El Principio de Proporcionalidad* (Mexico: Universidad Nacional Autônoma de México, 2007).

[160]See Miguel Carbonell, *El Principio de Proporcionalidad en el Estado Constitucional* (Bogotá: Universidad Externado de Colombia, 2007); Carlos Bernal Pulido, *El Principio de Proporcionalidad y los Derechos Humanos* (Madrid: Centro de Estúdios Políticos y Constitucionales, 2003).

[161]We will examine the development of this principle in Brazil in the next chapter.

[162]For an interesting analysis of the deployment of the principle of proportionality in Australia, showing the differences and possible conflicts in this reality, see Jeremy Kirk, "Constitutional Guarantees, Characterization and the Concept of Proportionality," *Melbourne University Law Review* 21, no. 1 (1997).

[163]In the United States, the deployment of a variant of the principle of proportionality, specifically balancing, can be seen mainly in four different scenarios, all of them related to the premise of the existence of a conflict between competing interests: (1) in the interpretation of the Eighth Amendment, as we could observe in the case *Ewing v. California*, (538 U.S., 11, 20, 2003) which sustained the existence of a proportionality principle in the Eighth Amendment applicable to non-capital sentences. In this opinion, the U.S. Supreme Court held that "the Eighth Amendment's prohibition of 'cruel and unusual punishments' expresses a broad and basic proportionality principle that takes into account all of the justifications for penal sanctions" (Justice Scalia, 126); (2) in the interpretation of the Fourth Amendment, as we can observe in *Tenessee v. Garner* (471 U.S. 1; 105, 1985), when the Court held that "to determine the constitutionality of a seizure 'we must balance the nature and quality of intrusion on the individual's Fourth Amendment interests against the importance of the governmental interests alleged to justify the intrusion'" (Opinion – Justice White). This case, particularly, shows how balancing deals with, on the one hand, governmental interests and, on the other, individual's private sphere; (3) in the interpretation of the Fourteenth Amendment, as we can observe in *City of Boerne v. Flores* (512 U.S. 507, 1997) or *Eldred v. Aschcroft* (537 U.S. 186, 218, 2003); (4) in the interpretation of the First Amendment (See *F.C.C.* v. *League of Women Voters* (468 U.S. 364, 1984)). A very critical and interesting analysis of balancing in the United States can be found in T. Alexander Aleinikoff, "Constitutional Law in the Age of Balancing," *Yale Law Journal* 96, no. 5 (April 1987): 943–1005. For a comparative study of methodologies adopted in Europe and the United States, including the deployment of a "balancing approach," see Daniel Halberstan, "Desperately Seeking Europe: On Comparative Methodoloy and the Conception of Rights," *International Journal of Constitutional Law* 5, no. 1 (2007): 166–182.

this matter.[164] There were also a clear spread of its promotion in different legal areas[165] and an improvement of its methodological comprehension, with a better systematization of its elements (suitability, necessity, and proportionality in its narrow sense or balancing), as well as a theoretical appropriation by constitutional scholarship.[166] It is certainly one of the most successful instruments already adopted by constitutional courts, and it has likewise radically transformed the constitutionalism.

The principle of proportionality, constructed as a dogmatic methodological structure that helps find the solution to a particular case and "rationally" systematizes the steps the judge has to observe in her duty, including how she has to deploy balancing, if necessary,[167] thanks much of its expansion to the contemporary transition to a casuistic jurisdiction in Germany[168] and in other countries.[169] In Germany, where the grounds and elements of this principle have been systematically developed, this transition implies the BVG's more activist approach.[170] This

---

[164]According to Gerald L. Neuman, the principle of proportionality in the United States does not exactly correspond to its usual concept, and it is not prominent in adjudication and doctrine. His words:

"The concept of proportionality does not lack paralells in U.S. constitutional law. Basically, it is a form of balancing of interests (*Güterabwägung*) common to both systems, and articulated with a tripartite structure. But balancing is not regarded in U.S. constitutional doctrine as an element of the rule of law, and it is not applied to interferences with all constitutional rights. Some degree of appropriateness (*Geeignetheit*) might be viewed as an aspect of nonarbitrariness required by the rule of law, but necessity and proportionality in the narrow sense are not."

"Moreover, this is not merely a peculiarity of *constitutional* doctrine. Even with regard to nonconstitutional debates about the rule of law in the United States, proportionality (or balancing) does not figure prominently as a feature. Procedural conceptions of the rule of law do not identify proportionality as an essential characteristic of law, and substantive conceptions of the rule of law may invoke human rights constraints without specifying proportionality as a necessary structural feature of rights" (Gerald L. Neuman, *Constitutional Conception of the Rule of Law and the Rechtsstaatsprinzip of the Grundgesetz*, http//papers.ssrn.com/paper.taf?abstract_id=195368 (accessed July 14, 2009)).

[165]As Schlink remarks, in the history of German constitutionalism, the principle of proportionality was applied, until mid-1950, particularly in the Administrative and "Police" law (*Polizeirecht*) and required only the examination of the legitimacy of the goal and the suitability and necessity of the means to realize it. Nowadays, its deployment reaches not only the constitutional and administrative law but also conflicts between organs, civil (especially with the *Drittwirkung* theory), criminal (particularly in the evaluation of the punishment) and European law. Moreover, it is also deployed not only when there is an excess of the interference with the private's sphere (*Übermaßverbot*), but also when the state remains passive and causes a severe encroachment on the individual (*Untermaßverbot*). See Bernhard Schlink, "Der Grundsatz der Verhältnismäßigkeit," in *Festschrift – 50 Jahre Bundesverfassungsgericht*, ed. Peter Badura and Horst Dreier (Tübingen: Mohr Siebeck, 2001): 445.

[166]See Schlink, "Der Grundsatz der Verhältnismäßigkeit."

[167]Balancing is theoretically understood as the third element of the principle of proportionality, whose deployment takes place after the legal provision under examination succeeds in the examination of suitability and necessity.

[168]See Bernhard Schlink, "Abschied von der Dogmatik: Verfassungsrechtssprechung und Verfassungsrechtswissenschaft im Wandel," *Merkur* 692 (December 2006).

[169]See, for example, Brazilian constitutional reality in the next chapter.

[170]See Bernhard Schlink, "The Journey into Activism," *Cardozo Law Review* 17 (1996).

consolidates the idea of being the guardian of not only the legal order but also of the social values[171] by using, as a communication towards society, arguments of policy, morality, economy, among others.[172] Normally, when a decision has to be made regarding the constitutionality of a legal statute, the court inquires whether it is legitimate to achieve a specific goal. There is, accordingly, the examination of means and goals based on the characteristics of a particular case and the possible encroachments the statute causes in an overall analysis of its effects in the society, which raises the need to deploy the principle of proportionality, and more specifically balancing, as a dogmatic method that seemingly best handles these tensions of constitutionalism. Indeed, if, on the one hand, we could remark that the BVG's workload is particularly notable in questions about rights of freedom[173] – and, in this case, the classic individualistic conception of opposition between private sphere and state action easily seems to call for this principle – on the other, its deployment is more and more omnipresent, first and foremost, in virtue of the premise that the basic rights of the Basic Law also demand a positive action concerning the BVG's protective duty towards society.

As a method that responds to this protective duty, the principle of proportionality, now comprising balancing in its triadic structure, is an outcome of this movement that led to the erection of different concepts and terms that could materialize rights as "principles of a total legal order whose normative content structures the system of rules as a whole."[174] Indeed, similar to this movement in the interpretation and application of principles, the principle of proportionality shifted historically from the mere examination of suitability and necessity[175] to

[171]See Schlink, "Abschied von der Dogmatik: Verfassungsrechtssprechung und Verfassungsrechtswissenschaft im Wandel."

[172]See Bernhard Schlink, "Open Justice in a Closed Legal System?," *Cardozo Law Review* 13 (1992): 1716.

[173]See Bernhard Schlink, "The Dynamics of Constitutional Adjudication," in *Habermas on Law and Democracy*, ed. Michel Rosenfeld and Andrew Arato (Berkeley, LA: University of California Press, 1998), 373.

[174]According to Habermas, the legal doctrine adopts nowadays this ideal of basic rights as principles of a total legal order in different perspectives:

"This specifically German doctrine of basic rights focuses primarily on a few key ideas. These include the 'reciprocal effect' or (*Wechselwirkung*) between ordinary legal statutes and fundamental rights (which remain inviolable only in their 'essential content' or *Wesengehalt*); the 'implicit limits on basic rights', which hold even for those basic individual rights, such as the guarantees of human dignity, that impose affirmative duties on the state (the so-called *sujektiv-öffentliche Rechte*); the 'radiating effect' (*Austrahlung*) of basic rights on all areas of law and their 'third-party effect' (*Drittwirkung*) on the horizontal rights and duties holding between private persons; the state's mandates and obligations to provide protection, which are tasks the Court derives from 'objective' legal character of basic rights as principles of the legal order; and finally, the 'dynamic protection of constitutional rights' and the links in procedural law between such rights and the 'objective' content of constitutional law" (Jürgen Habermas, *Between Facts and Norms* (Cambridge: Polity Press, 1996), 247).

[175]The principle of proportionality developed first in police and administrative law, in the form of the now labeled principle of necessity (*Grundsatz der Erforderlichkeit*) and also the principle of suitability (*Grundsatz der Geeignetheit*), at the end of 19th century, when the liberal idea that the

the integration of balancing (now called principle of proportionality in its narrow sense) into its inner core. In this regard, it appears as a key concept that seemingly serves as a mechanism that promotes unity and consistency in the process of decision-making.[176] It is a dogmatic framework that looks adequate for the complex dilemmas of contemporary constitutionalism and fulfills very well the aim to expand the BVG's protective activity to other dimensions than the traditional liberal concepts of individual freedom and equality. As a consequence, rather than being deemed an instrument to be deployed in strict areas such as police and administrative law with a narrower content and incidence, the principle of proportionality, with balancing,[177] reaches the most distinct legal areas. Moreover, it represents, as a result of its flexibility, a very robust instrument that consolidates the BVG's tendency to increase its influence and authority within the context of constitutional democracy. Briefly, the principle of proportionality, now with balancing, concurrently contributes to the growth of a casuistic jurisdiction – and the consequent amplification of the BVG's power and influence towards society – and expands itself because of this constitutional court's shift to activism.

The BVG's appearance, in 1951, radically transformed the comprehension of basic rights in Germany, and certainly one of the main transformations relates to the increasing deployment of the principle of proportionality, which supposedly "rationally" systematizes the practice of balancing in distinct legal matters. Unlike other constitutional realities, this principle is almost a requirement for the rightness of

---

state can only limitedly intervene in private freedom sphere according to a delimited goal gained force, and when the control over administrative acts by an administrative higher court could be more independently carried out. From this period onwards, the examination of the binary means/goals, in reference to the less harmful means to achieve a goal, began to be deployed, even though without the definition of a criterion to evaluate this intensity. After the Second World War, the principle of proportionality in its narrow sense, such as nowadays conceived, began to be used because of the new dilemmas derived from a more active role of judicial review, and many laws, particularly in the field of police law, started to describe it. Doctrine and the courts' precedents took, nonetheless, a long time before establishing a clear distinction among the different elements of the principle of proportionality in its broad sense. Furthermore, it is interesting to remark that all this evolution was practically not followed by judicial reflection or even by constitutional scholarship's criticism. See, for a comprehensive analysis of the historical development of the principle of proportionality, Lothar Hirschberg, *Der Grundsatz der Verhältnismäßigkeit* (Göttingen: Otto Schwartz & CO, 1981), 1–42.

[176]Habermas, *Between Facts and Norms*, 248.

[177]The terminology in constitutional doctrine and in the BVG is not uniform. The principle of proportionality can be deployed as a broader principle encompassing the principles of suitability, necessity and proportionality in its narrow sense, as here stressed, but it can also have a different meaning. Sometimes, the term "principle of proportionality" is applied simply to the principle of necessity or the principle of proportionality in its narrow sense. In other cases, the term *prohibition of excess* (*Übermaßverbot*) is used as synonym for the principle of proportionality in its narrow sense, principle of necessity, or to designate the principle of proportionality in its broad sense. Balancing, likewise, is not a harmonious term, but it is used more specifically as a correspondence to the principle of proportionality in its narrow sense. In this book, it will be used: the principle of proportionality in its broad sense encompassing the three elements (suitability, necessity, and proportionality in its narrow sense). This last one we will also call balancing.

the decision and must be deployed as a necessary parameter to deliver a minimum of consistency and "rationality" in legal reasoning. Judges rarely decide constitutional issues involving a conflict of basic rights without making reference to it, and the scholars follow the same predisposition to explore constitutional problems through the eyes of this methodology. In Law Schools, a constitutional problem is solved as a proportional-analysis problem, and most social and political issues are considered a problem of fundamental rights[178] whose solution demands a proportional investigation of means and goals. German constitutional culture, therefore, is nowadays mostly a proportionally-oriented culture centered on the BVG's decisions, which is regarded as the most popular and admired constitutional organ[179] and whose decisions are rarely criticized and opposed by constitutional scholarship.[180] It is a challenge, for this reason, to grasp how this interaction between the social expectations for the BVG's assumed duty to safeguard their values and its inclination to explore the terrain of politics can be regarded in the realm of a constitutional democracy. The question of the separation of powers evidently stands out, and the BVG's legitimacy, as a forum of discussion of political, economic, and social issues, becomes a primary concern. As a consequence, the question of whether the BVG can undertake a proportional analysis of social, political and economic values, as an activity of the very practice of judicial review, is also of central interest.

The BVG's posture as an activist court, which treats social, economic and political problems[181] as if they were part of its own field of responsibility and authority, is a sign of a controversial historical development that not only favored the disbelief in traditional politics but also claimed a new institution that could occupy the existing vacuum of legitimacy observed after the Second World War.[182] The BVG took also advantage of a certain crisis in Jurisprudence derived from the period of National Socialism[183] and, therefore, began to act without a consolidated scholarship that could develop the critique of its decisions and the progressive gain of authority in the different matters of social life. In this process, the reinforcement of the material judicial review based on the concept of legal rights as optimization requirements gave rise to the possibility of establishing a form of argumentation that clearly politicized decision-making, thereby increasing the power of the judiciary and its discretion, as Habermas suggests, "in a way that threatens to upset the equilibrium in the normative structure of the classical constitutional state at the cost of citizens' autonomy."[184] The BVG appeared as the institution that assumed this

[178] See Bernhard Schlink, "German Constitutional Culture in Transition," *Cardozo Law Review* 14 (1993): 729.

[179] See Schlink, "Abschied von der Dogmatik: Verfassungsrechtsprechung und Verfassungsrechtswissenschaft im Wandel, 1125."

[180] See Schlink, "German Constitutional Culture in Transition," 734.

[181] Ibid., 729.

[182] Ibid., 725.

[183] Ibid., 733.

[184] Habermas, *Between Facts and Norms*, 246.

role of social guard that might have put in jeopardy much of the principle of sovereignty of people, although, paradoxically, an expansion of its popular acceptance has followed this movement.[185] The constitutional reality in Germany is, accordingly, complex not only because of the particular configuration of a nation strongly challenged by the reconstruction of social values and the legal order in the postwar period, which brings to light the tortuous process of filling the institutional gap by an organ that was accredited to exercise this expected function, but also on account of the following social acceptance of this process, as if, apart from the consolidation of basic rights, the political and social order needed to be heteronomously conducted by a group of honest, skillful and prudent judges. This second aspect is even more intricate when we witness the prevalence of a constitutional scholarship that canonizes the BVG's decisions, instead of undertaking its duty of offsetting its authority and providing the critique.[186]

There is no possible reconstruction of the expansion of the deployment of the principle of proportionality, and balancing specifically, if not accompanied by the comprehension of these relevant aspects that transformed the BVG into the fundamental organ of German constitutional reality. The principle of proportionality and the continuous use of balancing, as a "rational" response to decision-making, are the reflex of the development of a constitutional court that descends from a period of crisis and assumes the role of centralizing the social and political discussions under its sphere of authority. It is the consequence of the BVG's progressive undertaking of the responsibility to represent and preserve German higher social values, to deal with many conflicts deriving from the dualism between law and politics,[187] as if any conflict of social life could be interpreted as a problem of basic rights,[188] and of its propensity to attend foremost to the present and future problems rather than focusing on the institutional development of the legal order as a whole.[189] It is thus the perfect instrument to embrace the vast authority that the BVG has taken on as its historical commitment to society and is likewise the bridge to promote an aura of methodological consistency and rationality in legal argumentation, which strengthens the justification for its growing influence and intervention in the public and private spheres.

---

[185]See Schlink, "Abschied von der Dogmatik: Verfassungsrechtsprechung und Verfassungsrechtswissenschaft im Wandel," 1125.

[186]See Schlink, "German Constitutional Culture in Transition," 734.

[187]According to Peter Häberle, this dualism can be seen – apart from the evident situations, such as the claim to the BVG's "self-restraint" or the appointment of the Justices to the constitutional court – in the practice of decision-making, as we can observe in the methods of interpretation followed by the question about the political consequences of the decision or in the investigation of the binding effects of constitutional decisions. There are also these dealings with law and politics in the admission of an appeal, in the definition of a principle, in the tactics and strategies used in the constitutional process, and in the specification of the intensity of the facts (See Peter Häberle, "Grundprobleme der Verfassungsgerichtsbarkeit," in *Verfassungsgerichtsbarkeit*, ed. Peter Häberle (Darmstadt: Wissenschaftliche Buchgesellschaft, 1976), 4–5).

[188]See Schlink, "German Constitutional Culture in Transition," 722.

[189]See Habermas, *Between Facts and Norms*, 246.

This chapter, above all, will focus on the process whereby the BVG gained this quality of a political activist constitutional court, and how it opened up the space for balancing in political matters. In this regard, it will provide an analysis of how the principle of proportionality, and balancing in particular, could be considered one of the foundations of German constitutionalism, and how it fits properly the purpose of expanding the BVG's activities to many different areas of social life by exposing some methodological and seemingly rational justification in this process. For this purpose, we will stress here four specific themes: (1) a brief introduction to the principle of proportionality, in order to examine its most well-known configuration: the triadic structure of suitability, necessity and proportionality in its narrow sense or balancing, which, in particular, will be the main focus of this investigation, even because it is balancing that is more intimately related to this BVG's shift to activism[190] (Sect. 2.2); (2) the historical analysis of the BVG's shift to politics after its institution in the middle of the postwar crisis, as a way to reveal that the dualism between law and politics that is at the core of its activities has an intimate relation to German institutional history (Sect. 2.3); (3) the transformations in constitutional dogmatics concerning the interpretation of basic rights, now transmuted into objective principles of a total legal order, which promoted serious outcomes in judicial review (Sect. 2.4); and (4) the constitutional scholarship reaction against this politicization of judicial review in Germany, particularly by reason of the increasing deployment of balancing, now "rationally" justified within the framework of the principle of proportionality, which can promote, as long as it weakens the consistency of the legal system through the objective structure of basic rights, the judicial exercise of arbitrary and discretionary rulings (Sect. 2.5).

2.2 Balancing Within the Triadic Framework of the Principle of Proportionality: A Brief Introduction

The principle of proportionality (*Verhältnismäßigkeitgrundsatz*), especially after having included balancing in its inner core, is a very persuasive instrument to resolve conflicts between basic rights and any other value the concept of constitution as a "concrete order of values" embraces. It is a technique of constitutional adjudication that, within this context of a value-based approach and the weakening of dogmatic reasoning, as we will shortly examine, seeks to set forth a systematized justification for decision-making. Strongly grounded in the BVG's activities, in the 1950s and 1960s, relevant dissertations provided a methodological comprehension of this principle and, while considering it to have a constitutional status, also

[190]This chapter will only briefly discuss the triadic structure of the principle of proportionality. For the central focus of this book refers to the rationality of balancing, particularly this element will be subject of more detailed analysis in the fourth chapter, using, for this purpose, Robert Alexy's *Theory of Constitutional Rights* as the main source.

examined the distinction among its elements.[191] In the following years, the commitment to promote systematization, a deeper examination of its characteristics, or its rational justification has been notorious.[192] Yet, much of this development has transformed constitutional scholarship into an activity shaped according to the BVG's decisions, especially because this court has continuously provided material, through its decisions, referring to the principle of proportionality. This aspect generates, consequently, a complex context, for the BVG, although expanding the deployment of this principle, has not truly contributed to set forth its basis nor presented a methodological structure of its elements. The BVG frequently uses different terminologies or mixes up the contents of its elements,[193] and most of the justifications for this principle – rule of law (*Rechtsstaatsprinzip*), "core content" of basic rights (*Wesengehaltsargument* of article 19 (2) of the Basic Law), for example – and its constitutional status are normally simply introduced without further discussion.[194] In addition, contrary to the role of scholarship, those incoherencies were not directly confronted – and when they were,[195] they were clearly

[191]Rupprecht von Krauss was the one who coined, in 1953, for the first time, the term principle of proportionality in the narrow sense (*Verhältnismäßigkeit im engeren Sinn*) by distinguishing it from the principle of necessity (even though then labeled principle of proportionality), which he identified with the principle of suitability. He mentioned that the proportionality in the narrow sense refers to a relationship between two or more measured quantities. For him, the new constitutional order is regarded as a proportional analysis order. He also attempted to treat this principle as a constitutional principle (See Rupprecht von Krauss, *Der Grundsatz der Verhältnismäßigkeit in seiner Bedeutung für die Notwendigkeit des Mittels in Verwaltungsrecht* (Hamburg: Appel, 1955)). In 1961, in turn, Peter Lerche developed a clear distinction, albeit connected, between the principle of necessity and the principle of proportionality. For him, they have distinct contents. Whereas the first refers to the premise that, among different possibilities that suitably can reach the goal, we have to choose the one that causes less encroachment on the private sphere, the second relates to the balancing between means and goals. He also sustained the constitutional status of the principle, as a consequence of the modern welfare state and the directive constitution (*dirigierende Verfassung*). See Peter Lerche, *Übermaß und Verfassungsrecht: zur Bindung des Gesetzgebers an die Grundsätze der Verhältnismäßigkeit und der Erforderlichkeit* (Goldbach: Keip, 1999).

[192]See Robert Alexy, *Theorie der Grundrechte* (Frankfurt a.M: Suhrkamp, 1994); Hans Hanau, *Der Grundsatz der Verhältnismäßigkeit als Schranke privater Gestaltungsmacht* (Tübingen: Mohr-Siebeck, 2004); Bernhard Schlink, *Abwägung im Verfassungsrecht* (Berlin: Duncker & Humblot, 1976). Schlink, nonetheless, concluded that balancing is not a rational response to adjudication. See item 2.5.

[193]See Bernhard Schlink, "Der Grundsatz der Verhältnismäßigkeit," in *Festschrift – 50 Jahre Bundesverfassungsgericht*, ed. Peter Badura and Horst Dreier (Tübingen: Mohr Siebeck, 2001): 446.

[194]Ibid., 446.

[195]Some authors have explicitly exposed this problematic situation in constitutional adjudication. See, for instance, Schlink, "Der Grundsatz der Verhältnismäßigkeit"; Helmut D. Fangmann, *Justiz gegen Demokratie: Entstehung – und Funktionsbedingungen der Verfassungsjustiz in Deutschland* (Frankfurt a.M; New York: Campus Verlag, 1979); Habermas, *Between Facts and Norms;* Walter Leisner, *Der Abwägungsstaat: Verhältnismäßigkeit als Gerechtigkeit?* (Berlin: Duncker & Humblot, 1997).

the minority's voice,[196] – resulting in a non-critical space where the BVG's expansionist purpose could more effortlessly happen.[197]

Nonetheless, by investigating the most prominent theories on the principle of proportionality, we can already find some connections that point out, if not with an expected coherence, at least a possible comprehension of its premises and elements. In this case, for instance, while Peter Lerche distinguishes the principle of proportionality (*Grundsatz der Verhältnismäßigkeit*) from the principle of necessity (*Grundsatz der Erforderlichkeit*) through an extensive analysis of both in the realm of a directive constitution (*dirigierende Verfassung*) – and here the principle of suitability (*Grundsatz der Geeignetheit*), even though introduced, is regarded as part of the principle of necessity and proportionality[198] – Lothar Hirschberg is very careful about setting forth the triadic division of the principle of proportionality (suitability, necessity, and proportionality in the narrow sense), although recognizing the difficulties of this division[199] and the problems these elements, especially the last-mentioned, raise.[200] By the same token, whereas Bernhard Schlink, aside from exposing the risks of confusion between law and politics, as well as of irrational rulings,[201] sets forth an extensive analysis of this principle by focusing on the suitability and necessity of the state intervention, the legitimacy of the goal, and the protection of citizen's minimal position,[202] criticizing, however, the

---

[196]See Hirschberg, *Der Grundsatz der Verhältnismäßigkeit*, 208.

[197]Indeed, there is an environment where the BVG's decisions are not really concerned with coming up with a systematized understanding of the principle of proportionality, its elements, and, foremost, its justification and *raison d'être*, and where the constitutional scholarship mostly upholds these characteristics, without entering into the most convoluted questions of this movement. For instance, how can balancing and the objective nature of basic rights behind it be compatible with the separation of powers? How can both be defended in a democratic regime where individuals respect each other as free and equals in their differences? Finally, how can both be compatible with the constitution? (See Schlink, "Der Grundsatz der Verhältnismäßigkeit," 447) Evidently, this leads to a vicious circle: while the BVG deploys this principle almost as a natural and evident premise in judicial review, and the constitutional scholarship endorses this development, the BVG can continue to act without discussing more deeply the grounds and the methodology of the principle of proportionality, and balancing in particular; while the constitutional scholarship basically becomes fashioned by the BVG's behavior, it can abstain from entering into those convoluted questions.

[198]See. Lerche, *Übermaß und Verfassungsrecht*, 76. Peter Lerche, in the preface of the second edition, mentions that, even though he examined the principle of suitability, his analysis was not so emphatic, and this could be deemed an omission. See Ibid., X.

[199]See Hirschberg, *Der Grundsatz der Verhältnismäßigkeit*, 59 ff.

[200]Ibid., 219 ff.

[201]See Schlink, "Der Grundsatz der Verhältnismäßigkeit," 461, 464–465.

[202]The protection of the citizen's minimal position varies from case to case, from area to area. For instance, in the case of property, the BVG pointed out the minimal position as what the personal job and effort can earn; in the area of profession, the BVG indicated the personal and economic existence; in the area of freedom of expression, the minimal position lies in the possibility of participation in the process of free communication, etc. In this realm, the question about the minimal position can lead to the protection of a minimal property or the protection of a certain role. See, for this purpose, Schlink, *Abwägung im Verfassungsrecht*, 193–195.

prevalent conception of balancing,[203] Robert Alexy, a notorious defender of the rationality[204] of this principle, and balancing in particular, presents a whole methodology for its deployment grounded in the classic triadic division of suitability, necessity, and proportionality in the narrow sense. Hence, it is from the points of contact among these authors and the BVG's practice that we can briefly introduce, more as an invitation to a future detailed analysis of balancing,[205] the triadic structure of the principle of proportionality.[206]

The principle of proportionality stems from the following premise: any concrete measure capable of affecting the private sphere has to be compatible with a proportional analysis of means and goals. It is, therefore, a principle that clearly stresses the protection of freedom and is founded upon the concept of limits of limits (*Schranken-Schranken*) of basic rights.[207] Accordingly, a radical difference occurs in the analysis of any act, in particular a state one: an act is not legitimate merely because it obeys the legal proviso (*Gesetzvorberhalt*), but also when it does not cause a disproportionate interference with the individual freedom (proportional legal proviso – *Vorbehalt des verhältnismäßigen Gesetzes*).[208] The legitimacy of an act then relies on an assessment of means and goals by observing the relationship between the act and the space of private autonomy. As a relational basis, the examination of proportionality demands that the space for action is also relative: there is no absolute definition of a measure to be employed, and, consequently, the existence of different possibilities of election of which measure is in a better proportional relationship with a goal becomes a condition for its deployment.

Those different possibilities of election, in turn, depend on a means/goals arrangement the judge will explore based on two main concerns reflecting the practicable proportional solutions: one stressing the empirical elements of this relationship, and the other concentrating upon the evaluation of the weight of two or more variables (a value, a legal principle, for instance) in conflict. Whereas the first lies in the idea of factual causality and in the experiences and "scientific"

---

[203]See Item 2.5.

[204]See Robert Alexy, "Balancing, Constitutional Review, and Representation," *International Journal of Constitutional Law* (Oxford University Press and New York School of Law) 3, no. 4 (2005): 577.

[205]See the fourth chapter.

[206]Since the purpose of this book relates to the rationality of balancing, the principles of suitability and necessity will be merely superficially discussed.

[207]The concept of limits of limits (*Schranken-Schranken*) in the realm of constitutional adjudication is linked to the principle of proportionality according to these premises: (1) the state's followed goal must be followed as such; (2) the state's appointed means must be appointed as such; (3) the deployment of the means must be suitable to reach the goal; (4) the deployment of the means must be necessary to reach the goal. A last criterion could be found in the realm of the principle of proportionality in the narrow sense: the intervention in the private sphere and the aimed goal must lie in a proportional basis between them. See, for this purpose, Bodo Pieroth and Bernhard Schlink, *Grundrechte: Staatsrecht II* (Heidelberg: C. F. Müller, 2006), 64–66.

[208]See Bernhard Schlink, "Freiheit durch Eingriffsabwehr – Rekonstruktion der klassischen Grundrechtsfunktion," *Europäische Grundrechte Zeitschrift* (N.P. Engel Verlag), 1984: 457–459.

knowledge about their natural or social environment,[209] the second in turn highlights the idea that every measure has a value the goal transmits to, which, expressed in a different way, corresponds to the hypothesis that it functions as a relational basis grounded in the weight those values have for a particular circumstance.[210] In this case, "means and goal are weighted together according to their 'proportion'."[211] The first assessment refers to the principles of suitability and necessity; the second, in its turn, leads to the most controversial and complex area where the principle of proportionality develops: balancing or the principle of proportionality in the narrow sense, which, for it is closely related to the BVG's shift to activism, will be the subject of a more detailed consideration. In addition, as long as balancing is where the defense of rationality of decision-making in this configuration of the BVG's activism appears as the most convoluted theme, the research here will place the emphasis on this matter and on the possible questions it raises, whose grounds will be directly confronted in the second part.

The principle of suitability (*Grundsatz der Geeignetheit* or *Zwecktauglichkeit*), the most elementary of them, indicates that, with a determined measure, the goal can be achieved, or, in other words, "the reality, that is, the hypothesis and theories we have about the reality, must allow the prognosis that, when the means are adopted, also the goal is reached."[212] For it is in the realm of prognosis, the principle of suitability is able to provide, better than the other elements of the principle of proportionality, an objective comprehension of the reality by bringing to light the knowledge of the past and present empirical facts as a means to determine whether a future result[213] is achievable.[214] The BVG has continuously

---

[209]See Hirschberg, *Der Grundsatz der Verhältnismäßigkeit*, 44.

[210]Lothar Hirschberg understands that, although most of the cases involving the deployment of the principle of proportionality in the narrow sense still lies in the relationship between means and goals, with the expansion of the areas the principle now reaches, we could conclude that there are some fewer cases in which this connection is no longer visualized. See Ibid., 45 ff.

[211]Lerche, *Übermaß und Verfassungsrecht*, 19, translation mine.

[212]Schlink, "Der Grundsatz der Verhältnismäβigkeit," 453, translation mine.

[213]Notwithstanding that it is an empirical examination, some difficulties could occur by reason of the very capacity to examine the facts. In this matter, for instance, we could point out how complicated it is to make a prognosis of whether a determined measure can indeed be suitable to achieve a certain result. In this regard, the analysis needs to take into account a very complex investigation of the possible capacity of prediction by whom practiced the activity at stake. The BVG has already indicated the difficulty of an accurate parliament's prediction as an aspect to be considered in the principle of suitability. According to it, it is necessary to observe whether "the legislator, in his view, could assume that the measure would be suitable to reach the stated objective, or whether his prognosis for the assessment of the economic-political connections was then appropriate and justifiable" (BVerfGE 38, 61 (1974) – *Leberpfennig*, translation mine).

[214]Schlink examines the elements of the principle of proportionality also by underlining the difference between prognoses and evaluations (*Bewertungen*). According to him, whereas prognoses refers to statements about the reality in the future based on the observation of the past and present, which allows to prove whether a determined knowledge is true or false (objective truth), evaluations correspond to subjective decisions concerning the analysis of the positive and negative consequences of a measure that are accepted or not by other evaluations. Whereas the first,

deployed it, usually by stating that a "measure is suitable, when with its help one can promote the desired consequence."[215]

Also in the context of factual examination, the principle of necessity (*Grundsatz der Erforderlichkeit*, *Notwendigkeit* or *Prinzip des geringsten Eingriffs*) has a greater weight in this analysis than the principle of suitability, inasmuch as "only what is suitable can also be necessary; what is necessary cannot be unsuitable."[216] In its basis, rather than inquiring whether, by observing the empirical circumstances, a determined measure can reach the goal, the principle of necessity relies on the quality of the interference with the private sphere. The question here resides in its quality grounded in the least encroachment on the private freedom. It is concerned with the deployment of a measure that, in comparison with others, is equally effective but encroaches less on the private freedom sphere, or, in other words, "the reality, that is, our hypothesis and theories must allow the further prognosis that, when the means is not adopted, the goal is not reached, either"[217]. The measure, therefore, in comparison with the possible others, has to cause minimum harm to the addressees. It has to be the mildest measure concerning the adverse intervention in the affected individuals' private status through the consideration of all factual circumstances involved in the case.[218] "Under several possible (= suitable to reach the goal) instruments must only be chosen the one that generates less drastic consequences."[219] The BVG, as well, has clearly deployed this principle – indeed, we could remark that this is the oldest understanding of the

---

therefore, refers to the legal rationality, the other is compatible with the rationality of politics. See, for this purpose, Schlink, "Der Grundsatz der Verhältnismäßigkeit," 455–456.

[215]See BVerfGE 33, 171 – *Honorarverteilung* (1972); BVerfGE 39, 210 – *Mühlenstrukturgesetz* (1975); BVerfGE 90, 145 – *Cannabis* (1994), translation mine.

[216]Bodo Pieroth and Bernhard Schlink, *Grundrechte: Staatsrecht II* (Heidelberg: C. F. Müller, 2006), 66, translation mine.

[217]Schlink, "Der Grundsatz der Verhältnismäßigkeit," 453, translation mine.

[218]Nonetheless, more than suitability, the principle of necessity also poses relevant questions. Bernhard Schlink, for instance, alleges that this principle can reach evaluative issues (not simply the prognosis of the mildest means in comparison with equal effective others), and, hence, on questions based on political arguments. Moreover, according to him, "to determine that a means is not necessary to reach a goal, because another mildest means exists, one must take into account the citizens' evaluation as his truth, not find his own evaluation" (Schlink, "Der Grundsatz der Verhältnismäßigkeit," 456, translation mine). There is, accordingly, a difficulty in the qualification of which is the mildest measure, for the reference – the citizen's evaluation – is not so predicable and can vary from context to context. After all, "different means burden different citizens differently" (Ibid., 457, translation mine). Similar thinking Lothar Hirschberg stresses when he mentions that, even though this issue is usually unproblematic, the definition of the mildest means causes relevant difficulties, especially when there is no previous definition of the addressees or the circle of addressees, and when, for different reasons, the situation brings about a conflict between the general and average consideration of the mildest measure and the particularities of the case (See Hirschberg, *Der Grundsatz der Verhältnismäßigkeit*, 69). True, in the definition of the mildest means, we could attempt to define an average parameter – and this is what should happen in the case of general rules – but even this investigation relies on empirical content that poses many difficulties.

[219]See Lerche, *Übermaß und Verfassungsrecht*, 19, translation mine.

principle of proportionality[220] – by stating that a means is necessary "if another equally effective but less noticeably restrictive means could not be selected."[221]

In any case, it is the third element, balancing or principle of proportionality in its narrow sense, that certainly raises the most complex doubts in this subject matter, especially in what refers to the theme of rationality. As the most visible expression among the three elements of the BVG's movement towards an active political role, owing to the fact that it comfortably absorbs the idea of constitution as an "order of values" (*Wertordnung*) embracing the totality of the legal order, as well as expresses a seemingly methodological justification for the assimilation of political arguments as a legitimate and correct component of constitutional adjudication, balancing (*Verhältnismäßigkeit im engeren Sinn* or *Abwägung*)[222] already appears marked by controversies and dilemmas. Unlike the previous elements, balancing stems from the conflictive basis of a universe of possible arguments to be deployed through its framework, insofar as, more than focusing on the empirical reality, it works in this margin of free appreciation in which a proportional assessment is carried out in abstract.[223] The purpose now is to establish an adequate, "reasonable"[224] and proportional relationship between two or more imaginable interests

---

[220]See note 175 *supra*.

[221]See BVerfGE 38, 281; BVerfGE 40, 71; BVerfGE 49, 24, translation mine.

[222]Some authors differentiate balancing or, more particularly, balancing of goods (*Güterabwägung*) from the principle of proportionality in its narrow sense. Since the focus here is on the question of rationality, particularly on Robert Alexy's approach, and the BVG's practice has not provided a clear distinction between them, both concepts will be identified. For an analysis of this distinction and the variations on this issue, see Hirschberg, *Der Grundsatz der Verhältnismäßigkeit,* 89 ff.

[223]Naturally, this conclusion does not mean that balancing is away from the concrete aspects of the case, but only that its focus is on the assessment of the conflictive goods or interests that the simple factual examination of which means is more suitable or necessary for a particular goal cannot solve. After all, every abstract examination, and more specifically the one relating to the elements of the case, is necessarily linked to the reality. The issue, therefore, relates to the focus of balancing.

[224]Usually the constitutional scholars and the BVG's decisions use the concepts of reasonableness (*Zumutbarkeit*) and proportionality (*Verhältnismäßigkeit*) as synonyms. An unreasonable act, therefore, can have the meaning of a disproportionate act (or more specifically, disproportionate in the narrow sense). Moreover, the concept of reasonableness – and thus of an unreasonable act – can be used as a concept in which any kind of injustice can be placed. However, sometimes constitutional scholarship differentiates them: the reasonableness refers to an absolute intervention in the private sphere, and, as such, an unreasonable act is unacceptable. A disproportionate act, on the other hand, is unacceptable only in a relational basis, that is, in the observance of the goal. This point of view, nonetheless, misses the central argument that an unacceptable act cannot be absolutely conceived, at least when the position of basic rights – and their deprivation in particular cases – remains without danger of being disrupted, but rather refers to a concrete situation where the goal and the intervention will be evaluated according to all the aspects of the situation. Furthermore, this issue leads to the debate on the absolute and relative meaning of the idea of an "essential core" (*Wesengehalt*) of basic rights. See, for this purpose, Schlink, "Der Grundsatz der Verhältnismäßigkeit," 451–453. For an analysis of the debate on the concept of "essential core" and its connection with the principle of proportionality, see Manfred Stelzer, *Das Wesesgehaltsargument und der Grundsatz der Verhältnismäßigkeit* (Wien; New York: Springer, 1991).

playing a special role in the examination of a case. Balancing, accordingly, lies in the capacity to define the best result in the assessment of two or more conflictive interests or goods, not in the factual control of the different means and their respective burdens caused to the addressees. In this respect, there is a balancing of goods or interests grounded in a relationship between means and goals that radically transforms the court's space of appreciation. This is the reason why it is a special instrument for its active role: more than the concern for the construction of a reliable and consistent system of rights built on established norms and guarantees,[225] on some methods and concepts framed by legal dogmatics, or even on the premise that "the court must decide each case in a way that preserves the coherence[226] of the legal order as a whole,"[227] we could argue that, with balancing, the focus can be easily transferred to a political argument of which value or interest offers, in a particular time, a better result for society (and, hence, based on the addressees' approval) from the elements of the case.[228]

In this balancing, for instance, there is, on the one hand, a legal right or a value the judge considers relevant; on the other, there is the sacrifice of another legal right or value. In this connection between *relevance* and *sacrifice*, balancing is employed in order to specify which legal right or value must prevail, which of them has more weight in a particular circumstance, and how proportionally they can thereby agree with each other. It furnishes a mechanism to evaluate how, in this appreciation of their relevance or weight, we can better achieve the best goal for society as a whole, without this meaning a total disregard for the other. This is why the judge attempts to create harmony between them, usually founded upon the idea of an "order of values" that can set up an adequate and proportional correspondence with the goal[229] as a means to avoid excesses: "The measure must not thereby excessively burden [the addressees]."[230]

[225]See Schlink, "Der Grundsatz der Verhältnismäßigkeit," 456.

[226]See the sixth chapter.

[227]Habermas, *Between Facts and Norms*, 237.

[228]Even though we sustain here the argument that balancing leads to the deployment of a political argument within the context of constitutional adjudication, one could argue that, notwithstanding that the realm of goods and interests is widely expanded in this situation, balancing must observe how every decision integrates the whole legal order, which refrains it from a political argumentation (See Hirschberg, *Der Grundsatz der Verhältnismäßigkeit,* 149). This conclusion, nonetheless, may not correspond to the BVG's practice, for more and more its decisions become a *case to case* analysis rather disconnected from this concern for keeping coherent the constitutional reasoning grounded in the idea of preserving the legal order as a whole and its institutional background. Besides, as we will further examine, in the structure of balancing lies normally a very serious concept of rationality that mixes up arguments of policy and arguments of principle. See, for this purpose, the second part.

[229]See BVerfGE 18, 353 (1965) – *Devisenbewirtschaftungsgesetz*; BVerfGE 35, 382 (1973) – *Ausländerausweisung*. BVerfGE 35, 202 (1973) – *Lebach.*

[230]BVerfGE 90, 145 (1994) – *Cannabis,* translation mine.

As regards the measurement of this excess, the judge must proceed to an assessment that will take into account, for example, other parameters than those the principles of suitability and necessity gave rise, such as the relevant economic and general interests at stake,[231] the superior primary social goods,[232] the cultural heritage of a community,[233] the social acceptance and effects of a good,[234] the interests of the subjects involved,[235] the efficiency of the process,[236] among others. Moreover, the judge can attempt to set up a material category[237] that can work as a higher standard[238] for the evaluation of a particular case, such as the Aristotelian concept of distributive[239] or commutative[240] justice[241] or a utilitarian

[231]See BVerfGE 18, 353 (1965) – *Devisenbewirtschaftungsgesetz.*

[232]See Pieroth and Schlink, *Grundrechte – Staatsrecht II*, 66

[233]See BVerfGE 93, 1 (1995) – *Kruzifix.*

[234]See BVerfGE 90, 145 (1994) – *Cannabis.*

[235]See BVerfGE 24, 119 (1968) – *Adoption I.*

[236]See BVerfGE 38, 105 (1974) – *Rechtsbeistand.*

[237]See Hanau, *Der Grundsatz der Verhältnismäßigkeit als Schranke privater Gestaltungsmacht*, 97.

[238]In the definition of a higher standard, we could remark the connection between balancing and the equality principle. Difficulties appear in the definition of a distinction between them. We could argue, for instance, that, whereas balancing refers to an assessment of goods or interests regarding a particular case, the equality principle, on the other hand, is linked to the premise of a comparison between cases in order to achieve equal solutions. Besides, they could provide distinct effects: whereas the court, in the assessment of a violation of the equality principle, could leave the solution open to the legislator (whether through the modification of the norm applied to the case, through the modification of the norm used as basis for comparison, or even through the modification of both), in the case of the deployment of balancing, the solution can only lead to the conclusion of whether the norm applied to the case is proportional and, thus, constitutional or not. According to this perspective, balancing provides an assessment of the individual case, which should walk in harmony with the equality principle grounded in a comparative perspective of cases that carry the idea of what is right for the particular context. This is the reason why "after the perception that the equality principle and the principle of proportionality (assessment of the individual case), like the most abstract legal approaches, contain the same elements merely disposed in a distinct arrangement, this observation [the idea that one can justify the other] can no longer cause surprise" (Hirschberg, *Der Grundsatz der Verhältnismäßigkeit,* 127, translation mine). Both principles, according to this point of view, are expression of a command of differentiation (*Differenzierungsgebot*): the principle of proportionality as a guarantee of this command in the individual case; the equality principle as a guarantee in the comparison of cases. In any circumstance, as Hirschberg remarks, the problem lies in the definition of the final point – the material value – that is behind both principles, and, moreover, this is where the danger of arbitrary rulings appears. In addition, there is always the difficulty in defining the extension of the influence of comparative cases as well as in establishing a proportion of the goods at issue. For a detailed analysis of this approach, see Ibid., 111–132.

[239]According to this principle, the goods must be proportionally distributed according to a prior defined criterion of differentiation.

[240]By abstracting the differences among the individuals, this principle stems from the contractual premise: to give everyone what belongs to him.

[241]For a defense of justice connected to the principle of proportionality, see Nils Jansen, *Die Struktur der Gerechtigkeit* (Baden-Baden: Nomos, 1998).

principle.[242] Different criteria, accordingly, can be brought to this assessment of means and goals in order to supply adjudication with a broader margin of appreciation and to apparently "correct" (balancing as a technique for correction of the limits of state's decisions)[243] the possible deviations of the state intervention not solved by the previous principles of suitability and necessity.[244]

At any rate, balancing, now embedded in the structural framework of the principle of proportionality, has this quality of allowing a *broader margin of appreciation* in adjudication while adapting perfectly for a practice oriented, more and more, towards activism. In this respect, the BVG embraced balancing as a powerful instrument to develop a more intervenient attitude towards the most distinct themes of social life, a characteristic that has a close connection with a controversial history that transformed it into the representative of the legal and social order within the context of a Germany recently emerging from the Second World War. This history, followed by a new interpretation of legal rights, originated from this shift to the assumption of a political duty towards the society – the idea of basic rights as objective principles of a total legal order, as we will examine shortly – is a fundamental premise to understand why balancing appeared as a natural characteristic of constitutional adjudication and, which is the focal point of this research, the "rational" response to the dilemmas arising therefrom. For this reason, the following sections will discuss this connection: why did balancing evolve so naturally from this movement, and why was it necessary to provide it with a "rational" aura through the stress on the principle of proportionality?[245]

2.3 The Bundesverfassungsgericht in the Postwar Crisis: The New Representative of the Legal and Social Order

In the realm of constitutionalism, where words resonate with different meanings, seldom a sentence comes into sight with a so accurate perception of a reality. Jutta Limbach, former BVG's Chief Justice, has correctly captured the paradox that surrounds the practice of constitutional adjudication by showing, on the one hand, the massive popular confidence in the BVG and, on the other, the crisis of democracy that might justify it. Her words, in the form of a question – "Does the unbroken great trust in the authority of constitutional jurisdiction indicate a

---

[242]According to the general guideline of this principle, an action is right if it promotes a useful benefit to the majority or happiness to the greatest number of people.

[243]See Leisner, *Der Abwägungsstaat: Verhältnismäßigkeit als Gerechtigkeit?*, 46.

[244]In the fourth chapter, a deeper analysis of balancing will be carried out through Alexy's *Theory of Constitutional Rights*.

[245]Particularly here, we use Robert Alexy's attempt to rationalize balancing through the stress on the principle of proportionality.

political mistrust of democracy?"[246] – might express more than they first seemed to aim at. They might reveal a movement that does not restrict itself to the contemporary reality, but is itself the consequence of a long process that began when the BVG was founded. The beginning of its activities, in 1951, is certainly the introduction of a new stage in German constitutionalism, but it is also the birth of a radical rearrangement of powers and authority, one that corresponds to the configuration of a constitutional court that concentrates on itself the judgment of the main questions of social life by interpreting them as problems of constitutional rights. This birth, as the sentence well exposes, might also feed itself on a certain mistrust in the face of democracy; it is a sequel to a period of crisis, when an institutional vacuum coexisted with a feeling for a necessary rapid reconstruction of the social and legal order. One institution had to assume this complex role that was already been drafted when it started to act, but it became more evident over the years: the court would not only bc thc guardian of thc Constitution; it would be, in truth, the guardian of the social values, a supposedly natural necessary duty that followed the expansion of the state's more intervenient attitude and the scope of providing benefits and services also through adjudication.[247]

The German Basic Law of 1949 was written under the feeling of reconstruction and of a radical institutional transformation. The introduction of an extensive catalogue of basic rights exactly at the beginning (articles 1-20) has much to do with a crusade against the authoritarianism that prevailed in the preceding years and the influence of a conflictive and complex situation that stemmed from the condition of a postwar Germany physically occupied by allied military forces.[248] In its formation and in the constitutionalism that has coexisted with it over the years, we can remark the influence of some relevant developments of the Weimar Republic.[249]

---

[246]Jutta Limbach, "The Effects of the Jurisdiction of the German Federal Constitutional Court," *European University Institute*, EUI Working Paper Law, 99/5, http://cadmus.iue.it/dspace/bitstream/ 1814/150/1/law99_5.pdf (accessed July 14, 2009), 22.

[247]See Habermas, *Between Facts and Norms*, 247.

[248]Although the American constitutionalism played an important role in the history of German Constitutionalism, Fangmann stresses that the establishment of a constitutional court with large competences was a German initiative. According to him, "the decisive initiative for the edification of a special judicial review comes, nonetheless, from the Germans." The establishment of judicial review, after all, was made in accordance with the will of the main political parties. Besides, notwithstanding that the influence of the United States was already strongly existent in the period of the Weimar Republic – and this, consequently, also influenced the period of the German Basic Law of 1949 – the idea of a centralized judicial review had no direct link with the American system. See Fangmann, *Justiz gegen Demokratie: Entstehungs – und Funktionsbedingungen der Verfassungsjustiz in Deutschland*, 226–27, translation mine.

[249]Helmut Fangmann stresses firmly this connection between both periods, especially in the realm of judicial review. According to him, "Despite all differences between Bonn and Weimar, one cannot deny that the Basic Law, the *Bundesverfassungsgericht* and the today predominant state law literature, in strong extent, assumed and continued the formal and functional change of Weimar's constitutional law. The introduced development in constitutional adjudication finds its continuation in the actual *Bundesverfassungsgericht*'s independence from the Constitution" (Ibid., 11, translation mine).

They are, among others: (1) the genesis of the material judicial review[250] of legal statutes and governmental acts[251] by a superior court that concentrated the exercise

[250]The existence of judicial review, with the power to invalidate legal statutes due to their material incompatibility with basic rights, is quite recent in Germany. Until the First World War, the decisions of the Supreme Court of the German Reich *(Reichsgericht)* were centered on the formal terms of a legal statute and did not exactly enter into the contents of the rule. The exceptions could be found in cases of ordinances issued by the government of the Reich, whereby, in virtue of not bringing conflict with the legislature and the monarch, the court could exercise a material judicial review or, which is particularly important here, the control over administrative and police acts (above all rights of freedom and property) by the administrative courts. This happened, above all, in the realm of the Prussian Higher Administrative Court (*Preußischen Oberverwaltungsgericht*) through the deployment of a preliminary version of the principle of proportionality (particularly the principle of necessity). In this case, we could remark the existence of a prelude to the later judicial review and the beginning of a comprehension of the Constitution as a superior law to be interpreted by courts in the control over administrative and police acts as well as already some relevant signs of the precursory deployment of the principle of proportionality. In the Weimar Republic, this situation changed somehow. First, the existence of a catalogue of rights in the Constitution of the German Reich of 1919, even though modifiable by ordinary statutes and initially deemed superfluous (due to the then existing idea of the supremacy of parliament, based on the democratic sovereignty, over the other powers), served as premise to introduce the material exam of the constitutionality of a legal statute. In this case, the establishment of the material review inverted the initial tendency to disregard basic rights as the basis of the legal order and promoted, instead, the conception of them as sacred and foundational for the German people. Second, there was a rich and notable development of the Jurisprudence and the state law doctrine, favored by the situation of a first historical democratic political context and all the crisis it brought about, which projected what Schlink called a "struggle over methods and aims" (See Bernhard Schlink and Arthur J. Jacobson, "Introduction – Constitutional Crises: The German and the American Experience," in *Weimar: a Jurisprudence of Crisis*, ed. Arthur J Jacobson and Bernhard Schlink (Berkeley, CA: Univeristy of California Press, 2000), 3) and also established the viewpoint of basic rights as the center of the legal order and a defensive parameter of the *status quo*. We could observe this movement in the texts of Rudolf Smend, Carl Schmitt, Heinrich Triepel, Erich Kaufmann, Gerhard Leibholz, among others. Third, there was the beginning of a movement towards the comprehension of basic rights not only as subjective rights but, mostly, as objective principles embodying social relationships and values, which could be adopted in judicial review as arguments to protect the bourgeoisie against state intrusions. Fourth, there was the understanding, derived from an anti-positivism and anti-parliamentarianism that flourished at that time, that basic rights are also to be observed by the parliament. Fifth, there was the expansion of the deployment of methods and criteria, especially a preliminary version of the principle of proportionality in administrative and police activities, which enhanced the instruments for judicial review. See, for this purpose, Michael Stolleis, "Judicial Review, Administrative Review, and Constitutional Review in the Weimar Republic," *Ratio Juris* 16, no. 2 (June 2003): 266–80; Schlink and Jacobson, *Weimar: a Jurisprudence of Crisis;* Hirschberg, *Der Grundsatz der Verhältnismässigkeit*; GUSY, Christoph. *Richterliches Prüfungsrecht: Eine verfassungsgeschichtliche Untersuchung*, 74–89.

[251]According to Michael Stolleis, notwithstanding that we could observe some developments in the realm of material judicial review in the Weimar Republic, at the beginning of this period, they could not be overestimated, because, apart from the fact of some legal provisions and courts were not implemented, there was not indeed a canon of review. Other particularities that mitigate a certain belief in the existence of a real material judicial review in the beginning of the Weimar Republic is the fact that neither the public opinion nor the public law were ready for this transformation. There was also a certain "anti-individualistic" mood that lessened the desire for

of constitutional review[252]; (2) the understanding of basic rights as the center of the legal order[253] and as a system of values[254] above the law to be used as instruments against state intrusion into the private sphere by means of constitutional review; (3) the deployment of methods and criteria for adjudication, chiefly a preliminary version of the principle of proportionality in the nowadays labeled principle of necessity (*Grundsatz der Erforderlichkeit*);[255] (4) the mistrust of politics[256] and of the legislature, for they could not rapidly deal with the social and economic problems of the time; (5) the struggle against the growing socialist discourse and

---

the existence of legally protected rights of individuals, and the conception that the law was the will of a higher authority still prevailed. However, these facts, after 1925, radically changed. The causes were: (1) the idea of a material judicial review could serve as a check upon the risk of a parliamentary absolutism, protecting hence the interests of the bourgeoisie; (2) the concept of basic rights, especially property and equality, as the bastion against the parliament and also as a system of values above the law; (3) the economic crises caused by the inflation of 1923–24 that threatened the republic, the property and the *status quo* of the bourgeoisie; (4) the disbelief in the legislature; (5) the political, academic and methodological expansion of an anti-positivism approach; (6) an anti-socialist and anti-parliamentary approach that could be defended by means of basic rights (See Stolleis, "Judicial Review, Administrative Review, and Constitutional Review in the Weimar Republic," 273–78). In any case, it was still missing the idea of basic rights as principles to be enforced by judicial review.

[252]This characteristic could be clearly observed in the creation of the previous Constitutional Courts of the *Länder* in the immediate period before the Basic Law of 1949. According to most Constitutions of the *Länder*, judicial review had to be carried out exclusively by these Constitutional Courts (*Verfassungsgerichten*). See, for this purpose, Fangmann, *Justiz gegen Demokratie: Entstehungs – und Funktionsbedingungen der Verfassungsjustiz in Deutschland*, 221.

[253]This idea of basic rights as the center of the legal order only gained supporters after 1924. Between 1919 and 1924, basic rights were still conceived more as political declarations without legal force. However, after the economic crisis of 1923–24 (among other factors, see note 251 *supra*), basic rights began to be regarded as inviolable and as a system of values unifying the legal order, as well as a mechanism to control the parliament (See Stolleis, "Judicial Review, Administrative Review, and Constitutional Review in the Weimar Republic," 273). In the practice of judicial review, this change could also be seen in 1924, when, for the first time, one court declared a law to be against the Constitution because of violation of basic rights. See Christoph Gusy, *Richteliches Prüfungsrecht: Eine verfassungsgeschchtliche Untersuchung* (Berlin: Duncker & Humblot, 1985), 82; Fangmann, *Justiz gegen Demokratie: Entstehungs – und Funktionsbedingungen der Verfassungsjustiz in Deutschland,* 10.

[254]See Stolleis, "Judicial Review, Administrative Review, and Constitutional Review in the Weimar Republic," 278. Schlink, however, instead of sustaining the dualism between values and rights, adopts the dualism between the concept of rights as subjective rights and one of rights as objective principles. For this purpose, see Schlink, "German Constitutional Culture in Transition," 711–736. This aspect, at any rate, will be shortly examined.

[255]According to Lothar Hirschberg, until the end of the Second World War, the principle of proportionality in *broad* sense was acknowledged only in the figure of the principle of necessity (*Grundsatz der Erforderlichkeit*) as well its counterpart principle of suitability (*Grundsatz der Geeignetheit*), although it was generically called the principle of proportionality. See Hirschberg, *Der Grundsatz der Verhältnismäßigkeit*, 14.

[256]See Stolleis, "Judicial Review, Administrative Review, and Constitutional Review in the Weimar Republic," 278; Schlink, "German Constitutional Culture in Transition," 724–725.

the strengthening of basic rights as a means to counter it[257] and appease social claims in a capitalist production system[258]; (6) the enfeeblement of parliament in favor of a broader field of judicial authority as a legitimate transformation[259]; (7) the coronation of the rule of law[260]; and (8) the demand for changes in the social order in conformity with the basic rights.[261] All these characteristics demonstrate that the resemblances between the Weimar Constitution of 1919 and the German Basic Law of 1949 did not limit to the coincidence of both being the consequence of a postwar crisis; they represent the sign that history somehow keeps alive in the memory the need for reconstruction. However, although these characteristics had a primary impact, by some means, during the framing of the Basic Law of 1949, evidently they gained a new shape, and many of them were reinforced given the condition of a Germany whose institutional consistency was, in 1949, still unsteady and which was striving to overcome the authoritarian past.

Among them, the idea of a centralized constitutional review, after the experience of this practice in the different *Länder* with their respective Constitutions from 1946 onwards,[262] achieved a central role in the framing of the Basic Law. This type of judicial review had two distinguishing marks: first, there was not a specific control regarding the binding of the courts to their respective Constitutions;[263] second, they were placed above any parliamentary constraint, as an opposition to parliamentary absolutism.[264] The idea that the constitutional court not only had to be the guardian of the Constitution ("*Hüter der Verfassung*") but also act as a defender against the risks of a new tragic end, as it happened in the Weimar Republic, helped design the concept of judicial review with a vast field of competences and justification. This juridification of the Constitution (*Verfassungsverrechtlichung*), accordingly, thanked much of its occurrence not only to a specific heritage from the Weimar Republic, when this process, despite progressively in course over the years,[265] revealed its impotence to deal with the increasing authoritarianism,[266] but also to the political frailness stemmed from its historical incapacity to prevent the advance

[257]See Stolleis, "Judicial Review, Administrative Review, and Constitutional Review in the Weimar Republic," 273.

[258]See Fangmann, *Justiz gegen Demokratie: Entstehungs – und Funktionsbedingungen der Verfassungsjustiz in Deutschland*, 233.

[259]Ibid., 10.

[260]Ibid., 8.

[261]See Schilnk, "German Constitutional Culture in Transition," 723.

[262]See Fangmann, *Justiz gegen Demokratie: Entstehungs –und Funktionsbedingungen der Verfassungsjustiz in Deutschland*, 220–221.

[263]Ibid., 222.

[264]Ibid., 222, 226.

[265]See notes 250 and 251 *supra*.

[266]See Stolleis, "Judicial Review, Administrative Review, and Constitutional Review in the Weimar Republic," 279.

of National Socialism and the dreadful consequences it gave rise to.[267] Traditional politics could not resolve the crisis that emerged at the end of the Weimar Republic and was responsible, in a sense, for allowing the transference of the idea of "Guardian of the Constitution" to a government already characterized by an authoritarian intent.[268]

The exercise of democracy was neither achieved by the traditional institutional background nor by the constitutional principles already present in the Weimar Constitution. Since the constitutional principles and the institutional background were severely effaced during the National Socialist regime, their reconstruction had to be as evident and immediate as possible. The trend already revealed in the Weimar Republic of a constitutional court came out as a suitable response to this task,[269] especially after the judiciary and the possibility of judicial review had been heavily undermined in the National Socialism. The BVG appeared as the appropriate institution to reconstruct the legal and social order, favored by a movement headed by jurists and judges who were before compelled by the authoritarian legality of Hitler's regime[270] and by the skepticism about the capacity of the Basic Law itself, an outcome of legislative activity, to avoid the return of authoritarianism. Besides, the still existing argument that judicial review contributed somehow to the dismantlement of the Weimar Republic converted into the conclusion that this outcome happened in virtue of the depoliticization of justice.[271] The introduction of judicial review with a political character, therefore, emerged as a natural response, and the idea of a higher court controlling the political system arose as an evident ineluctable movement.[272] The juridification of the Constitution became the juridification of politics.[273]

Indeed, the exercise of material judicial review by a constitutional court seemed to best handle the social claims to a better life than the capitalistic approach that prevailed in the political conscience and which – not only the politicians but also the capitalism itself – was firmly discredited after the end of National Socialism. The discredited political class was, after all, impotent to oppose to this juridification

---

[267]See Fangmann, *Justiz gegen Demokratie: Entstehungs – und Funktionsbedingungen der Verfassungsjustiz in Deutschland*, 224.

[268]See Carl Schmitt, *Der Hüter der Verfassung* (Tübingen: Mohr, 1931).

[269]See Fangmannm, *Justiz gegen Demokratie: Entstehungs – und Funktionsbedingungen der Verfassungsjustiz in Deutschland*, 230.

[270]See Gottfried Dietze, "Unconstitutional Constitutional Norms? Constitutional Develpment in Postwar Germany," *Virginia Law Review* 42, no. 1 (January 1956): 8.

[271]See Fangmann, *Justiz gegen Demokratie: Entstehungs – und Funktionsbedingungen der Verfassungsjustiz in Deutschland*, 233.

[272]See Heinz Laufer, "Politische Kontrolle durch Richtermacht," in *Verfassung, Verfassungsgerichtsbarkeit, Politik: Zur verfassungsrechtlichen und politischen Stellung und Funktion des Bundesverfassungsgerichts*, ed. Mehdi Tohidipur (Frankfurt a.M.: Suhrkamp, 1976), 94.

[273]See Fangmann, *Justiz gegen Demokratie: Entstehungs – und Funktionsbedingungen der Verfassungsjustiz in Deutschland*, 224.

of political issues.[274] The exclusion of an only parliamentary democracy[275] and, consequently, the construction of a strong constitutional court that seemingly represented a conciliatory solution between capital and social claims appeared as an adequate response to this dilemma. The BVG undertook the role that the bourgeoisie class represented by the parliament could no longer assume, and began to manage through its decisions a perspective of rights that seemed to more adequately respond to the social claims by interpreting them in a positive dimension towards collective benefits[276] and as objective principles to be deployed according to a proportional analysis based on the rule of optimization.[277] Nonetheless, this interpretation tended to perpetuate, in reality, the interests of the capital through other means, that is, property and freedom were protected as much as possible by the BVG's control over legislative activities. This means, in other words, that the encroachment on the sovereignty of parliament denounced the juridification of the social and economics order.[278] Even when the bourgeoisie was the majority in parliament, this process was not reverted and no fundamental conflict with the constitutional court appeared[279]: the BVG, in this case, acted as an ally, safeguarding thereby the guarantees of the capital; however, in doing so, it had also to reconcile them with some social claims, which became visible through the idea that its duty also embraced a positive protection of basic rights through demands for administrative regulation and governmental services.[280]

The reemergence of a natural-rights approach also in the realm of adjudication, as a reaction against the authoritarian years, helped this process and even promoted the idea that the existence of judicial review was protected under natural law, now transmuted into a positive dimension in the Basic Law of 1949.[281] Moreover, the stress on natural law widened the BVG's field of argumentation, as long as it could also have its origins in supra-positive grounds[282] as a mechanism against parliamentary democracy.[283] This vast field of argumentation complemented hence the already broad sphere of incidence of the BVG's decisions, which, due to a "super

[274]Ibid., 229.

[275]Ibid., 228.

[276]See Habermas, *Between Facts and Norms*, 247.

[277]See Schlink, "German Constitutional Culture in Transition"; Alexy, *Theorie der Grundrechte*.

[278]See Fangmann, *Justiz gegen Demokratie: Entstehungs – und Funktionsbedingungen der Verfassungsjustiz in Deutschland*, 237.

[279]Ibid., 237.

[280]See Schlink, "German Constitutional Culture in Transition," 720–721.

[281]See Dietze, "Unconstitutional Constitutional Norms? Constitutional Development in Postwar Germany," 8.

[282]See, for instance, the appearance of the doctrine of unconstitutional constitutional norms, which derives from the premise of the existence of supra-positive norms to be used in adjudication. For this purpose, see Dietze, "Unconstitutional Constitutional Norms? Constitutional Development in Postwar Germany."

[283]See Fangmann, Justiz *gegen Demokratie: Entstehungs – und Funktionsbedingungen der Verfassungsjustiz in Deutschland*, 231.

constitutional system"[284] where most citizens' political activities and governmental powers were bound to, tended to convert any discussion of social issues into a constitutional debate the BVG had to decide. The political, economic and social order entered almost unreservedly into the concept of constitutional order. Indeed, constitutional adjudication, as Helmut Fangmann correctly remarks, followed thereby less the logic of constitutional rights and more the logic of the relationship among political forces.[285] Briefly, the BVG was born in a scenario where the parliamentary democracy was not only discredited but also seen as a threat to democracy, and where an institutional and social crisis hastened the process of transferring the political discourse to constitutional adjudication, which, as a means to handle the crisis, demanded that the legal reasoning was also a political one. Politics and constitutional adjudication mixed with each other.

Although we could remark some political resistances to this BVG's political role –especially stemmed from the idea that the misuse of this power could bring back authoritarianism, as it happened in the Weimar Republic – the truth is that the conception of a constitutional court highly deprived of parliamentary control prevailed.[286] This resulted in the weakening of the parliament, which, more than being bound to the Basic Law, was considered constrained by the Basic Law according to how the BVG interpreted it and the conflicts it raised in different levels of social life.[287] The parliament, within this context, became a merely first interpretative instance of the Basic Law, submitted to the BVG's last interpretation.[288] The parliament itself assumed this secondary role, inasmuch as it, to avoid damages to its image, transferred much of the political discussion to the BVG.[289] Moreover, in many relevant issues, the parliament waited for a BVG's definition before enacting a law.[290] For it was – and still is – regarded as the guardian of the Constitiution, the prevalent idea has been that the BVG provides the last word, whose interpretation and extension lie in its own hands.[291] The depoliticization of

284Ibid., 238, translation mine ["*Superverfassungssystem*"].

285Ibid., 239.

286Ibid., 234–235.

287This characteristic differs from the initial period of the Weimar Republic, when the idea that the parliament is bound to the basic rights did not prevail, as long as the parliament could limit basic rights through infra-constitutional norms. See Dieter Grimm, "Die Entfältung der Freiheitsrechte," in *Das Bundesverfassungsgericht: Geschichte, Aufgabe, Rechtssprechung*, ed. Jutta Limbach (Heidelberg: C. F. Müller, 2000), 57.

288See Christian Hillgruber and Christophh Goos, *Verfassungsprozesrecht* (Heidelberg: C. F. Müller, 2004), 14.

289See Fangmann, *Justiz gegen Demokratie: Entstehung – und Funktionsbedingungen der Verfassungsjustiz in Deutschland*, 15.

290See, for instance, the recent example of the parliament and the federal government waiting for a BVG's definition of the possibility of online searches of individual's computer as stated by the Nordheim-Westphalia Constitutional Protective Law (*Verfassungsschutzgesetz*). See Elke Luise Barnstedt, "Judicial Activism in the Practice of German Federal Constitutional Court: Is the GFCC an Activist Court?," *Juridica International* II (2007): 38.

291See Hillgruber and Goos, *Verfassungsprozessrecht*,14.

the political struggle led to the juridification of politics. It is interesting to remark, besides, that this BVG's approach, with this vigorous power on the central issues of social and political life, was based on the standpoint that the principle of separation of powers, unlike that abstract conception of the past – which could not avoid the authoritarianism – demanded that judicial review be grounded in a centralized institution acting as the last interpreter.[292] Only by concentrating the judicial review, with vast competences to reach the most different claims of social life, could the risks of an authoritarian return be eroded.

The consequence of this process is that, as a means to achieve the goal of avoiding any authoritarian return and preserving freedom, the BVG, by concentrating the discussion of social claims on the concept of constitutional order, focused on looking at the present and future problems of society rather than considering the need to build a coherent system of rights by respecting their institutional history.[293] This characteristic led to the concept of legal rights that encompassed in itself an objective structure[294] with a value-oriented dimension that favored building an "unstructured argumentation field"[295] according to which the BVG's decisions could achieve its broad intents. Reinforcing legal rights, according to this perspective, gained a political dimension towards the present and future of society with an ambivalent axiological anatomy in which rights, now embodied in this "system of values," could be interpreted in a proportional basis. Furthermore, it helped, on the one hand, establish an intimate contact with the people, who, although not aware, except for rare important cases, of its activities,[296] strengthened the conviction of its responsibility towards the social life, particularly by also intervening in issues of moral, political and economic order. On the other, as a result of this conviction and link to society, it promoted its own popularity. Naturally, the BVG's popular confidence in its decisions does not evidently exempt it from a deeper analysis of its political arguments and its connection with the principle of separation of powers. Popularity, after all, does not necessarily mean democratic legitimacy. But these characteristics reveal that the BVG acts in the middle of a political responsibility towards the future and the belief, indeed existing, in the social acceptance of its decisions, which supposedly bestows on it legitimacy and provides it with a vast field of arguments as a way to not only avoid the return of authoritarianism, but also promote democracy by expressing, through decision-making, the highest standards and virtues of a society in need of rebuilding its conscience devastated by the authoritarian past.

---

[292]See Fangmann, *Justiz gegen Demokratie: Entstehungs – und Funktionsbedingungen der Verfassungsjustiz in Deutschland*, 236.

[293]See Habermas, *Between Facts and Norms*, 246.

[294]This subject will be examined in the next section.

[295]Schlink, "Freiheit durch Eingriffsabwehr – Rekonstruktion der klassischen Grundrechtsfunktion," 463, translation mine.

[296]See Schlink, "Abschied von der Dogmatik: Verfassungsrechtssprechung und Verfassungsrechtswissenschaft im Wandel," 1125.

2.4 The Bundesverfassungsgericht's Shift to Activism: From Subjective Rights to Objective Principles and the Consequences in Judicial Review

The beginning of the BVG's activities, as shown, was marked by a social crisis and the disbelief in traditional politics. This point of departure, over the years, proved to be a crucial aspect for the movement towards the axiological expansion of constitutional adjudication: it identifies with the following relativization of law through the progressive admission of arguments of morality and policy in adjudication to be deployed in a proportional analysis based on optimization rules. In doing so, the court set forth some already existing value-based concepts and some methodologies able to bear this specific role. Among them, we could point out the strengthening of the principle of proportionality, now presented as a vigorous system structurally encompassing balancing, which became the main mechanism to solve constitutional cases in German judicial review. This principle, and specifically balancing, at any rate, could only gain this vigorous feature insofar as the way legal rights were interpreted identified with the political opportunity of a court aspiring to expand its authority. This interpretation took place by the shift in legal reasoning from subjective rights to objective principles[297] or, in other words, from classical rights to "principles of a total legal order."[298]

This transition from the idea of basic rights as subjective rights to objective principles, according to which legal rights are no longer entitlements of the individual citizen[299] with a defensive character against state intrusions, but rather maxims encompassing the total legal order, is paradigmatic in German constitutionalism. Every interpretation, therefore, provides a proportional analysis of basic rights in accordance with the context and the legal possibilities. In this case, more than the simple relationship between state apparatus and the individual, they irradiate throughout the normative structure and different types of social relationships. Basic rights reach now any overpowering force[300] threatening society in general, even in private relations (*third-part effect* or *Drittwirkung*). They gain a material structure that is bound to the axiological conscience seemingly gathered and apprehended from society, resulting in the change of focus from developing a consistent and reliable system of legal norms and guarantees to the interest in providing responses to the present and future problems of society.[301] Adjudication, as a consequence, becomes a prospective task of finding a solution to a social

---

[297] See Schlink, "German Constitutional Culture in Transition."

[298] See Habermas, *Between Facts and Norms*, 247.

[299] See Schlink, "German Constitutional Culture in Transition," 713.

[300] See Erhard Denninger, "Freiheitsordnung – Wertordnung – Pflichtordnung," in *Verfassung, Verfassungsgerichtsbarkeit, Politik: Zur verfassungsrechtlichen und politischen Stellung und Funktion des Bundesverfassungsgerichts*, ed. Mehdi Tohidipur (Frankfurt a.M.: Suhrkamp, 1976), 168.

[301] See Habermas, *Between Facts and Norms*, 246.

problem with an interventionist and active attitude, without nonetheless being followed by a conceptually clarified and structured manner of argumentation that intends to reinforce a system of coherent legal norms.

With the idea of objective principles expressing an axiological order, the individual freedom relies on a close connection with the society and its institutions, as well on the need to provide social benefits through the concretization of basic rights. There is, therefore, an institutional (and no longer individualistic) comprehension of basic rights[302] as a means to promote their effectiveness through case to case evaluation. Behind this prospective basis, there is the definition of relevant essential concepts supporting the premise that basic rights are part of an axiological order, such as the idea of human dignity as the highest value[303] in opposition to the liberal conception of protecting individual freedom. This brings about a constitutional court's distinct approach: human dignity is preserved not only when the state activity towards individual freedom is limited, but also when this freedom is interpreted in conformity with the social moment and the total legal order. Or it brings the argument of "essential core" (*Wesensgehalt*), according to which basic rights cannot be encroached on their "essential content,"[304] one of the justifications for the deployment of balancing.[305] Both perspectives project the idea that every basic right has an objective structure, a *telos* that is linked to a general comprehension of society and its values.

These characteristics give rise to relevant consequences. First, basic rights obtain the character of optimization requirements. According to this approach, they become norms requiring that something be realized to the greatest extent possible in accordance with the factual and legal possibilities.[306] This means that, unlike the concept of subjective rights, in which we observe the idea that, in principle, the citizen is entitled to do something, with rare expressly and justified

---

[302]This is the terminology used by Peter Häberle. For him, the institutional comprehension of basic rights is connected to the claim to an active duty of judicial review to reinforce social values, thereby transforming individual freedom into an institutional one whereby norms and facts are understood according to their correlative relationships. See Peter Häberle, *Die Wesengehaltsgarantie des Artikel 19 Abs. 2 Grundgesetz: zugleich ein Beitrag zum Institutionellen Verständnis der Grundrechte und zur Lehre vom Gesetzesvorbehalt* (Heidelberg: C. F. Müller, 1983), 119.

[303]See BVerfGE 27, 1; BVerfGE 34, 269; BVerfGE 49, 24; BVerfGE 98, 169.

[304]This argument stems from the interpretation of art. 19 (2) of the Basic Law: "In no case may a basic right be infringed upon in its essential content." See BVerfGE 7, 377 (411) – *Apotheken-Urteil* (1958).

[305]The argument of "essential content" has two possible interpretations: the absolute one, according to which every basic right has a nucleus that cannot by any means be violated. The other – the relative one – links directly this argument to the deployment of the principle of proportionality, and particularly balancing. The BVG has already deployed both. See, for this purpose, Alexy, *Theorie der Grundrechte*, 267 ff; Stelzer, *Das Wesesgehaltsargument und der Grundsatz der Verhältnismäßigkeit*.

[306]See Alexy, *Theorie der Grundrechte*, 75.

exceptions, interpreting basic rights as optimization requirements leads to the submission of basic rights to what is factually and legally possible.[307]

Second, which is a direct outcome of the former, adjudication leaves aside the idea that individual freedom is, except in sparse situations, to be enforced and, instead, understand it, as well as other basic principles, as a value to be balanced according to variable degrees of satisfaction.[308] In other words, interpreting constitutional rights results in a proportional analysis of their weight in compliance with the characteristics of the context and possibilities the reality and the legal norms provide, which exposes the fact that every interpretation should follow some rules of optimization. Basic rights are no longer determinations linked to the civic practice of discourse and the protective instruments to be enforced by the judiciary against the administrative apparatus. Rather, they acquire a moralizing content that can, paradoxically, harmonize interests of this same administrative apparatus by expressing the abstract idea of representing the values shared by a community. More than appraising, accordingly, the conditions for the exercise of freedom, judicial review legitimates itself insofar as it successfully confirms the gathered values.[309] Values, after all, need to be permanently acknowledged, propagated, actualized, achieved and realized, thereby leading, rather than a negative status, to an "offensive struggling character"[310] with a duty of action in the foreground.[311]

Third, constitutional adjudication loses its connection with a dogmatic structure and establishes instead a flexible methodology able to absorb this new perspective, a role balancing, now with the aura of rationality the principle of proportionality aims to provide, properly exercises. With balancing, any basic principle can be relativized according to the relevance and intensity of goods or interests without the need to find, even when one attempted to,[312] a convincing criterion to compare and emphasize[313] the prevalence of a good or interest over a basic right. Indeed, as Bernhard Schlink mentions, even when a theory or dogmatics of basic rights treats them as values and goods, it cannot present any system or arrangement of values and goods.[314] By reason of this impasse, decision-making develops the fragile and

[307]See Schlink, "German Constitutional Culture in Transition," 714.

[308]See Alexy, *Theorie der Grundrechte*, 76.

[309]See Denninger, "Freiheitsordnung – Wertordnung – Pflichtordnung," 166–167.

[310]Ibid., 169, translation mine.

[311]Ibid., 169.

[312]See, for instance, Alexy, *Theorie der Grundrechte*. See the fourth chapter.

[313]See Schlink, "Freiheit durch Eingriffsabwehr – Rekonstruktion der klassischen Grundrechtsfunktion," 461.

[314]According to Bernhard Schlink:

"However, we can show that the construction of systems of interindividual preferences, when they should orient towards the interested and affected individuals, regularly miscarry. In decision-making and in the game theory, as well as in the normative economics, it is extensively proved that the enclosed interindividual value or utility units and standards, required for the construction of a system of interindividual preferences, fail. Certainly, the advantage and disadvantage of certain events or conditions for the individual and for the society can be, under financial and temporal tolerant concept of *judicial self-restraint* as a means to expose at least the apparent concern for the possible encroachments on other power spheres, without, evidently, this meaning that this concept could, for some reason, effectively promote a revision of the way the constitutional court acts.[315]

Fourth, basic rights become, for they are interpreted as values, a necessary instrument in the struggle of concurrent and divergent political-social forces, which attempt to evaluate them for their benefit, jeopardizing thereby the specific historical, political, and social conditions[316] in which basic rights should be deployed. In other words, judicial review, rather than assuming its role of "Guardian of the Constitution," becomes the producer of an ideological and political system of values.[317]

This process towards a more totalizing concept of basic rights and its consequences are outcomes of the BVG's longstanding practice, whereby not only was its active role extended, but also its influence on political grounds became an expected attitude in German constitutional reality. The movement has been gradual, but persistent: more and more the BVG has expressed itself as "forum for the treatment of social and political problems."[318] Some decisions can illustrate how judicial review gained this axiological and political dimension. They complement, accordingly, the perception the cases discussed in the first chapter – the *Crucifix* and *Cannabis*, in particular– brought about.

Already in 1958, the BVG, in the famous *Lüth* case (*Lüth-Urteil*), which referred to a call for a boycott against a film directed by a former Nazi moviemaker,[319] while establishing the central issues relating to the freedom of speech and its boundaries, provided the elements for the construction of a dogmatics of basic rights. According

viewpoints, measured and compared. But time and money are only insufficient arithmetical units and standards for individual and interindividual value and utility; money and time are for different individuals of distinct value and also for a balancing, perhaps, between freedom of expression and state security totally useless. In the examination of the proportionality in the narrow sense, only the examiner's subjectivity takes effect and leads to incidental evaluation of the deployment of basic rights as if they were more or less valuable" (Schlink, "Freiheit durch Eingriffsabwehr – Rekonstruktion der klassischen Grundrechtsfunktion," 462, translation mine).

[315]See Jürgen Seifert, "Verfassungsgerichtliche Selbstbeschränkung," in *Verfassung, Verfassungsgerichtsbarkeit, Politik: Zur verfassungsrechtlichen und politischen Stellung und Funktion des Bundesverfassungsgerichts*, ed. Mehdi Tohidipur (Frankfurt a.M.: Suhrkamp, 1976), 128.

[316]Ibid., 125–126.

[317]Ibid., 125.

[318]Schlink,"German Constitutional Culture in Transition," 729.

[319]BVerfGE 7, 198 (15.01.1958). This case involved the discussion of the claim to a boycott raised by Erich Lüth, a famous German movie critic, against a film directed by Veit Harlan, the well-known moviemaker of the film *Jud Süß*, a Nazi film that incited strong violence against the Jewish people. Harlan and the movie company, by sustaining moral damage grounded in the §826 BGB, filed a lawsuit against Lüth, which was judged in favor of the plaintiff by the State Court of Hamburg. Against this decision, Lüth filed a constitutional complaint (*Verfassungsbeschwerde*) grounded in the freedom of speech (art. 5 (I) 1 of the Basic Law) in the BVG, which reversed the decision in favor of Erich Lüth.

to the BVG, "the Constitution erects an objective system of values[320] in its section on basic rights, and thus expresses and reinforces the validity of the basic rights."[321] It set up likewise the notion that all basic rights, if they primarily protect the citizen from the state, "they also incorporate an objective scale of values which applies, as a matter of constitutional law, throughout the entire legal system."[322] Unlike the previous understanding that basic rights lie in a relationship between citizen's freedom and the state encroachment (not reaching thereby the relationship among distinct citizens, whose solution would rather derive from civil laws),[323] this case introduced the premise that basic rights have an objective structure embracing the totality of the legal order. With this decision, the BVG inaugurated the notion of constitution as an objective axiological order as well as stated, based on this premise, that basic rights bind the total legal system, including legal norms applicable only to private relationships (*third-part effect of basic rights* – "*Drittwirkung*" and "*Ausstrahlungswirkung*"). It also highlighted the need to balance those basic rights in reference to the particularities of the case: "There has to be a 'balance of interests'; the right that expressing an opinion must yield if its exercise infringes interests of another which have a superior claim to protection."[324] As a reference for future decisions, the *Lüth* case, certainly one of the most discussed decisions in German constitutionalism, allows us to understand how basic rights, previously understood merely as guarantees against the intrusion of state actions, became the center of the legal order together with the value-based approach carried out through the framework of balancing.

The following decisions continuously highlighted this character of the constitution as an order or a system of values,[325] and many of them established, as the primary value of this axiological system, the principle of human dignity. "In the

[320]The idea of an order of values (*Werteordnung*) is previous to the *Lüth* case (See, for instance, BVerfGE 5, 85 (1956); BVerfGE 6, 32 (1957)), but it gained a special treatment after this case, when a dogmatics of basic rights in this matter was clearly established.

[321]BVerfGE 7, 198. Translated by Institute for Transnational Law. The University of Texas at Austin. http://www.utexas.edu/law/academics/centers/transnational/work_new/german/case.php?id=1477 (accessed July 19, 2009).

[322]BVergGE 7, 198. Translated by Institute for Transnational Law.

[323]See Schlink, "German Constitutional Culture in Transition," 718.

[324]BVerfGE 7, 198. Translated by Institute for Transnational Law.

[325]See BVErfGE 12,1 (1960); BVerfGE 13, 46 (1961); BVerfGE 13, 97 (1961); BVerfGE 17, 232 (1964); BVerfGE 23, 191 (1968); BVerfGE 24, 367 (1968); BVerfGE 25, 256 (1968); BVerfGE 27, 1 (1969); BVerfGE 27, 18 (1969); BVerfGE 28, 243 (1970); BVerfGE 30, 1 (1970); BVerfGE 30, 173 (1971); BVerfGE 31, 58 (1971); BVerfGE 32, 98 (1971); BVerfGE 33, 23 (1972); BVerfGE 33, 1 (1972); BVerfGE 33, 303 (1972); BVerfGE 30, 173 (1973); BVerfGE 34, 269 (1973); BVerfGE 35, 79 (1973); BVerfGE 35, 202 (1973); BVerfGE 37, 271 (1974); BVerfGE 39, 1 (1975); BVerfGE 42, 95 (1976); BVerfGE 47, 327 (1978); BVerfGE 49, 24 (1978); BVerfGE 49, 89 (1978); BVerfGE 53, 30 (1979); BVerfGE 53, 366 (1980); BVerfGE 58, 208 (1981); BVerfGE 63, 131 (1983); BVerfGE 76, 1 (1987); BVerfGE 81, 278 (1990); BVerfGE 83, 130 (1990); BVerfGE 88, 203 (1993); BVerfGE 90, 145 (1994); BVerfGE 98, 169 (1998); BVerfGE 98, 265 (1998); BVerfGE 102, 347 (2000); BVerfGE 105, 279 (2002); BVerfGE 105, 313 (2002); BVerfGE 107, 104 (2003); BVerfGE 108, 282 (2003); BVerfGE 109, 279 (2004).

Basic Law's order of values, human dignity is the highest value"[326] and its primary point[327]. With this axiological order, grounded mostly in human dignity, the Constitution has to be interpreted as a unified order with a goal as a way to avoid contradictions among its particular norms.[328] The value-based approach to basic rights directly links, accordingly, to the structure of balancing (*Abwägung*), and any goal, value and prognosis the parliament envisages must be reviewed in order to verify whether its decisions and evaluations are offensive, invalid or whether they contradict the constitutional axiological standards.[329] The order of values, which binds the parliament, corresponds, at the same time, to a hierarchy of values, bringing about then the necessary balancing between basic rights.[330] In other words, the parliament is bound to a weighting of goods (*Güterabwägung*), whose grounds lie in an axiological framework the BVG has the monopoly to define.

This characteristic was reinforced especially after the Pharmacies case (*Apothekenurteil*),[331] in 1958, which upheld the premise that balancing is an indispensable instrument to be deployed in cases involving conflicts of basic rights. This case, whose subject referred to the freedom of profession, occurred in virtue of a pharmacist who had his claim to open a pharmacy in Bavaria denied. The argument was a state law establishing that the authorities could verify the necessity of public interest and the protection of the pharmaceutical market as a way to reject or accept the claim. The authorities understood that, in the market of that location, there were already enough pharmacies for the population. Based on this fact, the pharmacist filed a constitutional complaint in the BVG questioning the constitutionality of the Bavarian law. The court accepted the complaint and declared the unconstitutionality of the legal norm insofar as it infringed the freedom of profession defined in article 12 (I) of the Basic Law. In this decision, the court deployed clearly the principle of proportionality in its narrow sense by mentioning that any restriction on the freedom of profession could only exist as long as rational considerations demonstrate the correspondence to the collective will, requiring thereby a gradual evaluation of the intensity of the intervention. The BVG sustained that the requirements for the exercise of a profession could not be disproportionate concerning the goal and the regular accomplishment of the professional activity.

Moreover, because of the capacity of balancing to structurally encompass this BVG's political role in a justified way, this court could deploy it in the definition of primary social issues and as a basis for expressing the state protective duty towards society. We could observe this inclination to the definition of primary social issues with this objective perspective of basic rights in decisions such as the University case

---

[326]BVerfGE 27, 1 (1969), translation mine. See also BVerfGE 6, 32 (1957).

[327]See BVerfGE 35, 202 (1973); BVerfGE 98, 169 (1998).

[328]See BVerfGE 33, 23 (1972).

[329]See BVerfGE 53, 366 (1980).

[330]See BVerfGE 7, 198 (1958).

[331]BVerfGE 7, 377 (1958).

(*Horschule-Urteil*), in 1973.[332] In this case, the discussion resided in the constitutionality of the university reform, which altered the internal organization and the procedures of the university bodies, without bringing about nevertheless an effective encroachment on academic freedom, for the question was not exactly how professors should develop the contents of their researches and lessons.[333] The BVG, however, held the argument that scholarship had to be free from any parliamentary influence, including the field of university organization, structure and procedures. According to the decision, the Basic Law guarantees a protective free space against any state intervention, for there is a system of values of the Basic Law safeguarding the function of education (article 5 (3)).[334] Hence, any university reform referring to university procedures and structure must rely on the effective participation of its professors and bodies in the deliberation of the main issues of educational activity. As it is possible to infer, the BVG replaced the legislature's definitions by its interpretation of which values applied to education were more in conformity with society, extending thereby the meaning of academic freedom. This case is remarkable, because the BVG exposed how its decisions could encroach by some means upon the parliamentary and governmental space of deliberation in social matters through the definition, in a particular case, of what this constitutional axiological-objective order encompassed. Indeed, as Schlink mentions, "since this decision, there have been a host of additional decisions that test government institutions and procedures based on fundamental rights – fundamental rights as objective principles."[335]

The BVG's definition of primary social issues through the interpretation of basic rights as objective principles reaches a particular configuration when it refers to state protective duties, showing thereby the court's positive activism in the face of governmental activities. In these circumstances, the BVG determines that the state must safeguard a certain constitutional principle, regarded as weightier when balanced with other values at issue, either by means of administrative provisions, evaluation of administrative organs, or even the punishment of whomever acts contrary to this determination. One relevant example in this matter is the first decision on abortion (*Schwangerschaftsabbruch I*),[336] in 1975, when the BVG argued that, in the realm of basic rights, the state has a duty of protection (*staatliche*

---

[332]BVerfGE 35, 79 (1973).

[333]Indeed, as Bernhard Schlink remarks:

"Whatever the composition and procedures of university bodies might have been – whether, for example, the university senate consisted only of professors, or instead, equal numbers of professors, assistants, students, and technical personnel – the decisive factor regarding academic freedom was whether the individual professors could freely determine the subjects, methods, and goals of their research and lessons within the framework of their teaching duties. If they could not, if there were intrusions into their academic freedoms, the intrusions were not mitigated because they were decreed by a body composed solely of professors, as opposed to only one-third or one-fourth professors" (Schlink, "German Constitutional Culture in Transition," 719).

[334]See BVerfGE 35, 79 (1973).

[335]Schlink, "German Constitutional Culture in Transition," 719–720.

[336]See BVerfGE 39, 1 (1975).

Schutzpflicht), which, based on the principle of proportionality, and more specifically balancing, pointed to the safeguard of life. The court sustained that the then new §218a of the Criminal Code (*Strafgesetzbuch*), which prescribed a temporal exemption from punishment in virtue of abortion,[337] was void, for it contradicted the Basic Law's axiological order and the principle of human dignity as its central point. There could be no exemption if no special reason for the practice of abortion existed. For it declared the unconstitutionality of the norm, the court laid down, until new rules were defined, a transitional arrangement[338] in accordance with the decision's contents (*Urteilstenor*),[339] establishing, in any case, the parliament's duty to promulgate a new legislation protecting the fetus's life since the conception.

This case is paradigmatic, as long as it clearly replaced the parliamentary definition of exemption from the general criminal law of abortion by the BVG's interpretation founded on what it understood as included in the concept of an objective axiological order.[340] By sustaining that the Basic Law "must come down in favor of the precedence of the protection of life for the child *en ventre sa mère* over the right of the pregnant woman to self-determination,"[341] the BVG assumed a political role towards the definition of what would be more in conformity with society, even though the parliament had comprehended it differently. Besides, it also postulated a state's duty of punishment and a duty of protecting life from an objective interpretation of basic rights, whose collision was resolved by means of balancing. The dissenting opinion realized this problematic situation by stressing that, "for the first time in opinions of the Constitutional Court an objective value decision should function as a *duty* of the legislature to enact *penal norms*, therefore to postulate the strongest conceivable encroachment into the sphere of freedom of

[337]This norm stated that, if a doctor were responsible for the abortion procedure, there would be no crime if done before the first twelve weeks of conception, and if done later, it was still legal provided that it followed some requirements.

[338]Another interesting case in which the BVG laid down a transitional agreement can be seen in BVerfGE 99, 341 (1999) – *Testierausschluß Taubstummer*.

[339]See BVerfGE 39, 1 (204).

[340]It is interesting to remark that, in the dissenting opinion, the other members of the court, based on the concept of self-restraint, criticized the majority opinion. They even sustained that self-restraint applies when the court issues directives for the positive development of the social order through constitutional review. Their words:

"The authority of the Federal Constitutional Court to annul the decisions of the legislature demands sparing use, if an imbalance between the constitutional organs is to be avoided. The requirement of judicial self-restraint, which is designated as the "elixir of life" of the jurisprudence of the Federal Constitutional Court, is especially valid when involved is not a defense from overreaching by state power but rather the making, via constitutional judicial control, of provisions for the positive structuring of the social order for the legislature which is directly legitimatized by the people. The Federal Constitutional Court must not succumb to the temptation to take over for itself the function of a controlling organ and shall not in the long run endanger the authority to judicially review constitutionality" (BVerfGE 39,1. Translated by Robert E. Jonas and John D. Gorby. *The John Marshall Journal of Practice and Procedure* 9, 605 ff).

[341]BVerfGE 39,1. Translated by Robert E. Jonas and John D. Gorby. *The John Marshall Journal of Practice and Procedure* 9, 605 ff.

the citizen. This inverts the function of the fundamental rights into its contrary."[342] The decision, undeniably, pointed out a new BVG's approach and exposed how intricate is the notion of judicial self-restraint through the assumption of an objective axiological standpoint.[343]

Similar complexity regarding the duty of protection could be seen in the second decision on abortion in 1995,[344] in which the BVG stated that, if it is to accept a temporal exception, then, at least, the state must provide a compulsory counseling in favor of the continuity of pregnancy. In this judgment, the BVG held that every interruption of pregnancy is in principle against the law and, therefore, legally forbidden. Only few exceptions exist, when the damages result in such sacrifice of vital values that no one could rationally expect any other woman's behavior. In practical terms, the BVG pointed out that, if it is to accept a temporal exception of twelve weeks from the general criminal rule, some requirements have to be fulfilled. First, there must be a counseling with the purpose of fostering the continuity of pregnancy. Second, there must be an indication of the motives, established in the counseling rules, to interrupt the pregnancy and its verification by a third-party. Third, a doctor must carry out the procedure. Accordingly, if, on the one hand, the parliament has a discretionary margin to evaluate those exceptions, at least, on the other, as a reflex of the state protective duty, the state must provide the conditions to advise the woman of her responsibility and encourage the continuity of pregnancy.[345]

Through the argument of state protective duty – intimately connected to this premise of basic rights as objective principles of a total legal order – the BVG made itself more present in relevant questions of social life, for instance, in the definition of the boundaries of state supervision of broadcast transmissions;[346] the need for absolute safety in the regulation of atomic energy;[347] the protection of workers against injuries occurred at night jobs;[348] the defense of the precept of German reunification and the extension of judicial review to international treaties;[349] the

[342]BVerfGE 39,1. Translated by Robert E. Jonas and John D. Gorby. *The John Marshall Journal of Practice and Procedure* 9, 605 ff.

[343]It is interesting to verify that, especially in the main controversial cases, the dissenting opinion remarks the abuse of the boundaries defined by the principle of judicial self-restraint. See BVerfGE 39, 1 (1975); BVerfGE 114, 121 (2005).

[344]See BVerfGE, 88, 203 – *Schwangerschaftsabbruch II.*

[345]See BVerfGE, 88, 203 – *Schwangerschaftsabbruch II.*

[346]See BVerfGE 57, 295 – *Rundfunkentscheidung* (1981).

[347]See BVerfGE 49, 89 – *Kalkar I* (1978); BVerfGE 81, 310 – *Kalkar II* (1990), and BVerfGE 53, 30 – *Mülheim-Kärlich* (1979).

[348]See BVerfGE 85, 191 – *Nachtarbeitsverbot* (1991).

[349]In this decision, which refers to the *Basic Treaty* of 1972 (*Grundlagenvertrag*) between the Federal Republic of Germany (FRG) and the German Democratic Republic (GDR) concerning the recognition, for the first time, of their respective sovereignties, the BVG even indicated the constitutional boundaries and the scope the federal government and the parliament had to observe in future agreements, always with the purpose of making clear that, by no means, the precept of reunification could be jeopardized. See BVerfGE 36, 1 – *Grundlagenvertrag* (1973).

protection of human life against terrorist blackmail;[350] the supervision of children;[351] the protection of unborn children;[352] the prohibition of narcotics;[353] the financial support of private schools;[354] the need for an adequate regulation regarding the noise of airplanes[355] and road traffic;[356] life insurance;[357] the civil partnership of people of the same sex;[358] the admission of baking on Sundays;[359] the prohibition to shut down airplanes even when hijacked by terrorists with the purpose of being used as a weapon to kill more people,[360] among others. In many of these decisions, the connection of the protective duty with the concept of basic rights as objective principles was manifest, and the deployment of the principle of proportionality, and specially balancing, was used as a criterion to define how the different values at stake should be interpreted.

Naturally, this theme is intricate, for the movement towards the BVG's activism, grounded in the premise of being the Constitution an axiological order covering positive claims to benefits and services, lies in the subtle delimitation of the activity sphere between politics and adjudication. Not only by pushing protective and even punishment duties towards the parliament, including the "provisional arrangements," but also by calling the parliament to a modification of the law[361] are instruments the BVG can use, which reinforces the hesitation in the face of a possible encroachment on other power spheres. Although in many of these cases it is possible to agree with the BVG's necessary intervenient attitude, even because this action is the immediate outcome of the Basic Law, many of them derives nonetheless from an extensive interpretation of basic rights as though they could embrace any theme of social life.[362] This situation gains a special attention insofar as, more than ensuring basic rights in these circumstances, the court begins to lay down political functional decisions – regardless, in some cases, of the demanding additional budgetary funds to insure their effectiveness[363] – in order to attend to the

---

[350]See BVerfGE 46, 160 – *Schleyer* (1977).

[351]See BVerfGE 99, 216 – *Familienlastenausgleich II* (1998); BVerfGE 98, 265 – *Bayerisches Schwangerenhilfegesetz* (1998).

[352]See BVerfGE 86, 390 – *Schwangeren – und Familienhilfegesetz I* (1992).

[353]See BVerfGE 90, 145 – *Cannabis* (1995). See the first chapter.

[354]See BVerfGE 75, 40 – *Privatschulfinanzierung I* (1986).

[355]See BVerfGE 56, 54 – *Fluglärm* (1981).

[356]See BVerfGE 79, 174 – *Straßenverkehrslärm* (1988).

[357]See BVerfGE 114, 1 – *Schutzpflicht Lebensversicherung* (2004).

[358]See BVerfGE 105, 303 – *Lebenspartnerschaftsgesetz* (2002).

[359]See BVerfGE 87, 363 – *Sonntagsbackverbot* (1992).

[360]See BVerfGE 115, 118 – *Luftsicherheitsgesetz* (2005).

[361]See, for instance, the *Appellentscheidung*, according to which the BVG declares that a norm is still constitutional, but, if no modification in its contents is carried out by the parliament, it can declare its unconstitutionality. See, for example, BVerfGE 108, 82 – *Biologischer Vater* (2003).

[362]See, for instance, the *Cannabis* case examined the first chapter.

[363]According to Bernhard Schlink, even though the "*Bundesverfassungsgericht* has interpreted defensive rights to be service rights, and inferred from these rights entitlements to government

interests of the population through the identification of its moral conscience with the axiological order that, presumably, the Constitution carries with itself.

The critical question in the realm of separation of powers stands out as long as this movement results in the configuration of a court promoting a sort of "extra-juridical legality" (*außerrechtliche Gesetzmäßigkeit*),[364] whereby, rather than reinforcing constitutional norms, the court transforms any fact, any moral value, or any political claim into a problem of constitutional rights. In other words, the court assimilates axiological to deontological issues, thereby weakening the normative force of legal rights. The consequence is the risk of transmuting constitutional adjudication into a compromised solution, through which balancing, now "rationally" justified in the structure of the principle of proportionality, could result, however, in arbitrary and opportunistic opinions jeopardizing the constitutional order and favoring a particular interest. In fact, as Schlink mentions, "while methodologically convincing decisions still occur every now and again, there are many others that simply arise from the court's feel for what is indicated by social and political life – for what is accepted and 'fits' into the social and political landscape."[365] For this reason, the movement the BVG led to an objective standpoint of the Basic Law – the Basic Law as an objective axiological order – poses a very concerning problem for constitutional democracy, because: first, objective principles, unlike subjective rights as protection against state intrusion, can be balanced according to certain criteria embraced by this BVG's axiological interpretation of social standards; second, there is the loss of the possible control over the BVG's activities, since now the historical development of the legal system can be remodeled in accordance with a goal-oriented perspective the court aims to achieve through decision-making; third, the dialectical and healthy relationship between the distinct institutional powers results in their submission to the BVG's capacity to define the last word in any social matter; fourth, the court frees itself not only from the law but also from the construction of a consistent interpretation of the system of rights, inasmuch as the presumable resolution of these political and social problems demands of the judge the relativization of the usual constraints she should be aware of. Briefly, in this BVG's shift to political activism, Jutta Limbach's words – "Does the unbroken great trust in the authority of constitutional jurisdiction indicate a political mistrust of democracy?"[366] – seem remarkably accurate.

support and distribution of positions, means, and opportunities," it "has, consistently, dampened the practical consequences of its decisions on government. It has never demanded that the government release additional budgetary funds to cover these entitlements, but has only required equal distribution of already available means" (Schlink, "German Constitutional Culture in Transition," 721).

[364]See Leisner, Der *Abwägungsstaat: Verhältnismäßigkeit als Gerechtigkeit?*, 232.

[365]Schlink, "German Constitutional Culture in Transition," 729.

[366]Limbach, "The Effects of the Jurisdiction of the German Federal Constitutional Court," 22.

2.5 The Constitutional Scholarship Reaction Against the Bundesverfassungsgericht's Shift to Politics and the Irrationalism of Balancing

The BVG's shift to activism provides distinctive and interesting analyses. One could remark, for example, some benefits this movement promoted towards democracy. By observing the BVG's historical practice, one could argue, for instance, that the BVG has presented an active role regarding the consolidation of basic rights in German constitutional reality, or that, through a skeptical opinion, it is impossible to separate politics from basic rights, particularly in the realm of constitutional adjudication. Indeed, it could even be possible to sustain, as Peter Häberle did, that the function of judicial review is inherently political, as long as it promotes "policies through the interpretation of the Constitution,"[367] thereby diminishing the weight of the criticism concerning the BVG's politicization. Moreover, the stress could be projected into the political effects of every constitutional decision, insofar as "it is hence inevitable that such decisions – according to the facts of the case with different relevance – are bound to political consequences."[368] On the other hand, the conflict between law and politics could set forth a distinct approach by arguing that the BVG should be, at least in its nature, an organ of legal rights, not of politics, thereby subjugating political issues to the Constitution, which, nevertheless, the BVG has often pragmatically carried out by deploying the principle of judicial self-restraint as a condition for its own legitimacy.[369] Furthermore, it is possible to reason that, if the BVG makes politics, it does so just because political issues are normally brought to it.[370] From other perspective, the counterargument could be mostly factual: despite all the conflicts between law and politics, as well the BVG's shift to politics, its activity, in a sense, has respected the legislature's evaluative prerogative.[371] One could also argue that it has seemingly appeared to be self-confident enough to mediate between where jurisdiction ends and where politics begins,[372] or even has exercised the "last competence" in a better neutral

[367]Häberle, "Grundprobleme der Verfassungsgerichtsbarkeit," 4, translation mine.

[368]Winfrieed Steffani, "Verfassungsgerichtsbarkeit und Demokratischer Entscheidungsprozess," in *Verfassungsgerichtsbarkeit*, ed. Peter Häberle (Darmstadt: Wissenschaftliche Buchgesellschaft, 1976), 386, translation mine.

[369]See Thomas Clemens, "Das Bundesverfassungsgericht im Rechts – und Verfassungsstaat: Sein Verhältnis zur Politik und zum einfachen Recht; Entwicklungslinien seiner Rechtssprechung," in *Das Bundesverfassungsgericht: ein Gericht im Schnittpunkt von Recht und Politik*, ed. Michael Piazolo (Mainz-München: Hase & Koehler, 1995), 17–19.

[370]See Michael Piazolo, "Zur Mittlerrolle des Bundesverfassungsgerichts in der deutschen Verfassungsordnung – eine Einleitung," in *Das Bundesverfassungsgericht: ein Gericht im Schnittpunkt von Recht und Politik*, ed. Michael Piazolo (Mainz-München: Hase & Koehler, 1995), 10.

[371]See Schlink, "Der Grundsatz der Verhältnismäßigkeit," 465.

[372]See Piazolo, "Zur Mittlerrolle des Bundesverfassungsgerichts in der deutschen Verfassungsordnung – eine Einleitung," 11.

and independent way than other political instances, a solution that proved to be the most accepted model by the population.[373]

On the other hand, the scholarship's critique of this BVG's shift to political activism, rather than balancing the possible hazardous consequences for constitutional democracy with the identification of some positive benefits of this movement, could highlight that the main issue here lies in the depoliticization of the public sphere, which is the reversal of any democratic intent. Ingeborg Maus's association of the BVG with the "social superego of a fatherless society" is a very well-known reference in this subject matter.[374] Although it is possible to question her empirical reference to so incisively sustain that the BVG turned, in practice, into a "social superego of a fatherless society," leading thereby to the depoliticization of the public sphere, the demoralization of society and the moralization of justice, she provides an interesting analysis of how dangerous it is for democracy the politicization or moralization of constitutional court, and how, in a more radical

[373]See Ulrich Ramp, *Hüter der Verfassung oder Lenker der Politik* (München: Grin Verlag, 2003), 3.

[374]In Maus's interpretation of the German development of constitutional adjudication through the psychoanalytical concept of superego, the main problem in this way to politics is the inversion of the democratic procedure. The realm of freedom, which is the basic principle of self-government and sovereignty of people, becomes an outcome of decision-making. "The foregoing realm of individual freedom converts itself then into a case to case manufactured product of judicial decision activity" (Ingeborg Maus, "Justiz als gesellschaftliches Über-Ich: Zur Funktion von Rechtsprecung in der 'vaterlosen Gesellschaft," in *Sturtz der Göter? Vaterbildeer im 20. Jahrhundert*, ed. Werner Faulstisch and Gunter Grimm (Frankfurt a.M.: Suhrkamp, 1989), 128, translation mine). The popular confidence and the "infantilism" with respect to the expectancy that the BVG will correct the very procedure of public deliberation (Ibid., 129), as if it were the revelation of justice, demonstrate that the field of social mobilization is jeopardized by this need for an external control or, in better words, by this anxiety about what the superego will express. Public autonomy is undermined by a heteronymous definition of what is good for society, which delegates the pursuit of consensus to the court. The arbitrary quest for an "order of values" ("*Wertordnung*"), as the character of the Constitution, and for "morally enriched concepts" (Ibid., 142, translation mine) is a sign of this process of transference of social issues to the court by overcoming the public autonomy, thereby opening it to arbitrary and incoherent definitions of what is good for society. This could be observed in different adopted criteria, for instance, the ones based on efficiency – the principle of proportionality, as here examined, could be brought to this scenario – on social acceptance, on traditional values, among others that are not directly stated by the Constitution, whereas the immediate constitutional norms are simply disregarded. "The written constitutional guarantees of freedom are placed thereby under the reserve of unwritten idiosyncrasies of economic and political apparatus" (Ibid., 142, translation mine) What remains, in this context, is a judicial system acting in accordance with different interests but the subjective guarantees of the Constitution. The Constitution loses its connection with democracy as a document institutionalizing procedures and basic guarantees, which ensure the exercise of social and political mobilization, and becomes a moral text of values – a Bible or Koran – from which the BVG can deduce the correct values for society (Ibid., 131). The respect for the Constitution becomes a theology of the Constitution (*Grundgesetztheologie*) (Ibid., 143). Judicial review, accordingly, functions by this double-sided process: the social BVG's deification, for it claims this guardianship, and the deification of the Constitution, for it becomes an "order of values." For this purpose, see Ibid., 121–149.

perspective, it could result in the inversion of the democratic process as long as the constitutional court exercises this role of rebuilding a conscience devastated by the authoritarian past. Still, if Maus's conclusions make sense and German history somehow corroborates her thesis, they might, though, be overstating the dualism moralization/demoralization or politicization/depoliticization and leaving aside some relevant characteristics of German constitutional democracy. The conflict between law and politics in the realm of constitutional adjudication seems more complex, and the reality has proven to be not as tragic as Maus's conclusions appear to indicate.

On this score, this investigation will concentrate upon three German constitutional scholars who intimately examined this theme: Ernst-Wolfgang Böckenförde, Bernhard Schlink and Friedrich Müller. They set forth pertinent remarks in this matter and expose the transformations German constitutional culture has passed through and the dilemmas they raise. Their analyses introduce some critical aspects that will connect to the further investigation of the deployment of balancing as a "rational" solution to constitutional conflicts. They can be ordered in five primary aspects: (1) the absence of a rational safe basis and a methodological system able to constrain judicial decisionism; (2) the loss of the boundaries between the parliament and the judiciary, which reinforces the prior conclusion concerning the BVG's political role and poses the question of separation of powers; (3) the transformation of constitutional adjudication from a dogmatic orientation into a *case to case* examination; (4) the weakening of legal dogmatics and the constitutional scholarship's critical review of the BVG's decisions; and, (5) the relativization of legal rights without the counterpart of a coherent proviso and a self-binding criterion, as well as its functional subjection to an axiological pattern.

Ernst-Wolfgang Böckenförde places this movement towards the assimilation of values and legal norms in one of the forms of interpretation of basic rights. After having examined the liberal theory of basic rights, in which he stressed the state's limited authority against the absolute individual and social freedom sphere (defensive rights), and the institutional theory of basic rights, according to which the former subjective character is replaced by an objective principled order interpreting freedom according to goals, the axiological theory (*Werttheorie*) appears. This value-based interpretation of basic rights recalls some of the developments of the institutional theory – broader field for normalization and arrangement of the basic rights protection; orientation towards goals; objectivization of freedom – however, it embodies the integrative character of a community of values, which Böckenförde links to Rudolf Smend's integrative theory.[375] With this "material status," basic rights become objective norms, whose contents lie in the axiological fundamentals of a community, and are interpreted as a means to realize and confirm those

[375]Ernst-Wolfgang Böckenförde, *Staat, Verfassung, Demokratie: Studien zur Verfassungstheorie und zum Verfassungsrecht* (Frankfurt a.M.: Suhrkamp, 1991), 119. See also Rudolf Smend, *Verfassung und Verfassungsrecht* (Berlin: Duncker & Humblot, 1928).

values.[376] The consequences of this process are, on the one hand, the emancipation of legal methods and, on the other, the increasing investigation of the prevalent social values in a particular time.[377] Evidently, this process leads to the relativization of basic rights, insofar as they are conditioned by the evaluation of values they presumably express, which are interpreted according to a general definition of a temporal axiological conscience.[378] This evaluation, nonetheless, does not really seem to provide a practical way to solve problems in the realm of collision and limits of basic rights: "Because so far it is not evident neither a rational justification for values, nor mostly a rationally visible and discussible system of preferences for the definition of the values hierarchy, and thereby a constructed balancing of values."[379] In other words, every decision based on values as well as on a balance of values falls into the irrational definition of their hierarchy and obeys a different logic that loses – or even conceals – the connections with the existing arguments. Böckenforde's viewpoint, accordingly, identifies this form of interpretation of basic rights with the practice of judicial decisionism.[380]

This critique of the value-based concept of basic rights extends to the deployment of balancing. This instrument, based on the asymmetry that exists among the different undefined dimensions of basic rights, now reaching all the system of rights (*third-part effect of basic rights* – "*Drittwirkung*" and "*Ausstrahlungswirkung*"), makes the variable fields of freedom to be ordered in accordance with hierarchic relations and the characteristics of the particular case.[381] With this instrument, the concept of constitution as an axiological order stands out. The counterpart is that legal adjudication loses its link with legal dogmatics,[382] for it covers, with more flexibility and dynamics, this broader and relative function of basic rights. Adjudication, consequently, becomes a practice of giving meaning to basic rights in the form of 'case law' rather than a practice of interpreting the law, which assimilates it with the practice of legislation.[383] Naturally, this aspect causes a serious transformation in the relationship between the parliament and the constitutional court. They become closer: whereas the parliament weakens its normative labor and emerges in favor of the concretization of legal rights, constitutional adjudication, as well, weakens its interpretative labor in favor of the concretization of legal rights. There is no longer, accordingly, qualitative difference between both,[384] except

[376]Böckenforde, *Staat, Verfassung, Demokratie: Studien zur Verfassungstheorie und zum Verfassungsrecht*, 130.

[377]Ibid., 131.

[378]Ibid., 131–132.

[379]Ibid., 132, translation mine.

[380]Ibid., 133.

[381]See the fourth chapter.

[382]Böckenförde, *Staat, Verfassung, Demokratie: Studien zur Verfassungstheorie und zum Verfassungsrecht*, 185.

[383]Ibid., 186.

[384]Ibid., 189/190.

that the last word belongs to the BVG. The conclusion Böckenförde brings forth is the perception of a transition from a parliamentary constitutional state to a jurisdictional constitutional state.[385] The risks associated with the assumption of an objective axiological standpoint in the realm of basic rights, apart from the risks of decisionism, is the BVG's transformation into the "Areopagus of the constitution,"[386] which poses the question of whether this transition is democratic legitimate.

Böckenförde contrasts what is for him an inevitable process on the way to the BVG's prevalence with the alternative of a liberal understanding of constitutional rights, according to which the axiological standpoint is replaced by the idea that basic rights are interpreted based on the contemporary subjective perspective of the relationship between the state and the citizen.[387] With this approach, it does not mean that the objective understanding of basic rights is lost, but only that the parliament is no longer bound to political predefinitions of the constitutional court. Therefore, it can exercise its original function of defining the ethical and political principles in the usual "struggle over rights" as well as of getting closer to the civic practice of democratic participation in the realm of rights.[388] More than irradiating as objective principles throughout the totality of the legal order, the contents of basic rights are grounded in the area of individual rights, which ought to be preserved and developed with the appropriate support of legal dogmatics, promoting thereby a culture based on rights. Hence, every basic right is deployed and limited in the direct correlation between the citizen's and the state's apparatus and defined, rather than in reference to the abstract idea of a total normativeness, within the specific domain of the legal order. The immediate consequence of this thinking is that the parliament becomes more sovereign, for it is not bound to a previous judicial definition of social values embracing the totality of the legal order, and the democratic political process, unlike the idea of objective principles according to which this process is bypassed or replaced by the constitutional court's decisions, becomes central in legal foundation. Moreover, balancing reassumes its place as a reference to parliamentary activity,[389] not to constitutional adjudication. The doubt, however, is how the parliament would deal with this configuration, which makes the final question lie in the citizens' confidence: if they trust in the elected parliament, the emphasis is on the subjective rights; if they trust in the constitutional court, the stress is on the objective principles.[390]

Bernhard Schlink has similar apprehension about the advance of a value-based conception in the realm of basic rights when he remarks that this transition weakened the function of ensuring the individual freedom and strengthened the

[385]Ibid., 190.

[386]Ibid., 191.

[387]Ibid., 194.

[388]Ibid., 194.

[389]Ibid., 195.

[390]Ibid., 199.

idea that judicial review is mostly a "forum for the treatment of social and political problems."[391] Like Böckenförde, Schlink underlines, which is for him an irreversible and inexorable process,[392] the effects of a less consistent legal methodology following the BVG's activist approach, because it becomes less useful when confronted with the political and social problems interpreted as its domain of activity.[393] A radical transformation in German constitutional culture,[394] therefore, is in process nowadays. The association of the idea of basic rights as objective principles with the weakening of dogmatic studies provides, especially when we remark the BVG's undertaking of a political role, a turbulent combination that discloses the following significant movements:

1. The development of a casuistic interpretation of legal rights, without this meaning a coherent casuistic development of legal rights, and the loss of a self-binding mechanism. Adjudication turns into an activity based on a case to case examination, no longer directly linked to the deployment of dogmatic concepts and scholarly interpretation of a particular issue. Apart from leaving aside traditional and longstanding dogmatic concepts and a system supplying possible solutions to specific cases, this casuistic jurisdiction grounded in a teleological basis of objective principles embracing entirely the legal order has not assimilated the conceivable counterpart of the *common law stare decisis*[395] precedent orientation as a way to promote, at least, the apparent awareness of the need for coherence in decision-making.[396] Even though, evidently, none of the mechanisms – legal dogmatics or *stare decisis* – could promote a harmonious legal interpretation alone (in fact, they could even be easily eroded),[397] they seemingly indicate that a general analysis had previously been carried out in order to orient adjudication towards the equal interest of all. Insofar as the BVG

[391] Schlink, "German Constitutional Culture in Transition," 729.

[392] Schlink, "Abschied von der Dogmatik: Verfassungsrechtssprechung und Verfassungsrechtswissenschaft im Wandel," 1133.

[393] Schlink, "German Constitutional Culture in Transition," 729.

[394] As Schlink remarks, this is a process, whose outcomes are still in progress. See Schlink, "Abschied von der Dogmatik: Verfassungsrechtssprechung und Verfassungsrechtswissenschaft im Wandel," 1132–1133.

[395] The *stare decisis* ("to stand by things decided") means that prior court's decisions are to be regarded as precedent for future decisions. Even though it does not mean a strict observance of the past, the court should hold the previous decisions, and, if a modification reveals necessary, it should establish strong arguments, not disrupting thereby the harmony and the longstanding interpretation of a particular issue. In American constitutionalism, this mechanism can be clearly verified in the continuous attempt, in decision-making, to bring the main arguments of similar prior cases to the decision of a particular issue at stake.

[396] Bernhard Schlink brings a very interesting analysis of this loss of any mechanism of coherence in BVG's decisions. See, for this purpose Schlink, "Abschied von der Dogmatik: Verfassungsrechtssprechung und Verfassungsrechtswissenschaft im Wandel," 1132.

[397] Ibid., 1133.

is no longer guided by either criteria, the conclusion is that the BVG currently "frees itself from any self-binding mechanism (*Selbstbindung*)."[398]

2. The disruption of a hierarchical external binding criterion among the courts, for continuously the lower courts disobey the commands of the BVG's decisions, and, in turn, its decisions, since they lack a consistent justification, become subject to correction by the European courts.[399] This outcome can sound paradoxical: whereas the idea of principle of the total legal order endorses a more extensive range of incidence, it also weakens the power of this incidence, inasmuch as the arguments become less persuasive. "The Bundesverfassungsgericht's decisions acquire a moment of arbitrariness, which easily makes the other courts lay down their arbitrary wishes (*Belieben*) in the place of *Bundesverfassungsgericht*'s ones."[400]
3. The transformation of constitutional Jurisprudence, which converts more and more, due to the complexity to develop a systematized harmonious structure of reasoning in this *case to case* basis, into a task for experts by examining the judges' personalities.[401] It no longer assumes the "leading role in the development and transformation of constitutional law;"[402] it presupposes, instead, besides the intricate mission to bring out some connections from case analysis, the investigation of trends, judges' personalities and the politics involved in their activities.[403] Obviously, this is a serious modification of standpoint: rather than attempting to establish a conceptual methodological system to interpret rights, the accent now depends strongly on the investigation of the idiosyncrasies of judicial life. This is besides also a consequence of the political character adjudication acquires with the concept of basic rights as objective principles;
4. The loss of critique of the BVG's decisions and the consequent worship of its activities. Nowadays, the constitutional Jurisprudence "is entirely under the 'spell' of the Bundesverfassungsgericht,"[404] which means that its primary material derives directly from the observation of how the BVG decides a particular issue. This inverted substantially the role Jurisprudence plays,[405] insofar as it

[398] Ibid., 1132, translation mine ["*(Es) befreit sich von jeder Selbsbindung*"].

[399] Schlink points out some examples of this process, which, even though still exceptions, did not occur before. See, for this purpose, Ibid., 1125–1128.

[400] Schlink, "Abschied von der Dogmatik: Verfassungsrechtssprechung und Verfassungsrechtswissenschaft im Wandel," 1133, translation mine.

[401] Ibid., 1134.

[402] Schlink, "German Constitutional Culture in Transition," 730.

[403] Ibid., 727.

[404] Ibid., 730.

[405] Schlink compares the historical development of constitutional scholarship with other disciplines in Germany in order to show that a different critical approach could be adopted. According to him, "above all, however, it is the relationship between legal scholarship and decision making in other German legal disciplines that proves that the relationship which now exists between constitutional scholarship and the decisions of the Bundesverfassungsgericht could be different. Neither the decisions of the Bundesgerichtshof in civil and criminal matters nor those of the

functions in the sequence of the court definitions: "Constitutional scholarship thus thinks and works in the wake of the Bundesverfassungsgericht, rather than ahead of it."[406] A subservient constitutional Jurisprudence, hence, stands out, almost without exercising the indispensable critical role.[407] The BVG's canonization[408] becomes a reality, in which constitutional scholarship turns into a "sort of junior partner and thus participate[s] in its authority, instead of offsetting its authority as a critical opponent,"[409] sometimes even with clear personal interests.[410] Moreover, even though the BVG expresses activist and political purposes grounded in a *case to case* basis and in the idea of basic rights as objective principles, the constitutional theory still adopts the traditional way of legal dogmatics to interpret the BVG's decisions: "It reads and interprets these decisions, and their reasoning, as though they were codified law."[411]

In virtue of this situation, similar to Böckenförde, Schlink argues that the interpretation of basic rights as subjective rights should be adopted. For him, "a concept of fundamental rights that simply guarantees subjective rights by repealing state intrusions upon personal freedom is quite capable of dealing with those societal problems that must be adjudicated in court."[412] For this reason, the notion that balancing is indispensable, because it is a rational consequence of the interpretation of basic rights, might not be entirely true. After all, many questions could be answered by the deployment of the equality principle[413] or by the stress on the

---

Bundesverwaltungsgericht are canonized in a form comparable to the decisions of the Bundesverfassungsgericht" (Ibid., 731).

[406]Ibid., 730.

[407]Schlink even compares the German constitutional scholarship with that of the United States. According to him: "Theory, as other social, cultural, and intellectual disciplines teach, can maintain an extremely critical distance from practice. For example, in the American legal and constitutional order, the United States Supreme Court maintains a position similar to that of the Bundesverfassungsgericht in German society. Yet American constitutional scholarship challenges the Supreme Court more frontally, and if not less respectfully, than at least less gently" (Ibid., 731).

[408]Ibid., 731.

[409]Ibid., 734.

[410]Schlink also links this characteristic to the scholar's interests in positions on the BVG. According to him, "various constitutional scholars have acted as advisors or representatives in cases before the Bundesverfassungsgericht, as loyal compilers and systematizers of its decisions, even as possible candidates for future positions on the court. Constitutional scholarship would like to participate in power, and it realizes that the courtiers are rewarded for their service to the royal court by being allowed to influence it" (Ibid., 734).

[411]Ibid., 735.

[412]Ibid., 727.

[413]Bernhard Schlink mentions that, although the deployment of the equality principle still leaves questions unsolved and is not a real substitute for the examination brought by the principle of proportionality in the narrow sense – indeed, it can even lead to the principle of proportionality in the narrow sense –, the equality principle can answer fundamental questions that the principles of suitability and necessity could not achieve. See Schlink, "Der Grundsatz der Verhältnismäßigkeit," 459.

institutional development of basic rights grounded in the citizen's confidence in the continuance of legal rights and their position.[414]

In this regard, he remarks that, rather than proceeding to an assessment of private and public goods or interests (which is the characteristic of the traditional and prevalent concept of balancing), constitutional adjudication should revolve around a broader understanding of the principle of necessity (and also of suitability),[415] and focus thereby on the encroachment the measure caused rather than on the conflictive interests balancing brings to light. He is aware of the central questions the deployment of balancing poses within the context of the separation of powers, especially because of the political argument and its lastly subjective and "decisionist" character.[416] In particular in the context of the control over parliamentary activities, the deployment of balancing is, according to Schlink, a serious problem, for, besides the nonexistence of any constitutional authorization for that,[417] it leads to the confusion between law and politics. As he mentions, "constitutional adjudication is not an integral component of politics, but rather regarded as its balanced adversary, with a proper rationality and proper legitimacy. The political rationality of the ultimately subjective and "decisionist" evaluation and assessment bears a political legitimacy which is not assigned to the *Bundesverfassungsgericht*."[418]

The problem of rationality of balancing, therefore, implies the subversion of the domain where the definition of policies (relationship between the state and social interests, determination of resources, etc) should occur, as long as it is transferred

---

[414]Ibid., 460. See also Pieroth and Schlink, *Grundrechte – Staatsrecht II,* 67–68.

[415]Bernhard Schlink establishes, in his proposal for a method for balancing in the realm of dogmatics of basic rights (dogmatics of the social state), a critical position on the deployment of the principle of proportionality in the narrow sense as a balancing of particular and public goods and interests. His focus lies in the principles of suitability and necessity with larger amplitude and the continuous protection of citizen's minimal position, whose basis shapes a "method of balancing as dogmatics of basic rights" that works in the realm of rights of freedom. The central premises of his thinking are: (1) the idea that the balancing model is not a reification or a division between state and society, but rather a model for argumentation open to different arrangements between state and society; (2) the balancing model does not "deny the possibility of a conciliation between the individual and the society, the citizen and the state," but this conciliation is a "task of the political system"; (3) in the balancing model, the stress is, more than on the idea of an assessment of public and private interests and goods, on a wider reception of the principles of suitability and necessity (See Schlink, *Abwägung im Verfassungsrecht,* 219). Moreover, Schlink, as he points out in another text, thinks that, in the realm of constitutional adjudication, the idea that the deployment of the principle of proportionality in the narrow sense is indispensable is not entirely true (See Schlink, "Der Grundsatz der Verhältnismäßigkeit," 458–460).

[416]See Schlink, "Der Grundsatz der Verhältnismäßigkeit," 460–462.

[417]Schlink argues that, unlike the control over administrative and judiciary acts, there is no constitutional authorization for the control over parliamentary acts, especially when the BVG replaces parliamentary political rulings with its own. For this reason, Schlink defends that the proportionality control should focus on the analysis of the legitimacy of the relationship between means and goals as well as on the principles of suitability and necessity (Ibid., 462). A more detailed examination of a possible methodology in this area is found in Schlink, *Abwägung im Verfassungsgercht.*

[418]Schlink, "Der Grundsatz der Verhältnismäßigkeit," 462, translation mine.

from the parliament to the constitutional court without answering the primary question of where the authorization for the BVG's control over legislative acts stems from, as well as its capacity to replace them by its own interpretation.[419] In other words, insofar as balancing works in the field of political legitimacy and lacks a rational and methodological standpoint, there is no legitimate argument that could sustain why the constitutional court should deploy balancing.

However, since this movement seems irreversible, either because of the value-based approach or on account of the increasing deployment of the principle of proportionality (particularly balancing), at least, together with this BVG's political and activist posture, constitutional scholarship needs to play a critical role by "determinedly and consistently [placing] cases and decisions at the center of its work"[420] and by "[confronting] the political and ethical aspects of the decision."[421] Accordingly, not only should the BVG itself take advantage of a possible confrontation of its activities, but also the scholarship should see that its role is more than upholding the BVG's decision. Whereas the latter needs to reconfigure its responsibility towards the development of a discursive soil where democracy is practiced, the former must be open to critical review as well as avoid the confusion between law and politics. They do so by grasping that the legitimacy of constitutional review, instead of being linked to the aim to please the population, derives from the reliability of established legal norms and guarantees, that is, the quality of being expected and validated through decision-making.[422]

This is a serious outcome of this movement. When Schlink connects legitimacy to the reliability of established legal norms and guarantees, we could point out that the concept of basic rights as objective principles could, instead, link legitimacy to the capacity of the constitutional court to please the public and satisfy its wishes. The immediate consequence of this process is the risk of overestimating the values in constitutional adjudication. "Since the value-based decision precedes the legal guarantee, the value is the ground and the law is the consequence, and when the value-based decision is systematically shaped and hence relativized with other value-based decisions, then it can also only become this relativity in the legal guarantee."[423] In other words, legal adjudication becomes a relative area where any solution can be found, and where the defensive character of basic rights is suppressed insofar as it becomes a "mere function of an immanent value."[424] True, every legal interpretation will gather and uphold some social values, but the difference is how those values will be assimilated in constitutional adjudication. Inasmuch as legal rights become subservient to a functional character of values, the

[419]Ibid., 461.

[420]Schlink, "German Constitutional Culture in Transition," 735.

[421]Ibid., 735.

[422]See Schlink, "Der Grundsatz der Verhältnismäßigkeit," 456.

[423]Schlink, "Freiheit durch Eingriffsabwehr – Rekonstruktion der klassischen Grundrechtsfunktion," 464, translation mine.

[424]Ibid., 464, translation mine.

space of freedom is jeopardized. This can sound paradoxical: whereas freedom is institutionalized and objectivized, which should lead to a stronger defense of this principle, it is also relativized and weakened in favor of the functional character of a value. Schlink's analysis of the connection between the value-based approach in the realm of legal adjudication and the consequent relativism and functional subservience to values, above all by deploying balancing, therefore, emphasizes that, if there is a problem in the principle of separation of powers, there is also a structural incompatibility in this understanding.

Finally, Friedrich Müller directly examines this secondary role of basic rights that stems from an overestimation of values in constitutional adjudication. He, by stressing the problems of this assumption a system of values, is particularly incisive when he remarks that, apart from transforming and radically limiting legislature's activities and authority, and losing the suggested and recommended methodological procedure when constitutional rights are at stake, this approach "is not enough for the afforded and really satisfying requirements under the rule of law for a legal-objective controllable construction of the decision and the rational statement within the framework of the concretization of constitutional and infra-constitutional order."[425] He acknowledges that an objective axiological viewpoint inverts the idea that the normative force of basic rights relies much more, on the one hand, upon the investigation of the methodical, structural and theoretical standpoints of a normative domain, and, on the other, upon the framing of the relevant elements for the concretization of a practical procedure of creation of precedents rooted in the rule of law. Thus, it does not lie in a general concept embracing the total legal order with the implied deployment of balancing.[426] Although it is not the purpose here to extend this debate on Müller's methodological grounds,[427] he brings forth the following perception of primary importance: the assumption of this axiological viewpoint leads to an idea of constitution that has neither normative contour nor normative quality,[428] whose solution – balancing – cannot be rationally apprehended and does not provide any sufficient requirement for transparency and methodological and legal safety.[429] The combination of an axiological approach to basic rights and the deployment of balancing as a methodological solution to possible conflicts can undermine the very structure of basic rights and ignore the autonomy of legal guarantees and the possibilities of limitation of basic rights

---

[425]Friedrich Müller, *Juristische Methodik* (Berlin: Duncker & Humblot, 1995), 63, translation mine.

[426]Ibid., 63.

[427]Friedrich Müller introduces a hermeneutical structural legal methodology centered upon the concretization of norms, which can afford, according to him, a relevant analysis for the practical area of the activity of case-related concretizations as well as "complement the structural analysis of concretization procedures through a structural model of concretization." See Ibid., 284 ff, translation mine.

[428]Ibid., 67.

[429]Ibid., 67.

through the constitution.[430] In other words, the space for uncontrolled and arbitrary decision-making is open.

From Müller's view, we can remark that the focus is primarily on the constitution, the basic rights, not a vague material category or parameter taken from the abstract concept of constitution as an order of values. This is why he argues that "as far as the constitution, through direct normalization, above all, through a formal preference rule, makes clear a kind of 'higher relevance,' that is, so far as it brings to light a greater assessment of political or ethical origins through a formal binding preference norm, 'balancing of goods' is unnecessary."[431] Briefly, he sustains that, if the constitution already specifies a formal binding preference norm, there is no reason to proceed to balancing. Constitutional adjudication and, particularly, the constitution itself cannot rely on illegitimate and lastly irrational individual decisions.[432] Constitutional adjudication must, on the contrary, rely on the constitution, on the "objective legal extension of *a* basic right."[433] For, when it proceeds to balancing through the totalizing concept of an "order of values," it may overstep its limits.[434] It may lead to the admission of values that do not observe the conditions of validity of constitutional guarantees as well as threaten the political and historical development of basic rights.[435]

These approaches, by reflecting upon relevant aspects of this BVG's shift to political activism, integrate a group of critiques that play a special role when the rationality of balancing is at stake. They reveal how complicated the simple acceptance of the deployment of balancing with the value-based approach is. They show that there is a fundamental connection between rationality and separation of powers, between rationality and deontology of legal norms. These authors reveal, above all, an evident confidence in the constitution, and particularly in the basic rights, as powerful instruments to solve constitutional cases. In other words, the constitution, with its binding criteria, can provide the answers, transforming then balancing, which lacks binding criteria, into an instrument that is not as essential as it seems, and which carries its own potentiality to jeopardize the historical institutional development of legal rights and even promote injustice.[436] Rationality in constitutional adjudication, at least one that acknowledges its limits, accordingly, does not automatically lead to balancing nor can it encompass every type of argument as components of legal reasoning. There are serious effects when this boundary is overstepped, when legal rationality becomes a rationality of an order of values, a rationality of what is good for all or, in practical terms, when legal

[430]Ibid., 69.

[431]Ibid., 67, translation mine.

[432]Ibid., 68.

[433]Ibid., 68, translation mine.

[434]Ibid., 68.

[435]Ibid., 69.

[436]See the second part.

rationality turns into a rationality of balancing values. Legal rationality, in these cases, can become its opposite.

2.6 Final Words

This chapter had the primary purpose of contextualizing the debate on balancing through the investigation of the main characteristics of the recent German constitutionalism. Two characteristics emerge as primary aspects of this movement: the historical context that favored the erection of an activist constitutional court in political matters, which assumed as its authority the discussion of the present and future problems of society in a way that challenges the principle of separation of powers; and the development of a dogmatics of basic rights, which, rather than interpreting them as subjective rights in the idea of entitlements of individual citizens with a defensive character against state intrusions,[437] understand them as objective principles[438] with a proportional nature. While, in the historical examination, we could observe how progressively the BVG gained the characteristics of a constitutional court with an intervenient and activist approach, whose grounds could already be found in the circumstances of a Germany leaving a period of crisis and entering into a process of re-democratization, the transformations in German legal dogmatics showed how this political character could be operationalized in decision-making. These two characteristics, among the others we examined, favored the deployment of balancing, whose features adapted perfectly for the purpose of basic rights as objective principles and for the constitutional court's more intervenient and activist role towards the main problems of society.

If balancing, however, served as an adequate instrument for this new German constitutionalism, it needed to be systematized in rational grounds, even to provide a greater degree of legitimacy to the BVG's decisions. When balancing started to appear as an element of the principle of proportionality, before concentrated on the principles of suitability and necessity, it qualitatively gained the possibility of being presented in a structural framework that could, in theory, provide it with some rational standards for decision-making. Although this chapter did not deeply explore these rational standards – which will be done further[439] – it discussed how the principle of proportionality, with its triadic structure, assimilated balancing as an element that seems to best handle the dilemmas of these characteristics of an activist constitutional court. Indeed, with balancing in the structure of the principle of proportionality, the BVG obtained a justificatory methodology to do politics through decision-making. But, naturally, this movement is not exempt from the most incisive criticisms, and this chapter ended by focusing on some relevant

[437]See Schlink, "German Constitutional Culture in Transition," 713.

[438]See Ibid.

[439]See the fourth chapter.

German constitutional scholars – Ernst-Wolfgang Böckenförde, Bernhard Schlink and Friedrich Müller – in order to demonstrate that, even though this reality is somehow consolidated in German constitutional culture, it raises many doubts whether it is not indeed leading to a constitutional court that seriously threatens constitutional democracy, both because it strongly weakens the concern for keeping consistent the system of rights and because, in this way, it threatens the healthy relationship among the institutional powers.

While in this chapter we examined German constitutionalism, the next one will explore how this German constitutionalism has directly influenced other constitutional reality. In this respect, to examine Brazilian constitutionalism is especially interesting because of the possibility of extending the perception of how the constitutional courts' activism is closely associated with the increasing deployment of the flexible structure of balancing. As a recent democratic regime, with many particularities of a willingness to overcome the authoritarian past, the connections with German constitutionalism, in a sense, are very remarkable, not only in this dualism between law and politics but also in this attempt to "rationalize" decision-making. Particularly, it is very notable how Robert Alexy's *Theory of Constitutional Rights* and the BVG's decisions were used as primary sources for some relevant Brazilian constitutional court's decisions. For this reason, the next chapter will complement the empirical reference to see how balancing has a close connection with constitutional cultures where their constitutional courts play an activist role. Again, the emphasis is on history and on the aim to "rationalize" constitutional decisions. The emphasis is again on the concept of rationality this other constitutional reality, through balancing, seems to endorse.

Chapter 3
Balancing Within the Context of Brazilian Constitutionalism: The *Supremo Tribunal Federal's* Shift to Activism

Abstract As an interesting example of the current movement towards judicial activism, the Brazilian Federal Supreme Court (*Supremo Tribunal Federal*) reveals how a different constitutional reality can lead, by using similar methodologies and interpretations of basic rights, such as the principle of proportionality (and thus balancing) and the idea of subjective rights as objective principles of a total legal order, to comparable outcomes to those observed in Germany. Except for the untranslatable differences between both countries, it is possible to verify that, also in Brazil, there is a growing process of juridification of politics exactly after a period of authoritarianism and the rebirth of a constitutional democracy. This movement is also followed by the attempt to "rationalize" decision-making, providing thereby decisions that seem not only more legitimate but also the rational result of a careful interpretation of the "Guardian of the Constitution". The question, nevertheless, is how Brazilian democracy, which must preserve the principle of separation of powers, deals with this reality, and how the exercise of citizenship is preserved in this process.

3.1 Introduction

"The Constitutional Court exists to make the most rational decisions."[440] Gilmar Mendes's words, the then Chief Justice of the Supremo Tribunal Federal (STF), the highest and constitutional court in Brazil, seems relevant to the debate on the rationality of balancing. Especially because of the increasing deployment of this methodology in Brazilian constitutional culture in the last years and, above all, the parallel activism by this court, which has gradually centralized many social and political matters under its scrutiny, there could be no better connection with

[440] Gilmar Mendes, interview by Izabela Torres, "Entrevista – Gilmar Mendes," *Correio Braziliense*, Brasília (August 17, 2008), translation mine.

J.Z. Benvindo, *On the Limits of Constitutional Adjudication*,
DOI 10.1007/978-3-642-11434-2_3,

some of the conclusions drawn from the last chapter about the reality of German constitutional culture. Although any immediate comparison in this subject matter could lead to a simplification of the historical and legal differences between both countries, it is possible to outline some remarkable connections when balancing is at issue. The quest for rational decisions, on the one hand, and the constitutional court's activism, on the other, seem usually connected when this theme appears. This connection, indeed, makes the analysis of the institutional history a fundamental element to grasp why, also in Brazilian reality, the idea that the constitutional court is the "Guardian of the Constitution"[441] and, as such, acts as its last interpreter, extending thereby to its realm of authority the discussion of present and future problems of society, is real and widely accepted. The premise of being the "Guardian of the Constitution" has become, rather than a real concern for keeping consistent the system of rights, a justification for gradually transforming the STF into the center of the main political and social debates in Brazil. This reality, at first glance, might seem natural and inevitable, as Justice Mendes endorsed by remarking that "any polemic issue has a scheduled meeting with the Supremo Tribunal Federal."[442] On the other hand, nevertheless, it could reflect a serious sign of a relevant transformation in Brazilian constitutional culture that raises doubts whether this movement respects the principle of separation of powers and, as such, is legitimate or not.

This is why, by following the premises of the last chapter, when we examined the German constitutionalism as a means to conclude that the BVG has continuously acted as a "forum for the treatment of social and political problems,"[443] we can verify that the STF's reality also engenders rather similar problems and conclusions. Except for some important particularities derived from the individual system of rights and the institutional history of both countries, as well as the intensity of these developments, we can also remark that, more and more, the STF has become a casuistic court with an activist approach to the most variable themes of social life. This characteristic is carried out, in some relevant constitutional complaints, through the examination of the possible effects of the decision for the society as a whole and the observance of whether the expected goals are suitable, necessary, and proportional in the narrow sense. Briefly, the STF, highly influenced by German constitutionalism, increasingly deploys the principle of proportionality, and particularly balancing, as the sign of a "rational" justification for this shift to a more political and intervenient attitude in the institutional ground.

---

[441]See the second chapter.

[442]Gilmar Mendes, interview by Izabela Torres, "Entrevista – Gilmar Mendes," *Correio Braziliense*, Brasília (August 17, 2008), translation mine.

[443]Bernhard Schlink, "German Constitutional Culture in Transition," *Cardozo Law Review* 14 (1993): 729.

This "rationalization" of adjudication, and particularly judicial review – an intent Justice Mendes's words clearly expressed above – may serve as the counterpart of a movement that makes the STF a fundamental piece in the representation of the social will.[444] This movement is even more expressive as we notice the existence of a deficit of political representation of the Brazilian population in the parliament, causing thereby a mobilization towards the judiciary, especially the STF, in order to discuss the polemical and complex questions legislators[445] usually disregard. Therefore, the debate on shaping a constitutional court with the premise of being the guardian not only of the legal order but also of the social values, which brings forward the idea of subjective rights as objective principles and a methodology upholding this premise (balancing, for example), also raises, in Brazilian reality, the controversy about the dualism between law and politics in constitutional adjudication. There is also here the troublesome perception of a politicization of judicial review followed by the intent to justify this movement "rationally."

Nonetheless, this theme is rarely faced. In a degree more prominent than in Germany, which also suffers from the loss of the critique of the BVG's decisions, Brazilian constitutional scholarship basically conducts itself as a mere observer and worshiper of the STF's decisions. As Baracho Júnior remarks, "the reflection on the STF's activities is a scarce effort in Brazilian constitutional theory."[446] Most publications on the activities of this court merely reproduce the contents of its decisions, without providing simultaneously a critical review of those problematic themes regarding its legitimacy and the risks of this movement for the separation of powers. Moreover, if we examine the great majority of the edited books on the new methodologies the STF deploys, particularly the principle of proportionality, the same conclusion applies here: they reproduce the contents of those decisions and, simultaneously, make an examination, which is normally very brief and simplified, of Robert Alexy's *Theory of Constitutional Rights* (*Theorie der Grundrechte*),[447] as the main theoretical source to systematize the principle of proportionality, and balancing in particular. There is almost neither a critical analysis of the STF's decisions deploying this principle nor a rigorous critical review of Alexy's thinking on the themes of legitimacy and separation of powers. While the STF's decisions

---

444See Marcelo Andrade Cattoni de Oliveira, "O Projeto Constituinte de um Estado Democrático de Direito," in *15 Anos de Constituição: História e Vicissitudes*, ed. José Adércio Leite Sampaio (Belo Horizonte: Del Rey, 2004), 149.

445See Enzo Bello, "Neoconstitucionalismo, Democracia Deliberativa e a Atuação do STF," in *Perspectivas da Teoria Constitucional Contemporânea*, ed. José Ribas Vieira (Rio de Janeiro: Lumen Juris, 2007), 33.

446José Alfredo de Oliveira Baracho Júnior, "O Supremo Tribunal Federal e a Teoria Constitucional," in *15 Anos de Constituição: História e Vicissitudes*, ed. José Adércio Leite Sampaio (Belo Horizonte: Del Rey, 2004), 211, translation mine.

447Robert Alexy, *Theorie der Grundrechte* (Frankfurt a.M: Suhrkamp, 1994). See the next chapter.

become the new law, Alexy's thinking turns into the new methodological conception of truth in Brazilian constitutional reality.

This movement is evidently the consequence of an intense transformation of the STF's structure and nature after the promulgation of the Brazilian Federal Constitution in 1988, the following legal statutes regarding the procedures carried out by this court, and its decisions. The STF's particularity of not necessarily requiring an agreement among the majority of its members respecting the court's opinion, but merely an agreement with the decision itself, makes, at any rate, even more complicated to visualize a methodological tendency of this court. In the STF's judgments, only the Justice responsible for the case is obliged to justify his vote, if his opinion prevails, or another Justice, who manifests the contrary majority opinion, will assume this duty. However, in the most controversial and complex cases, it is very common that each one of the eleven Justices expresses his opinion, which, although leading to the same decision, can have a completely different justification and even incompatible premises. It is, by reason of this characteristic, almost impossible to account for the existence of a court's tendency. As a consequence, their personalities, idiosyncrasies and political positions become a real concern for most empirical investigations of those tendencies, a situation, however, that makes more difficult to visualize any prognostic of the future decisions and even more complicated the desire for a greater consistency of those decisions over the years.

Apart from this structural difficulty in achieving consistency in the STF's decisions, it is possible nonetheless to verify, among some of its members, the increasing attempt to "rationalize" the decisions through the deployment of well-known methodologies designed to solve the most controversial and complex cases, such as balancing. By "rationalizing" the decision through a determined methodology, the decision may achieve a greater degree of legitimacy. Still, on the other hand, this "rationalization" and even the quest for legitimacy do not appear to be concerned with the quest for normative consistency, which could be regarded as a paradox. The deployment of a methodology, in this regard, may be more in accordance with a certain premise of rationalization than indeed with the search for normative consistency. The problem of a lack of normative consistency refers, accordingly, not merely to the court's structure. The deficit of normative consistency emerges from the fact that the court gives the impression, in its decisions, that this problem is not really a chief concern.

It is difficult to argue likewise that the court is nowadays, de facto, deploying dogmatic concepts and scholarly interpretations of particular issues. What prevails is, more and more, albeit the difficulties in visualizing a real court's tendency, a casuistic jurisdiction without this meaning a coherent development of rights, for it is strongly grounded in teleological assumptions pointing to what is best according to a certain interpretation of the governmental will, or the political purpose of providing a solution that is best for the overall society. This reality, moreover, is even more noticeable through legal statutes that progressively provide the STF with the instruments to exercise its activism towards the most variable themes of social life, such as the expansion of the concentrated and abstract judicial review whereby this court has been regarded as a negative[448] and even positive legislator,[449] the introduction of the binding precedents (*Súmulas Vinculantes*), the modulation of the effects of the decision, among many others.

These are some signs of this institutional movement towards the increasing deployment of the principle of proportionality, and particularly balancing, in Brazilian constitutional reality. They are signs that interconnect with the STF's more activist approach, which gains space in the vacuum of legitimacy of a discredited parliament and of an almost inexistent scholarly critique of its decisions. It is interesting, besides, to observe that the shift to activism and the corresponding development of a value-based approach to adjudication – with the deployment of methodologies reinforcing arguments with a clear political intent – also in Brazil coexist with the evolution of the democratization process after a period of authoritarianism. If this is merely a coincidence, it is nevertheless arguable that judicial activism seems to be associated, at least in the constitutional realities here examined, with the challenge to establish a new institution that could catalyze the exercise of citizenship still not practiced enough by the overall society. Moreover, to legitimate and "rationally" justify this new position, inevitably leading to the undertaking of the role of defender of the values of this recent democratic period, it is necessary to provide methods that, in principle, can expose reflected arguments the population is able to widely accept. Decisions, for this reason, must be not only right but also correspond to the axiological parameters the society reasonably accepts. The court, consequently, appears to democratically anticipate the exercise of citizenship that, otherwise, other institutional channels, by expressing the political will and connecting directly to the social claims (parliament and government), must be representing.

Thus, this chapter begins with the perception that the first stage of any analysis of the increasing deployment of the principle of proportionality, and specially balancing, in Brazil is associated with the history of democratization after the military regime, which starts with the promulgation of the Federal Constitution of 1988 (Sect. 3.2). In this regard, a fundamental topic that is connected to the STF's activities is the advance of the abstract system of judicial review and the consequent weakening of its diffuse counterpart. Not only were many instruments created with the purpose of expanding this court's realm of activity, but also the STF assumed, during the consolidation of these new mechanisms, a more activist approach. Indeed, in the particular case of Brazilian reality, it is not possible to dissociate the investigation of the STF's shift to activism from this radical

---

[448]It is possible to associate the idea of negative legislator with Hans Kelsen's discussion of the constitutional courts' role, a simple consequence of his premise that adjudication is not qualitatively different from legislation, except for creating singular norms for the case. For this purpose, see Hans Kelsen, *Wer soll de Hüter der Verfassung sein?* (Berlin: Rotschild, 1931); Hans Kelsen, "Wesen und Entwicklung der Staatsgerichtsbarkeit." Berichte [der] Verhandlungen der Tagung der Deutschen Staatsrechtslehrer zu Wien am 23. und 24. April 1928, *Veröffentlichungen der Vereinigung der deutschen Staatsrechtslehrer* 5 (1929): 30–88, 117–123.

[449]See Bello, "Neoconstitucionalismo, Democracia Deliberativa e a Atuação do STF," 31 ff.

transformation in the mechanisms of judicial review. This is why we shall stress here this particularity of Brazilian constitutionalism, unlike we did in the case of Germany, where the judicial review of the Basic Law is historically a BVG's exclusive duty.[450] If this section refers to the analysis of this movement towards the STF's concentration of powers, the following one will stress which are the implications of this concentration for decision-making, in order to verify how activism and the deployment of balancing, now encompassed by the framework of the principle of proportionality, have been gradually, but intensively, verified (Sect. 3.3). The purpose here is to explore how the political discourse walked side by side with the expansion of this methodology, especially in a moment when the abstract system of judicial review was reinforced in Brazilian constitutional reality.

3.2 The Supremo Tribunal Federal in the Democratization Process: the Federal Constitution of 1988 and the Opening to Activism

It is not a simple enterprise to grasp how the STF, progressively, since the rebirth of the democratic period in 1988, has undertaken the role of defining many relevant political and social matters in Brazilian reality, and of justifying this new role with the interest in providing "rational" arguments in decision-making. After all, the STF did not become automatically, with the Federal Constitution of 1988, a constitutional court as the ones in continental Europe, for instance, the German BVG, which concentrates in its hands the judicial review. Indeed, it kept almost the same structure and authority of the previous years, as well as a system of judicial review rather inspired by the American model – even though in an eclectic and limited fashion[451] – according to which judicial review can be carried out by any judge in a *singular* and *concrete* case, and the STF, as the highest

[450]According to the German Basic Law: Art. 100 (1): "If a court concludes that a law on whose validity its decision depends is unconstitutional, the proceedings shall be stayed, and a decision shall be obtained from the Land court with jurisdiction over constitutional disputes where the constitution of a Land is held to be violated, or from the Federal Constitutional Court where this Basic Law is held to be violated. This provision shall also apply where the Basic Law is held to be violated by Land law and where a Land law is held to be incompatible with a federal law."

[451]For a detailed analysis of this American influence in the diffuse model of judicial review and the limits of its introduction in Brazilian pre-Republican reality, see Álvaro Ricardo de Souza Cruz, "Habermas, Ação Estratégica e Controle de Constitucionalidade," in *15 Anos de Constituição: História e Vicissitudes*, ed. José Adércio Leite Sampaio (Belo Horizonte: Del Rey, 2004), 219–280.

federal court of appeal, through the extraordinary appeal[452] (*Recurso Extraordinário*),[453] finally examines the constitutionality of the legal statute.[454] The so-called diffuse model of judicial review, traditionally present in Brazilian reality since the STF's origin in 1891[455] – not followed nevertheless by any similar mechanism as the American *stare decisis* – was kept almost intact in the new Constitution. The traditional and already experienced system, accordingly, weakened the claim, clearly observed in the debates before and during the National Constituent Assembly of 1987 and 1988,[456] to create a constitutional court as the ones in Europe. Moreover, it also limited any attempt to transform the STF's decisions into binding decisions to the other branches of the judiciary and the government. The premise of free appreciation of the constitutionality of a legal statute by all branches of the judiciary was regarded as a compromise with the longstanding characteristic of Brazilian constitutionalism, and mostly with the defense of democracy, as long as it could foster the dialogue among the distinct institutional actors and provide a more direct contact with the population. For this reason, if it is to investigate how the STF became a political activist court, the central aspect is to comprehend, first, how it instrumentally achieved this quality,

[452]The Extraordinary Appeal is the name of the constitutional appeal filed against a lower decision, based normally on a possible violation of the Constitution (art. 102, III, of the Constitution of 1988).

[453]The only possibility to extend, in the diffuse model, the effects of the decision to other similar cases occurs with the Senate's participation, which can suspend the execution of the legal statute. This mechanism has existed in Brazilian constitutionalism since the Constitution of 1934 (art. 90, IV). It is now established in art. 52, X, of the Federal Constitution of 1988. Gilmar Mendes provides a detailed analysis of this institute, who, nevertheless, as a supporter of the abstract and concentrated model of judicial review, understands it as still existing merely due to history, whose function, especially after the possibility of this court to suspend the efficacy of the law in the abstract judicial review, is only to give publicity to the STF's decisions. For him, the STF's own decisions have already the normative power to suspend the law, thereby conferring on the Senate solely the duty to publish the decision. For this purpose, see Gilmar Mendes, "O papel do Senado Federal no Controle de Constitucionalidade: um Caso Clássico de Mutação Constitucional," *Revista de Informação Legislativa*, no. 162 (April, June 2004): 149–168. Against this understanding, by presenting a favorable approach to the Senate's suspension of the legal effects, see Lenio Luiz Streck, Marcelo Andrade Cattoni de Oliveira and Martonio Mon'Alverne Barreto Lima, *A Nova Perspectiva do Supremo Tribunal Federal sobre o Controle Difuso: Mutação Constitucional e Limites da Legitimidade da Jurisdição Constitucional*, http://www.mundojuridico.adv.br/ sis_artigos/artigos.asp?codigo=912 (accessed July 7, 2009).

[454]For an investigation of the STF's history, and particular this influence of the American model of judicial review, see Lêda Rodrigues Boechat, *História do Supremo Tribunal Federal* (Rio de Janeiro: Civilização Brasileira, 1991); Oscar Vilhena Vieira, *Supremo Tribunal Federal: Jurisprudência Política* (São Paulo: Malheiros, 2002); Marcelo Paiva dos Santos, *A História Não Contada do Supremo Tribunal Federal* (Porto Alegre: Sergio Antonio Fabris, 2009); Gilmar Mendes, *Controle de Constitucionalidade: Aspectos Jurídicos e Políticos* (São Paulo: Saraiva, 2004).

[455]Art. 59, § 1o., *b* of the Constitution of 1891.

[456]See Oscar Dias Corrêa, "O 160o. Aniversário do STF e o Novo Texto Constitucional," *Arquivos do Ministério da Justiça*, no. 173 (1988): 67.

and, second, how it transported it to the practice of decision-making. Our concern in this section is with the first aspect.

This historical movement is very instructive to understand that the idea that the STF is the "Guardian of the Constitution," a characteristic already observed since the beginning of the Brazilian Republic in 1891,[457] was intimately associated, in the debates on the new Constitution, with the premise of ensuring the reliability of the diffuse system of judicial review. Nonetheless, this quality of exercising the protection of the Constitution by stating the unconstitutionality of a legal statute[458] in the ordinary diffuse system was intensively threatened by the establishment – already existent in a very limited fashion in the previous Constitutions,[459] though – of an abstract and concentrated model of judicial review. Since the Constitution of 1988, the traditional diffuse and concrete model has coexisted with a strong concentrated model of judicial review, which led, for instance, Justice Gilmar Mendes, one of the main supporters of this new mechanism, to say that "from 1988 onwards, however, there is only sense in thinking of a mixed system, if one is conscious that the basis of this system centers on the concentrated model."[460] Unlike the diffuse system, the

---

[457]See Oscar Dias Corrêa, *O Supremo Tribunal Federal, Corte Constitucional do Brasil* (Rio de Janeiro: Forense, 1987), 6.

[458]In Brazilian constitutional reality, it is possible the judicial review of a constitutional amendment, when it violates the Federal regime, the basic rights and guarantees, the separation of powers or the direct, secrete, universal and periodic vote, according to art. 60, §4 of the Constitution of 1988.

[459]The Constitution of 1934 (art. 12, § 2) introduced the possibility of the Attorney-General of the Republic (*Procurador Geral da República*), responsible at that time for judicially representing the interests of the federal government, to file an action (*Representação Interventiva*) directly in the Supremo Tribunal Federal in order to question any action or omission against the fundamental principles of the federative order (art. 7, I, *a* to *h*), which could lead to a federal intervention in the state. The Constitution of 1946 improved this model, inasmuch as the Attorney-General of the Republic could file the *Representação Interventiva* to question the constitutionality of state laws (art. 8) that infringed some *sensible principles* (republican representative system, separation of powers, municipal autonomy, judiciary's guarantees, periodicity of elections, etc, according to art. 7, VII). The declaration of unconstitutionality, nevertheless, did not necessarily lead to the intervention in the state, for the simple decision already had the power to suspend the effects of the state law. This model was then expanded in 1965, with the constitutional amendment 16/65, according to which exclusively the Attorney-General of the Republic could file an action to question the constitutionality of not only state laws but also federal legal statutes (art. 101, n. 1, *k*). Unlike the previous system, now the questions were not merely related to a certain violation of sensible principles involving a conflict between the Federal Union and a state, but rather the defense itself of the Constitution against unconstitutional laws. It is important to mention that this system, whose decisions had an *erga omnes* effect, was born with the purpose of reducing the excess of appeals in the STF, and not, in fact, of expanding the possibilities of protection of individual rights (See, for this purpose, Vieira, *Supremo Tribunal Federal: Jurisprudência Política*, 120–126; Gilmar Mendes, *Jurisdição Constitucional: Controle Abstrato de Normas no Brasil e na Alemanha* (São Paulo: Saraiva, 2004), 23–38 and 64-77; Cruz, "Habermas, Ação Estratégica e Controle de Constitucionalidade," 245–257).

[460]Mendes, *Jurisdição Constitucional*, XII, translation mine.

court's decisions in the abstract judicial review could immediately consider a determined legal statute void, and therefore without any *erga omnes* effect. Besides, there would be no specific conflict between subjects, but a direct analysis of the constitutionality or unconstitutionality of a certain legal norm through a direct action filed by some authorities and institutions.[461] Therefore, if the creation of a constitutional court as those of Europe did not become true, at least a "compromised solution"[462] took place, a solution that led to the expansion of the STF's activist approach.

The implementation of a mixed model of judicial review in Brazilian reality seemed to be in accordance with the democratic purpose of providing the instruments to effectively exercise the protection of the Constitution to the court. More than before, it was necessary that the court could undertake its role of "Guardian of the Constitution" and thus act directly against any practice that offended the constitutional and democratic system. Especially if we remark how the discussions preceding the promulgation of the Constitution of 1988 were civilly representative, democratically legitimate,[463] and printed a strong lack of confidence in the ordinary legislator and the government, we can conclude that the idea that it was indispensable to provide the constitutional court with the instruments to safeguard the constitutional principles and practices was natural. The Constitution of 1988 expressed, *de facto*, this feeling of a new political and legal era in Brazilian society, a democratic process that should be preserved over the years, both by establishing subject and social rights never before imagined in that reality (it has

[461]According to the Brazilian Constitution of 1988: "Article 103. The following may file an action of unconstitutionality and the declaratory actions of constitutionality:

I – the President of the Republic;

II – the Directing Board of the Federal Senate;

III – the Directing Board of the Chamber of Deputies;

IV – the Directing Board of a State Legislative Assembly or of the Legislative Chamber of the Federal District;

V – a State Governor or the Governor of the Federal District;

VI – the Attorney-General of the Republic;

VII – the Federal Council of the Brazilian Bar Association;

VIII – a political party represented in the National Congress;

IX – a confederation of labor unions or a professional association of a nationwide nature (...)"

[462]Mendes, *Jurisdição Constitucional*, 38, translation mine.

[463]See Menelick de Carvalho Netto's analysis of the participation of many and distinct civil organizations in the preparatory works for the elaboration of the new Constitution, which was not concentrated on some personalities, but rather on the direct mobilization and participation of the population. There was a strong receptivity to this participation by the legislators and the internal legal statute of the constituent process. According to Carvalho Netto, "it was from this process, profoundly democratic, that the Constitution earned its original legitimacy, resulting from an authentic manifestation of the constituent power, by reason of the employed process" (Menelick de Carvalho Netto, "A Revisão Constitucional e a Cidadania: a Legitimidade do Poder Constituinte que deu Origem à Constituição da República Federativa do Brasil de 1988 e as Potencialidades do Poder Revisional nela Previsto," *Revista do Ministério Público Estadual do Maranhão*, no. 9 (2002): 45, translation mine).

one of the most extensive bill of rights in the whole world)[464] and a comprehensive range of mechanisms protecting the access to the judiciary.[465] It was, in many aspects, a rupture with the authoritarian past, when both the parliament and the government were strongly discredited,[466] thereby giving rise to social demands that resulted in the introduction of many civil rights, even to protect this process. The judiciary in general and the STF in particular, was then a fundamental piece of these new dilemmas the Constitution of 1988 unfolded and could thereby act as real legitimate protector of democracy by making decisions that, while based on principles, were consistent throughout history and also externally rationally justified.[467] The STF, in other words, should be an institution that would protect this reality against any possible reemergence of authoritarianism by upholding the Constitution and its principles. It should embody this democratic reaction against the authoritarian past by strengthening and enforcing the democratic Constitution.

Yet, notwithstanding this ample democratic movement with the promulgation of the Constitution of 1988, the judiciary could not adequately assume its duty of a real protector of the legal order nor did the STF really act as the "Guardian of the Constitution." The judiciary in general, strongly influenced by the authoritarian period, usually adopted either a legalist or, on the contrary, creative and rather voluntaristic approach to legal interpretation as a means to deal with legal

---

[464]Gilmar Mendes, *New Challenges of Constitutional Adjudication in 21st Century: A Brazilian Perspective*, Lecture presented in Washington (US), October 10, 2008, http://www.stf.jus.br/arquivo/cms/noticiaArtigo Discurso/anexo/Jurisdicao_Constitucional_ no_Seculo_XXI_v__Ing.pdf (accessed July 14, 2009).

[465]The Brazilian Constitution provides nowadays many mechanisms to access the judiciary in order to guarantee civil rights and the democratic process (*Ação Civil Pública, Ação Popular, Mandado de Segurança, Mandado de Injunção, Habeas Corpus, Habeas Data, Ação Direita de Inconstitucionalidade, Ação Direta de Inconstitucionalidade por Omissão Ação Declaratória de Constitucionalidade, Arguição de Descumprimento de Preceito Fundamental*, among others). For a detailed analysis of these actions, see Gilmar Mendes, *Jurisdição Constitucional: Controle Abstrato de Normas no Brasil e na Alemanha* (São Paulo: Saraiva, 2004); Gilmar Mendes, *Arguição de Descumprimento de Preceito Fundamental* (São Paulo: Saraiva, 2007); Gilmar Mendes and Ives Gandra da Silva Martins, *Controle Concentrado de Constitucionalidade* (São Paulo: Saraiva, 2009); Gilmar Mendes, *Direitos Fundamentais e Controle de Constitucionalidade*, (São Paulo: Saraiva, 2004); André Ramos Tavares, *Arguição de Descumprimento de Preceito Fundamental* (São Paulo: Atlas, 2001); Motauri Ciochetti de Souza, *Ação Civil Pública* (São Paulo: Malheiros, 2003); Jose Adonis Callou de Araújo Sá, *Ação Civil Pública e Controle de Constitucionalidade* (São Paulo: Del Rey, 2002); Luzia Nunes Dadam, *Ação Popular – Controle Jurisdicional* (Rio de Janeiro: Lumen Juris, 2000); Celso Agricola Barbi, *Mandado de Segurança*, (Rio de Janeiro: Forense, 2008); Hely Lopes Meyrelles, *Mandado de Segurança* (São Paulo: Malheiros, 2008); Heráclito Antônio Mossin, *Habeas Corpus* (São Paulo: Manole, 2008); J. E. Carreira Alvim, *Habeas Data* (Rio de Janeiro: Forense, 2001).

[466]See Lênio Luiz Streck, *Entrevista ao Conjur: Lênio Streck Fala sobre o STF*, http://www.conjur.com.br/2009-mar-15/entrevista-lenio-streck-procurador-justica-rio-grande-sul (accessed July 14, 2009).

[467]See Jürgen Habermas, *Between Facts and Norms* (Cambridge: Polity Press, 1996), 199. This debate will be examined, with more details, in the sixth chapter.

authoritarianism.[468] In this scenario, the real concern for the construction of a consistent system of rights did not materialize. The quest for consistency in decision-making was usually confronted, right after the promulgation of the Constitution of 1988, with an authoritarian past that could not furnish many parameters for decision-making in the new democratic regime, especially if we consider the extensive range of basic rights, on the one hand, and the urgent needs of a society still marked by social inequalities, on the other. Moreover, this lack of consistency also derived from the incapacity of many judges to cope, on the one hand, with the new demands of the democratic regime, and, on the other, with a real understanding and knowledge of how constitutionalism and democracy should be interconnected with each other in their interpretation of legal statutes, a complex task if we consider the inertial effect of a practice framed by a past aimed at having judges literally acting in conformity with the governmental will.[469] The authoritarian legacy, which, as Oscar Vilhena Vieira remarks, was characterized by a "great silence of the Supremo Tribunal Federal and the tribunals in general in working in pursuit of the reconstruction of the *rule of law* (*Estado de Direito*) and democracy,"[470] weakened the space for an active construction of a consistent system of rights, for it carried with itself a deficit of constitutional-democratic practice and a deficit of constitutional-democratic knowledge.

The conjunction of both deficits established the grounds for the judiciary's gradual movement towards activism, without being followed, nonetheless, by a serious interest in providing and constructing a real consistent system of rights. Against the legalist and passive approach of the previous years, when the military regime subverted the judiciary, and also on account of "visible deficiencies in the elaboration and implementation of public policies necessary to make fundamental rights effective"[471] in the new constitutional order, the judiciary, especially the lower courts, progressively undertook, as their realm of responsibility, the duty to implement those public policies through decision-making. The judiciary's autonomy and the guarantees the Constitution of 1988 created gave more instruments and naturally freedom to the exercise of this more intervenient attitude in the definition of policies also. This is why Lênio Streck remarked that "the judges (and the scholarship is also guilty), who now should apply the Constitution and filter the

---

[468]See Streck, *Entrevista ao Conjur: Lênio Streck fala sobre o STF*.

[469]Indeed, during the military regime, not only were the constitutional guarantees of the judiciary suspended, but also some crucial legal statutes, such as the Institutional Acts (which were authoritarian legal statues the government directly enacted usually restricting the exercise of basic rights), were not under the judiciary's authority. Besides, no rarely were judges, and even STF's Justices, automatically compulsorily retired. This circumstance promoted a judiciary's silence, and the STF, after these confrontations, no longer opposed any resistance to the military regime. See, for this purpose, Vieira, *Supremo Tribunal Federal: Jurisprudência Política*, 123–126.

[470]Ibid., 123, translation mine.

[471]Mendes, *New Challenges of Constitutional Adjudication in the 21st Century: a Brazilian Perspective*.

bad laws, that is, those unconstitutional ones, started to think they knew more than the constitutional framer. We moved, accordingly, from stagnation to activism, understood as the replacement of law by the judge's subjective judgments."[472]

Indeed, many judges coming from the authoritarian period, right after the Constitution of 1988, felt, for the first time, that they could assume a more activist posture and, contrary to the old times, an activist posture against the government. In addition, more and more, the political and social actors began to file lawsuits concerning their basic rights (which are meticulously detailed and defendable through numerous instruments in the new Constitution) and, in an exponential manner, lawsuits against governmental policies and economic programs. This two-sided movement created frequent situations where the judiciary intervened directly in the definition of public policies by ordering immediate governmental action, even in the economic area.[473] Furthermore, if we investigate the increasing number of problematic economic plans and the enactment of *Provisional Measures* (*Medidas Provisórias*)[474] by the government, many of them of questionable constitutionality, then the number of lawsuits and, consequently, judicial activism seriously gained an impulse. This activism, besides, since the lower courts, could already promote a decision, albeit appealable, stating the unconstitutionality of a certain legal statute, and therefore its invalidity for all effects in that singular case.

---

[472]See Streck, *Entrevista ao Conjur: Lênio Streck fala sobre o STF*.

[473]See the examples of judicial order to raise the pension of retirees paid by the Brazilian Social Security Service (*Instituto Nacional de Seguridade Social*) in 1992, whose disobedience caused the imprisonment of the president of that institution; the prohibition of the Brazilian Central Bank (*Banco Central do Brasil*) to put into practice a program of financial reorganization framed to prevent a crisis of confidence in Brazilian banking industry; the imprisonment of the President of the Brazilian Central Bank, in 1991, by virtue of his refusal to suspend the dissolution of a brokerage firm; the order to reduce the monetary correction of incoming tax penalties, according to the technical definition deemed correct by the federal revenue service; the judicial liberation of the blocked Cruzados Novos (Brazilian currency at that time) during the government of President Fernando Collor de Mello, among others. For an accurate analysis, see Marcus Faro de Castro, "The Courts, Law and Democracy in Brazil," *International Social Science Journal*, no. 152 (June 1997): 241–252; Vieira, *Supremo Tribunal Federal: Jurisprudência Política*, 135 ff.

[474]This reality is even more evident, if we remark that, in Brazil, there is the *Provisional Measure* (*Medida Provisória*), which, although being a mechanism for the government's legal production to be used merely in situations of relevance and urgency, has been widely employed in many other circumstances with the parliament's collaboration. The *Provisional Measure*, nonetheless, does not pass through the procedures where a more accurate debate on the constitutionality of its contents takes place, as it happens with legal statutes. Indeed, in Brazilian history after the Constitution of 1988, the use of *Provisional Measures* has become a generalized mechanism of legislation, which has often been passively accepted by the parliament, and not really controlled by the STF in what refers to its constitutional requirements of relevance and urgency (art. 62 of the Constitution of 1988). See, for this purpose, Vieira, *Supremo Tribunal Federal: Jurisprudência Política*. 219. For a critical analysis of the institutional disrespect for the due process of law and the constitutional process within this context, see Marcelo Andrade Cattoni de Oliveira, "Devido Processo Legislativo e Controje Jurisdicional de Constitucionalidade no Brasil," in *Jurisdição Constitucional e Direitos Fundamentais*, ed. José Adércio Leite Sampaio (Belo Horizonte: Del Rey, 2003).

In the STF, however, this movement started very gradually, but it has especially intensified in the last years. As shown, this court already inherited from the previous period a structure and a composition that were marked by a strong silence on the abuses of the military regime, and one could even argue that it acted submitted to this governmental power.[475] There was not, accordingly, an immediate connection between that democratic movement in the constituent process and this court's practice, although its authority for judicial review had increased. In reality, in the years immediately following the Constitution of 1988, the STF had a very timid activity in distinct areas of basic rights and, in many cases, even created legal interpretations and precedents to avoid entering into some sensible areas that could expose a judicial encroachment upon the other powers. For instance, mostly grounded in an economic standpoint, the STF never held the claim to federal intervention when a particular state did not honor the individual's alimony credits (*precatórios alimentícios*) or even transformed the *Writ of Injunction*[476](*Mandado de Injunção*), a powerful instrument to make effective an individual right in case of legal omission,[477] and the *Direct Action of Unconstitutionality due to Omission*

---

[475]See Vieira, *Supremo Tribunal Federal: Jurisprudência Política*, 125; Cruz, "Habermas, Ação Estratégica e Controle de Constitucionalidade," 251–257.

[476]The *Writ of Injunction* is a constitutional writ whose purpose is to preserve the exercise of subjective rights and freedoms, as well as prerogatives inherent to nationality, sovereignty and citizenship, in case of a lack of regulatory norm for this purpose. The court, at the beginning of its activities after the Constitution of 1988, had a very restrictive attitude in this matter by stating that, if the writ were granted, the effect was merely an order to the legislator to take the appropriate measures, without any sanction (See MI n. 107, published on 08.02.1991). This understanding, nonetheless, began to change over the years. In the judgment of the MI n. 283 (published on 10.02.1992), the court determined a deadline to correct the omission caused by the legislature's delay, establishing besides the sanction of considering the legal rights, now denied due to omission, automatically granted. Similar understanding happened in the analysis of the MI n. 232 (published on 03.27.2002). The most radical innovation, nonetheless, would happen only in 2007, when the STF, in the judgment of the MI n. 670 (published on 10.31.2008) and MI n. 708 (published on 10.31.2008), determined that, by reason of an omission regarding the regulation of public servants' strike, the Law n. 7.783/89, applicable to strikes in the private sector, should be extended to the public domain where appropriate. As Gilmar Mendes remarks "the Court, moving away from the course initially followed of attaining to declare the existence of legislative omission and issuing a specific regulating norm, without any commitment to the exercise of a legislative function, began to accept the possibility of provisory regulation by the judiciary itself" (Gilmar Mendes, Constitutional Jurisdiction in Brazil: The Problem of Unconstitutional Legislative Omission, http://www.stf.jus.br/arquivo/cms /noticiaArtigoDiscurso/ anexo/Omisao_Legislativa_ v__Ing.pdf (accessed July 7, 2009), 12).

[477]According to Gilmar Mendes, "the injunctive writ is granted based on the constitution whenever there is a lack of implementing rule that makes it impossible to exercise constitutional rights and freedoms, as well as prerogatives that are inherent to nationality, sovereignty and citizenship. Thus, the injunctive writ must be aimed at *non-compliance with the constitutional duty to legislate*, which in some way may affect rights that are ensured by the constitution (*lack of a regulatory norm that makes it impossible to exercise constitutional rights and freedoms and prerogatives that are inherent to sovereignty and citizenship*). Such omissions may have either an *absolute or total* character or be *partial* in nature" (Gilmar Mendes, *Controlling Constitutionality in Brazil*, Lecture

(*Ação Direta de Inconstitucionalidade por Omissão* – ADIo),[478] into almost useless instruments.[479] Also in the realm of abstract judicial review, this court was very restrained and, in a very slow fashion,[480] judged lawsuits in this area. As Marcus Faro de Castro remarks, "the Supreme Court [seemed] relatively impervious to pressures for the expansion of judicial power"[481] and "largely refrained from resorting to available remedies in order to expand its power by the cumulative articulation of substantive doctrine in case law."[482] On the other hand, since the STF is likewise a court of appeal, the argument referring to the large amount of constitutional complaints – exponentially growing over the years – gained relevance. As a result of several unconstitutional legal statutes and different economic plans after the Constitution of 1988, the STF spent most of its effort judging questions regarding these matters,[483] which brought about a real crisis in its capacity to protect the Constitution from its violations and mostly diminished its potentiality to safeguard the new individual and social rights that were enhanced or introduced in the new constitutional model.

While there was a strong expansion of lawsuits in the judiciary, we could thus picture two simultaneous developments: first, a growing activism, particularly in the lower courts, which directly confronted many political intents and projects, and also superior judicial decisions, without being followed, however, by a real concern for the consistency of the system of rights; second, a rather restrained and inexpressive STF, which did not act as a real "Guardian of the Constitution" and even created mechanisms to make some constitutional rights ineffective, either by virtue of the

presented at Harvard Law School: http://www.stf.jus.br/arquivo/cms/noticiaArtigo Discurso/anexo/Controle_ de_Consti tucionalidade_v__Ing.pdf (accessed July 14, 2009), 4).

[478]The purpose of this *Direct Action of Unconstitutionality due to Omission* (ADIo), an instrument of the abstract system of judicial review, is to make constitutional norms effective, informing thereby the appropriate authority to adopt the necessary measures and, in case of the government, to do it within thirty days. Right after the Constitution of 1988, the STF, when it favorably decided the case, merely notified the responsible organ of the decision and the need to provide the required measure as a means to overcome the omission (See ADI 2520, published on 03.15.2002; ADI 2525, published on 04.05.2002; ADI-MC 267, published on 05.19.1990; ADI-MC 1458, published on 09.20.1996). Moreover, in case the legislative procedure had already started, even when during many years (*inertia deliberandi*), the precedent was that ADIo, in these cases, could not be filed (See ADI 2.495 – published on 08.02.2002). This thinking changed recently. Now, with the STF's more active approach, this court has understood that, in some specific cases, it can establish a reasonable term for the parliament to supply the omission (See, for this purpose, ADI 3.682, published on 09.06.2007).

[479]See, for this purpose, the decisions MI n. 107; MI n. 232; MI n. 283; MI n. 419–9.

[480]Indeed, between 1988 and 1992, 113 *Direct Actions of Unconstitutionality* (*Ações Diretas de Inconstitucionalidade – ADI*) were filed by political parties, but only 6 were judged at the beginning of 1993. For this purpose, see Castro, "The Courts, Law and Democracy in Brazil," 245.

[481]Ibid., 246.

[482]Ibid., 243.

[483]Many of these appeals are the *Agravo de Instrumento*, which seeks only the discussion of whether the STF should judge the question, after the lower court had decided that it was not a constitutional matter, and therefore not on the STF's competence.

legacy of the authoritarian past, the still complacent attitude towards the government, or its complex characteristic of not being merely a constitutional court but also a court of appeal bound to judge an increasing number of constitutional complaints. There would be no other better political context to revitalize the frustrated proposal to transform the STF into a constitutional court as those of Europe that appeared during the Constituent Assembly of 1987 and 1988. There were: first, political motivation (it was necessary to stop the lower courts' growing activism, insofar as they were disturbing the politics in the different areas of social life); second, legal justification (it was indispensable to make coherent decisions and promote a consistent system of rights, on account of the lower courts' continuous inobservance of superior decisions, particularly when we remark the lack of any similar mechanism as the American *stare decisis*); and, third, the argument of efficacy (it was imperative to reduce the number of lawsuits in the whole judiciary and, above all, constitutional complaints in the STF). The lower courts' activism had to be controlled while the STF's authority had to be expanded to its real possibilities. Within the context where the STF was not totally yet detached from a certain complacency towards the government's will, this expansion of authority did not seem to be a danger to the exercise of politics by the other constitutional powers. In fact, it sounded more as a solution to many of the inconveniences of a widespread activist but inconsistent and institutionally unstable[484] attitude of the judiciary in general. Nonetheless, if this gave the impression of a necessary and reasonable answer, it also opened up the space for the installation of a progressive activism in the STF, which was not yet a serious actor in the new constitutional democracy and was popularly discredited in virtue of its unproductive capacity to handle the main questions of constitutionalism.

It is not simple to outline a specific date to visualize the beginning of this transformation, but some signs point to 1993 as the year when the STF gradually started to acquire some characteristics that would make it resembles an European constitutional court by concentrating in its hands the judicial review. As seen, there were already at least those three justifications that promoted a political context supporting the framing of instruments to concentrate judicial review. Besides, there was the process of constitutional revision originally introduced in the Constitution of 1988,[485] which had to take place five years after its promulgation. The year of 1993, accordingly, tied up a short experience of five years of judicial review in the new constitutional model, which was causing some conflicts with the government, with the possibility of constitutional revision, which relieved the burden of rigid formal requirements in comparison with the ordinary form of constitutional

---

[484]Marcus Faro de Castro argues that this judiciary, even though shifting to a more active and audacious attitude, did not lead to the improvement of a stable judicial power from the institutional standpoint, inasmuch as this shift was not followed by a clear definition of the judiciary's institutional role regarding its political participation in this new democratic system. For this purpose, see Castro, "The Courts, Law and Democracy in Brazil," 244 ff.

[485]According to art. 3 of the *Temporary Constitutional Provisions Act*, "the revision of the constitution shall be effected after five years as of its promulgation, by the vote of the absolute majority of the members of the National Congress in a unicameral session."

amendment. The conservative political parties took advantage of the constitutional revision to alleviate the so-called excesses of the democratic reaction during the Constituent Assembly of 1987 and 1988, and, within this context, the creation of the constitutional amendment establishing the *Declaratory Action of Constitutionality* (*Ação Declaratória de Constitucionalidade* – ADC), with serious outcomes in the realm of judicial review, materialized. The widespread possibility of the exercise of judicial review by the whole judiciary through the diffuse system appeared to be an excessive constitutional permission that, while still the ordinary and longstanding procedure, was not the most compatible with the new democratic regime, according to this view. The conservative thinking, by using those three justifications above, sustained that it was necessary to redesign the system of judicial review as a more adequate mechanism for democracy.

In reality, however, by diminishing the relevance of the diffuse system of judicial review – which, more than being the result of a longstanding tradition that places upfront the citizenship as its main reason,[486] has a more direct participation of the citizens and a widespread discussion of the constitutional rights through the different branches of the judiciary – one could say that this transition might have carried with itself "an antidemocratic prejudgment of not bestowing on the citizen the possibility of unmaking through its own initiative what was the legislator's work."[487] Indeed,

[486]Menelick de Carvalho Netto develops a critical analysis of the abstract system of judicial review, which, in his opinion, goes in the opposite direction of a constitutional experience already solidified in Brazilian reality. His words: "(. . .) I would like to highlight another challenge, not less serious, even though of internal origin: the importation through legal means of typical premises of the concentrated or Austrian judicial review. Our premises are of a tradition much older and also better in terms of experience and constitutional living than the German one, extremely more sophisticated and much more effective as a guarantee of the idea of concrete freedom and equalities. The basic principles of the diffuse judicial review are put in jeopardy, which constitute our heritage of more than one hundred years, a heritage expressing the comprehension of the Constitution as everyone's authorial work. The diffuse judicial review makes everyone of us an authorized interpreter of the Constitution, insofar as it did not authorize the legislature nor any other power to violate basic rights, and in which the constitutional matter, for it always relates to the basic rights of all of us, has recognized itself the authority for discussion, investigation and decision of this issue by any judge in any concrete case whatsoever appearing to him. It is important to remark the tremendous effort Peter Häberle endeavors to be able to affirm the existence of an open community of interpreters of the German Constitution, which, for us, is a premise, a basic point of departure for more than one hundred years. It is clear that it is no longer possible the artificiality of the Kelsenian standpoint, absolutely overcome, as Prof. Lênio Streck sustained. The authority in charge to apply the Constitution cannot do whatever he wants from the constitutional text; there are boundaries, which are intersubjectively shared, and the greatest guarantee of any constitution calls citizenship, a live and active citizenship, careful of its rights" (Menelick de Carvalho Netto, "A Hermenêutica Constitucional e os Desafios Postos aos Direitos Fundamentais, " in *Jurisdição Constitucional e Direitos Fundamentais*, ed. José Adércio Leite Sampaio (Belo Horizonte: Del Rey, 2003), 163, translation mine). Marcelo Cattoni de Oliveira expresses similar point of view. See Marcelo Andrade Cattoni de Oliveira, *Direito Processual Constitucional* (Belo Horizonte: Mandamentos, 2001), 212 ff.

[487]Paulo Bonavides, *Curso de Direito Constitucional* (São Paulo: Malheiros, 1994), 278, translation mine.

if we examine carefully, behind the enthusiasm for the abstract system of judicial review and those justifications, there were many other interests in play. It was not actually the greatest democratic intent, but rather a political purpose that had two relevant targets: to convince the STF of the need to expand its authority and expressivity – a purpose this court strongly aimed to achieve, as some of its activities already testified to[488] – as if the government were promoting its institutional improvement and prestige, while selling the image of modernity to the constitutional scholarship;[489] and to reduce the conflicts with the judiciary, which were causing disturbance to the government, while influencing more directly the STF either with political, axiological, economic justifications or arguments of efficacy, such as the one related to the reduction of lawsuits.[490] More than a mechanism to guarantee the Constitution and democracy – an argument we could use, for instance, in the case of other abstract actions originally present in the Constitution of 1988, such as the *Direct Unconstitutional Suit (ADI)*[491] and the *Direct Unconstitutional Suit due to Omission* (ADIo), but not exactly in the case of the original text of the *Direct Action of Constitutionality (ADC)*[492] – the abstract system of judicial review was strengthened

[488]The year of 1993 also points out the STF's more intervenient attitude towards the other powers. Marcus Faro de Castro stresses an event referring to a provisional order issued by Justice Marco Aurélio de Mello, after an action filed by the left wing parties, which ordered the immediate suspension of the legislative works concerning the process of constitutional reform. Although the other Justices revoked afterwards this provisional order, this attitude already reaffirmed its authority and had, accordingly, "sent a message across to politicians that the Court was a political power not to be underrated" (Castro, "The Courts, Law and Democracy in Brazil," 248). This development of a more intervenient attitude towards the other powers was then reestablished, with more frequency, in other relevant opportunities, many of them also in the economic area, such as the exclusion through a STF's decision of the salary raise ordered by the President Itamar Franco to the employees of public companies; the judgment related to the ex-President Fernando Collor de Mello's impeachment; the decision invalidating the governmental *Provisional Measure* (*Medida Provisória*) that prohibited indexation of the contracts after the creation of the *Plano Real*, which provided a general stabilization of the Brazilian currency and attacked the inflation; the decision provisionally suspending the legislative activity, which discussed the reform of the Social Security System, among others. For this purpose, see Ibid., 247 ff.

[489]See Cruz, "Habermas, Ação Estratégica e Controle de Constitucionalidade," 237.

[490]See Ibid., 237.

[491]According to Streck, Cattoni de Oliveira and Lima: "The ADI was the way the original framer had found to also involve the organized civil society in the protection of the Constitution. We can prove the objectivity of this assertion by reading the roll of actively legitimate actors to file the ADI: we can find in art. 103 of the Constitution of the Republic both representatives of the state and the society. Accordingly, we can remark the democratic-participative keynote of the Constitution, for the very Constitution does not embrace the society without its connections with the state and vice-versa" (Streck, Cattoni de Oliveira and Lima, *A Nova Perspectiva do Supremo Tribunal sobre o Controle Difuso: Mutação Constitucional e Limites da Legitimidade da Jurisdição Constitucional*, 6, translation mine).

[492]Indeed, if we could correctly say that the Constitution of 1988, in its original text, introduced the possibility of the abstract judicial review by means of a lawsuit (*Direct Unconstitutional Suit – ADI*) able to be filed by different authorities and sectors of civil society (art. 103 of the Constitution of 1988) – and this could represent an expansion of democratic participation in the realm of constitutional adjudication – the constitutional amendment n. 3/93, which introduced the

to facilitate the governability. More than the purpose of reinforcing the system of rights, this movement provided, more than ever, the means to do politics in constitutional adjudication. The abstract system of judicial review had to be redesigned to more flexibly conjoin law and politics.

In this scenario, the introduction of the *Declaratory Action of Constitutionality* with some new elements for judicial review – by the way, an unknown instrument in many other constitutional realities, particularly because of the risks of canonizing its interpretation over the years as well as expanding even more the constitutional court's authority[493] – was a viable and useful instrument for this goal. Created by the constitutional amendment 3/93, within the context of a fiscal reform proposed by the federal government,[494] it introduced, as one of its main characteristics, the fact that its definitive decision stating the constitutionality of a legal statute had a binding effect to the other branches of the government and the judiciary.[495] This

---

Declaratory Action of Constitutionality (ADC), established, however, that solely the President of the Republic, the Directing Board of the Federal Senate, the Directing Board of the Chamber of Deputies, and the General-Attorney of the Republic could file it. The other authorities and institutions of civil society – as the Federal Council of the Brazilian Bar Association and the confederations of labor unions or professional associations of a nationwide nature, for instance – were then excluded. This situation only changed after the constitutional amendment n. 45 in 2004, which identified the competent authorities and institutions of civil society of the ADC with those of the ADI (art. 103 of the Constitution of 1988, with the text of the amendment n. 45/2004).

[493]It is interesting to remark that, in many consolidated systems of judicial review as in Germany (see, for this purpose, BVerfGE 40, 88 (93ff)), Portugal and Spain, the declaration of constitutionality does not cause the binding effect, thereby not preventing someone from filing an action questioning the constitutionality of a legal norm before declared constitutional. There are many reasons for this understanding: (1) the declaration of constitutionality would make the open and variable contents of the constitutional principles static and rigid, causing therefore an impediment to its evolutive constitutional interpretation; (2) this declaration would confer on the constitutional court an uncontrollable power to infallibly decide on the constitutionality of a legal norm, turning then into an irresponsible arbiter of the constitution and an owner, instead of a serve, of the constitution; (3) a wrong decision in this matter could have the same value of a constitutional norm that was used as a parameter and could only be corrected by means of a constitutional amendment. For this purpose, see Streck, Cattoni de Oliveira and Lima, *A Nova Perspectiva do Supremo Tribunal sobre o Controle Difuso: Mutação Constittucional e Limites da Legitimidade da Jurisdição Constitucional*, 14–15.

[494]According to Álvaro Ricardo de Souza Cruz, "it is curious to remark that, once more, the institute of judicial review, born to safeguard the basic rights and the supremacy of the Constitution, comes out within a context of arbitrariness to respond to the fiscal interests of the federal government" (Cruz, "Habermas, Ação Estratégica e Controle de Constitucionalidade," 261, translation mine).

[495]The original text of art. 102, §2, of the Constitution of 1998 did not establish the binding effect for the Direct Unconstitutionality Suit. It was the constitutional amendment n. 45/2002 that instituted it. Its actual text is this one below: "Art. 102 (...)

§2 – Final decisions on judgments, pronounced by the Supreme Federal Court, in direct actions of unconstitutionality and in declaratory constitutionality suits, shall have force against all, as well as a binding effect, as regards the other bodies of the Judicial Power, as well as direct and indirect public administration, at federal, states and municipalities levels."

action, which only some authorities could file,[496] intended, as one of its purposes, to avoid that lower courts would not apply or judge unconstitutional certain federal legal statutes, intervening thereby directly in the implementation of federal public policies (in the case, primarily, policies related to the fiscal intents of the government), while restricting the number of extraordinary appeals (*Recurso Extraordinário*) to the court. It was, accordingly, a suitable instrument to put into action the purpose of concentrating judicial review in the STF's hands as well of controlling the lower courts' activism. This instrument, moreover, could avoid that many governmental measures – some of which very unpopular and certainly representing most of the lawsuits – had distinct results in the different branches of the judiciary. Hence, the STF, once the action was filed by one of those authorities, could state the constitutionality of the legal statute, and consequently all other judges and the government became obliged to obey its contents, suspending, in any case, the diffuse control of constitutionality carried out in the lower courts.[497]

The creation of the *Declaratory Action of Constitutionality* was a major aspect in this process of concentration of powers. It instituted the possibility of suspending the exercise of the diffuse judicial review – which is, as a matter of fact, the only way a common citizen can file a suit questioning the constitutionality of a legal statute – while transforming the abstract judicial review into a viable mechanism for the STF's endorsement of governmental policies, no longer questionable by the lower courts. This is why one could say that "the *Declaratory Action of Constitutionality* [opened] the space for the installation of a cooperation between the government and the Supremo Tribunal Federal."[498] If the constitutionality of a determined legal statute had been strongly questioned by the lower courts, the STF's manifestation in this procedure could immediately alter this scenario by stating its constitutionality. It was a direct relationship between those authorities and the STF, no longer disturbed by any manifestation from below, and ultimately any formal manifestation from the common citizen.

In any case, if the introduction of the *Direct Action of Constitutionality* in 1993 already started to redesign the configuration of judicial review in Brazil, it was the enactment of the Laws n. 9.882/99 and 9.868/99[499] in 1999 that transformed it radically. The first law regulates the *Petitions for Non-Compliance of a Fundamental Precept* (*Arguição de Descumprimento de Preceito Fundamental* – ADPF), which already existed in the original text of the Constitution of 1988,[500] despite not regulated so far. This constitutional action, which, according to its subsidiary nature, enlarged the possibilities of the abstract judicial review to other areas not

---

[496] See note 461 *supra*.

[497] A very interesting example is the famous *Apagão* case (ADC 9 – DF) further examined.

[498] Vieira, *Supremo Tribunal Federal: Jurisprudência Política*, 136, translation mine.

[499] The Laws 9.882/90 (art. 6, §1) and 9.868/90 (art. 9, §1) introduced, in any case, an interesting mechanism in the abstract system of judicial review: the *amicus curiae*, that is, the Rapporteur Justice can solicit the presence of representatives of civil society and experts of a determined subject matter to manifest their opinion in a public audience.

[500] Art. 102, §1 of the Constitution of 1988.

before embraced by the *Direct Unconstitutionality Suit* (*ADI*) or the *Declaratory Action of Constitutionality* (ADC),[501] brought two mechanisms for the concentration of powers in the hands of the STF: first, the possibility that, through a provisional order, all lower courts and judges have to suspend the analysis of the lawsuits or the effects of already-made decisions (if still appealable), as long as the issue they pose relates to the one of the *Petition for Non-Compliance of a Fundamental Precept*;[502] and, second, the modulation of the effects of the decision, by introducing a mechanism totally unknown in Brazilian constitutional culture,[503] that is, the possibility of unconstitutional but still effective legal norms, if reasons of legal security or exceptional social interest demand it.[504] The second law, in turn, expanded the range of the *Direct Unconstitutionality Suit* (ADI) and the *Declaratory Action of Constitutionality* (ADC) by also establishing the possibility of suspension of any lower courts' judgment involving a legal statute under the STF's scrutiny by reason of an ADI or an ADC,[505] and the modulation of effects, in a similar fashion as the Law n. 9.882/90.[506] Therefore, with these two legal statutes, the Brazilian constitutional reality introduced more one mechanism to centralize judicial review under the STF's authority, which now was able to suspend any judgment carried out in the lower courts as long as its subject had a mere *relation* to the issue discussed in the ADPF, ADC or ADI, and, secondly, created a mechanism seemingly incompatible with a longstanding constitutional tradition that placed the Constitution in a prevalent position over any unconstitutional legal statute, thereby nullifying *ex tunc* any of the effects it produced.

Considering some aspects of German constitutional tradition,[507] the STF now had the authority to decide, with binding force and *erga omnes* efficacy, when the effects of the decision should take place according to its convenience, grounded in the *legal security* and the *exceptional social interest*.[508] The STF could use these two values – extremely fluid, by the way – as a justification to keep in force, for a determinate period, a legal statute whose contents were deemed unconstitutional. In

---

[501]For instance, the judicial review of municipal legal statutes, pre-constitutional legal statutes still valid, and even an act practiced by the government that infringes a fundamental precept of the Constitution (in this case, the concept of fundamental precept has been created by case law).

[502]Art. 5, §3 of the Law n. 9.882/99.

[503]In these cases, the decision must be made by two-thirds of the STF's Justices.

[504]Art. 11 of the Law n. 9.882/99.

[505]Art. 21 of the Law n. 9.868/99.

[506]Art. 27 of the Law n. 9.868/99.

[507]See the extensive analysis Gilmar Mendes carried out on the modulation of effects – which he clearly endorses (See Mendes, *Jurisdição Constitucional*, 271) – when he examined the German constitutional model. For this purpose, see Ibid., 196–321. It is interesting to remark that Mendes himself understands that the modulation of effects is a direct consequence of the political character of judicial review. According to him, the more political the decision is, the more it demands the modulation of effects. See Ibid., 197.

[508]See, for this purpose, ADI 2.240 (published on 08.03.2007); ADI 3.682 (published on 09.06.2007); ADI 1.351 (published on 06.29.2007).

this case, based, for instance, on the economic outcomes of the decision, the STF could simply state that, even though the legal norm was unconstitutional, it would be ineffective only after a certain date.[509] With this instrument, this court gained the authority to balance constitutional rights with an axiological standpoint. There would be no other better institutional connection with balancing.[510] Indeed,

> To recognize the unconstitutionality and allow that a legal norm remains enforceable, in a temporal or indefinite manner, in the local space or in the whole national territory, is to make a balancing judgment on what one understands as the "best". This means to enter into the semantic field of values, of the opinion of what is politically or economically adequate. In other words, law and politics are confounded with each other.[511]

Notwithstanding this reality pointing out the confusion between law and politics in the realm of the abstract judicial review, there would not be much time until a similar approach also occurred in its diffuse counterpart. The lower courts' remaining space of action was then jeopardized by mechanisms that restrained further their already disrupted capacity to exercise judicial review. Moreover, this concentration of powers achieved other areas not directly related to judicial review. In 2004, with the constitutional amendment n. 45,[512] the Brazilian constitutional reality introduced the *binding precedent*[513](*Súmula Vinculante*), also justified by those three arguments above (stop the lower courts' activism, provide coherent decisions, and reduce the number of constitutional complaints in the STF).[514] With this instrument,

[509]Imagine, for instance, the possibility of a determinate tax being considered unconstitutional, but the STF states that it will be ineffective only in two years, allowing thereby that the government continues to enforce this unconstitutional law on the citizens for more two years. For this purpose, the STF simply says that, if it decided otherwise, there could be a serious economic outcome for the whole society.

[510]See, for instance, the clear connection between the modulation of effects and the deployment of balancing according to the Brazilian constitutional scholarship. See, for this purpose, Daniel Sarmento, "Eficácia Temporal do Controle de Constitucionalidade: O Princípio da Proporcionalidade e a Ponderação de Interesses das Leis," *Revista do Direito Administrativo* (Renovar), no. 212 (April-June 1998): 27–40.

[511]Cristiano Paixão and Paulo Henrique Blair Oliveira, "O Julgamento das Células-Tronco: Ponderação contra a Constituição," *Constituição e Democracia*, June 2008: 17, translation mine.

[512]The constitutional amendment n. 45/2004, in any case, introduced some interesting mechanisms, such as the expansion of the legitimate actors to file the *Declaratory Action of Constitutionality* (ADC), thereby identifying them with those of the *Direct Unconstitutionality Suit* (ADI) (art. 103 of the Constitution of 1988), and the institution of the *National Council of Justice*. This organ, composed of representatives of the judiciary, public prosecutors, Brazilian Bar Association and civil society, is responsible for supervising the judiciary's administrative and financial activities (art. 103-B of the Constitution of 1988).

[513]The binding precedent can be created after the approval of two-thirds of the members of the STF, that is, eight judges (art. 103-A, *caput*, of the Constitution of 1988).

[514]See, for instance, Justice Gilmar Mendes's justification: "This instrument [the binding precendent] plays an obvious role in stabilizing expectations and in reducing the overload of cases in the judiciary in general and particularly in the Federal Supreme Court" (Gilmar Mendes, *Judicial Reform as a Fundamental Element to Ensure Legal Security to Foreign Investments in Brazil*, New York (US) Lecture presented at the Council of Americas, http://www.stf.jus.br/

any binding precedent the STF issues becomes a rule the lower courts must obey.[515] Furthermore, the same amendment, with similar motivation,[516] created the *General Repercussion Requirement*;[517] that is, in order for an extraordinary appeal to be examined, it must pass through an analytical filter of its economic, political, social and legal repercussions[518] beyond the subjective interests involved.[519] In this circumstance, whenever the court understands that there is no general repercussion, any other suit treating identical matter is immediately rejected in the future.

The first innovation, clearly concentrating even more powers in the hands of the STF, while seemingly interested in establishing a consistent interpretation of legal rights, is nevertheless problematic within the context of separation of powers, and specially the disruption of the primary characteristics of the diffuse system of judicial review as developed in Brazil. Unlike the American model on which it was based,[520] however, it is founded upon the elaboration of general prospective thesis concerning a certain subject matter, and not the features presented in a particular case, which ought to be compared with those of future similar cases *incidentally*. True, the basis for the construction of a thesis arises from concrete cases, but, since the binding precedent is issued, it becomes a general clause applicable to future controversies as if it were a law, not longer necessarily demanding a review of the specifications of the original cases. It gains the quality

arquivo/cms/noticiaArtigoDiscurso/anexo/Reforma_do_Sistema_Judiciario_no_Brasil_v_Ing.pdf (accessed July 14, 2009), 5).

515Two interesting examples occurred when the STF, by using the institute of the *binding precedent*, limited the nepotism in the three powers and restricted the use of handcuffs by the police when arresting a suspect of a crime (Binding Precedent n. 13 and 11, respectively).

516According to Gilmar Mendes: "As it contributes to a drastic reduction in the number of cases that reach the Court, as well as to limiting the subject of decisions to constitutional questions of an objective nature, the new requirement of general repercussion for extraordinary appeals opens up promising prospects for constitutional jurisdiction in Brazil, especially as to the Federal Supreme Court assuming the typical role of a truly constitutional court" (Mendes, *Judicial Reform as a Fundamental Element to Ensure Legal Security to Foreign Investment in Brazil*, 8).

517Art. 102, §3, of the Constitution of 1988.

518See art. 543-A, §1, of the Brazilian Civil Procedural Code.

519The STF can only dismiss the constitutional appeal (*Recurso Extraordinário*), in any case, after the decision taken by two-thirds of its members, that is, eight judges (art. 102, §3, of the Constitution of 1988). This decision will serve as a parameter for the future ones with identical matter, which can be dismissed *a limine* (art. 543-A, §5, of the Brazilian Civil Procedural Code).

520In the American model, nonetheless, the *stare decisis* derives from a longstanding tradition and not from any law or written rule. Besides, it is intimately connected to the common law system, which has as its focal point the case and its particularities. In Brazil, on the other hand, the binding precedent is a consequence of a legal determination, which has no connection whatsoever with the tradition of the Brazilian legal system (grounded in the roman-germanic model), and, instead of focusing on cases, it leads to the framing of general theses binding the different branches of the judiciary and the government. The American *stare decisis* is, therefore, *contextually bound;* the Brazilian *binding precedent* is, in turn, decontextualized. This fundamental difference, however, might have not been observed by some who defend its implementation, such as STF's Justice Gilmar Mendes, according to whom the binding precedent is similar "to what occurs in Anglo-American law."

of universality.[521] As a consequence, while in the American system the judge has to make distinctions and comparisons with the leading precedent in order to contextualize the arguments of her decision with the peculiarities of the case, in Brazil, the judge can simply justify her decision by stating that it is in accordance with a general thesis, in a similar way as she does when she argues that her decision complies with a certain legal statute. This is why one could remark that, in Brazilian judicial system, there is the "power being exercised without checks and balances, all because the binding precedent turns, in practice, from individual norms – valid for each case – into general norms with *erga omnes* validity."[522] Accordingly, it could either radically restrain the lower courts' activity (they ignore, for instance, the particularities of their case by thinking the general thesis applies thereto) or simply be innocuous, when the lower courts act in the opposite direction.

By constructing a general thesis as a binding precedent, the STF began to jeopardize the possibility of a greater dialogue among the different legal interpreters – and came close to the activity of legislation,[523] and even of the constituent power.[524] As a direct consequence of the attempt to control the diffuse system of judicial review through forms and effects derived from the abstract model, this constitutional amendment engendered a discourse similar to that of lawmaking, disconnected somehow from the context of a case and its features.[525] In summary, it endangered the discursive feature of the diffuse model by binding the lower courts to a precedent that is merely a general thesis not substantially different

[521]For this reason, Justice Mendes's words saying that, contrary to the objective cases, "the binding precedent *is derived from decisions that were in principle made when dealing with concrete cases, under the incidental model*" (Mendes, "Controlling Constitutionality in Brazil," 13) are not wrong, but he failed to verify that, as soon as it is created, the binding precedent becomes an abstract and objective thesis without any effective control of other powers (always the last word is STF's) and without the possibility of directly questioning of its contents by a common citizen.

[522]Lênio Luiz Streck, *O Fahrenheit Sumular no Brasil: o Controle Panóptico da Justiça*, http://leniostreck.com.br/index2.php?option=com_docman&task=doc_view&gid=17&Itemid=40 (accessed July 14, 2009), 4, translation mine.

[523]See Vieira, *Supremo Tribunal Federal: Jurisprudência Política*, 224–225.

[524]Indeed, to modify or revoke a binding precedent, unless the STF does not implement it, it is necessary three-fifths of the congressmen's votes, in both the Senate and the Deputy Chamber, in two rounds. Another possibility is a *Direct Unconstitutionality Suit* (ADI) filed by the authorities of art. 103 of the Constitution.

[525]According to Streck, Cattoni de Oliveira and Lima: "By intending to accept complaints against its *thesis* and not against the decisions from cases (remark that we are dealing with the diffuse judicial review, whose *ratio* is the examination of concrete cases and prejudicial questions), the Supremo Tribunal Federal diverts the legal discussion to the discourses of justification (*Begründungsdiskurs*), elaborated in a decontextualized way. They become "concepts without things." And this is metaphysics, if we use a language relevant to philosophical hermeneutics." (Streck, Cattoni de Oliveira and Lima, *A Nova Perspectiva do Supremo Tribunal sobre o Controle Difuso: Mutação Constittucional e Limites da Legitimidade da Jurisdição Constitucional*, 22, 26, translation mine.)

from a legal norm,[526] and not a concrete case able to foster a greater dialogue based on the distinctions and similarities of every new case in comparison with the precedent.[527]

On the other hand, the second innovation synthesized the spirit springing from the movement towards the STF's expansion and accumulation of powers. With the introduction of the *General Repercussion Requirement*, the STF became responsible for deciding not only a constitutional matter emerging from the analysis of a particular case, but also whether it is politically, economically or socially relevant enough to engender its scrutiny. Hence, even if a case deals with a violation of constitutional nature, it can then be dismissed by reason of its economic or political irrelevance. Now, any appeal to the STF relies on the discretionary evaluation of other values that might have a prevalent position according to the Justice's opinion

---

[526]Moreover, with this understanding, as already the STF clearly exposed (See RCL 4335-5 – Justice Eros Grau's opinion), it put at risk the institute of the Senate's participation in the suspension of the legal execution deemed unconstitutional by a STF's definitive decision (art. 52, X, of the Constitution of 1988), an interesting mechanism existing since 1934 that aims to bring to judicial review the representatives of the population. The elaboration of a binding precedent in the diffuse system of judicial review, nevertheless, creates a general rule – it refers to a *thesis*, not an immediate case, after all – without the involvement of any political organ more directly connected to the citizens. Insofar as the STF can declare the *erga omnes* and binding effect in its decisions, now in the realm of the diffuse system of judicial review, the Federal Senate, within this context, becomes merely responsible for the publicity of that court's decisions (See, for this purpose, Mendes, "O papel do Senado Federal no Controle de Constitucionalidade: um Caso Clássico de Mutação Constitucional").

[527]Furthermore, with the introduction of the binding precedent as a general thesis, the diffuse system of judicial review shifted progressively to the idea that it could no longer have, as its ratio, the case and its features, but rather the simple judgment of a thesis, capable of being enforced, in the hypothesis of the lower courts' or the government's disobedience, through a direct complaint (*Reclamação*) filed in the STF. With this instrument, this court can then straightaway revoke the administrative act or annul the lower decision, determining thereby that the lower court makes another conclusion (Art. 103, §3, of the Constitution of 1988). This institute, nonetheless, seems to go in the opposite direction of the premise that, in the diffuse system of judicial review, as the Constitution clearly specifies (Art. 103, III, *a*, *b*, *c*, *d* of the Constitution of 1988), "the result of the Supremo Tribunal Federal's activity (. . .) is never the judgment of a thesis, and this activity does not result in a theory, but a decision" (Streck, Cattoni de Oliveira e Lima, *A Nova Perspectiva do Supremo Tribunal sobre o Controle Difuso: Mutação Constittucional e Limites da Legitimidade da Jurisdição Constitucional*, 4). Yet, from that moment onwards, the STF could establish a certain interpretation of a legal statute and immediately impose it – as a general prospective thesis – on all other different procedures taken place in every branch of the judiciary and the government. It could even revoke an already-made decision, if still appealable, as well as modulate its effects (See RE 197.917, published on 05.07.2007; HC 82.959, published on 09.01.2006; MS 26.602, published on 10.17.2008), based on reasons of legal security or exceptional public interest. Ultimately, therefore, the binding precedent, in the way as such defined, radically altered the structure of the diffuse system of judicial review, its inner core, which could lead one to sustain that it subverted the constitutional principles of due process of law, ample defense and contradictory, as long as it immediately excluded the claim of those who did not participate in the very process of decision-making that affects them (See Streck, Cattoni de Oliveira e Lima, *A Nova Perspectiva do Supremo Tribunal sobre o Controle Difuso: Mutação Constittucional e Limites da Legitimidade da Jurisdição Constitucional*, 7).

over the constitutional matter itself. The STF, with this instrument, confirmed its position of not merely the "Guardian of the Constitution"; it has also become the guardian of the political, economic and social order.

It is discernible that many of these developments converged upon the erection of a constitutional court that, more than constructing a reliable and consistent system of rights, became responsible for balancing those legal rights with other values emerging from a particular case, even to frame its analysis according to the premise of a general repercussion it inspires. On the one hand, the court expanded its possibilities to exercise judicial review through the abstract model; on the other, it transposed many of these features into the diffuse model of judicial review, with the consequent disfiguration of its main attributes. It also received the authorization for the evident value-based interpretation of basic rights through many of these changes (the modulation of effects, the binding effect and the general repercussion requirement, just to cite some), as a necessary response to the quest for reducing the number of constitutional complaints and likewise providing a more adequate consistency in the system of rights. The growing abstract system of judicial review and the concomitant deterioration of its diffuse counterpart have been, accordingly, deeply connected to the relativization of the authoritative nature of constitutional principles, which can ultimately modulate a constitutional right into what is regarded as a necessary response to the aspirations of society.

The constitutional court, for this reason, can have a more generalist approach than the restrictive space the diffuse model of judicial review provides. Similarly, it can assume the so-desired active space in the new Brazilian democratic reality by focusing not merely on the subjective rights of a particular situation, but rather on the general consequences a decision could bring about in a prospective basis for all. The constitutional rights turned into objective principles embracing the totality of the legal order,[528] and the constitutional court, in turn, became the guardian of this broad content. As a "Guardian of the Constitution," now shaped by this objective approach, the STF undertook an activism that is not merely directed to "protecting the constitution and fundamental rights,"[529] as Justice Mendes describes it, but also to exercising somehow the role of a positive and negative legislator, or even a "permanent constituent power,"[530] transforming thereby the Constitution and its principles into a "concrete order of values" to be managed by the court towards the interests of all.

Ironically, what began as collaboration among the different powers to prevent the lower courts' activism promoted the higher court's activism, now with a much greater impact. In this shift from a discrete and inexpressive constitutional court to an activist one denoting the figure of a guardian of the political, economic and social order, the previous analysis of the intimate connection between the

---

[528]See the second chapter.

[529]Mendes, *New Challenges of Constitutional Adjudication in the 21st Century: a Brazilian Perspective*, 2.

[530]Cattoni de Oliveira, "Jurisdição e Hermenêutica Constitucional no Estado Democrático de Direito: um Ensaio de Teoria da Interpretação Enquanto Teoria Discursiva da Argumentação Jurídica de Aplicação," 385.

government and the STF regarding their convergence on interests – on the one hand, stop the lower courts' activism; and, on the other, expand the authority and influence of this court in Brazilian democratic reality – would inevitably enter into conflict. The expansion of influence, after all, is not so fraternal as to allow a longstanding symbiosis of interests among the different powers, especially when we remark that, behind this movement, the exercise and definition of politics are at issue. Frequently since then this court has been deemed responsible for creating norms filling the legislative vacuum,[531] causing thereby a disturbance to the relationship with the other powers.[532] "The judiciary, here and there, in the face of the legislature's omission, is really legislating,"[533] said Garibaldi Alves, the then president of the Federal Senate. Yet, what should be a serious concern (after all, the debate on the separation of powers is central to constitutionalism) is normally envisaged as the sign of a natural and irreversible development. In fact, the STF itself, at least according to Justice Gilmar Mendes' words, has not given the impression of being concerned with the undertaking of this legislative function, as long as this court has merely made provisional decisions[534] while the functionality of the parliament is affected, based on what he called a cooperative model, "according to which one attempts to make the decision functional but, at the same time, expects that the legislature reacts."[535] Everything, in his opinion, seems a natural evolution of Brazilian democracy and constitutionalism, and the consequence of a healthy relationship between constitutional adjudication and the

[531]See, for example, what the then Brazilian Minister of Justice, Tarso Genro, said about the binding precedents concerning the restriction on nepotism in the three powers and the use of handcuffs by the police: "if the STF reaches the realm of creation of norms, I would say a little without precedents here in the country, this means that there are legal vacuums that have to be filled. And the Supremo has been doing it" (*Folha de S. Paulo*, Interview by Lucas Ferraz, Brasil (August 28, .2008), translation mine).

[532]See, for instance, the Federal Deputy Arnaldo Faria de Sá's words (PTB-SP), which, although praising the contents of the decision, mentioned: "The STF did the homework in place of the Chamber of Deputies. Congratulations to the judges." One can reach the same conclusion through the Federal Deputy Henrique Alves's words (PMDB-RN): "This is a very important discussion and the Chamber of Deputies, with its own legs, should approve a constitutional amendment prohibiting this practice" (*Folha de São Paulo*, Brasil, (August 21, 2008), translation mine).

[533]*Folha de São Paulo*, Brasil, (August 26, 2008), translation mine.

[534]We can observe this circumstance either in the cases of *binding precedents* or in the *Writ of Injunction*, whereby the STF undertakes the responsibility to establish provisional regulations until the parliament or the government enacts the regulatory measure. See MI 670 and MI 708 (both published on 10.31.2008).

[535]According to Gilmar Mendes: "(. . .) the functionality of the Congress is affected, and therefore the Court is going to make decisions, albeit provisional ones"; "The model suggested is of cooperation, in which one tries to make the decision functional but, at the same time, expects that the legislature reacts."

Besides, he even says that, in order to promote the legislature's reaction, it would even be possible to discuss mechanisms such as the salary cut of the congressmen or even threaten them with a criminal lawsuit. (*Folha de S. Paulo*, Interview by Andreza Matais, Brasil, (September 13, 2008), translation mine).

democratic legislator, making possible thereby the development of an "open and pluralistic society, based on principles and fundamental values."[536]

Paradoxically, nevertheless, the construction of this "open and pluralistic society, based on principles and fundamental values," led progressively to the common citizen's loss of the space for questioning the constitutionality of a legal norm. The court's activism, now equipped with legal instruments allowing the axiological shaping of the Constitution, diminished the relevance of the citizen's participation in the debates on constitutional matters. The independence of the judiciary in general, and the STF in particular, became the dependence of the lower courts on the STF's statements – a characteristic that also affected the other powers – and somehow into its independence from the common citizen. Now it is the STF, as a "forum for the treatment of social and political problems,"[537] that "monologically" represents this "open and pluralistic society," whose "principles and fundamental values" are shaped according to its viewpoint. The "open and pluralistic society," as such qualified as a desirable goal of democracy, became *a contrario sensu* the justification for a practice that closed even more the possibility of exercising a dialogue between the judiciary and the common citizen. Instead of introducing mechanisms able to foster a greater interaction with the other powers and expand the dialogue with the possible affected ones by each decision, building thereby, in a self-correcting learning process, a consistent system of rights, the formula for the crisis directed to the definition of parameters that restricted the debates on constitutionality, without achieving, nevertheless, the desire for establishing a consistent system of rights. In this regard, while the concentration of powers and activism were walking side by side, more and more, in the STF's practice, decision-making and policy-making were coming closer. Through tortuous paths, the STF would then realize the dream it could not entirely achieve at the time of the Constituent Assembly of 1987 and 1988: it could resemble a constitutional court as the ones in Europe.

3.3 Balancing in the Decisions of the Supremo Tribunal Federal: The Quest for Rationality in Decision-Making

It is not a mere coincidence that the strengthening of balancing and the idea of subjective rights as objective principles occur side by side with the expansion of the constitutional courts' activist posture. In the last chapter, we could outline this connection by verifying how the BVG and balancing are closely associated with a history leading to placing this court in the central arena of Germany's most fundamental social and political matters. In this chapter, in turn, we can already

[536]Mendes, *New Challenges of Constitutional Adjudication in the 21st Century: a Brazilian Perspective*, 10.
[537]Schlink, "German Constitutional Culture in Transition," 729.

foresee that the STF's gradual increase in influence has also an intimate relationship with this methodology. When we remark the existence of a progressive deployment of techniques such as the modulation of effects based on values of legal security and relevant social interest, the binding precedent with a general prospective character, the general repercussion requirement linking every claim to social, economic, political values, among others, it is not difficult to infer that Brazilian constitutional culture has progressively accepted a more flexible interpretation of the Constitution as a means to assimilate those values into the core of its constitutional practice. In this respect, it is relevant that this purpose of undertaking, as its realm of authority, the definition of social, economic and political matters reveals likewise a theoretical ground. The history pointing to the STF's activism needs to converge upon a theory able to explain that this movement is not simply a random effect of a society aiming to accomplish the most desirable values of democracy. Judicial activism must be justified and legitimized *rationally*.

As the "Guardian of the Constitution," the STF seems to hold those values, necessary for the exercise of democracy and citizenship, with a rational argumentation in its decisions, differing thereby from the normal and ordinary policy-making of other powers. Insofar as, by holding those values to shape constitutional rights, the STF's activity connects, more and more, constitutional principles to arguments of policy, strategically oriented to what is good for all, the fundamental difference we could find between this court and the other political powers would hypothetically lie in its technical capacity to provide answers with a rational justificatory force. The use of a methodology, therefore, that could reveal this spectrum of rationality becomes a requisite for the legitimacy of the political character of this court's activities. Whereas the other powers would legitimize themselves through periodic elections and representation, the constitutional court would achieve equivalent quality by means of a rational activity and by deploying a method revealing this quality. No one could then think of a violation of the principle of separation of powers, according to this view, if the judge could justify those values, although shaping the constitutional principles in a particular decision, in a rational basis. If the classical principle of separation of powers is outdated, albeit still necessary,[538] this results from the perception derived from the relationship between constitutional adjudication and rationality, and hence constitutional adjudication and rational methodology. Rationality and method are the words that open up the space for and give reason to activism.

Balancing appears within this context of conjunction of activism and the quest for a rational justification. One could even assert that balancing, especially now in the framework of the principle of proportionality, is deemed a sufficient argument to legitimize the new constitutional courts' activist and political role,[539] and also to justify the concentration of powers in their hands. Indeed, one could also sustain that

[538]See Mendes, "O papel do Senado Federal no Controle de Constitucionalidade: um Caso Clássico de Mutação Constitucional," 155.

[539]See Cruz, "Habermas, Ação Estratégica e Controle de Constitucionalidade," 266.

those different instruments (modulation of effects, binding effect, general repercussion requirement, just to cite some) would be grounded in the premise that constitutional principles are "optimization requirements," which could be weighted according to the factual and legal circumstances.[540] Yet, if, on the one hand, this approach could be generally identified with those practices and this process of concentration of powers in the court's hands, it is also possible, on the other, to specifically establish how it has appeared in this court's decisions. What must be verified, accordingly, is how, step by step, the deployment of this methodology emerged side by side with the arrival of the STF's new approach. The aim is to identify points of contact with a history indicating a more active and political attitude and its methodological justification for this movement arising from balancing.

As many other constitutional realities, as we could observe in Germany,[541] however, the deployment of balancing, even as an element of the principle of proportionality, has been characterized by a strong deficiency of systematization and scholarship's critical review. Neither the precedents nor scholarly analyses have promoted a reliable comprehension of this methodology,[542] which poses a serious problem in identifying it in many decisions as well as a significant lack of critical evaluation of its consequences for Brazilian constitutionalism. Still, some signs converge on the understanding that the Brazilian courts have increasingly deployed this methodology as a fundamental premise in the hypothesis of a collision of constitutional principles or values, endowed with a constitutional status. Moreover, there is likewise the attempt, albeit still incipient, to systematize it. Step by step, the principle of proportionality, thereby balancing, has become the fundamental argument of many decisions and, by the same token, the primary subject of many scholarly texts,[543] most of them, nonetheless, still expressing a great fascination with this methodology without developing a more comprehensive critical analysis of it and its deployment in constitutional adjudication.

---

[540]See Cattoni de Oliveira, "Jurisdição Constitucional e Hermenêutica," 399.

[541]See the second chapter.

[542]See, for this purpose, Silva's analysis of the usual confusions the Brazilian scholarship makes in the investigation of the principle of proportionality in: Luís Virgílio Afonso da Silva, "O Proporcional e o Razoável," *Revista dos Tribunais*, no. 798 (April 1992): 23–50.

[543]See, for this purpose, Suzana de Toledo Barros, *O Princípio da Proporcionalidade e o Controle de Constitucionalidade das Leis Restritivas de Direitos Fundamentais*, (Brasília: Brasília Jurídica, 2003); Helenilson Cunha Pontes, *O Princípio da Proporcionalidade e o Direito Tributário* (São Paulo: Dialética, 2000); Wilson Antônio Steinmetz, *Colisão de Direitos Fundamentais e Princípio da Proporcionalidade* (Porto Alegre: Livraria do Advogado, 2001); Paulo Arminio Tavares Buechele, *O Princípio da Proporcionalidade e a Interpretação da Constituição* (Rio de Janeiro: Renovar, 1999); Luis Roberto Barroso, *Interpretação e Aplicação da Constituição: Fundamentos de uma Dogmática Constitucional Transformadora* (São Paulo: Saraiva, 1998); Humberto Ávila, *Teoria dos Princípios: da Definição à Aplicação dos Princípios Jurídicos* (São Paulo: Malheiros, 2008); Alexandre Araújo Costa, *O Princípio da Proporcionalidade na Jurisprudência do STF* (Brasília: Thesaurus, 2008); Raquel Denise Stumm, *Princípio da Proporcionalidade no Direito Constitucional Brasileiro* (Porto Alegre: Livraria do Advogado, 1995); Luís Virgílio Afonso da Silva. "O Proporcional e o Razoável"; Luís Virgílio Afonso da Silva, *Interpretação Constitucional* (São Paulo: Malheiros, 2005).

Furthermore, Brazilian constitutional scholars have sustained that the principle of proportionality, and particularly balancing, is not new in Brazilian constitutional reality, usually connecting it to the beginning of the STF's debates on reasonability and substantive due process of law.[544] In this circumstance, the idea of proportionality would be similar to what happened in the American constitutionalism, where the discussion is more connected to the requirement that all state acts must be reasonable to achieve the goal, without, nonetheless, presenting a methodology "rationally"[545] and structurally built under an ordered set of examinations such as the suitability, the necessity and the proportionality in the narrow sense or balancing. This assimilation, nonetheless, is not exempt from serious critiques,[546] either because the idea of reasonability has another origin, or because it has a different structure and form of reasoning.[547] In any case, it is interesting to observe this shift in Brazilian constitutionalism: if the STF's first manifestations against parliament's and government's arbitrariness and excesses were founded mostly upon the American concept of reasonability and substantive due process of law, then, with its expansion and concentration of powers, this court progressively deployed balancing with the so-called "rational" methodology of the principle of proportionality according to the framework as normally German scholarship discuss and the BVG deploys it. The evolution of judicial review of governmental and legislative acts, accordingly, followed somehow the expectation for a "rational" response – not easily found in the American idea of reasonability – which should better agree with the STF's new role.

Consistent with this premise that reasonability and proportionality are not synonyms – although presenting a theoretical connection as long as both refer to the control over governmental and legislative acts – the investigation here will discuss some cases where we can observe this transition in order to demonstrate how, step by step, the STF, however erratically, has changed the focus to the German

---

[544]See, for instance, Gilmar Mendes, "O Princípio da Proporcionalidade na Jurisprudência do Supremo Tribunal Federal," in *Revista Diálogo Jurídico* I, no. 5 (2001); Suzana de Toledo Barros, *O Princípio da Proporcionalidade e o Controle de Constitucionalidade das Leis Restritivas de Direitos Fundamentais* (Brasília: Brasília Jurídica, 2003); Luis Roberto Barroso, "Os Princípios da Razoabilidade e da Proporcionalidade no Direito Constitucional," *Revista dos Tribunais – Cadernos de Direito Constitucional e Ciência Pollítica* 23 (1998): 65-79; Alexandre Araújo Costa, *O Princípio da Proporcionalidade na Jurisprudência do STF*.

[545]Silva, "O Proporcional e o Razoável," 30.

[546]See, for instance, Humberto Ávila, *Teoria dos Princípios: da Definição à Aplicação dos Princípios Jurídicos* (São Paulo: Malheiros, 2008), 151–179.

[547]According to Luís Virgílio Afonso da Silva: "The rule of proportionality in the control over restrictive laws to basic rights arose from the German Constitutional Court's practice and is not a simple agenda that vaguely suggests that the state acts must be reasonable, nor a simple analysis of the relationship means-goal. In the way the German constitutional practice developed it, it has a rationally defined *structure*, with independent sub-elements – the analysis of *suitability*, *necessity* and *proportionality in the narrow sense* – which are deployed in a predefined order, and which confer on the rule of proportionality a singularity that differentiates it *clearly* from the mere exigency of reasonability (Silva, "O Proporcional e o Razoável," 30, translation mine)".

model, quiet temporally coinciding with the consolidation of the abstract system of judicial review and the weakening of its diffuse counterpart. With the STF's expansion of influence and power, the German model of the principle of proportionality appeared to best handle the presumption Justice Gilmar Mendes drafted by stating that "the Constitutional Court exists to make the most rational decisions."[548] In this respect, as we have done so far, the analysis will concentrate on the period after the promulgation of the Federal Constitution of 1988, even though relevant signs of this movement, still characterized nonetheless by the idea of reasonability, could somehow be found in the previous years.[549] By the same token, as long as the main question of this research relates to the rationality of balancing, the analysis will stress how the deployment of balancing has continuously embodied the idea that subjective rights are objective principles of a total legal order, leading therefore to a practice of decision-making very close to lawmaking.

Unlike the previous authoritarian years, when the STF's decisions were normally characterized by a strong formalism, especially with regard to the judicial review of governmental and parliamentary acts,[550] with the new democratic

[548]Gilmar Mendes, interview by Izabela Torres, "Entrevista – Gilmar Mendes," *Correio Braziliense*, Brasília (August 17, 2008).

[549]Some interesting examples: RE n. 18.331 (published on 11.08.1951); RE 16.912 (published on 06.28.1968); HC 45.232 (published on 06.17.1968); RP 930 (published on 08.02.1977); RP 1.054 (published on 06.29.1984).

[550]During the period of the military regime, the manifestations concerning the deployment of the principle of reasonability and balancing were very diluted by the formalist character of constitutional interpretation that prevailed in those years, and, when they appeared, they were still marked by an unsystematic approach that could, indeed, denature the characteristics of the idea of reasonability. Apart from some very rare manifestations (see note 488 *supra*), the STF, even after receiving the authorization to evaluate the constitutionality of federal legal statutes in the abstract system of judicial review, would not undertake a more activist approach. The military dictatorship could not live with the idea of having a constitutional court with sufficient powers to invalidate legal statues, many of them directly originated from the military government. Therefore, the STF, by normally using formal arguments, behaved with a certain complacency and even supported some government's arbitrary practices. The development of a value-based account of constitutional principles to be balanced with other values emerging from the economic, social and political order was, even though presented in some court's opinions (many of them coming from the minority of its Justices), a premise that contrasted substantially with the prevalent judiciary's and scholars' legal thinking. It is from this period the idea that the judiciary could not control the merits of the administrative act, and above all, the legislative ones, a principle that, if existing in different legal realities, achieved an expressive passive character towards many of the illegalities and arbitrariness the dictatorship yielded, leaving then the common citizen, in many cases, without any legal protection. When the STF attempted to react against this situation, the military regime created barriers that strongly intimidated its activities. It was necessary thus that this court thought and acted in accordance with the military regime, and nothing better for this purpose than establishing a strict and formal comprehension of the principle of separation of powers and, as Alexandre Araújo Costa mentions, "an almost mythical respect to the so-called *administrative merit* and the legislative discretion, ideas well aligned with an authoritarian regime where the governmental axiological choices could not even be contested by the society, and much less annulled by the judiciary" (Costa, *O Princípio da Proporcionalidade na Jurisprudência do STF*, 93–94, translation mine).

regime, the STF could finally play a more active role and indeed attempt to function as the "Guardian of the Constitution." Within this context, the debates on *due process of law* and the material review of governmental and legislative acts, before strongly restricted, appeared as a justification for judicial review. The idea of reasonability of the governmental and legislative acts, used nevertheless in many cases as a simple self-reinforcing argument[551] or interpreted as a complement to the equality principle without serious methodological justification,[552] became a central concern for constitutional adjudication. The so-called principle of reasonability somehow progressively received a better configuration in the STF's decisions and started to appear more frequently in the Justices' opinions.[553] However, the

---

[551]Reasonability (or proportionality) is normally, in these cases, used to express the idea of an existing common sense of a determined matter. See, for this purpose, RE 192.568 (published on 09.13.1996), REED 199.066 (published on 08.011997), ADIN 1.326 (published on 09.26.2007); AGRAG 203.186 (published on 06.12.1998); ADI-MC 2273 (published on 05.25.2003); ADI 2019 (published on 06.21.2002); RE 453.740-1 (published on 02.28.2007); RE 197.917 (published on 05.05.2004).

[552]In this case, the idea of reasonability or proportionality (some already with the triadic structure of the German doctrine) is instrumentally introduced as a criterion of validity to evaluate discriminatory treatments, as though the discrimination established by the legal statute or the administrative act not only violates the equality principle but is also unreasonable to achieve a certain goal. See, for instance, ADIn 1.326 (published on 09.26.1997); AGRRE 205.535 (published on 08.14.1998); RE 184.635 (published on 05.04.2001); RE 176.479 (published on 09.05.1997); ADInMC 1.753 (published on 06.12.1998); ADI-MC 2317 (published on 03.23.2001); ADI 3522 (published on 05.12.2006); RE 140.889 (published on 12.15.2000); RE 150.455 (published on 05.05.1999); ADI 1.040 (published on 04.01.2005); ADI 1.351 (08.05.2005).

[553]In the following years, the STF's decisions began to provide a better shape to the principle of reasonability and present it explicitly as a central argument. In 1993, the STF introduced the so-called principle of proportionality, still mixed up, however, with many other arguments, as a justification to provisionally suspend a law (Law n. 10.248/93 of the state of Paraná) that determined that all gas (LPG) cylinders had to be weighted before being sold in favor of the consumer (ADI-MC 855-2, published on 10.01.1993). Based on a neutral technical report saying that this measure would be ineffective and would not provide any real benefit to the consumer, the STF deployed the principle of proportionality – specifically here suitability and necessity, even though these elements were not explicitly discussed – in order to suspend the law by virtue of its non-suitability to achieve the goal and the existence of other means less harmful to the consumer's rights. In the same year, the STF judged the reasonability of the Complementary Law n. 75/93, which required the lapse of two years after the graduation as a legal bachelor to participate in public contests created to select public prosecutors, deciding in favor of the reasonability of the measure. Other cases that also deployed, in a certain sense, the idea of reasonability can be found in the AGRAG 153.493 (published on 02.25.1994), which stated that the constitutional model of monetary correction of public debts was not suitable; the ADIn n. 966 and ADIn n. 958 (published on 08.25.1995), which discussed the constitutionality of the Federal Law n. 8.713/93, responsible for restricting the possibility of minor political parties to indicate a candidate for President, whose decision had some opinions founded on the reasonability of the restriction (Justice Sepúlveda Pertence connected the concept of reasonability to that of moderation; Justice Moreira Alves connected it to the *due process of law*) as a principle to control legal norms restricting constitutional principles. See also ADInMC 1.158 (published on 05.26.1995), which discussed the constitutionality of a legal norm of the State of Amazonas that extended a benefit related to vacations to retired public officers; the HC 76.060 (published on 05.15.1998), in which Justice

American constitutionalism was still the main source, especially by associating the concept of reasonability with the fundamentals of *due process of law* and the *abuse of the legislative function*, no longer merely examined in its formal perspective but also in its material dimension. Since the American constitutionalism's primary characteristic is the ongoing construction of precedents from case to case without a rigid and methodical definition of criteria and formulas, the STF seemed to follow, in this matter, a similar recipe. However, it did not provide real precedents, either because there was not a court's unified opinion, but a compilation of different arguments of each one of the Justices in the most complex cases, or because the very idea of reasonability, in this multiplicity of arguments, gained different meanings,[554] although the association with the *substantive due process of law* was the most convincing.

In this complexity of different arguments and a still unsystematic comprehension of reasonability, notwithstanding the evident increasing movement towards the control over governmental and legislative acts based on the *substantive due process of law*, balancing (not yet, however, as a third and subsidiary element of the principle of proportionality), gained force in Brazilian constitutionalism. The idea of shaping constitutional rights with other values evolved naturally from the role the STF progressively played in the new political order. Yet, this impulse towards the material review of governmental and legislative acts, despite the prevalent justification based on the substantive due process of law, was still deprived of a better comprehension of what the premise of being the "Guardian of the Constitution" really meant. The STF did not have longstanding experience of actively seeking coherence in its decisions, above all in the realm of material review of governmental and legislative acts, a practice, after all, this court very scarcely engaged in during the previous authoritarian period. Moreover, constitutional scholarship, which could theoretically provide this court with dogmatic concepts, relevant discussions of the interconnection between constitutionalism and democracy, and critical review of its decisions, was far from representing a solid voice, mainly in the debates on constitutional law. In reaction against the self-restraint posture of the previous times, the new STF, gradually emerging, would become a more active court, carrying with itself, nevertheless, a deficit of constitutional-democratic practice and a deficit of constitutional-democratic knowledge. It was an opportune scenario for balancing (activism, deficit

Sepúlveda Pertence proceeded to the analysis of proportionality as a means to decide that the father could not be compelled to make an investigation of paternity (DNA test), if there were already other sufficient evidences indicating who was the father; the ADI-MC 2667 (published on 03.12.2004), concerning the possibility of issuing a high school degree before finishing the last grade, in case of approval in the vestibular (Brazilian access exam to the University), when Justice Celso de Mello connected this principle to the abuse of the legislative function and to the substantive due process of law; the RE 319.556-5 (published on 03.12.2002), in which Justice Sepúlveda Pertence connected the principle of proportionality to the substantive due process of law in order to sustain the principle, in criminal law, of "minimal intervention."

[554]See notes 551 and 552 *supra*.

of coherence and constitutional knowledge, and absence of critical review), still not logically and structurally shaped by a methodology such as the one German constitutionalism contributes,[555] but undeniably present.

On the other hand, in many opportunities in which this court could decide in favor of a consistent interpretation of the legal system, enforcing it against arbitrary governmental and legislative unconstitutional provisions, it did not offer any resistance, using, moreover, the idea of balancing as a means to justify a noticeable political decision. Balancing, therefore, appeared as an argument either to a more intervenient attitude with regard to the other institutional powers or, rather, to a lenient validation of evident unconstitutional measures. In this hypothesis, this court frequently conducted itself timidly and passively vis-à-vis the other powers, guaranteeing hence the economic and political space of governability. This characteristic was especially visible in virtue of the expansion of the lower courts' activism and the growing number of lawsuits, many of them arising from the vast constitutional rights of the new democratic regime. In these circumstances, the STF's deployment of balancing was often used as a form of justification for arguments of policy in its decisions and as a mechanism in favor of governability. This scenario is remarkable: as long as the constitutional court's activism was not rooted in a consistent interpretation of the legal system, but instead in the ample realm of possibilities this broader space of interpretation balancing brought forth, paradoxically, this activism could also turn into a real passive approach to the protection of the Constitution. It was then politically active and constitutionally passive.

Two cases are paradigmatic of this posture regarding the governmental acts. In favor of governability and against the lower courts' activism, for instance, the STF did not state, in the abstract system of judicial review, the unconstitutionality of the Provisional Measure (*Medida Provisória*) n. 173/90,[556] which prohibited preliminary verdicts and immediate executions of provisional sentences in some areas related to economic and fiscal matters enacted during the Fernando Collor de Mello's Presidency. Rather, it transferred this discussion to each case brought to the judiciary through the diffuse system of judicial review. The main argument, proposed by Justice Sepúlveda Pertence, was that the legal norm was not

[555] These cases demonstrate, therefore, that balancing appears in the most different contexts, and not only as a necessary third element of the principle of proportionality, normally deployed after the principles of suitability and necessity. The premise that a basic right can be restricted depending on how relevant the realization of another value is reveals a usual practice in the construction of political arguments by constitutional courts. The German scholarship's methodological construction and the BVG's practice attempt to provide a "rational" aura for decision that is, fundamentally, open to the relativization of a basic right with other value (economic, sociological, political). This is its greatest danger and the main reason to associate it with the constitutional court's discretionary power.

[556] See ADInMC 223 (published on 06.29.1990).

unconstitutional; it could at the most be unreasonable, demanding thereby the examination of each case in concrete.[557] In a more restrictive way, Justice Moreira Alves, by using the fundamentals of due process of law and some historical examples, argued that the restriction on the lower courts' provisional verdicts was not totally unreasonable, especially in the circumstances of an economic plan of emergency and of an economic crisis, and thereby he could not declare the unconstitutionality of the legal norm in abstract.[558] In this case, the principle of judicial access, defined in article 5, XXXV, of the Constitution of 1988, was balanced somehow with the economic and political interests of not having any provisional verdict or the execution of a provisional sentence from the lower courts that could, as Justice Sydney Sanches mentioned, disrupt the "economic-political plan that, if it [had] imperfections from the very human elaboration, it still [had] the noble purpose to attempt a return to the economic and social stability and a restart of development."[559]

On the one hand, there was the principle of judicial access; on the other, the economic value that was deemed weightier for the interests of the overall society through a teleological approach. This case shows that this court assumed somehow the role of providing answers that were, according to the Justices' discretionary perception, the best economic solution to society and governability, leaving aside the primary concern for keeping consistent the system of rights. Rather than reinforcing this legal system, and particularly connecting the principle of judicial access to the democratic character of the new constitutional reality, the decision was directed mostly to the discretionary evaluation of what was best for the nation and for governability according to the Justices' interpretation of an economic matter, shaping hence the concept of reasonability. Weightier than the principle of judicial access were the interests of the nation, even though this could be regarded as a "very hard decision," either "under the legal or the political aspect."[560] Balancing, therefore, appeared within the context of a quest for justifying through constitutional adjudication a political program, using thereby arguments of policy with a seemingly legal approach.

Another interesting example was the famous "*Apagão*" case, ADC 9,[561] judged in 2001. In this case, the government filed a declaratory action of constitutionality in the STF with the purpose of sustaining the constitutionality of the energy policy.[562]

---

[557] In this case, Justice Moreira Alves mentioned that, in order to evaluate the reasonability of the measure, the abstract system of judicial review was not adequate (See ADInMC 223, published on 06.29.1990).

[558] Ibid.

[559] See ADInMC 223 (published on 06.29.1990), translation mine.

[560] Ibid., translation mine.

[561] Published on 04.23.2004.

[562] It is interesting to verify that Luís Virgílio Afonso da Silva, one of Brazilian most notorious and expert scholars in the principle of proportionality, when examined specifically this case, concluded that a correct decision would point out the unconstitutionality of the Provisional Measure, because the restrictions it created were clearly unnecessary (there were alternatives less harmful to the citizen's consumer rights, after all). See, for this purpose, Silva, "O Proporcional e o Razoável,"

The case related to the enactment of the Provisional Measure n. 2148/2001, responsible for establishing mechanisms to face the energy crisis of that time, such as rules to reduce the energy consumption (definition of goals for energy consumption, surcharge for the users who did not accomplish those goals, and suspension of energy supply in the repeated situations). In the judgment of this *Declaratory Action of Constitutionality*, the STF deployed balancing as a way to decide that suspending the energy supply was "less serious than the collapse of the system," and that the "balancing procedure, as an analysis of efficiency (...) imposes that one applies the cost-benefit relation."[563] In this decision, the court balanced some values, such as the "minimal social solidarity,"[564] the "popular reaction,"[565] and even the fact that "Brazilian people understood the urgent measures that had to be taken in the face of the seriousness of the situation which the country [went] through,"[566] with the constitutional principles at stake.

This case is a patent example of the deployment of balancing to provide a legal justification for a political program, an evident association of balancing with the idea of efficiency and cost-benefit, and a serious demonstration of how governmental goals can obtain a higher degree of relevance than the legal system through balancing, hence jeopardizing the constitutional principles at stake, above all the consumer protection.[567] This case, furthermore, also indicates how the government used the STF as a partner not only to validate an evident unconstitutional measure in favor of a political or economic goal, but also to consolidate the abstract system of judicial review as a mechanism to avoid direct citizens' lawsuits in this matter. Balancing, on the one hand, and the abstract system of judicial review, on the other, shaped a perfect marriage for policy-making in constitutional adjudication.

Naturally, the deployment of balancing also took place in some relevant situations in which the STF managed to work more specifically with constitutional principles in a more coherent fashion. A relevant precedent is the HC 71.373, judged in 1994, which discussed the right of a child to know her genetic origin, leading thus to balance human dignity and the inviolability of human body with the child's right to know her genetic origin and the minimal sacrifice to the father's physical integrity.[568] Another interesting case is the ADI

39–41. This conclusion also demonstrates that balancing, as practiced in this case, not necessarily occurs after a systematical investigation of the suitability and necessity, even though the German doctrine (Alexy in particular) and Silva, in Brazil, mention that balancing is a procedure to carry out after the other two (suitability and necessity), given its subsidiary nature. Besides, as it will be shortly examined, it is not by reason of strictly deploying this methodology that the discretionary character of this principle disappears.

[563] ADC 9 (published on 04.23.2004). Judge Neri da Silveira's opinion, translation mine.

[564] ADC 9 (published on 04.23.2004). Judge Ellen Gracie's opinion, translation mine.

[565] ADC 9 (published on 04.23.2004). Judge Sydney Sanches' opinion, translation mine.

[566] ADC 9 (published on 04.23.2004). Judge Maurício Corrêa's opinion, translation mine.

[567] Art. 5, XXXII, of the Federal Constitution of 1988.

[568] The HC 71.373 (published on 11.22.1996), a *habeas corpus*, judged in 1994, referred to the obligation of a father to proceed to an investigation of paternity (DNA test). After the lower courts

1.511-7,[569] in which the court, already anticipating the debate on the principle of proportionality,[570] discussed the constitutionality of the law that determined the submission of every undergraduate student, before receiving the university degree, to a national exam (*Provão*) in order to evaluate the quality of the superior education in Brazil. The court understood that the measure was reasonable, proportional, and, although the constitutional principle of providing education with quality[571] was not directly mentioned, we could say that balancing took place between this principle and the individual burden – considered minimal – the student would suffer for that purpose.

Considering the right to health and life, similarly, the STF used balancing in a case related to a law of the state of Rio de Janeiro that determined progressive discounts to elderly people in buying medicines, stating thereby that those principles were more relevant than the eventual and partial financial loss a determined commercial company would suffer.[572] It is interesting to mention that, in this

had determined that, in favor of the child's right to know her genetic origin, the father had to submit himself to this test, the STF granted the writ to the father. After Justice Francisco Resek's opinion, according to whom "the sacrifice imposed on the plaintiff's physical integrity is ridiculous when confronted with the child's interests as well as with the certainty the expert evidence can provide to the judge's decision," the court, in a very tight result oriented by Justice Marco Aurélio's opinion, stated that the principle of human dignity was a fundamental principle that should prevail in this situation. The court did not use the word "balancing" in this case, but there was already, at least implicitly, the discussion of which principle was more relevant to the circumstances of the case: human dignity, intangibility of the human body and the principle of legality (there was no law compelling someone to be submitted to a DNA test in these cases, for instance), on the one hand, or the child's interest in knowing her original ancestry, and the "ridiculous sacrifice imposed on the corporal inviolability," on the other. It is, besides, important to notice that the discussion of the "minimal sacrifice" the father would suffer was a primary value for balancing (translation mine).

In 1998, in a similar case, HC 76.060 (published on 05.15.1998), the court was again encouraged to balance comparable constitutional principles. The difference, nonetheless, was that now the plaintiff, the registered father by reason of marriage, asked to revoke a lower courts' decision, which compelled him to proceed to the investigation of paternity, even when there were already evidences of the child's DNA tests, as well as his mother's and the third-party's who defined himself as the real father. In this case, Justice Sepúlveda Pertence, who was one of the dissenting opinions in the previous case, and clearly mentioning that this was a case of collision of principles leading to the evaluation of their weight, argued that it was against the principle of proportionality to compel someone when there are already enough evidences proving who was the father.

[569]Published on 06.06.2003.

[570]This case is particularly remarkable for it carries out an analysis of what the principle of proportionality is (indicating, for instance, how this debate occurs in Germany and Spain), a doctrinal differentiation between proportionality and reasonability (even though this difference was afterwards not clearly developed according to the features of the case), and an investigation of the evolution of this principle (reasonability and proportionality used as a synonym). See ADI 1.51107 (published on 06.06.2003).

[571]Art. 205 and 206, VII of the Constitution of 1988.

[572]See Justice Ellen Gracie's opinion. ADI 2.435 (published on 10.31.2003).

decision, Justice Marco Aurélio, as a dissenting opinion, also founded on the proportionality, asserted that, provided that the age was the only criterion (and not the beneficiary's financial condition), the measure was disproportionate. He balanced those rights with the excessive intervention in the economic field and in the private autonomy, and concluded that this measure would adversely affect the whole society, causing then the rise of prices of medicines in a period when economic stability was a goal.[573] Another situation that stimulated interesting discussions on balancing in the situation of illegal evidences was the HC 71.373,[574] also in 1994, when the court balanced the principle of inviolability of domicile,[575] on the one hand, and the general interests in the efficacy of criminal repression, on the other[576]; the HC 80.948,[577] judged in 2001, which is particularly remarkable for its debate on the inappropriateness of balancing to relativize the constitutional guarantee of inadmissibility of illegal evidences[578] (showing, therefore, the risks of balancing to constitutional adjudication)[579]; and the HC 80.949,[580] in which Justice Sepúlveda Pertence examined the problem of bringing to Brazilian reality the German theory of the principle of proportionality in cases of illegal evidences, pointing out, besides, that, as long as there is the supremacy of basic rights, when in conflict with the interest in finding the real truth in the criminal prosecution, there is no need to deploy balancing.[581]

---

[573]See Justice Marco Aurélio's opinion. ADI 2.435 (published on 10.31.2003).

[574]Published on 11.22.1996.

[575]Art. 5, XI, of the Federal Constitution of 1988.

[576]This case is very interesting, because it directly examined the principle of proportionality and the collision of legal principles and interests in play. Justice Sepúlveda Pertence argued that the principle of proportionality can be used as a means to relativize the principle of inviolability of domicile, by balancing it with the general interests in the efficacy of criminal repression. The result, nonetheless, which seemed to consider the evidence illegal, because there was no previous judicial authorization for that, was in the other direction. Justice Sepúlveda Pertence, who expressed the court's opinion, understood that, insofar as there was no evidence of plaintiff's opposition to the government agents in the case, the collected evidence was therefore valid for all effects (See HC 71.373, published on 11.22.1996).

[577]Published on 12.19.2001.

[578]Art. 5, LVI, of the Federal Constitution of 1988.

[579]This was one of the most interesting analyses of the risk of balancing to the constitutional principles and guarantees. According to Justice Celso de Mello, based on some relevant views of Brazilian scholars, in the case of evidence collected through illegal means, it is a serious risk to admit them with support of the principle of proportionality. See, for this purpose, Justice Celso de Mello's opinion in the HC 80948 (published on 12.19.2001).

[580]Published on 12.14.2001.

[581]Justice Sepúlveda Pertence even says that the STF is the Guardian of the Constitution and not the Guardian of the prisons, as if it were its duty to relativize the guarantee of the inadmissibility of illegal evidences in favor of the real truth of the criminal prosecution. See this interesting analysis in Justice Sepúlveda Pertence's opinion in the HC 80949 (published on 12.14.2001). See also HC 87927 (published on 06.23.2006); HC 90232 (published on 03.02.2007).

The shift to the deployment of the principle of proportionality, and particularly balancing, with the triadic framework as normally discussed in Germany[582] was, accordingly, being already suggested in those decisions, but in a very unsteady and scattered way. Moreover, it followed, at least temporally – which is remarkable to notice – the movement towards the emphasis on the abstract system of judicial review. It appeared, here and there, at the beginning, more as a suggestion, usually linked to the debate on substantive due process of law, without expressing the systematic framework German scholarship normally associates with this principle. However, apart from the evident influence of the American constitutionalism, marked less by some methodological approach and formulas and much more by a case to case discussion, this movement towards the abstract system of judicial review and the consequent concentration of powers in the hands of the STF would point out the attempt to rationalize what was indeed reasonability (or proportionality). The criteria German scholarship framed[583] and the BVG somehow deployed in many cases,[584] founded on the triadic structure of suitability, necessity and proportionality in the narrow sense (or balancing), arrived thus as a serious subject for consideration and a relevant source to "rationalize" constitutional decision-making, providing thereby supposedly more legitimate decisions within the context of a new comprehension of the STF's role.[585]

Yet, it was only from 2003 onwards that the principle of proportionality clearly appeared in the STF's decisions with the triadic framework.[586] The STF's Justice

[582]See the previous chapter.

[583]See the previous chapter.

[584]See the first and second chapters.

[585]This influence, nonetheless, is complicated to be verified. Different Justices have different forms of justifying the principle of proportionality, some still recalling the debates on American *substantive due process of law*, and others connecting it more directly to the German constitutional scholarship, or mixing up both views. In any approach, however, it is possible to observe that the idea of proportionality, and balancing in particular, became more and more frequent. From cases where the idea of balancing was not explicitly mentioned, but obviously inferred, to cases where terms such as "weight," "proportionality" and the triadic structure appeared, what is relevant to observe is that the principle of proportionality, and especially balancing, accompanied progressively the STF's decisions and has become a fundamental criterion for constitutional adjudication.

[586]A transitory example of this tendency was the judgment of the *Direct Unconstitutionality Suit*, ADI-MC 2.213 (published on 04.23.2004), in 2002, which exposed, despite the decision declaring the constitutionality of the legal statute, one of the most extensive analysis of the possibility of judicial review of the motives – relevance and urgency (art. 62 of the Constitution of 1988) – for the enactment of Provisional Measures (*Medidas Provisórias*). This case refers to the Provisional Measure n. 2.027-38/2000 (later 2.183-56/2001), responsible for the introduction of some serious restrictions on the program of land reform, especially motivated by some invasions of rural properties by organized civil movements. The STF carried out the principle of proportionality, *indirectly* deploying the triadic structure – for it held the Advocate-General of the Union's arguments grounded in this triadic framework – while, simultaneously, associating it with the premise of reasonability and substantive due process of law as a fundamental category for the limitation of excesses practiced by the government in this area.

who systematically introduced this debate, especially founded upon Robert Alexy's *Theory of Constitutional Rights* (*Theorie der Grundrechte*),[587] was Gilmar Mendes, then recently appointed to the court. The particularity of having carried out his doctoral studies in Germany, where he investigated the differences between the abstract system of judicial review of both constitutional realities[588] and also entered into contact with German constitutional scholarship, such as the idea of basic rights as objective principles of the total legal order or the treatment of the constitution as a "concrete order of values," for instance, were certainly chief aspects for the turning point within this context. Along with his explicit intentions to expand the abstract system of judicial review in the STF's decisions, as well as extend some of its characteristics to the diffuse system,[589] Justice Mendes headed the movement towards exploring systematically the principle of proportionality.

One of the cases that exposed, initially, this turning point was the IF 2.915-5,[590] related to a claim to federal intervention in the state of São Paulo as a result of not having paid the alimony credits (*precatórios alimentícios*) resulting from a condemnation in another procedure. After Justice Marco Aurélio's favorable opinion to the plaintiffs by stating that the state of São Paulo's continuous disobedience of judicial decisions offended the primacy of the judiciary and the certainty of the value of judicial decisions,[591] therefore ordering the federal intervention, Justice Gilmar Mendes applied the principle of proportionality to contradict this understanding. According to him, it was necessary to examine the legitimacy of the federal intervention from its "conformity with the constitutional principle of proportionality."[592] After having established the constitutional nature of this principle, he associated it, along with the substantive due process of law, with German

---

[587]See the next chapter.

[588]See Gilmar Mendes, *Die abstrakte Normenkontrolle vor dem Bundesverfassungsgericht und vor dem brasilianischen Supremo Tribunal Federal* (Berlin: Duncker & Humblot, 1991). In Portuguese: Gilmar Mendes, *Jurisdição Constitucional: Controle Abstrato de Normas no Brasil e na Alemanha* (São Paulo: Saraiva, 2004).

[589]See, for instance, Justice Gilmar Mendes's extensive analysis of the necessity of modulation of effects also in the diffuse system of judicial review (RE 197.917/SP – published on 05.07.2004).

[590]Published in 11.28.2003. See also IF 298 (published on 02.27.2004); IF 444 (published on 11.14.2003); IF 2194 (published on 06.20.2003); IF 1690 (published on 05.20.2003); IF 1466 (published on 06.20.2003); IF 470 (published on 06.20.2003); IF 237 (published on 05.20.2003); IF 139 (published on 05.23.2003); IF 449 (published on 08.29.2003); IF 2257(08.01.2003); IF 1952 (published on 08.01.2003); IF 1317 (published on 08.01.2003); IF 492 (published on 08.01.2003); IF 317 (published on 08.01.2003); IF 171 (published on 08.01.2003); IF 3578 (published on 08.22.2003); IF 3292 (published on 08.29.2003); IF 2973 (published on 08.29.2003); IF 3601 (published on 08.22.2003); IF 3046 (published on 08.22.2003); IF 2975 (published on 08.22.2003); IF 2909 (published on 08.22.2003); IF 2805 (published on 08.22.2003); IF 2737 (published on 08.22.2003); IF 2127 (published on 08.22.2003); IF 164 (published in 11.14.2003).

[591]See IF 2.915-5 (published on 11.28.2003). Justice Marco Aurélio's opinion.

[592]Ibid., Justice Gilmar Mendes's opinion, translation mine.

concepts such as "limits of limits" (*Schranke der Shranke*)[593] and "prohibition of excess" (*Übermaßverbot*). Besides, he, based on Robert Alexy's words, connected this principle to the "essential core of basic rights conceived in the relative way"[594] and to the collision of constitutional principles, goods and values. He even discussed Alexy's distinction between rules and principles[595] by mentioning, in this last case, that the conflict is resolved by "balancing the relative weight of each one of the norms theoretically applicable and able to justify decisions in opposite directions."[596] Therefore, unlike the previous cases, which we could at most only indirectly observe the deployment of the triadic framework, now Justice Mendes differentiates the three elements (suitability, necessity and proportionality in the narrow sense), explains each one of them, and remarks that this investigation can be carried out by using as example the German recent experience.[597]

This explanation, nonetheless, which could impress due to its dogmatic basis, had an evident purpose: to proceed to balancing as a means to introduce the concept of the "reserve of the financially possible" as a value to be weighted with the constitutional norm determining the payment of the alimony credits in those circumstances. First, Justice Mendes mentioned that one could not disregard the state economic limitations (payment of public officers, investments, debts, etc.). Second, he introduced another axiological argument: the state of São Paulo had undertaken a great effort to honor its judicial debts, exposing, for this purpose, a detailed analysis of facts and numbers. Third, he equated this obligation originated from a constitutional norm[598] with other "multiple obligations of identical hierarchy,"[599] which could bring about the inefficacy of other constitutional norms, such as the ones related to health and education. With these arguments, he then deployed the principle of proportionality by stating that, first, the intervention would not even

[593]The idea that constitutional rights have limits according to the interests of the overall society and other constitutional rights, bringing about conflicts in particular situations, is what is behind the concept of "limits of limits" (*Schranke der Schranke*). See, for this purpose, Bodo Pieroth and Bernhard Schlink, *Grundrechte: Staatsrecht II* (Heidelberg: C. F. Müller, 2006), 51ff.

[594]IF 2.915-5 (published on 11.28.2003). Justice Gilmar Mendes's opinion, translation mine.

[595]See the next chapter.

[596]Ibid., Justice Gilmar Mendes's opinion, translation mine.

[597]According to Justice Gilmar Mendes: "(...) The deployment of the principle of proportionality in cases such as the present one, in which there is the claim to the Federal Union's activity in the realm of federal state's autonomy, is admitted by the German law. For this purpose, Bruno Schmidt-Bleibtreu and Franz Klein remark, in commentary on art. 37 of the Basic Law, that "the means of federal execution (*'Bundeszwang'*) are established by the Constitution, by the federal laws and by the principle of proportionality" ("Die Mittel des Bundeszwanges werden durch das Grundgesetz, die Bundesgesetze und das Prinzip der Verhältnismäßigkeit bestimmt," *Kommentar zum Grundgesetz*, 9a. ed., Luchterhand, 795). (Ibid., Justice Gilmar Mendes's opinion, translation mine)."

[598]Art. 100 of the Federal Constitution of 1988.

[599]IF 2.915-5 (published on 11.28.2003). Justice Gilmar Mendes's opinion, translation mine.

pass through the principle of suitability, for that measure would have limits of constitutional nature emerging from other state obligations, thereby not being suitable to achieve its goal; second, it was not necessary, either, because it would be less harmful if the plaintiff waited for the government's normal activity, which was working progressively and in good faith to honor its obligations; and third, it was not proportional in the narrow sense, for the goal – the payment of alimony credits – was not weightier than the onus this measure would cause to the *whole society*. By using the premise of a "relation of conditional precedence,"[600] Justice Mendes, finally, reached the conclusion that the principle of state autonomy and the state interest in not having its public services threatened should prevail over the constitutional principle of protection of judicial decisions and the constitutional norms concerning the right to receive alimony credits in those circumstances.[601] This would be, moreover, the right interpretation, the one that was not a "simplistic reading of the constitutional text."[602]

This is a leading case nowadays in every new lawsuit related to alimony credits the Brazilian federal states must pay in virtue of a judicial condemnation, one that transformed the mechanism of federal intervention into a rhetoric and ineffective constitutional measure. This is also a leading case to comprehend how balancing, deployed within the triadic framework, emerged as an instrument to shape constitutional rights with other values (social, political, economic, etc.), conditioning them hence to a teleological evaluation of weight that has much of the court's discretionary view. The turning point of balancing towards this systematized deployment according to a methodological and logical framework, at first glance, would supposedly provide a better rational and legitimate response to the dilemmas arising from the new constitutional model, and particularly the new approach the STF would undertake from that period onwards. Yet, it was not this methodological quality that changed the perception that balancing, no longer unsystematically applied, is closely connected to the use of discretionary power in constitutional adjudication. The "rational" comprehension of balancing, except for being framed in accordance with the *Weight Formula*,[603]

[600]Ibid., Justice Gilmar Mendes's opinion, translation mine.

[601]It is interesting to verify that, in other opportunities, this court understood that the argument of the "reserve of the financially possible" could not be used against a constitutional norm, demonstrating thereby how this concept is extremely malleable according to the judge's discretionary viewpoint. See, for instance, the case in which it was discussed an appeal against a lower decision that determined the ample and unrestricted access of all children up to six years old to crèches in the city of Santo André. The STF, in this case, clearly stated that the argument of the "financially possible" – except when there are motives objectively assessed (not presenting, though, any explanation of which motives they were) – could not be used as a justification for not observing the duty established by the Constitution (See RE-AgR 410.715. published on 02.03.2006).

[602]IF 2.915-5. (published on 11.28.2003). Justice Gilmar Mendes's opinion, translation mine.

[603]See the next chapter.

did not seem essentially different from its unstructured deployment of the previous years.[604]

[604]After this case, many other interesting debates took place in the STF that led to the deployment of the principle of proportionality with the triadic dogmatic structure. A very well known example is the HC 82.424 (published on 03.19.2004), the *Ellwanger case* previously examined (see the first chapter), judged in 2003, where this constitutional court balanced the freedom of speech, on the one hand, with the principle of human dignity and equality, on the other, with distinct and irreconcilable approaches by each of its Justices, particularly Gilmar Mendes and Marco Aurélio, the former by centering more on the constitutional principles in play, and the latter by using balancing to insert a naturalistic justification tied to a semantic approach. Another example was the ADI 3.324 (published on 08.05.2005), judged in 2004, which discussed the possibility of transference of civil or military public officers and his or her dependents from private to public universities, according to the Law n. 9536/97. In this case, Justice Gilmar Mendes, founded upon the equality principle – using, for this end, Alexy's *Theory of Constitutional Rights* – sustained that this privilege had no "sufficient reasons" (translation mine) for the discrimination. Afterwards, he proceeded to the deployment of balancing, in which the equality principle, the university autonomy and the economic argument of Brazilian public universities' budgetary limits were introduced after a theoretical analysis of the "thinking of the possible" (translation mine). For this purpose, Justice Gilmar Mendes entered into the debate with authors such as Gustavo Zagrebelsky and Peter Häberle, providing thereby the basis for the premise of the "legal thinking of the possible," as well as bringing out the arguments to conclude that the obligatory transferences have limits on the public universities' budgets.

Moreover, in what refers to the possibility of modulation of effects, balancing became a very powerful instrument, even to justify it in the diffuse system of judicial review, whose possibility was not legally defined. We can cite two relevant cases in this matter:

(1) The HC 82.959 (published on 09.01.2006), in criminal law. In this case, the principle of proportionality with this framework appeared, – as a dissenting opinion, though – in Justice Gilmar Mendes's extensive opinion. He contradicted the then prevailing precedent, which stated that, according to the Law n. 8.072/90, in case of heinous crimes, there was no possibility of progression of the regime of the punishment, and therefore the criminal would remain in prison during the whole period of the conviction. The main argument was that this law, by abstractly establishing the impossibility of this progression, violated the constitutional principle of individualization of punishment. For this purpose, he presented initially the doctrine of the "essential core of fundamental rights," its divergences (absolute and relative theory), based, above all, on German scholarship and practice, and then began to develop the connection between the idea of "essential core of fundamental rights" and the principle of proportionality. With these premises in mind, he sustained that this legal norm offended the "essential core of the principle of individualization of punishment." Through the examination of other legal statutes, he concluded that the prohibition of progression of regime was neither suitable nor necessary, for there were other mechanisms equally effective and less harmful to the rights of the criminal. However, if we could consider this decision a right one, at the end, Justice Gilmar Mendes proceeded to the modulation of its effects, according to balancing, extending through decision-making thereby a possibility that existed only in the abstract system of judicial review. Founded upon this presumption, he declared the unconstitutionality of some articles of the law n. 8072/90 but defended the effect *ex nunc* of the decision, for it could cause serious repercussions for the civil, procedural and criminal area. Balancing appeared within the context of placing a value – the *serious repercussions for the civil, procedural and criminal area* – to be weighted in favor of a practice (the modulation of effects in the diffuse system of judicial review) that was neither historically nor even legally authorized. See also HC 85692/RJ (published on 09.02.2005); HC 85687/RS (published on 08.05.2005).

(2) The ADI 2240 (published on 08.03.2007). This case referred to the analysis of the constitutionality of the Law n. 7.619/2000 of the state of Bahia, which created the municipality

This perception reveals itself more clearly when we verify the deployment of balancing to justify this increasing STF's shift to political activism. In this respect, a very remarkable case is the ADI 3.112,[605] judged in 2007, in which Justice Gilmar Mendes, as a dissenting opinion, deployed the principle of proportionality to declare the unconstitutionality of article 21 of the Law n. 10.826/2003, concerning the *Disarmament Act* (*Estatuto do Desarmamento*). This legal norm stated that crimes of illegal possession of fire weapon of restricted use, illegal commerce of fire weapons and international traffic of fire weapons were not subject to parole, with or without bail. Justice Gilmar Mendes deployed the principle of proportionality, using, for this end, Robert Alexy's thinking, as his main source, and five BVG's cases.[606] With respect to balancing, he exposed clearly that this procedure leads to political arguments by arguing that the court examines whether the legal intervention in the basic rights "keeps a relation to the proportionality of the established goals for the criminal policy."[607] The court should deploy balancing, accordingly,

---

of Luís Eduardo Magalhães in a municipal electoral year. After Justice Eros Grau's opinion, which declared the constitutionality of that legal norm and thus the dismissal of the action, Justice Gilmar Mendes affirmed that this was a situation that required the observation of principles as optimization requirements, leading inevitably to the technique of balancing. In the circumstances, there was the principle of legal security, for the municipality had been already factually established for more than six years, and the principle of nullity of unconstitutional laws. Within this context, Justice Mendes initiated a serious analysis of why this last principle could be mitigated in Brazilian constitutional reality. Considering comparative law and examples, even from the United States (*Linkletter v. Walker* (381 U.S. 618), judged in 1965), he demonstrated that, on account of practical reasons, one could shape this principle according to the particular circumstances, founded on legal security, equity or exceptional public interest. This was why, in complex state of affairs, the STF would have to deploy balancing between the legal security, the exceptional social interest, on the one hand, and the principle of nullity of unconstitutional laws, on the other. For this reason, Justice Mendes argued that the law that created the municipality of Luiz Eduardo Magalhães was unconstitutional, but not void, establishing in the sequence a period of twenty four months for the state legislator to appreciate again this matter. He used then balancing to reduce the normative force of the constitutional principle of nullity of unconstitutional laws. Justice Sepúlveda Pertence, although afterwards having agreed with the balanced solution, on the other hand, did not see this shift as a normal one (and also Justice Marco Aurélio), expressing, in this regard, his serious apprehension about what was occurring at the court in this shift to the modulation of effects of the declaration of unconstitutionality. His words: "I am not able yet, with all the excuses to the Washington Court – case Linkletter and others – to remain calm and pose, in this very difficult living of the diffuse and the direct judicial review, more this problem of temporal modulation of the effects of the declaration in theory of the unconstitutionality of a law. Particularly, in a legal system as ours, where the guarantee of the vested rights, the perfect legal act and res judicata against the law is an expressed and intangible text of the Constitution (. . .) Now, legal acts are perfected and rights are vested by the non-incidence of an unconstitutional law. Therefore, where there are vested rights, I cannot really accompany the possibility of the court to project to the future the initial term of the unconstitutionality of a law" (translation mine).

[605]Published in 10.28.2007.

[606]*Mitbestimmungsgesetz* (BVerfGE 50, 290); *Lagerung Chemischer Waffen* (BVerfGE 77, 170); *Mühlenstrukturgesetz* (BVerfGE 39, 210), *Cannabis* (BVerfGE 90, 145), and *Apothekenurteil* (BVerfGE 7, 377).

[607]Ibid., translation mine.

as an instrument to place side by side a constitutional principle and a policy as a way to evaluate whether it is a proportional measure for the promotion of the social welfare.[608] In this regard, after having discussed many other interesting arguments,[609] he defended that this measure was disproportionate, excessive, insofar as for more serious crimes, such as murder, parole was possible in some circumstances.[610] Justice Marco Aurélio manifested similar understanding.[611]

This case draws attention particularly because Justice Gilmar Mendes, founded upon the disproportionality of the measure, considered an important aspect of a political program that aimed to drastically reduce the use of fire weapons in Brazilian society to be unconstitutional, which resulted from many discussions that took place in the parliament. Indeed, Justice Sepúlveda Pertence argued that the Brazilian Constitution, in its article 5, LXVI,[612] conferred on the legislator the authority to define which crimes would be subject to bail, with only the exceptions of article 5, XLIII.[613] Only in absolute unreasonable cases could, therefore, the court judge the legislative definition unconstitutional in this matter. He even said that there was "this real fever to transform the Supremo Tribunal itself into a real court of appeal of the National Congress' judgment of reasonability."[614] Justice Eros Grau, by the same token, sustained that "there is no sense in taking decisions outside the Constitution, according to the criterion of reasonability or the proportionality."[615]

If this case already seriously indicates this STF's shift to political activism, despite some manifest resistances, the ADI n. 3510[616] brought it to the maximum point. In this case, regarding the use of embryonic cells for research, a relevant debate took place between Justices Marco Aurélio and Gilmar Mendes that can well illustrate how far the deployment of balancing is able to sustain a political discretionary opinion. Whereas the former denied the possibility of the court to supply the

---

[608]In this case, Justice Gilmar Mendes also makes reference to the discussion of the prohibition of excess (*Übermassverbot*) and the protection against deficient legal protection (*Untermassverbot*). See ADI 3.112 (published on 10.28.2007). Justice Gilmar Mendes's opinion.

[609]For instance, he associated this problem with the debate on the prevalence of the principle of innocence (art. 5, LVII) and the constitutional norm that requires justification for every type of prison (art. 5, LXI).

[610]In any case, after his opinion, Justice Gilmar Mendes manifested the difficulty in verifying whether, in these circumstances, the legislator exceeded its legislative function. See ADI 3.112 (published on 10.28.2007). Justice Gilmar Mendes's opinion.

[611]See ADI 3.112 (published on 10.28.2007).

[612]"Art. 5, LXVI – no one shall be taken to prison or held therein, when the law admits release on own recognizance, subject or not to bail."

[613]"Art. 5, XLIII – the practice of torture, the illicit traffic of narcotics and related drugs, as well as terrorism, and crimes defined as heinous crimes shall be considered by law as non-bailable and not subject to grace or amnesty, and their principals, agents and those who omit themselves while being able to avoid such crimes shall be held liable."

[614]Ibid., Justice Sepúlveda Pertence's opinion, translation mine.

[615]Ibid., Justice Eros Grau's opinion, translation mine.

[616]Published on 05.28.2008.

legislative omission through recommendations, remarked that "strange opinions to the law cannot themselves prevail," and mentioned the risks of "relativizing, under the viewpoint of convenience, the realm of evaluation of the legislator elected by the people,"[617] the latter defended the court's political role by arguing that "important decisions of the contemporary societies have been decided not by the people's representatives joined together in parliament, but by constitutional courts."[618] Justice Gilmar Mendes even asserted that, in cases like that one, the STF turns into a "house of commons, as the parliament,"[619] where the "multiple social claims and the political, ethical and religious pluralism find refuge in the debates procedurally and argumentatively organized through previously established norms,"[620] such as the public hearings, the *amicus curiae* intervention, and the participation of society through different civil organizations during the procedure.[621]

The court, as long as it exists to "make the most rational decisions,"[622] has no problem with the parliament in this matter. In Justice Mendes's view, relevant matters as the one related to the exam of embryonic cells for research could not lead to the conclusion that they "would be best decided by majoritarian institutions, and that they would have therefore a greater democratic legitimacy;"[623] constitutional adjudication is a matter that has usually many "tasks that transcend the boundaries of the legal matter and involve moral, political and religious arguments."[624] Hence, inasmuch as the legal statute under scrutiny, according to his point of view, was insufficient for the necessary protection of the embryo, based on the principle of proportionality (as a prohibition against insufficient protection),[625] he, as positive legislator, remarked: the creation of a Central Ethical Committee[626] in this matter was indispensable.[627] Every research using

[617]ADI n. 3510 (published on 28.05.2008). Justice Marco Aurélio's opinion, translation mine.

[618]Ibid., Justice Gilmar Mendes's opinion, translation mine.

[619]Ibid., Justice Gilmar Mendes's opinion, translation mine

[620]Ibid., Justice Gilmar Mendes's opinion, translation mine.

[621]Ibid., Justice Gilmar Mendes's opinion.

[622]Gilmar Mendes, interview by Izabela Torres, "Entrevista – Gilmar Mendes," *Correio Braziliense*, Brasília (August 17, 2008).

[623]ADI n. 3510 (published on 05.28.2008). Justice Gilmar Mendes's opinion, translation mine.

[624]Ibid., Justice Gilmar Mendes's opinion, translation mine.

[625]Justice Gilmar Mendes used the doctrine of Claus-Wilhelm Canaris to associate the idea of prohibition of excess with the prohibition of insufficient protection, deploying, in the sequence, the principle of proportionality.

[626]Justice Gilmar Mendes's opinion resembles the BVG's opinion in the second case of abortion (BVerfGE, 88, 203 – *Schwangerschaftsabbruch II*), in which that court pointed out that the temporal exception of twelve weeks to the general criminal rule could only occur if some requirements were previously filled, establishing besides the need of counseling with the purpose of fostering the continuity of pregnancy. With this case, therefore, it is possible to observe how influenced by the BVG the STF has progressively been, not only in the deployment of similar methodologies but also in this shift to a more active approach. See the second chapter.

[627]Ibid., Justice Gilmar Mendes's opinion.

embryonic cells would, accordingly, previously require the authorization and approval of this Committee.[628]

This case is particularly interesting because it reveals how the deployment of balancing based on the debate on human dignity led to different results, some Justices using this argument to protect the embryo, others to safeguard the interests of the beneficiaries of the research on embryonic cells. In any case, what is particularly remarkable is that five Justices of this court,[629] even though not declaring the unconstitutionality of the legal provision, attempted to establish either some safeguards or even some norms, as a condition for considering those legal provisions constitutional. While Justice Gilmar Mendes sustained the requirement of a Central Ethical Committee, Justice Menezes Direito, in a more extensive list, created six norms starting from the inspection by the Ministry for Health, passing through the prohibition to destroy the embryo, and reaching, we could say, the definition of a crime.[630] There were no more limits indeed to judicial review, for there was no more difference, both qualitatively and democratically, between its activity and that one of parliament. In reality, this assimilation to the legislative function would provide a comparative advantage: the STF, in the political realm, could make, with balancing, decisions rationally justified.

The criterion of rationality appeared then as a differential force to legitimate an activity that was, step by step, getting closer to legislation. The STF, whose involvement in the construction of the democratic regime was very timid and self-restraint at the beginning of this period, would transform itself gradually into an activist court. The "Guardian of the Constitution," a function it seemingly was not achieving because of its passive approach, would only supposedly be suitably exercised as long as it built a new perspective, one that called for this political attitude towards a variety of themes of social life. The practice of citizenship, increasingly jeopardized by means of the disruption of the diffuse system of judicial review, was now reestablished from above: the STF, more than ever, in this new activist approach, would undertake this enterprise, justified by its capacity to provide more rational decisions.

It could hence legitimate itself thanks to its aptitude and talent for discovering answers originated from longstanding debates, reflected arguments, and rationally and methodologically structured reasons. Now not only would citizens but also their elected representatives – who should exercise the political function of evaluating the different values at stake in lawmaking – have their decisions reviewed by each one of the court's members, according to their discretionary but seemingly rational axiological choices. The capacity to provide "the most rational decisions" would justify its political role, and balancing, especially now in the framework of the

628 Ibid., Justice Gilmar Mendes's opinion.

629 Justices Gilmar Mendes, Eros Grau, Ricardo Lewandowski, Menezes Direito and Cezar Peluso.

630 According to Justice Menezes Direito, if the institutions responsible for the research, once having submitted their projects to the Ministry for Health, had their projected approved in violation of the presented recommendations, this would configure a crime according to art. 24 of the Law n. 11.105/2005. See Ibid., Justice Menezes Direito's opinion.

principle of proportionality, would accomplish the goal of marrying policy with rationality in decision-making. The pathway to power had to be justified both politically, as a necessary rearrangement of the principle of separation of powers, and rationally, as a reorganization of an institution that could suitably promote rational responses to society. If, on the one hand, we could then argue that this new STF was born from the inevitable and natural tendency to concentrate powers, on the other, it could certainly achieve it due to the strength of its seemingly rational motivation. The problem of legitimacy, accordingly, shifts to a debate on rationality.

3.4 Final Words

This chapter aimed to explore the Brazilian constitutional culture, as a means to verify that, in another reality, with a distinct historical context and untranslatable particularities, some connections with German constitutionalism could be found. By the same token, through an analysis that sought to prove that balancing and constitutional court's activism walk side by side, this chapter revealed that, also in Brazil, the constitutional court's progressive shift to activism raises similar apprehensions about the possible encroachment on the principle of separation of powers. Initially, we examined the historical background of a constitutional court emerging as the "Guardian of the Constitution" after a period of military regime and authoritarianism, which should then undertake the role of protecting the Constitution against the possibility of any return of the authoritarian past. In a rather comparable perspective with Germany, being the "Guardian of the Constitution" changed progressively into the idea of protecting social values through a broader comprehension of basic rights. The theory of basic rights as objective principles of a total legal order or the idea that the Constitution is a "concrete order of values" became, more and more, a reality in Brazilian constitutional culture, and the STF's decisions likewise welcomed this interpretation of basic rights. Naturally, as we examined, this was not an immediate outcome of this process of democratization, even because, at the beginning of this period, the STF was somehow very timid and self-restrained in the review of the governmental and parliamentary acts. Still, as long as many legal instruments were created to expand the mechanisms of abstract judicial review, and the court gradually extended these instruments to other areas not in fact legally established, the STF assumed finally the quality of a "forum for the treatment of social and political problems."[631]

If this historical background favored this STF's shift to activism, it also favored the deployment of balancing. The second part of this chapter was oriented towards discussing this relationship between, on the one hand, the dualism between law and politics, and, on the other, the deployment of balancing. In this regard, after some case analysis, a relevant perception was that, with this deployment of balancing and

[631]Schlink, "German Constitutional Culture in Transition," 729.

the growing activism, there was a movement towards the rationalization of this methodology. Particularly, balancing transformed into an element of the principle of proportionality, in a similar fashion to what happened in Germany. Also in Brazil, accordingly, the idea of rationality of balancing came out as a justification for the legitimacy of constitutional adjudication.

For this reason, it is remarkable how the discourse on rationality, both in Brazil and Germany, is connected to the idea of providing a methodological response to the dilemmas of the new constitutional order. After all, by discussing it, the question of legitimacy and rightness of constitutional adjudication seems to happen as a natural consequence. In this respect, the search for a rational justification for balancing is particularly notable, whose most well-known and influential theoretical defense will be examined in the following chapters. After the investigation of the empirical grounds where balancing appeared as a rational solution to the main dilemmas of constitutional adjudication, and a mechanism that can provide correctness and legitimacy when filled with arguments and by following some "rational standards," the next step is to investigate how constitutional scholarship has developed a methodological comprehension of balancing that could account for this movement and prove how rational and indispensable it is for constitutional adjudication. The second part, for this reason, will explore these "rational standards," in order to verify how they could account, or not, for a rationality that seems more adequate within the context of indeterminacy of law. Balancing, for this reason, will be challenged in its very structural justification through the stress on a *concept of limited rationality*, gradually unfolded, one that knows that "the most hateful and unconstitutional attempt to privatize the public" – as if constitutional adjudication were the expression of judges' private viewpoint – is the "egoistic possession and the normative annihilation of the constitution."[632]

[632]Menelick de Carvalho Netto, "A Contribuição do Direito Administrativo Enfocado da Ótica do Administrado para uma Reflexão acerca dos Fundamentos do Controle de Constitucionalidade das Leis no Brasil: Um Pequeno Exercício de Teoria da Constituição," *Forum*, March 2001: 26, translation mine.

Part II
The Debate on the Rationality of Balancing

Chapter 4
The Aim to Rationalize Balancing Within the Context of Constitutional Courts' Activism

Abstract As long as balancing appears as a fundamental instrument of this new constitutionalism where the constitutional court plays the role of "Guardian of the Constitution" through the interpretation of subjective rights as though they were objective principles of the total legal order, the aim to rationalize it appears as an immediate consequence. Rationality, accordingly, relates to the purpose of providing decisions that could best fulfill the exigency of legitimacy. In this regard, Robert Alexy's *Special Case Thesis* and *Theory of Constitutional Rights* are clear examples of this connection, first, between morality and law, and, second, between balancing and rationality. Since, for Alexy, legal discourse is a special case of general practical discourse, and balancing, through the *Weight Formula* and the definition of preference relations, can provide rationality in decision-making, the main question is how to apply, in constitutional adjudication, this vast field of argumentation according to the premise of "unity of practical reason" without sacrificing the consistency of the system of rights. The relationship between constitutional court's activism, particularly the *Bundesverfassungsgericht,* and balancing seems to adequately links this dilemma to Alexy's premises. Yet, it also reveals that this debate on rationality, as Alexy's defends it, should be challenged by new questions.

4.1 Introduction: The Quest for a Systematization and Rationalization of Balancing

The constitutional court's shift to activism, as we examined in the realities of Germany[633] and Brazil,[634] is intimately connected to this movement towards the expansion of the idea of basic rights as objective principles and the deployment of

---

[633]See the second chapter.

[634]See the third chapter.

J.Z. Benvindo, *On the Limits of Constitutional Adjudication*,
DOI 10.1007/978-3-642-11434-2_4, © Springer-Verlag Berlin Heidelberg 2010

balancing. This movement can be interpreted according to two major points of view: one that stresses the loss of legal dogmatics, the coherence and self-binding criteria in a more active and popular exercise of judicial review, with the consequent loss of a rational basis (for the lack of methodological grounds); and the one that underlines that, since constitutional adjudication inevitably deals with principles – and not only with rules - the procedure is to balance principles in accordance with their weight, which corresponds to a rational methodology for deploying basic rights in constitutional cases. Briefly, whereas the first considers this movement a loss for the constitutional culture with the risks of more and more personal arbitrary rulings, the second understands it as a necessary step in the direction of the rationalization of constitutional adjudication. In other words, whereas the first standpoint interprets it as a shift to irrationalism, subjectivism and decisionism, the second aims to justify rationally this constitutional court's activism. From other perspective, insofar as the situation is of an increasing casuism, whereas the first underlines that the constitutional court's necessary role is to carry out a critical review of constitutional decisions, and expose the relevant incompatibilities and illogical grounds of their contents while constructing the basis for a coherent development of legal adjudication, the second is more oriented to accepting the rationality of the decision, provided that the procedure of balancing is deployed.

The first standpoint has some of its grounds discussed in the first part,[635] which could already raise the concern for the increasing deployment of balancing in this constitutional courts' shift to activism. Indeed, it provided some relevant perceptions to the diagnosis, the symptoms, as well as posed chief questions in this subject matter. It represents, nonetheless, a dissenting opinion where the belief in the rationality of balancing prevails. Perhaps because of its very possibility to set up a methodological framework for balancing (as if the problem of rationality of decision-making were a question of defining a formal structure), perhaps due to the argument surrendering to the reality (balancing, after all, is so inserted into German constitutional culture that the solution is to accept it how it is), or perhaps because the concept of legal rationality is still a methodological rationality bearing an objectivistic self-comprehension of science and technique, the fact is that most German scholars strongly defend the rationality of balancing.

In this respect, the usual impasses arising from the connection between rationality and separation of powers as well as between rationality and deontology of legal rights are mitigated by the argument that balancing itself does not provide the solution to the case, nor is responsible for the quality of the argument that is placed in its framework, but rather is simply a formal structure that eases the complicated tasks of constitutional adjudication. Balancing, accordingly, is basically an analytical and structural framework;[636] a formal principle, whose contents are defined

[635]See the second chapter.

[636]See Robert Alexy, *Theorie der Grundrechte* (Frankfurt a.M: Suhrkamp, 1994), 32 ff.

in each decision;[637] a formal structure shaping and coupling the contents;[638] an instrument for a formal, "metatheoretically" neutral and clear conception of justice;[639] or even a mere permanent form, always correct and valid, in which only the contents can vary when it is concretized in every decision.[640] By arguing that it is a formal structure, the question regarding the material category, the last argument guiding balancing, is transported to the development of the contents, which are flexible enough to be adapted for different circumstances and, mostly, to different justifications. The question of rationality, consequently, does not face directly those issues, for they are concerned with the material contents of the argument placed in the structural framework of balancing, or, when it does, it attempts to deal with it either by presenting more formulas and general rules or by introducing some decisions as examples.[641]

Within this context, rationality becomes a question of capacity to provide a formal and universal structure where arguments and justifications can be correctly placed and a proportional analysis carried out, without this leading to the insurmountable dilemma of categorically establishing material contents defining how the proportion needs to be defined. In other words, the rationality of balancing, according to this view, is concerned with the capacity to deploy a formal structure, which, because of this quality, gains universality, and to place arguments in this structure. If it is not possible to point out a material methodological or normative status of the demanded evaluation,[642] at least a formal framework can be rationally conceived in which different views establishing a rule for balancing can manifest

---

[637]See Lothar Hirschberg, *Der Grundsatz der Verhältnismäßigkeit* (Göttingen: Otto Schwartz & CO, 1981), 77.

[638]See Peter Lerche, *Übermaß und Verfassungsrecht: zur Bindung des Gesetzgebers an die Grundsätze der Verhältnismäßigkeit und der Erforderlichkeit* (Goldbach: Keip, 1999), 224.

[639]See Nils Jansen, *Die Struktur der Gerechtigkeit* (Baden-Baden: Nomos, 1998), 162.

[640]See Theodor Lenkner, *Der rechtfertigende Notstand: Zur Problematik der Notstandregelung im Entwurf eins Strafgestzbuches* (Tübingen: Mohr-Siebeck, 1965), 156.

[641]Robert Alexy attempts to link rationality and correctness to the formal structure of balancing. See Robert Alexy, "On Balancing and Subsumption – A Structural Comparison," *Ratio Juris* 16, no. 4 (December 2003); Alexy, *Theorie der Grundrechte*. The question of democratic legitimacy of balancing is also examined through the premise of the "weight formula" in Robert Alexy, "Balancing, Constitutional Review, and Representation," *International Journal of Constitutional Law* (Oxford University Press and New York School of Law) 3, no. 4 (2005); Robert Alexy, "Law and Correctness," in *Law and Opinion at the End of the Millenium: Current Legal Problems* (Oxford: Oxford University Press, 1988).

[642]Schlink indicates the problems of balancing. He introduces some relevant questions that demonstrate this complex situation: "How should the evaluation of goals and means be found? Is the parameter found in the principle [*balancing*] it demands? Are the evaluations, accordingly, correct, when and because the BVG finds it? Do correct evaluations simply understand themselves from themselves? Can they obtain methodical reliability and normative enforceability in another way? Or must they be particularly exempt from evaluation, because they are unreliable and non-binding? See Bernhard Schlink, "Der Grundsatz der Verhältnismäßigkeit," in *Festschrift – 50 Jahre Bundesverfassungsgericht* (Tübingen: Mohr Siebeck, 2001): 454, translation mine.

themselves.[643] The question of rationality becomes, above all, a question of methodology.

As regards this view, we will focus now on one of the most prominent theories concerning the defense of the deployment and the rationality of balancing. This theory is Robert Alexy's *Theory of Constitutional Rights* (*Theorie der Grundrechte*)[644] as well as his other writings. After having briefly introduced some different approaches to balancing,[645] the choice for Robert Alexy's theory has several relevant reasons. First, it is one of the most widespread contemporary influential theories with respect to the deployment of the principle of proportionality, and balancing in particular. Indeed, Alexy is certainly one of the most well-known authors in the discussion of legal reasoning. His *Theory* is, as Augustín José Menéndez and Erik Oddvar Eriksen describe, "one of the most authoritative general expositions of German fundamental rights law,"[646] and its adoption as a fundamental reference book in this matter is worldwide. Second, it is one of the most consistent defenses of the rationality of balancing and a very interesting and instigating thinking to be confronted with the investigation of the next chapters. Third, as one of the most notable defenders of the rationality of balancing, particularly by sustaining it as a formal framework, Robert Alexy is a necessary reference to observe how scholarship deals with the contemporary transformations in German and Brazilian constitutional cultures. Finally, by critically reviewing Alexy's premises, especially because he sustains his arguments directly by connecting them to the BVG's decision, we are somehow developing a critical analysis of the BVG's - and, in a sense, STF's - constitutional practices.

Against most of the above critiques, Alexy attempts to expose how balancing can provide not only a rational response to constitutional dilemmas but also rightness and legitimacy. He is a strong exponent of how constitutional adjudication can be structured in some models and schemas of reasoning founded upon a methodological comprehension of the principle of proportionality, and especially balancing. This is why Robert Alexy's theory can be regarded as one of the most refined examples of the new constitutionalism in Germany, one that attempts to cope with the expansion of a casuistic Jurisprudence and with the idea of objective principles embracing the totality of the legal order by bringing forward a structural analysis of possible responses to constitutional adjudication. Moreover, Alexy reproduces and systematizes many of the developments of the BVG's practice, and reveals many of the characteristics this court displays through its decisions. Accordingly, the attack on Alexy's approach reaches the practice of constitutional courts and vice-versa, which links it to the examination of the developments of

[643]Lothar Hirschberg indicates many examples of different BVG's viewpoints in the deployment of balancing. See Hirschbert, *Der Grundsatz der Verhältnismäßigkeit*, 78–83.

[644]See Alexy, *Theorie der Grundrechte*.

[645]See the second chapter (Sect. 2.5).

[646]Agustín José Menéndez and Erik Oddvar Eriksen, "Introduction," in *Arguing Fundamental Rights*, ed. Agustín José Menéndez and Erik Oddvar Eriksen (Dordrecht: Springer, 2006), 4.

German[647] and Brazilian[648] constitutionalisms and the central questions they raise. Furthermore, it will guide our analysis as a reference source to question the rationality that is behind balancing as well as gradually help us unfold the concept of limited rationality in constitutional adjudication. For this purpose, this chapter will introduce the foundations of his thinking and his advocacy of the rationality of balancing, which are intimately related to his Special Case Thesis (*Sonderfallthese*), whereby he constructs his defense of the "unity of practical reason" (Sect. 4.2). Afterwards, we will explore with more details the framing of his rational approach to the question of indeterminacy of law through the emphasis on balancing within his structural framework, a debate that will enter directly into his *Theory of Constitutional Rights* (*Theorie der Grundrechte*) (Sect. 4.3). In the following chapters, at any rate, some ramifications of his premises will be discussed,[649] bringing thereby into light the main features of this powerful thinking that has intensively influenced not only constitutional decision-making but also how constitutional scholarship deals with its own constitutionalism.

4.2 Robert Alexy's Special Case Thesis (*Sonderfallthese*)

In the debates on legal reasoning, Robert Alexy's Special Case Thesis (*Sonderfallthese*), introduced in his book *Theory of Legal Argumentation* (*Theorie der juristischen Argumentation*),[650] echoes as a very controversial view regarding the connection between law and morality. His main premises are that legal discourses, first, address practical questions, that is, they are concerned with what is prohibited, permitted, obligatory; second, raise the claim to correctness, which needs to be discussed and decided; and, third, have their decisions limited to the context of a legal framework and a certain valid legal order.[651] The first two premises demonstrate that legal discourses are part of the general practical discourses, and the third justifies why they are special. They also indicate that his theory could be categorized, in principle, as a discourse theory, for it is grounded in the premise of a discourse rationality – which seems to recall Jürgen Habermas's considerations on *communicative action*[652] - by focusing on the speech acts involving claims to truth and correctness, and thus leading to an exigency of justification.

---

647 See the second chapter.

648 See the third chapter.

649 Particularly, the question of how Alexy defends the correctness, legitimacy, and coherence of legal reasoning through balancing will be more directly examined in the next chapters.

650 See Robert Alexy, *Theorie der juristischen Argumentation: Die Theorie des rationalen Diskurses als Theorie der juristischen Begründung* (Frankfurt a.M.: Suhrkamp, 1989).

651 Ibid., 263.

652 See the sixth chapter.

According to this view, in legal reasoning every discourse must be justified based on some rationality rules - such as the idea that all potential participants have the right to raise claims in equal conditions, all potential participants must justify their claims, unless they explain why they refuse to do so, and no participant can have the right to participation suppressed by any type of coercion, whether internal or external to the discourse, among others[653] - and thus guaranteed through procedures of intersubjective participation. The different arguments are then evaluated according to rules guaranteeing the conditions for a procedure allowing the rational acceptability of the best argument. Nevertheless, although these rules apply here, legal discourse is special because they are a type of rational justification in the realm of a valid legal system,[654] which, unlike general practical discourse, prompts the need for a decision that will have to deal with the restrictions of space, knowledge and time in a way that can provide rationality as much as possible. Indeed, legal discourses complement general practical discourses, inasmuch as, through legislative procedure, general decisions are made, and, through adjudication, singular decisions come to light and be deemed rational as a result of a procedure in which different arguments are examined in compliance with pre-established norms. The claim to correctness – and thus its justifiability - for this reason, in Alexy's view, is not concerned with what is "absolutely correct," that is, what is universally accepted as correct, but rather with "what is correct within the framework and on the basis of a validly prevailing legal order."[655] It takes place in the boundaries of a legal order, and, although it is "bound to statutes and to precedents,"[656] it has also to "observe the system of law elaborated by legal dogmatics."[657]

In any case – and this is where the main controversies arise – it seems that these institutional and authoritative boundaries of legal discourses are more flexible than the words above give the impression of. In Alexy's opinion, the similarity and specialty of legal discourses in comparison with general practical discourses, at any rate, is justified in a broader sense by the idea that they express the "unity of practical reason"[658] and, as such, have to be deployed in decision-making as

[653]Alexy develops a very detailed explanation of his *Theorie der juristischen Argumentation* in order to expose a general theory of general rational practical discourse, where he examines: possible discourse theories, justification of practical rules of discourse, forms and rules of general practical discourse (basic rules, rationality rules, forms of arguments, rules of justification, etc), and limits of general practical discourse. In this explanation, he defines the basic lines for the comprehension of his theory as a discourse theory and opens up the debate to examine how and why legal discourses are a special case of general practical discourse (*Sonderfallthese*). See Alexy, *Theorie der juristischen Argumentation*, 221–257.

[654]Ibid., 264.

[655]Robert Alexy, "The Special Case Thesis," *Ratio Juris* 12, no. 4 (December 1999): 375.

[656]Ibid., 375.

[657]Ibid., 375.

[658]Ibid., 383.

though the claim to legal correctness necessarily implied the claim to moral correctness.[659] Accordingly, if, on the one hand, he says that legal reasoning must take into account the institutional background of norms and procedures, on the other, "legal reasoning remains deeply connected with what can be called the free, discursive, or ideal side of law;"[660] that is, it demands more than what is authoritatively solely determined;[661] "it must be free to a certain degree."[662] The stress on institutional grounds is highly diluted by a certain conditioning of legal reasoning to a broader sense of practical reason, even though recognizing the specialty of legal discourses. Evidently, what this broader sense – this "unity of practical reason" – means in the practice of decision-making is not of unchallenging answers. But Alexy's words appear to be sure of their adequacy. This is true in that, whereas the *Theory of Legal Argumentation (Theorie der juristischen Argumentation)* introduces and explains why legal discourses are a special case of general practical discourses, which leads to the idea of "unity of practical reason," his *Theory of Constitutional Rights* (*Theorie der Grundrechte*) operationalizes this premise into adjudication.

As regards this "unity of practical reason," the different types of arguments – institutional and non-institutional (moral arguments, founded upon universal considerations of what is equally good for all; ethical-political arguments, grounded in self-understandings and traditions of a collectivity; and pragmatic arguments based on means/goals relations of interests and compromises)[663] - give rise to an integrative theory,[664] according to which these arguments are gathered not simply as an addition, but rather in conformity with a complementary,[665] systematical and rational evaluation of their strength to legal reasoning.[666] The combination of legal and other types of practical discourses, which are "combined at all levels

[659]Robert Alexy, *The Argument from Injustice: A Reply to Legal Positivism* (Oxford: Clarendon Press, 2002), 77.

[660]Alexy, "The Special Case Thesis," 375.

[661]Ibid., 375.

[662]Ibid., 375.

[663]Ibid., 379.

[664]Alexy calls his theory founded upon the special case thesis "integrative theory," according to which rational legal arguments are combined with general practical arguments at all levels (See Ibid., 380).

[665]Alexy remarks that "general practical discourse is not a simple mix or combination but a systematically necessary connection expressing the substantial unity of practical reason. This is the basis of the special case thesis" (Ibid., 379).

[666]Alexy constructs a system of priority relations between the elements of general practical discourse founded on the following rule: the good (ethical-political arguments grounded in self-understandings and traditions of a collectivity) prevails over the suitable (pragmatic arguments founded on means/goals relations of interests and compromises), and the just (moral considerations of what is equally good for all) prevails over the good. He knows, nonetheless, that this rule is complex, especially in the realm of just and good, for the "just is permeated by the good" (Ibid., 379).

and applied jointly,"[667] therefore, is made with some criteria from mathematic and economic[668] models and regarded as a condition for the correctness and coherence of legal discourse. Hence, Alexy's legal theory (*Sonderfallthese*) exposes that his concept of legal discourse springs from the assumption that legal arguments can be controlled by, and recreated in, general practical arguments - a condition, besides, that transforms balancing into an effective mechanism to evaluate them, and an indispensable instrument to provide correctness and coherence.

The consequence of his integrative theory founded upon the *Sonderfallthese* is the conclusion that the system of rights cannot itself provide the answers to legal reasoning. Instead, legal reasoning stems from the premise that, even though legal discourses work with the claim to correctness applied to a particular framework and system of rights, it is integrated with other general practical arguments,[669] for there is a "unity of practical reason."[670] It has to embrace them as a totality originating from the legal and social order. What matters to legal reasoning is that these general practical arguments, as a unity, are institutionalized, regardless of whether this institutionalization takes place in legislation or adjudication: "The legal system of the democratic constitutional state is an attempt to institutionalize practical reason."[671] Practical reason not only justifies the legal system but is also an essential part of legitimate procedures of will-formation and a demanding source for correctness,[672] for they supplement, permeate, and even control legal discourses:[673] "General practical arguments have to float through all institutions if the roots of these institutions in practical reason shall not be cut off."[674] The float of these general practical arguments through the institutions, according to Alexy's special case thesis, is enough to expose how similarly strong they can be for legal reasoning in the process of balancing, and how this process can coherently house them without this meaning a change of their non-institutional character.[675]

---

[667]Ibid., 380.

[668]Alexy acknowledges that "the conceptualization of the principles as optimization commands does indeed lead to the incorporation of criteria of economic rationality into the law" (Robert Alexy, "Jürgen Habermas's Theory of Legal Discourse," in *Habermas on Law and Democracy: Critical Exchanges*, ed. Michel Rosenfeld and Andrew Arato (Berkeley, CA: University of California Press, 1998), 229).

[669]This is why Alexy can only view the idea of basic rights as substantiated by a certain morality and not by the logic of norms: "The basic rights 'strict priority,' as far as it exists, is substantiated morally, rather than by the logic of norms" (Ibid., 228).

[670]Alexy, "The Special Case Thesis," 383.

[671]Ibid., 383.

[672]Ibid., 383–384.

[673]See Alexy, "Jürgen Habermas's Theory of Legal Discourse," 232.

[674]Alexy, "The Special Case Thesis," 384.

[675]In Alexy's opinion, what matters is that these non-institutional arguments remain arguments: "(...) as long as they remain arguments they retain what is essential for this kind of argument: their free and non-institutional character" (Ibid.).

Since they are arguments, they can be "embedded, integrated, and specified as much as one wants."[676] Accordingly, it is not problematic that an argument in legal adjudication stresses a collective goal instead of an individual right by balancing the distinct general practical discourses (institutional and non-institutional). What is really relevant at the end is that a proportional analysis of them is carried out, that is, "no matter whether an institutional right is restricted in favor of collective goods or of other persons' individual rights, the restriction is necessarily prohibited and violates the right unless it is suitable, necessary, and proportional in the narrower sense."[677] For this reason, unless a collective good causes a disproportionate harm to the individual right, the decision considering the collective good weightier than the individual right can be coherent. Indeed, for Alexy, notwithstanding that there is the danger of collective goods causing undue restrictions on individual rights, the possibility that a teleological argument prevails over a deontological one should not be banned.[678] What could count in favor of individual rights is merely the fact that they can have what Alexy labels *prima facie* priorities,[679] which only through balancing will be brought to light. In this respect, balancing is the rational response to the indeterminacy of law in the vast world of the "unity of practical reason." This is where Alexy's structural theory appears as a response to the so-strived quest for the rationality of balancing.

4.3 The Quest for the Rationality of Balancing: The Core of Robert Alexy's Theory of Constitutional Rights

The quest for the rationality of balancing and its inexorability within the context of the indeterminacy of law could characterize the central words of Robert Alexy's *Theory of Constitutional Rights*. As a clear example of a constitutional analysis that connects to many of the developments of German constitutional transition to a case to case perspective and activism, as well as the idea of principles with an objective nature embracing the totality of the legal order, Robert Alexy's account is based on the premise of developing a theory with a rational purpose, which can be considered a *structural theory* with an analytical-normative (for it is concerned with the correction of the decision)[680] and some degree of systematic-conceptual clarification[681] in the realm of constitutional rights. Contrary to other approaches that

[676]Ibid.

[677]Alexy, "Jürgen Habermas's Theory of Legal Discourse," 230.

[678]Ibid.

[679]Ibid.

[680]Robert Alexy, *Theorie der Grundrechte* (Frankfurt a.M.: Suhrkamp, 1994), 32.

[681]Ibid., 33–34.

are, according to him, either one-dimensional or too abstract, which Alexy calls "one-point theories" or "combined theories,"[682] a structural theory can provide constitutional dogmatics with clarity, which, in his words, "is an elementary requirement for the rationality of any domain of knowledge."[683] Therefore, rather than attempting to discover a last argument or even the most elementary foundations of legal rights - which could lead to the prior questions here examined - a structural theory devotes itself to supplying a contingent framework of reasoning in constitutional adjudication that corresponds to a formal core able to be discovered behind the employed substantive arguments. With its deployment, not only are many of the material questions previously discussed here placed in another domain of discussion, but mainly an analytical and controlled appropriation of value-judgments[684] emerging from this "unity of practical reason" can be carried out by, in his words, "taking away all political rhetoric and the vacillating struggle of world-views," as well as by providing "instruments which hold out the promise of a fruitful development of existing legal conceptual research."[685] A structural theory, accordingly, becomes a necessary mechanism for developing constitutional law within the context of radical transformations in constitutionalism. It is the new methodological response to the dilemmas that appear in this manifest consolidation of an activist and casuistic constitutional court, and the possible rationalization and systematization of a seemingly coherent model for constitutional adjudication in the realm of a value-based approach.

As a structural theory, which detaches itself from the material discussion that occurs when arguments are inserted into this framework, it does not intend to establish the solution to the case, but rather specify how, structurally speaking, a decision in the field of constitutional rights can be made. It is, for this reason, a system that, although recognizing the multidimensionality of legal theory[686] and all the criticisms a logical approach receives,[687] sustains that only a systematic-conceptual consideration based on the rules of analytical logic is capable of promoting a "rational control of all indispensable evaluations in Jurisprudence and of a methodological controlled use of empirical knowledge."[688] Alexy's *Theory of Constitutional Rights*

[682]According to Robert Alexy, one-point theories are the those that attempt to "derive all constitutional rights from a basic thesis" (Ibid., 30, translation mine) carrying thereby the problem of being very abstract or insufficient to the complexities of the contemporary constitutionalism. Combined theories, on the other hand, "forms the basis of the BVG's jurisdiction" (Ibid., 30, translation mine), which means, in other words, theories that have many different perspectives as premises. The objection, which Alexy presents, is that they "cannot provide any guidance to legal decision-making and justification, but simply represent a collection of highly abstract *topoi*, which one can adopt at will" (Ibid., 31, translation mine).

[683]Ibid., 32, translation mine.

[684]Ibid., 38.

[685]Ibid., 38, translation mine.

[686]Ibid., 37.

[687]Ibid.

[688]Ibid., 38, translation mine.

and its examination of how the BVG deploys the principle of proportionality, and particularly balancing, therefore, clearly aims to establish a structure that can, by acknowledging the relevance of analytical logic, guarantee a "rational control" of how cases are decided by constitutional courts through the consideration of all their features, whose knowledge can also be incidentally and methodologically controlled. Indeed, in his words, "without a conceptual-systematic exposition of the law, Jurisprudence is not possible as a rational discipline."[689] Rationality and methodology are thus intimately related in this purpose of controlling empirical knowledge and constitutional evaluations.

The primary premise of this analytical structure is the principle of proportionality, and mainly balancing, which Robert Alexy introduces in the third chapter of his *Theory of Constitutional Rights.* This principle, with balancing - for it represents the method that can somehow control the empirical knowledge and the constitutional courts' evaluations when there is a collision between constitutional rights - furnishes a practical source for a concept of rationality that deposits in some abstract rules and formal schemas, even though deriving from practical examinations of the courts' activities, the answer to achieve clarity and methodological control in legal discourse, and thus rationality itself. This analytical structure provides the indispensable mechanism to solve the problems of constitutional rights. Its formal basis, when correctly applied, shapes the substantive contents in a way that can be logically inferred that they were rationally justified.

In summary, the rationality of decision-making within this context relates to: first, the relative safety and stability of the procedure (the deployment of all its correlated and concatenated steps) – this is the formal parameter; second, the decision is reached through arguments – this is the substantive aspect. In a certain way, the substantive aspect is not exactly Alexy's focus,[690] but it becomes a subject of interest insofar as arguments comply with the analytical criteria afforded by the principle of proportionality. These criteria allow to establish that those arguments were inferred by means of an adequate apprehension of the relevant facts and of a clarified evaluation of constitutional rights in a singular case. Guarantee, control, and rationality: these seem to be the core of this systematic-conceptual theory. Besides, notwithstanding that Alexy emphasizes that his theory does not furnish definitive solutions to constitutional cases – and neither obviously could it, for cases are always singular - he transforms the principle of proportionality into the basis of a method that aims to provide *correctness* and *coherence*[691] to a particular decision.

---

689 Ibid., 37–38, translation mine.

690 Ibid., 32–38. At any rate, Alexy develops an investigation of the arguments that can be applied according to this formal structure in his book *Theorie der justischen Argumentation: Die Theorie des rationalen Diskurses als Theorie der juristischen Begründung* (283–348) based on three groups: rules of positive law, empirical statements, premises that are neither empirical statements nor rules of positive law (Ibid., 283).

691 We will investigate more directly the claim to correctness and coherence in the next chapters.

The correctness and coherence, which are also the manifestation of *rationality*, are closely associated with the conviction that the application of constitutional rights can be *rationally controlled and guaranteed* by an abstract and analytically structured method that places arguments in logical grounds.

The central argument leading to the conclusion that the principle of proportionality (and especially balancing) is the most elementary premise of his structural theory stems from the distinction between two classes of norms: rules and principles. It is by reason of this distinction that the foundations of his structural theory are presented. Indeed, although Alexy understands that no constitutional norm can in reality result in pure rules or principles,[692] he points out that this is a necessary differentiation to solve the most essential problems of constitutional theory, particularly the issues on the limits and roles of basic rights and the solution to their conflicts.[693] It is also the starting-point for the theme of rationality and its limits in the domain of constitutional adjudication.[694] As two classes of norms, and hence as "basic deontic expressions of command,"[695] principles and rules distinguish one from the other by reason of their quality: whereas principles are "norms which require that something be realized to the greatest extent possible given the legal and factual possibilities,"[696] and therefore have the character of optimization requirements with the dimension of satisfaction in different degrees according to the factual and legal possibilities,[697] rules, on the contrary, are "norms which can be always either fulfilled or not,"[698] which means, in other words, that they contain "established determinations in the field of the factually and legally possible."[699]

[692]Robert Alexy, although differentiating both dimensions, examines the possibility of norms resulting in principles and rules, which he calls "double aspect constitutional norms" (Alexy, *Theorie der Grundrechte,* 124, translation mine). Considering this analysis, Alexy sustains that "it is inadequate to conceive of constitutional legal norms either merely as rules or merely as principles. An adequate model, on the other hand, associates both rules and principles with the provisions of the constitution." (Ibid., 125, translation mine) We can better verify this aspect, for instance, when we insert into a principle a limitation clause that transforms it into a rule. Alexy indicates the example of the principle of freedom of artistic activity, which can become a rule when, for instance, the provision that guarantees the freedom of artistic activity obtains the following prescription: "state interference in activities belonging to the artistic domain is prohibited, unless it is necessary to satisfy competing principles of constitutional degree (whether protecting the constitutional rights of others or collective goods), which in the circumstances of the case take precedence over the principle of artistic freedom" (Ibid., 123, translation mine) In this case, nonetheless, insofar as the limitation clause expressly makes reference to competing principles, it is not a pure rule, but rather what he calls "double aspect constitutional rights norms."

[693]Ibid., 71.

[694]Ibid.

[695]Ibid., 72, translation mine.

[696]Ibid., 75.

[697]Ibid., 76.

[698]Ibid., translation mine.

[699]Ibid., translation mine.

Both, for this reason, have different mechanisms when conflicts occur. In the case of rules, the conflict can be resolved by two ways: one of the rules is declared invalid or an exception is established in the contents of one of the rules.[700] There is no gradation between the validity of rules, and the resolution of the conflict usually relies on such maxims as the *lex posteriori derogat legi priori* or *lex specialis derogat legi generali,* among others.[701] Furthermore, the resolution is found in the realm of validity. In the case of principles, on the other hand, the resolution lies not in the realm of validity, but in the domain of weight.[702] Every principle has a weight that is measured in accordance with the particularities of the case, which will lead to the deployment of balancing as the instrument to define which principle is weightier in a particular circumstance.

The nature of principles, therefore, implies necessarily the principle of proportionality;[703] this principle "logically follows from the nature of principles; therefore, it can be deduced from them."[704] Its justification lies, basically, in the very nature of principles.[705] There cannot be *rational* adjudication in this area without following the principle of proportionality. As a result of this procedure, none of the principles at issue will be deemed invalid and thus excised from the legal system, but simply a proportional harmonization between them takes place. In this analysis, one sets up a "conditional relation of precedence between the principles in the light of the circumstances of the case;"[706] that is, they will need to be regarded as conditioned by factual (the principles of suitability and necessity) and legal (the principle of proportionality in its narrow sense or balancing) features. Moreover, as optimization requirements, the principles, contrary to the rules, do not have the ability to define how they should be applied, and thus they are not definite, but have a *prima facie* character[707] that requires the specification of conditions of precedence. These conditions of precedence can shape the *Law of Competing Principles*, examined shortly, which will serve as a rule linking the legal consequences of a principle to its precedence over the other.[708]

---

[700]Ibid., 77.

[701]Ibid., 78.

[702]Ibid., 79.

[703]Ibid., 100.

[704]Ibid., translation mine.

[705]This conclusion, nonetheless, as Alexy points out, does not exclude other usual justifications for balancing, such as the rule of law or concepts of justice, but it serves as a justification that derives directly from the structural framework of his theory.

[706]Alexy, *Theorie der Grundrechte,* 81, translation mine.

[707]Alexy examines some possible exceptions in which rules could be regarded as also encompassing a *prima facie* character, and principles, in turn, a definite character. According to him, however, both norms could be identified: "The fact that rules, by enfeebling their definitive character, do not obtain the same prima facie character as principles is only one side of the coin. The other side is that principles, by strengthening their prima facie character, do not obtain the same prima facie character as rules either." (Ibid., 89, translation mine).

[708]Ibid., 104.

The idea of principles as optimization requirements is at the core of Alexy's theory.[709] With this flexible structure, principles achieve a vast possibility of contents and can better situate in the discussion of the "unity of practical reason." Indeed, Alexy stresses that "principles can be related either to collective goods or to individual rights,"[710] which shows that, in their basis, the idea of subjective rights, as previously examined,[711] is no longer the focus in his view of constitutional adjudication. Principles, with this objective nature, are relative and hence do not have an absolute character.[712] In this case, more than highlighting the protective function of subjective rights, the analysis gains a greater extent by focusing on the collective interests, the communitarian wills in adjudication, which are embraced, as a unity, by the concept of principles. Alexy, in fact, thinks that it is necessary that this goal and the collective value-based structure become part of the idea of principles. In opposition to opinions that remark the need to focus on subjective rights,[713] such as Ronald Dworkin's,[714] Alexy aims to set forth an extensive meaning of principles, for "it is neither necessary nor convenient to tie the concept of a principle to that of an individual right."[715] For him, principles also entail political arguments, policies that are placed in the structure of balancing, to the extent they encompass the idea of "what is good for all." This approach appears, in his words, to be more suitable,[716] although he remarks that every interpretation must begin with the constitutional text and not depart therefrom except in special cases. This means that the constitutional provisions cannot be deprived of their enforceability (*Verbindlichkeit*).[717] In his account, the broad meaning of principles does not contradict the deontology of constitution, a conclusion that might be, as this research demonstrates, not entirely correct.[718]

---

[709]Indeed, in the *Postscript* of the English version of his *Theorie der Grundrechte*, he remarks that "the central thesis of this book is that regardless of their more or less precise formulation, constitutional rights are principles and that principles are optimization requirements" (Robert Alexy, "Postscript", in *A Theory of Constitutional Rights* (Oxford: Oxford University Press, 2002), 388).

[710]Alexy, *Theorie der Grundrechte,* 94, translation mine.

[711]See the second chapter.

[712]The only complex analysis of this statement would be found in the case of the principle of human dignity, which, according to Alexy, cannot, in a more conclusive way, be deemed absolute. See, for this purpose, Alexy, *Theorie der Grundrechte,* 97.

[713]See Ernst-Wolfgang Böckenförde's, Bernhard Schlink's and Friedrich Müller's critique of the assumption of a value-based approach to constitutional adjudication in the second chapter.

[714]Alexy, *Theorie der Grundrechte*, 99.

[715]Ibid., 99, translation mine.

[716]Ibid.

[717]Ibid., 106.

[718]Indeed, as previously examined, the idea of principles embracing the totality of the legal order relates to the expansion of a political rationality in the realm of constitutional adjudication. This aspect causes serious outcomes in constitutional democracy, among them, the weakening of the deontology of basic rights and the confusion between law and politics. These conclusions, which will be more deeply examined in the following chapters, reveal that the flexible structure of principles as optimization requirements does not agree with the idea of a strong constitution,

Principles, which are not limited to the notion of subjective rights insofar as they also contain collective interests that "could be used above all as reasons against *prima facie* constitutional rights, but also as reasons for them,"[719] are the central elements of balancing. For they can have different origins, his concept of principles resembles that of values. In truth, Alexy, in this matter, remarks that "there is a lot of room in the spacious world of principles,"[720] which can be a legal provision protecting a subjective right, a collective interest derived from the constitution (democracy, rule of law, social state, etc) or a value with no direct origin in the constitution (a social tradition, a communitarian practice, public interests, etc), just to cite some. He emphasizes this approximation: "The graduated satisfaction of principles corresponds to the graduated realization of values."[721] That there is no solid difference between both can be seen in the idea that any formulation of usually employed values when balancing is at issue – we can observe it in the idea of constitution as an "order of values" – can be reformulated in terms of principles and principles or maxims in terms of values without loss of meaning.[722]

Accordingly, even though he recognizes that values have an axiological nature founded upon the idea of good, and principles, in turn, have a deontological character grounded in the notion of command (the "Ought"),[723] at the end, principles and values are assimilated. Principles, indeed, according to this view, are connected to the application of evaluative criteria in conflict through balancing as a means to define what is best in a particular situation. Still, evaluative criteria are what Alexy calls values.[724] Therefore, they, although differentiating in the premise of deontology and axiology, are not, in practice, deemed distinct in the realm of balancing. Alexy even argues that "the structural distinction between rules and principles is also found on the axiological level."[725] Moreover, inasmuch as the best mechanism in constitutional adjudication when there is a conflict, in his mind, is to proceed to comparative value judgments, balancing appears as the inevitable solution either to an axiological or deontological perspective. It is interesting, in this aspect, to observe that, even though clearly assimilating both, he shows his preference for a model of principles founded upon the argument that it "always

whose contents are not confounded with collective interests, and, therefore, cannot be, since they are established through institutional procedures of democratic participation, included in the concept of legal principle to be balanced in particular circumstances. Deontology is not compatible with the teleological character of desirable goods represented by the preferences of a communitarian or social will, as though they were similar in balancing to legal norms originated by longstanding and democratic institutional procedures of will formation.

[719]Alexy, *Theorie der Grundrechte,* 118, translation mine.

[720]Ibid., 120, translation mine.

[721]Ibid., 125, translation mine.

[722]Ibid., 125, translation mine.

[723]Ibid., 127.

[724]Ibid., 130.

[725]Ibid., 131, translation mine.

expresses the obligatory character of law clearly."[726] The problem of this twofold character of norms becomes, hence, a problem of simple clarity. Insofar as, in his view, the transition from one to the other is simple and acceptable,[727] and both have conceptually the same structure,[728] deontology and axiology, at the end, seem to become two concepts without practical difference.

The distinction between rules and principles, therefore, by assuming the character of principles as similar to that of values, inevitably transforms balancing into a natural solution to constitutional adjudication. Balancing seems to be an adequate mechanism when the variables at issue are in the realm of preferences of goods shared by a collectivity, which can result in the estimation of their degrees in accordance with some intersubjectively shared interests through a flexible case to case perspective. In other respects, this approach seems to more reasonably correspond to the idea of principles of a total legal order and the BVG's more activist and political approach. This is why Robert Alexy uses his methodological defense of balancing as a clear response to the critiques against the idea of a value-based approach.[729] Whereas the critiques, as some we previously discussed,[730] attack this theory on account of the lack of rational justification, Alexy, through the systematization of balancing, argues that it can promote rationality in this matter; whereas the critiques sustain the imminent risk of subjectivism and decisionism in constitutional adjudication, Alexy asserts that, by centering on balancing, it is possible to supply adjudication with an instrument to rationally control the decision, as well as to bring about correctness and coherence. Therefore, the idea of principles assimilated to values is reasonable and can, as a matter of fact, be defended, since it is followed by the consequent and inevitable deployment of balancing in the way he systematizes it. For Alexy, the critiques against balancing are indeed incorrect "as long as the conclusion is that balancing is a non-rational or irrational procedure."[731]

Nonetheless, the quest for the rationality of balancing seems much more complex. Alexy recognizes that the simple deployment of this procedure does not mean that the judge attained a rational solution. For this reason, it is necessary to investigate the structure of this mechanism. Every balancing leads to a statement of preferences in which one principle is regarded as precedent over the other given determined conditions. It is in this aspect that the idea of the *Law of Competing Principles* appears: given the conditions C^1, the principle P^1 takes precedence over the principle P^2, which results in the legal consequences R of P^1, or, in other words, a rule is established requiring that, under the conditions C^1, the legal consequences of P^1 must be R.[732] Hence, there is, based on the factual and legal possibilities, the

[726]Ibid., 133, translation mine.

[727]Ibid.

[728]Ibid., 134.

[729]Ibid., 138.

[730]See the second chapter.

[731]Alexy, *Theorie der Grundrechte,* 143, translation mine.

[732]Ibid., 143.

definition of a preferential statement that will lead to a rule: P^1 takes precedence over P^2 given the conditions C^1; under the conditions C^1, the consequences of P^1 must be R. This preferential statement, however, needs to be established not merely intuitively, but rather by distinguishing the mental process that leads to the determination of the preferential statement and the justification.[733] Unlike the simple creation of a preferential statement, a rational approach to balancing demands this distinction: every statement of preference must be necessarily justified. For this justification, Alexy introduces the most elementary rule that reveals the rationality of this procedure: the *Law of Balancing*.

According to Alexy, balancing, in a justified way, applies the following formula: "*the greater the degree of non-satisfaction of, or detriment to, one principle, the greater must be the importance of satisfying the other*,"[734] which leads to three stages of analysis: first, the establishment of the degree of non-satisfaction of the first principle; second, the importance of satisfying the competing principle; and, third, the satisfaction of the latter must justify the non-satisfaction of the former.[735] This structure is the major reference to the defense of the rationality of this procedure, and it is through the explanation of how we can achieve an optimum point between principles in collision that it is possible to sustain many arguments in this respect. Alexy argues that, with the law of balancing, we can justify the preferential statement by exposing the degrees of intensity of non-satisfaction of, or detriment to, one principle and the importance of satisfying the other. It provides thereby a justification insofar as it sets forth a requirement linking the degree of non-satisfaction of one principle and the degree of importance of the other.[736] It establishes a conditional preferential statement serving as a rule for decision-making founded upon the aim to optimize a principle.[737] It provides, for this reason, a justification that is not, in Alexy's view, the result of a "matter of all or nothing."[738] The issue here is not, after all, a problem of validity, but of setting up the most adequate weight of a principle in accordance with the particularities of the case.

Balancing takes into account "which of the interests having *equal degree in the abstract* [has] the *greater weight* in the *concrete case*."[739] They are thus *relativized* to achieve the most adequate situation with respect to the interference with the private sphere. There is a harmonization, a "practical concordance" between them founded upon a relation of conditional preference[740] that indicates, according to the

[733]Ibid., 144.

[734]Ibid., 146, translation mine.

[735]Robert Alexy, "On Balancing and Subsumption – A Structural Comparison", *Ratio Juris* 16, no. 4 (December 2003): 433–449.

[736]Alexy, *Theorie der Grundrechte*, 149.

[737]Ibid., 151.

[738]Ibid., 152, translation mine.

[739]Ibid., 80, translation mine.

[740]The collision between principles is resolved by establishing a statement of *conditional* preference when the judge examines the particularities of the case, indicating thereby the *conditions* under which it is possible to define the precedence of a principle over the other (See Ibid., 83).

singularities of the case, why one principle is preferred over the other.[741] At the end, what is achieved is a rule constructed by means of the process of optimization of the principles at issue, which synthesizes the *Law of Competing Principles*: "If a principle P^1 takes priority over the principle P^2 under conditions C: (P^1 **P** P^2) C, and if P^1 under conditions C implies legal effect R, then a rule, which has 'a relatively high degree of concretization,'[742] is valid that comprises C as the operative fact and R as the legal effect: $C \rightarrow R$."[743] In other words, a rule, which contains the conditions of priority obtained from the particular case, is formulated from each process of balancing, and this results in a system of *prima facie* priorities between principles that will serve as a parameter for decision-making. Notwithstanding that this system, obviously, does not mean that definite answers are reached, at least it creates a certain order in the realm of principles[744] and serves as a relevant parameter and subject of analysis for legal dogmatics.[745] Moreover, it supplies constitutional adjudication with criteria insofar as it "ties the Law of Balancing to the general theory of rational legal argumentation."[746] It is, for this reason, not a "null formula" (*Leerformel*), but rather a mechanism that says what "has to be rationally justified"[747] and, as such, has a universal character.[748] It is a formal structure through which a theory of legal reasoning can be carried out.

Yet, we can further explore this formal structure and specify its central aspects. The justification of a preferential statement through the *Law of Balancing* leads necessarily to the *Weight Formula*, normally associated with a triadic scale. Alexy understands that, by linking the explanation of the Weight Formula to the Law of Balancing, he can prove the rationality of balancing. Indeed, Alexy introduced a more detailed explanation of the Weight Formula as a reaction against a certain disbelief in this quality of balancing. In response to Jürgen Habermas and Bernhard Schlink, who, in his opinion, are the two most prominent representatives of a skeptical view of the rationality of balancing in legal reasoning,[749] he explains, with details, what the *Weight Formula* is and how it applies to the context of balancing. A specification of his framework, accordingly, would ratify his position:

[741]According to Alexy, the result of balancing is a statement of conditional preference, but this result is followed by a justification (which differentiates it from the model of an intuitive definition of preferences leading to subjectivism and uncontrollable results). He sustains that balancing is rational when the statement of conditional preference can be rationally justified (See, for this purpose, Ibid., 144). This culminates in his discussion of legal argumentation as a particular case of general practical argumentation.

[742]Ibid., 153, translation mine.

[743]Robert Alexy, "On the Structure of Legal Principles," *Ratio Juris* 12, no. 4 (September 2000): 297. See also Alexy, *Theorie der Grundrechte*, 83.

[744]Robert Alexy, "Sistema Jurídico, Princípios Jurídicos y Razón Práctica," *Doxa* 5 (1998): 148.

[745]Alexy, *Theorie der Grundrechte*, 153.

[746]Ibid., 152, translation mine.

[747]Ibid., translation mine.

[748]Ibid.

[749]Alexy, "On Balancing and Subsumption: A Structural Comparison," 436.

"Habermas and Schlink would be right if there were no structure making it possible for one to construct balancing as a rational form of argumentation."[750] However, since this structure exists, their opinions are not correct; the structure, after all, can promote rationality: "If it were not possible to make rational judgments about, first, intensity of interference, second, degrees of importance, and, third, their relationship to each other, then the objection raised by Habermas and Schlink would be justified. Everything turns, then, on the possibility of such judgments."[751]

Through the introduction of the degrees "light," "moderate" and "serious" (the triadic scale) to the analysis of the intensity of interference with one principle and its importance in comparison with another, Alexy endeavors to demonstrate that, apart from the analysis of some BVG's examples,[752] those objections fail in their aim to expose the irrationality of balancing. The degrees "light" *(l)*, "moderate" *(m)*, and "serious" *(s)* are inserted into the *Law of Balancing* as degrees of "non-satisfaction of, or detriment to, one principle and the importance of satisfying another,"[753] but they can also be considered "in terms of the 'intensity of interference'."[754] In this respect, Alexy constructs a system that has two levels: the concrete one concerning the intensity of interference, and the abstract one respecting the abstract weight of principles in relation to others. He defends that it is possible, previously to the case, to sustain that a principle has a higher abstract weight than others: "many constitutional principles do not differ in their abstract weight. Some, however, do."[755] In the examination of this abstract weight, he takes into account, for instance, the different legal sources from which principles were established, social values,[756] earlier decisions and, regardless of the case, defines how one principle is weightier than the other.[757] The abstract weight, accordingly, provides the link with other elements that,

[750]Ibid.

[751]Ibid., 437.

[752]Alexy, in order to demonstrate the rational character of this structure, examines some BVG's important decisions. He usually discusses two cases for this purpose: the *Tobacco case* (BVerfGE, 95, 179) and the *Titanic case* (BVerfGE, vol. 86, 1). According to him, "the Tobacco and Titanic Judgments show that rational judgments about degrees of intensity and importance are possible at least in some cases" (Ibid., 439).

[753]Ibid., 440.

[754]Ibid.

[755]Ibid.

[756]Carlos Bernal Pulido, "The Rationality of Balancing," *Archiv für Rechts- und Sozialphilosophie* 92, no. 2 (2006): 202.

[757]This might be metaphysical standpoint in Robert Alexy's theory. Indeed, there is no satisfactory explanation why one principle has a higher abstract weight than another. His conclusions are quite intriguing, especially when, for instance, he, before any case, concludes that one principle is abstractly weightier than another, as when he remarks that "the right to life, for instance, has a higher abstract weight than the general freedom of action" (Alexy, "On Balancing and Subsumption: A Structural Comparison," 440). It is possible to observe in this Alexy's conclusion that some categories are previously assumed even before the concrete aspects of a particular case are examined. However, how can the weight of a principle be abstractly measured, detached from the concrete aspects of the case? What does exactly an abstract weigh mean? Laura Clérico

according to this approach, are not associated with the case. Principles can nonetheless have equal abstract weights. As a consequence, balancing deals with two different scenarios: the principles have equal abstract weights (this is, according to Alexy, what usually occurs)[758] and thus we can disregard this variable in balancing; or they have different abstract weights, which influence somehow this process.

In the first situation, balancing takes two steps: the examination of the intensity of interference with one principle and the investigation of the importance of satisfying the other principle. These two situations, as Alexy argues, can be summarized by explaining that "the concept of concrete importance of P^1 is identical [to] the concept of the intensity of interference with P^2 by omitting the interference with P^1,"[759] which means, in other words, that balancing functions by comparing two situations: an actual and real interference with one principle and the "intensity of the hypothetical interference that would be inevitable if the actual interference were omitted."[760] The evaluation carried out in this scenario takes place by measuring these two situations with the triadic scale: 'light', 'medium', and 'serious'. Each interference with one principle is thus graduated in order to find the optimum point in this relationship between principles, which is the basis for specifying the *Weight Formula*. Despite that, Alexy remarks the difficulties in adopting a numeric scale in this process[761] and shows the possibility of creating a "formula which expresses the weight of a principle under the circumstances of the case to be decided."[762] The *Weight Formula,* in this context where the abstract weights of principles are equivalent, demonstrates the relative weight a principle

---

introduces a possible definition of the abstract weight through three different criteria: (1) the force of the interests in play; (2) the weight of the principle in comparison with other principles; (3) the earlier decisions (Laura Clérico, *Die Struktur der Verhältnismäßigkeit* (Baden-Baden: Nomos, 2001), 178/179). However, although these criteria are regarded as some parameters to define the abstract weight of a principle, at the end, it seems that a set of principles is construed as if some principles could be regarded as *superprinciples* and the others as principles of minor relevance. Again, there might be sort of metaphysical standpoint guiding the process of balancing whenever there is a difference in this abstract weight of principles. Still, Alexy sustains here the rationality through the specification of more criteria. These are, after all, categories, in agreement with his point of view, that the judge must accept as a way to provide objectivity and logical constitution to balancing.

[758]Alexy mentions that "the Law of Balancing names as the first object of balancing only the intensity of interference. This shows that it is shaped for the situation in which the abstract weights are equal, that is, they play no role at all" (Alexy, "On Balancing and Subsumption: A Structural Comparison," 440).

[759]Ibid., 441.

[760]Ibid.

[761]Alexy remarks that "graduation in terms of light, moderate or serious is often difficult enough as it is. In some cases one can just barely distinguish light and serious, and in some cases even that seems impossible. Legal scales can thus only work with relatively crude divisions, and not even that in all cases" (Ibid., 443). According to him, the nature of constitution brings about this complexity: "In the end, it is the nature of constitutional law which sets limits to fitness of graduation and altogether excludes the applicability of any infinitesimal scale" (Ibid., 443–444).

[762]Ibid., 444.

has according to the particularities of the case, and it can be illustrated (despite reservations)[763] by quotients and numbers.[764]

The second scenario in turn refers to the existence of a distinct abstract weight between principles. The foregoing explanation gains thereby a new variable, which can obviously influence the result of balancing. Moreover, as well as in the concrete interference, the triadic scale also applies here. Yet, the weight formula, to be complete, needs a third variable, which refers to the "reliability of the empirical assumptions concerning what the measure in question means for the non-realization of P^1 and the realization of P^2 under the circumstances of the concrete case."[765] This is what he calls the *second Law of Balancing*: "The more heavily an interference with a constitutional right weights, the greater must be the certainty of its underlying premises."[766] The focus now is on the certainty and quality of the premises with respect to the empirical investigation and how they relate to the balancing of principles (unlike the first *Law of Balancing*, which is linked to the "substantive importance of the reasons underlying the interference").[767] As well as the first two variables, the triadic scale can also apply here. At the end, a complete weight formula works with three distinct variables: the concrete interference with one principle (which also refers to the concrete importance of a principle), the abstract weight of a principle, and the reliability of the empirical premises. These three variables are expressed by a mathematical formula that has the two principles in collision:[768] one as the numerator and the other as the denominator. The result of this equation gives the relative weight of both with reference to the case. More importantly, it provides, according to this approach, even when the courts do not explicitly apply it, a rational justification for the decision: "The Weight Formula can then be used to infer those values

763 Alexy, although having observed the difficulties in this numerical understanding of the weight formula, show at least his interest in doing so. His words:

"Now one can only talk about quotients in the presence of numbers, which is not the case in any direct sense with balancing. So concrete weight can only really be defined as a quotient in a numerical model which illustrates the structure of balancing. In legal argumentation it is only analogous to a quotient. But the analogy is an interesting one" (Ibid., 444).

764 In his attempt to illustrate how the weight formula functions, Alexy applies different criteria "for allocating numbers to the three values of the triadic model" (Ibid.). First, he introduces the geometric sequence, then the nine classes of double-triadic model, which can be geometrically and arithmetically represented (See, for this purpose, Ibid., 444–446).

765 Ibid., 446.

766 Ibid.

767 Ibid.

768 The Weight Formula becomes much more complex when it involves more than two principles in collision. Alexy, with the same aim to deploy a rational justification for constitutional adjudication, examines this more complex configuration of the Weight Formula in Robert Alexy, "Die Gewichtsformel", in *Gedächtnisschrift für Jürgen Sonnenschein*, ed. Joachim Jickely, Peter Kreutz and Dieter Reuter (Berlin: de Gruyter, 2003), 771–792.

which have not been determined."[769] Indeed, constitutional cases can be expressed through this mechanism.[770]

Robert Alexy's intent, through the explanation of the *Law of Balancing* and its elements, such as the *Weight Formula,* results in a very detailed complex of models and mechanisms that attempt to provide, as a dogmatic structure, a rational methodology for decision-making, particularly in the realm of constitutional adjudication. It is notable his aim to provide not obviously definitive solutions to the cases, but at least a formal parameter that could guide decision-making. This formal dogmatic structure, as he points out, is essential to bring about rationality in this process. Albeit formal, he mentions that "this cannot diminish the value of identifying the kind and the form of the premises which are *necessary* in order to justify the result."[771] The rational justification, as a consequence, derives from the correct deployment of this balancing structure, from following, as best as possible, its distinct but concatenated and correlated levels (degrees of interference, importance

---

[769]Alexy, "On Balancing and Subsumption: A Structural Comparison," 447.

[770]See, for example, Alexy's application of the Weight Formula to the *Cannabis case*, examined in the first chapter. It is clear that, according to this approach, the decision can be justified in rational patterns, for it could be grounded in accordance with the Weight Formula:

"The Cannabis Judgment of the Federal Constitutional Court offers an example. Whether the legislature is allowed to prohibit cannabis products depends mainly on whether the interference with constitutionally protected liberty caused by the prohibition is suitable and necessary to combat the dangers associated with the drug. If criminal prohibition were not suitable or not necessary, it would be definitively prohibited on account of constitutional rights. The court explicitly states that the legislature's empirical premises were uncertain. It considered adequate that the empirical assumptions of the legislature were "maintainable" (*BVerfGE* vol. 90, 145, 182). This can be grasped by the Weight Formula in the following way: *Ii* stands for the interference with the constitutionally protected liberty caused by the prohibition of cannabis products. *Ij* represents the losses caused on the side of collective goods, especially public health, if cannabis products were not prohibited. The abstract weights of the colliding principles *Pi* and *Pj* shall be considered as equal, which allows one to neglect them. If cannabis products are prohibited, the interference with *Pi* must be considered as certain. The value of *Ri* is therefore 2(o) = 1. *Rj* stands in our case for the reliability of the empirical assumption of the legislator that the prohibition of cannabis products was necessary in order to avoid dangers for collective goods, especially public health. The Courts classes *Rj* as "maintainable," that is, as *p*. If one presupposes the simple triadic model, *Rj* receives by this explicitly the value 2(−1) = ½. From this and the fact that the Court considered prohibition of cannabis products as constitutional, it follows that the interference with *Pi* is not of the highest degree. Its highest possible value is 2, that is *m*. This becomes clear by putting the following values into the Weight Formula: 1= 2 × ¼ × ½. *Rj* must be because the Court explicitly assumes this degree of reliability. *Ri* must be 1, because interference in case of prohibition is certain. *Wi,j* must not be more than 1, for if it exceeds 1 the prohibition would be unconstitutional. The Court, however, declares the prohibition constitutional. In this constellation the highest possible value which *Ii* can achieve is 2, that is, moderate, because *Ij* cannot achieve in the simple triadic model a higher value than 4, that is, *s*. *This demonstrates that the Weight Formula allows one to grasp the interplay between the six elements which are relevant in order to determine the concrete weight of a principle in case of a collision of two principles.*" (Alexy, "On Balancing and Subsumption: A Structural Comparison," 447–448, emphasis mine)

[771]Ibid., 448, emphasis mine.

of abstract weights and degrees of reliability).[772] It is through its application that what has to be rationally justified is defined.[773]

4.4 Final Words

This chapter had the purpose of investigating one of the most influential and well-known defenses of the rationality of balancing, one that has strong connections with the characteristics of constitutional courts' shift to activism, as we discussed in the first part. Robert Alexy's *Theory of Constitutional Rights,* as a continuation of the project initiated in his *Special Case Thesis,* demonstrated how far German scholarship has developed methodologies and "rational standards" to account for the possibility of rationally justifying balancing, and thereby the constitutional courts' shift to activism. This chapter, for this reason, focused initially on his *Special Case Thesis* in order to explain what Alexy's defense of the "unity of practical reason"[774] means, and how he ties up legal correctness with moral correctness,[775] as an integrative theory.[776] The consequences of his *Special Case Thesis,* such as the fact that there is no problem that collective goals prevail over constitutional guarantees, since balancing is used, or that the system of rights cannot provide itself the answers to legal reasoning, demanding thereby the appeal to the generality of practical reason, whose institutionalization can occur either in lawmaking or decision-making, were hence subject of consideration in this chapter.

Yet, it was the following investigation of his *Theory of Constitutional Rights* that allowed us to visualize how this "unity of practical reason" could be operationalized in decision-making, especially through balancing. By examining the main characteristics of his structural and analytical theory, as a formal framework behind the substantive arguments that could guarantee the "rational control" of knowledge in decision-making, we could discuss how he carries out the distinction between rules and principles. Besides, we verified how he reaches the consequent conclusion that the principle of proportionality, with balancing, follows from the very nature of principles as long as they are interpreted as optimization requirements. This, ultimately, assimilates them to values. Since the justificatory strength of his analytical theory stems from some "rational standards," this chapter ended by stressing, in the structure of balancing, his construction of the *Law of Competing Principles* and the *Law of Balancing,* which led, finally, to the *Weight Formula.* These "rational standards" could then prove how balancing is rational, and how it is an

---

[772]Ibid.

[773]See Alexy, *Theorie der Grundrechte*, 152.

[774]Alexy, "The Special Case Thesis," 383, translation mine.

[775]Alexy, *The Argument from Injustice: A Reply to Legal Positivism*, 77.

[776]See Alexy, "The Special Case Thesis," 380.

indispensable instrument for constitutional adjudication. Rationality in decision-making becomes then, above all, a question of abstract criteria and formulas filled with arguments.

Consistent with this conclusion, we can observe that it is not possible to grasp Robert Alexy's thinking detached from the time when it appeared. His *Special Case Thesis* and *Theory of Constitutional Rights* are closely connected to the radical transformation in German constitutional culture, to the movement of the BVG towards a more activist and political attitude, as well as to the expansion of a concept of subjective rights as objective principles embracing the totality of the legal order.[777] While the first theoretically justifies why legal reasoning can be controlled by, and recreated in, general practical arguments, as the BVG's practice – as well as STF's – has experienced,[778] the second appears as a methodological response to this new constitutionalism. The need to provide a method as the condition for rationality in constitutional adjudication appears to be the answer that constitutional scholarship had so long strived for within the context of a case to case Jurisprudence; it systematizes, and gives orientation to, this background of new dilemmas and new challenges. His theory, we could say, is thereby an attempt to *rationally justify* this new constitutionalism. His main concern seems to be that of providing a *rational* justification for the way the BVG decides. The legal reasoning in constitutional adjudication, within the new context of German constitutionalism, to be rational, should be a proportional reasoning following some general rules able to embrace this "unity of practical reason." Yet, in the realm of a case to case Jurisprudence, of a constitutional scholarship so linked to the belief in the rationality of balancing, of the BVG's - and, as we examined, also the Brazilian STF's[779] - political and activist role, and of a dimension of basic rights as an objective comprehension of basic rights, it is therefore not totally purposeless to question: is it really necessary to be so absorbed by balancing to discuss constitutionalism?

This question, evidently, demands a more complex debate. It stems from the question of which rationality is behind this structural framework, and which rationality derives from the attempt to justify, in methodological grounds, the practice of a constitutional court grounded in the idea of constitution as an "order of values." The next chapters will probe this problem more directly and deeply, and will expose how complex the confidence that some abstract rules could provide a rational ground for deploying arguments in constitutional adjudication is. They will show that, perhaps, behind this belief, there is a metaphysical standpoint, whose consequences can be serious for constitutional democracy. This word, metaphysics, might sound, at this time, complex and even mysterious, but it opens up the fascinating world where the discussion of a *concept of limited rationality* will take place. In the

[777]See the second chapter.

[778]See the second and third chapters.

[779]See the last chapter.

next chapters, this debate on rationality will investigate Jacques Derrida's deconstruction and Jürgen Habermas's proceduralism, revealing thereby that it is possible to think of reason in constitutional adjudication according to another perspective. In summary, the concept of limited rationality will challenge the rationality of balancing, especially in the way Robert Alexy defends it, in order to vindicate that it is not necessary to be so absorbed by balancing to discuss constitutionalism, and that other rationality, in this respect, is possible and necessary.

Chapter 5
When *Différance* Comes to Light: Balancing Within the Context of Deconstruction

Abstract The aim to rationalize balancing seems to follow this movement towards the juridification of politics. Robert Alexy's defense of rationality is a relevant source to grasp how this process could be legitimately justified according to some predetermined rules and formulas. Yet, his theory poses some necessary questions and possibly leads to the question of whether there is a metaphysical standpoint behind his central premises. In this respect, to challenge Alexy's premises with Jacques Derrida's complex, fascinating and powerful philosophy is an interesting and instigating theme, insofar as his deconstruction leads to an incessant questioning of all our beliefs and certainties. There cannot be a metaphysical standpoint behind our activities, for this results in the forgetfulness of the other's otherness, and there cannot be a *logos* behind the dualism between law and justice, for this culminates in the practice of violence with ground, and thus injustice. For this reason, it is necessary to verify whether Alexy's claims to correctness, rationality and legitimacy are not metaphysically justified, and, if they are, which are the consequences they bring about to constitutional democracy. Particularly, as long as the principle of separation of powers is a fundamental issue here, the problem of the legitimacy of balancing through Alexy's idea of an "argumentative representation" must face the question of "who are the people?" in order to show the risks of a possible construction of a substantive comprehension of democracy in this process. It is here where *iterability, undecidability, autoimmunity,* and *responsibility* in the negotiation between constitutionalism and democracy and between law and justice demonstrate their critical potential towards the other's otherness, and hence towards doing justice to the case.

5.1 Introduction

After having discussed the main arguments of Robert Alexy's thinking concerning his defense of balancing through a theory that could rationalize the way constitutional courts decide cases, such as the German *Bundesverfassungsgericht* and the

J.Z. Benvindo, *On the Limits of Constitutional Adjudication*,
DOI 10.1007/978-3-642-11434-2_5, © Springer-Verlag Berlin Heidelberg 2010

Brazilian *Supremo Tribunal Federal*,[780] it is not purposeless that we begin the critical investigation of his premises with Jacques Derrida's deconstruction. With his discourse on deconstruction (*déconstruction*) and *différance*, and also with a different logic of writing – "a text of pleasure (. . .) Pleasure and play (*jouissance et jouer*)",[781] – Derrida opens a new world for reflection, a world where many of our basis and beliefs are rigorously challenged. Indeed, as a critical heir of the Heideggerian and Nietzschean thinkings, he carries the purpose of questioning and combatting the hegemonic metaphysics that is present in the grounds of Western philosophy, but now radicalizes it by stressing the otherness. In this respect, it seems that a relevant message to the dogmatic problem here focused on appears: to what extent could we understand that balancing, as we have discussed here so far both empirically[782] and theoretically,[783] is not the sign of metaphysics? Besides, to what extent can this metaphysics put the other, as the reflex of Derrida's stress on otherness, in jeopardy?

Derrida's philosophy, since it is clearly marked by this intent to disclose and undercut metaphysics, emerges, therefore, as a very interesting and intriguing source for this purpose. It opens up the possibility to exercise the critique, which now, more than ever, gains an interminable and challenging character. It reaches the core of many of the assurances guiding human reasoning and actions by showing how they are marked by a metaphysical standpoint. It also exposes how this metaphysics can be the sign of an identity, as well as its outcomes. It centers on the premise that we cannot forget the unconditionality of the other, of the absolute singularity, words that give rise to a framework powerful enough to sustain the hypothesis that balancing can become the sign of a metaphysics that goes in the opposite direction of constitutional democracy.

In this respect, the question of justice is central in this strike against metaphysics, and this is the realm where justice arises as *deconstruction*; after all, according to Derrida, "deconstruction is justice."[784] For deconstruction aims to disclose and undercut metaphysics, it questions all types of presuppositions, including the ones behind Alexy's defense of the rationality of balancing, even though "this questioning of foundations is neither foundationalist nor anti-foundationalist."[785] Here the first counter-metaphysical thinking[786] challenges the empirical and theoretical

[780]See the second and third chapters.

[781]Gary John Percesepe, *Future(s) of Philosophy: The Marginal Thinking of Jacques Derrida* (New York: Lang, 1989), 1.

[782]In this respect, we remark the BVG's and STF's shift to activism, as examined in the second and third chapters.

[783]In this respect, the emphasis is on Robert Alexy's *Theory of Constitutional Rights*. See the last chapter.

[784]Jacques Derrida, "Force of Law: The 'Mystical Foundation of Authority'", *Cardozo Law Review* 11 (1990), 945.

[785]Ibid., 931.

[786]In the next chapter, the procedimentalist approach will challenge this metaphysical thinking, as the second counter-metaphysical source.

examination of the last chapters. Similarly to the hermeneutical debate in philosophy – this continuous project of disclosing the metaphysics – the analysis now has to radicalize the perception that the deployment of methods and criteria in adjudication, specifically in the context of balancing, can also mean the denial of *différance* and, accordingly, justice. This is particularly an interesting theme, inasmuch as we can study the problem from a philosophical tradition that somehow inherits from Nietzsche and Heidegger, and which, as Richard Rorty suggests, "show[s] how the creation of new discourses can enlarge the realm of possibility."[787] The opening to new possibilities is, as a matter of fact, what deconstruction proposes when justice is brought into question or, in Derrida's words, "for in the end, where will deconstruction find its force, its movement or its motivation if not in this always unsatisfied appeal, beyond the determinations of what we call, in determined contexts, justice, the possibility of justice?"[788] It is, therefore, a very particular look into the problem, which will guide the discussion of how far the deployment of some methods and criteria, especially when we observe the expansion of the political influence of constitutional courts, if not followed by this message *différance* brings out, can become an expression of injustice, that is, the closeness of the realm of possibility.

Furthermore, this study will confront the problem of a metaphysical appropriation of constitutional adjudication with the question of legitimacy. This brings into discussion one of Jacques Derrida's most intriguing and interesting articles, one that exposes the faith behind the process of foundation. His *Declarations of Independence*[789] is not only a direct attack on the basis of our beliefs in institutional legitimacy but mainly a reflection on the foundations of constitutionalism. His provocative manner of dealing with the act of signing the Declaration of Independence poses the question of whom "signs, and with what so-called proper name, the declarative act that founds an institution"[790] This is, certainly, one of the main points to understand that his philosophy radicalizes the very basis of many of the supports grounding the democratic parameters. What does, after all, legitimate an institution? Again, it seems that this question needs to face a new challenge, now reinforced by a strong problematization of the way we interpret our public sphere. When, therefore, the institutional legitimacy is at stake – and this brings into debate, evidently, the legitimacy of constitutional courts – it is also necessary to observe this founding moment, this special violent act of foundation.[791] Maybe there is much to be revealed in this process and, principally, much of a complementary

[787]Richard Rorty, *Truth and Progress: Philosophical Papers*, Vol. 3 (Cambridge: Cabridge University Press, 1998), 310.

[788]Derrida, "Force of Law," 957.

[789]Jacques Derrida, "Declarations of Independence," in *Negotiations: Interventions and Interviews, 1971–2001* (Stanford, CA: Stanford University Press, 2002), 46–54.

[790]Ibid., 47.

[791]See Derrida, "Force of Law," 973–1045 (Part Two). See also Walter Benjamin, "Zur Kritik der Gewalt," in *Zur Kritik der Gewalt und andere Aufsätze* (Frankfurt a. M.: Suhrkamp, 1965).

discourse on the relevance of understanding the institutional history as an indispensable criterion to criticize the way courts act.

This chapter is a first analysis of the consequences of a metaphysical thinking in the realm of constitutional adjudication in a constitutional democracy. It will introduce some of the elements and premises that will guide throughout this book the perception of how methods and criteria, particularly balancing, can bear much of a logocentric meaning and which are the outcomes this characteristic brings about. Notwithstanding the complexities and criticisms regarding the application of his philosophy to the practical realm,[792] as we can observe in the context of constitutional rights, Jacques Derrida's philosophy can offer a different and relevant insight into the problem of how methods and criteria can express this so-called *metaphysics of presence*. His combat against the closeness of the *a priori* (this logocentrism) that serves as a guidance for human actions and his aim to disclose its totalizing effects for democracy are a serious point that can help disclose, step by step, the *concept of limited rationality* that will challenge the so defended rationality of balancing. This is by no means a remote and disconnected debate nor transforms the researched problem into an abstract problem. In spite of the fact that the Derridian philosophy is complex and has some difficulties when transported to the institutional debate, especially when a very specific issue, as the one here of legal reasoning, is the subject of discussion, it is possible, from its fascinating opening to this other philosophical dimension, to acknowledge that, no matter what, adjudication ought to be an act of responsibility. This is not an obvious conclusion: responsibility relies on what Derrida calls deconstruction. How is decision-making a responsible act? What does deconstruction, for this reason, have to do with the problem of interpreting legal principles and, specifically, balancing?

This could be considered a chapter of challenges. First, the disclosure of Derrida's philosophy is itself full of tortuous paths and difficulties, whose words will, step by step, expand the realm of possibilities in this democratic discussion. Second, the application of this thinking to democracy, particularly to the issue investigated here of constitutional court's reasoning, is not of simple development – indeed, this might be the most challenging – but is the worthiest, insofar as it brings much of this necessary abstraction into the practical world, where, according to Derrida, it is to see a "democracy that comes,"[793] never reached, though. In any

---

[792]This application of Derrida's philosophy to practical problems is, nonetheless, contested by some approaches. Jürgen Habermas, for instance, sustains that the Derridian philosophy loses its seriousness, and also its suitability for the *praxis,* when he mentions that "the linguistic contextualist approach, imbued with life philosophy, is insensitive to the factual strength of the counterfactual" (Jürgen Habermas, *Der philosophische Diskurs der Moderne: Zwölf Vorlesungen* (Frankfurt a.M.: Suhrkamp, 1985), 242, translation mine). Richard Rorty, in turn, remarks that "Heidegger's and Derrida's only relevance to the quest for social justice is that, like the Romantic poets before them, they make more vivid and concrete our sense of what human life might be like in a democratic utopia – a utopia in which the quest for autonomy is impeded as little as possible by social institutions" (Rorty, *Truth and Progress: Philosophical Papers, Vol. 3,* 310).

[793]See Jacques Derrida, *Rogues: Two Essays on Reason* (Stanford, CA: Stanford University Press, 2005), 86.

case, despite the natural abstractions Derrida's philosophy bears, this chapter, ultimately, centers on a practical and empirical problem, for the philosophical approach reaches the democratic realm, particularly the theme of separation of powers, which is connected to the institutional legitimacy. How could, after all, this debate bring new insights into the quest for a democratic comprehension of the role constitutional courts should have? How could this attempt to unmask metaphysics be related to the way constitutional courts justify their decisions?

Consistent with these premises, this chapter will be divided into two main sections. The first (Sect. 5.2) will, initially, center on Jacques Derrida's philosophical thinking (Sect. 5.2.1). The purpose here is to introduce and investigate the premises and concepts that shape the background of this intriguing but also powerful philosophy, as well as expose how it opens up the possibility to establish the critique of metaphysics through the stress on *différance*. Even though the discussion here is more abstract and complex, it is a fundamental premise to grasp what exactly Derrida means when he thinks of *différance,* a concept that has an intimate relationship with the purpose of unfolding the *concept of limited rationality*. Afterwards, it will begin to explore the political extension of Derrida's philosophy. In this matter, it will, at first, examine what Derrida calls *democracy to come,* as a "call for a militant and interminable political critique"[794] (Sect. 5.2.2); then, it will examine the negotiation between constitutionalism and democracy, as a central argument to understand the problematics of legitimacy (Sect. 5.2.3); finally, it will enter, more directly, into the specific realm of adjudication by stressing the negotiation between law and justice (Sect. 5.2.4). In the first section, accordingly, the philosophical and political premises of this deconstructionist approach will open up the possibility to begin to understand how they will lead to the dogmatic problem examined in the second section.

In the second section (Sect. 5.3), on the other hand, the purpose is to apply those premises to the dogmatic problem. Here the study will focus on many of Robert Alexy's considerations on the activity of constitutional adjudication, especially his defense of the rationality, correctness and legitimacy of constitutional courts' decisions when they deploy balancing. The intention is to confront balancing and Alexy's *Theory of Constitutional Rights* (*Theorie der Grundrechte*) with Derrida's philosophical premises. First, we will analyze the claim to correctness and the claim to rationality that are normally associated with balancing, and reveal how they transform themselves into a *logos of correctness* and a *logos of rationality* (Sect. 5.3.2) The *logos of correctness-rationality,* in turn, will lead to the final analysis of this chapter, which will deal with the most central question regarding the consequences the deployment of methods and criteria such as balancing could cause: the debate on legitimacy. The focus, in this section, is to demonstrate how a metaphysical standpoint can culminate in a *logos of legitimacy,* thereby jeopardizing the principle of separation of powers. In this section, we will stress two different approaches: the discussion of the most elementary question regarding

[794] Ibid.

legitimacy, which Derrida suggests in his text *Declarations of Independence:* "Who are the people?" (Sect. 5.3.3.1); and how balancing and the belief in the "argumentative representation" Robert Alexy presents as an argument to account for the legitimacy of constitutional courts can erode the basis of a negotiation between constitutionalism and democracy through a logocentric approach (Sect. 5.3.3.2). Finally, we will show that, as long as a metaphysical standpoint places itself behind the negotiation between constitutionalism and democracy, the space to establish a substantive conception of democracy is open, which is the opposite of the dynamics towards the other's otherness that *différance* consists of, sustaining then the practice of violence without legitimacy.

5.2 Différance and the Political-Legal Realm of Deconstruction

5.2.1 Jacques Derrida and Différance

In an interview entitled *Politics and Friendship*, Jacques Derrida announced the basis of his thinking when applied to politics: "For the present, to me, democracy is the place of a negotiation or compromise between the field of forces as it exists or presents itself currently (...) and this 'democracy to come.'"[795] This paradox, which expresses the dynamics of a memory that is incapable of entirely recollecting and gathering the past, for history is not coherent and linear, at the same time it has to deal with a future as an opening to the other, is at the core of his philosophy. Every situation is a new situation, and the particularity of a reality can never again be properly and entirely remembered. There is no safe place for conscience; we can never find it in a sort of a harmonious identity. Rather, what exists is an unlimited play of signs, which informs his intention to expose the absence of transcendental signified: "One could call play the absence of transcendental signified as limitlessness of play, that is, as the destruction[796] (*ébranlement*) of the ontotheology and the

[795]Jacques Derrida, "Politics and Friendship," in *Negotiations: Interventions and Interviews, 1971–2001* (Stanford, CA: Stanford University Press, 2005), 180.

[796]The purpose of unmasking and undercutting metaphysics Derrida suggests, which continues somehow the Heidegger's project, does not mean, as further examined, that he projects a kind of world without metaphysics. This would lead to an affirmation of a future-present and close the realm of possibility that the emphasis on *différance* sets forth. What he seeks is to disclose this metaphysics of presence existing in the Western philosophy and the consequences it brings about. Moreover, at the same time, he seeks to bring forth a thinking that is concerned with the other's otherness, which means the incessantly opening to interpretability and invention mediated by language. This is, nonetheless, in its finite character, metaphysics. Richard J. Bernstein understands that it is a usual erroneous interpretation of Derrida, when one calls his theory a project that nails down the effective destruction of metaphysics. This would mean, however, the opposite of *différance* as an incessantly opening to the future, albeit never a future-present. Instead, for Derrida, "we are never simply 'inside' or 'outside' metaphysics":

metaphysics of presence."[797] These words bring to light the perception that "determining the something is conceptually not possible, insofar as *différance* withdraws from the idea of a founding substantiality."[798] It is towards *différance*[799] that Derrida reveals much of his thinking. It means this lack; this emptiness we must articulate, but never fill, for it would lead to a new form of identity. Instead of fundamentals, centers towards which the history of Western philosophy gravitated,[800] Derrida sustains that there is nothing else for conscience than language, expressed by traces, metaphors, which, in turn, show the intrinsic relation to the other, as long as the conscience is woven into the narrative of *différance*:

> It is not here, therefore, the question of a constituted difference, but rather, before all determination of content, of the pure movement which produces the difference. *The (pure) trace is différance*. It does not depend on any sensible plenitude, audible or visible, phonic or graphic. It is, on the contrary, the condition of it. Although it does not exist, although it is never a being-present outside of all plenitude, its possibility is by rights anterior to all that one calls sign (signified/signifier, content/expression, etc), concept or operation, motor or sensory.[801]

Derrida is acutely aware that we cannot question or shake traditional ethical and political claims without at the same time also drawing upon these traditional claims. The very dichotomy of 'inside-outside' is also deconstructed. We are never simply 'inside' or 'outside' metaphysics. Derrida has been read – I think seriously misread – as if he were advocating a total rupture with metaphysics, as if some apocalyptic event might occur that would once and for all release us from the metaphysical exigency. But he mocks the very idea of such an apocalyptic happening. He tells us that 'the idea that we might be able to get outside of metaphysics has always struck me as naïve', and that 'we cannot really say that we are 'locked into' or 'condemned to' metaphysics, for we are, strictly speaking, neither inside nor outside' (Richard J. Bernstein, "An Allegory of Modernity/Postmodernity: Habermas and Derrida," in *The Derrida-Habermas Reader*, ed. Lasse Thomassen (Edinburgh: Edinburgh University Press, 2006), 81).

797 Jacques Derrida, *De la Grammatologie* (Paris: Les Éditions de Minuit, 1967), 74, translation mine.

798 Toni Tholen, *Erfahrung und Interpretation: Der Streit zwischen Hermeneutik und Dekonstruktion* (Heidelberg: Universitätsverlag C. Winter, 1999), 25, translation mine.

799 The use of the term *différance* instead of *différence* reveals much of Derrida's thinking, for it exposes the difference in the very concept of difference. According to Geoffrey Bennington, "it is a good Derrida's word: the difference between "différence" and "différance" is only noted in the writing, which takes then a certain revenge on the speech by obliging it to take as reference its own written trace, as if it wanted, for example, in the course of a conference, to say that difference" (Geoffrey Bennington, "Derridabase," in *Jacques Derrida* (Paris: Éditions de Seuil, 1991), 70, translation mine). Indeed, this connects the term *différance* to his analysis of the marginalization of the writing throughout the Western philosophy. For a better comprehension of this discussion, see Jacques Derrida, *L'écriture et la Différence* (Paris: Éditions de Seuil, 1979) and *De la Grammatologie* (Paris: Les Éditions de Minuit, 1967).

800 According to Derrida, he "since then had doubtless to start thinking that there was no center, that the center could not be thought in the form of a present Being, that the center had no natural place, that it was not a fixed place, but a function, a form of no-place where it would be thrown itself towards the infiniteness of signs replacements" (Derrida, *L'Écriture et la Différence*, 411, translation mine).

801 Derrida, *De la Grammatologie*, 92, translation mine.

These words – "*the* (*pure*) *trace is différance*" – reveals the complexity of this concept, for it cannot be confounded with a new form of safe place for conscience, nor can we consider it a complete denial of any kind of reference. Indeed, "Derrida does not deny the reference but only seeks to destroy the semantic determined by the traditional metaphysics of presence."[802] His philosophical heritage was, for this reason, essential for this purpose. The Heideggerian hermeneutics[803] led him into the discussion of the determination of Being (*sein*) as presence throughout the history of Western philosophy, as well as opened up the possibility to think of the Being (*sein*)[804] as an opening, lack or absence.[805] It also encouraged him, for

---

[802]Miroslav Milovic, *Comunidade da Diferença* (Ijuí, RS; Rio de Janeiro: Unijuí; Relume Dumará, 2004), 103, translation mine.

[803]The presence of Heideggerian philosophy in Derrida's thinking is so intense that he himself acknowledges that "nothing of what [I] attempted would be possible without the opening of Heideggerian questions" (Jacques Derrida, *Positions* (Paris: Minuit, 1972), 18, translation mine). This is, besides, what his critics usually remark, as Rodolphe Gasché, who mentions that, despite the influence and referred indispensability of Heidegger's questions and also his critical reflection on his philosophy, "even this criticism, including Derrida's 'disseminative gesture', is made, at least to a certain degree, in Heideggerian language" (Rodolphe Gasché, *Inventions of Difference: on Jacques Derrida* (Cambridge, MA: Harvard University Press, 1994), 78). Richard Rorty, in turn, sustains that "Derrida's books are just what you need if you have been impressed and burdened by Heideggerian language but want to avoid describing yourself in terms of it" (Rorty, *Truth and Progress*, 307). Notwithstanding the acknowledgement of Heidegger's influence on Derrida's philosophy, there are critics, however, that emphasize that Derrida is no Heideggerian, even in works as *Of Grammatology*. This is Joshua Kates's interpretation: "Nevertheless, granting all this, even at this moment of perigee in respect to Derrida's and Heidegger's orbits, Derrida, I would argue, is still no Heideggerian: especially when it comes to these issues of the epoch and the totality of metaphysics – to themes that can be summed up as falling under the heading of history" (Joshua Kates, *Essential History: Jacques Derrida and the Development of Deconstruction* (Evanston, Il: Northwestern University Press, 2005), 160).

[804]See Martin Heidegger, *Sein und Zeit* (Tübingen: Neomarius, 1949). See also Martin Heidegger, *Identität und Differenz* (Pfullingen: Neske, 1957), in which Heidegger brings to the discussion the *Being as difference.* Rodolphe Gasché's interesting analysis of this matter shows, however, that, although Heidegger had thematized *difference,* he still interpreted it as only the ontological difference between Being (*sein*) and beings (*seined*), which is a vestige of a metaphysics of presence in his thinking. See, for this purpose, Gasché, *Inventions of Difference: On Jacques Derrida,* 100–101.

[805]The Heideggerian thinking expressed in his book *Sein und Zeit* (*Being and Time*) opens up a new dimension in Western philosophy. By radicalizing the Husserlian phenomenology, which, in turn, attacked the Kantian "thing itself" by stressing the phenomenon and disrupting any reference to an essence conditioning the world, Heidegger sustains that the Being was forgotten by philosophy. Here, Husserl's message that the conscience is not a thing, but an *act,* is renewed with distinct contours by the discussion of *Dasein.* The Being, on the contrary, is revealed as an expression of the time, and the difference appears as the consecration of this existential discovery. It comes out as a possibility of self-comprehension, as a project of comprehension whereby the beings opens itself in its very possibility. Therefore, the Being can only be comprehended in its possibility, which is opened over time. Consistent with this view, the Being is no longer a structure, a thing itself, a *logos*, but rather it appears as an existential condition. Instead of an essentialism, the stress now is on the particular. As Habermas mentions, Heidegger intends to "comprehend the very world-projected subjectivity as Being-in-the-world, as a singular *Dasein* that finds itself in the

Heidegger did not find *différance* in his philosophy,[806] to continue this project of disclosing and undercutting metaphysics, although recognizing its insurmountable presence in reality, for the language is metaphysical.[807] Against the solitary character of Heidegger's *Dasein,* Derrida constructs a thinking strongly associated with this perception of the metaphysical structure of language that is the basis for setting forth an approach to the otherness: "Insofar as such logocentrism is not absolutely absent in Heideggerian thinking,"[808] the relationship between metaphysics and language emerged as a new standpoint to overcome this solitude.

The message in the book *Of Grammatology* (*De la Grammatologie*) is clear: the language is metaphysical. He aims to assert that the metaphysics of presence is at the core of any conception of language and signification. Intentionally, he goes directly to the empirical language in its different expressions, speech and writing, by bringing much of Saussure's study of signs. In this matter, the difference between the signifier and the signified becomes the basis of his opening to *différance.* According to him, the signified, which represents the articulated concept, the reference, has always been prioritized by Western philosophy, whereas the signifier, the way we can achieve this articulation, the "movement of the language,"[809] on the other hand, has been underestimated. However, it is exactly in this primacy

facts of a historical contour, which, however, needs not forfeit its transcendental spontaneity" (Jürgen Habermas, *Nachmetaphysisches Denken: philosphische Aufsätze* (Frankfurt a.M.: Suhrkamp, 1988), 49, translation mine). It is a new moment for the philosophical thinking, insofar as the hermeneutic circle, the temporality radicalizes now the traditional resting place of conscience. "Instead of the transcendental distinction between *constituens* and *constitutum,* another one appears: the ontological difference between the projection of the world, which opens up the horizon for possible meetings in the world, and that one which factually occurs inside of it" (Ibid., 49, translation mine). See, for this purpose, Heidegger, *Sein und Zeit.*

[806]Although Heidegger adopted the term *Differenz* (See Heidegger, *Identität und Differenz*) and questioned the continuous search for a rational basis – which becomes a forgetfulness of the Being, according to this perspective – it seems that he still held the metaphysics of presence in this reinforcement of the value of the Being. Indeed, notwithstanding that Heidegger demonstrated that the comprehension of the Being is self-comprehension and an opening to a spectrum of comprehensibility and possibilities, he did not escape from a certain reference towards the *Dasein* in his hermeneutics. As Miroslav Milovic remarks: "the Heideggerian difference seems more a reified difference determining – we could say – the places for the appearance of the authentic" (Miroslav Milovic, "A Impossibilidade da Democracia," *Anais do Congresso Nacional do Compedi* 14 (Florianópolis: Fundação Boiteux, 2005), 259, translation mine). By the same token, Habermas remarks that "the problem of intersubjectivity becomes insoluble under the accepted premises of a *Dasein,* which can only in loneliness authentically be projected into its possibilities" (Habermas, *Nachmetaphysiches Denken,* 50, translation mine).

[807]As Derrida remarks: "The system of language associated with the phonetic-alphabetic writing is the one in which the logocentric metaphysics, by determining the sense of the Being as presence, has been produced" (Derrida, *De la Grammatologie,* 64, translation mine).

[808]Ibid., 23, translation mine.

[809]Ibid., 16, translation mine.

of the signified that the metaphysics of presence, in its different configurations, has been revealed.[810]

They forgot thereby the comprehension that "there is no signified that escapes, eventually falling into it, from the play of referenced signifiers constituting the language."[811] Accordingly, the signified requires the signifier; it needs it to make the language move, as a condition to be articulated. The connection between signified, as the expression of the same, the identity, and the signifier, as the other, is, by that means, visible in the free play of signs: the signifier is in play with the signified. It is crucial thus to "put in evidence the systematic and historical solidarity of concepts and gestures of thinking usually believed as possible to be innocently separated."[812] This means, consequently, the absence of a transcendental meaning and the emphasis on the unlimited play of signs, on this empty space that "requires a field of infinite substitutability, where each signified could, in turn, become signifier, for nothing forbids thus the permutation of all terms without exemption any longer, nothing stops the play anymore."[813] We can convert a signified, as a consequence, into a signifier and vice-versa in a play that has no boundaries, no origin nor end, but only a continuous and fluid process of re-signification. It is therefore a play of continuous substitution, of presence and absence, which, however, will never express the real thing, but only the metaphors, for it would otherwise mean a new form of identity and presence. The dynamics between signified and signifier, besides, leads Derrida to the perception that it "must be thought as a signifying trace,"[814] which, consequently, allows the possibility to think of *différance*.

This linguistic turning point that occupies much of Derrida's philosophy provides the requisite for going further than Heidegger's hermeneutics, for it, from this perception of signified/signifier, identity/other, transposes the *Dasein*'s solitude concerned with a sort of historicism into a thematic closer to the otherness and the possibilities language provides. Unlike the discussion of Being and beings Heidegger raised, Derrida sustains a difference that is no longer the ontological difference – the focus shifts from the presence of Heidegger's Being to the unlimited play of signs. Accordingly, at this point, it seems that he takes Nietzsche as a crucial reference. In this matter, as Françoise Dastur argues, Nietzsche provides

---

[810]For instance, the metaphysics of objectivity that we observe in classical philosophies, such as in Plato and Aristotle; the metaphysics of subjectivity in the Kantian transcendental conscience and the Hegelian phenomenology of Spirit; the metaphysical solipsism still verified in Husserl's phenomenology and Heidegger's ontological difference, only to cite some. All of them expressed a sort of *logos* that was not undercut, not overcome by the dimension of *différance*.

[811]Derrida, *De la Grammatologie*, 16, translation mine.

[812]Ibid., 25, translation mine.

[813]Françoise Dastur, *Philosophie et Différence* (Paris: Les Éditions de la Transparence, 2004), 113, translation mine.

[814]Kates, *Essential History: Jacques Derrida and the Development of Deconstruction*, 166.

Derrida with a better support, because, for him, "the conscience is only the effect and not the cause of vital forces, and these are not present realities, but only pure differences."[815] In order to reach the dynamic dualism between absence/presence, identity/other, which refers to his *différance*, the Nietzschean perception of Being as a result of a difference of forces led him to a deconstructive approach centered on the notion of traces.[816]

However, not only this emphasis appears on the play of forces here, but also – and this is crucial to the further investigation – *différance* has a bearing on an active and creative movement. The play, after all, is an "invitation to an active interpretation."[817] Instead of fundamentals or any principle reducing the history, the accent is now directed to creation, to invention by exposing this difference of forces in the very structure of language. In place of presence, a sort of active nihilism[818] takes place, which ruptures and fragments, although not totally – for it would lead to a sort of presence – the differences. This is one of the central aspects distinguishing Derrida from Heidegger: he escapes from the metaphysical enclosure of Being: "It is probably that what Nietzsche intended to write and what resists to the Heideggerian reading: the difference in its active movement – which is comprised, without exhausting it, in the concept of *différance* – and which not only precedes metaphysics but also overflows the Being's thinking."[819] The free plays of signs – *différance* as the lack of any external reference guiding and governing these plays – acts, therefore, as an attack on any sort of positivism, any *logos*: we have to acknowledge the signified within the context of a linguistic interaction, where, instead of those references, the traces appear naked of this traditionally existing metaphysic: "*Différance* does not resist to the appropriation, it does not impose on itself an exterior limit."[820]

For traces only obtain a signification to the extent that they are inserted into an unlimited play of signs, where linguistic interaction puts in parentheses the reference, the substantiality behind the context, then their identity is only attained with

---

[815]Dastur, *Philosophie et Différence,* 109, translation mine.

[816]According to Derrida, "this deconstruction of presence realizes itself through the deconstruction of conscience, therefore through the irreducible notion of traces (*Spur*), as it appears in the Nietzschean as well as in Freudian discourse" (Derrida, *De la Grammatologie,* 103, translation mine).

[817]Dastur, *Philosophie et Différence,* 114, translation mine.

[818]This term is used by Françoise Dastur to describe Derrida's emphasis on free play of signs. According to her, "(...) the lack of the presence, in place of being experienced as a defect, should be an invitation to an active interpretation. This doubtless means, if we retranslate it in Nietzschean terms, that we should be capable of transforming the passive nihilism, which is a denial of life, into an active nihilism, which is invention and creation, and as such free of all nostalgia and all hope in an afterlife (*au-delà*) that would be that of the full presence" (Ibid., 114, translation mine).

[819]Derrida, *De la Grammatologie,* 206, translation mine.

[820]Ibid., translation mine.

respect to the other. Every presence, for this reason, presupposes its own absence: "No element is never nowhere present (nor simply absent), there are only traces."[821] The consequence of this consideration is that we cannot recognize the same, the identity without its own difference. This is, besides, the very possibility of language: "In each element is not 'present' but the other element, the 'absence', which must present, so as to render the language possible, this otherness *as* otherness."[822] This paradox, this dynamics of presence-in-difference and difference-in-presence is the basis of creation, of movement. After all, *différance* cannot be deemed absolute, but relative, whereby its non-placement shows its *asymmetrical* and indefinite interpretability towards the relationship between the same and the other, without this meaning the radicalization of one or another, for it would cause the end of the play and thus the language itself.

Similarly, it is the requisite to avoid comprehending *différance* as a new form of *logos* guiding, as a safe place and guarantee, the language, as a criterion whereby we could control the unlimited play of signs. If we understand it in this way, *différance* would get mixed up with a category of fundamentalism or transcendental signified. The emptiness would be filled by the presence. D*ifférance* could turn into an expression of an identity. The illusion of an entity responsible for saying the last word, a new type of God, would cease the dynamics, the traces, regarded as a movement towards creation and questioning. However, *différance* is the condition,[823] the opening to the possibility, although being itself impossible. Albeit its nonexistence, its impossibility, it "renders possible the opposition of presence and absence."[824] It is hence the possibility of language, the articulation of signs,[825] this continuous and fluid play of signifiers not followed by a corresponding metaphysical signified. It is the disclosure of deconstruction, the "producing causality of differences,"[826] and an infinite interpretation. Therefore, to seek a signified for *différance* is beyond question, since it would fill it with presence. The filling of *différance* is the closeness of the possibility; it is the cessation of its creative movement; it is the effacement of otherness. In sum, it is the denial of deconstruction.

As "difference in its active movement,"[827] *différance* shows its coordination with the context, with the empirical world of language. It is not, in view of that, an abstract entity kept out of reality. It is not a thing in itself, not a presence, but always an endless and active process among the different forces: "*Différance* is

---

[821]Bennigton, "Derridabase," 74, translation mine.

[822]Ibid., translation mine.

[823]See Derrida, *De la Grammatologie*, 92.

[824]Ibid., 206, translation mine.

[825]See Ibid., 92.

[826]Pierre Chassard, *Derrida: La Destruction du Monde* (Brussel: Mengal, 2004), 86, translation mine.

[827]Derrida, *De la Grammatologie,* 206, translation mine.

the non-full, non-simple 'origin', the structured and *different* (*différant*) origin of differences."[828] Despite its impossibility, its nonexistence, its non-essence, *différance* is the "movement according to which the language (...) reconstructs itself *historically* as system on the field of differences."[829] There is no *différance* detached from the relations among the differences, from the singularity of a context. In this debate, Derrida assumes that *différance* is a condition for connecting the differences that express linguistically in time and space, as a condition of signification: "In a language, in the language system, there is nothing but differences."[830] This historical and spatial aspect relates, accordingly, to traces, for they reveal this fragmentary character of language, whose construction and development cannot be guided by an original essence – there is no internal pure signified. Rather, what exists is an unlimited play of traces, which, temporally and spatially, projects an endless field of possibilities and an infinite realm of interpretations.

Consistent with this view, Derrida's philosophy provides the basis to account for a comprehension of otherness. What matters is not a sort of solipsism and egoistic presence, but rather the unlimited play of signs, the other's otherness. D*ifférance,* as the impossible, is the incessant question of the other, of the different, in contrast to the metaphysics that thinks of an identity or even reifies the difference.[831] A form of presence cannot thereby jeopardize the lack provided by stressing the unlimited play of signs, this endless process of re-signification. Indeed, this stress on the play is the very itinerary towards otherness. "The play is the disruption of presence."[832] It is always a dynamic process of absence and presence: *différance* is the eternal possibility of play. The philosophy, as a consequence, has to deal with this "joyful affirmation of world-play,"[833] with this irreducible otherness to come. "The philosophy needs the sensibility towards the different; otherwise it will just repeat the forms of identity, and thus close the possibilities of the 'new', the 'spontaneous', the 'authentic' in history."[834] It is an important message: there is no safe play,[835] no guarantee; the play itself, in its own insecurity, in its own fragility, is the condition of otherness. *Différance* is this ongoing non-existing possibility, this always to come.

---

[828]Jacques Derrida, *Marges de la Philosophie* (Paris: Les Éditions de Minuit, 1972), 47, translation mine.

[829]Chassard, *Derrida: La Destruction du Monde,* 84, translation mine.

[830]Derrida, *Marges de la Philosophie,* 47, translation mine.

[831]In his *Sein und Zeit* (*Being and Time*), we could still consider the difference as still connected to the metaphysical structure of the Being in the project of finding the "authentic." See note 806 *supra*.

[832]Derrida, *L'Écriture et la Différence,* 426, translation mine.

[833]Ibid., 427, translation mine.

[834]Milovic, *Comunidade da Diferença,* 130, translation mine.

[835]See Derrida, *L'Écriture et la Différence*, 427.

5.2.2 *Différance and Constitutional Democracy: The Democracy to Come*

The continuous possibility, this opening towards the singularity of the other, "the event, the singularity of the event: this is what *différance* is about."[836] This is also why *différance,* first woven in the intricate structure of language, is pervaded by a practical attitude. If *différance* is concerned with the margins, this overflow of boundaries, there is no space in Derrida's philosophy for stipulating a limit between theoretical and practical issues. *Différance,* therefore, is embedded in political-legal issues. As Bernstein remarks, "all deconstruction for Derrida is always and also political."[837] This shift to practical problems, nonetheless, can sound, from the standpoint of Derrida's premises, to a large extent, generic and abstract. It is not hard to find, after all, criticisms placing Derrida's philosophy in a realm of a certain rhetorical and unproductive social critique,[838] or considering his texts disconnected from an "empirical research into the structural dynamics of society and politics."[839] Indeed, it seems that this shift to practical problems from this linguistic debate is largely abstract, and the stress on otherness can hardly be applied to political-legal issues without showing its theoretical fragility and insufficiency.

At any rate, although it is important to acknowledge beforehand this tortuous transposition of his philosophy into practical and institutional problems, such as democracy and constitutionalism, we could say, on the contrary, that his thinking, since the beginning, was rather immersed in political matters. His biography doubtless pushed him into the debate on otherness, and, even though his initial works were more concerned with the philosophical purpose of disclosing and undercutting the metaphysics of presence, there were already strong messages we could apply to practical problems. The political accent of his last works was a clear demonstration of this connection. Derrida himself acknowledges that he "expected to be able to articulate [his] work of deconstruction with a renewed concept of politics."[840] *Différance,* now immersed in the political realm, becomes a word of action. By emphasizing the otherness through the dimension of democracy, Derrida writes an invaluable book, *The Politics of Friendship* (*Politiques de l'Amitié*),[841] which announces his perspective of a "*democracy to come.*" Along with the active movement that *différance* projects towards the other, *democracy to come,* as he himself sustains in another text, "does indeed translate or call for a militant and

---

[836]Jacques Derrida, "The Deconstruction of Actuality," in *Negotiations: Interventions and Interviews, 1971–2001* (Stanford, CA: Stanford University Press, 2002), 93.

[837]Bernstein, "An Allegory of Modernity/Postmodernity: Habermas and Derrida," 83.

[838]See Habermas, *Der philosophische Diskurs der Moderne*, 219/46.

[839]Bernstein, "An Allegory of Modernity/Postmodernity: Habermas and Derrida," 93.

[840]Jacques Derrida, "Entretien par Jerôme-Alexandre Nielsberg – Penseur de l'événement," *L'Humanité* 28 (January 2004), translation mine.

[841]Jacques Derrida, *The Politics of Friendship* (London, New York: Verso, 2005).

interminable political critique."[842] It seems that he designed a distinct and powerful theoretical approach to account for political action; nevertheless, this did not mean the reformulation of his previous works. Although now inserted into practical reflection, we can also examine the debates on politics through the dimension of metaphysics and his attempt to disclose and undercut it. Politics, public institutions, democratic regimes, constitutional principles, for instance, can express the sign of presence, defend different types of identity, fill the absence, using a sort of discourse that reifies the movement towards the other.

When, for example, we investigate an institution, we can register that it undermines this *democracy to come* insofar as it closes the possibility of otherness. The opening brought by this coming of the other is closed by placing a *logos,* an identity, behind the context, as if it could govern and guide somehow the destiny of society. As Derrida says, "nondemocratic systems are above all systems that *close* and *close themselves off* from this coming of the other."[843] More than ever, here the particularity of a context gains relevance in our reflections. D*ifférance*, as this infinite possibility, this dynamics towards the other – without forgetting nonetheless the insurmountable metaphysics in reality, for language is metaphysical – is examined from the singularities of a political context. It has thereby a practical and significant application: the disclosure of metaphysics is a crucial step on the way to a new concept of politics directed towards the other:

> For democracy remains to come; this is its essence in so far as it remains: not only will it remain indefinitely perfectible, hence always insufficient and future, but, belonging to the time of the promise, it will always remain, in each of its future times, to come: even when there is democracy, it never exists, it is never present, it remains the theme of a non-presentable concept.[844]

Consistent with this premise, the initial quotation we introduced in the last section – "for the present, to me, democracy is the place of a negotiation or compromise between the field of forces as it exists or presents itself currently (...) and this 'democracy to come"[845] – seems now much more understandable. *Democracy to come* is associated with the previously discussed unlimited play of signs. The traces, the forces empirically existing in the public realm, have to deal continuously with this promise of a democracy to come. This is where Derrida's stress on negotiation appears. Negotiation requires this act of responsibility towards the future in a present that aims, as an obligation, to protect the other's otherness. It is carried out in the interaction between the reality, with its differential forces, and the *to come*. Notwithstanding that *democracy to come* is not confounded with the negotiation, there is no sense at all in thinking of this *to come* without the empirical process of play of forces where negotiations take place. There is no *democracy to come* that is, at a certain time, considered apart from the singularity of a context.

---

842 Derrida, *Rogues: Two Essays on Reason*, 86.

843 Derrida, "Politics and Friendship,"182.

844 Derrida, *The Politics of Friendship*, 306.

845 Derrida, "Politics and Friendship,"180.

It is not an abstract entity behind reality, but rather the condition itself of a dynamics of possibilities manifesting in reality: "It makes possible the impossible, the coming of the other, the invention of the future."[846] It is a future, nonetheless, which is projected without erasing the past, even though incompletely recollected and gathered as fragments. This is why the process of negotiation is violent: it has incessantly to deal with momentary acts and decisions that are insufficient to the other's otherness. In sum, *democracy to come,* in Derrida's philosophy, is the possibility of overcoming the repetition of the same, and is also what makes negotiation a tense moment: it "operates in the very place of threat, where one must [*il faut*] with vigilance venture as far as possible into what appears threatening and at the same time maintain a minimum of security – and also an internal security not to be carried away by this threat."[847]

Accordingly, democracy refers to this realm of possibilities opened up by the process of negotiation, in every singularity of a reality, where the future is stressed by this *democracy to come* and where the past, never entirely recollected, offers a minimum of security: "Thus, there is a feeling of duty – a respect for the law."[848] This is why, for Derrida, the institutional history plays a central role in the democratic field. However, on the other hand, the stress on the past does not mean the denial of the "impossibility of stopping,"[849] of this "fatigue, of this without-rest, this enervating mobility preventing one from ever stopping"[850] *democracy to come* brings to this process, for it would mean the closeness of the realm of possibilities that is necessary for democracy. He is not, for this reason, a radical philosopher that intends to disrupt or replace the institutional background, but rather he attempts to set forth a critical thought that is concerned with the possibility of its always-necessary improvement without establishing a timeline of progress.

If we could sustain a sort of radicalism in his philosophy, this would be related to his effort to disclose and undercut the metaphysics of presence also in the political ground. Nonetheless, it is precisely this "radicalism" that denies the very possibility of establishing an entire political rupture to the extent that this could lead to a form of identity, a form of metaphysical presence. A new sort of presence would fill the lack following the previous presence, and thereby the free play of traces where democracy should take place would be interrupted. There would be no more negotiation, not at least a negotiation that acknowledges its own impurity. If democracy "is what it is only in *différance* by which it defers itself and differs

---

[846] Elisabeth Rothemberg, "Introduction," in *Negotiations: Interventions and Interviews, 1971–2001* (Stanford, CA: Stanford University Press, 2002), 5.

[847] Jacques Derrida, "Negotiations," in *Negotiations: Interventions and Interviews, 1971–2001* (Stanford, CA: Stanford, 2002), 17.

[848] Ibid., 13.

[849] Ibid.

[850] Ibid.

from itself,"[851] and thus "interminable in its incompletion beyond all determinate forms of completion,"[852] the negotiation that takes place in its grounds cannot be pure. This is why it has to negotiate, or the "thing would be even more impure."[853] The opening to a realm of possibilities is what Derrida always defended in politics, not the replacement or disruption of the past through a future-present conceived as a kind of pure entity negotiation would, at a certain time, accomplish. Deconstruction, after all, does not avoid tradition, but it does attack the closeness of otherness that a past or a future-present could bring forth. Hence, if negotiation is "none other than deconstruction itself,"[854] then it is not destruction: "If deconstruction were a destruction, nothing would be possible any longer."[855] Instead, it is a continuum of invention and reinterpretation, which, although "[involving] the structures' or the constructa, the things constructed that make life or existence possible,"[856] it is critical of it, for it does not see them as stabilized over time.

Democracy is thus marked by *iterability*, according to which we cannot presently take the past as a simple repetition, but as part of a process of reinterpretation and invention of the context to produce something new. The invention, which is characteristic of the democratic realm, is thereby to produce iterability.[857] Therefore, democracy cannot represent a sign of identity, but the play of traces that manifests this negotiation between the present and the *to come* with no guarantees regarding the future. Iterability exposes the negotiation that takes place between the impossible and the present as to render possible the democracy itself. For deconstruction is "the experience of impossible,"[858] the institutional and cultural heritage must embody its own perspective of deconstruction, as iterability reveals it, in order to avoid the closeness of the indispensable play where democracy manifests itself. As the impossible, accordingly, *democracy to come* is the political side of *différance*,[859] the other that is necessary in the process of negotiation where iterability

[851] Derrida, *Rogues: Two Essays on Reason*, 38.

[852] Ibid.

[853] Derrida, "Negotiations," 14.

[854] According to Derrida, "So negotiation is constantly under way, the negotiation which is none other than deconstruction itself" (Ibid., 16).

[855] Ibid.

[856] Ibid.

[857] According to Derrida, "To invent is to produce iterability and the machine for reproduction, simulation and simulacrum." (Jacques Derrida, *Psyché: Inventions de l'Autre* (Paris: Galilée, 1987), translation mine).

[858] Jacques Derrida, "As If It Were Possible," in *Negotiations: Interventions and Interviews, 1971–2001* (Stanford, CA: Stanford University Press, 2002), 352.

[859] According to Derrida:

"What announced itself thus as 'différance' had this singular quality: that it simultaneously welcomed, but without dialectical facility, the same and the other, the economy of analogy – the same only deferred, relayed, delayed – *and* the rupture of all analogy, absolute heterology. Yet one could also, in this context, retreat this question of *différance* as a question of legacy. The legacy would consist here in remaining faithful to what is received (...), while breaking with the particular figure of what is received" (Derrida,"As if It Were Possible," 352).

reveals itself. It makes possible the interminable opening: "The impossibility, the possible as impossible, is always bound to an irreducible divisibility that affects the very essence of the possible,"[860] after all. This *to make possible* that is characteristic of deconstruction, hence, deals incessantly with the past, never entirely gathered and recollected, as a way to construct what will also be the subject of deconstruction. If it is to have respect for the past, for an institutional tradition, on the one hand, it is indispensable to acknowledge that it also has to be deconstructed, on the other – this is where negotiation is taken into account: "One cannot imagine oneself alive renouncing all consciousness, all presence, all ethics of language: and yet this is precisely what must be deconstructed."[861]

In any case, a very important aspect we must stress in order to avoid misunderstandings concerning Derrida's philosophy is that negotiation is not a place for prophetic or progressive guarantees. *Democracy to come,* although linked to this dynamics towards the other, does not indicate a better future than the present; it is not a future-present: "The 'to-come' not only points out to the promise but suggests that democracy will never exist, in the sense of a present existence: not because it will be deferred but because it will always remain aporetic in its structure."[862] For this reason, this *democracy to come* seems to be an impossible project, "a democracy not to be identified with any of its existing institutional forms."[863] In negotiation, therefore, what really matters is that the realm of possibilities remains open. Were there, otherwise, this progressive term, as if the future were better than the present, then it would be necessary that we created a rule or a criterion to determine how better a reality is in a certain time. It is not a regulative idea in the Kantian sense, either.[864] For Derrida, after all, "there is no general law, there is no general

[860]Ibid., 358–359.

[861]Derrida, "Negotiations," 16.

[862]Derrida, *Rogues: Two Essays on Reason*, 86.

[863]Bernstein, "An Allegory of Modernity/Postmodernity: Habermas and Derrida," 87.

[864]As Simon Critchley correctly remarks by stressing the non-regulative idea behind its concept: "Democracy-to-come is much easier to describe in negative rather than positive terms. Recalling the deconstruction of the idea of presence in his earlier work, Derrida is particularly anxious to distinguish democracy-to-come from any idea of a *future* democracy, where the future would be a modality of presence, namely the not-yet-presence. Democracy-to-come is *not* to be confused with the living present of liberal democracy, lauded as the end of history by Fukuyama, but *neither* is it a regulative idea or an idea in the Kantian sense; *nor* is it even a utopia, insofar as all these conceptions understand the future as modality of presence. For Derrida, and this is something particular clear in *Spectres of Marx,* it is a question of linking democracy-to-come to the messianic experience of the *here* and *now* (*l'ici-maintenant*), without which justice would be meaningless. Namely, what was described above as 'the universal dimension of experience' that 'belongs to all languages'. So, the thought here is that the experience of justice as the here and now is the *à venir* of democracy. In other words, the temporality of democracy is *advent,* it is futural, but it is arrival happening *now,* it happens – and one thinks of Benjamin – as the messianic now blasting through the continuum of the present" (Simon Critchley, "Frankfurt Improptu – Remarks on Derrida and Habermas," in *The Derrida-Habermas Reader*, ed. Lasse Thomassen (Edinburgh: Edinburgh University Press, 2006). 108).

rule for negotiation."[865] The existence of a rule in the field of negotiation would mean placing an identity in what supposed to be a free play of signs. In sum, this free play of signs, this free play of forces that is so central to Derrida's philosophy would not be so freely conceived to the extent that it would be submitted to a sort of presence. Furthermore, it would contrast with one of the main characteristics of *différance:* a certain essentialism would take over the very singularity of the context. Derrida himself mentions that "there are only contexts, and this is why deconstructive negotiation cannot produce general rules, 'methods.'"[866] When, nonetheless, we take into account a prophetic or progressive term, rather than the accent on this healthy scenario where deconstruction takes place "*here and now*,"[867] an emphasis on criteria – a kind of *logos* – is made. The consequence is to put democracy in jeopardy. If democracy is this "place of a negotiation or compromise between the field of forces as it exists or presents itself currently (...) and this 'democracy to come,'"[868] then it stops manifesting itself insofar as this negotiation is closed by an identity, a presence, a *logos*, a general method filling the *to come* of an event.

Yet, this risk of interpreting Derrida as a philosopher of a certain prophetic and progressive timeline when democracy is at stake is not only erroneous, but also adopting him as a defender of the past as the reference denies the very basis of *différance* and deconstruction. After all, the *to come* does not mean something that is temporal, but instead the time itself. As Caputo concluded, "the *to come* enjoy[s] an 'infinite' structure that never takes a finite form."[869] Therefore, thinking of democracy as a timeline, where there is the past as a self-correcting learning process (although offering security), does not correspond to the fragmentary emphasis on traces Derrida so intensively worked with. This fragmentary character, nonetheless, does not mean that we cannot "distinguish between better and worse political efforts to show fidelity to the unconditional."[870] It is not relativism. It simply indicates that, although the past offers a minimum of security, what remains are the traces, which can always be deconstructed. The concept of self-correcting

In this respect, Derrida remarks that "in speaking of an unconditional injunction or of a singular urgency, in invoking a *here and now* that dos not await an indefinitely remote future assigned by some regulative idea, one is not necessarily pointing to the future of a democracy that is going to come or that must come or even a democracy that *is* the future" (Derrida, *Rogues: Two Essays on Reason*, 90).

[865]Derrida, "Negotiations," 17.

[866]Ibid.

[867]Derrida, *Rogues: Two Essays on Reason*, 90.
Derrida, "Negotiations," 17.

[868]Derrida, "Politics and Friendship," 180.

[869]John D. Caputo, "L'Idée Même de L'à Venir," in *La Démocratie à Venir: Autour de Jacques Derrida*, ed. Marie-Louise Mallet (Paris: Galigée, 2004), 302, translation mine.

[870]Bonig Honnig, "Dead Rights, Live Futures: On Habermas's Attempt to Reconcile Constitutionalism and Democracy," in *The Derrida-Habermas Reader*, ed. Lasse Thomassen (Edinburgh: Edinburgh University Press, 2006), 169.

learning process prints an idea of progress that is absent in Derrida's approach. We cannot interpret Derrida's philosophy as establishing conditions of a progress, but instead as an uneasy and infinite movement towards the other's otherness, which means undercutting the metaphysics of presence in this process. There is, for this reason, a promise in the very concept of democracy and also a paradox at the core of this negotiation between the present and the infinite promise. *Democracy to come*, contrary to the finite present, "always remains inaccessible, not just as a regulating ideal but also because it is structured like a promise and like a relation to otherness, because it never possesses the identifiable form of the presence or of the presence to self."[871] By stressing this inaccessible and open future, Derrida sustains the respect for all singularities, for the heterogeneous, for the other:

> Everything is at play in this paradox that I cannot develop here: Singularity is never *present*. It *presents itself* only in losing or undoubling itself in iterability, thus in the mark and the generality or ideality that, moreover (threat or luck), will allow later for a calculated negotiation between the presentable and the nonpresentable, the subject and a-subjective singularity, rights and justice beyond rights and ethics, and perhaps even beyond politics (...) The here-now indicates that this is not simply a question of utopia. There is a constant and concrete renewal of the democratic promise as there is of the relation to the other as such, of the relation to infinite distance, incalculable heterogeneity, etc.[872]

The stress on this connection between the present field of forces and this promise sets forth the demanding theme of responsibility. If it is to sustain a responsible decision, it means that it always works with the 'perhaps' in this process of negotiation:[873] "But this 'perhaps' would be that of the possible *and* the impossible *at the same time*, of the possible *as* impossible."[874] There are no guarantees, certitudes when we have to make a responsible decision. The opening brought by the *to come* irrupts as a deconstruction, but not destruction, of any dogmatic certitude that could guide somehow our decision. The negotiation between this promise and the present field of forces leads to a responsible decision to the extent that it settles the aporia of *undecidability* in its basis: "To think responsibility calls indeed for an absolutely unprecedented responsibility, one that cannot look back upon given assurances, but that is attentive to and responsive to the structural features of undecidability characteristic of responsibility itself."[875] Inasmuch as history is never linear and thus cannot be entirely gathered and recollected, and the future is marked by a promise full of uncertainties, the singularity of the context addressed to the other's otherness keeps alive the imperative of incessantly questioning the origins, certitudes, memories that are at the core of deconstruction. As Derrida remarks, "responsibility without limits"[876] relates to this philosophical and

---

[871]Derrida, "Politics and Friendship," 180.

[872]Ibid.

[873]See Gasché, *Inventions of Difference*, 228.

[874]Derrida, "As if It Were Possible," 344.

[875]Gasché, *Inventions of Difference*, 228.

[876]See Derrida, "Force of Law," 955.

cultural heritage the subject receives – "the task of an interpretative memory is at the heart of deconstruction"[877] – but, on the other hand, this responsibility is concerned with a "heritage that is at the same time the heritage of an imperative or of a sheaf of injunctions."[878]

Hence, it is not a recollection of the past, but rather the impulse towards an ongoing reinvention by questioning these origins, these dogmatic certitudes. We receive the heritage, and this encompasses the responsibility itself as a critical interrogation of all concepts and references that are part of this legacy. This unwearied interpretability and reinvention, which is the very heritage and the character of this responsibility, accordingly, is an aporetic experience as long as a selection of what was inherited takes place by promoting the reinterpretation of the tradition according to the particularity of a context, to the singularity of the other's otherness: "To be responsible in the name of heritage is to be governed by the need to intervene in what we received, and thereby restarting in a singular and new way this heritage itself."[879] Thinking democracy entails this responsible act towards the otherness, which means deconstructing the assurances and rules (as a *logos*) that could guide decision-making: all is construed in this contextual negotiation that never forgets the inherent undecidability linked to the impossible experience of the *to come*: "I *think* this impossibility, and it is there that I *think* what my responsibility should be, which is to say, infinity."[880]

As a reflex of the discussion of *différance*, democracy requires, for this reason, a non-fundamentalist, yet non-arbitrary (since not voluntaristic),[881] process of decision-making. The unlimited play of signs, the traces, the metaphors are here transported to the political dimension of the inevitability of a future surrounded by uncertainty, and even the risk of disrupting the negotiation between *democracy to come* and the present field of forces through a form of essentialism or dogmatic structure. The risk is always present. Derrida argues that "without the possibility of radical evil, of perjury, and of absolute crime, there is no responsibility, no freedom, no decision."[882] This is, indeed, the immediate consequence of the opening *différance* brings forward, but, on the other hand, it is this opening that also guarantees the very functioning of democracy. There is no democracy without the threat that is at the heart itself of its inventive and interpretative character, or, in Derrida's words, "democracy protects itself and maintains itself precisely by

---

[877]Ibid.

[878]Ibid.

[879]Rodolphe Gasché, "L'Étrange Concept de Responsabilité," in *La Démocratie à Venir: Autour de Jacques Derrida*, ed. Marie-Louise Mallet (Paris: Galilée, 2004), 364, translation mine.

[880]Jacques Derrida, "Performative Powerlessness – A Response to Simon Critchley," in *The Derrida-Habermas Reader*, ed. Lasse Thomassen (Edinburgh: Edinburgh University Press', 2006), 113–114.

[881]See Derrida's analysis of Carl Schmitt in his book *The Politics of Friendship*. A very interesting debate on this subject can also be found in Critchley, "Frankfurt Improptu – Remarks on Derrida and Habermas," 98–109.

[882]Derrida, The Politics of Friendship, 219.

limiting and threatening itself."[883] For there are no guarantees, what accompanies democracy is merely the 'perhaps'. And what accompanies the responsible decision is this eternal undecidability, which opens it up to the other and reveals this always-indispensable singularity of a context: "There is no other decision than this one: decision in the matter and form of the undecidable."[884] Democracy functions thus in this field of forces, differential forces, where a responsible negotiation is carried out through iterability, which, in turn, is marked by the heritage, the *to come*, and its threat. Democracy is marked by *undecidability*.

5.2.3 *The to Come in the Negotiation Between Constitutionalism and Democracy*

The previous arguments and concepts have enormous effects when we examine, more particularly, the unlimited play of traces and the promise of the *to come* within the context of the relationship between constitutionalism and democracy. The previous discussion of *différance* as this dynamical opening of presence/absence now reaches the compromise between constitutionalism and democracy. As introduced at the core of the concept of responsibility, the undecidability is unavoidable if we think of the contradictory structure embedded in the concept of *democracy to come*. The negotiation between the *to come* and the field of forces as they currently present themselves – this incalculable, impossible and nondeductible aspect of the *to come* in contrast to the reality in any responsible decision – after all, is behind Derrida's political thought: "It is on the basis of this undeconstructible infinite responsibility that one is propelled into moral and political problems, into the realm of decision."[885] When, therefore, Derrida highlights the opening as an ongoing reinterpretation and reinvention, he acknowledges, in the political realm, the gap that exists between the presence and the absence as the necessary condition of democracy itself. As an attack on the identity derived from the shortcomings of the movement towards otherness, democracy needs its other as a means to remain open to the future. Democracy needs constitutionalism. As well as the responsibility and the negotiation, this relationship with the other is woven into insurmountable undecidability. It is the infinite play of democracy and constitutionalism that shows the opening to the other. It is a play, besides, that remains open as the coming of the 'perhaps', as a sign of its always-remained undecidability.

The stress on *différance* ensures that the connection between democracy and constitutionalism is always deferred, for *différance* can never mean the destruction of the existing gap between the same and the other. Democracy, as a consequence, cannot be confounded with constitutionalism and vice-versa, because this would

[883] Derrida, *Rogues: Two Essays on Reason*, 36.

[884] Derrida, *The Politics of Friendship*, 219.

[885] Critchley, "Frankfurt Improptu – Remarks on Derrida and Habermas," 105.

stop the play and even interrupt the dialectical movement towards the other's otherness. In sum, the destruction of this interaction would result in interrupting the dynamics of time, for there would be no *to come* any longer, the infinite time itself. Accordingly, the unlimited play of traces, in the political realm, can only be preserved insofar as a continuum in the negotiation takes place by never disrupting this complex and irreducible interaction between both. Under other circumstances, the filling of this lack – the absence of transcendental signified – would bear a metaphysical thought at the heart of the political field, and thereby jeopardize democracy. Politics possessed by metaphysics is the closeness of the realm of possibility, the disruption of the coming of the impossible, and thereby the denial of democracy: "When a democracy attempts to close off this openness to the other (which is the very condition of democracy) then democracy commits suicide."[886] Metaphysics breaks up the movement towards otherness, and thus ceases the indispensable interaction between democracy and constitutionalism. This connection, for this reason, must be intense, complex, dynamical, but, above all, remain open. Whereas democracy evokes the sovereignty of people, who have to endlessly reinterpret and shape the constitution, establish its conditions and bestows its legitimacy, constitutionalism defines how the exercise of this sovereignty will take place by submitting it to the rule of law and basic rights. In order to remain open to the other, "constitutionalism must be at once iterable by (because the ground of) and alterable by (because the product of) democracy;"[887] "the condition of possibility of constitutionalism (namely, democracy) is simultaneously its condition of impossibility and limit (and *vice versa* from the viewpoint of democracy),"[888] and this means that one cannot assume the other's place, or it would simply interrupt the democratic time marked by iterability.

The mediation between democracy and constitutionalism, which is pervaded by undecidability, sets forth an instigating structure that shapes the functioning of constitutional democracies. There are no simple resolutions in this process: the threat encompassing undecidability is a necessary motor for this interaction between both and a condition to avoid that one overcomes the other. The gap between them is to remain unfilled, but it is also to remain dynamical and aware of the responsible call for the other. The otherness points out that one cannot be the other nor jeopardize the other, but, on the other hand, cannot subsist without the other. Since the signifier does not exist without the signified and vice-versa (as the same presupposes the other), democracy and constitutionalism are inseparable, although not confounding. Both function by negotiating the conditionality of a reality and the unconditionally of the *to come*, as an opening towards the other and an endless process of interpretability and reinvention. The undecidability of the

[886] Mark Dooley and Liam Kavanagh, *The Philosophy of Derrida* (Stocksfield: Acumen, 2007).

[887] Lasse Thomassen, "'A Bizarre, Even Opaque Practice': Habermas on Constitutionalism and Democracy," in *The Derrida-Habermas Reader*, ed. Lasse Thomassen (Edinburgh: Edinburgh University Press, 2006), 185.

[888] Ibid., 186.

process, therefore, marks their constitutive and asymmetrical interaction as a process of limiting and enabling their very functioning. Both are, therefore, the condition of possibility of the other and both presuppose and constitute the other. Since they are characterized by undecidability, there is neither origin nor end in this process: what remains is simply an infinite regress.[889]

The conclusion that democracy and constitutionalism are inseparable and not confounding, yet permeated by undecidability, is a very strong argument to understand how we can examine *différance* within the political realm. It is a play between the same and the other, the presence and the absence, the signified and the signifier, but never the prevalence of one over the other. It is the same message: there cannot be a transcendental signified behind politics, for there would be no movement towards the other and legitimate transformation. The *logocentrism* means an attack on the free play of traces where constitutional democracy should develop. The question, nevertheless, is how it develops in practice. How can we reinforce constitutional democracy in every new context? In reality, constitutional democracy is marked by a set of responsible decisions that take place through iterability. The differential forces and the promise of the *to come* are negotiated in order to fight against the identities that undermine the very possibility of thinking of otherness. The negotiation addresses to the other; it opens up the possibility of deconstruction; it leads to the autoimmune character of constitutional democracy, that is, its capacity of self-critique and perfectibility[890] in every new event. On the one hand, constitutional democracy requires the calculable, the differential forces, the institutional history; on the other, it demands the promise, the incalculable, the absolute singularity of the other's otherness, the *democracy to come*. In this process, political decisions must transform realities, as an urgent call against identities, *logocentrisms*. Decisions have to combat metaphysics. Hence, it is a practical and intervenient call, for "the thinking of *différance* is also, therefore, a thinking of urgency."[891] Constitutional democracy is characterized by its capacity of action (otherwise it would imply inaction and lead to metaphysics) by a set of decisions

[889] Lassen Thomassen sums up, in a very clear manner, this perception in the following passage:

"The relationship between constitutionalism and democracy is not one of either internality or externality, either mutually enabling conditions or limits. You cannot have one without the other, yet they stand at a slight distance from one another. It is not a distance that can be measured, or a gap that can be closed, though. This lack (or lag), the slight but infinite distance between constitutionalism and democracy, cannot be recuperated; it is constitutive. This is what makes constitutional democracy [go] around. Without the undecidability, constitutional democracy would not work. Without democracy as its condition of possibility, constitutionalism would not be properly constitutional, yet democracy, at the very moment it makes constitutionalism possible, also limits it. But, if democracy needs constitutionalism (and *vice versa*), then democracy cannot repair a lack in constitutionalism, because democracy will itself be lacking as a result of this incompleteness in constitutionalism. We cannot escape the vicious circularity and infinite regress" (Ibid., 186).

[890] Derrida, *Rogues: Two Essays on Reason*, 87.

[891] Derrida, "The Deconstruction of Actuality," 93.

that, to be responsible, thinks of otherness: "We cannot escape responsibility, decision, and choice. They are thrust upon us by the other."[892]

This interaction leading to practical interventions, therefore, is pervaded by a set of decisions that should reflect iterability and *différance*. Constitutional democracy functions through an "active and interminable critique"[893] that affects dogmas, certitudes, prejudices, identities, in order to reinvent them in conformity with the singularity of the event, of the other's otherness. Every decision in this process, as a consequence, for it has to deal with the complex interaction between constitutionalism and democracy, is inevitably a "madness."[894] It is not a total destruction of the heritage, but it is not its simple reproduction, either. More particularly, constitutional democracy is neither simply the sovereignty of people nor the constitution and the institutional history it follows. It cannot be a reconciliation of both, either. Still, we must negotiate. This has to be done by taking into account the inevitable threat of setting one up as a form of presence, instead of seeing the constitutive gap embedded in their compromise towards the other's otherness. This is why constitutional democracy is a very anguishing process. It is marked by a set of decisions that have to abruptly interrupt the negotiation taking place between the conditional of a moment and the unconditional of the *to come*. This must done in such way that democracy and constitutionalism are mediated without putting each other in jeopardy. A responsible decision, in this field, is a decision that takes into consideration the risks involved in this mediation according to the singularity of the context, for there is always uncertainty and undecidability, while keeping alive the irreducible incommensurability of the other's otherness: "No code can close the gap or diminish the undecidability that confronts us in making an ethical-political decision or choice."[895]

Moreover, the responsible call – this urgent call constitutional democracy expresses in practice – knows the threats that are at the heart of this negotiation. There is no responsible decision that is unaware of the risks of negotiation, thereby of the *aporia of undecidability*, and it must be made by focusing on the particularity of the context: "There must be decision, there must be absolute risk, and thereby there must be the undecidable."[896] The negotiation between democracy and constitutionalism, for this reason, is embedded in this play where responsible and urgent decisions are made, and where the constitutive gap between them remains open and unfilled. In this realm, the question about the legitimacy of a political transformation and the institutional practice gains a relevant focus of analysis: the play between constitutionalism and democracy, as long as not filled by any *logos* and carried out only through iterability, leads to responsible decisions in the political arena, and this is the premise to evaluate the legitimacy of a certain practice.

---

[892]Bernstein, "An Allegory of Modernity/Postmodernity: Habermas and Derrida," 85.

[893]Derrida, *Rogues: Two Essays on Reason*, 86.

[894]See Derrida, "Force of Law," 967.

[895]Bernstein, "An Allegory of Modernity/Postmodernity: Habermas and Derrida," 84.

[896]Derrida, "Negotiations," 31.

Likewise, this is the realm where constitutional democracy connects to decision-making. In this matter, the premises of the negotiation between constitutionalism and democracy imply their counterpart in adjudication. Constitutional democracy is also characterized by a set of decisions in the field of legal reasoning. In this matter, there is here negotiation, which is carried out between the dichotomy of law and justice. A responsible decision in adjudication, which deals with the differential forces in reality and the *to come*, now justice to come, might also be marked by iterability and the absence of transcendental signified. It is a more particular subject, but, like all Derrida's concepts, it is identified by this message of *différance*: in adjudication, there must be also the purpose of disclosing and undercutting metaphysics.

5.2.4 Différance Within the Context of Decision-Making: The Negotiation Between Law and Justice and the First Insight into Legitimacy

From the standpoint of this negotiation between democracy and constitutionalism, two relevant questions, nonetheless, still need to be faced. First, even though the gap between them is to remain open and unfilled, it is not yet clear how this opening to the other's otherness extends to the institutional discussion of legal rights, and more particularly the way legal rights are interpreted in decision-making. Second, constitutionalism and democracy lead to the question of legitimacy, and thereby how the *to come* of democracy keeps alive a continuous project of legitimation.[897] In this matter, the accent on negotiation reaches Derrida's analysis of the American Declaration of Independence[898] in order to show how law, at the precise moment of foundation, excluded the other's otherness, as a power against the other,[899] and how this violence is reproduced throughout history. Both questions are interconnected. The extension to the debate on adjudication is the counterpart of the foregoing negotiation between constitutionalism and democracy, which is now expressed through the negotiation between law and justice. The discussion of legitimacy, in turn, although initially constructed in the political realm of the negotiation between constitutionalism and democracy, extends to the decision-making, for the debate on legitimacy can also be visualized in the negotiation between law and justice. More specifically, the question here is to see how a decision, even though practicing violence (every law, after all, practices violence, since it cannot entirely render justice to the other), can be legitimate. Legitimacy and violence seem to be a tense

[897]We shall examine this discussion further in order to show how methods and criteria, such as balancing, can become a sort of *logocentrism* that fills the gap between constitutionalism and democracy.

[898]See, for this purpose, Derrida, "Declarations of Independence," 46–54.

[899]See Milovic, "A Impossibilidade da Democracia," 259.

paradox that occurs in the realm of adjudication, which connects the second question to the first: law, as an expression of power, as a sign of force and violence, is taken by its always-present capacity to bring about injustice. The outcomes of this discussion in the particular debate on decision-making are central: there is no decision without the risk of failure, of the threat of injustice, and it is thus, paradoxically, the very condition of the possibility of justice. This conclusion connects, in turn, the second question to the first: the very condition of justice, albeit never reached – for justice is, like democracy, a *to come* – is the very condition of legitimacy.

Moreover, the previously discussed opening we assumed in the relationship between democracy and constitutionalism reaches the opening of the process of adjudication, here expressed by the negotiation between law and justice. The foregoing conclusions, accordingly, apply here. This negotiation, as the previous one, is to remain open and unfilled; it is to keep alive the endless potentiality of interpretation and transformation. At the same time, this opening carries its own threat of closeness, and, more specifically, the closeness of the realm of possibility in adjudication. When democracy and constitutionalism are at stake, and one of them is put in jeopardy, the play is interrupted by this closeness and thus constitutional democracy commits suicide. When decision-making, inherently violent, does not take into account the play between law and justice and settles any sort of metaphysical thought as a justifying *logos*, there is no responsibility, and, hence, it commits injustice. On the side of the question about legitimacy, the same reasoning applies. As well as in the negotiation between constitutionalism and democracy, in the negotiation between law and justice, there should be autoimmunity,[900] as the right to self-critique and perfectibility,[901] and iterability, but now more directed to adjudication. Whereas, in the first case, a presence of *logos* would cause the attack on the right to criticize publicly everything that is brought up into the play, in the second, it would cause the practice, in a singular case, of violence without legitimacy. In any case, the absence of autoimmunity and iterability would disrupt the democratic legitimacy of a certain activity and expand the possibility of

[900]Derrida develops a very instigating analysis of democracy and its autoimmune character in his book *Rogues: Two Essays on Reason* in which he remarks that, by assuming that every identity presupposes the other and thus cannot be entirely filled, autoimmunity is part of democracy as a means to keep it open to the realm of possibilities, including the threat of the closeness of the play. According to him:

"If an event worthy of this name is to arrive or happen, it must, beyond all mastery, affect a passivity. It must touch an exposed vulnerability, one without absolute immunity, without indemnity; it must touch this vulnerability in its finitude and in a nonhorizontal fashion, there where it is not yet or is already no longer possible to face or face up the unforeseeability of the other. In this regard, autoimmunity is not an absolute ill or evil. It enables an exposure to the other, to *what* and to *who* comes – which means that it must remain incalculable. Without autoimmunity, with absolute immunity, nothing would ever happen or arrive; we would no longer wait, await, or expect, no longer expect one another, or expect any event" (Derrida, *Rogues: Two Essays on Reason,* 152).

[901]Ibid., 86.

establishing through violence (no more legitimate) an identity in the political arena of public mobilization or in the legal space of decision-making. Besides, in any case, the forgetfulness of the other (constitutionalism as the other of democracy, and vice-versa; law as the other of justice, and vice-versa) would preclude deconstruction. This would lead to the loss of the possibility of surmounting the same structures, the identity. Indeed, the logical consequence of forgetting the other is the reinforcement of the same. This is why if, instead of deconstruction, a *logos* were affirmed, then there would be no possibility of the 'new', of democracy, of justice. After all, "deconstruction is justice;"[902] deconstruction is a requirement for legitimacy. Were there no deconstruction, and thus autoimmunity and iterability, there would be, instead, violence without legitimacy, and the law, applied to a context of simple repetition without legitimacy, could become a simple process of calculation or identification of a particular will, which would be, definitely, neither just nor democratic.

Hence, the interaction between democracy and constitutionalism leads to the debate on legal rights, on the negotiation between law and justice. If it is to think of legal rights under constitutional democracy, then we cannot forget this irreducible gap that is to remain open and unfilled by any type of logocentrism. Law, for it is marked by its inherent violence, needs at least to be legitimate, and thereby let deconstruction do its role. Adjudication, in turn, needs to make deconstruction possible. If decision-making can never render presently justice,[903] for it would stop the play, then it needs at least to be legitimate and, for that, take into account "the institutive act of a constitution that establishes what one calls in French *l'état de droit*."[904] On the contrary, if we assume a *logos* as the referential guidance for decision-making, then not only arises the practice of violence without legitimacy, but also the decision becomes a reified discourse of reproduction of an identity. As a consequence, this identity, instead of taking into account *l'état de droit*, becomes a reference in itself, an essentialism against the singularity the context provides, thereby putting in jeopardy constitutional democracy.

Yet, how can we observe the existence of this identity and logocentrism in the realm of constitutional rights, particularly in constitutional adjudication? How can we, after all, remark that the negotiation between law and justice (as the other facet of the negotiation between constitutionalism and democracy) is characterized, in a singular situation, by a metaphysical standpoint? How can a responsible decision be attained in this field? The title of Derrida's book that explores this complex negotiation between law and justice already provokes some important questions. *Force of Law* (*Force de Loi*), originally called *Deconstruction and the Possibility of Justice*,[905] represents the sign of a strong perception of the necessary call for justice

---

[902] Derrida, "Force of Law," 945.

[903] Ibid., 961.

[904] Ibid., 963.

[905] Jacques Derrida, "Deconstruction and the Possibility of Justice," *Cardozo Law Review* II, no. 5–6 (July–August 1990).

when deconstruction is at stake. In its contents, the message – "deconstruction is justice"[906] – seems to call everyone's attention given the intriguing connection it brings forth: on the one hand, there is the word *deconstruction* with all the implications and complexities it raises; on the other, there is the word *justice,* as a direct link with the discussion of law and its force. Both are involved in a mutual correspondence with *différance. Force of Law* introduces the previously discussed problem of *différance* and deconstruction into the debate on legal rights and makes the link with decision-making. It also elucidates why, to undercut metaphysics, a certain *logos* cannot guide the negotiation between law and justice. The message is clear: there cannot be a metaphysical ground in the concept of justice. Indeed, justice means the effacement of any transcendental signified in the realm of legal rights.

When Derrida sustains that "law (*droit*) is essentially deconstructible, whether because it is founded, constructed on interpretable and transformable textual strata (and that is the history of the law (*droit*), its possible and necessary transformation, sometimes its amelioration), or because its ultimate foundation is by definition unfounded,"[907] he is simply stressing this process of negotiation between the conditional of a reality and the unconditional of the *to come,* now a *justice to come.* This negotiation, similar to the one between constitutionalism and democracy, needs also to take into account the desire to keep the past, traditions, identities alive, while simultaneously enacting deconstruction as a means to open them up to the threat the unconditional poses. Law and justice connect to each other as an opening to a non-future-present that will draw attention to the singularity of the context as the expression of justice. As constitutionalism cannot be reduced to democracy and vice-versa, so justice cannot be reduced to law, and vice-versa. They presuppose and constitute each other and can never be entirely reconciled:

> Justice can never be reduced to law, to calculative reason, to lawful distribution, to the norms and rules that condition law, as evidenced by its history and its ongoing transformations, by its recourse to coercive force, its recourse to a power or might that, as Kant with the greatest rigor, is inscribed and justified in the purest concept of law or right (. . .) The interruption of a certain unbinding opens the free space of the relationship to the incalculable singularity of the other. It is there that justice exceeds the law but at the same time motivates the movement, the history, and becoming of juridical rationality, indeed the relationship between law and reason, as well as everything that, in modernity, will have linked the history of law to the history of critical reason. The heterogeneity between justice and law does not exclude but, on the contrary, calls for their inseparability: there can be no justice without an appeal to juridical determinations and to the force of law; and there can be no becoming, no transformation, history, or perfectibility of law without an appeal to a justice that will nonetheless always exceeds it.[908]

This passage seems to set forth similar premises as the ones hitherto examined, but now centered on adjudication. Whereas the law refers to the real state of legal

---

[906] Derrida, "Force of Law," 945.

[907] Ibid., 943.

[908] Derrida, *Rogues: Two Essays on Reason*, 149–150.

history, concrete norms and institutions, justice appears as the incalculable, the impossible, the *to come*. It is their interaction that permits to initiate the movement towards the other, thereby not reducing the free play of traces to pre-established rules determining exactly how to proceed. We can now better visualize, therefore, the answer to the above questions. In the realm of the negotiation between law and justice, a responsible decision, while respecting the institutional history, the law and its movement, opens up the possibility to deconstruct them, which means neither destruction[909] nor an inertial movement. Moreover, since law is inherently violent, and therefore enforceable, consequently, justice needs paradoxically to be connected to the force of law. Justice must appeal to juridical determinations.[910] In this matter, we should stress a very important message here, for we will further apply it to the dogmatic problem here investigated: if there were not this call, justice would become a mere *moralizing principle*. The demand for the other would become a *logos* that could justify any practice. Justice would become a *transcendental meaning*.

This relationship between justice and law is consistent with the perception that one of the conditions to avoid the use of legal discourse as a moralizing one is to regard justice not as a *logos* behind the context, but rather as an opening to the context. There is no justice that does not critically take into consideration the institutional history and deconstruct it, and this is, indeed, the very condition to keep alive the force of law. The call for justice embodies the negotiation with the law: "*Both* calculation *and* the incalculable *are necessary*."[911] There is, consequently, a very specific correlation between both: if the emphasis is placed on one of them to the detriment of the other, there is no more play, no more negotiation, and thus a metaphysical thought is affirmed. First, the metaphysics of justice would mean the forgetfulness of the institutional history and force of law, and, since justice relates to the respect for the singularity of a context, there would be, in fact, no justice. Second, the metaphysics of law would express the indifference to the other, and thus indicate a simple repetition of the same structures, as if the law were a ground in itself. Both are prejudicial to constitutional democracy: they represent the other facet of the disruption of the play between democracy and constitutionalism. If there is no negotiation between law and justice, there will be no negotiation between constitutionalism and democracy. As democracy without constitutionalism would open up the space for the voice of majority, even when disregarding minorities, so justice without law would open up the space for practicing violence without legitimacy. As constitutionalism without democracy would suffer from a crisis of legitimacy, so law without justice would suffer from the incapacity to question its own basis, and thus repeat the same. Hence, there is no justice without law; there is no democracy without constitutionalism.

---

[909] See Derrida, "Negotiations," 16.

[910] Ibid.

[911] Derrida, *Rogues: Two Essays on Reason*, 150.

For it is to keep open the negotiation between law and justice – similarly to the negotiation between constitutionalism and democracy – decisions must be aware of the call for responsibility and also the threat embedded in this process. The risk surrounding decision-making is always present. Yet, the risk becomes reality when the gap between constitutionalism and democracy is closed, and subsequently the gap between law and justice. On the one hand, justice stops being the opening to the other and becomes the enclosure of the other and the justification of the same. In this case, law, by following this *transcendental signified* that justice embodies, is applied with no further justification but the ground in itself of this transcendental signified. As a consequence, the play where constitutional democracy takes place and where justice and law ought to be negotiated is dominated by a *logos*, a *substantiality* that outlines the itineraries of an apparent democracy, for there is no more critique, no more autoimmunity, no more undecidability. The fragile character that is embedded in the negotiation between constitutionalism and democracy, and law and justice – which is, nonetheless, the condition of self-critique and perfectibility – is now replaced by the safe place the *logos* sets forth, a *moralizing logos*, which uses the force against democracy, uses the force against justice, uses the force without legitimacy.

This argument is a serious one: against the metaphysics, justice cannot become a *moralizing principle*, for this would mean, in decision-making, the denial of the institutional history and the inherent enforceability of law. Legal interpretation, as a consequence, would not have the minimum of necessary security[912] to confront the threat of the *to come*, which is now possessed by a certain substantiality. Without law, there would be no justice, but rather substantiality in its place. Without justice, there would be no law, but rather a certain ruling without legitimacy. In addition, every law, to be law, needs to have the *enforceability*, this ability to be applied with force: "There is no such thing as law (*droit*) that doesn't imply *in itself, a priori, in the analytic structure of its concept*, the possibility of being 'enforced,' applied by force."[913] This means that law has, in its structure, the ability to be applied, even when the addressee does not intend to obey it. However, since law can *enforce* someone, it must do it in a legitimate way. To be legitimate, the decision needs to keep unfilled the gap between law and justice (and thus the gap between constitutionalism and democracy), and therefore justify itself through iterability. The play between law and justice, on this account, is carried out not from a substantiality, which has to be attacked, but from its own performative process where different forces interact with one another and gain new signification in conformity with a particular context.[914]

---

[912]Derrida, "Negotiations," 17.

[913]Derrida, "Force of Law," 925.

[914]According to Derrida:

"A first precaution against the risks of substantialism or irrationalism that I just evoked involves the differential character of force. For me, it is always a question of differential force, of difference as difference of force, of force as *différance* (*différance* is a force *différée-différante*), of the relation between force and norm, force and signification, performative force, illocutionary or

The legitimate use of force appears, for this reason, connected to *différance,* with its performative construction according to the time, but it also embodies the paradoxes and conflicts of this process. It refers to history. On the other hand, it is exactly this connection with *différance* that does not limit it to history. The force must be open to the other. The force must be connected to justice; as well, justice must be connected to the force: "If justice is not necessarily law (*droit*) or the law, it cannot become justice or *de jure* except by holding force or rather by appealing to force from its first moment, from its first word."[915] That is to say, the force, to be legitimate, needs to be in a play with justice and vice-versa, and thereby decision-making, although enforcing the law, needs always to regard the other. It has to maintain this double bind in negotiation, as a condition to reach the responsible act, the responsible decision. A responsible decision knows that the "law is essentially deconstructible,"[916] which means that it is open to the *to come,* and thus "constructed on interpretable and transformable textual strata."[917] A responsible decision is connected to institutional history and all complexities and tensions it gives rise to, but, at the same time, it does not establish a last argument in this process, because it knows that the ultimate foundation of law, which is the very justice, is by definition unfounded.[918] The foundation of law is its own absence of foundation. Besides, this absence of foundation and the deconstructible character of law make deconstruction possible.[919] This relationship between the deconstructibility of law and the undeconstructibility of justice, since it is the very deconstruction,[920] is the mark of the responsible negotiation between law and justice in decision-making.

The responsible decision, which is, in sum, the expression of deconstruction in a particular case, has, as a consequence, to face three important aporias. First, there is what Derrida calls *épokhè and rule,*[921] which stresses the singularity of each case through the premise that, in decision-making, there is, on the one hand, the need to follow the law, and, on the other, the indispensability of recreating and reinventing the law in conformity with the singularities of the case: "For a decision to be just and responsible, it must, in its proper moment, if there is one, be both regulated and without regulation."[922] The process of decision-making cannot simply rely on the law, its history and its inherent enforceability, but must also reinvent it with reference to the case, "it must conserve the law and also destroy it or suspend it enough to have to reinvent it in each case, rejustify it, at least reinvent it in the

perlocutionary force, of persuasive and rhetorical force, of affirmation by signature, but also and especially of all the paradoxical situations in which the greatest force and the greatest weakness strangely enough exchange places. And that is the wholly history" (Ibid., 929).

915 Ibid., 935.

916 Ibid., 943.

917 Ibid.

918 Ibid.

919 Ibid., 945.

920 Ibid.

921 Ibid., 960.

922 Ibid., 961.

reaffirmation and the new and free confirmation of its principle."[923] There cannot be, therefore, an essentialism defining a standard to be followed when a case is to be decided, for every interpretation is always unique.[924]

Second, there is the *ghost of the undecidable.*[925] Indeed, to do justice, albeit never presently possible, the decision must be made and needs to have in its basis the 'perhaps' of undecidability. This is a condition for attacking the metaphysics of presence: "Its ghostliness deconstructs from within any assurance of presence, any certitude or any supposed criteriology that would assure us of the justice of a decision, in truth of the very event of a decision."[926] Instead of these assurances, there is only an infinite and irreducible idea of the other's otherness, an idea that is not a transcendental signified, but simply points to a "desire for justice,"[927] to the singularity of the other, as "the very movement of deconstruction at work in law and the history of law."[928]

Third, as a consequence of the premise that the negotiation between law and justice will inevitably have to face with decisions, for there is always in this process a thinking of urgency,[929] in every decision, we cannot escape from the "urgency that obstructs the horizon of knowledge."[930] The negotiation between law and justice, accordingly, is not marked by a regulative idea, by a "horizon of expectation,"[931] for justice must be rendered immediately; decision-making is "always a finite moment of urgency and precipitation."[932] On the one hand, it is impossible to entirely gather and recollect the past, the institutional history, the law; on the other, justice is presently unattainable. For it deals with this interruption in space and time, "the instant of decision is a madness."[933] It has, at its core, dissymmetry, because of the very character of differential forces, and violence,[934] because the *to come* cannot be entirely reached. Although justice is a *to come,* "the experience of absolute alterity,"[935] it is also the very condition of history.[936] The absence of a transcendental signified, in the realm of the negotiation between law and justice, is, consequently, the very condition of legitimate transformations.

---

[923]Ibid.

[924]Ibid.

[925]Ibid., 963.

[926]Ibid., 965.

[927]Ibid.

[928]Ibid.

[929]Derrida, "The Deconstruction of Actuality," 93.

[930]Derrida, "Force of Law," 967.

[931]Ibid., 969.

[932]Ibid., 967.

[933]Ibid.

[934]Ibid., 969.

[935]Ibid., 971.

[936]Ibid.

Briefly, since the negotiation between law and justice lies in this realm of the calculable and the incalculable,[937] decision-making should be characterized simply by iterability, which encompasses the complexities and tensions of a past and a never reachable future. Decision-making should let deconstruction do its role. The relationship between the incalculable and the calculable, the *épokhè and the rule,* and the urgency that makes the double bind of law and justice the condition of the legitimate transformations of history are the core of Derrida's deconstructionist approach in the realm of adjudication. By examining how he understands decision-making in his book *Force of Law*, Derrida applied much of his previous concepts from his study of language; however, they gain now a political and legal meaning and become a very powerful premise to understand and criticize how politics and legal adjudication take place. More directly, they become a relevant and powerful premise to develop the dogmatic problem of this research. This is what the next section proposes: the connection of these premises with the dogmatic problem of the rationality of balancing.

5.3 Balancing Within the Context of Différance

5.3.1 Introduction

Force of Law introduces the double bind of justice and law as the expression of deconstruction, and hence the call for *différance.* It has the interesting ability to encourage the critique of the most central aspects and beliefs embedded in the debates on legal rights, and it has the potential to criticize the institutions and how they operate the negotiation between law and justice, constitutionalism and democracy. Particularly, it permits to develop a critical analysis of the way constitutional courts act. It is not a direct attack, though. As said previously, Derrida has not been immediately involved in the specific questions of decision-making, despite the fact that we could observe some analysis of this when he discusses the three aporia (*épokhè and rule, the ghost of undecidable, the urgency that obstructs the horizon of knowledge*). Still, he seems to miss a deeper investigation of the institutional background, which is remarkably significant when we seek to go inside the structure of a methodology of decision-making and study its consequences for constitutional democracy. It is possible, nonetheless, to extend his thinking to the dogmatic problem at issue. Constitutional rights, after all, are at the center of the debates on constitutional democracy, and the constitutional courts' growing influence and power – and how they proceed in their function – are not detached from the argumentation on the relationship between law and justice. Yet, how can we identify this debate with the critique of the defense of the rationality of balancing as Robert Alexy particularly interprets it?

---

[937]Ibid.

From the previous investigation, we can already conclude that, for every case, every decision *requires* a singular interpretation, the complexities and insurmountable dilemmas of constitutional adjudication inevitably disrupts the aura of safety and relative stability criteria and formulas aim to bring about. The reconstructive task that is at the core of deconstruction, when extended to constitutional adjudication, cannot simply be shaped or guided by an analytical framework based on logical considerations, as we observed in Robert Alexy's *Theory of Constitutional Rights* (*Theorie der Grundrechte*).[938] This is the very disruption of the responsible call for reconstruction and reinvention that takes place within the context of a negotiation between law and justice. For there is always negotiation, the free play of traces is marked by an inevitable unpredictability no formal parameter can apprehend. The problem of rationality, accordingly, may be much more complex. Since general rules do not apprehend deconstruction, every attempt to achieve rationality simply through this connection of arguments with methods and formulas loses the most intimate counter-metaphysical confrontation: for there are only traces,[939] whose signification is only obtained because that they are inserted into an unlimited play of signs in every new context, and where linguistic interaction puts in parentheses any reference, any method behind the context, constitutional adjudication also needs to put in parentheses any method, any formula that could define its itinerary. Hence, *methods and criteria must be deconstructed.*

When we investigate balancing, we shall raise the following question: is balancing part of the play? As such, does it let deconstruction do its role, or, instead, does it control the negotiation between law and justice, and thus disrupt the play? The second question stems from the first: if balancing disrupts the play, how does this *logos* operate? There are two suggestions for tackling these problems, which are the hypotheses here: the first is the perception, as we will shortly examine, that balancing, when carried out with no limits, is not part of the play; rather, it controls from outside the negotiation between law and justice and, hence, disrupts the play; the second refers to the construction of a *logos* based on two essential presuppositions: (a) the *logos* that operates from a concept of rationality connected to the promotion of correctness by deploying an analytical framework: we will call this *logos of correctness-rationality* (Sect. 5.3.2); and (b) the *logos* that emanates from the belief that this analytical framework, as long as it is filled with correct arguments accepted by rational persons,[940] bestows legitimacy through the idea of a "true argumentative representation,"[941] regardless of the possibility of jeopardizing the enforceability of law: we will call this *logos of legitimacy* (Sect. 5.3.3).

---

938 See the fourth chapter.

939 Bennington, "Derridabase," 74.

940 See Robert Alexy, "Balancing, Constitutional Review, and Representation," *International Journal of Constitutional Law* (Oxford University Press and New York School of Law) 3, no. 4 (2005): 580.

941 Ibid., 579.

5.3.2 Balancing and the Logos of Correctness-Rationality

5.3.2.1 Previous Considerations

The first suggestion is connected to the perception that balancing is not part of the play, but rather attempts to control it. As we examined in the last chapter, based on Robert Alexy's thinking, there is the conviction that an abstract and analytically structured method that places arguments in logical grounds can rationally control and guarantee the correctness of legal interpretation. The question is whether an analytical framework that is itself not part of the play can involve the continuous negotiation between law and justice. In this case, balancing, instead of being deconstructed – and, as such, has no insurmountable origin nor a pattern behind its traces – presents itself as a natural consequence of any premise, which is not deconstructed either. It is not necessary to go much further to conclude that balancing, embedded in the framework of the principle of proportionality, as Alexy defines it, develops in conformity with a premise that is not deconstructed but only introduced as a natural result of the differentiation between rules and principles.[942] The premise here is the character of principles as optimization requirements. The character of principles implies the principle of proportionality and vice-versa.[943] In turn, principles, as optimization requirements, demand that they be realized as great as possible according to their legal and factual possibilities.[944] This is the structure of a circular system that justifies itself by itself. Nothing can overcome the premise that principles are optimization requirements, as well as nothing can overcome the presupposition that the character of principles and the principle of proportionality, and hence balancing, are *necessarily* associated. For we cannot overcome these presuppositions, they configure a *logos*: a *logos* that is simply presented as the condition of constitutional argumentation.

In addition, apart from this natural presentation of the character of principles as optimization requirements and the unavoidability of the principle of proportionality, Alexy even defends that this procedure can be individualistically carried out.[945] This seemingly means: whereas arguments are to be extended to discourse,[946] the procedure itself can remain monological.[947] Since it is not projected into the realm of negotiation, but, instead, the negotiation is controlled by this formal system molded "monologically," consequently, a monological perspective shapes the different gathered arguments to decision-making. The arguments, with their constitutional complexities and tensions, are placed, as judgments and propositions,[948] in the

---

[942]See Robert Alexy, *Theorie der Grundrechte* (Frankfurt a.M.: Suhrkamp, 1994), 71–74.

[943]See Ibid., 100.

[944]Ibid., 75.

[945]See Robert Alexy, *Recht, Vernunft, Diskurs* (Frankfurt a.M.: Suhrkamp, 1995), 104.

[946]As we will shortly see, even this discursive character turns into a monologue.

[947]See the last chapter.

[948]See Alexy, "Balancing, Constitutional Review, and Representation," 577.

logical model of balancing, which is exempt, however, from being part of the play, and, hence, it is not confronted but with itself. Accordingly, provided that the negotiation is controlled from outside, the gap between law and justice that is to remain open and unfilled is, nonetheless, closed and filled by a *logos* guiding how the negotiation has to take place. The free play of traces is thus interrupted by a certain presence that is not part of this free play of traces. This is why the singularity of the context, even when we sustain that there is no possibility to think of balancing detached from cases, is covered by a certain essentialism. It is an essence attempting to foresee or synthesize a contingency that, although impossible to be entirely gathered in the moment of decision-making, cannot be, without affecting the play between law and justice, "*monologically*" apprehended.

This monologue, despite the discursive character Alexy attempts to defend,[949] shapes the core of a construction that intends to provide *rationality* and *correctness* to constitutional adjudication. It is behind its grounds that it is possible to control the evaluations of constitutional rights and the empirical knowledge the judiciary apprehends. Ultimately, two metaphysical perspectives guide the negotiation between law and justice: first, the metaphysics that sustains that an analytic structure is indispensable to achieve correctness and rationality in this process; second, the metaphysics that directly concludes that principles have to be regarded as optimization requirements, as well as its consequent conclusion that the principle of proportionality (and thereby balancing) is indispensable for constitutional adjudication. Both metaphysics close the anatomy of a system that understands that rationality implies control and guarantee of arguments through the establishment of some formal boundaries, and both shape the two elements of the *logos of correctness-rationality* above indicated.

These two metaphysics will be, therefore, the focus of this section. Both will demonstrate how the grounds for the defense of the indispensability of this structural-analytical framework derive from a logocentric approach. Indeed, the purpose is to see how distant this procedure and the theory supporting it are from the *concept of limited rationality*. The first metaphysics relates to the eloquent defense of rationality and the link with the claim to correctness, which Alexy establishes in his structural-analytical framework. The confrontation here is with the very basis of the belief that a procedure can provide rationality and correctness to decision-making, or, better, how it can rationally control and guarantee the correctness of a decision by means of an analytically structured method that places arguments in logical grounds. This is also the space to show how the deconstructionist approach would see this problem differently. In this regard, the analysis will be carried out by didactically separating them into the *logos of correctness* (Sect. 5.3.2.2.1) and the *logos of rationality* (Sect. 5.3.2.2.2). The second refers to the natural conclusion

---

[949]See Ibid., 577; Alexy, *Theorie der Grundrechte*, 498–501; Alexy, *Recht, Vernunft, Diskurs: Studien zur Rechtsphilosophie*; Robert Alexy, "Discourse Theory and Human Rights," *Ratio Juris* 9, no. 3 (August 2007): 209–35.; Robert Alexy, *Theorie der juristischen Argumentation: Die Theorie des rationalen Diskurses als Theorie der juristischen Begründung* (Frankfurt a.M.: Suhrkamp, 1989).

that principles are optimization requirements, which leads automatically to the indispensability of the principle of proportionality, and balancing consequently. Here, there is also a central issue we shall investigate, which will connect the *logos of correctness-rationality* to the *logos of legitimacy:* for principles are optimization requirements, they become moralizing principles and, as such, disrupt the play between constitutionalism and democracy. These two metaphysics close the structure of a monologue that is at the core of the *logos of correctness-rationality* and open up the space to configure the *logos of legitimacy.*

5.3.2.2 The First Metaphysics of the Logos of Correctness-Rationality: The Rationality and Correctness of the Analytical-Structural Framework

The Claim to Correctness as a Logos of Correctness

The first investigation is concerned with the expected correctness this procedure is able to provide. In the basis of Alexy's *Special Case Thesis,*[950] his purpose is clearly to connect legal arguments to moral ones. By extending this premise to constitutional adjudication, he attempts to demonstrate how the deployment of balancing associates rationality[951] with correctness, as an "internal justification."[952] According to him, the claim to correctness, indispensable to judicial review,[953] confirms the link between law and morality.[954] Indeed, he understands that "discursive constitutionalism, as a whole, is an enterprise of institutionalizing reason and correctness."[955] We could consider it a plausible argument: reason and correctness, in constitutionalism, are associated with institutionalized discourse. Despite that, a problem appears in this field as a consequence of the prior perception that Alexy's defense of balancing is not followed by placing this procedure in negotiation. The evaluation of correctness, given that it refers to a "rational" structure that can be "monologically" carried out, can also become a monologue. In this perspective, a

[950]See the last chapter (Sect. 4.2).

[951]See Alexy, *Theorie der Grundrechte*, 37–38.

[952]According to Alexy, the *internal justification* relates to the connection between the justification and the premises, that is, whether the decision resulted from a justification that logically followed the premises (See Alexy, *Theorie der juristischen Argumentation,* 273). It refers to the formal-analytical structure of legal justification, which guarantees its universality. We can call it the "rule and form of the formal justice" (Ibid., 280, translation mine). The *internal justification* differs from the *external justification,* which relates to the justification of the premises that are adopted in the *internal justification*, and which, according to Alexy, can have different origins: "(1) rules of positive law, (2) empirical statements, and (3) premises that are neither empirical statements nor rules of positive law." They are thus the contents inserted into the logical system of justification.

[953]Alexy, "Balancing, Constitutional Review, and Representation," 574.

[954]See Robert Alexy, "Law and Correctness," in *Law and Opinion at the End of the Millenium: Current Legal Problems*, ed. Michael D. A. Freeman (Oxford: Oxford University Press, 1988), 221.

[955]Alexy, "Balancing, Constitutional Review, and Representation," 581.

judge can define, provided that she applies this structural framework, whether the argument is correct or not or, in other words, since for Alexy legal judgment is likewise a moral judgment, good or bad.[956]

We could infer the correctness of the decision, ultimately, from simply observing whether the interference was justified or not, whether there was a justification of the presented premises ("external justification") within the criteria balancing provides ("internal justification"): "Judgments about proportionality raise, as do all judgments, a claim to correctness, and this claim is backed by judgments about degrees of intensity as reasons."[957] He links correctness to an inferential system that is implicit in the structure of balancing, which will lead to an "internal justification" constructed with help of the logic of the Weight Formula.[958] This formula, in turn, transfers the correctness of the arguments to a "judgment expressing the ruling of the court."[959] Yet, attached to this possible monological construction of the "internal justification," the discursive character that could still remain in the realm of the "external justification," of the premises inserted into those criteria and projected into an expectation of discursive acceptability,[960] might not resolve this monological impasse. In this matter, in which the claim to correctness could be connected to legitimacy through linguistic interaction,[961] there is still another relevant difficulty: behind its apparent argumentative character, there are some moral standards that legal argumentation must follow.[962] Here it appears that, even in this discursive dimension, there is a *logos* controlling the negotiation between law and justice.

---

[956]We can observe this characteristic in Alexy's defense of his theory when he, by observing the decision of the Titanic case (*BverfGE*, 86, 1) with respect to the inconvenience of an interference with the freedom of the press, affirmed that "this is an argument, and it is not a bad argument" (Robert Alexy, "Constitutional Rights, Balancing, and Rationality," *Ratio Juris* 16, no. 2 (June 2003): 139); or when he, by recognizing the intensive interference with the plaintiff's freedom of personality on account of the word "cripple," mentioned that "this is, first, argument, and, second, a good argument" (Ibid., 139). Neither the graduation of the interference (which leads to the result of the decision) nor the result was achieved by linguistic interaction; they both stemmed from the judge's capacity to place herself in the plaintiff's situation and understand whether the interference was light, moderate or intense. As we will investigate through Habermas's approach, the validity claim – associated with the claim to correctness –, in this situation, is not projected into a discursive interaction but stems from an observation made by the judge's conscience. Obviously, the judge has the capacity to express whether the argument is good or not (this is, besides, a condition of any critique), but, democratically speaking, any judgment based on validity claims needs to be projected into an intersubjective perspective. This refers to the external legitimacy of any discourse.

[957]Ibid., 139.

[958]See Alexy, "Balancing, Constitutional Review, and Representation," 575. See also the last chapter.

[959]Robert Alexy, "Discourse Theory and Fundamental Rights," in *Arguing Fundamental Rights*, ed. Agustín José Menéndez and Erik Oddvar Eriksen (Dordrecht: Springer, 2006), 27.

[960]See Alexy, "Law and Correctness," 208.

[961]Ibid.

[962]Ibid., 217.

It is necessary, in this matter, to briefly examine Alexy's explanation of the relationship between law and correctness.[963] According to Alexy, the claim to correctness is associated with three elements: first, the assertion of correctness, which means that "institutional acts are always connected to the non-institutional act of *asserting* that the legal act is substantially and procedurally correct;"[964] second, the guarantee of justifiability, derived from the perception that the claim to correctness "not only [accepts] a general obligation to justification on principle; it also maintains that this obligation is complied with or can be met;"[965] third, the expectation of acceptance of correctness, which is "the *expectation* that all addressees of the claim will accept the legal act as correct as long as they take the standpoint of the respective legal system and so long as they are reasonable."[966] These three elements represent the basis for the defense of, first, the need to connect law to correctness[967] and, second, the need to connect law to morality.[968]

Accordingly, correctness is shaped *morally*. This brings about two outcomes. Correctness is not the result of a negotiation between, on the one hand, law and its enforceability, and, on the other, justice, which, while opening up the law to invention, works with history[969] and the calculable.[970] Rather, correctness, in addition to legal standards, stems from traditional and collective values, utilitarian considerations, etc.[971] Indeed, the first element above is a clear example of how Alexy argues that legal reasoning requires the non-institutional act of asserting whether the legal act is substantially and procedurally correct. In other words, the legal system demands an assessment from outside of the legal parameters; it is necessary to establish not only a simple material correspondence between law and morality, but also their necessary conceptual connections.[972] Besides, as long as he points out this indispensable connection, he reveals an understanding that inherits a tradition that interprets the constitution as a "concrete order of values." Both characteristics allow reasoning, for instance, that justice is related to distribution and balance[973] (and, thus, balancing is a requisite for justice), and that other non-legal criteria are essential to shape the correctness in legal discourses. Alexy even says that, given the incapacity of law to provide correctness alone, "only recourse to standards other than legal standards is available, such as general reflections on utility, traditional and common ideas of what is good and evil, as well as principles

---

963 See Ibid., 205–221.

964 Ibid., 208.

965 Ibid.

966 Ibid.

967 Ibid., 209.

968 Ibid., 221.

969 Derrida, "Force of Law," 929.

970 Ibid., 971.

971 See Alexy, "Law and Correctness," 216.

972 Ibid., 221.

973 Ibid., 211.

of justice. In short, utility, custom, and morality."[974] In constitutional adjudication, accordingly, balancing cannot avoid the recourse to moral, political, utilitarian, or traditional arguments.[975] They are thus a requirement for correctness.

The second outcome, which is associated with the former, is the *equivalence* of moral discourses to legal discourses in the realm of balancing, which leads to the discussion of how judges apprehend these vast normative and factual parameters and apply them to the case, and, more importantly, how they manage these contents in a legitimate way. Indeed, as Eriksen points out, "the problem that emerges here pertains to the limits of the law and those of the judge's competence. How can we know that judges are right when they do not merely apply politically laid down laws and reasons but make use of their own discretionary know-how and value-base?"[976] This serious question leads, moreover, to a crucial conclusion: even though Alexy sustains the discursive character of correctness by establishing a reference to the expectation of addressees' acceptance, this expectation becomes ultimately conditioned by this moral apprehension of legal discourse.

Insofar as legal discourse is considered a special case of general practical discourse, and given that the procedure, with its rules and criteria, can be "monologically" carried out, at the end, the claim to correctness loses this connection with the external world. We could observe this result by examining how Alexy deals with the third element above in conjunction with the first one. The third element refers to the addressees of the claim, who will accept it or not as correct based on the legal system and its reasonableness. These addressees, nonetheless, have to be more than those connected to the legal act,[977] inasmuch as every institutional act is linked to non-institutional ones. In other words, every legal argument is linked to moral arguments, which will *assert* whether the legal act is substantively and procedurally correct or not. Since this connection with morality is a requirement for correction and its evaluation, the non-institutional addressees must be considered likewise a condition for correctness. In sum, we can only infer the correctness of the decision, along with the inferential system, when there is an expectation that the addressees in general, this external forum, also accept the correctness of the claim. Alexy mentions that there are, in this field, two addressees: the institutional ones, who are the "circle of addressees of the respective legal acts,"[978] and the non-institutional ones, who are "everyone who takes the point of view of a participant in the respective legal system."[979] The correctness in law refers to these two groups,

---

[974]Ibid., 216.

[975]Evidently, in adjudication, these different types of arguments are also relevant for the decision, but, as we will shortly examine, they cannot be balanced with legal arguments as sources of equivalent weight, for this can disrupt the enforceability of law and, mostly, the very premise of separation of powers.

[976]Erik Oddvar Ericksen, "Democratic or Jurist-Made Law?," in *Arguing Fundamental Rights*, ed. Agustín José Menéndez and Erik Oddvar Eriksen (Dordrecht: Springer, 2006), 83.

[977]Alexy, "Law and Correctness," 207.

[978]Ibid.

[979]Ibid.

who will give a certain universal character to the law within a legal system.[980] This is the required external forum for the claim to correctness.

Now, a simple modification of focus can demonstrate how the discursive character is ultimately lost in this perspective. Instead of the addressees, the judge is considered. The judge, when she decides a case according to these premises, has to obey the "rational" analytical framework and expect that the addressees, not only the institutional, but also the "non-institutional" ones *assert* and *accept* the claim to correctness. Whenever a decision is made, the judge must focus on these two dimensions and justify (the second element above) the claim to correctness. She has to infer which are the values (traditions, utilitarian considerations, collective goals) those institutional and non-institutional addressees would accept beside the legal aspects of the case. She must associate values and the satisfaction of the addressees (this forum) with the legal issues involved and place them in a framework of preferential conditioning.[981] She could, for instance, sustain that a traditional value – a religious belief, for example – is weightier than the religious freedom of an individual, for it is secularly shared by a collectivity, and the addressees could regard this argument as correct, insofar as it satisfies their moral values[982] (the *Crucifix case* could be an example here). By the same token, she could sustain that racism, in virtue of a historical tradition, only applies to certain groups and place this argument inside the analytical framework of balancing (the *Ellwanger case* could be an example here).[983]

All these examples demonstrate that, through this structure of the claim to correctness, it is possible to justify the decision internally as well as externally, and that the result was reached through arguments (and, thus, based on discourse rationality). It is here however where we identify a very clear *logos*. Despite his statement on the existence of the *Law of Competing Principles* that could restrain

---

[980]Ibid.

[981]See the last chapter.

[982]Alexy attempts, in response to his critiques, to sustain that his *theory of legal discourse as a special case of general practical discourse* respects the institutional background. He remarks that "what is correct in a legal system essentially depends on what is authoritatively or institutionally fixed and what fits into it. It must not contract the authoritative and coherence with the whole. If one wants to express this in a short formula, it can be said that legal argumentation is bound to statutes and to precedents as to observe the system of law elaborated by legal dogmatics" (Alexy, "The Special Case Thesis," 375). However, although bringing this defense of coherence and institutional background, at the end, the focus is on how different arguments can fit into the analytical-structure of reasoning in which balancing is employed. It suffices that non-institutional arguments be integrated in this structure: "General practical arguments have to float through all institutions if the roots of these institutions in practical reason shall not be cut off. General practical arguments are non-institutional arguments. General non-institutional arguments floating through institutions may be embedded, integrated, and specified as much as one wants, as long as they remain arguments they retain what is essential for this kind of argument: their free and non-institutional character" (Ibid., 384). We will examine this issue in the next chapter.

[983]See the first chapter.

judge's discretion,[984] it seems that, behind this construction, it is the judge's conscience that ultimately defines which is this group (much broader than the institutional one) and which are the values this group would *assert* and *accept* as necessary to be balanced in order to provide correctness to the case. She defines who justifies the claim and what can be justified. Hence, "it follows from the constraints of a legal discourse that only judicial arguments are allowed and that it is not the participants but the judge who has the last word."[985] At the end, there is nothing but a monological structure whereby the judge says the last word, and whose criteria for correctness are nothing but simple criteria for judicial discretion.

We can then verify the *logos of correctness*: the correctness of a decision is based not only on the monological structure of the procedure ("internal justification"); even the arguments placed inside this procedure are "monologically" shaped ("external justification"). The discursive approach that is the basis of the claim to correctness reveals that it is, in reality, an open territory where the legal enforceability could be jeopardized by any other value (traditional, collective, utilitarian, economic, etc.), justified because it satisfies the external forum's expectancy, and reasonably inferred from an analytical-structural framework capable of rationalizing and clarifying the argumentation. We could, nevertheless, adopt both justifications (internal and external) and the expectancy of addressees' acceptance as premises to *control* the negotiation between law and justice in the realm of adjudication. They both seem plausible: the analytical-structural framework provides a very consistent, verifiable, useful, and effective methodological basis for reasoning by establishing the necessary steps a judge ought to follow in order to reach the right conclusion. The external forum, in turn, brings to this framework and reasoning the discursive parameter, which establishes the premises for legitimation and the "conditions of a true argumentative representation."[986] Still, as previously examined, although the claim to correctness is indispensable in this matter,[987] both foundations of Alexy's theory seem to transform this claim into a *logos of correctness*, a *logos* that lies in the need of law for a prior moral ground,[988] revealed "monologically," though. The fact that all legal discourses, to be correct, are "in

[984]See the last chapter.

[985]Eriksen, "Democratic or Jurist-Made Law?," 84.

[986]Alexy, "Balancing, Constitutional Review, and Representation," 579.

[987]See Alexy, "Law and Correctness," 209.

[988]This is also the basis of Alexy's critique of theories based on coherence, as we can observe in Klaus Günther's *Der Sinn für Angemessenheit: Anwendungsdiskurse in Moral und Recht* (Frankfurt a.M.: Suhrkamp, 1988). For the claim to legal correctness requires a claim to moral correctness, there is no possibility to think that the legal system alone can provide the answers in the realm of constitutional adjudication. As we will investigate in the next chapter, for Alexy, "radical coherence theories do not put up with the doubtlessly reasonable and correct thesis that systematical completeness and systematical connection are essential criteria of rationality and correctness. They assert further that coherence is a sufficient and, indeed, the only criterion in hard cases" (See Alexy, "Law and Correctness," 217). See also his critique of coherence in Robert Alexy, "Coherence and Argumentation or the Genuine Twin Criterialess Super Criterion," in *On Coherence Theory of Law*, ed. Aulis et al. Aarnio (Lund: Juristförlaget I Lund, 1998), 47.

need of a prior supplementation"[989] casts some doubts on how and by whom this supplementation is implemented as well as leads to the ontological question: what is this "prior supplementation?" The claim to correctness becomes a *logos of correctness*.

The Claim to Rationality as the Logos of Rationality and the Opening to The Claim to the Other's Otherness

The *logos of correctness* does not appear isolated. At its core there is the belief that the structural framework can provide rationality. A correct decision is, as before mentioned, a rational decision.[990] This is why the *logos of correctness* is linked to the *logos of rationality*. Both are interconnected in this construction of a theory whose purpose lies in furnishing "rational justified answers to constitutional questions."[991] More particularly, in the realm of balancing, both are interconnected by the premise that, since there are arguments,[992] we can defend the rationality of this procedure. Rationality, according to this approach, derives from the deployment of a structural system that "improves" judges' capacity to find a solution to the case. Through arguments placed in this analytical framework, it seems that the judge reached a transparent and justifiable conclusion and deeply carried out a discussion before making the decision. Rationality is associated with the judge's capacity to correctly place the arguments in the structure of the principle of proportionality, and balancing in particular. Rationality, after all, means, according to this approach, the ability to clarify and coherently establish the arguments through some criteria of this analytical framework.[993]

A simple look into any of Robert Alexy's texts allows us to conclude that his thinking is followed by this interest in proving rationality to legal reasoning through balancing. In his *Theory of Legal Argumentation,*[994] it is there, right in the preface, his aim to promote the comprehension of what is a rational legal argumentation and which is its extent[995] through an emphasis on the rational practical discourse. In his *Theory of Constitutional Rights,* rationality is directly associated with the capacity to provide conceptual and analytical clarity,[996] which can be reached through methods and criteria,[997] and by opening up the legal system to the system of

989 Alexy, "Law and Correctness," 217.

990 See the last chapter.

991 See Alexy, *Theorie der Grundrechte,* 18, translation mine.

992 See Alexy, "Balancing, Constitutional Review, and Representation," 577.

993 See Alexy, *Theorie der Grundrechte,* 27.

994 See Alexy, *Theorie der juristischen Argumentation.*

995 See Ibid., 15.

996 See ALEXY, *Theorie der Grundrechte,* 32.

997 According to Alexy, balancing, for example, says what has to be rationally justified. See Alexy, *Theorie der Grundrechte,* 152.

morality[998] (a reflex of his conception of legal discourse as a special case of general practical discourse). In different texts, he has stressed this rational quality of his theory and of balancing by responding to critiques,[999] and his thought has reverberated through his followers.[1000] Yet, notwithstanding all his persuasive effort to bring rationality to discussion, there is always the feeling that something deeper is still missing. Indeed, there is always the impression that, to defend rationality, the answer must be the defense of more methods and criteria,[1001] as if everything, to be rational, needed to be expressed by formulas. Truc, hc rcmarks that "definite determination is not necessary for the acceptatibility of a criterion,"[1002] which means that criteria need not be precise or establish some kind of calculus.[1003] However, he seems to hold a concept of rationality that is not limited, as if methods and criteria, as long as they are filled with arguments, could solve the problem of legal adjudication. He expresses, for instance, a certain belief that principles are beforehand firm and clear[1004]; he sustains the premise that the constitution can deny a certain epistemic discretion, as if it could define, in advance, the terms to be adopted[1005]; he even seems to believe in the possibility of assuring the truth of

---

[998]Ibid., 19.

[999]See Robert Alexy's following texts: "The Special Case Thesis"; "Balancing, Constitutional Review, and Representation"; "On Balancing and Subsumption. A Structural Comparison"; "Discourse Theory and Fundamental Rights"; "Coherence and Argumentation or the Genuine Twin Criterialess Super Criterion"; "Postscript," in *A Theory of Constitutional Rights* (Oxford: Oxford University Press, 2002), 388–425; *Recht, Vernunft, Diskurs* (Frankfurt a.M.: Suhrkamp, 1995).

[1000]See Carlos Bernal Pulido, "The Rationality of Balancing," *Archiv für Rechts – und Sozialphilosophie* 92, no. 2 (2006); Carlos Bernal Pulido, "On Alexy's Weight Formula," in *Arguing Fundamental Rights*, ed. Agustín José Menéndez and Erik Oddvar Eriksen (Dordrecht: Springer, 2006); Laura Clérico, *Die Struktur der Verhältnismäßigkeit* (Baden-Baden: Nomos, 2001); Luís Virgílio Afonso da Silva, "O Proporcional e o Razoável," *Revista dos Tribunais*, no. 798 (April 1992): 23–50; Virgílio Afonso da Silva, *Grundrechte und gesetzgeberische Spielräume* (Baden-Baden: Nomos, 2003) [In Portuguese: Silva,. *Direitos fundamentais: Conteúdo Essencial, Restrições e Eficácia* (São Paulo: Malheiros, 2009)].

[1001]We can see, for example, Alexy's response to Jürgen Habermas's critique. The debate is centered on demonstrating, through formulas and some BVG's decisions, the rationality he defends. See Alexy, "Balancing, Constitutional Review, and Representation"; Alexy, "Constitutional Rights, Balancing, and Rationality."

[1002]Alexy, "Coherence and Argumentation or the Genuine Twin Criterialess Super Criterion," 47.

[1003]Ibid., 47. Alexy calls these criteria, as the coherence, "criterialess criteria of rationality" ("*Kriterienlose Rationalitätskriterien*").

[1004]Alexy, "Postscript," 404.

[1005]Alexy says:

"Epistemic discretion in balancing gives rise to particular problems on account of its proximity to structural discretion in balancing, which can also be found in the decision just discussed (...) If the legislature is permitted to base its interferences with constitutional rights on uncertain premises, then it is possible that the protection afforded by constitutional rights will be refused on the basis of false assumptions, even though constitutional rights have in reality been breached. *Constitutional rights would offer more protection if the legislature were to be refused an epistemic discretion*" (Ibid., 416, emphasis mine).

empirical premises.[1006] This might show that, although rationality, for him, is associated with arguments, there is a conception of truth that is rather not subject to critical review. There is a *logos*. In other words, there might be, behind these criteria broadly introduced in his different texts, arguments that are not part of the play. This is why, perhaps, the answer is not exactly in developing methods, criteria, formulas to provide rationality to constitutional adjudication. Perhaps, rationality resides in the deconstruction of any criterion.

Derrida is aware of the problems of a certain accent on rationality. In his book *Rogues: Two Essays on Reason,* when he mentioned the term *reason,* it was followed by a very careful explanation full of adjectives denoting its risks. He mentioned that, in order to attribute a meaning, "the most difficult, least mediocre, least moderate meaning,"[1007] to the word *reasonable,* "this well-worn, indeed long-discredited, word,"[1008] it would mean the "reasoned and considered wager of a transaction between these two apparently irreconcilable exigencies of reason, between calculation and the incalculable."[1009] Indeed, what could be considered the spectrum of rationality is properly deconstruction, the negotiation that takes place between the calculation or conditionality and the unconditional uncalculability.[1010] It is not, accordingly, any *logos,* but the destruction of all-possible transcendental signified, even in the very concept of rationality. If it is to sustain a "rational deconstruction,"[1011] it simply means that nothing can rely on a last argument.

Therefore, concepts of rationality, in order to sustain a possible rationality, must be deconstructed. What really matters is that the negotiations between constitutionalism and democracy, law and justice, remain open and unfilled in order to render possible deconstruction. The free play of traces cannot be guided by a certain *logos* establishing the parameters according to which decisions are made, especially when this *logos* abstains from being deconstructed. Rather than traditional claims to rationality, which rely on some last arguments, a rational deconstruction goes further by stressing the otherness, by emphasizing that there is no other without the experience of the impossible.[1012] Still, what do these abstract words mean in this particular issue? Besides, what do they contribute to the construction of a *concept of limited rationality*? First, they undercut the pretension that establishing general rules can solve somehow the problem of legal adjudication: "For a deconstructive operation *possibility* would rather be the danger, the danger of becoming

[1006]According to Alexy, "(. . .) the legislature could only interfere in any way with constitutional rights on the basis of empirical premises the truth of which was assured" (Ibid., 417).

[1007]Derrida, *Rogues: Two Essays on Reason*, 151.

[1008]Ibid.

[1009]Ibid.

[1010]Ibid., 150.

[1011]Ibid., 151.

[1012]Derrida, "Force of Law," 981.

an available set of rule-governed procedures, methods, accessible approaches."[1013] It places, therefore, in legal adjudication the imperative of doubt, which is the fist step towards the deconstruction of a methodology. In order words, by stressing the otherness and the experience of the impossible, methodology, rather than controlling the play and being confronted only with itself, becomes part of the play and is thus deconstructed. Second, they reveal that a responsible decision ought to center on the singularity of the case, while simultaneously opening it up to deconstruction. This means, in the particular situation of constitutional adjudication, a respect for the "calculable," the history by stressing the enforceability of law, while simultaneously reinventing it as a means to render, albeit factually impossible, justice to the other. Applied to the problem here, it means that a responsible decision, in the play between law and justice, cannot surrender law to moral standards, for it loses its *enforceability,* but has to understand the tensions and complexities of the negotiation between law and justice; that is, it has to reinforce the law, its history, while reinventing it in favor of the other, of the singularity of the case.

Robert Alexy's account of the rationality of balancing through his structural framework seems, nevertheless, to have a distinct concern, even to achieve its universality: regardless of the contents of the argument,[1014] the analytical framework can control how constitutional courts apply them as a means to guarantee correctness and rationality to the process. However, his structural framework seems to dilute the tensions and complexities arising from a responsible negotiation. The analytical framework, in order to sustain its rationality, must submit these tensions to this identity: it must control the gap between law and justice as a means to orient the judgment to a realm of clarification, coherence and non-contradiction.[1015] This is its very condition of rationality. Arguments are rationalized insofar as they become part of a logical system. Rationality, according to this approach, relies on a negotiation controlled by a certain *logos*. There is, as a consequence, the prevalence of the need to deploy a methodology holding this *logos* over the concern for a responsible decision. The *claim to rationality*, therefore, turns into a *logos of rationality*.

Despite that, when we defend the deconstructionist approach to constitutional adjudication, it does not follow that criteria are dispensable for decision-making: they also have their role. Yet, they cannot become a general abstract rule behind the singularity of the case as though they guided the negotiation between law and justice in decision-making: they must also be deconstructed. The existence of a general rule in the field of negotiation means placing an identity where there should

[1013]Ibid.

[1014]Kai Möller stresses, for example, that "for Alexy's enterprise to succeed, his theory must have the potential to be applied fruitfully to different substantive theories of constitutional rights – to a socialist perspective, a libertarian perspective, or a liberal perspective. I call this the 'framework character" of Alexy's theory. Kai Möller, "Balancing and the Structure of Constitutional Rights," *International Journal of Constitutional Law* 5, no. 3 (2007): 458).

[1015]See Alexy, *Theorie der Grundrechte*, 27.

be a free play of traces. A sort of essentialism, substantiality stifles the particularism by transferring the concrete problem of legal rights to the abstract. It is exactly here that *différance* is denied: "There are only contexts, and this is why deconstructive negotiation cannot produce general rules, 'methods'".[1016] This is why a responsible decision, instead of concentrating on methodologies and formulas into which the arguments regarding the case are inserted, centers on the *context*, on the *singularity of the case,* because it knows that every knowledge is nothing but a fallible knowledge. Only the context, the particular, the place where the play comes about, matters to *différance*. Its complexities and tensions, its history, the applicable norms matter to a responsible decision. On the one hand, it takes into account the desire to keep the past, traditions, identities, tensions alive. Responsibility works with history. Moreover, responsibility works with the calculable, the institutionalized norms, for the "incalculable justice requires us to calculate."[1017] On the other hand, all of this minimum of internal security[1018] must simultaneously enact deconstruction as a means to open it up to the threat of the unconditional. After all, "this opening of context is not saturated by the determination of a context."[1019] The history, the institutionalized norms have to be open to deconstruction, to a process of reconstruction and reinvention, and this is why the unconditional, the *to come,* must be always present in this interaction. There is thus no argument that is out of the play, and there is no responsible decision closing the gap between law and justice. If it has to claim rationality, it has to let deconstruction operate.

Decision-making ought to be a moment of reinvention, recreation, reconstruction, a moment in which deconstruction plays its role. The past, the history, the institutionalized norms enter inevitably into negotiation, but they enter as an argument that, as any other possible reference, cannot abstain from being part of the play. History and the calculable are always invited to an active interpretation,[1020] which does not reduce them to a sort of controlled empirical knowledge and normative evaluation gathered and "rationalized" by formulas. Indeed, this analytical framework goes against the very character of deconstruction: "Deconstruction cannot be captured or reduced to a formula."[1021] It is not by placing an analytical structure into which arguments are inserted that we can conclude that, to a certain extent, now the empirical knowledge and the constitutional rights can be clearly gathered, and thus legal reasoning happens in a safe environment for decision. Analytical criteria do not furnish internal safety to rights, nor even to decision-making. Indeed, if there is any safety, it lies in the premise that every

[1016]Ibid., 17.

[1017]Derrida, "Force of Law," 971.

[1018]Derrida, "Negotiations," 17.

[1019]Ibid.

[1020]Dastur, *Philosophie et Différence*, 114.

[1021]Bernstein, "An Allegory of Modernity/Postmodernity: Habermas and Derrida," 81.

decision rends the time and defies dialectics;[1022] "it is a madness."[1023] Moreover, although it is important to have a feeling of duty and respect for the law,[1024] the necessary interaction with the *to come* culminates in the perception that there is no other way to keep internally consistent the system of rights than the recognition of the singularity of the other, that is, the interaction between the constative and performative, the iterability. The consequence is that, if it is to keep alive the interaction between law and justice, as a reflex, besides, of the one between democracy and constitutionalism, every new decision has to be justified as a consequence of a play, of keeping alive the history, the institutions, the norms, while simultaneously reinventing them. It is as if the founding moment of law were somehow perpetuated in this reinvention, which is the prerequisite for the responsible decision: while keeping alive this "founding moment of law," it is reinvented according to the circumstances of the particular case. Every reinterpretation of the "very foundations of the law" must be taken "such as they had previously been calculated and delimited."[1025]

Deconstruction, as a consequence, does not result in relativism. It just highlights that every case is a singular case adjusted to deconstruction,[1026] which, in turn, operates in this realm of iterability. There is, on the contrary, relativism, when the play relies on the deployment of a methodology that becomes a *logos* behind the negotiation, for it depends merely on the judge's capacity to correctly insert her arguments into those formal parameters. However, at the end, when she finds the solution, the rationality and correctness are linked to this monological framework, which is not subject to further critical review.[1027] It is thus not the absence of general rules the reason for relativism; it can be, paradoxically, the very existence of them. For every knowledge is nothing but a fallible knowledge, it is not criteria or methods that will transform any claim to rationality into a claim against relativism. Hence, the *logos of rationality* gives rise to relativism.

The first metaphysics of the *logos of correctness-rationality* is clearly disclosed: there is, indeed, a lack of reflection on the basis of the formal procedure, but also the automatic conclusion that this analytical framework can provide rationality and correctness to adjudication lies in a belief that will ultimately reveal its relativism and monological nature. Alexy's structural analysis seems to lie in a metaphysical standpoint able to be operated by judicial discretion. Consequently, Alexy's claim to rationality appears to be premised on, first, a semantic approach of an observer embodied by judge's conscience and, second, a relativized claim to correctness. Accordingly, if rationalization associates with the capacity to control the empirical

[1022] Derrida, "Force of Law," 967.

[1023] Ibid.

[1024] Derrida, "Negotiations," 13.

[1025] Derrida, "Force of Law," 971.

[1026] Derrida, "Negotiations," 17.

[1027] See the discussion of the *logos of correctness* in the last section.

knowledge and the evaluations in adjudication[1028] by applying a method that orients the judgment to a realm of clarification, coherence and non-contradiction,[1029] then there might be a paradox in its basis: at the end, the relevance of a principle over another and how the conditions of priority are settled are defined by judicial discretion.[1030] This discretion says the last word without being controlled: it controls the negotiation by deploying a *logos* that might not take into account the other. The *claim to rationality*, as *logos of rationality*, is therefore the denial of the other's otherness.

5.3.2.3 The Second Metaphysics of the Logos of Correctness-Rationality: Principles as Optimization Requirements, Principles as Moralizing Principles

The second metaphysics of the *logos of correctness-rationality* is concerned with the premise that we must regard principles as optimization requirements and the consequent conclusion that balancing is unavoidable[1031] when constitutional rights are at stake. It implies, in conformity with this view, that principles must be optimized to the greatest extent as possible with respect to the factual and legal possibilities. The metaphysics here lies in the unquestionable acceptance of these premises. As the structural foundation of Alexy's *Theory of Constitutional Rights*, they are simply introduced as the consequence of the difference between rules and principles.[1032] There is no further discussion of why principles must be necessarily conceived with this character. As a result, the automatic deduction that we must proceed, when there is a collision of principles, to the *rule of collision* based on balancing,[1033] for they are optimization requirements, is also metaphysical. It is not on account of the German BVG's decisions that we can conclude that we should incontrovertibly regard principles in this way. Although largely applied, there are important cases of collision of principles whose decisions did not deploy balancing.[1034] Besides, in many other constitutional realities, principles are deemed

[1028]See Alexy, *Theorie der Grundrechte*, 38.

[1029]See Ibid., 27.

[1030]See Massimo La Torre, "Nine Critiques to Alexy's Theory of Fundamental Rights," in *Arguing Fundamental Rights*, ed. Agustín José Menéndez and Erik Oddvar Eriksen (Dordrecht: Springer, 2006), 59.

[1031]According to Alexy, "balancing is not an alternative to argumentation but an indispensable form of rational practical discourse" (Alexy, "Constitutional Rights, Balancing, and Rationality," 131).

[1032]See Alexy, *Theorie der Grundrechte*, 71–104.

[1033]Ibid., 79–80.

[1034]Indeed, there are some cases to which the BVG did not apply balancing as a consequence of the nature of principles as optimization requirements. One important example is brought by Kai Möller:

differently,[1035] and other forms of justification are used. It is likewise known that there are different possibilities to apply principles that do not necessarily establish balancing as an indispensable procedure.[1036] Therefore, the questions remain: why do principles have to correspond to varying degrees of satisfaction? Why do they have to be acknowledged by this teleological attribute of being realized to the greatest extent as possible? Why is balancing a natural consequence of the nature of principles?

There is no answer: the very nature of them responds to these questions;[1037] the BVG's practice reinforces those conclusions.[1038] The character of principles leads automatically to the deployment of the principle of proportionality, and

"(. . .) Alexy does not come close to a full examination of the German jurisprudence: on a closer look, the claim that all constitutional rights are principles qua balancing norms cannot be sustained."

"This becomes particularly clear in the case of the right to human dignity, protected in article 1 (1) BL. According to German constitutional theory, any interference with the right to dignity is prohibited: there is no balancing. Thus, there is at least one right that is not open to balancing and, therefore, not a principle in the Alexian sense. Alexy tries to resist this conclusion by arguing that the right to dignity is a rule in the sense that interference with it cannot be justified, but that its language is to open that courts can always do the necessary balancing beforehand, when determining whether something pertains to human dignity. This means for him that human dignity is, in truth, a principle (a balancing norm) just like other constitutional rights."

"(. . .) However, there have been developments in German jurisprudence since 1985 that show the limits of Alexy's approach."

"One of the laws passed in Germany as a response to the terrorist attacks of September 11, 2001, is the Aviation Security Act [*Luftsicherheitsgesetz* of January 11 2005, Bundesgesetzblatt 2005 I], allowing the government to shoot down planes that had been hijacked and were likely to be used as terrorist weapons. The FCC has declared this unconstitutional and a violation of human dignity, arguing that shooting down a plane with innocent passengers aboard violates the human dignity of those passengers [*BVERG*, 1 *BvR* 357/05 of February 15, 2006]. There was no balancing involved. Even in a case where it was clear that many more people would die if the plane were not shot down, the destruction of the plane would still be impermissible. The concern here is not with the merits of this or any specific judgment, doctrine, or conception of rights, but only to demonstrate that *even in the jurisprudence of the FCC, which is the material on which Alexy builds his theory, there are cases where the Court obviously rejects a balancing approach.* Moreover, these are cases which Alexy's theory has no capacity to explain, other than by conceding that some constitutional rights are "rules," or, as I prefer to call them, balancing-free norms." (Möller, "Balancing and the Structure of Constitutional Rights," 465–466, emphasis mine).

[1035]In Brazilian reality, for instance, only recently the principle of proportionality and balancing in particular, similarly to their configuration in Germany, has been deployed. The history of STF's decisions is marked by distinct approaches to the interpretation of constitutional rights that are not directed linked to the perspective of principles as optimization requirements. In the United States, as largely known, the Supreme Court has historically based its interpretation on a variety of arguments, and, in many of its cases, did not understand principles with this optimization perspective.

[1036]In the next chapter, we will investigate other different approaches in order to show that balancing, from the premise that principles are optimization requirements, is not uncontroverted.

[1037]Alexy derives the consequence that principles are optimization requirements by differentiating them from rules. See Alexy, *Theorie der Grundrechte*, 75.

[1038]See Alexy, "Balancing, Constitutional Review, and Representation," 572.

vice-versa.[1039] Its justification is rather related to concepts such as the "nature of fundamental rights"[1040] or "essential core guarantee" (*Wesensgehaltgarantie*) of article 19(2) of the German Basic Law.[1041] In this hypothesis, the argument is that this "essential core" can be evaluated by and is embedded in this principle.[1042] It is evident the circularity functioning from a metaphysical standpoint.

The critique in this matter may seem, at first glance, purposeless. One may argue that there is no possibility to define a methodology in the realm of constitutional rights without previously establishing some standards (taken, for instance, from the court's arguments). Moreover, one could argue that any attempt to investigate beyond those standards would inevitably lead to an eternal search for a justification, which, ultimately, would also have to be pre-established as a standard in similar patterns. Indeed, it may seem problematic to foresee any other perspective for decision-making not lying in those criteria. Even so, perhaps these counterarguments might be centered on a belief that places emphasis on the need to design a system of rules that can, to a certain extent, assist judgments, instead of striving for overcoming some criteria whose deployment has been a success in the practice of courts of justice. The ongoing effectiveness of those standards legitimates by some means its adoption, and the studies stemming from this perception need hence to demonstrate how they develop in practice. The priority, according to this approach, is not to investigate how far it is possible to account for a criterion, but mostly how far a criterion is able to yield effective results in the realm of constitutional rights, and how it can be seen as a rational pattern for decision-making.

Yet, the establishment of some prior standards (principles as optimization requirements and the indispensability of balancing, for instance) brings about some relevant consequences. Although no one could question that every mechanism to reach a decision needs to be effective and assist decision-making, we could sustain that it is not by focusing on its capacity to facilitate the reach of the result that they produce right decisions. This conclusion is remarkably relevant, for the basis of the procedure (prior standards simply introduced) can be the origin for the justification of any result in adjudication. If there is a final argument that is not confronted, the decision can be justified based on a non-confronted argument. For instance, the "essential core guarantee" (*Wesensgehaltgarantie*), which, for Alexy, accounts for the principle of proportionality, could indeed, instead of guaranteeing a constitutional right, be the argument to relativize it in favor of a traditional value, a policy. The last word in this matter would be nothing other than the very inevitability of the principle of proportionality, thereby balancing,[1043] but it could

[1039]See Alexy, *Theorie der Grundrechte*, 100.

[1040]Ibid.

[1041]Ibid., 272.

[1042]See La Torre, "Nine Critiques to Alexy's Theory of Fundamental Rights," 59–60.

[1043]See Alexy, *Theorie der Grundrechte*, 272.

lead to the subjugation of a constitutional principle to the opposite of any guarantee. Rather, the only guarantee, in this account, is a relative one:[1044] the relativism of balancing itself. Therefore, it is not purposeless the attempt to undercut these premises.

It is in this matter that we confront those counterarguments above. The opposite perspective of the emphasis on criteria rather than the justification for criteria does not mean the emphasis on justification for criteria rather than criteria, nor culminates in more criteria. It simply understands that this distinction is meaningless, for it assumes that every knowledge – and that one obtained with help of criteria – is a fallible knowledge. As a consequence, the problem is not the deployment or not of some prior standards shaping a methodology, not even when it is filled with arguments, for every case is always singular. In virtue of the singularity of the case, it takes as premise that there is no method or pattern that can reduce it without the risk of putting further stress on one of the two sides of the negotiation, law or justice. Moreover, since there is this risk – and this is why every decision is a "madness," a "finite moment of urgency and precipitation"[1045] – the decision cannot be guided by a *logos;* the decision cannot ultimately be the utterance of a monologue.

In what follows, if it is to use criteria, and if these criteria can provide some support for decision-making, then they must be part of the play: they have to be deconstructed. If there is a *logos* instead, especially when this *logos,* by controlling the empirical knowledge and legal evaluations, is in charge of judicial discretion, the inevitable reduction of the particularities of the case – inevitable, because every decision "cannot furnish itself with infinite information and the unlimited knowledge of conditions, rules or hypothetical imperatives that could justify it"[1046] – becomes a moment of violence *with ground.* In turn, this *ground,* for it descends from criteria and arguments managed according to the judge's conscience, can disrupt the play between law and justice, and hence between constitutionalism and democracy. The first play leads to the question about the risk of transmuting decision-making into a reproduction of calculative reason or into a moralizing discourse; the second one culminates in the issue referring to the danger of transforming decision-making into a practice that takes into account the need to preserve the institutional history of legal rights but forgets its claim to legitimacy, or that stresses the claim to legitimacy but leaves aside the institutional history.

It is possible to argue that Alexy's defense of balancing within the context of collision between constitutional rights can be ultimately used as a justification for the practice of violence *with ground.* There is a *logocentrism* that operates by controlling the empirical knowledge and the legal evaluations, and this *logocentrism* is managed by judicial discretion, as previously examined. Accordingly, the *logos* disrupts the interaction between law and justice inasmuch as, from outside,

---

[1044] Ibid.

[1045] Derrida, "Force of Law," 967.

[1046] Ibid.

the decision can easily stress one of the sides of this play. By stressing law or justice through a methodology that is not part of the play, the danger, in practice, can be of two sorts: first, the repetition of the same, for the "calculable" is not confronted with the *threat* of the *to come;* second, the transformation of constitutional rights into moralizing principles, for the *to come* becomes a reified discourse that jeopardizes the *enforceability* of law. Both are a sign of a monologue, of the forgetfulness of the other, of a *logos.* The first destroys *iterability,* to the extent that there is no reinvention or recreation, but rather the maintenance of an identity that is not contradicted but by further affirming its identity. The second disrupts the *institutional history* and the *force of law,* considering that legal principles are balanced with any other argument the idea of a *moralizing principle* could embrace, and it would not promote justice, either, not only because it would be presently impossible, but because the singularity of the context would be covered by an essentialism.

The second effect can be gathered from the concept of principles as optimization requirements. As a consequence of the previous discussion of its metaphysical character when we examined the *logos of correctness*, the focus now is directed to the effects of this concept. According to Alexy, "principles are norms commanding that something be realized to the highest degree that is actually and legally possible."[1047] He establishes thereby a teleological ground in the concept of principles, one examined factually (suitability and necessity), and the other normatively (balancing). Specifically balancing is employed according to the definition of degrees of non-satisfaction of one principle and the importance of satisfying the other,[1048] based, in the most complex configuration of the Weight Formula, on three variables: first, the degree of importance (and the intensity of interference); second, the abstract weight; third, the reliability of empirical assumptions. These three variables are measured in conformity with a triadic scale (or other numerical parameter).[1049] At the end, we obtain a relativized and proportional arrangement of principles.[1050] Moreover, he constructs, as before mentioned, the *special case thesis*,[1051] according to which legal discourse is a special case of general practical discourse. Considering this connection, he links the claim to legal correctness to the claim to moral correctness[1052] by sustaining that "the legal system as such cannot produce completeness and coherence,"[1053] and hence needs moral arguments in its grounds. Both characteristics reveal a very important aspect of Alexy's theory, which is somehow connected to the BVG's history:[1054] the teleological character of principles and insertion of legal rights into the realm of moral-practical discourse

[1047] Alexy, "On the Structure of Legal Principles," 295. See the last chapter.

[1048] See Alexy, "On Balancing and Subsumption: A Structural Comparison," 440.

[1049] For instance, the double-triadic model. See Ibid., 445.

[1050] See the last chapter.

[1051] See Alexy, *Theorie der juristischen Argumentation,* 263ff. See the last chapter.

[1052] See Alexy, "Law and Correctness," 217.

[1053] Ibid.

[1054] See the second chapter.

demonstrate that Alexy might not clearly distinguish legal rights, with their *force*, their deontological force, from general values.[1055]

For there is no more practical difference between values and principles, and if the procedure is carried out "monologically" as a *logos*, there is no more force, but simply a *moralizing principle*. In addition, if there is no more *force of law*, there is no more play, there is no more promise of justice. "There can be no justice without an appeal to juridical determinations and to the force of law."[1056] After all, deconstruction is ultimately a differential force; it refers to history.[1057] The loss of force and the transformation of principles into *moralizing principles* expose that, regardless of the structural character Alexy sustains for his theory,[1058] his thinking might, in reality, fall into a substantive approach. A certain moral parameter behind the collision of principles needs to be established in order to promote a solution to balancing. Behind balancing, there is a moral argument defining which principle has to take priority over the other. "The outcome of our moral argument then dictates what is possible."[1059] This is why, behind all this rational, structural, formal, analytical framework, the configuration of a *moralizing principle* is the opening to judicial discretion. His structural theory, which he considers essential for a theory of constitutional rights,[1060] becomes, indeed, what he attempted to avoid. He says that a constitutional theory cannot rest in some "superficial grounds of general assumptions,"[1061] but the premises he takes (principles are optimization requirements, and, consequently, the principle of proportionality is indispensable for constitutional adjudication), and his aim to bind law to morality[1062] transforms his theory into a substantive one. The link of his structural theory (based on the premises above) with his concept of legal discourse as a special case of a practical discourse is not capable of preventing him from falling into abstractions,[1063] for instance, which moral argument dictates what is possible and who are the addressees the judge expects will *assert* and *accept* the correctness of the decision. Therefore, the need for a prior supplementation of law[1064] corresponding to those addressees' interests (institutional and non-institutional) is a mystery that can jeopardize the *law*, the *force of law*.

Derrida is aware of this danger of a moralizing principle. He knows that, in the play between law and justice, one cannot be reduced to the other: they are inseparable. In *Force of Law*, he intensively carried out the discussion of the

[1055]See the last chapter.

[1056]Derrida, *Rogues: Two Essays on Reason*,150.

[1057]Derrida, "Force of Law," 929.

[1058]See Alexy, *Theorie der Grundrechte*, 32–38.

[1059]Möller, "Balancing and the Structure of Constitutional Rights," 460.

[1060]See Alexy, *Theorie der Grundrechte*, 32.

[1061]Ibid., 31, translation mine.

[1062]See Alexy, "Law and Corretness," 217.

[1063]See Alexy, *Theorie der Grundrechte*, 29.

[1064]See Alexy, "Law and Correctness," 217.

enforceability of law, and a crucial aspect he mentioned is that, when deconstruction is under consideration, this enforceability is one of the aspects that demonstrate that his theory works against "risks of substantialism or irrationalism."[1065] Paradoxically, therefore, it seems that, by transforming principles into moralizing principles, Alexy's theory expands the risks of irrationalism and substantialism behind the structural framework he developed. Besides, since deconstruction is not destruction[1066] – it is, on the contrary, the "wholly history"[1067] – when principles are bound to some superior moral standards, to some moralizing principles that can be deemed weightier in a singular case, it seems that also the rational guarantee he intended to set up is threatened, for there is, in this situation, no more deconstruction, and consequently, no more history. After all, deconstruction, as we repeatedly argued, requires that law *enforces;* otherwise, justice, as the other side of negotiation, becomes a *logos*. It is necessary this *enforceability,* accordingly, as a fight against the risk of "giving authorization to violent, unjust, arbitrary force."[1068] The purpose of revealing and undercutting the metaphysics of presence in the realm of constitutional adjudication, although knowing its insurmountable reality, is a struggle against the use of arbitrary force that is driven in the opposite direction of this "whole history," of *iterability*. When, however, a metaphysical standpoint prevails, as we can observe in this loss of the *enforceability* of law that balancing can cause, the consequence is that the struggle against the use of arbitrary force surrenders itself to some value, which, since it is ultimately "monologically" gathered, can even be employed as an argument to the use of arbitrary force.

The opening to the use of arbitrary force, within this context, can be better visualized by the simple reasoning derived from the negotiation between law and justice. Derrida remarks that "it cannot become justice legitimately or *de jure* except by withholding force or appealing to force from its first moment, from its first word."[1069] This shows that justice calls for the force of law and, while interacting with it, transforms contexts, produces the "new," but it does so by keeping alive the "mystical foundation of the authority of law;"[1070] that is, it conserves the law while opening it up to the otherness, to invention, to recreation. It produces *iterability*. Nonetheless, when there is no force, the promise of the moment of foundation, the institutional history is put in jeopardy. The law loses its structural call for repetition, the promise that iterability inscribes in the negotiation. The law is no longer deconstructed; it is applied according to a *logos* that transforms it into a *moralizing principle*. It is no longer the play towards the other's otherness; it is the moralizing principle towards satisfying the *logos*. Justice, as the opening to the other, becomes a *logos,* as the enclosure of the other. Law, by following this

[1065]Derrida, "Force of Law," 929.

[1066]Derrida, "Negotiations," 16.

[1067]Derrida, "Force of Law," 929.

[1068]Ibid.

[1069]Ibid., 935.

[1070]See Derrida, "Force of Law," 943.

transcendental signified, can be applied with no other justification but what the *logos* defines. This may be paradoxical: whereas deconstruction runs in this interaction between the constative and performative, and, as such, is conservative of the law while opening it up to the other, the *logos,* since it is not part of the play, while keeping the law as a means of its satisfaction, can merely surrender it to a particular interest (the judge's conscience, for instance) and, as such, close it to the other and account for the "same." In other words, whereas *deconstruction* creates the "new," while producing *iterability,* the *logos* creates the "same," while closing the law to the other and establishing an essentialism, instead.

It is possible to see it when the *logos* defines, in a decision, as a moralizing principle, what is good for everyone in place of what is due to the case. A general perspective transcending the case overcomes its singularity, transforming thereby decision-making into a political program of satisfying the interests of a community or particular interests.[1071] Balancing, while working with values and principles with identical criteria of evaluation[1072] and application, allows that this essentialism materializes. Principles can become *moralizing principles,* that is, norms without their *force,* without their institutional history, their performative construction over the years, their paradoxes and complexities. As moralizing principles, they do not have enough power to combat other values that, although possibly applicable to the case, would not be, if the negotiation between law and justice had taken place, instead. Besides, inasmuch as these moralizing principles lose their force, they lose the realm of the possibility opened by justice. The double bind character of the interaction between law and justice is disrupted.

Furthermore, the transformation of principles, as optimization requirements, into *moralizing principles* can paradoxically lead to a repetition of the "same," for the "calculable," as long as it is not confronted with the *threat* of the *to come,* is submitted to the control of these *moralizing principles.* This is the first effect we previously mentioned, and it derives from the double bind character of the mediation between law and justice. As well as there is no law, with its *force,* the *force of law,* when there is no justice, there is no justice if there is no *force of law.* The consequence of this reasoning is that, if, instead of *justice,* this opening towards the other, there is a *transcendental meaning,* a certain *substantiality,* as the closeness to the other, then the law is not reinvented, or better, legitimately reinvented; the law is not deconstructed. This is a very serious consequence for constitutional democracy: when there is no play, when this double bind is lost, there is no deconstruction, and, consequently, the "same" remains. It remains, however, not as a conservative violence reinvented by the *to come,* a *justice to come,* which takes into account the

---

[1071]See next chapter, where we explore this characteristic regarding the distinction Ronald Dworkin sets forth between *arguments of policies* and *arguments of principles.* See also the first part, where we examined the BVG's and STF's shift to activism, and thus to the deployment of political arguments in decision-making.

[1072]This is one more example of how Alexy seems to confuse values with principles. According to him, "the application of evaluative criteria that have to be balanced with each other corresponds to the application of *principles.*" See Alexy, *Theorie der Grundrechte,* 131.

other's otherness, the singularity of the context, and only the context. Still, the "same" remains as a conservative violence *with ground,* which expresses a certain essentialism, a certain confusion between norms without force and norms with force. The creation, if there is one (and this is clearly possible when the double bind is disrupted), is not in the realm of the performative force of law; it does not improve it in this dimension of the "differential character of force,"[1073] of the "force as *différance*,"[1074] of the "relation between force and norm, force and signification."[1075] It does not improve it with regard to its history.[1076] Rather, it disturbs the history of law, the history of the *force of law*. It is thus not invention, not iterability; it is, instead, the affirmation of an identity, the sign of a repetition, the conservative violence with no history. In sum, a violence of law with no *différance*.

The disruption of the *force of the law,* the loss of its improvement throughout history, the end of iterability are the outcomes of a *logos* from which the perspective of principles as optimization requirements seems to result. The double bind character of law and justice is affected; there is no more interaction between the constative and performative. Since the "mystical foundation of the authority of law" is replaced by the violence with ground, the space for the use of "violent, unjust, arbitrary force" is open.[1077] Here it seems that the elementary conclusions to grasp why balancing can surrender the negotiation between constitutionalism and democracy to a *moral principle,* to a *logos* arise. Here it seems that the reason to conclude why balancing can distort the principle of separation of powers manifests itself. If there is no play between law and justice, if there is no play between constitutionalism and democracy, the *logos of correctness-rationality,* by attacking the first play, also attacks the second one. Briefly, the *logos of correctness-rationality* culminates in the *logos of legitimacy*.

5.3.3 *Balancing and the Logos of Legitimacy*

5.3.3.1 The Elementary Question of Legitimacy: Who Are the People?

The *logos of correctness-rationality* already mirrored the primary aspects to foresee why balancing, sustained by the premise that principles are optimization requirements, can transform itself into a *logos of legitimacy*. All the previous conclusions apply directly to the investigation here: the claim to rationality is, indeed, a *logos of rationality;* the claim to correctness, as well, is a *logos of correctness*. They both

[1073]Derrida, "Force of Law," 929.

[1074]Ibid.

[1075]Ibid.

[1076]See Ibid.

[1077]Ibid.

revealed that Alexy's theory on balancing not only seems to derive from unjustified premises, for they are only introduced and not projected into the play, but mostly, his fundamentals can result in disrupting the play. We could remark serious consequences for constitutional adjudication from the transformation of principles into moralizing principles. Likewise, the conviction that methods and criteria can realize a controlled evaluation of empirical knowledge and legal assessments, since argumentation occurs, demonstrates a certain semantic approach, for the discursive character of his theory might culminate in a simple utterance of judge's conscience and also a relativism in its basis. These outcomes end in the focus we will concentrate on this context of the *logos of legitimacy*: at the core of Alexy's structural theory and defense of balancing, there might be a substantive conception of democracy. The *logos of legitimacy* guides the negotiation between constitutionalism and democracy, and this undermines the autoimmunity, a condition for the self-critique and perfectibility of democracy, to the extent that, at the end, what remains is the use of force against democracy, use of force without legitimacy.

A passage of Robert Alexy's article *The Special Case Thesis,* when he objects to Ulfrid Neumann's critique of the monological character of the *Theory of Constitutional Rights*, can be a sign of this perception: "Even if one agrees with Neumann that the accused should have the right to discuss all relevant legal questions of his case with the judge, *one cannot deny that it is the court which has to decide and argue in the last instance*."[1078] Indeed, although the court has obviously to decide, and thus uphold or deny the claim with arguments, the question regarding the last word is central when legitimacy is under discussion. Who has the last word when legal rights are at stake? As previously examined, much of the basis of Alexy's defense of correctness and rationality in adjudication is intimately associated with the premise that a public forum (institutional and non-institutional) has to evaluate and assert the result of a decision. Nonetheless, we showed that, behind this structural analytical schema, there is practically nothing but the judge's conscience defining which is this forum and which are the values this group would assert and accept as necessary to be balanced in order to provide correctness to the case. It is therefore not difficult to conclude why Alexy's words concerning the *last word* would point to the court. The discursive character applied to balancing, since it is filled with arguments, becomes a monologue of this institution. Besides, we discussed how the concept of the legal discourse as a special case of general practical discourse,[1079] connected to this *logos of correctness,* promotes the conditions to transform constitutional principles into *moralizing principles* to be managed by the court. Both characteristics shape the critical scenario where the constitutional court's activity, with its power to invalidate the acts of parliament, poses the question about its legitimacy. How can constitutional decisions, in this scenario, be legitimate?

[1078] Alexy, "The Special Case Thesis," 377, emphasis mine.

[1079] See Alexy, *Theorie der juristischen Argumentation*, 259–360.

Robert Alexy defends the legitimacy of constitutional courts based on what he calls "argumentative representation."[1080] Unlike the parliament, which is decisional and argumentative,[1081] the constitutional court represents the people in a "purely argumentative"[1082] fashion. Marked by the claim to correctness,[1083] this representation establishes an ideal that links decision-making to discourse as a means to have the assessment and acceptance of the general addressees. This ideal allows establishing what is a good or bad argument, connecting, then, decision-making to "what people really think."[1084] The rationality of the process (the internal justification) associated with the arguments (the external justification) provides the background to find objectivity to a certain degree in the evaluation of the arguments (and thus their quality in balancing).[1085] Nonetheless, these objectivity and rationality alone are not enough to bestow legitimacy: "The existence of good or plausible arguments is enough for deliberation or reflection, but not for representation."[1086] It is here that appears the bridge between the court and the population: "It is necessary that the court not only claim its arguments are the arguments of the people; a sufficient number of people must, at least in the long run, accept these arguments for reasons of correctness."[1087] The court's decision, to be legitimate, requires that it represents the people's aspiration and also that they could be able to assert and accept it. These are, consequently, the two conditions for argumentative representation: "(1) the existence of sound or correct arguments, and (2) the existence of rational persons, that is, persons who are able and willing to accept sound or correct arguments for the reason that they are sound or correct."[1088] The argumentative character of this representation helps define why judicial review is legitimate and why it can even have priority over the acts of the parliament by invalidating them: since the court institutionalizes discourse and follows those two conditions, it achieves legitimacy. In reality, without this relationship of the court with arguments and the people (the institutional and non-institutional addressees), there would be no space for instituting a deliberative democracy.

It is unnecessary to develop here again the prior discussion of the *logos of correctness*, which is closely connected to this perception that constitutional decisions have to satisfy the population and need to be in accordance with "what people

---

[1080]Alexy, "Balancing, Constitutional Review, and Representation," 578.

[1081]According to Alexy, the parliament works with the concepts of election and majority rule. However, to achieve the configuration of a deliberative democracy, it must develop according to arguments, which makes this representation "volitional or decisional as well as argumentative and deliberative" (Ibid., 579).

[1082]Ibid.

[1083]Ibid.

[1084]Ibid., 580.

[1085]Ibid.

[1086]Ibid.

[1087]Ibid.

[1088]Ibid.

really think."[1089] We showed that, ultimately, the defense of the discursive character of this argumentative representation, in Alexy's view, could become a monologue of the court, which erodes the foundation of its argumentative representation. Besides the claim to correctness, we discussed not only why the claim to the rationality of the methodology is insufficient to furnish the so desired objectivity to the decision – for every knowledge is nothing but a fallible knowledge – but also why the claim to legitimacy, by associating the decision with the aspirations of society, reveals its unsustainable justification. This happens because the definition of this "jury" (institutional and non-institutional) and the values that will be gathered for balancing are not controlled but by the court, which will "argue and decide in last instance."[1090] This conclusion is even more evident when we observe that principles, according to this view, can become *moralizing principles,* with the loss of their institutional force as long as they are balanced in equal patterns with any other value. This brings the question of whether this argumentative representation really represents people or, instead, uses this "argumentative" character to conceal what, in reality, occurs: a discretionary decision founded upon value-base standards.

Indeed, this leads to a deeper discussion of the meaning of legitimacy when constitutional courts decide cases. Does it refer to satisfying society's aspirations in accordance with "what people really think," or, rather, legitimacy, in judicial review, has a different configuration when compared to that of the parliament? By observing the court's activity according to this point of view, with its tendency to comply with society's aspirations, the conclusion is that its activity does not clearly differentiate itself from that of the parliament: both have to represent people by deciding cases (constitutional court) or by enacting laws (parliament) concerning the interests of society. Perhaps, the only crucial difference is the instrument involved: decision, in the first case; laws, in the second. Moreover, by following this approach, argumentative representation has priority even over representation based on election,[1091] for it has the power to invalidate the laws. Still, the question is still unanswered: why do constitutional courts have to behave as if they were political institutions? Does argumentative representation mean political representation?

These questions lead to the nuclear dimension of the debate on legitimacy. The confusion of the constitutional courts' activity with that of the parliament involves a complex and serious investigation of the origins of legitimacy, more specifically how an institution gained the authority to act in the name of a society. It is a discussion that goes beyond the prior examination of the *logos of correctness-rationality* directly affecting the *logos of legitimacy*. Derrida's philosophy, in this matter, can provide a very intriguing interpretation of how legitimacy is connected to the act of foundation, and how it has implications for the way an institution operates in a constitutional democracy. Now, more than before, the question refers

---

[1089] Ibid.

[1090] Alexy, "The Special Case Thesis," 377.

[1091] See Alexy, "Balancing, Constitutional Review, and Representation," 580.

to which force is legitimate in a constitutional democracy, and how it should function without becoming a *logos,* thereby practicing violence with ground. More particularly, the issue here relates to the doubt whether constitutional courts, insofar as they act by representing the society and its interests and aspirations, keep the gap between constitutionalism and democracy unfilled or, instead, establish themselves as the conductor of a *logos* that jeopardizes democracy by transforming it into a substantive democracy.

It is a serious question to understand why the considerations on argumentative representation, according to Alexy' parameters sustaining the inevitable relationship between constitutional rights and balancing, although seemingly discursively justified, opens up the possibilities to affirm a model of substantive democracy. It is an issue that is closely related to the confusion of roles between the constitutional courts and the parliament, a confusion that can jeopardize the principle of separation of powers. This is also the background to comprehend why the model of argumentative representation, by following those parameters, might well delineate the basis of a justification for the use of force, although apparently democratically justified, against democracy, thereby using the force without legitimacy.

We previously explored why the play between law and justice is nothing but the other facet of the play between constitutionalism and democracy. The stress on moralizing principles, the belief in the indispensability of balancing (as well as the nature of principles as optimization requirements), and the conviction that a decision must satisfy the interests of people in general (institutional and non-institutional) led to the conclusion that the claim to correctness and the claim to rationality became the *logos of correctness-rationality.* This *logos,* at any rate, opens up the discussion of the *logos of legitimacy.* Since the *logos of correctness-rationality* culminates in the possibility of using the argument of justice as a transcendental signified that could justify judicial discretion or, on the contrary, apply the law as a means to repeat the same (for there would be no legitimate innovation without the demand for the other), the *logos of legitimacy* could also reflect the use of arbitrary force (violence with ground) and the repetition of the same structures. However, the *logos of legitimacy,* although directly reflecting the *logos of correctness-rationality,* could also point to a more specific conflict. In this realm, the negotiation between constitutionalism and democracy would result in two major effects when disrupted by a *logos* behind the context, a *logos* that invalidates the opening deconstruction sets forth: first, there could be the use of force as a means to satisfy the voice of the majority, and thus disregard minorities (democracy without constitutionalism); second, the use of law against democracy (constitutionalism without democracy), and, therefore, the use of law without legitimacy. The first is intimately connected to the previous discussion of the transformation of principles into moralizing principles and the link of correctness with the assessment and acceptance of the society in general. The second requires a deeper investigation of legitimacy itself. More specifically, it demands the investigation of the question of how an institution could be deemed legitimate.

In his text *Declarations of Independence,* Jacques Derrida poses a relevant question when legitimacy is at issue: "Who signs, and with what so-called proper name, the declarative act that founds an institution."[1092] When the negotiation between constitutionalism and democracy is under analysis, particularly by stressing the activity of an institution, we must argue how that institution could be part of this negotiation without disrupting the play. In order to grasp the problematics of the relationship between the institution and the play, we must investigate properly how and in whose name was it founded. Moreover, if the institution inaugurates a new order – and, hence, strongly shakes the play between constitutionalism and democracy – we shall explore how this new order shapes this play. In this matter, it comes to light the question of how legitimate or democratic is this new order. In *Declaration of Independence,* Derrida stresses the founding act of a new order: the independence of the United States of America. Despite that, his investigation opens up the realm of possibilities to analyze likewise the very democratic character of any institution, for he entered into the most basic and complex domain of the question of legitimacy: the domain where nothing but the very history (with its violence and faith)[1093] could account for the establishment of a new interaction between constitutionalism and democracy. In fact, he entered into the most elementary question of legitimacy: who are the people?

This questioning refers to democracy, to the other side of the play with constitutionalism, for democracy is directly linked to the sovereignty of people. This also gives rise to the premise to discuss whether an institution, in its act of foundation, followed the democratic premise. This is the first issue: the connection between the act of an institutional foundation and the democratic premise regarding the sovereignty of people. The second issue, in turn, derives from the first: how is the democratic premise preserved throughout history in the activities of an institution? In other words, how is the link with sovereignty of people kept alive over time? This second issue, as we will shortly examine, requires that the democratic premise, to be preserved, does not become a *logos.* This danger is real insofar as the democratic premise loses its contact with constitutionalism. Indeed, this is the first effect we previously mentioned (democracy without constitutionalism). On the other hand, constitutionalism can also become a *logos* to the extent that it loses this contact with the democratic premise over time. This is the second effect: constitutionalism without democracy. Therefore, there are two possible aspects here to investigate

---

[1092]Derrida, "Declarations of Independence," 47.

[1093]Derrida acknowledges that every new order is full of history, "every signature finds itself thus affected" (Derrida, "Declarations of Independence," 49). He remarks that there is no pure founding moment, since it is always already affected by iterability (See Derrida, "Force of Law," 997). Yet, this revolutionary moment, notwithstanding the suspension of the law, is the very history of law. His words:

"This moment of suspense, this *épokhè,* this founding or revolutionary moment of law is, in law, an instance of non-law. But it is also the whole history of law. *This moment always takes place and never takes place in a presence.* It is a moment in which the foundations of law remain suspended in the void or over the abyss, suspended by a pure performative act that would not have to answer to or before anyone. The supposed subject of this pure performative would no longer be before the law, or rather he would be before a law not yet determined, before the law as before a law not yet existing, a law yet to come, *encore devant et devant venir*" (Ibid., 993).

springing from the *logos of legitimacy*: the *logos of democracy* and the *logos of constitutionalism*.

Both issues refer to both *logos*. The question of who signs and whose name the declarative act that founds an institution is in is central to this debate. Derrida's question is oriented towards the *Declarations of Independence*, but his words reach a much broader intent. The signature that founds an institution, a new order, is marked by what Derrida calls, in his *Force of Law*, the "mystical foundation of authority;"[1094] that is, in the moment an institution is inaugurated, there is always a performative act, a promise that is confounded with the constative. On the one hand, for example, in the case of the *Declaration of Independence*, there was the constative act derived from the affirmation that the Declaration was made in the name of the "laws of nature and in the name of God"[1095] – this was the declarative truth behind the signature. On the other, there was the performative utterance concerning the promise of these truths to be self-evident. As Benhabib mentions, "for if it is the Laws of Nature and Nature's God which show these truths to be 'self-evident', then their self-evidence should be apparent to all."[1096] It is in this matter that a first problem appears, which is central to legitimacy: how could one account for this self-evidence, if not based on arbitrary force or on a metaphysical faith? How, after all, could one argue that the laws of nature, in the way they were institutionalized, are self-evident to all without a minimum of violence (meta or physical), especially when a new order is being established? The "mystical foundation of authority" is indeed characterized by this confusion: we could not consider the founded institution legal or illegal in its founding moment;[1097] however, it must give, although practicing some violence, an air of legality and legitimacy through the signature, as if it were the expression of self-evident laws, which agree with the people represented by the signature. Here, the central question of legitimacy is born: if the self-evidence derives from God, from where originates the signature? In better words, who are the people represented by the signature?

Derrida acknowledges that the Declaration inevitably oversteps the boundaries of the constative: it has an intentional structure that goes beyond the act itself: "The signature maintains a link with the instituting act, as an act of language and an act of writing, a link that has absolutely nothing of the empirical accident about it."[1098] It is thus much more than the event itself. It reaches the performative; it keeps itself alive in the uttering performative. The performative must always remember the signature; it has to appeal to the constative.[1099] As Derrida writes, "the founding act

---

[1094]See Ibid., 943.

[1095]See Derrida, "Declarations of Independence," 51.

[1096]Seyla Benhabib, "Democracy and Difference: Reflections on the Metapolitics of Lyotard and Derrida," in *The Derrida-Habermas Reader*, ed. Lasse Thomassen (Edinburgh: Edinburgh University Press, 2006), 133.

[1097]See Derrida, "Force of Law," 943.

[1098]Derrida, "Declarations of Independence," 47.

[1099]See Benhabib, "Democracy and Difference: Reflections on the Metapolitics of Lyotard and Derrida," 134.

of an institution – the act of achieve as well as the act as performance – *must maintain within itself the signature*."[1100] The signature is the event that brings forth the complexity of legitimacy. Since it has to maintain within itself the signature, the question of what this signature really represents is fundamental. In this matter, Derrida shows the circularity that exists at the core of every act of foundation. The question of who signs such acts leads to an insurmountable search for a justification, but, as he remarks in the very *mystical* character of this moment, the "origin of authority, the foundation or ground, the position of the law can't by definition rest on anything but themselves, they are themselves a violence without ground."[1101] In the *Declarations of Independence,* he strongly stresses this aspect: on the one hand, we could argue that Thomas Jefferson signed the Declaration of Independence; however, he was only representing, and thereby could not properly be the signer. He was only the "draftsman"; he wrote but could not sign.[1102] He was representing the "representatives who have delegated to him the task of drawing what they knew *they* wanted to say."[1103] However, these representatives, who have the duty to approve, revise, correct, ratify the project,[1104] were also obviously representing. Although they, by right, signed, they did not sign only for themselves but also for the others.[1105] These others are ultimately the "good people," who are the signers of the founding act.

Nevertheless, this conclusion only exposes the problematics of the origin of legitimacy. After all, who gives the right to the people to sign a document for their own freedom? As Derrida points out: "Is it that the good people have already freed themselves in fact, and are only stating the fact of this emancipation in the Declaration? Or is it rather that they free themselves at the instant of and by the signature of this Declaration?"[1106] There is a knot in this issue that we cannot easily undo. On the one hand, the Declaration brings the authority of the people and expresses their sovereignty, but it only brings them through the premise of the existence of this sovereignty, which bestows legitimacy and authority on the Declaration. The relation between constitutionalism and democracy seems, within this context, introduced by circularity that shows that one necessarily demands the other. The legitimacy of the founding act is linked to its constitution in conformity with the will of the people, but, reversely, the sovereignty of people gains authority through the founding act: "The signature invents the signer."[1107]

In the founding moment, the connection between constitutionalism and democracy refers to this insurmountable problematics that occurs between the constative

[1100]Derrida, "Declarations of Independence," 48.

[1101]Derrida, "Force of Law," 943.

[1102]See Derrida, "Declarations of Independence," 48.

[1103]Ibid.

[1104]Ibid.

[1105]Ibid.

[1106]Ibid., 49.

[1107]Ibid.

and the performative,[1108] which is, besides, the very condition of iterability. This is the bridge between the founding moment and history. This is also the premise to comprehend how deconstruction relates to the negotiation between constitutionalism and democracy throughout history. In the basis of this negotiation, there are no grounds, no foundation, but only traces, linguistic interaction; there is only history, violence and faith. Accordingly, albeit without grounds, this moment cannot be pure, as if nothing from the past remained. Even in the moment of foundation, there is iterability, which inscribes the repetition in this new beginning[1109]: "There is no more a pure foundation of a pure position of law (. . .) Position is already iterability, a call for self-conserving repetition."[1110] Nonetheless, we could not find the grounds in constitutionalism, for there is no constitution, nor could we find it in democracy, for democracy is institutionalized by the constitution and is paradoxically required to frame the constitution. There are only facts, full of complexities and contradictions; there are only traces. However, this is also why history gains relevance. For the founding moment is a mystical moment, what remains then is history, with its tensions and linguistic interactions, but also the promise, the *to come,* as a call for critique, as a "weapon aimed at the enemies of democracy."[1111] Briefly, a principle of political legitimation[1112] is established as iterability, based on which institutional history and deconstruction play a crucial role in order to protect constitutional democracy. This is the paradox: "Iterability requires the origin to repeat itself originally, to alter itself as to have the value of origin, that is, to conserve itself."[1113]

This history and the *to come,* the signature and the invention: they are at the core of a responsible negotiation between constitutionalism and democracy. Both interact in their own fragility, in their own risk of being disrupted. Still, this fragility is

---

[1108]Ibid.

[1109]See Derrida, "Force of Law," 997.

[1110]Ibid.

[1111]Derrida, *Rogues: Two Essays on Reason*, 86.

[1112]Although Seyla Benhabib argues that Derrida's philosophy does not exactly point out a new principle of political legitimation in history, for, according to this philosophy, "appeals to humanity and morality appear all too indefensible" (Benhabib, "Democracy and Difference: Reflections on the Metapolitics of Lyotard and Derrida," 143), Derrida does not diminish the value of history, its tensions and institutions as indispensable elements to constitutional democracy. He only does not see here the need to transform them into a sort of moralizing principles or a regulative idea that overcomes the singularity of the context. This is why, when he introduces the concept of iterability, he establishes at the core of this "calculable" the opening to the other, as a stress on the singularity of the context. This is, obviously, a complex question, but it does not mean that Derrida is only, as Benhabib's words seem to demonstrate, examining the American Declaration of Independence as if it were "harboring the conflation of the performative with the constative and the normative" (Ibid., 143). This tense relationship between the constative and performative is marked by deconstruction, which is not, as before mentioned, destruction, but rather the "whole history" (Derrida, "Force of Law." 929). Therefore, there is, indeed, a principle of political legitimacy, which is not a moralizing principle, but rather the iterability, which is shaped by the negotiation between the "calculable" and the *to come.*

[1113]Derrida, "Force of Law," 1007–1009.

the very condition for iterability, for the call for the urgency of the other, for opening the history to the "new," for opening the history to the *here and now*[1114] of a promise brought by *to come*, a *democracy to come,* which, albeit unattainable, calls for "a militant and interminable political critique."[1115] First, the maintenance of the signature, of this mystical moment in the performative inscribes the rule of law in the history, and therefore shapes the institutional history. The maintenance of the signature over time – and, consequently, the rule of law – is the condition for preserving the interaction between constitutionalism and democracy. The signature, nonetheless – and this is the second aspect to examine – does not sustain itself alone: it is continuously reinvented. The negotiation between constitutionalism and democracy is thus a space for invention, and this is paradoxically the condition for maintaining the signature. It is a space for respecting the history, the linguistic traces manifested diachronically, but also is the space where the *to come* opens up the history to self-critique and perfectibility, as a means to preserve the responsible negotiation between constitutionalism and democracy. This is why Derrida remarks that:

> (...) 'democracy to come' takes into account the absolute and intrinsic historicity of the only system that welcomes in itself, in its very concept, that expression of autoimmunity called the right to self-critique and perfectibility. Democracy is the only system, the only constitutional paradigm, in which, in principle, one has or assumes the right to criticize everything publicly, including the idea of democracy, its concept, its history, and its name. Including the idea of the constitutional paradigm and the absolute authority of law. It is thus the only paradigm that is universalizable, whence its chance and its fragility. But in order for this historicity – unique among all political systems – to be complete, it must be freed not only from the Idea in the Kantian sense but from all teleology, all onto-theo-teleology.[1116]

The history of constitutional democracy is, therefore, a history of self-critique and perfectibility; the signature needs thus to be continuously reinvented. Yet, this reinvention occurs by stressing the singularity of the context, not by setting up beforehand a *telos,* nor by justifying the invention through an appeal to an essence or superior abstraction, such as a metaphysical concept. In order for history to be democratic and constitutionally bound, it cannot be conditioned by a *logos,* for it would disrupt the play between constitutionalism and democracy. Moreover, it is the very history with its opening to the other, to self-critique and perfectibility, that makes the negotiation between constitutionalism and democracy possible. Therefore, it is this negotiation that supports and feeds the institutional history and gives rise to iterability, but, on the other hand, it is this institutional history through iterability that supports and feeds the negotiation. Iterability is thus an inscription of the possibility of repetition in the very reinvention of the act[1117] throughout history. It is the reinterpretation of the signature in accordance with the particularities of every event, every context, as if the founding moment were somehow perpetuated

---

[1114]Derrida, *Rogues: Two Essays on Reason*, 85.

[1115]Ibid., 86.

[1116]Ibid., 86–87.

[1117]See Derrida, "Force of Law," 997.

in the singular situation. It is the opening of the signature to self-critique and perfectibility over time.

The unanswerability of the previous questions regarding the signature, the founding moment, is crucial to the interminable reinvention of new contexts through iterability. This is the nuclear characteristic of deconstruction. For this "violence without ground," this "coup de force"[1118] refers to the founding act and legitimacy, it is essential that undecidability, as this link with the signature and the opening to the other,[1119] be inscribed at the heart of the interaction between both over time. As Derrida sustains, "the constitution and the law of your country somehow guarantee the signature,"[1120] and this is the space where iterability appears: the "simulacrum of the instant,"[1121] when the signer gives himself the right to sign,[1122] when the air of legality is built, is somehow repeated, but also reinvented, throughout history. Derrida's words concerning the American Declaration of Independence point to this process: the constative of the signature in the name of the laws of nature and in name of God must resound through new contexts, as if the founding moment were guided by an entity guaranteeing the goodness of people,[1123] who founded laws and rights in a legitimate way. Performative utterances appeal to this constative utterance,[1124] to this "vibrant act of faith."[1125] The promise requires the facts, even if they are, in reality, an act of faith that accounts for the signature. There is a mystical character, therefore, linking the *to be* with the *ought to be*,[1126] as a condition to give meaning and effect to the Declaration.[1127] The circularity, which is the basis of the act of institutional foundation, hence, is not resolved but by appealing to an act of faith – the God as the ultimate signer – and this is once more why Derrida calls it the "mystical foundation of authority," which resounds by means of iterability throughout history.

We can now more adequately explore the two issues earlier introduced. The first issue is concerned with the connection between the act of foundation and the democratic premise. The second, in turn, refers to the maintenance of this connection throughout history. From Derrida's analysis in *Declarations of Independence*, it is clear that the relationship between the act of foundation and the sovereignty of people is characterized by an insurmountable circularity, and, if it is to overcome

---

[1118]See Derrida, "Declarations of Independence," 50.

[1119]Undecidability, examined in Derrida's second aporia of his *Force of Law*, is, according to him, "the experience of that which, though heterogeneous, foreign to the other of the calculable and the rule, is still obliged – it is of obligation that we must speak – to give itself up to the impossible decision, while taking account of law and rules" (Derrida, "Force of Law," 963).

[1120]Derrida, "Declarations of Independence," 50.

[1121]Ibid., 51.

[1122]Ibid., 50.

[1123]Ibid., 51.

[1124]Ibid.

[1125]Ibid., 52.

[1126]Ibid., 51.

[1127]Ibid., 52.

the unanswerability of the question of who signs the act of foundation, the recourse will, inevitably, fall into an act of faith. The same reasoning applies to the debate on the constituent power and the constituted power: there is also here this circularity in the very beginning of a new institutional order. This is why "in a constitutional democracy, there is no final seat of sovereignty."[1128] Even the people, who are nominally sovereign, must submit to rules that are constantly interpreted, re-appropriated and contested.[1129] However, it is this undecidability, this unanswerability that opens up the possibility for deconstruction, inasmuch as it inscribes the conflict that occurs between the constative and the performative, the reality and the promise, in the very realm of constitutional democracy. In truth, the "mystical foundation of authority" becomes an indispensable element to sustain the negotiation between constitutionalism and democracy, as if it rendered visible that, ultimately, there cannot be a *logos*, but rather the understanding that the ultimate foundation of an institution is not founded.[1130]

Therefore, we can examine the first issue as if the connection between the act of foundation and the democratic premise were, in reality, a condition for deconstruction, to the extent that it opens up the possibility to overcome a *logocentric* approach by stressing the "violence without ground" of the act of foundation. In turn, the second issue brings to discussion iterability and autoimmunity. The first refers to the assumption of a position, a "call for self-conservative violence,"[1131] while a performative process takes where different forces interact with one another, and where the context gains new significance by embodying the paradoxes and conflicts that appear throughout history. Iterability is therefore connected to differential force[1132] in order to produce something new. The second, in turn, is concerned with the right to self-critique and perfectibility of democracy.[1133] Whereas the first issue inscribes the void, the emptiness, the absence of foundation in the origin of the negotiation between constitutionalism and democracy, without this meaning a pure foundation, for the signature is already affected,[1134] the second issue stamps at the heart of constitutional democracy the respect for the institutional history – the signature must remain – while opening it up to reinvention in favor of the singularity of the other through iterability and autoimmunity.

However, notwithstanding that we can now verify both issues, it is still necessary to address why history, this signature, must be reinvented according to the singularity of the other. Particularly in law, the differential forces towards the other have

1128 Benhabib, "Democracy and Difference: Reflections on the Metapolitics of Lyotard and Derrida," 140.

1129 Ibid., 141.

1130 See Derrida, "Force of Law," 943.

1131 Ibid., 997.

1132 Ibid., 925.

1133 See Derrida, *Rogues: Two Essays on Reason*, 87.

1134 See Derrida, "Declarations of Independence," 49. On the impossibility of pure foundation, see Derrida, "Force of Law," 997.

to deal with the inherent violence of law. There is violence not only in its foundational moment ("violence without ground"), in the constituent power, but also in the one that echoes in every new context where the signature must remain. Jacques Derrida's study of Walter Benjamin's *Critique of Violence* (*Zur Kritik der Gewalt*) in the second part of his *Force of Law* elucidates how the dynamics of the founding act and its preserving effects over time configure two sorts of violence: a violence – the "mystical foundation of authority" – which is called "founding violence, the one that institutes and positions law,"[1135] and the conservative violence, that is, "the one that maintains, confirms, insures the permanence and enforceability of law."[1136] Both support the basis of the history of constitutional democracy, which develops in accordance with this inevitable presence of violence in the very symbolical order of law.[1137] This happens because law has the elementary interest in monopolizing the violence in order to protect the law, as well as the power to provide the means to guarantee, although always threatened by the law, the validity of the performative.[1138] Accordingly, founding and conservative violences correspond to the other side of the relationship between the constative and the performative. The founding moment practices violence, as we can observe in the American Declaration of Independence, to the extent that there is the practice of exclusion of the other, especially some groups who could not exercise their rights.[1139] As Miroslav Milovic remarks, "the institutionalization of the power does not articulate this iterability in the performative part and make it visible only in the constative one, which then, only apparently, speaks in the name of people and democracy."[1140] There are, indeed, in the foundation of law, two "we the people," one in the constative and the other in the performative, which is a paradox, for, whereas the first one includes, the second one excludes the other, establishes the violence within the law. Nonetheless, this violence – now conservative violence – remains throughout history. For the signature must remain, albeit reinvented, a performative violence occurs in the realm of the very interpretation of law,[1141] and here, similarly to the founding violence, the exclusion of the other also occurs insofar as every interpretation of law will have to deal with choices, thereby exposing the impossibility of responding to every other in every context.

[1135] Derrida, "Force of Law," 981.

[1136] Ibid.

[1137] Ibid., 983.

[1138] Ibid., 985.

[1139] In the particular case of the United States, we can observe this aspect in the exclusion of Black American slaves and American Indians, who, as Bernhabib remarks, "are included in the second 'we', in the we to whom the law of the land applies, but they have no voice in the articulation of the law of the land" (Benhabib, "Democracy and Difference: Reflections on the Metapolitics of Lyotard and Derrida," 136).

[1140] Milovic, "A Impossibilidade da Democracia," 259, translation mine.

[1141] Derrida, "Force of Law," 995.

Accordingly, legitimacy is marked by two essential conclusions. First, ultimately, legitimacy cannot be founded, for there is circularity in the founding act, which practices violence without ground. Since the founding moment is a myth, in turn, institutional history gains relevance. Second, legitimacy throughout history is linked to the perpetuation of this founding violence in the conservative violence; yet, this violence is only legitimate inasmuch as it lets deconstruction do its role; that is, instead of being founded, it develops merely according to iterability. In conclusion, there is the respect for the signature, but this signature is only legitimated over time whenever there is a differential force towards the other and whenever *no logos,* no *transcendental signified* destroys the non-foundational character of iterability. The negotiation between constitutionalism and democracy, as a consequence, while maintaining the signature, needs the *to come* in order to keep alive a continuous project of legitimation and to overcome the violence of law as the power against the other.[1142] It needs the undecidable as a means to protect constitutional democracy from its original potentiality to self-destruction. It has thus to overcome the appeal to a *logos,* to a *substantiality*, and transforms itself solely through iterability, where the gap between constitutionalism and democracy remains unfilled and open to an interminable interpretation towards the other. Institutional history, rule of law, on the one hand, and opening to the other, the right to criticize everything publicly, even the very history, on the other, are hence the conditions to overcome the identity, the repetition of the same. They are the conditions for the legitimacy of an institution, which, albeit its insurmountable violence, in a constitutional democracy, needs always to be a legitimate institution.

5.3.3.2 The Logos of Legitimacy in the Structure of Balancing

Derrida's words – "iterability precludes the possibility of pure or great founders, initiators, lawmakers"[1143] – can be the message to begin investigating, more directly, how balancing can lead to the disruption of the deontology of legal principles and the principle of separation of powers, which is, indeed, the consequence of the *logos of legitimacy*. From the previous investigation, we can remark two important problems in this discussion. First, the *logos of legitimacy* can be revealed when, instead of iterability, balancing promotes the formation of a belief in the abstract, which coordinates and guides every sort of argument from outside. In this case, the deontology of legal principles is distorted in favor of a conviction that a teleological approach to satisfying collective wills is legitimate, inasmuch as it conducts the decision towards what the people really desire.[1144] Consequently, adjudication becomes the realization of policies. The question here is whether, by

[1142] Milovic, "A Impossibilidade da Democracia," 259.

[1143] Derrida, "Force of Law," 1009.

[1144] See the cases discussed in the first chapter and the analysis of the development of the German BVG and the Brazilian STF in the second and third chapters, respectively.

affirming this abstract, this transcendental signified, there might not be an implicit authoritarianism. More particularly, the discussion aims to understand why, by precluding iterability, there is the formation of an institution as the great founder of law, which will act as a supreme instance, as the one that "has to decide and argue in the last instance."[1145] Moreover, by acting as the supreme instance, this institution forgets not only the signature but also the opening to the other, jeopardizing thereby the singularity of the context of judgment in favor of an essence derived from establishing an origin where there is circularity and undecidability.

This aspect already culminates in the second issue. We can disclose the *logos of legitimacy* when the autoimmunity is jeopardized by the deployment of balancing, for, instead of promoting the perfectibility and the self-critique of law by reinforcing it and opening it up to deconstruction, there is the subversion of law by subjugating it to a monological utterance expressing the ruling of the court.[1146] In this case, even if we could account for the critique and perfectibility of court's decisions, we could sustain, instead, that, by transforming legal principles into moralizing principles, there is the loss of the effectiveness of this critique. After all, the critique operates out of the realm of the essential core of the violence practiced by law, or, as Derrida argues, "these attacks against violence lack pertinence and effectiveness because they remain alien to the juridical essence of violence, to the *Rechtsordnung,* the order of law (*droit*)."[1147] Accordingly, the autoimmunity requires that the critique, in order to keep the negotiation between constitutionalism and democracy active, cannot become the expression of a logocentrism, as if a moral standard were superior and the definer of the legal order and its correctness. Rather, autoimmunity requires that the critique develops in the play between constitutionalism and democracy, not by appealing to an external entity, for this is not pertinent to – and effective for – the legitimate transformation of law: "An effective critique must lay the blame on the body of *droit* itself, in its head and in its members, in the laws and the particular usages that laws adopt under protection of its power (*Macht*)."[1148] In other words, the negotiation between constitutionalism and democracy, which is characterized by the very inevitable threat of its disruption, perfects itself by reinforcing, in the system of law, the conditions of its self-critique through the attack on any metaphysical standpoint and the emphasis on the transformation of law through deconstruction, which are the very condition for the perfectibility of law.

Nonetheless, when we observe the defense of balancing as Robert Alexy describes it and his argument of "argumentative representation"[1149] of constitutional courts, we could present three relevant arguments to demonstrate why and how, in the basis of his theoretical construction, even though derived from the

[1145]Alexy, "The Special Case Thesis," 377.

[1146]See Alexy, "Discourse Theory and Fundamental Rights," 27.

[1147]Derrida, "Force of Law," 1003.

[1148]Ibid.

[1149]Alexy, "Balancing, Constitutional Review, and Representation," 578.

observation of constitutional courts' activities, iterability and autoimmunity can be jeopardized. First, by establishing a teleological approach tied up with the assessment and acceptance of the general addressees of the law (institutional and non-institutional), Alexy does not seem to draw attention to the undecidability and unanswerability of the founding act of an institution, which, through the signature, characterize the institutional history of law. Indeed, by arguing that decision-making has to express "what people really think,"[1150] he establishes, at the heart of legitimacy, a final point, as an essence that guides adjudication, without being however in a tense relationship with constitutionalism.

The sovereignty of people becomes a *logos,* insofar as it is not followed by an emphasis on the institutional history and legal certainty achieved over time. In other words, the sovereignty of people is not accompanied by the signature, by constitutionalism, with its inherent enforceable character. This is particularly evident to the extent that, according to Alexy's view, there is the assimilation of legal legitimacy to moral acceptance through the premise that legal discourse is a special case of practical discourse.[1151] What exists in the negotiation between constitutionalism and democracy, rather than iterability, is the appeal to a last justification relying on a "value Jurisprudence" or on a "concrete order of values" whereby legal principles can be transformed into moralizing principles. In this quality, they are evaluated in accordance with abstract rules of intensity and efficiency, whose basis is the metaphysical assumption that principles must be necessarily interpreted as optimization requirements.

Briefly, a *logos of democracy* is visualized here, for adjudication must satisfy collective's goals without comprehending the *force of law* expressed by a responsible negotiation between constitutionalism and democracy throughout history. The *logos of democracy* jeopardizes the premise that democracy requires constitutionalism. It demands that, in order to preserve the *force of law,* the reinvention of law through the appeal to the sovereignty of the people is, simultaneously, an appeal to the signature, to the legal principles and norms that have been historically framed and accepted as legitimate by a process of self-critique and perfectibility brought by autoimmunity. In other words, democracy and constitutionalism, as mutually co-original and presupposed, imply that adjudication cannot be reduced to the activity of balancing directed towards satisfying collective goals, for it loses, by diminishing the *force of law* and establishing an answer to legitimacy through the premise of "what people really think," the link with constitutionalism, as well as it opens up the possibility to jeopardize the principle of separation of powers.

Indeed, inasmuch as the constitutional court acts as if its duty were to interpret the law according to "what people really think," deploying thereby methods that point out a teleological solution to the case, its activity is assimilated to that of the parliament. However, in this case, decisions involving constitutional principles are balanced with moral standards without the guarantees of anti-majoritarian rules,

---

[1150]Ibid., 580.

[1151]See the last chapter.

which democratic constitutions introduce in order to avoid the elimination of the voice of minorities by the majority represented by the parliament. Besides, these political decisions are made without the control exercised by periodical elections. The legitimacy of the parliament is connected to the observance of "procedural conditions for the democratic genesis of legal statutes"[1152] as well as the will of people preceding the legislative act. There is, therefore, in the parliament's activity, like any democratic institution, the need for a responsible negotiation between constitutionalism and democracy, which is in truth the condition for its legitimacy. In order to pursue policies and enact laws, the parliament is bound to strict constitutional procedures that set up its connection with the signature. It is, thus, tied up with an institutional history that establishes the mechanisms for enacting laws and pursuing policies. Still, it is also bound to the current people's will preceding the legislative act.[1153] This is why it has to negotiate in this tense realm of constitutionalism and democracy.

The judiciary as well, to be legitimate, must rely on this link with a previous system of rights, must sustain its *enforceability* to the current interpretation of law, as a means to reinvent it through deconstruction. Adjudication is not expressed by pursuing policies, but by strengthening the system of rights in coordination with the differential forces that manifest themselves in the case and all the characteristics and norms extracted therefrom.[1154] Hence, the judiciary legitimately acts whenever it is open to *différance,* as a "movement according to which the language (...) reconstructs itself *historically* as system upon the field of differences." In other words, it acts legitimately whenever it takes in advance the historical development of the negotiation between constitutionalism and democracy as a way to render possible deconstruction, that is, to reinvent the system of rights in conformity with the singularity of the case – an activity, besides, that is marked by undecidability and "infinite regress." The constitutional court's legitimacy, for that reason, cannot simply rely on the observance of "what people really think" ("the existence of rational persons, that is, persons who are able and willing to accept sound or correct arguments for the reason that they are sound or correct")[1155] and on the "existence of sound or correct arguments."[1156] There is a crucial element that is missing in this formulation: the *enforceability* of law is a requirement for the very preservation of democracy, for it sets forth the premise that, although constitutionalism calls for this link with the sovereignty of people, people can only be sovereign – always as an unattainable possibility, since there is no origin, but only undecidability – if they call for constitutionalism, and hence, paradoxically, for the protection of the

---

[1152]Jürgen Habermas, *Between Facts and Norms* (Cambridge: Polity Press, 1996), 263.

[1153]See Ibid., 262.

[1154]We will develop another way of examining this problem in the next chapter through the differentiation between discourses of justification and discourses of application.

[1155]Alexy, "Balancing, Constitutional Review, and Representation," 580.

[1156]Ibid.

singularity of each individual.[1157] In other words, the "argumentative representation"[1158] Robert Alexy defends can threaten the negotiation between constitutionalism and democracy, inasmuch as it might: first, transform the sovereignty of people into a *logos of democracy,* because it is not followed by the mutual co-original and presupposed constitutionalism and its inherent enforceability (founding and conservative violence); second, assimilate, by placing a methodology that converts enforceable rights into teleological preferable interests, the constitutional courts' activity to that of parliament, without being followed by previously enacted counter-majoritarian procedures that are typical of legislation, and without being directly controlled through elections by the addressees of the decision; third, by sustaining the democratic principle through the "argumentative representation" in conformity with "what people really think," paradoxically promote the disruption of democracy, as long as the protection of the singularity of the other is a condition for the sovereignty of people[1159]; fourth, forget the undecidable character of the negotiation between constitutionalism and democracy, since, by undermining the enforceability of law, it establishes a last point in the question of legitimacy, an origin where there would be only differential force; finally, jeopardize the principle of separation of powers, because the discursive character of decision-making lies ultimately in an abstract balancing of interests and values that operates without iterability.

The *logos of democracy* leads to the *logos of constitutionalism,* which is the second argument to understand why and how Alexy's approach to balancing can put in peril the iterability and autoimmunity of constitutional democracy. For the *enforceability* of law is relativized by means of a methodology that transforms principles into moralizing principles, the critique of the legal system becomes a moral critique of law, which threatens the autoimmunity. Inasmuch as autoimmunity is concerned with "the right to self-critique and perfectibility,"[1160] and since moral attacks on the violence of law "lack pertinence and effectiveness because they remain alien to the juridical essence of violence, to the *Rechtsordnung,* the order of law (*droit*),"[1161] the conclusion is that the legitimate transformation of law must be carried out not from outside, as if a *logos* were commanding the development of law, but from its own internal capacity of perfectibility through iterability.

[1157]As Lasse Thomassen argues: "Constitutionalism is supposed to protect the singularity of each individual, but must itself be mediated by democracy" (Thomassen, "A Bizarre, Even Opaque Practice: Habermas on Constitutionalism and Democracym," 180).

[1158]Alexy, "Balancing, Constitutional Review, and Representation," 578.

[1159]In the next chapter, we will explore this question through the discussion of the mutual and presupposed relationship between private and public autonomy. In constitutional democracy, both must be, despite their tense character, continuously reinforced. This is, besides, one of the constitutional courts' roles, as Habermas remarks: "The constitutional court should keep watch over just that system of rights that makes citizens' private and public autonomy possible." (Habermas, *Between Facts and Norms,* 263).

[1160]Derrida, *Rogues: Two Essays on Reason*, 87.

[1161]Derrida, "Force of Law," 1003.

There can be no metaphysical standpoint at the heart of constitutionalism, since the very history of law is reinvented according to the particularity of the context, according to the inevitable threat of the 'perhaps', which inscribes the "infinite regress" and circularity in its relationship with democracy. There can be no transformation of constitutionalism into a "concrete order of values," whose conflicts are decided in conformity with preference relations, and which is based on the metaphysical standpoint of the inevitability of balancing and the character of principles as optimization requirements. Instead, constitutionalism involves a process of adjudication that strengthens legal principles through an invention that, although criticizing its dogmatic certitudes and origins (for this heritage cannot become a *logos*), does not appeal to a sort of external substantiality. This means that it stresses the endless and active differential forces the relationship between constitutionalism and democracy promotes. In virtue of the undecidability of this relationship, which is the reflex of the *to come*, a *democracy to come*, constitutionalism and democracy reinforce and perfect themselves (since one always presupposes the other) by establishing, within the negotiation itself, the conditions for self-critique and the purpose of undercutting metaphysics. This is why the *logos of constitutionalism*, although capable of transforming the law, does not transform it legitimately, for a logocentric critique of law blurs the possibility of improvement of law through iterability. The consequence is that, insofar as the self-critique is threatened, the transformation of law can ultimately occur through the voice of an institution as the great founder of law, as a supreme instance – the signature, after all, blurs – which subjugates the law to a monological utterance. Moreover, since there is no self-critique, but a superior monologue, there is a certain reproduction of an identity. Instead of a process of perfectibility of law through self-critique, there is the reinforcement of a *logos* through the manifestation of an identity.

The analysis of both arguments, which brings forth the conclusion on the existence of the *logos of democracy* and the *logos of constitutionalism*, demonstrates that they point to the same problematic discussion: for democracy presupposes constitutionalism and vice-versa, the *logos of democracy* presupposes the *logos of constitutionalism* and vice-versa. On the one hand, the *logos of democracy* indicates the loss of the signature by stressing the will of people; nonetheless, the will of people will only be sovereign by linking it to the signature. On the other hand, the *logos of constitutionalism* indicates the metaphysical appeal to a certain substantiality in the negotiation between constitutionalism and democracy – for instance, by transforming the constitution into a "concrete order of values," whose conflicts (at least the ones involving principles) need necessarily to be solved by deploying balancing –, which brings about a moral apprehension of law and, consequently, the disruption of the very process of self-critique and perfectibility of law. In this case, a certain moral identity overcomes the framing and perfectibility of law, surpasses the conditions for self-critique, which, to be democratic, must lead this moral identity to also be part of the play, that is, to be deconstructed. With the *logos of democracy*, there is thus neither constitutionalism nor democracy; with the *logos of constitutionalism*, there is neither democracy nor constitutionalism. Both are embedded in the *logos of legitimacy*.

Nonetheless, the most serious question in this matter – which is the third argument to understand why Alexy's approach can undermine iterability and autoimmunity – is the perception that, behind this logocentrism, there might be the endorsement of a substantive conception of democracy. For the self-critique is threatened by the placement of a *logos,* which subjugates the critique of law to a realization of moral values, the negotiation between constitutionalism and democracy is no longer characterized by the conjunction of differential forces – *différance* as a movement towards the other is the word here – but is controlled from outside by a moral standard the judiciary has to accomplish in its decisions. Democracy, in this case, expresses the community's morality, which the judge has to express through decision-making. This morality may consist of different essentialist arguments, of an identity that can jeopardize the individual, the particular, going thereby in the contrary direction of *différance* and, more specifically, the gap that ought to exist in the negotiation between constitutionalism and democracy. The reaffirmation of a certain essentialism goes in the opposite trajectory of the philosophical discussions brought by hermeneutics, which stresses the particular as the condition for the novelty.

Indeed, in a radical perspective,[1162] the essentialism can establish the conditions for the erosion of constitutional democracy, insofar as the other, the singularity of the other is kept out of the play. Instead of self-critique, a *transcendental signified* coordinates the way institutions should operate by establishing an ideology of the collectivity in place of the particular.[1163] Self-critique is replaced by the metaphysical stress on the people, on "what people really think."[1164] Naturally, Alexy, as a constitutionalist, is aware of the problems this emphasis on the people can bring about, but we could not deny that an important characteristic of his thinking seems to be this appeal to some sort of metaphysical ground that could support, as the last

[1162]In a more radical perspective, which is certainly not Alexy's one, we can mention the problems of a substantive conception of democracy. The history, as a matter of fact – as we can observe in discourses of identity, nationalisms, which have inspired many wars and regimes – is plenty of examples of how the assumption of a substantive content behind democracy can reveal this erosion of the singular. By the same token, the philosophy has many examples that expressed the essentialism by reinforcing the unity of people in opposition to the particular (we could indicate, for instance, Rudolf Smend's substantive concept of politics in the construction of a content that would integrate the community (See Rudolf Smend, *Verfassung und Verfassungsrecht* (Berlin: Duncker & Humblot, 1928)) or Carl Schmitt's homogenizing and unifying concept of people through the dichotomy between friend and enemy (See Carl Schmitt, *Verfassungslehre* (Berlin: Duncker & Humblot, 1957)).

[1163]Another possible and interesting investigation in this matter we see in Hannah Arendt's concept of "banalization of evil," in which she stresses the death of politics, when the individuals, instead of actively participating in the construction of the public sphere through arguments, are manipulated as instruments for the exercise of power. In this case, without the critique of this form of domination, the totalitarian ideology transforms the individuals into a collectivity under the argument of a controlling power. See Hannah Arendt, *Was ist Politik?: Fragmente aus dem Nachlaß* (München: Piper, 1993); Hannah Arendt, *Eichmann in Jerusalem: ein Bericht von der Banalität des Bösen* (Leipzig: Reclam, 1990).

[1164]Alexy, "Balancing, Constitutional Review and Representation," 580.

point, his premises. What, therefore, we must stress is that, no matter how this metaphysics expresses itself, it affects deeply the play between constitutionalism and democracy by disrupting what the call for *différance* sets forth: a movement towards the other's otherness, the singularity, which balancing and all this theoretical background, even though paradoxically sustaining the relevance of the case, can, in fact, as previously verified, deny. When there is a *logos*, a *transcendental signified* behind the play between constitutionalism and democracy, after all, there is no *différance*, no movement towards the other's otherness, but the establishment of an identity, which configures, according to Heidegger's words, a metaphysics of presence.

It is an irony, therefore, to conclude that Alexy was right when he, in advance, introduced the possible critiques of his thinking and, more particularly, of his defense of balancing, as the instrument to ensure the functioning of constitutionalism in the realm of constitutional courts. When he mentioned that we could argue that his position with respect to the "argumentative representation" of constitutional courts would lead to a field of "no limits and no control,"[1165] where "legitimation of everything is possible,"[1166] this is indeed a true perception of the problems his theory raises. It is not by appealing, as he usually does, to the BVG's practice as a means to demonstrate how he is right[1167] that these questionings are solved, for the practice can also be metaphysical (as we can observe, for example, in the cases here examined – *Crucifix, Cannabis,* and *Ellwanger*[1168]-, and also in many characteristics of the BVG's and STF's shift to activism),[1169] nor can they be justified by calling for the two premises he presented (the existence of sound or correct arguments and the existence of rational persons who are able and willing to accept those arguments),[1170] because they, in conformity with the three arguments previously studied, can configure the *logos of legitimacy*. Although his defense that "constitutional review as argument or discourse does not allow for everything,"[1171] especially on account of its connection with the will of people,[1172] the reality is that it seems that he could not escape from a logocentric approach by reflecting a substantive conception of democracy, which makes his theory and his understanding

[1165] Ibid.

[1166] Ibid.

[1167] Alexy sustains that the examples he brought forward regarding some BVG's decisions demonstrate how those objections are not strong enough to disturb his theory: "The analysis of the examples presented above shows that rational argument and, thereby, objectivity is possible in constitutional argumentation to a considerable degree. It shows, too, that the existence of cases in which the arguments lead to a stalemate represents no danger at all for constitutional review (Ibid.).

[1168] This metaphysics will be better verified when, through the concept of limited rationality, we will critically reexamine these cases in the eighth chapter.

[1169] See the first part.

[1170] Alexy, "Balancing, Constitutional Review and Representation," 580.

[1171] Ibid.

[1172] Ibid.

of the practice of judicial review a fertile example of a possible identity that can blur *différance* as a movement towards the other's otherness. What remains is a monological representation expressing the "ideal values"[1173] of society. What remains is metaphysics, identity, which can be reinforced and reproduced by way of balancing, an instrument that can, by ultimately sustaining a *logos,* disrupt the play between constitutionalism and democracy, and thus deny *différance.* Constitutional courts become the supreme instance of the ideals of society, the voice of a "concrete order of values." Nonetheless, we shall raise the question: in constitutional democracies, as the ones we worked with in the first part,[1174] is it the role we expect from constitutional courts the defense of collective ideals?

5.4 Final Words

This chapter had the intent to explore one of the most intriguing recent philosophies – Jacques Derrida's deconstruction – and apply it directly to the dogmatic problem of this research, which, in turn, could open up the premises to begin unfolding the concept of limited rationality. Initially by introducing some of its primary concepts – *différance, iterability, autoimmunity, responsibility, negotiation,* among others – it, in the sequence, extended these premises to the political and legal grounds, as a means to reveal that Derrida's philosophy "does indeed translate or call for a militant and interminable political critique."[1175] By discussing *différance,* as the impossible, as the asymmetrical *to come,* and looking into the political characteristics of his thinking, we could verify the double bind of constitutionalism and democracy, and the undecidability that is embedded in the negotiation between both. By reaching the legal realm, we could conclude how *différance* relates to the quest for justice to the other's otherness, while demonstrating that there is no justice without the force of law, which, in the same way, is also characterized by undecidability, the urgency of taking a decision *here and now,* and the need to reinstitute the law in conformity with the singularities of the case. The examination of each of these double binds and the stress on *différance* provided then a powerful edifice to start questioning the premises of the defense of the rationality of balancing, as Robert Alexy justifies it, exercising therefore Derrida's intent to disclose and undercut metaphysics.

The second section of this chapter was oriented to demonstrating that Alexy's premises seem to be marked by a logocentric – or metaphysical – approach. It concentrated on the analysis of three central aspects: the claim to correctness, the claim to rationality, and, finally, the claim to legitimacy. Initially, we demonstrated why, in the basis of Alexy's framework, it seems that there is no questioning of the

---

[1173]Ibid., 579.

[1174]See the second and third chapters.

[1175]Derrida, *Rogues: Two Essays on Reason,* 86.

main premises that led to his conclusion of the indispensability of balancing in decision-making, as well as it seems that the argument that principles are optimization requirements is also exempt from critical review. Not only by showing that Alexy's framework seems to lead to a monological perspective towards the interpretation and application of legal rights, but also by demonstrating that his claim to correctness can, indeed, even when justified by the *Law of Competing Principles* and by the connection between *internal* and *external* justification, result in the judge's discretion, this study concluded that there is a *logos of correctness* in Alexy's claim to correctness. Yet, as long as the claim to correctness is related to the claim to rationality, we could verify that this quest for setting forth an analytical framework in order to defend the rationality of balancing – as if rationality were a question of a methodology able to control the empirical knowledge and orient decision-making to clarification – might not be adequate to grasp the complexities and dilemmas of constitutional adjudication, such as the concern for keeping consistent the system of rights and the quest for doing justice to the other by stressing the singularities of the case. Finally, this research ended by showing, through the question posed by Derrida in his text *Declarations of Independence*[1176] – "who are the people?" – that Alexy's idea of "argumentative representation"[1177] to account for the legitimacy of constitutional courts in this shift to activism and deployment of balancing might, at the end, reveal a *logos of legitimacy*, expressed by a substantive comprehension of democracy in his emphasis on a practice of adjudication oriented to "what people really think."[1178]

Through the application of Derrida's philosophy to the dogmatic problem of this research, accordingly, many nuances of this debate on reason appeared, and the critical review of Alexy's premises, which are connected to the practice of constitutional courts as previously examined,[1179] could be then carried out. This is the context where Derrida's approach reveals its power: it is an active movement, an active interpretability that never stops, because it knows, beforehand, that metaphysics, although always to be confronted, always exists. There cannot be simple conformism to this situation, either: albeit always existing, we must interminably act to disclose and undercut it, because deconstruction knows the outcomes metaphysics causes in reality and, particularly here, in constitutional democracy. This is why deconstruction opens up the possibility to a new project; it opens up the possibility to a different behavior of constitutional courts, one that is committed, even though aware of its unattainableness, to the unconditionality of the other. Perhaps, the message of deconstructionism could be summarized in the words of Martin Morris: "The thrust of deconstructive critique is to transform the experience of the encounter between the self and other such that new institutions appropriate to

[1176] Derrida, "Declarations of Independence."

[1177] Alexy, "Balancing, Constitutional Review, and Representation," 578.

[1178] Ibid., 580.

[1179] See the first part.

such experience can emerge."[1180] Perhaps, what is missing in this movement is a responsible call for constitutional adjudication. Perhaps, what is missing is simply the 'perhaps'.

This chapter ends with the message that *différance*, as the struggle against the metaphysics of presence, is also a struggle against the metaphysics in constitutional adjudication and in constitutional democracy. It also ends with the perception that, even though Derrida's philosophy could, in principle, sound abstract enough to come up against the problem here investigated, it offers, in reality, more than simple intriguing words. However, it is necessary to go further in the critique of this dogmatic problem; it is necessary to complement those conclusions with different outlooks. They can have different perspectives, but they can also open up the space for new possibilities to exercise the interminable critique, one that shapes the *concept of limited rationality*. The next chapter reflects this intent. It might expose that with metaphysics "no thing and no one, nothing *other* and thus *nothing*, arrives and happens."

[1180]Martin Morris, "Deliberation and Deconstruction: The Condition of Post-National Democracy," in *The Derrida-Habermas Reader*, ed. Lasse Thomassen (Edinburgh: Edinburgh University Press, 2006), 249.

Chapter 6
When Procedures Towards Mutual Understanding Come to Light: Balancing Within the Context of Proceduralism

Abstract Robert Alexy's defense of the rationality of balancing is associated with the premise that this procedure is indispensable in the current constitutionalism, for the very nature of principles, interpreted as optimization requirements, leads necessarily to the exigency of deploying it. However, insofar as it seems that there is a metaphysical standpoint behind Alexy's view, the question shifts to the need to develop a possible alternative to balancing, one that best reconciles the exigency of rational justification with legal certainty. In this regard, the claim to coherence may be the point of a departure of this discussion, as long as, by placing Alexy's idea of coherence side by side with Klaus Günther's differentiation between discourses of justification and discourses of application, Ronald Dworkin's integrity in law, and Jürgen Habermas's proceduralist approach, it is possible to provide a robust response to the indeterminacy of law within the context of complex, post-conventional and plural societies. Moreover, through this response, justified in the premise of a dualism between facts and norms in decision-making, the metaphysics embedded in Alexy's thinking can be disclosed, and the problems emerging from the deployment of balancing through an axiological viewpoint, attacked.

6.1 Introduction

In the last chapter, Jacques Derrida's philosophy applied to politics and law guided the examination of three central themes regarding the development of balancing and of a value-based dimension of legal adjudication: the claim to correctness, the claim to rationality and the claim to legitimacy. Their investigation led to the conclusion of how each of these claims, particularly by stressing Robert Alexy's *Theory of Constitutional Rights*, can express a metaphysical standpoint in the realm of decision-making and which consequences we can, from this perspective, achieve. Indeed, it inscribed the attitude of a quest for disclosing and undercutting metaphysics in the realm of constitutional adjudication. His deconstruction of attack on logocentrism, as well as the insurmountable dichotomies (constitutionalism-democracy;

J.Z. Benvindo, *On the Limits of Constitutional Adjudication*,
DOI 10.1007/978-3-642-11434-2_6, © Springer-Verlag Berlin Heidelberg 2010

law-justice) embedded in constitutional democracies and legal adjudication, respectively, revealed that his emphasis on the gap, on the "void" is, in truth, an emphasis on the continuous opening the context brings forth, and thereby a strike against the authority. For this reason, it fitted accurately for the critique of constitutional courts, particularly when they transform subjective rights into objective principles embracing the totality of the legal order, to the extent that it showed how this exercise of adjudication could become the sign of a certain authoritarianism. It is now, nonetheless, necessary to address the question of how this debate can operate in the reality of decision-making with the presentation of a possible alternative to balancing, which, by following the intent to disclose and undercut metaphysics, connects, in a complementary fashion, to the previous debate.

For this purpose, this chapter will begin with a challenge: it will examine the metaphysics embedded in the structure of balancing, as Alexy sustains it, and enter into the debate on an alternative to balancing with a distinct and sometimes untranslatable philosophical language in comparison with the previous chapter. It is no longer deconstruction that will be focused on here, but rather the basis of a proceduralist approach. It is hence not an investigation that is primarily founded upon Derrida's premises, but instead one that dialogues with a tradition that has a Kantian influence, particularly his structure of two worlds (facts and norms), but now discursively remodeled. Specifically, this chapter will focus on Klaus Günther's, Ronald Dworkin's and, more emphatically, Jürgen Habermas's proposals for the problem of indeterminacy of law. While drawing attention to the premises of the proceduralist account might sound, in a sense, contradictory with the previous analysis, on the other hand, it complements it with another look into the issue of legal application as well as with the purpose of disclosing and undercutting metaphysics. The central interest – which will orient the thesis of this investigation, concerned with unfolding the *concept of limited rationality* in legal reasoning – is that, despite their insurmountable divergences, both deconstruction and proceduralism complement each other in the analysis of legal adjudication in the realm of constitutional democracy.[1181]

In the specific field of this investigation, they can demonstrate, through distinct angles, how balancing can be metaphysical – and hence make a diagnosis of the metaphysics in institutional practices – as well as already indicate the steps towards an appropriate account of the practice of decision-making, providing thereby a therapy rooted in the purpose of disclosing and undercutting metaphysics. For this last purpose in particular, the therapy, the procedural model seems to attack the problem more directly, although, as discussed in the last chapter, we could already see many relevant suggestions in Derrida's deconstruction. This opens up, therefore, a possible dialogue between both accounts. In any case, the proceduralism goes direct to the point and provides a response to the indeterminacy of law that is neither founded upon the idea of balancing as a proportional analysis of constitutional principles or values nor grounded in the idea that rationality, at least one that

[1181]See the next chapter.

acknowledges its boundaries, is the result of the deployment of an abstract methodology. Instead, it draws attention to procedures of mutual understanding as the basis of a post-metaphysical thinking in the realm of constitutional democracies, which reaches the very practice of legal adjudication.

While the language of this chapter differs from the previous one, it also opens up the analysis of other possible and relevant standpoints, and this discloses, step by step, the connections between deconstruction and proceduralism, which will then be more directly focused on in the next part. Insofar as the purpose now is to introduce this new elegant and relevant language, the chapter will center on the premise that is intimately related to the issue of indeterminacy of law: it will start with the debate on coherence. First, it will expose how Robert Alexy envisages coherence and how he connects this claim to the deployment of balancing (Sect. 6.2). Against this background, which recovers some of the developments examined in the previous chapters,[1182] but now centered on the discussion of coherence, we will examine a viable response to balancing in the realm of indeterminacy of law. The purpose, at this moment, is to present, step by step, the foundations of a legal theory that will culminate in the proceduralist account, which will serve as a post-metaphysical counterargument to balancing, regarded, particularly by Robert Alexy, as the indispensable mechanism for providing coherence and rationality in the realm of legal adjudication (Sect. 6.3). Accordingly, the first analysis will concentrate on the debate on the discourses of justification and discourses of application, which is at the core of Klaus Günther's legal theory (Sect. 6.3.2), as a first proposal for the indeterminacy of law that understands the tense and complementary character of both discourses in constitutional democracies. A second investigation will concentrate on Ronald Dworkin's premises, which, albeit very similar to Klaus Günther's, focus on the themes of integrity and the "single right answer," primary concepts for the construction of a proceduralist account (Sect. 6.3.3). Finally, we will explore Jürgen Habermas's proceduralist account (Sect. 6.3.4), first by examining how he conceives of a post-metaphysical thinking lying in the tension between facts and norms through discourse (Set. 6.3.4.2), and then by showing how this tension is transported to the debate on legal adjudication (Sect. 6.3.4.3).

By exposing the construction of the proceduralist proposal applied to the problem of indeterminacy of law and legal adjudication, which sets forth a robust response to the generalized idea that balancing is the mechanism that provides rationality and coherence in legal adjudication, it is possible to outline an immediate critique of many of the metaphysical assumptions surrounding balancing, and particularly Robert Alexy's theoretical interpretation of this practice. This critical investigation will be carried out by underlining four major topics: (1) the construction of an axiological content in the structure of principles as the point of departure for balancing (Sect. 6.4.2); (2) the confusion between discourses of justification and discourses of application, which reveals the loss of the tension between facts and norms in the realm

[1182]Particularly, Alexy's claim to correctness examined in the last chapter.

of legal adjudication, resulting in the lack of protection of minorities (Sect. 6.4.3); (3) the relativization and misunderstanding of the "single right answer," whose character of regulative idea is put in jeopardy (Sect. 6.4.4); and (4) the rationality approach that is embedded Alexy's theoretical interpretation of balancing (Sect. 6.4.5). All these sections will corroborate the thesis that balancing, practiced as a proportional application of constitutional principles or values, which gains an objective nature embracing the totality of the legal and social order, does not promote coherence nor rationality – at least not one that acknowledges its boundaries – in decision-making and can be structurally metaphysical, for it is rooted in some material contents that are not transformed into arguments subject to critical review.

This chapter will complement the previous one with a new viewpoint of the metaphysics or logocentrism embedded in the practical and theoretical justification of balancing, while addressing a possible post-metaphysical response to the problem of indeterminacy of law. It introduces new elements, a new language and a new perception of the problem, reinforcing thereby the conclusion that is the focus of this thesis: in the context of legal adjudication, the metaphysical assumption of balancing can convert into a serious problem for constitutional democracy as long as it subverts the principle of separation of powers. In addition, this chapter will bring to light, to the extent that it aims to disclose and undercut metaphysics, how distinct philosophical traditions can dialogue with each other when they face practical dilemmas of social life. This is the perception this chapter aims to reveal: either by means of a deconstructionist or a proceduralist approach, which are two central philosophical standpoints applied nowadays to legal studies, the conclusions on the problem here investigated are intimately connected. It is, for this reason, the exercise of critique by acknowledging that, more than stressing the insurmountable differences between those views, they must dialogue with each other as a means to materialize an effective thinking that has an immediate intervenient attitude towards the world and, more specifically, towards the institutional practices of constitutional adjudication.

6.2 The Claim to Coherence in Robert Alexy's View: When Rights Lapse into General Practical Discourse

As the consequence of his Special Case Thesis (*Sonderfallthese*),[1183] Robert Alexy emphasizes balancing as the viable response to operationalize the "unity of practical reason"[1184] in constitutional adjudication. Still, in the realm of a multiplicity of arguments expressing this unity in legal argumentation, the question of whether and how balancing will promote a rational response necessarily implies the claim to coherence. The claim to correctness, whose metaphysical and ultimate monological

[1183]See the fourth chapter.

[1184]Alexy, "The Special Case Thesis," *Ratio Juris* 12, no. 4 (December 1999): 383.

character was already exposed in the last chapter,[1185] gains, with the debate on coherence, a new complement to the debate on rationality: it implies the investigation of how different arguments can be jointly related without contradicting the system of rights. After all, from Alexy's view, these arguments, to be correct, "must not contradict the authoritative and cohere with the whole."[1186] In other words, their correction is closely related to the maintenance of a coherent response. Still, what could be a coherent response for Alexy in this realm of "unity of practical reason"?

Coherence, in Robert Alexy's view, appears as a necessary criterion for achieving rationality in decision-making. It "implies the claim to justifiability,"[1187] which is the paramount aspect of discourse rationality. In its basis, there is the idea of creating a "cluster of arguments" consisting of "chains of arguments, cumulations of arguments, and arguments against counterarguments which, again, appear in chains of arguments, cumulations of arguments, and arguments against counterarguments, and so on."[1188] In other words, it expresses the capacity to link distinct arguments, which, as long as they are consistent, represent a justification for decision-making. Since adjudication is based on discourse rationality, and discourse rationality is also associated with this capacity, it is a duty of adjudication to make, in the complexity of distinct arguments, coherent decisions. For this purpose, a theory of principles must be connected to the claim to coherence,[1189] which, from Alexy's view, necessarily implies the deployment of his structural framework and the idea that principles are optimization requirements. Coherence, according to Alexy, leads inevitably to balancing.

The implications of this premise can be inferred from the analysis developed in the former chapters: coherence, insofar as it results in balancing, also demands the assumption of a value-based perspective in the realm of adjudication. It works with the idea that legal reasoning inevitably comprises "moral principles valid solely by their moral substantiality."[1190] A coherent reasoning, as a consequence, must be comprehensive, that is, it has to encompass as many different arguments as possible; it must have an "all-embracing"[1191] character. Whether it is an institutionalized practice, an ethically shared tradition, or a moral value, a coherent decision must deal with these variable reasons in a complementary way. For this goal, they are weighted as a means to find the best solution in a particular circumstance. A coherent solution, when these different reasons come into play, is therefore a weighting-based solution. Indeed, according to Alexy, "weighting is the most

---

[1185]See last chapter.

[1186]Robert Alexy, "The Special Case Thesis," 375.

[1187]Robert Alexy, Robert Alexy, "Coherence and Argumentation or the Genuine Twin Criterialess Super Criterion," in *On Coherence Theory of Law*, ed. Aulis et al. Aarnio (Lund: Juristförlaget I Lund, 1998), 43.

[1188]Ibid., 44.

[1189]Ibid., 45.

[1190]Ibid., 46.

[1191]Ibid.

important method for achieving coherence."[1192] By the same token, coherence embodies, for it gives rise to balancing and has this "all-embracing" character, the idea of indeterminacy in the very basis of legal reasoning. As "an elementary postulate of rationality,"[1193] coherence indicates that legal reasoning is also marked by indeterminate criteria of rationality, which "contain neither precise rules nor some kind of calculus or algorithm which definitely determine the solution."[1194] It is, for this reason, a species of "criterialess criteria of rationality (*kriterienlose Rationalitätskriterien*)."

Since it inscribes this indeterminacy in legal reasoning, the claim to coherence demands that decision relies on a flexible structure that can house the distinct possible arguments in order to define preference relations among them in accordance with the particularities of the case. With balancing, this indeterminacy, inherent to legal reasoning, is confronted with the possibility of establishing some rules settling how the preferable argument should be interpreted in a specific situation. This transforms the incommensurable values that appear in this debate into commensurable ones as well as solve the problem of value pluralism.[1195] This is why, for Alexy, the claim to coherence, insofar as it culminates in balancing, is not a "value-free guide."[1196] Rather, it is a condition for a rational value-based evaluation: it, by encompassing all types of reasons, conducts legal reasoning to balancing, which, in turn, creates concrete preference relations among the arguments as a means to make them commensurable and rationally related. There is no coherence without fixing preference relations, as well as there is no rationality in legal reasoning without balancing.

In this framework that places coherence as a "super criterion"[1197] for legal reasoning, and inasmuch as it culminates in conditioning it to balancing, two other elements are, nevertheless, deemed necessary for its completion. First, Alexy points out the discourse rationality, which, along with coherence, "[constitutes] a genuine twin super criterion."[1198] Second, he indicates the history, the "real powers vivid in history comprising the needs, interests, self-interpretations and aspirations of individuals and groups, that is to history's anthropological and sociological dimension."[1199] All these elements are part of legal reasoning and demonstrate that it has to deal with more than a simple reference to the institutional grounds where adjudication takes place. By referring to discourse rationality, Alexy links his debate

[1192]Ibid.

[1193]Robert Alexy, Robert Alexy, "Jürgen Habermas's Theory of Legal Discourse," in *Habermas on Law and Democracy: Critical Exchanges*, ed. Michel Rosenfeld and Andrew Arato (Berkeley, CA: University of California Press, 1998), 227.

[1194]Alexy, "Coherence and Argumentation or the Genuine Criterialess Super Criterion," 47.

[1195]Ibid.

[1196]Ibid.

[1197]Ibid.

[1198]Ibid., 48.

[1199]Ibid.

on coherence to his premise of legal discourse as a special case of general practical discourse (*Sonderfallthese*). By alluding to the "real powers vivid in history," he connects it to the idea of legal reasoning embracing the totality of the legal and social order. Both are interconnected in the thesis that the claim to coherence is pervaded by the need to gather moral values, traditional and collective standards, utilitarian considerations, self-understandings and pragmatic evaluations of interests and compromises, and by a call for a procedure of argumentation whereby balancing is used as a means to establish among them preference relations.

According to Alexy, this procedural model[1200] is this process of integrating general practical discourse, thereby achieving coherence through balancing. In his opinion, the procedural way contrasts with other approaches that attempt to promote coherence in decision-making, particularly with what he calls a coherentist model, which, in its most radical configuration, would point out the idea of legal holism.[1201] In this model, as Alexy describes it, the legal framework would furnish beforehand the answers, which only need to be discovered.[1202] From Alexy's view, nevertheless, an idea of legal holism, although seemingly perfect, "is not realizable,"[1203] and even if one defends it, inevitably she would fail to demonstrate how it does not rely on the appeal to general practical reasons not previously inserted into the system of rights. Legal arguments are necessarily "in need of supplementation beforehand,"[1204] and only through this supplementation can one achieve coherence: "Just as norms cannot apply themselves, a legal system as such cannot produce coherence."[1205] In addition, the idea that coherence relies merely on the legal system would not solve the problem of a rational application of law:[1206] "Just as much as rules are unable to apply themselves, a system cannot itself create the right answer."[1207] An integrative procedure through balancing is therefore indispensable to promote coherence, and, insofar as it culminates in the right answer through arguments, it also yields rationality.

[1200]Alexy, "The Special Case Thesis," 383.

[1201]Ibid.

[1202]Alexy suggests that Habermas's proceduralist theory could be, in principle, an example of a coherentist model, for he defends that legal discourses must come from the legal framework (Ibid.). Nonetheless, Habermas does not exclude from argumentation other possible arguments and reasons that do not stem from the legal system, but instead defends the priority of arguments of justice over arguments of good, a characteristic that appears to contradict Alexy's view. Besides, as we will shortly investigate, Alexy does not seem to grasp the real meaning of the tension between form and content that is the basis of Habermas's proceduralist theory, which makes Alexy believe that Habermas defends a perfect coherent model instead of a tense model directed towards coherence, never totally achieved, though. It lies at the core of this misunderstanding Alexy's non-separation between discourses of justification and discourses of application. See Sect. 6.4.3. *infra*.

[1203]Alexy, "The Special Case Thesis," 383.

[1204]Ibid.

[1205]Ibid.

[1206]Alexy, "Jürgen Habermas's Theory of Legal Discourse," 227.

[1207]Ibid.

The question we can pose in this subject is how this integration would differ from that of the parliament. Alexy's view seems to consider that the duty of adjudication is also of reviewing the reasons why a statute was enacted. On the one hand, If in the democratic process of lawmaking different arguments were taken into account, which Alexy categorizes into three different groups (institutional or authoritative reasons; ethical–political reasons founded upon traditional and self-understandings of a collectivity; pragmatic reasons based on means/goals analysis of interests and compromises),[1208] on the other, adjudication has to reexamine them: "If legal argumentation is to connect with what has been decided in the democratic process, it has to consider all three kinds of reasons presupposed by or connected with its results."[1209] In other words, adjudication will have to deal with the validity claim of the enacted norms and, for that, review their pertinence in contrast to other values deemed relevant to the decision by focusing on their results to society. This explains why, for Alexy, "every application discourse includes a discourse of justification."[1210] In decision-making, in compliance with this point of view, the judge will have to evaluate whether the appropriate norms to the case can also be justified in a broader perspective, that is, in the general interests of all those affected by the norm. Its validity is conditioned, in adjudication, by a teleological analysis of its capacity to achieve a result that not only is pertinent to the case but can also be justified in a further general analysis of the interests of society This is why balancing is essential to provide coherence: it places arguments in a structure of teleological analysis of what is good for a particular social self-comprehension.

6.3 The Post-Metaphysical Response to Balancing as an Indispensable Instrument for Coherence: The Coherence and the Single Right Answer Within Democratic Procedures of Opinion – and Will Formation

6.3.1 Introduction

Robert Alexy's focus on balancing as a response to the indeterminacy of law and as an indispensable instrument for coherence in legal reasoning lies in the premise of the "unity of practical reason," whose outcomes we can verify in the almost indistinctness of the way reasons enter into argumentation in lawmaking and in decision-making, in discourses of justification and in discourses of application. A proportional analysis of means and goals through a general and broad perspective of what is good for a particular society transforms then into the rational mechanism for

[1208]Alexy, "The Special Case Thesis," 377–378.

[1209]Ibid., 377.

[1210]Alexy, "Jürgen Habermas's Theory of Legal Discourse," 231.

the problem of indeterminacy of law. The judge, accordingly, must work on reasons, regardless of their quality (institutional, ethical–political or pragmatic), and deploy balancing as a means to justify, in a rational basis, the decision. Even though this mechanism seems plausible, in the last chapter, through Derrida's deconstruction, its metaphysical standpoint and how this metaphysics echoes in the questions of legal rightness, rationality and legitimacy was already proved. For this reason, if the claim to coherence is a necessary step to face the challenge of the indeterminacy of law, it might not simply rely on the idea of balancing with a proportional analysis. We must then search for a post-metaphysical response to the indeterminacy of law.

A viable post-metaphysical response to the indeterminacy of law resides in three primary aspects: first, the clear differentiation between discourses of justification and discourses of application; second, the quest for the single right answer; and third, the stress on procedures of opinion – and will formation directed towards mutual understanding. It makes explicit that it is not only conceivable but also essential that the indeterminacy of law is challenged by another form of reasoning that is not simply balancing in which legal arguments lapse into general practical arguments. It is necessary rather to establish the clear priority of arguments of justice over arguments of good as a condition for a post-metaphysical thinking, for justice is envisaged within democratic procedures directed towards mutual agreement. Klaus Günther's, Ronald Dworkin's and Jürgen Habermas's proposals are strong examples of this perception. They, in a complementary manner, provide a distinct and more convincing answer to the indeterminacy of law than Robert Alexy's theory rooted in balancing.

For they are somehow concerned with constructing a thinking that does not lie in metaphysical assumptions, they can overcome many of the unanswered problematic issues we can still find in Alexy's view. Moreover, even though they do not originate from the same philosophical tradition of Derrida's thinking, since they come from a Kantian tradition discursively remodeled, they, by some means, complement it to the extent that they provide a more institutionally-related analysis of the problem of legal application in cases of collision of principles. For now, the purpose is to explore how Günther's, Dworkin's and Habermas's views can confront the premise of balancing as Alexy, clearly based on the BVG's practice, develops it. The intention is to disclose, from each of these theories, the grounds for an alternative to balancing that challenges metaphysics while showing the boundaries of the rationality in legal reasoning.

6.3.2 Klaus Günther's View: Coherence Through the Distinction Between Discourses of Justification and Discourses of Application

Within the context of a discourse theory of legal reasoning, Klaus Günther's proposal of a distinction between discourses of justification and discourses of application appears as a powerful counterargument to Robert Alexy's premise that legal reasoning, to be coherent, necessarily demands balancing. Moreover, it reveals that Alexy's theory might suffer from the mistake of mixing up these discourses, thereby undermining the firewall that separates decision-making from lawmaking. In Günther's approach, legal reasoning cannot simply lapse into general practical discourse by means of balancing, which should deliver a rational mechanism to the incommensurability of values. Rather, two different discourses take place in legal reasoning, which are connected to two distinguishable moments with their own premises: one that refers to the validity of norms, and one that is concerned with their application by taking into account all the particularities of a certain situation. The first embraces an idealized universal principle (U), according to which valid norms are those that take into account the interests of all those possibly affected by the norm, whereas the second in turn centers on the factual features of the case and on all the norms that can be applied thereto. With this separation, Günther argues that the justification of norms cannot be occupied with the insurmountable dilemmas arising from the practical feasibility of the validity claim, whereas the application of norms is exonerated from considering previously all other possible situations to which the norm could be applied. Besides, with this premise, Günther indicates that the discourses of justification, rather than being complemented by new singular justifications – as we can observe in Alexy's idea that legal discourse is a special case of general practical discourse – are supplemented by discourses of application, which have a distinct nature.

The starting point of Klaus Günther's thesis is that a particular moral judgment, deemed correct in a particular circumstance, is not necessarily compatible with principles that, in other circumstances, we would accept as valid.[1211] What is appropriate to a particular situation does not combine with what is universally accepted as a principle by all affected people. Moreover, the differentiation between discourses of justification and discourses of application corresponds better to the premise that knowledge is fallible, without this meaning the renunciation of the ideal of a "perfect norm." It is, for this reason, necessary to demonstrate that the "distinction between justification and application of moral norms is possible and makes sense."[1212]

According to Klaus Günther, the discourse of justification of a norm is relevant only to the norm, regardless of its possible application to each one of the situations.[1213] It is concerned with its validity by considering *in abstracto* all the possible interests in conformity with the actual circumstances. For this reason, it implies a weak version of the principle of universalization (U),[1214] according to

[1211] Klaus Günther, Ein normativer Begriff der Kohärenz für eine Theorie der juristischen Argumentation, *Rechtstheorie,* Vol. 20 (Berlin: Duncker & Humblot, 1989): 165.

[1212] Klaus Günther, *Der Sinn für Angemessenheit: Anwendungsdiskurse in Moral und Recht* (Frankfurt a.M.: Suhrkamp, 1988), 21, translation mine.

[1213] Ibid., 55.

[1214] We can remark the connection of Klaus Günther's theory with the grounds of communicative rationality, as Jürgen Habermas develops it, in these discourses of justifications. They originate from the idea of a reciprocal consideration of the interests of each individual and should be which a norm is valid "when the consequences and the side effects of its general observance for the interests of each individual, under the same circumstances, could be accepted by them all."[1215] For this premise, it takes into account all the existing characteristic signs in different situations to which the norm could be applied and artificially suspends the possibility of relevant distinctive signs.[1216] The purpose, in this matter, is to idealistically foresee an agreement on the universal acceptability of the norm by artificially presupposing that the cases to which the norm could be applied are kept unchanged (*ceteris paribus* clause). This is why the principle of universalization cannot be deployed in its stronger configuration,[1217] since it does not stem from the premise of an idealization that comprises and previews *all* the possible situations of normative application. Insofar as the knowledge is fallible and the time is finite, the discourse of justification must take into account that it is impossible to foresee all the situations of application. In this case, it would lead to what Günther calls a "perfect norm,"[1218] that is, a norm that would be not only valid but beforehand already appropriate to each situation susceptible of having it applied.[1219] All the cases of its application would be already established in the discussion of its validity: "The appropriateness of its application would belong to the signification of its validity."[1220] Indeed, there would be a possible confusion between discourses of justification and discourses of application.[1221]

performed according to certain idealized conditions of argumentation that will link their validity to the strength of the arguments used to justify them. The validity claims should derive from the premises of a non-coercive practice of communication (such as the ideas of free and equal participation of all subjects of the discourse, the reciprocal presupposition of rationality among the participants, the requirement of justification of validity claims through arguments), which points out an idealized rational consensus. Accordingly, the validity claims operate in the realm of an ideal community of communication where a non-coercive rational consensus is presupposed. Still, on the other hand, this idealization is established according to a weak form of the principle of universalization (U), resulting from the standpoint that knowledge is fallible, and hence the validity claims must be in connection, without being confused with, the facts. It its thus a type of weak transcendentalization, a hypothetical reference comprising the counterfactual conditions of reciprocal understanding, which, nevertheless, will be confronted with the characters of a particular situation when one deploys it.

[1215]Günther, *Der Sinn für Angemessenheit: Anwendungsdiskurse in Moral und Recht*, 53, translation mine.

[1216]Ibid., 266.

[1217]A stronger configuration of the principle of universalization (U) can be as such described:

"A norm is valid, and, in any case, appropriate, if the consequences and the side effects of the general observance of this norm for the interests of each individual, in each particular situation, could be accepted by them all." (Ibid., 50, translation mine).

[1218]Günther, "Ein Normativer Begriff der Kohärenz für eine Theorie der Juristischen Argumentation," 167.

[1219]See Günther, *Der Sinn für Angemessenheit: Anwendungsdiskurse in Moral und Recht*, 50.

[1220]Günther, "Ein Normativer Begriff der Kohärenz für eine Theorie der Juristischen Argumentation," 167, translation mine.

[1221]Günther, *Der Sinn für Angemessenheit: Anwendungsdiskurse in Moral und Recht*, 51.

Since our knowledge cannot reach all possible situations of normative application, the principle (U) must be regarded as an open principle relying, in a tense way, on the historical and empirical knowledge whereby its application is conditioned.[1222] It comprises merely the consequences and side effects that can previously result from the general normative observance[1223] as well as anticipates some imaginable outcomes as long as they can be generalized according to a criterion of similar traces of potential relevant cases of application. It furnishes, for this reason, *prima facie* applicable valid norms,[1224] which demands a specification and a proper adaptation to the circumstances in discourses of application. There is no problem, accordingly, if all the possible situations were not previously considered: "The consideration of additional contexts does not force a secret revision of the claim to validity of a legitimate norm."[1225] In reality, this is one of the main reasons why both discourses – justification and application – must be separated. This is also the condition for the openness of the norm: "This openness of the moral principle to the contents of a norm (which is generalized because it is hypothetically set and *as such* examined for its compatibility with everyone's interests) does guarantee that no content remains excluded *a limine*."[1226] Moreover, discourses of justification do not define whether a norm is applicable to a particular circumstance and, if not, attempt to reevaluate its validity, because this would invert the sequence from justification to application.[1227] In discourses of justification, as Günther remarks, the question is not whether it is correct to apply a norm to a certain reality, nor to establish criteria for its appropriateness,[1228] but rather merely "question which are the consequences that would expectedly result for our interests, as if it were applied to each one of the situations"[1229]. Briefly, the "validity refers only to the question of whether, *as a rule,* the norm is in our common interests."[1230]

The discourses of application in turn assume beforehand the normative validity and center instead on the particularities of a determined reality. It is concerned not with the validity of the norm, but rather with its appropriateness to a particular situation. By doing so, it conducts legal reasoning to a dependence on the facts and possible reinterpretations of the valid norms applicable to a certain case. Both discourses are connected by the perception that, whereas the reality is fundamental

[1222]Ibid.

[1223]Ibid., 52.

[1224]Günther, "Ein Normativer Begriff der Kohärenz für eine Theorie der Juristischen Argumentation," 171.

[1225]Klaus Günther, "The Idea of Impartiality and the Functional Determinacy of Law," *Northwestern University Law Review* 83, no. 1 & 2 (1989): 165.

[1226]Ibid., 159.

[1227]Günther, *Der Sinn für Angemessenheit: Anwendungsdiskurse in Moral und Recht*, 53.

[1228]Günther, "The Idea of Impartiality and the Functional Determinacy of the Law," 159.

[1229]Günther, *Der Sinn für Angemessenheit: Anwendungsdiskurse in Moral und Recht*, 54, translation mine.

[1230]Ibid., translation mine.

for legal reasoning – and this can be observed in discourses of justification (the fallibility of knowledge in validity claims) and, more incisively, in discourses of application (by assuming idealistically some preconditions for the exercise of non-coercive communication) – this reality is reflexively and critically gathered according to a claim to mutual understanding.[1231] This is why the normative validity and the facts are distinguishable, albeit complementary; this is why the discourses of justification and the discourses of application express two complementary, but not confusable, moments of legal reasoning.

The immediate consequence of this thinking is that, even though one could point out a conflict of norms in a certain reality – and, in this case, the discourse of application is considered – the norms remain valid. It is evident, however, that, in situations of normative application, the interpreter may have to face the dilemma of verifying that the facts in play are in contradiction to the *prima facie* character of the norm, and thus think that this norm is no longer valid. As mentioned, these are two distinguishable moments of legal reasoning: "The decision about the validity of a norm does not imply any decision regarding its appropriateness in a situation, and vice-versa."[1232]

When the interpreter applies a norm, instead of presenting the reasons why it should be observed by everyone as a rule, considering the circumstances, consequences and side effects of the norm, she examines all the special characteristics of a reality and evaluates whether and how this norm should be observed in this singular situation.[1233] The norm, in this case, has to be observed by all people involved in an event by confronting its contents with the assembly of the particular data she can extract from that specific space and time. Hence, the discourse of application refers exclusively to one circumstance, not to all others possible:[1234] "The vehicle for accepting this judgment is a discursive determination of what is the most appropriate principle to guide action in a specific circumstance."[1235] The different norms, valid in abstract, when applied, will be evaluated in accordance with their appropriateness to the specific reality.

The example Günther sets forth[1236] of a conflict between the norms – "one must keep promises" – and "help the other, if he is in a situation of necessity" is paradigmatic: although both could be considered valid, because they express a general interest of observance, they could conflict with each other in a certain

---

[1231]This characteristic, which is clearly observed in Günther's approach, will be examined further, when we will carry out an investigation of Jürgen Habermas's communicative action applied to legal reasoning.

[1232]Günther, *Der Sinn für Angemessenheit: Anwendungsdiskurse in Moral und Recht*, 55, translation mine.

[1233]Ibid.

[1234]Ibid., 56.

[1235]Jeffery Smith, "Justifying and Applying Moral Principles," *The Journal of Value Inquiry* 40 (2006): 404.

[1236]See Günther, "Ein Normativer Begriff der Kohärenz für eine Theorie der Juristischen Argumentation," 168.

reality, resulting in the need to decide which one is more appropriate to that particular event.[1237] The decision about the appropriate norm, for this reason, will reveal the insufficiency of the discourse of justification, as long as it will have to continuously face cases where its universal claim to observance of reciprocal interests under unchanged circumstances seems no longer justified. Nonetheless, the distinctiveness of the discourse of application is exactly its capacity to reveal some relevant unexpected areas of normative application that were not beforehand established and confront them with the circumstances that justify its validity according to the principle of universalization (U). In discourses of application, those valid norms function as *prima facie* reasons for a particular normative statement.[1238] This is the complementary aspect of discourses of application, which will inevitably lead to a tense moment between the idealization of possible circumstances to which the norm is applied and the concrete and conflictive reality, whose facts demand more than pre-established arguments.

In what refers to the complementary relationship between both discourses, Klaus Günther ascribes to them an impartiality principle, which, in its formal sense, embodies the idea that, in the same conditions of application, we must apply the same norm.[1239] In order for this principle to be perfect, all the possible circumstances to which we could apply the norm should be already foreseen. It is, therefore, the case of a "perfect norm," when we can preview the acceptance of all those affected, considering the consequences and side effects of the norm, but also all the situations to which it could be applied. In these circumstances, the impartiality principle would culminate in the identification of discourses of justification with discourses of application, which, in other words, would lead to the deployment of a strong configuration of the principle of universalization (U). Günther, in spite of this conclusion, sees that the impartiality principle is compatible with a weak version of the principle of universalization (U), if, instead of being thematized in only one act,[1240] the impartiality principle were comprehended in a complementary relationship, without mixing them up, between the justification and application of a norm. Since the weak version of the principle of universalization (U)

[1237]Günther examines the famous Kantian example of the lie. If a political fugitive enters into his classroom and hides himself under Kant's desk, and right afterwards the police officers ask Kant whether he saw the fugitive, he must decide whether he follows the moral principle of saying the truth or helps the fugitive by concealing the fact in order to save his life. These two moral principles – "do not lie" and "help the other, in case of necessity" - contain a controversial dilemma of morality. Both norms, consequently, could not foresee all the situations of their application, even though no one would say that they are invalid. Insofar as every situation is a new situation, with distinct signs, and the knowledge is limited, this dilemma could only be solved by, initially, distinguishing the discourses of application from those of justification. See, for this purpose, Günther, *Der Sinn für Angemessenheit: Anwendungsdiskurse in Moral und Recht.*, 47 ff.

[1238]See Günther, "Ein Normativer Begriff der Kohärenz für eine Theorie der Juristischen Argumentation," 172.

[1239]Günther, "The Idea of Impartiality and the Functional Determinacy of the Law," 164.

[1240]See Günther, *Der Sinn für Angemessenheit: Anwendungsdiskurse in Moral und Recht*, 55.

would partially exhaust[1241] the impartiality principle, it must be complemented by an independent and impartial discourse of application,[1242] that is, a discourse that takes into account all the characteristic signs of a determined case, in which the interpreter will confront the idealistic claim to validity with a description of all the aspects of the concrete reality. Only after these two moments do the *prima facie* reasons of normative statements, selected in compliance with the particularities of the case, gain the quality of definitive[1243] ones. Impartiality, in this condition, is connected to the interpreter's capacity to confront the *prima facie* applicable norms with a detailed and integral description of all the characters of a singular case[1244]. It is achieved by two concatenated and complementary steps:

> (. . .) Both embody, respectively, a determinate aspect of the idea of impartiality: the claim that the consequences and the side effects for the interests of each individual should be accepted by all them together operationalizes the universal-reciprocal sense of impartiality, whereas, complementary to it, the claim to consider, in a particular situation of application, all the characteristics operationalizes the applicative sense. As we combine both aspects with each other, we approximate ourselves of the complete sense of impartiality, as if it were through bifurcated ways.[1245]

The complementary character of both discourses that result in the deployment of the impartiality principle through a ramified way indicates their dependency on each other: "Only their combination in historical and social processes fulfills the sense of impartiality."[1246] From another viewpoint, the achievement of the impartiality principle in two concatenated steps leads to the indirect achievement of the "perfect norm," which works as a regulative idea: "We do not anticipate, in a given moment, all the circumstances of each situation of particular application, but rather, in each situation of corresponding application, we anticipate, in a determined moment, all the circumstances."[1247] This ideal of a "perfect norm" brings about a

---

[1241]Ibid.

[1242]Ibid., 27.

[1243]See Günther, "Ein Normativer Begriff der Kohärenz für eine Theorie der Juristischen Argumentation," 172.

[1244]Since Günther's argument concerning legal reasoning lies in a hermeneutical basis, the question of how this description of the facts will be carried out in relation to *prima facie* applicable norms becomes irrelevant. According to Günther: "We can only determine which norm is appropriate in a situation, if the participants of discourse referred the applicable *prima facie* norms to a complete description of the situation. Here is useless to ask whether the participants of the discourse precede first to a complete description of the situation and then to all applicable *prima facie* norms or whether the description of the situation is particularly only shown "in light of" a pre-comprehension of the possible applicable norms. The problem of the hermeneutical circle can remain open." (Ibid., 175, translation mine).

[1245]Günther, *Der Sinn für Angemessenheit: Anwendungsdiskurse in Moral und Recht*, 56, translation mine.

[1246]Günther, "The Idea of Impartiality and the Functional Determinacy of the Law," 165.

[1247]Günther, "Ein Normativer Begriff der Kohärenz für eine Theorie der Juristischen Argumentation," 172, translation mine.

dynamical process of legal reasoning: on the one hand, the *prima facie* applicable norms work as reasons for a particular judgment;[1248] however, on the other, their *prima facie* character makes this judgment dependent on the reciprocal argumentation that exposes, in comparison with possible other *prima facie* applicable norms, their compatibility with, and appropriateness to, a singular reality. This tense combination between validity and appropriateness demonstrates that legal reasoning is necessarily dialogical, and also that legal reasoning cannot be simply one or the other kind of discourse, whether justification or application.

Besides, legal reasoning demands that the interpreter sets forth strong reasons to justify why a determined norm must be applied in place of other also *prima facie* applicable norms; that is, it requires a strong argument inside the structure of the legal framework. The interpreter must justify why, even when there are many other *prima facie* applicable norms, just one of them is more appropriate to the case. These two discourses embody, for this reason, a structure of thinking that reveals: first, the idea of rational discourse by means of a general acceptance of a norm, considering its consequences and side effects under equal circumstances – a primary condition we take into account in discourses of application; and, second, the definite establishment of the reasons why this norm, before *prima facie* applicable norm, is the one selected for that individual circumstance. The prerequisite for a norm, after all, to be applied to a certain case is that it could at least be justified in a general and ideal dimension of reciprocal interests, that is, it is valid: "What is in contradiction with our rational interest should not even be deployed as *prima facie* reason in a discourse of application."[1249] The *prima facie* applicable norms and, of course, their reasons are presupposed to be valid in discourses of application,[1250] as if they were an abstract equality claim used for the exercise of critique in these very discourses.[1251] This is why, as mentioned, the simple fact that they are valid does not mean that they are appropriate to that reality. That fictional premise of presupposing the maintenance of equal circumstances in discourses of justifications will unavoidably be confronted with the characteristic signs of a determined situation.

Consistent with these premises, Günther develops a logic of appropriateness argumentation (*Logik der Angemessenheitsargumentation*)[1252] founded upon what he calls *coherence,* strictly tied up with the tense and complementary relationship between discourses of justification and discourses of application, between the ideal premises of validity and the reality shaping the appropriateness of a norm. The problem in this matter is to develop a system according to which it is possible to bring forth right answers in the realm of a collision of different norms, whose

[1248]Ibid., 173.

[1249]Ibid., 174, translation mine.

[1250]Ibid., 175.

[1251]Günther, Der *Sinn für Angemessenheit: Anwendungsdiskurse in Moral und Recht*, 259.

[1252]Ibid., 287.

contents, for being "indeterminate[1253] in their references to situations,"[1254] are marked by the "need of *additional* specifications in the individual case."[1255] The deployment of the impartiality principle, in the context of discourses of application, results, after all, in the inevitable collision of norms, particularly because, by examining all the aspects of a determinate circumstance, these aspects can be relevant according to different perspectives.[1256] Normative collision, for this reason, refers to the appropriateness, not to the validity of norms.[1257] Naturally, there will only be a collision of norms whenever the characteristic signs of a certain situation contradict the general foreseen circumstances in which the *prima facie* applicable norms appear, thereby leading to a conflict in concrete. The *prima facie* applicable norms, which serve as a regulative idea but are insufficiently appropriate to a certain reality, result in collision.[1258] On the other hand, the characteristic signs of this situation can only be deemed relevant as long as they are confronted with the abstract and general *prima facie* reasons of discourses of justification, which will require, moreover, the presentation of arguments to justify the relevance of those signs in comparison with the others. The immediate outcome is the need for a complete description of the situation, at least implicitly.[1259] Indeed, this is a requirement to overcome, in concrete, the original indeterminacy of norms resulting from the "division of labor between justification and application."[1260]

The structure of reasoning in discourses of application, for this reason, according to Günther, will be carried out through two levels: first, the complete description of the situation, and, second, the normative coherence. The first level relates to the need to select, in obedience to the impartiality principle, the relevant data in a singular circumstance as well as to justify this selection by comparing it to the other also present data. Its starting point is the question *"why do you have to refer to these data and not to the others?"*[1261], which will demand the presentation of arguments justifying the selection of a specific datum or, on the contrary, reasons explaining

---

[1253]Günther examines the indeterminacy of norms according to his model of legal reasoning founded upon a distinction between discourses of justification and discourses of application. The indeterminacy of norms, which can be observed in the definition of *prima facie* norms, must be complemented by the discourse of application, which will evaluate those norms in conformity with the particularities of the situation. The indeterminacy, which also links to the collision of norms, is a characteristic of legal norms and proves why both discourses must be distinguished, for the indeterminacy, inherent to the legal system, does not harm the validity of the norm, but rather exposes how this validity only makes sense, in the practice of adjudication, insofar as it is complemented by discourses of application.

[1254]Jürgen Habermas, *Between Facts and Norms* (Cambridge: Polity Press, 1996), 217.

[1255]Ibid.

[1256]Günther, *Der Sinn für Angemessenheit: Anwendungsdiskurse in Moral und Recht*, 258.

[1257]Ibid., 267.

[1258]Ibid., 280.

[1259]Ibid., 287.

[1260]Habermas, *Between Facts and Norms*, 219.

[1261]Günther, *Der Sinn für Angemessenheit: Anwendungsdiskurse in Moral und Recht*, 288, translation mine.

why the other data are also relevant:[1262] "The selectivity of the interpretation demands justification and has also justificatory capacity within the discourses of application."[1263] When we ask, deny or confirm why a determined character is relevant and not the others, we have already taken into account a complete description of the facts as a means to justify our assertion. By the same token, the other possible *prima facie* applicable norms were already inserted into the interpretation process, as well as the variations of their terminological meanings, which will be related to the characteristic signs of the situation.[1264] Indeed, as Günther sustains, "we are obliged, in practical discourses, to integrally exercise the variations of meaning, which *are possible in a situation,* if we do not intend to infringe the principle of impartial application."[1265] For this, a complete description of the situation is indispensable.[1266]

Nonetheless, in case of collision of norms, one could still argue that, because a previous selection of the relevant signs is carried out, the result is that the decision necessarily will favor the appropriateness of a norm to the detriment of the others in a certain reality. Again, the question "*why these facts and not the others*?" appears, since, were other facts considered, the appropriate norm could be different. Günther, despite this, demonstrates that, even in this situation, there will be the need to justify the reasons for the selection, whether because the proponent must maintain her assertion against critiques- and, for this, must present reasons that will attack the other characteristic signs- or because the opponent will introduce other signs to demonstrate why the original assertion is incorrect:[1267] "The justification of the affirmation of relevance is under the presupposition of a complete situational description."[1268] In other words, as a means to justify why a norm and not the others is appropriate to a determinate circumstance, it will prevail solely the best

[1262]Günter ascribes to the statements about these data the requirement of truth: the participants of the argumentation must agree that the facts described and the reasons connected to them really exist, as a condition for the necessary integral description of the situation (Ibid., 289).

[1263]Günther, "Ein Normativer Begriff der Kohärenz für eine Theorie der Juristischen Argumentation," 177, translation mine.

[1264]Günther sustains that, for an impartial application of norms, the variations of normative meaning need to be related to the characteristic signs of the situation, provided that "different variations of meaning of normative terms can, therefore, deny the affirmation of relevance of a situational characteristic sign, or can confirm or bind them to other characteristic signs." (Günther, Der Sinn für Angemessenheit: Anwendungsdiskurse in Moral und Recht, 293, translation mine).

[1265]Ibid., 294, translation mine.

[1266]According to Günther, "exactly because a selection of this characteristic sign of the factual reality, and not that other sign, is always bound to the determination of a meaning, this selecting decision should be justified in view of all the other characteristic situational signs." (Ibid., 295, translation mine). Moreover, "the principle of normative impartial application implies that, in this case, the norm is to be applied after having exhausted all the possibilities of meanings that could be obtained by a complete description of the situation (Ibid., translation mine).

[1267]Ibid., 296.

[1268]Ibid., translation mine.

argument, whose persuasive force will rely on its capacity to confront it with the integral description of the facts and their relation to the other *prima facie* applicable norms.

The second level, in turn, refers to the coherence of norms. In this case, it is assumed beforehand the premise of the need for integral description of the characteristic signs. For this reason, although centering on the reasons why a norm is more appropriate, it is still in the realm of discourses of application. Günther's thesis is that a coherent interpretation of law will not lead to the confusion between discourses of justification and discourses of application, whose conditions remain separated from each other.[1269] Moreover, it will not rely on material conceptions of an axiological point of view establishing preference relations between the norms to solve the problem of their collision, as Günther sees in Alexy's *Theory of Constitutional Rights*. In this case, what is appropriate becomes what is good in a particular situation, resulting in the danger of withdrawing the material criterion, introduced in the structure of reasoning, from the discourses of application.[1270] The claim to appropriateness, if it intends to observe the impartiality principle, must deal with all type of arguments, including those possible and implicit material ones that correspond to the structure of legal reasoning, through a methodological procedure that is not dependent on "material criteria."[1271] It has to take into account the integrity of the characteristic signs of the facts, examine the reasons for considering them relevant by confronting them with the general and abstract *prima facie* applicable norms (and thus with their presupposed unchanged circumstances), and critically reflect upon the other norms and their variations of meaning, thereby including the relevant facts that are significant to them. A coherent application of norms lies hence in an integral critical review of the possible applicable and valid norms and in their variations of meaning, which will require an integral gathering of the characteristic signs of the situation. To be coherent in discourses of application, an integral comprehension of the impartiality principle, therefore, stands. It has, in conformity with Günther's approach, to observe the following criteria:

1. A norm Nx is appropriate in the situation Sx, if it is compatible with all the other variations of meaning NBn and all the norms Nn, and if the validity of each individual variation of meaning and each individual norm in a discourse of justification can be justified.[1272]
2. A norm Nx is appropriately applicable in Sx, if it is compatible with all the other applicable norms Nl that belongs to a way of life Lx and can be justified in discourses of justification (the same applies to the variations of meanings).[1273]

[1269] Ibid., 302.

[1270] Ibid., 301.

[1271] Ibid., 302.

[1272] Ibid., 324, translation mine.

[1273] Ibid., 324–25, translation mine.

The primary aspect we can observe in Günther's viewpoint is that his criterion of coherence becomes as regulative idea of discourses of normative application that is grounded, first, in the integral gathering of the characteristic signs of a determinate situation, and, second, in the observance of all the possible *prima facie* applicable norms – which must be, in any case, valid – and their variations of meaning. Nonetheless, to the extent that it is impossible to know, in a particular situation, which are all the valid and *prima facie* applicable norms, for the knowledge is fallible, the solution is to take into account the already proved valid norms we can link to the relevant characteristic signs.[1274] This facet of the complementary connection between justification and application will give rise to the connection with paradigms, ways of life, history, and, in the particular case of legal reasoning, with the historical and institutional development of rights, especially because of the institutionalization of their reasons in lawmaking. A coherent application of norms must consider all the *prima facie* valid and applicable norms (and their variations of meaning); however, for this end, there must be an integral description of the situation, and thus the link with the facts, the history is unavoidable, even to open up the interpretation to always-new possibilities and unpredicted circumstances. Indeed, this is a consequence of the very indeterminacy of norms, resulting in the fact that arguments of justification must be necessarily complemented by arguments of application, thereby embracing the facts at stake. An equilibrium – which is tense and hermeneutically open – between these two levels should therefore be reached.

It is important to notice, in any case, that Günther's account stresses that this tense relationship between normative validity and facts does not induce the binding to a certain fact, for there is no confusion between both discourse, nor to a predetermined established order, since the reference is always the singular case.[1275] As previously mentioned, the impartiality principle will manifest, in two steps, the regulative idea of a "perfect norm," which, on the one hand, calls for the general and ideal acceptance of a norm, considering its consequences and side effects under equal circumstances; on the other, requires the definite establishment of the reasons why this norm ought to be applied to a certain circumstance. Only valid norms, after all, can be considered in discourses of application: "The justification of a particular judgment – says Günther – pounces on a *valid* norm, whose general recognizability no one will seriously put in doubt."[1276] This premise brings forward the possibility of critically reflecting upon the facts, insofar as, to be relevant, their characteristic signs will be confronted with the regulative idea of a "perfect norm," thereby reaching the conditions of rational discourse. In other words, this tense moment and critical reflection that should follow it give rise to

[1274]Ibid., 304.

[1275]Ibid., 307.

[1276]Günther, "Ein normativer Begriff der Kohärenz für eine Theorie der juristischen Argumentation," 178, translation mine.

the submission of any argument used in discourses of application to the examination of their respect for the reciprocity principle of effective observance of valid norms.[1277]

This condition of normative validity, accordingly, serves as a regulative idea to evaluate whether the argument, stemmed from the appreciation of the relevant characteristic signs of a reality and the *prima facie* applicable norms, does not indeed violate this presupposition. In the specific realm of legal reasoning, this presupposition lies in the consideration of each participant of the discourse as legal subjects with equal rights of participation, that is, as citizens. Nonetheless, these facts can also be subject to critical review insofar as their emerging norms are also confronted with the other side of the impartiality principle, that is, as long as their contents – and the general and abstract facts sustaining their *prima facie* character – are placed side by side with the complete description of the reality.[1278] The impartiality principle in the domain of validity and in the domain of appropriateness, therefore, introduces a powerful presupposition for the critical review of the very history, its paradigms, but also of all the norms and principles springing therefrom, since the established order will ultimately face the peremptory challenge of promoting coherent responses to a singular circumstance.

By extending this presupposition to the practice of adjudication, the conclusion is that legal reasoning claims that judges observe the legal valid positive norms, enacted through institutional procedures guaranteeing the exercise of citizenship, as a means to impartially apply the correct one according to the characteristic signs of the case. The discourses of justification are tied up with institutionalized procedures that make possible the consideration of all individual interests, even though in an abstract and generic form, as we can remark in the practice of lawmaking, whereas the discourses of application, in turn, operate through institutionalized procedures

[1277]Günther examines the distinction between moral and legal norms by stressing the effective validity of the principle of reciprocity based on the observance of the norm by all people affected as well as on the requirement of a decision, for they are under restrictions of time and fallible knowledge. According to him:

"The only sense of this 'right' consists in making possible the effective validity of the reciprocity principle. Only under these strict premises the Kantian equivalence between law and the prerogative of reciprocal coercion can be justified. The law constitutes a relationship between virtual participants of discourse, whose mutual claim is the effective observance of valid norms. Therefore, they recognize each other reciprocally as legal subjects." (Günther, *Der Sinn für Angemessenheit: Anwendungsdiskurse in Moral und Recht,* 315, translation mine). Moreover, "general and singular legal norms have, therefore, to stem from discourses able to be concluded through a decision. Hence, unlike the practical discourse, they are under conditions of narrow time and incomplete knowledge" (Ibid., 316, translation mine).

[1278]According to Klaus Günther:

"(. . .) The paradigms must be always criticizable regardless of each form of live by considering two aspects: the validity of individual norms, when they, in light of changed interest positions, can no longer keep the reciprocity of the consideration of interests, and the coherent relationship among the individual norms, if they, by reason of the generalized description of the situation that serve as their basis, are no longer compatible with a complete description of the situation." (Günther, "Ein normativer Begriff der Kohärenz für eine Theorie der juristischen Argumentation" 183, translation mine).

that make possible the consideration of all the characteristic signs of a situation.[1279] A coherent decision, accordingly, needs to justify, by taking into account all the *prima facie* applicable norms, their variations of meaning, and their specific correspondence with the facts, why a norm is more appropriate than the others. This activity is not confounded, for this reason, with the definition of preference relations nor with balancing of interests or goods. The primary issue here is to justify the meanings of each norm and confront them with the situation as way to define the appropriate one to that circumstance. The decision results not from balancing according to a proportional analysis, but from the "best theory of all the applicable principles."[1280] Günther is very direct in this distinction: "The norm that offers that relation of priorities should not appear as the optimal realization of the concurring goals in reference to all the possibilities, but rather as the optimal exhaustion of the normative meaning of both principles under the consideration of all circumstances."[1281] Hence, the impartial decision, by means of a complementary relationship between discourses of justification and discourses of application, contrasts with a decision grounded in balancing, for the question is not how to proportionally measure each principle, but instead how to confront them with an impartial consideration of the reality as a means to achieve the *only* appropriate norm to this reality.

Indeed, Günther sees, in this complementary model, the only way to, first, resolve the apparent paradoxes of positive norms, as they emerge from their indeterminacy; second, reach an agreement between their potential modifications and the validity claim based on the general acceptance of the norm (as well as their variations of meaning); and, third, reconcile the selection of *prima facie* applicable norms with the impartial application.[1282] Only the clear distinction between justification and application, now institutionalized within the legitimate procedure of decision-making, can offer a coherent response that preserves the impartiality principle. In truth, this is the condition for simultaneously presupposing, on the one hand, a coherent counterfactual system lying in the ideal dimension of validity and, on the other, the possibility to reach, in each case, the impartial and *single right answer*.[1283]

Naturally, by reason of the conflictive relationship between justification and application, which brings about the indeterminate character of legal norms, the practice of adjudication gains a very broad space of activity. It will be, after all, responsible for making this connection of the valid norms with the situation, many of them previously undefined and unexpected. Still, Günther does not regard this

[1279]Günther, *Der Sinn für Angemessenheit: Anwendungsdiskurse in Moral und Recht*, 337.

[1280]Günther, "Ein normativer Begriff der Kohärenz für eine Theorie der juristischen Argumentation," 179, translation mine.

[1281]Ibid., translation mine.

[1282]Günther, *Der Sinn für Angemessenheit: Anwendungsdiskurse in Moral und Recht*, 337.

[1283]Günther, "Ein normativer Begriff der Kohärenz für eine Theorie der juristischen Argumentation," 182.

characteristic of a possible expansion of discourses of application as a surprising movement.[1284] It rather reflects the characteristics of a post-conventional society,[1285] that is, a society governed by principles accepted by their citizens – and, as such, not controlled from outside – relying thus merely on internal procedures of rational communication. It is a society always open to new unpredicted situations, and thus demanding new discourses of application. However, discourses of application are not enough. In order to preserve the firewall between the parliament and the judiciary, these discourses of application ought to hold the idea of impartiality, therefore stemming from the premise of observing valid norms and from the claim to a coherent reasoning that will lead to the *single right answer*.

6.3.3 Ronald Dworkin's View: Integrity in Legal Reasoning and the Claim to the Single Right Answer as a Response to Coherence

It is not purposeless that Klaus Günther ends his book *The Sense of Appropriateness: Application Discourses in Moral and Law* (*Der Sinn fur Angemessenheit: Anwendungsdiskurse in Moral und Recht*) making direct reference to Ronald Dworkin's theory of legal reasoning. There is an intimate connection between Klaus Günther's approach and Ronald Dworkin's focus on the principle of integrity in adjudication. Günther himself remarks that "Dworkin's theory of the coherent interpretation of principles is the one that comes closer to the model of

---

[1284]Günther, *Der Sinn für Angemessenheit: Anwendungsdiskurse in Moral und Recht*, 338.

[1285]This term – post-conventional – is attributed to Lawrence Kohlberg's theory of moral development based on three distinct levels of moral development, each one divided in two stages. The last stage – the post-conventional – refers to a level where society is regulated by universal principles accepted by the individuals, which become the primary sources for the self-comprehension of a society. According to Kohlberg, at the post-conventional level:

"Stage 6: *The Universal-ethical-principle orientation.* Right is defined by the decision of conscience in accord with self-chosen *ethical principles* appealing to logical comprehensiveness, universality, and consistency. These principles are abstract and ethical (the Golden Rule, the categorical imperative); they are not concrete moral rules like the Ten Commandments. At heart, these are universal principles of *justice,* of the *reciprocity* and *equality* of human *rights,* and of respect for the dignity of human beings as *individual persons*" (Lawrence Kohlberg, "The Claim to Moral Adequacy of a Highest Stage of Moral Judgment," *The Journal of Philosophy* 70, no. 8 (1973): 632).

Klaus Günther sees, in the post-conventional level, particularly in the sixth stage, a fundamental source for developing his distinction between discourses of justification and discourses of application. For him, only in the sixth stage the individuals comprehend the general principles as the representation of procedural normative conditions of communication, which everyone must accept to achieve a mutual agreement. In his view, at this moment, every norm must virtually link to all the characteristic signs as a means to become an appropriate norm capable of being accepted by all affected persons (see Günther, *Der Sinn für Angemessenheit: Anwendungsdiskurse in Moral und Recht*, 174).

argumentation proposed here."[1286] Whereas Klaus Günther underlines the distinction between discourses of justification and discourses of application, Ronald Dworkin stresses, on the one hand, the counterfactual premise of a community of principles where individuals behave with equal concern and respect and, on the other, the principle of integrity in adjudication, according to which every case must be judged in accordance with a coherent interpretation of all applicable principles of a political community, thereby demanding the consideration of all relevant circumstances of the case. Dworkin even uses the term "sense of appropriateness"[1287] to indicate this interconnection between the exhaustive investigation of the valid principles of this "community of principles" and the concern for the singularities of a determinate situation. There is, accordingly, a direct connection between Dworkin's principle of integrity and Klaus Günther's principle of impartiality. Their theories encompass a very clear perception that a coherent response, in the realm of legal reasoning, will lead to the single right answer by carrying out a comprehensive and exhaustive reflection upon the valid principles emerging from a political community, and by regarding the case as a singular case demanding the consideration of all its special features. Similarly, their theories contrast with Alexy's perception that coherence necessarily leads to balancing with an optimization character behind.

Ronald Dworkin sums up his thinking on adjudication at the very beginning of the seventh chapter of his book *Law's Empire*, where he sustains that "the adjudicative principle of integrity instructs judges to identify legal rights and duties, so far as possible, on the assumption that they were all created by a single author – the community personified – expressing a coherent conception of justice and fairness."[1288] From these words, we can immediately conclude that Dworkin seeks to investigate the realm of adjudication as founded upon a deontological standpoint by, first, assuming beforehand the existence of valid norms originated from a political community and, second, rationally reconstructing them as a means to provide the right answer.

The counterfactual presupposition of Dworkin's theory resides in the idea of a community personified to which he ascribes the requirement of "equal concern and respect," that is, the reciprocal-universal presupposition that all affected persons have equal right to participate in the decisions of their community, and thus accept the other as a legitimate member of, and a contributor to, these debates. It is, as Günther correctly sustains, the sense of impartiality in the context of normative justification or, more directly, a "rule of argumentation in practical discourses."[1289] It embodies the premise of a community committed to principles, which is not

[1286]Günther, "Ein normativer Begriff der Kohärenz für eine Theorie der juristischen Argumentation," 190, translation mine.

[1287]See Ronald Dworkin, *Taking Rights Seriously* (Cambridge, MA: Harvard University Press, 1978), 40.

[1288]Ronald Dworkin, *Law's Empire* (Cambridge, MA: Harvard University Press, 1987), 225.

[1289]Günther, *Der Sinn für Angemessenheit: Anwendungsdiskurse in Moral und Recht*, 350–351, translation mine.

merely governed by past political laid down norms and decisions, but also by an autonomous practice of decision-making in which people begin exploring new possibilities of application of those norms to new and unpredicted circumstances. This is what shapes Dworkin's concept of integrity, as if it, on the one hand, "[expanded] and [deepened] the role individual citizens can play in developing the public standards of their community"[1290] and, on the other, established an intersubjective practice of argumentation through an exchange of demands among individuals as a means to promote a coherent conception of "citizens' moral and political lives."[1291]

Integrity is thus a political virtue that has a twofold character: first, it embraces a cooperative and solidary relationship among individuals, who act as authors of the law, and whose interpretation needs to be coherent as if it were a narrative both critically following past decisions and preparing the terrain for the future ones; second, it sets up a commitment to principles with an integrative feature by transforming each individual into a citizen who will act fairly and justly. In other words, it means that individuals will act according to practices and procedures that preserve the conditions of equal participation and influence of all members in the community's deliberations,[1292] while pursuing what is due to this community. In a broader sense, it expresses the impartiality principle of treating like cases alike,[1293] of always coherently justifying each decision in accordance with an extensive comprehension of all principles, as if there were one voice acting "in a principled and coherent manner towards all citizens,"[1294] as well as enforcing and extending "to everyone the substantive standards of justice or fairness it uses for some."[1295] While, therefore, establishing a responsibility towards each one in relation to the other through the ideas of fairness and justice, it also raises the claim to coherence as an inherent character of the practice of citizenship. Integrity, as Günther remarks, "binds self-rule to a coherent scheme of principles, which applies to the authors of the law as well as to the addressees."[1296] Indeed, the condition for the practice of citizenship – and thus for the concern and respect for the other – lies in this search for coherence in all domains of community's decisions, as if it were a promise in the name of fraternity.[1297]

When this premise of integrity is transferred to the institutional realm, it reveals two distinct but complementary perspectives: the integrity in legislation, as an ideal of politics expressed by the requirement of enacting coherent norms in accordance

[1290]Dworkin, *Law's Empire,* 189.

[1291]Ibid., 189.

[1292]Ibid., 164–165.

[1293]Ibid., 165.

[1294]Ibid.

[1295]Ibid.

[1296]Klaus Günther, "Legal Adjudication and Democracy: Some Remarks on Dworkin and Habermas," *European Journal of Philosophy* 3, no. 1 (April 1995): 45.

[1297]See Dworkin, *Law's Empire,* 214.

with an extensive set of moral values springing from the community of principles,[1298] and the integrity in adjudication, as an ideal of legal reasoning that instructs judges to decide cases by interpreting the legal norms and precedents in a coherent way. In a comprehensive manner, this ideal will link adjudication to the requirement of coherence that, since legislation, has been already pursued, thus binding decision-making to the moral and political standards unfolded by legislation in its search for coherence. Finally, therefore, the claim to coherence in adjudication will reveal its "interconnection between societal solidarity and an intersubjectivist concept of law"[1299] from which the premise of "equal concern and respect" reveals its justificatory force. This link with moral and political standards, however, is made through the intermediation of the democratic process of legislation, which must institutionally hold the premise of self-rule procedures of discourse rationality, and hence preserve the idea of "equal concern and respect," even though in a broader sense of the welfare of the community.

For this reason, adjudication is in a different position from legislation.[1300] Whereas the last leads to political decisions "made through the operation of some political process designed to produce an accurate expression of the different interests that should be taken into account,"[1301] that is, with legitimate procedures of rational decision that guarantee the space of discourse to all groups interested in manifesting their claims, as we can see in the safeguards for minority representation, adjudication leads to decisions rooted in principles and, as such, intended to establish an individual right.[1302] The "equal concern and respect" that is the basis of Dworkin's principle of integrity is therefore translated into two dimensions. In the realm of legislation, arguments of policy are the primary focus, for the question here is to make decisions that satisfy the community. In this case, the control of "equal concern and respect" is achieved through procedures for integrating the minority into the debates, on the one hand, and through mechanisms for controlling the legislature's activity, such as the elections or pressure groups, on the other: "The political system of representative democracy may work only indifferently in this respect [consideration of all interests], but it works better than a system that allows nonelected judges, who have no mail bag or lobbyists or pressure groups, to compromise competing interests in their chambers."[1303] In the realm of adjudication, in turn, the focus is on arguments of principles, for now the focus is not on satisfying the collectivity, but rather on making a decision that indicates and justifies whether the legal right claimed by an individual is appropriate to the case based on a comprehensive and exhaustive interpretation of the legal norms

[1298]Ibid., 176.

[1299]Günther, *Der Sinn für Angemessenheit: Anwendungsdiskurse in Moral und Recht*, 353, translation mine.

[1300]Ibid., 244.

[1301]Dworkin, *Taking Rights Seriously*, 85.

[1302]Ibid., 90.

[1303]Ibid., 85.

and precedents. The judge's constraints, therefore, reside in the observance of the democratic procedure of legal rights development and in the claim to coherently improve it by applying the best interpretation possible to the case at issue.

Indeed, this is one of the major subject matters Dworkin continuously reinforces in his texts: legal norms cannot be confounded with policies, for this confusion erodes the firewall between legislation and adjudication. For this purpose, Dworkin distinguishes two classes of legal norms: rules and principles, whose differences are merely logical,[1304] but fundamental to the comprehension of his theory. Besides, as mentioned, he sets up the requirement of not confusing principles with policies. In the first case, Dworkin remarks that whereas rules are "applicable in an all-or-nothing fashion"[1305] with some exceptions,[1306] and therefore are valid or not insofar as the facts are given, principles, on the other hand, operate according to the dimension of their weight or importance,[1307] and hence either have an unspecified validity claim or are constrained by other general conditions.[1308] The first operates with an "if clause" that already foresees the conditions of applicability: if the facts occur, the rule must be applied, and vice-versa; the second, in turn, does not have this quality of an "if clause." Consequently, if two rules are apparently in conflict, the solution lies in some parameters to identify the one's precedence over the other, for only one can be valid. On the contrary, a conflict of principles will not lead to their invalidity, but merely to their inappropriateness to a particular case. Principles, even though mutually excluding and in conflict, can paradoxically not be contradictory.

This dichotomy seems very close to Robert Alexy's[1309] – he himself points out this influence;[1310] however, principles here do not have an optimization structure,[1311] for Dworkin's stress on their deontology clearly requires their distinction from policies. Moreover, Alexy constructs this dichotomy based on logical-structural differences between principles and rules, not on the purpose of building a principle-based interpretative theory of law as an objection to theories leading to discretionary decisions.[1312] For this reason, Dworkin's theory centers on the premise that judges must focus on the individual case, on the context of application,

---

[1304]See Ibid., 24.

[1305]Ibid.

[1306]According to Dworkin: "The rule might have exceptions, but if it does then it is inaccurate and incomplete to state the rule so simply, without enumerating the exceptions. In theory, at least, the exceptions could all be listed, and the more of them that there are, the more complete is the statement of the rule" (Ibid., 25).

[1307]Ibid., 26.

[1308]See Habermas, *Between Facts and Norms*, 208.

[1309]See the fourth chapter.

[1310]See Robert Alexy, *Theorie der Grundrechte* (Frankfurt a.M: Suhrkamp, 1994), 79.

[1311]See Ibid., 75ff.

[1312]See Dworkin's attack on positivist theories such as H. L. A. Hart's in Dworkin, *Taking Rights Seriously*, 14–80.

as a "political responsibility"[1313] towards the individual, for they must justify their decision "within a political theory that also justifies the other decisions that [they propose] to make."[1314] They must thus justify their decision in conformity with the premise of "equal concern and respect;" that is, the individual cannot be taken by surprise, as if the decision were dissociated from a comprehensive interpretation of the legal framework and precedents.

The distinction between rules and principles, accordingly, demands the comprehension of the realm of principles, since their opening is constrained by a process of interpretation calling for integrity in law. Rather than norms of gradual application whose solution to their conflicts relies on some methodological hierarchic schemas, as we can observe in Alexy's Weight Formula, Dworkin links principles to a learning process that is connected to the institutional development of rights, whose application will demand the examination of the features of the case, not a general consideration of what is best or useful for the community as a whole. This is why the distinction between rules and principles can only be adequately grasped when followed by this demarcation of the realm of judicial activity. Since judges must address themselves to arguments of principles, their concern is not to safeguard an economic, political or social situation desired by the community as a whole, but rather to confirm and reinforce, as much as the available knowledge permits, the individual's expectancy of being treated and respected as equal in rights and duties by the legal institutions, whose legitimacy, as a matter of fact, lies in this observance.

Naturally, these different types of arguments appear in legal reasoning and are necessary for the legitimacy of decision-making, but they are translated into the code of rights and, as legal norms, interpreted as trumps against policies, to the extent that "justice is in the end a matter of individual right, and not independently a matter of public good."[1315] This practice is then case-oriented towards the rights and duties the individual is entitled to claim, not goal-oriented towards the wishes and interests of the community, as if it were committed to providing their improvement. Integrity recognizes that adjudication works with arguments of principles, which have a counter-majoritarian dimension and prevalence over policies,[1316] so far as they apply to the particular case as a way to preserve an individual right even when the consequences could harm somehow the community as a whole.[1317] At the core of Dworkin's thinking, there is a clear perception that, by emphasizing

---

[1313]Dworkin, *Taking Rights Seriously*, 87.

[1314]Ibid.

[1315]Ronald Dworkin, *A Matter of Principle* (Cambridge, MA: Harvard University Press, 1985), 32.

[1316]This perspective contrasts directly with many BVG's and STF's decisions, as examined in the first part of this book.

[1317]According to Dworkin,

"(. . .) Arguments of policy try to show that the community would be better off, on the whole, if a particular program were pursued. They are, in that special sense, goal-based argument. Arguments of principle claim, on the contrary, that particular programs must be carried out or abandoned because of their impact on particular people, even if the community as a whole is in some way principles instead of policies, adjudication connects to the exercise of citizenship, and thus to the virtues of justice, fairness and due process[1318] that are at the core of rational discourse. Even when a decision could prejudice the community, it is indispensable to keep alive, at least as a promise – and hence as a regulative idea in adjudication – the conditions of rational discourse or what he calls the "forum of principles," where everyone's claims will be steadily and seriously considered.[1319] This represents, after all, a respect for the most elementary premise of democracy, that is, all legal institutions must treat people as equals.

With this binary configuration – integrity in legislation and integrity in adjudication – Dworkin establishes an all-embracing purpose to bring out right decisions that preserve the ideal of a community of principles, so far as, in the last instance, when adjudication is brought to activity, the judge assumes, as much as she can, the posture of both grasping all the applicable principles to the case and making a decision that preserves a coherent interpretation of them. As a consequence, adjudication is carried out through the requirement of guaranteeing and strengthening, in the best way possible, the ideal of "equal concern and respect." The judge must interpret the law as if it were integrally in compliance with the community of principles, and, since the community of principles functions by preserving the conditions of rational discourse, the judge must interpret the law in the best way possible to reinforce, in reality, these ideal conditions. In other words, the judge must, as much as she can, have a posture of grasping all the applicable valid norms to a determinate case and all its features, and then interpret them as though all these norms derived from a community of principles whose basis lies in the consideration of all persons as equals in rights with solidary forms of sociability. Integrity associates the deontological nature of rights with the claim to coherence leading to the right answer, one that, while examining all the features of the case, follows the impartiality principle of treating like cases alike.[1320]

The adjudicative principle of integrity, as Dworkin remarks, instructs "our judges, so far as this is possible, to treat our present system of public standards as expressing and respecting a coherent set of principles, and, to that end, to interpret these standards to find implicit standards between and beneath the explicit ones."[1321] It is an exhaustive practice of both back-and-forward elements, which, from the

---

worse off in consequence. Arguments of principle are right-based" (Dworkin, *A Matter of Principle*, 2–3).

[1318]According to Dworkin, apart from fairness (existence of "procedures and practices that give all citizens more or less equal influence in the decisions that govern them") and justice (concern for the outcomes of the decisions rather than the procedure, as if the results were distributed in the most just form possible), there is the procedural due process, which "is a matter of the rights procedures for judging whether some citizen has violated laws laid down by the political procedures" (Dworkin, *Law's Empire*, 164–165).

[1319]Ibid., 32.

[1320]It is interesting to observe that, in their general lines, Dworkin's conclusions are very similar to Günther's.

[1321]Dworkin, *Law's Empire*, 217.

specific features of a singular case, looks into the all-valid applicable principles as if it were an "unfolding political narrative."[1322] It is, for this reason, intimately connected to history. However, the fact that "history matters in law as integrity"[1323] does not cause, similarly to the discussion of Günther's approach, the consequence of binding judicial activity to the past, as if it were a conception of truth not subject to further review. Rather, since Dworkin's theory of law is concerned with the tension between the regulative idea of integrity, on the one hand, and the application of norms by preserving and reinforcing, as much as possible, the conditions of a coherent interpretation of principles, on the other, history must necessarily be submitted to critical reconstruction. Dworkin himself stresses that history is relevant to the extent that it raises the requisite of justifiability – "history matters because that scheme of principle must justify the standing as well as the content of these past decisions"[1324] – but also because it makes the association of legal reasoning with the conditions of rational discourse, for history is reconstructed in the spirit of unfolding what Habermas calls "traces of practical reason."[1325]

Throughout history, we can observe, first, the practice of rational discourse through arguments that appear in legal argumentation and which will uncover, at the end, the principles of fairness and justice, and, second, the signs of a learning process that, although bringing a certain stability and predictability to legal reasoning, also demonstrate that history is only the tip of the iceberg: learning processes mean, after all, revisiting the past with the eyes of reconstruction towards the future. As an element of the learning process, the past can bear an ideological dimension, and hence, with such quality, a practice against discourse rationality; it can express a metaphysical standpoint that is not subject to further critical review. Dworkin's viewpoint, for this reason, assumes the past in a reconstructive manner, for his hermeneutics has already absorbed the critique of a communitarian position towards the application of legal norms.[1326] He knows that an interpretation that is

---

[1322]Ibid., 225.

[1323]Ibid., 227.

[1324]Ibid.

[1325]Habermas, *Between Facts and Norms*, 203.

[1326]The critique, in this matter, refers to theories that indicate the prudence – or *phrónesis,* as it stems from Aristotelian philosophy – as a response to the application of norms. However, in a post-conventional level, these theories suffer from the problem of confusing discourses of justification with discourses of application or, in other words, normative validity with facts. The critique, consequently, operates in the same level of the facts, thereby limiting itself to the context of application, which can nevertheless be ideological and metaphysically oriented against the possibility of carrying out the reflexive critique. For a detailed critique of the prudence in the realm of rights, see Günther, *Der Sinn für Angemessenheit: Anwendungsdiskurse in Moral und Recht*, 216–254 ('Phrónesis' als Beispiel kontexgebundener Anwendung). Against the appeal to *phrónesis* by Hans-Georg Gadamer, see Jürgen Habermas, "Zu Gadamers 'Wahrheit und Methode'," in *Hermeneutik und Ideologiekritik*, ed. Karl-Otto Apel (Frankfurt a.M.: Suhrkamp, 1971), 45–56, as well as the reply by Hans-Georg Gadamer, "Rethorik, Hermeneutik und Ideologiekritk: Metakritische Erorterungen zu 'Wahrheit und Methode'," in *Hermeneutik und Ideologiekritik*, ed. Karl-Otto Apel (Frankfurt a.M: Suhrkamp, 1971), 57–82.

limited to the context of application, as if the normative validity were confounded with the facts, fails to face the dilemmas of a pluralist and post-conventional society, because it loses the regulative idea the justificatory premise of a community of principles, and hence integrity, brings to light, which serves in turn as a starting point to criticize the facts. The conditions of critical review, in this case, operate in the same level of the features of the case, and therefore do not carry the regulative idea of integrity, which presupposes a personified community that, according to Dworkin, works as "special kind of entity distinct from the actual people who are its citizens."[1327] Briefly, the community of principles, to which Dworkin ascribes a justificatory character, and the moment of application of norms, when one gathers the facts in their plenitude, are distinct.

Inasmuch as the past is reconstructed and cannot provide safe answers to the most complex dilemmas of legal reasoning, the consequence is that the judge works in the realm of a strong indeterminacy of law. Dworkin's theory, nevertheless, in this reconstructive perspective of the past, takes the past instances of legal interpretation and attempts to apply the best interpretation of law possible as a response to this indeterminacy. Like Günther, Dworkin is aware of the risks of the belief in a norm that, insofar as it is not determinate in all its possibilities of application, can be relativized by reason of other interests, as if the context of application were rather relieved of the justification brought by the ideal of a community of principles. Indeed, he develops a powerful critique of what he calls conventionalism, whose source lies in theories such as legal positivisms in the way of Hans Kelsen's and H.L.A. Hart's approaches,[1328] because they, by concluding that, in some cases, the judge cannot rely on any source to decide, lead to a solution based on a discretionary choice.[1329] Dworkin sees also the risk of theories that, grounded in the indeterminacy of law, promote a goal-oriented perspective of legal reasoning. In this case,

---

[1327]Dworkin, *Law's Empire,* 168.

[1328]See Dworkin, *Taking Rights Seriously*, 1–80.

[1329]Dworkin differentiates integrity from two other types of discourse in legal reasoning: conventionalism and pragmatism. Briefly, conventionalism relies on social conventions defining what law is, which the judge must discover. However, since, in many cases, these conventions may not apply, the judge has to proceed, for she has no more source to act, to a discretionary solution. According to Dworkin:

"(...) Conventionalism explains how the content of past political decisions can be made explicit and noncontestable. It makes law dependent on distinct social conventions it designates as legal conventions, in particular on conventions about which institutions should have power to make law and how. Every complex political community, conventionalism insists, has such conventions (...)."

"Second, conventionalism corrects the popular layman's view that there is always law to enforce. Law by convention is never complete, because new issues constantly arise that have not been settled one way or the other by whatever institutions have conventional authority to decide them. So conventionalists add this proviso to their account of legal practice. 'Judges must decide such novel cases as best they can, but by hypothesis no party has any right to win flowing from past collective decisions – no party has a *legal* right to win – because the only rights of that character are those established by convention. So the decision a judge must make in hard cases is discretionary in this strong sense: it is left open by the correct understanding of past decisions.

due to the impossibility of the norm to provide answers to all circumstances, the solution must then reside in the pursuit of objectives that are best for society as a whole. Pragmatism,[1330] as Dworkin categorizes these theories, transforms the context of application oriented towards to the interests of society as the only justificatory premise for legal reasoning, and hence confuses rights with policies. Both conventionalism and pragmatism fail to establish a justificatory premise for legal reasoning with a reconstructive dimension of history as well as undermine the firewall between legislation and adjudication. Both force confusion between normative validity – and its justificatory claim – and facts, as though the indeterminacy of law could only be solved by judicial activism founded lastly upon the judge's discretion.

Consistent with the integrity standpoint, the indeterminacy of law stems not from the structure of law itself, but rather from the incapacity of the judge to deploy the best interpretation possible.[1331] As a consequence, the response to this problem is to endeavor, according to the available knowledge, to provide the best possible interpretation of law, one that reconciles the claim to legal certainty with the claim to legitimacy,[1332] bringing thereby a vaster comprehension of all principles and precedents, as justificatory premises, than any other possible. This theory, in Dworkin's view, is the one that aims to guarantee in reality, as much as possible, the counterfactual presupposition of a community of principles, as if the legal framework were structured in a coherent way by preserving the principles of justice, fairness and procedural due process.[1333]

In the practical realm, this would mean that judges should "enforce these in the fresh case that comes before them, so that each person's situation is fair and just according to the same standards."[1334] They must virtually connect their decisions to

A judge must find some other kind of justification beyond law's warrant, beyond any requirement of consistency with decisions made in the past, to support what he then does." (Ibid., 114–115).

[1330]The pragmatism standpoint ascribes an instrumental-teleological perspective to legal reasoning, as if the law were an instrument to achieve certain goals, even when against the texts and procedures. In this case, rather than being oriented by principles, adjudication functions by promoting policies to reach an objective. It is not, as conventionalism, a certain reference to the past, insofar as there is no character of historical continuity, but a practice looking to the future by transforming the law into a mere instrument to reach a determined result. According to Dworkin,

"The pragmatist takes a skeptical attitude towards the assumption we are assuming is embodied in the concept of law: he denies that past political decisions in themselves provide any justification for either using or withholding the state's coercive power. He finds the necessary justification for coercion in the justice or efficiency or some other contemporary virtue of the coercive decision itself, as and when it is made by judges, and he adds that consistency with any past legislative or judicial decision does not in principle contribute to the justice or virtue of any present one. If judges are guided by this advice, he believes, then unless they make great mistakes, the coercion they direct will make the community's future brighter, liberated from the dead hand of the past and the fetish of consistency for its own sake" (Ibid., 151).

[1331]See Habermas, *Between Facts and Norms*, 214.

[1332]Ibid., 211.

[1333]Dworkin, *Law's Empire*, 243.

[1334]Ibid.

all applicable principles by deploying the counterfactual presupposition of treating the individual with equal concern and respect, and place themselves as if they were the other of the discourse, the one and the other litigant, in order to see the case from all viewpoints. This procedure shows that the indeterminacy of law is confronted with a reconstructive practice of law that cannot, beforehand, promote any guarantees: the rationality is thus limited. More than the law itself, it is the procedure of interpreting it as integrity that matters to face the challenges of its indeterminacy, bringing, at least to that particular case, certainty, with all its hermeneutical fragility and fallibility, and rightness. This is the condition of a practice that takes rights seriously, for it deals with the dilemmas of adjudication by pensively considering the tension that occurs in the relationship between the counterfactual premise of integrity and the facts, and by providing answers that are in the expectancy of being the result of a coherent interpretation of all principles. Still, how can a judge apply this best theory to the practice of adjudication?

Dworkin, in this matter, sustains that the judge must thus personify the community of principles and, from this premise, interpret and apply the law. After all, law is, according to integrity, a reconstructive practice guided by principles. The judge, for this reason, must justify why a determinate norm, in comparison with others, is the one that provides the right answer. This justification, in turn, is made through an integral interpretation of the internal principles of the legal framework, including also moral and political principles placed in this framework, which, ultimately, associates adjudication with the community's legal and constitutional morality already manifested – but now interpretatively reconstructed – in the foregoing legal tradition. In the first stage, the judge must proceed to the test of fit, according to which she will refer to the institutional history of law and its practices as a means to select the best interpretation among the possible others in a way that preserves the consistency of the legal system. This test, nevertheless, especially when it does not provide an answer because of several eligible interpretations, must be complemented by the test of justification,[1335] according to which the judge "must choose between eligible interpretations by asking which shows the community's structure of institutions and decisions, its public standard as a whole, in a better light from the standpoint of political morality."[1336]

Dworkin also explains this twofold process in metaphorical way. By personifying the community of principles, the judge assimilates the regulative premise that orients the reconstruction of the past instances of legal interpretation, while inserting into legal reasoning the claim to coherence of the actual answer. This coherence is manifested in the idea of a "chain of law," as if the actual decision were like a result of an interpretation assimilated to the activity of writing a chain novel by

[1335]Evidently, this is merely an analytical structure that facilitates the comprehension of his stress on integrity. Dworkin remarks, in this matter, that "the distinction between the two dimensions is less crucial or profound than it might seem. It is a useful analytical device that helps us give structure to any interpreter's working theory or style" (Ibid., 231).

[1336]Ibid., 256.

different authors.[1337] Each author, in this case, must write her "chapter" as if it were in accordance with preceding ones and construct a coherent story therefrom; she is compelled both by the past and by the need to reveal the right answer to the new situation that preserves the integrity of the novel. There is, accordingly, coercion from the institutional history and the very form of law. This limits the judge's activity, without undermining her creativity manifested by her capacity to discover the law in a manner that is not political (at least in the sense of promoting polices, and not in the sense of political responsibility, for this she must have)[1338] but based on principles, since "judges are in a very different position from legislators."[1339] She is also constrained by the need to improve the quality of the novel; she must then "judge which of these eligible readings makes the work in progress best, all things considered,"[1340] and consider this reading when she is confronted with similar cases in the future. She has, accordingly, "responsibilities within the institution and to the institution."[1341] She must endeavor to provide the best legal interpretation possible, one that both coherently connects the individual decision to the foregoing legal tradition and improves this learning from the past by reconstructing it in a way that protects the integrity of the legal system. In this circumstance, the focus is on the principle of treating the other with "equal concern and respect," the basis of a community of principles and the legitimate source of law.

The personification of the community of principles, and hence integrity, is expressed by the fictional personage Dworkin calls Hercules, "an imaginary judge of superhuman intellectual power and patience who accepts law as integrity."[1342] Hercules not only has a comprehensive knowledge of all the valid legal principles and other possible justificatory arguments (and their connections with the legal framework) but also aims to make decisions that, while interpretatively reconstructing the institutional history, preserve and enforce the premises of fairness, justice and procedural due process in practice. Therefore, the judge Hercules, who embodies the regulative idea of integrity in adjudication, "reconciles the rationally reconstructed decisions of the past with the claim to rational acceptability in the present, it reconciles history with justice."[1343] He takes rights seriously by expressing the process of interpreting law as a way to deliver certainty and rightness

---

[1337] Ibid., 228–232.

[1338] According to Dworkin, "judicial decisions are political decisions, at least in the broad sense that attracts the doctrine of political responsibility;" that is, "an argument of principle can supply justification for a particular decision, under the doctrine of responsibility only if the principle cited can be shown to be consistent with earlier decisions not recanted, and with decisions that the institution is prepared to make in the hypothetical circumstances" (Dworkin, *Taking Rights Seriously,* 88).

[1339] Dworkin, *Law's Empire*, 244.

[1340] Ibid., 231.

[1341] Paul Gaffney, *Ronald Dworkin on Law as Integrity: Rights as Principles of Adjudication* (Lewiston, NY: The Edwin Mellen Press, 1996), 163.

[1342] Dworkin, *Law's Empire,* 239.

[1343] Habermas, *Between Facts and Norms*, 213.

to the case and also, so far as he works in the realm of arguments of principle, to make decisions that do not encroach upon legislature's responsibilities. He expresses the reconstructive character of adjudication by means of his posture of respecting and applying to the case the integrity in law, as if it were, in Günther's words, a "principle for appropriateness argumentation."[1344] This means that legal certainty only comes through a procedure able to provide the expectancy of treating like cases alike by means of a coherent and extensive reading of all the legal order and an impartial consideration of the features of the case. The "equal concern and respect," represented by the impartiality principle of treating like cases alike, becomes the immediate goal of adjudication, and the premise of "equal concern and respect" in a broader sense, as the equality and freedom of participation or a "forum of principles," becomes the mediate end of this practice. Hercules is the metaphor, at the end of this chain, of a discourse rationality that preserves the exercise of citizenship.

Needless to say, Hercules cannot be compared to any real judge, since the reality proves that judges are much more limited in their knowledge and sensitivity to grasp all the legal norms and history. Still, as a regulative idea, he operates as a counterfactual premise in the validity field of the argument in order to provide a criterion of critical review of adjudication. Moreover, it works as a promise of increasingly promoting, although acknowledging the risk of regression, the conditions through adjudication of the exercise of citizenship, reaching thereby the equal consideration and respect for all individuals. This is why Dworkin says that, notwithstanding that "no actual judge could compose anything approaching a full interpretation of all of his community's law at once,"[1345] we cannot deny "an actual judge can imitate Hercules in a limited way."[1346] The tension between the counterfactual Hercules and the reality of judge's limited knowledge, especially in the scenario of more complexities and dilemmas, can never be eliminated, to the extent that it is a requirement of this learning process, as though the past could be reflexively reconstructed as an opening towards the future. In other words, the personification of the community of principles in the figure of Hercules represents the admission of the fallibility of knowledge, of a limited rationality, inasmuch as it reveals that the right response to a reality cannot simply derive from the application of seemingly methodological schemas. Rather, it stems from a posture of continuously problematizing this very reality, of learning from this process of growing difficulties in the realm of decision-making, as a means to open it up to new dilemmas and complexities, always presupposing, though, a coherent interpretation and application of this knowledge. It is thus the perception that adjudication, in a complex and post-conventional society, is indeed the "specific functional aspect of the establishment, consolidation, development and reproduction of the legal

---

[1344]Günther, *Der Sinn für Angemessenheit: Anwendungsdiskurse in Moral und Recht*, 353, translation mine.

[1345]Dworkin, *Law's Empire*, 245.

[1346]Ibid., 245.

certainty as well as of the 'constitutional feeling' and feeling of justice – the only feeling that can adequately ensure the solidity of the legal order under the democratic rule of law."[1347]

Consistent with this premise, the answer in adjudication is right insofar as it embodies this tension and understands that the application of law must not only carry out an exhaustive understanding of the legal framework- and, in a broader sense, of the political morality of the community- but also grasp all the features of the case. By assuming Günther's terminology, it must complement the justificatory force of principles with the application conditions of an impartial consideration of the characteristics of a determined circumstance. It is right, because it stems from an integral analysis of the valid principles and possible interpretations, all the circumstances of the case considered. The rightness derives from a justified assertion, given the available knowledge, corresponding to the best interpretation possible, one that arises from the integrity criterion of treating like cases alike. This is why the right answer in the case of application of law is the "one right answer,"[1348] for it appears as a regulative idea, a counterfactual premise, manifested in a reconstructive argumentation founded upon principles, which consists in the concept of integrity. Every case must be interpreted in a coherent way with the valid principles, as if there were only one right answer to the case.

As a counterfactual premise, evidently this does not mean that judges cannot have different views when deciding a case or even that there cannot be, due to the complexity of the facts, different forms to interpret the rightness of a decision.[1349] The fact that the case admits one right answer does not lead to the conclusion that it must comply with everyone's viewpoint. Thinking differently would lose the tense relationship between the reciprocal-universal principle of treating like cases alike, on the one hand, and the reality and all its features, on the other, or, by using Günther's terminology, confuse the discourses of justification with the discourses of application. It is rather a posture of respecting the institutional history, the everyday experience, while, based on the features of the case and on the deontological force of legal rights, perfecting the law towards the future. In other words, it is, first, an attack on the discretionary power of judges, an attack on the authority, as though the indeterminacy of law could only be solved according to judge's conscience, and, second, an individual guarantee of being treated with equal right and respect, even when this protection upsets the interests of the majority. After all,

[1347]Menelick de Carvalho Netto, *Pragmatic Requirements of Legal Interpretation under the Paradigm of the Democratic Rule of Law* (Athens: Paper presented at the VII World Congress of the International Association of Constitutional Law, 11–15 June 2007): 19.

[1348]See Dworkin, *A Matter of Principle*, 119–145.

[1349]According to Dworkin, "(...) it is no part of this theory that any mechanical procedure exists for demonstrating what the rights of parties are in hard cases. On the contrary, the argument supposes that reasonable lawyers and judges will often disagree about legal rights, just as citizens and statesmen disagree about political rights. This chapter describes the questions that judges and lawyers must put to themselves, but it dos not guarantee that they will all give these questions the same answer" (Dworkin, *Taking Rights Seriously*, 81).

what finally matters is that the grounds of legal argumentation are taken seriously. The answer is right as a result of a rational procedure that safeguards not only the normative coherence of the legal system but also its appropriate application to the singular case. The belief in the right answer is, besides, the belief in the premise that the legal system can provide the answer, given that it is justifiably and coherently achieved through a rational procedure of argumentation that transforms the other into the primary focus of its activity.

6.3.4 *Jürgen Habermas's View: Between Facts and Norms Within Democratic Procedures of Opinion – and Will Formation*

6.3.4.1 Introduction

The final words of the last section might sound, for a respectful understanding of Ronald Dworkin's theory, an extensive interpretation of what, indeed, Dworkin sought to explain when he stressed the integrity in law and the community of principles founded upon "equal concern and respect" as the source of law. The idea of the *other* as the primary focus of adjudication, taken by a rational procedure of argumentation, might refer to a procedural understanding of democratic forms of participation that is more than Dworkin projected from his texts. Frank Michelman, for instance, for whom the judge Hercules, because of his loneliness and excessive heroism, lacks dialogue and thus misses the pluralist character of adjudication,[1350] sees in this procedural emphasis the most characteristic difference between Ronald Dworkin and Jürgen Habermas. For him, "on Dworkin's conception, 'democracy' points not to a procedure but to a state of affairs – points to government treating 'all members of the community, as individuals, with equal concern – and respect."[1351] By the same token, Klaus Günther, while identifying the intimate connection between both authors in the premise of 'interpretation and shaping' of a 'system of rights', remarks that, "in contrast to Dworkin, Habermas links interpretation to a *procedure* with certain qualifications."[1352] Habermas sees that Dworkin's concept of integrity in adjudication, even though marked by some characteristics of communicative dimension[1353] is, nevertheless, excessively centered on the figure of the

[1350] See Frank Michelman, "The Supreme Court 1985 Term - Foreword: Traces of Self-Government," *Harvard Law Review* 100 (1987): 76.

[1351] Frank Michelman, "Democracy and Positive Liberty," *Boston Review*, November 1996, http://bostonreview.net/BR21.5/michelman.html (accessed July 15, 2009).

[1352] Günther, "Legal Adjudication and Democracy," 46.

[1353] For instance, the idea of law as a means of social interaction, the practice of argumentation demanding of each participant to assume the angle of the other, the connection with paradigms and institutional history in order to reduce complexities in adjudication, and the stress on the conditions of the exercise of citizenship, such as the premise of "equal concern and respect," among others (See Habermas, *Between Facts and Norms,* 222–223).

judge Hercules, which raises "some initial doubts about the tenability of this *monological* approach."[1354]

Yet, what does Habermas have to suggest to overcome this monological approach he still observes in Dworkin's theory of law? What is this procedural emphasis that appears to be the primary focus to confront the metaphysical idealizations Habermas could indicate in the realm of law as integrity? Since Dworkin's premises are at the core of Habermas's proceduralist approach to adjudication – complemented, in any case, by Klaus Günther's distinction between discourses of justification and discourses of application – what can we point out as distinctive in this debate? These questions are not simple: they reach the structure of Habermas's theory of law and demonstrate how Habermas critically appropriated Dworkin's and Günther's premises to develop his proceduralist account of legal adjudication. In this matter, three relevant aspects must be stressed for this investigation: (1) how the communicative assumptions Habermas uses reaches the intervenient practices in the world; (2) how the communicative assumptions Habermas uses as a post-metaphysical response to a theory of law can be implemented in the relationship between constitutionalism and democracy; and, (3) how the communicative assumptions Habermas uses as a post-metaphysical response to a theory of law can be implemented in the relationship between justification and application of norms, and hence as a response to the legitimacy and the indeterminacy of law in the realm of adjudication. These three aspects are closely connected, and indeed the analysis of one naturally results in the other. Didactically differentiated, nevertheless, they will expose that his proceduralist approach is nothing other than the radicalization of the realm of validity and the conception of rationality. This is a radicalization that arises from a post-metaphysical purpose and ends in the stress on the other.

6.3.4.2 Communicative Action as an Intervenient Attitude in the World

Michel Rosenfeld sums up one of the main concerns of present theories of law: how in contemporary pluralist societies can we reconcile law with legitimacy without sacrificing either democracy or justice?[1355] The main quest here, accordingly, is for a theory that can, concurrently, overcome the residual arbitrariness of lawmaking while maintaining a neutral perspective towards communitarian conceptions of good.[1356] A viable response would be a proceduralist theory,[1357] so far as its chief

[1354]Ibid., 222.

[1355]Michel Rosenfeld, "Can Rights, Democracy, and Justice be Reconciled through Discourse Theory? Reflections on Habermas's Proceduralist Paradigm of Law," in *Habermas on Law and Democracy*, ed. Michel Rosenfeld and Andrew Arato (Berkeley, CA: University of California Press, 1998), 84.

[1356]Ibid., 83.

[1357]Michel Rosenfeld, in any case, is critical of Habermas's proceduralist approach and does not think that Habermas's theory can indeed overcome this dilemma. (See Ibid.)

purpose is to set forth a post-metaphysical thinking that could overcome legal theories that are not able to justify their main premises, as if they were *given* without a further review. In other words, a proceduralist approach would attempt to establish that even the most elementary argument we could use to justify our claims must necessarily be subject to critical review, in order to render possible the absence of any substantive final point no one could question. Instead of a final argument from which the legitimacy of law, its interpretation and application should derive, the accent is on the *procedure* whose grounds lie in an intersubjective practice of understanding. Besides, since language is the self-reflexive parameter of excellence, after all "no one individually disposes over an intersubjectively shared language"[1358] and "no single participant is capable of controlling the course and dynamics of the interpenetrating process of *mutual* understanding and *self*-understanding,"[1359] this intersubjective practice of understanding could overcome metaphysics, albeit never factually totally achievable, inasmuch as it inherently embodies the premise of any raised argument being vindicated and redeemed from the other's perspective.

In this respect, Habermas's pragmatic approach could be presented as a proceduralist response in the search for a post-metaphysical theory of law, for he clearly aims to provide: first, a thinking centered on practices of mutual understanding, as the source of law, and thus the communication as the ground for institutional legitimacy as long as it develops in accordance with procedures in which the other is inevitably considered; and, second, a thinking focused *exclusively* on these procedures, and not on any other substantiality that could guide the birth, development, interpretation and application of law. In the context of interpretation and application of law, the main characteristic of Habermas's approach resides in this stress on mutual understanding from which no one could point out a final fundament establishing how law ought to be interpreted and applied, but rather simply procedures preserving the conditions for a rational communication towards rights.

In order to preserve these procedures, from Habermas's view, we have necessarily to stand on some ideal presuppositions, which are inevitable expressions and conditions of the very communicative rationality, provided that, at the end, "no dispute about validity claim is beyond rational argumentation by the participants involved."[1360] Therefore, when it is to apply the law, rather than appealing to a last argument, in many cases founded on a certain communitarian conception of good, the interpreter must then attempt to apply the law by preserving and strengthening the conditions of rational communication in the empirical world. It is the procedure of rational communication that, carried out in a tense way between inevitable

[1358] Jürgen Habermas, "How to Respond the Ethical Question," in *The Derrida-Habermas Reader*, ed. Lasse Thomassen (Edinburgh: Edinburgh University Press, 2006), 122.

[1359] Ibid.

[1360] Richard Bernstein, "Introduction," in *Habermas and Modernity*, ed. Richard Bernstein (Cambridge, MA: The MIT Press, 1988), 19.

idealizations of discourse and the reality, promotes rightness in adjudication, and not the recourse to a substantive justification, even when it is presumably in accordance with the communitarian will.

In the basis of Habermas's thinking, there are some premises that expose how the communicative rationality comes to light. First, he establishes that no one could think of a theory of law "monologically," that is to say, if it is to sustain a post-metaphysical legal theory, especially on account of complex and post-conventional societies, it is indispensable to affirm the intersubjectivity, for communication relies on reciprocal discourse: "This rationality is inscribed in the linguistic telos of mutual understanding and forms an ensemble of conditions that both enable and limit."[1361] However, this mutual understanding only occurs because everyone who dialogues will inevitably depend on some presuppositions, or, otherwise, the communication would be unthinkable: "Whoever makes use of a natural language in order to come to an understanding with an addressee about something in the world is required to take a performative attitude and commit herself to certain presuppositions."[1362] In this matter, Habermas recalls the Kantian structure of two worlds (facts and norms) but sets up, in place of the solipsistic and metaphysical structure we could still observe in the Kantian transcendental (or constitutive) subject,[1363] the dialogue, which is the reflexive premise of any validity claim raised by the participants of the discourse. One particular source, Karl-Otto Apel's reflections on the communities of communication,[1364] also contributed to this quest for a dialogical approach to the structure of two worlds, particularly by his account of the tense relationship between the ideal community of communication, as the realm of justification or validation of any argument, and the real community of communication, as the spatiotemporally existing community.

From this viewpoint, if an argument is presented, the presupposition of the *other* of discourse already indicates that the argument demands more than this simple presentation; it must be rather subject to a rational reflection based on a dialogue as a means to be accepted. Moreover, if someone raises a validity claim in this dialogue, she already foresees that the other can contradict it. This is why the performative understanding of the Kantian separation between facts and norms (counterfactual presuppositions) has an operative importance: while some presuppositions are established as regulative ideas of communication, they also model a process of understanding and organizing the coordination of actions in practice.[1365] The establishment of communicative presuppositions in the realm of validity is, for

[1361]Habermas, *Between Facts and Norms*, 4.

[1362]Ibid., 4.

[1363]For an accurate analysis of this metaphysical assumption of Kant's transcendental subject, see Jürgen Habermas, *Kommunikatives Handeln und detranszendentalisierte Vernunft* (Stuttgart: Philipp Reclam, 2001); Miroslav Milovic, *Filosofia da Comunicação: Para uma Crítica da Modernidade* (Brasília: Plano, 2002), 49–120.

[1364]See Karl-Otto Apel, *Transformation der Philosophie: Das Apriori der Kommunikationsgemeinschaft*, Vol. 2 (Frankfurt a.M.: Suhrkamp, 1976).

[1365]Habermas, *Kommunikatives Handeln und detranszendentalisierte Vernunft*,11.

this reason, closely tied to the factual practice of understanding. This connection, nonetheless, although not bringing about a confusion between the two fields, is tense – a tension, in truth, that is indispensable for the very dynamic process of understanding.

These counterfactual presuppositions are, for Habermas, neutral: they have no substantive contents but simply indicate that communicative rationality is viable, without setting up any determination of how each participant should act.[1366] Since they, instead of embracing contents, sustain a reflexive basis for practices of cooperative discourse, they represent the key-aspect of a post-metaphysical thinking in Habermas's viewpoint. Indeed, these counterfactual presuppositions can be summarized in four major items, all of them operated in a performative way: first, the common presupposition of an independent world of all existing objects, that is, an objective world; second, the reciprocal presupposition of rationality and liability (*Zurechnungfähigkeit*) among individuals; third, the unconditionality of validity claims beyond contexts; fourth, the validity, required for argumentation, able to consolidate a reason that leaves aside a purifying perspective and situates itself in the world ("argumentative presuppositions full of claims the participants hold in order to decentralize their interpretive perspectives").[1367] These presuppositions, on the one hand, set up the existence of a validity field not confusable with the real world; on the other, they specify the conditions of rational communication leading to an intervention in the real world. They demonstrate that every validity claim calls for some idealizations that, albeit beyond the context, only exist in virtue of the context. No one could introduce a validity claim seeking an understanding without assuming, beforehand, that the other "participants pursue their illocutionary goals without reservation, that they tie their agreement to the intersubjective recognition of criticizable validity claims, and that they are ready to take on the obligations resulting from consensus and relevant for further interaction."[1368] Evidently, these idealizations represent a perfect realm of communication not identifiable with any reality, but, paradoxically, they are indispensable for sustaining the communication as the source of validity of an argument in any reality.

That being the case, although these presuppositions can sound very abstract, they are only assumed in virtue of concrete activities towards dialogue and, from a distinct perspective, do not obviously mean that this ample and unrestricted discursive environment occurs in practice. Habermas discernibly knows that, in the real world, communication is usually subject to ideological and strategic practices, manipulative interests, and is thus never perfect as the ideal presuppositions indicate. Indeed, it is by reason of this imperfect environment of communication that the separation between these two fields, facts and norms, becomes necessary, as a means to demonstrate that the justification of any argument must be made so

[1366]Habermas, *Between Facts and Norms*, 4.

[1367]Habermas, *Kommunikatives Handeln und detranszendentalisierte Vernunft*, 12–13, translation mine.

[1368]Habermas, *Between Facts and Norms*, 4.

intensively as if its reasons were continuously projected into an unrestricted debate. Briefly, it points out how possible it is to articulate, in reality, the conditions of a reciprocal understanding where reason unveils itself. Furthermore, by projecting a field beyond the context, the justification can operate in a level that is not restrained by the context of application. In other words, the normative validity is connected but not limited to the facts, and hence the critical review of an argument must take place even when the facts – for example, an argument that is in accordance with a communitarian prevailing conception of good – seem to agree with it, for the facts themselves can be ideological, that is, operate against the practices of mutual understanding. The ideal presuppositions, in this case, function then as motor against ideological practices, because they bring an impulse towards the enhancement of communication in reality. By showing how a perfect communication would be, they indicate how the imperfections of the real world could be overcome, although never totally, by means of *procedures* of reciprocal understanding.

The rationality of the deliberation arising from a validity claim is thus referred to procedures through which, at the end, the argument is accepted in virtue of being publicly accepted for its good reasons, and not because of a manipulative or strategic interest coercively limiting the communicative action. The speakers and the listeners are, therefore, free "only in virtue of being subject to the *binding power* of the reasons that they offer to one another, and take from one another."[1369] This is the basis for constructing a cognitive theory lying in the ideal of achieving the best argument, one that results from the reflexive continuation of actions oriented towards reaching mutual understanding, excluding thereby all other motives except the cooperative search for the truth.[1370] The best argument is the one that can be justified in an ideal procedure in which all the participants are involved in the deliberation of its contents without being compelled to this end. Its validation, besides, can only be attained so far as all the possible participants involved in this rational deliberation can achieve consensus by taking into account all the side effects and consequences of its general observance.[1371] The validation of the argument, for this reason, is closely connected to the idea of impartiality or, in Habermas' words, to the discourse principle reflecting the "symmetrical relations of recognition built into communicatively structured forms of life in general."[1372]

The quest for the best argument, for this reason, becomes a regulative idea we must anticipate as a means to promote, in reality, actions towards mutual understanding. For this purpose, we must transform any validity claim into an argument able to be criticized, vindicated and redeemed because of its own quality, and not in virtue of any "*external sources of validity,* since the sphere of validity

---

[1369]Habermas, "How to Respond to the Ethical Question," 123.

[1370]See Jürgen Habermas, *The Theory of Communicative Action: Reason and the Rationalization of Society*, Vol. I (Boston: Beacon Press, 1985), 25.

[1371]See Habermas, *Between Facts and Norms,* 108–109.

[1372]Ibid., 109.

is – conceptually is – identical with the sphere of human speech."[1373] The justification of any validity claim takes place solely by means of argumentation.[1374] This is the condition of a post-metaphysical thinking, because other types of justification, as a strategic one regarding the efficacy of results,[1375] might rely on premises that are not subject to critical review. The quest for the best argument results, from this premise, in an action towards intersubjective agreement in practice, for, as shown, validity only makes sense in reference to an interaction within the practical world; it is, in truth, dependent on situational contexts:[1376] "About this formal world assumption [the objective world where validity operates], the communication about something in the world converges on a practical intervention in the world."[1377] The tension between facts and norms leads thus to an intervenient action in the world to transform it through communication, as an emancipation from the prevailing ideological[1378] structures of society.

Hence, Habermas's approach can be regarded as a rational reconstructive project with an intervenient attitude in the real world. First, it develops according to a neutral dimension of the conditions of rational communication. After all, insofar as they are conceived as *a priori* transcendental presuppositions only to be "detranscendentalized" in reality, as to make possible the insertion of the socialized subjects into the life world as well as the convergence between the speech and the action,[1379] these presuppositions are regulative ideas in need of being mediated with the empirical world.[1380] In this case, we could call them justificatory premises of weak "transcendentalization", either because they do not point or embrace any content, as we could think of some superior moral norms above the discourse itself,[1381] or because, for Habermas, "norms cannot be, at the same time, normative-transcendental conditions of the possibility and validity of argumentative discourses *and* morally substantive basic principles of a normative moral discourse."[1382] Moreover, these *a priori*

[1373]Albrecht Wellmer, "Reason, Utopia, and the Dialectic of Enlightenment," in *Habermas and Modernity*, ed. Richard J Bernstein (Cambridge, MA: The MIT Press, 1988), 53.

[1374]Ibid., 54.

[1375]See Habermas, *The Theory of Communicative Action*. Vol. I., 285ff.

[1376]See Ibid., 279.

[1377]Habermas, *Kommunikatives Handeln und detranszendentralisierte Vernunft*, 17, translation mine.

[1378]It is ideological because of its capacity to hinder actions towards mutual understanding.

[1379]See Habermas, *Kommunikatives Handeln und detranszendentralisierte Vernunft*, 16.

[1380]See Milovic, *Filosofia da Comunicação: Para uma Crítica da Modernidade*, 280.

[1381]See, for example, Karl-Otto Apel, for whom it is not possible to sustain a morally neutral discourse, as Habermas defends, for moral and legal principles stem from the presupposition of moral contents based on the equality of rights of all participants involved in the discourse. For him, the morality in legal discourses is unavoidable; otherwise, the ethical imperative of discursive co-responsibility for the consequences of all would be undermined. See, for this purpose, Karl-Otto Apel, *Auseinandersetzungen in Erprobung des transzendentalpragmatischen Ansatzes* (Frankfurt a.M.: Suhrkamp, 1998).

[1382]Marcel Niquet, *Moralität und Befolgunsgültigeit: Prolegomena zu einer realistischen Diskurstheorie der Moral* (Würzburg: Königshausen & Neumann, 2002), 82, translation mine.

transcendental conditions do not determine patterns of how to act in certain actions, for the communicative rationality is "not a subjective capacity that would tell actors what they *ought* to do."[1383] What ultimately matters are the procedures whose norms behind – those presupposed conditions – are simply assumed as a requirement for the validation of the argument, as "normative principles reflexively founded."[1384] What matters is simply the unavoidability of the discursive form, whose expansion must be always targeted. In Habermas's words, "a set of unavoidable idealizations forms the counterfactual basis of an actual practice of reaching understanding, a practice that can critically turn against its own results and thus *transcend* itself."[1385] They are a set of norms that show how Habermas's focus is on how to expand the *procedures* of rational communication, as a request for continuously extending, in practice, the possibilities of reflexive deliberation on all validity claims. As Milovic remarks, "it is necessary to seek its intersubjective basis in these real conditions, which could possibly lead us from the world of strategies to the world of mutual recognition and respect."[1386]

6.3.4.3 Communicative Action in the Relationship Between Constitutionalism and Democracy

Under these premises, we can understand how Habermas's proceduralist account could reconcile democracy with constitutionalism, more particularly, how it could deal with the residual arbitrariness of lawmaking while maintaining a neutral perspective towards communitarian conceptions of good. On the one hand, it is clear how his thinking assumes the communicative transcendental presuppositions in a weak and neutral way. There is not, consequently, in the context of modern law, any previous submission to any kind of morality or to any conception of good. The *a priori* conditions of communication achieve a universalist character, simply because, in their absence, there would be no communication, at least the one oriented towards mutual understanding. Besides, since rational communication is, in his view, the condition for social transformation, the denial of them would mean the very impossibility of mutual mobilization towards any communicatively rational reconstruction of social structures and institutions. There is not, in the realm of law, any submission of an argument to a determinate social *ethos,* a certain conception of good of a particular group, for normative validity does not confuse with the facts. Accordingly, notwithstanding that the tradition, the history, the past are essential to the very development of legal rights, they appear in the condition of arguments and, as such, able to be subject to critical scrutiny and reconstruction. The transcendental conditions of communication, therefore, serve as an operative

[1383] Habermas, *Between Facts and Norms*, 4.

[1384] Milovic, *Filosofia da Comunicação,* 230, translation mine.

[1385] Habermas, *Between Facts and Norms*, 4.

[1386] Milovic, *Filosofia da Comunicação*, 251, translation mine.

reflexive basis to evaluate whether a determinate *ethos* really holds a communicative character or whether it operates grounded in a metaphysical assumption leading to actions oriented to success and, as such, to practices not rooted in the purpose of expanding the communicative rationality.

In addition, the non-submission of law to morality and to an *ethos* demonstrates that, for Habermas, the areas where modern law operates are much more complex – they are plural, fragmented, multifaceted – and hence it is no longer possible to observe a fusion of facts and norms, especially because every consensus on the most distinguished functional spheres (economic, political, and so on) is fragile, spatiotemporally limited, and normally coordinated by strategic actions. Therefore, law becomes a necessary instrument of social integration by means of communicative action and, above all, an instrument, if it is to be post-metaphysically conceived, whose grounds rely solely on *procedures* oriented towards mutual understanding. True, each one of those types of discourse – legal, moral or ethical, for instance – represents constitutive forms of integration and, as such, can be under an extensive critical scrutiny of a process of societal rationalization. Yet, positive law has an integrative attribute the other forms of discourse cannot achieve by themselves: it brings forward a form of integration that, instead of relying on the perspective of the participants of the discourse and on their cognitively indeterminate and motivationally unreliable results,[1387] operates according to independent institutional standards responsible for stabilizing behavioral expectations by fixing orientations followed by sanctions.[1388] Briefly, it releases individuals from moral and ethical obligations, for now rather than stemming from procedures of opinion- and will-formation, norms arise from "collectively decisions of authorities who make and apply the law."[1389]

Yet, law does not legitimately isolate itself from the other types of integration: there is no legitimate law, albeit its autonomy, without the observance of basic moral principles: "In virtue of the legitimacy components of legal validity, positive law has a reference to morality inscribed within it,"[1390] and this can be verified in the fact that law and morality share contents and refer to common social problems.[1391] Moreover, law cannot evidently be conceived dissociated from its historical developments and its ethical background. In brief, the comprehension of law in contemporary societies, while involving a neutral, autonomous, but complementary perspective towards morality and communitarian conceptions of good, must, on the other hand, establish on the *procedures* of mutual understanding the basis for supplying this lack of substantive justification law now encompasses. Hence, procedures – and not any substantiality, for they are always projected into

[1387]Ibid., 257.

[1388]See Habermas, *Between Facts and Norms*, 37.

[1389]See Jürgen Habermas, *The Inclusion of the Other: Studies in Political Theory* (Cambridge, MA: The MIT Press, 1998), 256.

[1390]Ibid., 106.

[1391]Ibid.

those procedures – give reason to the law, which, as a consequence, derives from forms of social participation where each individual can act. Law, for this reason, integrates in a legitimate way so far as it relies on procedures of intersubjective participation and not on any authority or argument whose justifications cannot be submitted to dialogical scrutiny. This is how the idea of still existing arbitrariness in lawmaking can be attacked by a neutral approach to communitarian conceptions of good and morality: the arbitrariness, as an ideological or strategic action, must be suppressed by means of procedures of mutual understanding that give rise – and are likewise internal to – the law.

The integration law brought about develops by focusing on procedures of mutual understanding rooted in the tense structure of facts and norms. In the realm of complex societies, this tense structure can be expressed by the continuous demand for conciliating, without detracting one from the other,[1392] the public and private autonomy. It is based on this complementary and tense conciliation that Habermas constructs the chief elements of his legal theory. Law must protect the autonomy of all equally, even to guarantee its own procedural character, but also must "prove its legitimacy under this aspect of securing freedom."[1393] On the one hand, law must provide a "stable social environment in which persons can form their own identities as members of different traditions and can strategically pursue their own interests as individuals,"[1394] but, on the other, "laws must issue from a discursive process that makes them rationally acceptable for persons oriented toward reaching an understanding on the basis of validity claims."[1395]

Private autonomy, therefore, must be in an internal reciprocal relationship with public autonomy: whereas individuals must have their freedom and equality guaranteed, there must be rights of political participation as a means to deliver to all individuals the condition of rational authors of norms. This is where Habermas links the integrative character of law to the process of democratic legitimation: the communicative rationality, whose processes and actions must react against the submission of society to forms of strategic domination, gains in strength insofar as it connects to the 'form of law', which, in turn, demands this interaction between private and public autonomy, between individual freedom and the reciprocal construction and enforcement of law. This relationship results in a set of abstract rights individuals must beforehand recognize as a means to legitimately regulate their practices through the law, such as the right to public participation in the procedures of law formation. In this case, the 'form of law' would shape the rational discourse.[1396] On the other hand, the rational discourse shapes the legal form

---

1392 See Ibid., 257.

1393 Ibid.

1394 William Regh, "Translator's Introduction," in *Between Facts and Norms* (Cambridge), xix.

1395 Ibid., xix.

1396 See Günther, "Legal Adjudication and Democracy: Some Remarks on Dworkin and Habermas," 46.

inasmuch as it guarantees equal rights to private autonomy, to the legal enforcement of rights and to equal membership in a legal community.[1397] Both mutually set up the conditions for integrating society while inserting each one of its members into institutional processes of legal deliberation and formation, and hence attaching communicative rationality, now expressed in the legal form, to democratic procedures of discourse.

The residual arbitrariness of lawmaking is thus combated by means of the expansion of democratic procedures of discourse in law, which is the institutional synonym for the attack of communicative rationality on ideological forms of social integration. Accordingly, the discourse principle, which is the abstract and neutral support for communicative rationality and "already presupposes that practical questions can be judged impartially and decided rationally,"[1398] gains now the form and quality brought by the legal form and, as such, the designation of democratic principle. As Habermas remarks, "the principle of democracy results from a corresponding specification [the norms can be justified if and *only* if equal consideration is given to the interests of all those who are possibly involved] for those action norms that appear in the legal form."[1399] With the democratic principle, norms are justified, and thus valid, as a result of institutionalized procedures that preserve the right of equal participation of each person in a "process of legislation whose communicative presuppositions are guaranteed to begin with."[1400] It is based on this that legal norms can be legitimately enacted. Besides, it is through the legal form of the principle of democracy that communicative rationality can, more efficiently, enhance rational practices and actions towards mutual understanding in complex societies. By intensifying the practice of democracy, not only is every citizen qualified as a rational author of legal norms, but also there is the institutionalization of the conditions for widening and deepening the social communication through an interaction between private and public autonomy:

> (...) The discourse principle is intended to assume the shape of democracy only by way of legal institutionalization. The principle of democracy is what then confers legitimating force on the legislative process. The key idea is that the principle of democracy derives from the interpenetration of the discourse principle and the legal form. I understand this interpenetration as a *logical genesis of rights,* which one can reconstruct in a stepwise fashion. One begins by applying the discourse principle to the general right to liberties – a right constitutive for the legal form as such – and ends by legally institutionalizing the conditions for a discursive exercise of political autonomy. By means of this political autonomy, the private autonomy that was at first abstractly posited can retroactively assume an elaborated legal shape. Hence the principle of democracy can only appear as the heart of a *system* of rights. The logical genesis of these rights comprises a circular process in which the legal code, or legal form, and the mechanism for producing legitimate law – hence the democratic principle – are *co-originally* constituted.[1401]

[1397]Ibid.

[1398]Habermas, *Between Facts and Norms*, 109.

[1399]Ibid., 108.

[1400]Ibid., 110.

[1401]Ibid., 121–122.

Rational discourse, now qualified by the legal form, is what shapes – and gives movement to – the system of rights through procedures in which the other is necessarily presupposed. Under this premise, there is no legitimate law and institution without being submitted to the scrutiny of the citizens who, freely and rationally, accept them under ideal conditions of public dialogue. We cannot think of legitimacy of law and institutions without rational communication. Still – and this is the primary issue to understand how Habermas links rational dialogue to constitutional democracy – we cannot think of rational dialogue, now democratic principle, without a system of individual rights that guarantees to each individual the free and equal space of participation in public debates, and vice-versa. The principle of democracy and the system of individual rights are mutually complementary and co-originally connected. In other words, democracy is co-original with constitutionalism, which is a reflex of the co-originality thesis regarding the public and private autonomy.[1402] One implies necessarily the other, for, while individuals exercise their public autonomy and are regarded, in this quality, as the sovereigns of law, they can only do so if freedom and equality are preserved through the medium of law. The popular sovereignty that corresponds to democracy as it develops according to rational discourse can only be exercised if individual's freedom and equality are guaranteed, but, on the other hand, these individual rights can be democratically safeguarded only if they stem from a practice of mutual understanding by way of institutional procedures of opinion- and will-formation.

On the one hand, "citizens can make an *appropriate* use of their public autonomy, as guaranteed by political rights, only if they are sufficiently independent in virtue of an equally protected private autonomy in their life conduct,"[1403] but, on the other, "members of society actually enjoy their equal private autonomy to an equal extent – that is, equally distributed individual liberties have 'equal value' for them – only if as citizens they make an appropriate use of their political autonomy."[1404] Communicative rationality inscribed in the legal form of the principle of democracy is therefore in mutual dependency on constitutionalism, for, whereas democracy is grounded in the popular sovereignty and in those procedures guaranteeing public autonomy, constitutionalism, by preserving the individual autonomy (for instance, by introducing mechanisms for minority participation), protects democracy, and vice-versa: "Just as autonomy is not mere freedom, so popular sovereignty is not mere majoritarianism."[1405] Democracy and constitutionalism, for

---

[1402]See Jürgen Habermas, "Constitutional Democracy: a Paradoxical Union of Contradictory Principles," *Political Theory* 29, no. 6 (December 2001): 767.

[1403]Ibid.

[1404]Ibid.

[1405]Bonig Honnig, "Dead Rights, Live Futures: On Habermas's Attempt to Reconcile Constitutionalism and Democracy," in *The Derrida-Habermas Reader*, ed. Lasse Thomassen (Edinburgh: Edinburgh University Press, 2006), 162.

this reason, are mutually implied and presupposed.[1406] The system of rights – and, consequently, all institutions bearing procedures of democratic participation –, while being a form of social integration, is only in accordance with the premises of communicative action inasmuch as it depends on practices of rational dialogue.

Habermas, with this insight into the circularity of this relationship,[1407] brings his previous analysis of the tension between facts and norms to the core of constitutional democracy and builds a powerful background to evaluate how democratically legitimate an institutional practice is. The question is how an institution or a determinate legal practice can reconcile, although knowing their insurmountable differences, private with public autonomy, constitutionalism with democracy. In a more abstract dimension, this would mean how the tension between facts and norms could be best arranged to enhance communicative actions within the lifeworld. Communicative actions within this institutional background would have to deal both with the need to be under the conditions of ideal acceptability and with the demand for being, in reality, accepted by the participants. In more practical words, this would imply that democratic actions have to handle both the need to be legitimately justified and the demand for preserving the conditions of democratic participation and free and equal will formation, which correspond to the very system of rights. Besides, insofar as communicative action is a dialogical process, for Habermas wants to expand it into different social practices – after all the weak transcendental pragmatic conditions must be "detranscedentalized" in reality – the principle of democracy should be also progressively extended to institutional procedures.

Still, how could this circularity provide, in the long run, a broadening and deepening of rational communication? Again, the tension between facts and norms, now expressed by the circularity of constitutionalism and democracy, is what responds to this dialogical process and practices of rational dialogue. However, as a dialogical process, this circularity must be connected to a diachronic perception of historical acquisitions that fortified the relationship between both. For this purpose, Habermas sustains the idea of a self-correcting learning process, in which this circularity can lead to more rational communication over the years, to a better relationship between constitutionalism and democracy, because every constitutional democracy has inherently a "performative meaning [that] remains the implicit but stable point of reference."[1408] This performative meaning implies the very history of the system of rights, which, in Habermas's account, makes the citizens "heirs to a founding generation, carrying on with the common project,"[1409]

[1406]See Lasse Thomassen, "'A Bizarre, Even Opaque Practice': Habermas on Constitutionalism and Democracy," in *The Derrida-Habermas Reader*, ed. Lasse Thomassen (Edinburgh: Edinburgh University Press, 2006), 179.

[1407]It is interesting to verify how this understanding connects somehow to Derrida's debate on the circularity between constitutionalism and democracy. See the last chapter.

[1408]Habermas, "Constitutional Democracy: a Paradoxical Union of Contradictory Principles," 776.

[1409]Jürgen Habermas, "On Law and Disagreement: Some Comments on 'Interpretative Pluralism,'" *Ratio Juris* 16, no. 2 (June 2003): 193.

while simultaneously foreseeing their possible active and critical position when confronted with this heritage: "Any citizen can put herself at any time in the shoes of a framer and check whether, and to what extent, the established practices and regulations of democratic deliberation and decision-making meet at present the required conditions for legitimacy-conferring procedures."[1410]

The practices within – and because of – the system of rights provide, for this reason, a "normative perspective from which later generations can critically appropriate the constitutional mission and its history."[1411] Every new procedure and, particularly, every new law and legal decision have to be deemed, in the background of communicative rationality, a fallible continuation of the founding constitutional event to be understood in the long run as a self-correcting learning process.[1412] In addition, if the validation of an argument is always referred to continuous tensions, with the facts throughout history – and, as such, as a necessary regulative idea that leaves the discourse open – the legitimacy of any institutional practice is also marked by this continuous and open, even though "not immune to contingent interruptions and historical regressions,"[1413] practice of rational discourse. Legitimacy, for this reason, is connected to a pedagogical process that, progressively, although not immune to endemic disagreement,[1414] achieves some "stable points of reference," which are, however, always subject to review and critical scrutiny.

For this reason, we can say that constitutional democracy is, on the one hand, historically connected to a past and present that brought and brings about some stability both in the social integration and in the mechanisms for exercising private and public autonomy, while attempting, in the long run, to promote more and more procedures of rational dialogue within the legal institutions, as if there were a promise, albeit never realizable, to reconcile constitutionalism with democracy. The tension between facts and norms is now transported to the tense and co-original character of constitutional democracy and demonstrates that, for Habermas, what matters is not to be bound to a certain reality, for it can be ideological and strategically oriented, but rather how this reality – and thus the institutional practices under the law – is critically gathered and reinterpreted by each new generation as a means to deliver both a consistent and legitimate system of rights. There is, accordingly, a prospective conception – similar to the discussion of the tension between the weak and neutral transcendental conditions of communication and the reality leading to new contexts of growing rational dialogue – of the relationship between constitutionalism and democracy that will give rise to new

[1410]Habermas, "Constitutional Democracy: a Paradoxical Union of Contradictory Principles," 776.

[1411]Habermas, "On Law and Disagreement: Some Comments on 'Interpretative Pluralism,'" 193.

[1412]Habermas, "Constitutional Democracy: a Paradoxical Union of Contradictory Principles," 774.

[1413]Ibid.

[1414]See Habermas, "On Law and Disagreement: Some Comments on 'Interpretative Pluralism,'" 189.

contexts where public and private autonomy can, more and more, reconcile with each other.

Consistent with these premises, the different institutional procedures are examined according to how they manage this relationship between constitutionalism and democracy. The process of lawmaking, even though dealing in a certain unlimited way with different reasons (ethical, pragmatic, normative, etc), is constrained by "democratic procedures designed for the justification of norms."[1415] This preserves the connection with the sovereignty of people, which in its turn works in circularity with constitutionalism. As Habermas remarks, "this principle [the principle of popular sovereignty] forms the hinge between the system of rights and the construction of a constitutional democracy."[1416] It follows that lawmaking is, ultimately, dependent on rational discourse shaped by the "form of law": it must be a reflex of the sovereignty of people, but, for that, it must respect the "generative grammar"[1417] the system of rights is for any type of institutionalized rational discourse. No political authority, for this reason, despite being able to sustain the most different reasons to justify a law, is exempt from following rules and principles the citizens gave themselves in the form of democratic procedures of will-formation. Legitimate legislation cannot undermine constitutionalism (and the acquisitions it achieved throughout history in a self-correcting learning process) and cannot proceed in the opposite direction of democracy: every new law must be justified in the premises of institutionalized rational discourse and, for that, must preserve the exercise of freedom and equality of all citizens.

As any other institutional democratic practice, moreover, lawmaking must point out the promise of conciliation between constitutionalism and democracy, as a regulative idea and as a quest for the correct solution. Otherwise, the deliberative character inscribed in the political will-formation can degenerate: "Political disputes would forfeit their deliberative character and degenerate into purely strategic struggles for power if participants do not assume – to be sure, fallibilistically, in the awareness that we can always err – that controversial political and legal problems have a correct solution."[1418] If, on the one hand, we cannot deny that, in reality, lawmaking is marked by continuous disagreements and disputes, on the other, we cannot refuse, if it is to sustain its legitimate character, the idea that consensus must be achieved – as an idealistic goal – by means of rational dialogue. It is simply the application of the previous analysis of the tension between facts and norms pointing out a prospective standpoint, although never entirely reached, of a rational consensus, but now channeled into legally binding institutional procedures of deliberation.

[1415]Habermas, *Between Facts and Norms,* 192.

[1416]Ibid., 169.

[1417]See Günther, "Legal Adjudication and Democracy: Some Remarks on Dworkin and Habermas," 47.

[1418]Jürgen Habermas, "Reply to Symposium Participants, Benjamin N. Cardozo School of Law," in *Habermas on Law and Democracy*, ed. Michel Rosenfeld and Andre Arato (Berkeley, CA: University of California Press, 1998), 396.

In any case, communicative rationality expressed by the democratic principle, which, in turn, is mutually dependent on constitutionalism and increasingly verified through mechanisms of self-correcting learning process, has also its correspondence in other forms of institutional communication. Legal adjudication is likewise carried out through ongoing procedures of shaping and interpreting the abstract system of rights.[1419] Nevertheless, unlike lawmaking, it has a very limited range of reasons for it must observe the "*principle of binding the judiciary to existing law.*"[1420] Consequently, the tension between facts and norms gains now new contours, notwithstanding that it can be interpreted as a specification of those complementary but also tense relationships between private and public autonomy, between constitutionalism and democracy. This can be verified by means of the relationship between the need for internal consistency of the system of rights, on the one hand, and for external rational justification of every decision, on the other. That is to say, legal decisions must correspond to the expectancy of observance of precedents and the existing legal statutes, and, at the same time, satisfy the presupposition of being communicatively rationally grounded. They are legitimately made in virtue of being justified in the light of rules and principles accepted by the participants of democratic procedures of opinion- and will-formation. Moreover, they are so, because they are oriented towards the right answer, which can be symbolized by the ideal of conciliation of law and justice (which is the specification to the case of the circularity and co-original tension between constitutionalism and democracy). Consistent with this viewpoint, legal adjudication is part of an institutional chain that begins with constitution-making, pass through lawmaking, and arrives at the need to apply the law to a particular circumstance, each one of them expressing those tensions, which, in a more abstract dimension, refers to the relationship between facts and norms. In any of these processes, rational dialogue is required as a means of legitimation, whereas the abstract system of rights is assumed as an "*idealized internal reference point* to the members of a society who conceive of themselves as authors and addressees of equal rights."[1421] The rational discourse (as the source of legitimacy) and the system of rights (as the prerequisite for certainty) are, due to their mutual dependency, the two premises legal adjudication, as any other institutional process in this chain, must rely on.

6.3.4.4 Communicative Action in the Relationship Between Justification and Application

Properly conceived, legal adjudication, in Habermas's account, legitimates itself through procedures of rational dialogue. In the realm of discourses of application,

---

[1419]See Günther, "Legal Adjudication and Democracy: Some Remarks on Dworkin and Habermas," 47.

[1420]Habermas, *Between Facts and Norms*, 172.

[1421]Ibid., 47.

this rational dialogue appears marked by the tension between safeguarding certainty or the expectation that the principles and legal statutes will be respected and enforced (facts) and the legitimacy of decision (normative validity). As part of a set of institutionalized procedures – constitution-making, lawmaking, and decision-making – that will ultimately characterize what Habermas calls "principle of guaranteeing legal protection for each individual,"[1422] legal adjudication already assumes the tension between constitutionalism and democracy inscribed in the discourses of justification taken place in legislation. It is hence bound to valid norms, whose discourses of justification, even to be qualified as legitimate, rely on the practice of rational dialogue. The judge is thus constrained by impartially valid legal norms, on which she must focus to avoid crossing the "'red line' that marks the division of powers between courts and legislation."[1423] Indeed, for Habermas, it is clear that the rationality of adjudication relies first on the legitimacy of legal statutes,[1424] i.e. it already expects that rational dialogue was seriously developed by the time of lawmaking: "The legal discourse of the judge should be confined to the set of reasons that legislators either in fact put forward or at least could have mobilized for the parliamentary justification of that norm."[1425] For this reason, she must not reopen the discourse of justification, for the reasons once therein used "play a different role when the courts, with an eye to the coherence of the legal system as a whole, employ them in a discourse of application aimed at decisions consistent over time."[1426]

In what follows, the tension between facts and norms in the realm of legal adjudication has a very particular feature that is embedded in this relationship between discourses of justification and discourses of application. Habermas knows that, in this specific issue, his premises of communicative rationality are seriously challenged, for valid norms raise, when applied to certain cases, the problem of legal indeterminacy. In these situations, the tension between guaranteeing the consistency of the system of rights and the legitimacy of decision-making that must orient decision-making can be jeopardized. This occurs either by decisions that are based on a certain substantiality not submitted to critical scrutiny (and, as such, metaphysically grounded) or decisions that reopen the discourses of justification, and thus jeopardize the institutional chain (constitution-making, law-making and decision-making) that corresponds to the observance of the principle of separation of powers. As a response to this problem, which is central to many legal theories, the quest for both consistency of law and legitimacy must assume, in Habermas's standpoint, the prospective search for the "single right answer," one that would idealistically conciliate both tense premises of discourses of application. The "single right answer," in turn, is accepted as the result of a procedure founded

[1422]Ibid., 172.

[1423]Ibid., 447.

[1424]Ibid., 238.

[1425]Jürgen Habermas, "A Short Reply," *Ratio Juris* 12, no. 4 (December 1999): 447.

[1426]Habermas, *Between Facts and Norms*, 192.

on rational discourse, because, for Habermas – and this is how he sustains the procedural account of his legal theory –, the problem of indeterminacy of law in legal adjudication can only be solved inasmuch as rational discourse is embedded in the validity of the decision, and thus no metaphysical point of reference – instead, only the procedure leading to the "single right answer" – founds the decision. Habermas, in sum, aims to construct a post-metaphysical theory of law whose basis lies in the procedural exercise of rational discourse.

By assuming this proceduralist account of legal adjudication, Habermas saw in Ronald Dworkin's theory of law a relevant and convincing proposal for treating the problem of consistency of law and external justification (legitimacy) in the quest for the "single right answer" in decision-making. With his premises, rationality and indeterminacy of law could find a solution rooted in a reconstructive perspective of legal principles, and thus internally originated from the system of rights. Dworkin's theory, nonetheless, only needed to be adapted for a proceduralist account: "Dworkin's proposals for a theory-guided, constructive interpretation of law could be defended on a proceduralist reading that transposes the idealizing demands on a theory formation into the idealizing content of the necessary pragmatic presuppositions of legal discourse."[1427] In this matter, Klaus Günther's separation between discourses of justification and discourses of application and his more explicit connection with a theory of rational procedural argumentation could assist in this purpose. From both theories, Habermas carefully and discernibly selected what he considered suitable for a procedural understanding of law, while rejecting, in his viewpoint, their potential regressive aspects.

From Dworkin, he upheld the values of a theory that aims to link legal coherency to legal legitimacy through the premise that "duties can only be 'trumped' by duties and rights by rights,"[1428] and hence as a response to the indeterminacy of law through the application of the best theory possible. He also seems to have incorporated the comprehension of coherence as a work-in-progress rather than something presupposed in the very nature of principles.[1429] On the other hand, he criticized the solipsistic and metaphysical character of Hercules: "The critique of Dworkin's solipsistic theory of law must begin at the *same* level and, in the shape of a *theory of legal argumentation,* ground the procedural principles that henceforth bear the brunt of the ideal demands previously directed at Hercules."[1430] In this matter, he remarked that Dworkin's approach, while underlining the exercise of citizenship as the source of legitimate law and judicial activity, is still dependent on the premise of judge's privileged position, whose knowledge and virtue qualify him to represent and secure the integrity of the legal community.[1431]

[1427]Ibid., 238.

[1428]Habermas, *The Inclusion of the Other: Studies in Political Theory*, 28.

[1429]See John P. McCormick, "Habermas' Discourse Theory of Law: Bridging Anglo-American and Continental Legal Traditions?," *The Modern Law Review* 60, no. 5 (September 1997): 740.

[1430]Habermas, *Between Facts and Norms*, 225.

[1431]Ibid., 222.

These premises appeared to be in conflict with a procedural interpretation of law,[1432] even though we could point out some of its crucial signs in Dworkin's premises of a solidary community rooted in reflexive forms of communicative action ("equal right to liberty as founded in the right to equal communicative freedom")[1433] and in his concern for legal paradigms reducing complexities of the practice of adjudication[1434]. Moreover, Habermas also attempted to relieve the burden of some Dworkin's idealizations, as the "normative self-understanding of constitutional orders"[1435] in the concept of "forum of principles," shortly examined, since they give the impression of tying up the validity field with a metaphysical assumption. From Günther, in turn, he absorbed the distinction between discourses of justification and discourses of application, as an adequate answer to the indeterminacy of law, and to construct a normative concept of coherence, while adding some more precision to Dworkin's attempt to differentiate legal validity from validity of decision-making. Despite that, he rejected Günther's thesis that the rationality of legal norms stems directly from moral norms,[1436] as if it were not possible to abdicate from the practical reason.[1437]

On that account, Habermas sustains two relevant steps for constructing a procedural theory of adjudication from both theories. First, the idealizations we could observe in Dworkin's Hercules and in his solipsistic character must be translated into "ideal demands on a *cooperative procedure of theory formation*,"[1438] which corresponds, in other words, to the tension between the regulative ideals of validity claims leading to the right answer, on the one hand, and the fallibility of adjudication[1439] derived from the very fallibility of knowledge, on the other. The regulative ideas we could remark in Dworkin's Hercules are now transferred to the pragmatic conditions

---

[1432]According to Habermas, the judge's knowledge, rooted especially in some professionally proven standards and procedural principles, is subject to a rational reconstruction, as any valid claim. In his words, "the mere fact that a hardly homogenous professional class legitimates itself is not sufficient to demonstrate the validity (*gültig*) of the very procedural principles that ground validity within the system (*geltungsbegründenden*). Procedural principles that secure the validity of the outcome of a procedurally fair decision-making practice require internal justification" (Ibid., 225).

[1433]Ibid., 223.

[1434]Ibid.

[1435]Ibid., 216.

[1436]Unlike Günther - for whom the specificity of legal norms is revealed in the transition from justification to application discourses, when the justificatory reciprocal-universal principle (U) referring to the acceptability of the argument is complemented by the impartial consideration of all the features of the circumstance – Habermas, as formerly investigated, understands that legal norms do not rely on the morality, but, instead, both have a co-original and complementary mutual relationship (Ibid., 104–118), particularly because, in legal discourses, what is primarily behind is the principle of democracy, which, instead of being grounded in a universal-reciprocal basis, "is tailored to legal norms" (Ibid., 111).

[1437]See Günther, *Der Sinn für Angemessenheit: Anwendungsdiskurse in Moral und Recht*, 9.

[1438]Ibid., 226.

[1439]Ibid.

of discourse, as though the premise of "equal concern and respect" could be expanded, in reality, by means of communicative action. This is how Habermas defends that, even in judge's conscience referring to discourses of application, it is possible to verify an intersubjective dimension in the validity field, insofar as the judge ultimately relies on the pragmatic presuppositions of communication when she has to interpret and apply the law. This characteristic can be observed in the connection between the parties' perspectives and the pragmatic conditions of communication that stand behind valid norms in discourses of justification.[1440] This is where we can see that interpretations of legal statutes and principles characterized by the quest for coherence in legal adjudication are necessarily connected to communicative action, "whose socio-ontological constitution allows the perspectives of the participants and the perspectives of uninvolved members of the community (represented by an impartial judge) to be transformed into one another."[1441]

Besides, the judge's supernatural powers must be replaced by anchoring those normative ideals to an "open society of interprets of the constitution."[1442] Rather than focusing thus on the judge's supernatural powers, we must emphasize the deliberative procedures founded on pragmatic presuppositions, and thus transpose Dworkin's idealizations into "the idealizing content of the necessary pragmatic presuppositions of legal discourse."[1443] In this matter, Günther's thesis based on the distinction between justification and application discourses serves as a relevant support to bring to light a proceduralist perspective to Dworkin's idea of a "forum of principles." The solution of the case, for this reason, instead of relying on the "intellectual resources of an idealized judge,"[1444] must then derive from "pragmatic constraints that distinguish the discourse of application from the discourse of justification."[1445]

[1440]Ibid., 229.

[1441]Ibid.

[1442]Habermas takes this term from Peter Häberle. For this purpose, see Peter Häberle, *Verfassung als öffentlicher Prozeß: Materialen zu einer Verfassungstheorie der offenen Gesellschaft* (Berlin: Duncker & Humblot, 1978).

[1443]Habermas, *Between Facts and Norms*, 238.

[1444]In any case, we cannot deny that, in Dworkin's stress on the "forum of principles," there are already some relevant signs of a rational discourse. After all, this forum is based on the idea of a compromise among individuals towards the institutional construction of their principles, which could be understood by way of procedures of rational dialogue in which each individual is considered a participant with equal rights and not submitted to coercive practices restricting her freedom of expression. The premise of "equal concern and respect" somehow embodies, after all, this thinking, especially in virtue of the presupposition that every participant involved in this discourse will take seriously every validity claim therein raised. It demonstrates that integrity stems from the premise of a compromise among citizens through common principles – and thus as a "forum of principles" - as their own constitutive character, and hence as participants whose freedom and equality are safeguarded.

[1445]Jacques Lenoble, "Law and Undecidability: Toward a New Vision of the Proceduralization of Law," in *Habermas on Law and Democracy*, ed. Michel Rosenfeld and Andrew Arato (Berkeley, CA: University of California Press, 1998), 64.

In this respect, Habermas seeks to relieve the burden Dworkin creates for this compromise through principles among individuals as the basis of integrity. He remarks that Dworkin's view sustains that the judge must justify her decision by reflecting on the obligation, originated from the act of framing the constitution, to retain the "integrity of a life in common,"[1446] a compromise that seems excessively idealized or expresses "a false idealization."[1447] He understands that "constitutional practice could thus deceive itself in a manner fraught with consequences, burdening institutions with wholly unsolvable tasks."[1448] The idea of a "forum of principles," therefore, ought to be released from a certain "normative self-understanding of constitutional orders"[1449] in this compromise among individuals Habermas still sees in Dworkin's legal theory. Indeed, as shown, for Habermas, the idealizations – where we should place the "forum of principles" – occur by the fact of the inevitability of the discursive form grounded in pragmatic presuppositions of argumentation of normative content. In the realm of normative validity, we could not set up an idealization bearing a strong idea of compromise among individuals, but rather only the pragmatic conditions of argumentation no one could deny. This is, after all, what corresponds to Habermas's thesis of a morally neutral discourse and the perception that his idealizations are weak, for they only exist to be "detranscendentalized" in the practice of lifeworld. Consequently, Habermas's emphasis on the inevitable presuppositions of rational discourse demonstrates that even this premise – the judge has an obligation to interpret law as if all citizens were committed to maintaining integrity of a life in common according to a "self-understanding of constitutional order" – is subject to rational reconstruction. Rather than this emphasis on a "normative self-understanding of constitutional order," Habermas is concerned with the procedure and the inevitable presuppositions of rational discourse. This is why the regulative idea of a "forum of principles" must be "immediately tailored to the rationality problem facing the adjudication process."[1450]

This "forum of principles" could be then translated into Habermas's concept of community of communication, in its idealized form, where the conditions of rational communication and the validity of any claim take place. More specifically, it could be translated into Günther's separation between discourses of justification and discourses of application. However, unlike Günther, this would happen not through the insertion of legal discourses into a special case of moral discourses – since the discourse is morally neutral – but rather as a specific forum regulated by the principle of democracy where the norms are valid because all the possibly

[1446] Habermas, *Between Facts and Norms*, 216.

[1447] Ibid.

[1448] Ibid.

[1449] Ibid.

[1450] Ibid.

affected persons could agree as participants in rational discourses through the democratic procedure of legislation: "The system of rights and constitutional principles are certainly indebted to practical reason, but they are due in the first instance to the special shape this reason assumes in the principle of democracy."[1451] Habermas clearly builds his account of a counterfactual "forum of principles" as the domain of validity, which has to be "detranscendentalized" in effective practices of rational dialogue. Under these conditions, the discourses of justification, in which the process of legislation takes place, must assume this counterfactual "forum of principles" as the source of law, and hence as the condition of validation of any law. The "forum of principles" would correspond, de facto, to the very dependency of lawmaking on the sovereignty of people, but, since this forum is regulated by principles, particularly freedom and equality, it also means the dependency of law on the system of rights and, more particularly, on the constitution. Besides, the "forum of principles" would also indicate a prospective idea of rational communication in the ongoing search for conciliation between democracy and constitutionalism, which are the two primary debates lawmaking must pursue.

Similarly, the "forum of principles" would mean the realm of validity of discourses of application, as its condition of legitimacy, which, in reality, would be consistent with the fact that every legal decision must rely on principles enacted by means of democratic procedures of lawmaking. This connects it to the practice of rational communication taken place in discourses of justification, while giving reasons to apply and interpret valid norms according to an impartial consideration of all features of the case. In this situation, we can indicate the practice of rational dialogue inasmuch as the decision must be, on the one hand, just to the parties involved (and, for that, express the expectancy of treating like cases alike) and consistent with the law (and thus with the existing rational discourse), on the other. Moreover, the "forum of principles" would also imply the prospective idea of rational communication in the ongoing process of seeking the "single right answer," one that reconciles consistency of law with legitimacy.

In harmony with these premises, in discourses of justification, the "forum of principles," while idealized, is factually considered by way of democratic procedures of legislation and the observance of the form of law represented by the constitution. Discourses of application, in turn, are connected to the "forum of principles" by means of discourses of justification, since they already assume the valid legal principles therein enacted. For the same reason, this connection is reached through the interpretation and application of these valid legal principles in a way that they correspond to the parties' expectancy of having their case treated like cases alike (justice) based on an impartial consideration of all the features of the case. As a consequence, discourses of application reveal their practice of communicative rationality in the search for the appropriateness – and no longer

[1451] Ibid., 206.

the validity – of legal principles to the case. Both are, accordingly, linked to the "forum of principles," as the source of legitimacy, but in different ways. Whereas the first relates to the validity of general norms, the second is concerned with the appropriateness of some *selected* norms to the particular case. This is why "the validity of general norm does not yet guarantee justice in the individual case,"[1452] i.e. the validity of a legal norm does not correspond immediately to its appropriateness to a particular reality. The validity – and hence the legitimacy – of discourses of justification is directly tied up with rational procedures of legislation; the validity of discourses of application, in turn, is directly tied up with rational procedures oriented towards pursuing the "single right answer" by stressing the appropriateness of some selected norms to the case. The "forum of principles," under these circumstances, has the counterfactual character of a "supreme tribunal of all rights and claims,"[1453] as the regulative idea of communication in the realm of validity, which must be continuously "detranscendentalized" in the practice of rational discussion taken place in legislation (justification) and legal adjudication (application). This provides a legitimate response to the indeterminacy of law in the realm of separation of powers. For that, the tension between facts (these both procedures) and normative validity (the "forum of principles") resounds, within the context of constitutional democracy, in the factual tension between discourses of justification and discourses of application, as two distinguishable but complementary forms of communicative action.

Furthermore, Habermas stresses the deontological character of rights, either through Günther's requirement that the application of norm demands a coherent interpretation of all valid *prima facie* applicable principles[1454] or Dworkin's emphasis on integrity, also demanding an extensive and exhaustive interpretation of all valid applicable norms.[1455] Both authors remark the need for legitimate procedures translating the vast contents of values, principles, interests of social life into the code of rights, and, from this stance, the construction of a coherent interpretation of law resulting in the single right answer. This is why Habermas

[1452]Ibid., 217.

[1453]Habermas, *Kommunikatives Handeln und detranzendentalisiert Vernunft,* 14, translation mine.

[1454]As shown, Günther presupposes that the practice of adjudication calls for the observance of legal valid positive norms, enacted through institutional procedures ensuring the exercise of citizenship, in order to impartially apply the right one in conformity with the features of the case. In this realm, whereas discourses of justification link to the practice of lawmaking, discourses of application refer to the consideration of all the characteristic signs of a particular situation, and hence identify with adjudication. Any confusion between these two discourses would undermine the primary deontological force of law and the firewall between parliament and adjudication. See Sect. 6.3.2. *supra.*

[1455]As shown, Dworkin, by emphasizing the premise of "equal concern and respect," establishes the counterfactual requirement of treating like cases alike according to an integral comprehension of the legal framework. Consequently, deontology here is protected inasmuch as the application of law is made in accordance with a reconstructive procedure of back-and forward interpretations in order to achieve the best interpretation possible, one that will not put in jeopardy the individual's expectancy of being treated with justice by all legal institutions. See Sect. 6.3.3. *supra.*

mentions that, "as deontological, rights contain a moment of inviolability that refers to a rational dimension and encourages the search for 'single right' answers based on principles."[1456] Besides, Habermas explicitly defines, in this advocacy of deontology, the idea of rights as trumps against policies that so intensively Dworkin developed, which could directly function as a critique of theories that sustain coherence through balancing founded upon the concept of principles as optimization requirements,[1457] as we can observe in Robert Alexy's approach.

In summary, both Dworkin's and Günther's approaches help construct an intersubjective comprehension of the realm of validity, one of Habermas's primary focuses to overcome the metaphysics in adjudication. For Habermas, a procedural rationality "locates the properties constitutive of a decision's validity not only in the logicosemantic dimension of constructing arguments and connecting statements but also in the pragmatic dimension of the justification process itself."[1458] The premises of rational discourse are already set up in the domain of legal validity and serve as an intersubjective counterfactual regulative idea that prevails over any possible solipsistic understanding in the realm of normative justification. The achievement of the "single right answer" that both Günther and Dworkin stress, according to the Habermasian model, is directly tied up with the conditions of rational communication, and therefore with the acceptability of the good reasons presented in a procedure of justification that is "*carried out* with arguments."[1459] The "single right answer" is, consequently, not simply the factual conclusion of examining different arguments and defining which one must prevail, but also the conclusion of a procedure whose arguments can be justified insofar as they preserve the conditions of rational communication and can be subject to rational assessments through principles. Furthermore, to the extent that the discourses of application, similarly to Günther and Dworkin's viewpoints, are constrained by the legal form and are part of an institutional chain regulated by rational discourse, the "single right answer" must be the outcome of an operation within the framework of legal valid norms in a way that reconciles legitimacy with coherence. These both requisites lead, in turn, to the pragmatic conditions of rational discourse. In order for an answer to be right, the argument must be valid, and validity demands necessarily dialogue.

The quest for the "right answer" is also a quest for bringing out reasons that, without disregarding the risk of regression, favor a construction of a coherent system whose basis is immersed in a nonstop critique. The decision, therefore, is right because it is intimately connected to a coherent system carried out according to valid reasons: "Under favorable conditions, we bring argumentation to a de facto conclusion only when the reasons solidify against the horizon of unproblematic

---

[1456] Habermas, *Between Facts and Norms.*

[1457] Ibid., 209.

[1458] Ibid., 226.

[1459] Ibid.

background assumptions into such a coherent whole that an uncoerced agreement on the acceptability of the disputed validity claim emerges."[1460] The coherence of the system is thus grounded in this purpose of expanding, in reality, the conditions of rational agreement in a way that, continuously, each decision could be justified as if it were the result of these agreements, even though never entirely factually reached. It is for this reason a dynamic coherence, inherently submitted to critique, for the tense relationship between facts and norms, application and justification of norms will always remain. In any case, there must be the possibility of learning from this process, as if the knowledge, albeit its fallibility, were anchored to an ongoing debate that preserves the presuppositions of communicative rationality, and hence the opening to the consideration of all arguments. At the same time, a reconstructive selection of the institutional history, whose criterion is the *acceptability* of the argument, is carried out by focusing on the achievements and by correcting and learning from the mistakes of the past.

There is, after all, a reconstruction of the rational past decisions with the claim to rational acceptability in the present.[1461] Every argument is thus submitted to a rational reconstruction: "These pragmatic process conditions ideally ensure that all the relevant reasons and information available for a given issue at a particular time are in no way suppressed, that is, that they can develop their inherent force for rational motivation."[1462] Accordingly, it is the process of rational discourse that provides coherence and leads to the "single right answer": idealistically, since the pragmatic presuppositions of rational discourse serve as a regulative idea encompassing a non-coercive debate on reasons, as a cooperative quest for the truth,[1463] and the consideration of all the possible reasons for a determinate issue; factually, because the "right answer" is achieved by an ongoing procedure that attempts to implement these normative presuppositions in reality, and hence project them into an interminable critique by means of a rational reconstruction. Both dimensions – normative validity, idealistically considered, and facts – correspond to a reasoning that, notwithstanding its tense character, aims to attain the right answer communicatively. This is why, for Habermas, coherence cannot be employed in conformity with purely semantic characterizations, but it instead refers always to pragmatic presuppositions of the argumentative process.[1464] By the same token, this is why the rightness of a decision is nothing other than the satisfaction of the "communicative conditions of argumentation that make impartial judgment possible."[1465]

[1460]Ibid., 227.

[1461]Ibid., 213.

[1462]Ibid., 227.

[1463]See Habermas, *The Theory of Communicative Action*. Vol. I., 25.

[1464]Habermas, *Between Facts and Norms*, 229.

[1465]Ibid., 230.

The decision must be, due to its rightness and claim to coherence, a result of a discursive procedure in the realm of courtroom proceedings.[1466]

If, on the one hand, this thinking could lead to even more indeterminacy, especially because it inherently embraces the "ripple effect argument"[1467] – i.e. every new decision demands a rational reconstruction of the past decisions, and hence creates a ripple in the coherent system of rights – on the other, it indicates that communicative action critically reconstructs history and learns from this ongoing process of decision-making. This shapes distinct "paradigmatic legal [understandings]"[1468] that serve as reducers of complexities of the very practice of decision-making. Decision-making and paradigmatic legal understandings mutually shape each other according to communicative action, as a condition of a post-metaphysical thinking in the realm of legal adjudication and as a reflex of Habermas's account that every legal decision and every legal paradigm must be carried out *procedurally*. What matters – a characteristic always present in Habermas's approach – is that the indeterminacy of law is ultimately challenged by the fact that the conditions for reaching understanding are continuously reinforced in society, as to render possible an ongoing process of revalidation of the validity of the legal system.

This is why, according to the proceduralist perspective, law is an opening towards the future, as if every new interpretation, simultaneously and paradoxically, closed – for it provides some stability – and opened the system of rights – because, in the quest for the "single right answer," that interpretation is continuously challenged by even newer interpretations and revalidation processes. Law is thus only the tip of the iceberg, which brings more and more complexities, opening and even risks of regression in this self-correcting learning process. Legal adjudication must deal with this opening also by knowing that, by practicing and enhancing rational communication in reality, the indeterminacy of law will be temporally "resolved" only as a *determining-in-process*. If legal adjudication must deal with the need for an immediate response to a particular case, it only does so post-metaphysically by knowing that the decision, while solving the case, brings even more complexities to the system of rights: the facts, after all, did and will not achieve the normative validity of discourse. There is rather a simple reshaping of the tension between facts and norms.

[1466]Habermas makes a description of the implementation of pragmatic presuppositions in the practice of courtroom proceedings at the end of the fifth chapter of his *Between Facts and Norms*, where he, by indicating the different phases of the procedure, sustains finally that the "court must decide each case in a way that preserves the coherence of the legal order as whole," words similar to Dworkin's. Similarly, he says that "the institutionalized self-reflection of law promotes individual legal protection from two points of view, that of achieving justice in the individual case and that of consistency in the application and further development of law," both in need of developing the conditions of rational discourse. "Procedural law does not regulate normative-legal discourse as such but secures, in the temporal, social, and substantive dimensions, the institutional framework that *clears the way for* processes of communication governed by the logic of application discourses." See Ibid., 235–237.

[1467]Ibid., 219.

[1468]Ibid., 220.

6.4 The Metaphysics of Balancing from the Perspective of the Proceduralist Account

6.4.1 Introduction

From Habermas's proceduralist account, Ronald Dworkin's theory of law as integrity, and Klaus Günther's distinction between discourses of justification and discourses of application, it is possible to outline a consistent response to Robert Alexy's argument that coherence necessarily implies balancing with that broad meaning stemming from his defense of the "unity of practical reason"[1469] in the realm of legal adjudication. If based on any of those previous theories, the response to the indeterminacy of law, within the context of constitutional democracy, cannot simply rely on the deployment of balancing, whereby legal and any type of practical discourse are "combined at all levels and applied jointly"[1470] according to some criteria inspired by mathematical and economic models.[1471] There is a clear differentiation between discourses of justification and discourses of application, between arguments of policy and arguments of principles, which transforms any attempt to unify practical arguments – as though legal arguments could be controlled by, and recreated in, the broad realm of practical discourses through balancing – into a serious affront to the principle of separation of powers and, consequently, to constitutional democracy. Besides, insofar as the claim to a post-metaphysical proceduralist account of law is assumed, the loss of this firewall between both justification and application discourses could result in the loss of the exercise of communicative rationality and, accordingly, in a metaphysical legal reasoning.

The investigation of those previous theories, for these reasons and also because they brought a distinct response to the indeterminacy of law, gave rise to doubt, first, whether balancing is necessarily a condition for providing coherence and correctness in the realm of legal adjudication (there are other ways to resolve this problem, after all), and, second, whether balancing can be regarded as a legitimate procedure in the application of law. More specifically, those theories can increase the skepticism about the so-called rationality of balancing Robert Alexy defends as a response to the indeterminacy of law. Indeed, they could even demonstrate how Alexy's account of a procedural discourse theory – which, in this quality, should promote rational argumentation – might lead to the monological exercise of authority – in this case, the authority of the judge who must apply the law. For facts and norms are confounded with each other by way of balancing – after all, with Alexy's integrative thinking, there is practically no need for a distinction between justification and application of norms[1472] – the judge can ultimately rely solely on her own

[1469] Alexy, "The Special Case Thesis," 383.

[1470] Ibid., 380.

[1471] Alexy, "Jürgen Habermas's Theory of Legal Discourse," 229.

[1472] Ibid., 231.

capacity to integrate the different practical reasons by establishing preference relations among them. Fundamentally, the firewall between lawmaking and decision-making that Günther, Dworkin and Habermas so fiercely attempted to preserve in their theories for the problem of indeterminacy of law in post-conventional societies becomes a mere abstract reference.

The problem that arises from this investigation is doubtless connected to a possible metaphysical and ultimately solipsistic character of Alexy's concept of rationality, which causes the direct outcome of a criticizable perception of legitimacy in his theory of constitutional rights. By accepting the arguments of the proceduralist approach to legal reasoning, it is feasible to conclude that, despite Alexy's defense of a discourse rationality in the deployment of balancing, his rationality is anchored to the belief in some implicit material criteria[1473] that are not subject to critical scrutiny in the validity field. In order to understand more deeply this perception, which later led to his conclusion that "every application discourse includes a discourse of justification,"[1474] it is necessary, however, to verify that the primary fundamentals of his thinking – principles are optimization requirements, the nature of principles necessarily leads to balancing, and balancing is indispensable to achieve rationality, coherence and correctness in legal reasoning – can be the sign of a metaphysical comprehension of discourse rationality, which, paradoxically, Alexy himself defends.

Habermas, in the quest for a post-metaphysical thinking also in the legal domain, drew special attention to this problem. Indeed, this issue is so crucial in his opinion that he even says that Alexy's dissertation was responsible for encouraging him to "extend discourse theory, which was originally developed for morality, to law and the constitutional state."[1475] He certainly saw there a substantial material to demonstrate how an implicit metaphysics remains in Alexy's theory, and, from that, he could visualize that another response had to be provided to the problem of indeterminacy of law in constitutional democracies in order to avoid its serious outcomes in reality, a task he managed to do by dialoguing with Ronald Dworkin and Klaus

---

[1473]According to Klaus Günther:

"The criterion, according to which we orient ourselves when weighting norms in collision may not, on its part, have a predetermined material content which gives priority to certain normative viewpoints over others. Alexy's conception of principles as optimization requirements already drew our attention to the danger that can arise when, for example, a model of values is projected onto a theory of norm structure. The decision on an appropriate norm is thus reduced to a decision for a relative better circumstance, which is also the optimum in the singular situation. The problem thereby indicated consists in the danger of already introducing, when determining the structure of argumentation, the material criteria, which should themselves be the subject matter of an appropriateness argumentation. A procedural concept of appropriateness, or a procedural application of norms, would, however, refrain from using such implicit material criteria. If appropriateness is to consist in the consideration of all features of a situation, then the *method* of considering may not, for its part, be determined by material criteria" ("Günther, *Der Sinn für Angemenssenheit: Anwendungsdiskurse in Moral und Recht,*" 301, translation mine).

[1474]Alexy, "Jürgen Habermas's Theory of Legal Discourse," 231.

[1475]Habermas, "Reply to Symposium Participants, Benjamin N. Cardozo School of Law," 428.

Günther's legal theories. There are some well-defined critical points Habermas elucidates in Alexy's legal theory for this perception, some of which we discussed in the previous chapter but now are examined through another philosophical language. They descend, in essence, albeit his attempt to distinguish deontology from axiology,[1476] from Alexy's confusion between principles and values in his structural theory.[1477] We could point out, as we did,[1478] this characteristic as stemming from a direct influence of the German value-based approach to objective principles embracing the totality of the legal order in constitutional adjudication, as the BVG's practice reveals. According to Habermas, Alexy's defense of the rationality of balancing by means of the Weight Formula relies on an axiological comprehension of principles, for balancing, in the way described, is necessarily deployed according to preference relations that are inherent to values, and its outcomes are teleological-oriented decisions: "Because no value can claim to have an inherently unconditional priority over other values, this weighting operation transforms the interpretation of established law into the business of *realizing values* by giving them concrete shape in relation to specific cases."[1479] Alexy himself is explicit in this correspondence between principles and values: "The problem of the preference relations between principles corresponds to the problem of the values hierarchy."[1480]

In Habermas's opinion, by assuming an axiological structure, nonetheless, legal norms lose their universal character of regulating a matter in the equal interest of all and "enter into a configuration with other values to comprise a symbolic order

[1476]See Alexy, *Theorie der Grundrechte*, 133.

[1477]As formerly investigated, Alexy's claims to correction and coherence are strictly dependent on axiological assumptions, which, as a part of his conception of the "unity of practical reason" (Alexy, "Jürgen Habermas's Theory of Legal Discourse," 233), transform the law, except for its institutional character, into a practical argument as any other, with similar strength to be measured with preference relations. The institutional quality of law becomes then a minor detail, provided that, at the end, law is not only supplemented or permeated but can also be "controlled by general practical arguments" (Ibid., 232). This conclusion is evident when we observe that Alexy adds to the binary deontological code of law – command/prohibit – the axiological concepts of necessity, impossibility and possibility (Robert Alexy, "Postscript," in *A Theory of Constitutional Rights* (Oxford: Oxford University Press, 2002), 393), as though they were not really different (Ibid., 393–394), transforming hence the constitutional framework into a "problem of the existence of a sphere of the constitutionally merely possible" (Ibid., 394). Accordingly, the application of law has to be justified based on moral, ethical and political goals, even when they constrain an individual right, for what really matters is whether the constraint is suitable, necessary, and proportional in the narrower sense (Alexy, "Jürgen Habermas's Theory of Legal Discourse," 230). For Alexy, the problem is not whether deontology or teleology prevails, but, instead, whether we can achieve preference relations in a broader sense of practical reason (Ibid., 230), which is an obvious axiological operation.

[1478]See the second chapter.

[1479]Habermas, *Between Facts and Norms*, 254.

[1480]Robert Alexy, "Sistema Jurídico, Princípios Jurídicos y Razón Práctica," *Doxa* 5 (1998): 145, translation mine.

expressing the identity and form of life of a particular community."[1481] Indeed, the definition of preference relations is directly connected to the need to measure the value in conformity with how a group of individuals qualifies it in comparison with possible others. When this premise is transferred to the structure of principles in collision through hierarchic relations, the consequence is that they lose their binding and deontological character and gain, instead, the quality of a gradual enforceability, which empties out their legal normative structure. After all, legal norms cannot rely on "more or less" observance of their contents, given that they have a "binarily coded obligation character of behavioral expectations"[1482] that will correspond to one's compliance or not with the law, and not a gradual duty as though each context could originate a relative obligation to the legal prescription. If we subvert this normative character of legal norms, then law loses its enforceable character and its priority over axiological points of view. It is not the values and social interests that are then translated into, and shaped by, the system of rights, but rather it is the system of rights that is translated into, and shaped by, the values and social interests. Inasmuch as this translation or shaping undermines the priority of the system of rights over axiological viewpoints and, anyhow, cannot be justified but by customary standards, a serious problem of rationality arises.[1483]

The axiological premises of Alexy's thinking explain why he sets up, as the primary basis of adjudication, not the *prima facie* valid norms, but the *prima facie* preference relations – after all, "law is ambiguous and cannot be strictly kept anyway"[1484] – which are defined by the very practice of adjudication. In other words, it is adjudication reinforcing its own discourse based on the idea of a "unity of practical reason,"[1485] and not adjudication relying on an integral interpretation of the system of rights, which links it to legislation and, ultimately, to the sovereignty of people. Legal decisions, based on interpretations of the abstract and broad content of practical reason and on preference relations – also defined by adjudication – are what shape an ongoing practice of decision-making. It is axiological interpretations reproducing axiological interpretations, while transforming

[1481]Habermas, *Between Facts and Norms*, 256.

[1482]Habermas, "Reply to Symposium Participants, Benjamin N. Cardozo School of Law," 429.

[1483]As Habermas argues:

"As soon as rights are transformed into goods or values in any individual case, each must compete with the others at the same level of priority. Every value is inherently just as particular as every other, whereas norms owe their validity to a universalization test (...) True, not every right will win out over every collective good in the justifications of concrete decisions. But a right will not prevail only when the priority of a collective good over a corresponding norm can itself be justified in the light of higher norms or principles. Because norms and principles, in virtue of their deontological character, can claim to be *universally binding* and not just *specially preferred*, they possess a greater justificatory force than values. Values must be brought into a transitive order with other values case to case. Because there are no rational standards for this, weighting takes place either arbitrarily or unreflectively, according to customary standards and hierarchies" (Habermas, *Between Facts and Norms*, 259).

[1484]Alexy, "Jürgen Habermas's Theory of Legal Discourse," 230.

[1485]Ibid., 233.

democratic institutional enacted laws into one argument among many others. By jeopardizing this institutional link with sovereignty of people, we can verify that the space for monological and ideological exercise of authority through adjudication is, therefore, open: "By thus assimilating ought-statements to evaluations, one opens the way to legitimating broad discretionary powers."[1486] Properly speaking, we can conclude that discourse rationality in law, necessarily rooted in the democratic principle, as Habermas point out,[1487] might be lacking in Alexy's legal theory.

This is why Alexy's claim to coherence has nothing to do with Habermas's claim to coherence. The coherent decision, in Alexy's view, is connected to a teleological satisfaction of general interests through the definition of preference relations rather than a response to a validity claim that attempts to reconcile legitimacy with coherence, now understood in conformity with a proceduralist comprehension of the legal system: "The court's judgment is then *itself* a value judgment that more or less adequately reflects a form of life articulating itself in the framework of a concrete order of values."[1488] Accordingly, the claim to the "single right answer" inscribed in the structure of the proceduralist account of legal adjudication turns into the claim to what is good for a determined community, for axiological assessments, by means of preference relations, refer to teleological comparative analyses of attractive goods of a certain *ethos*. A legal decision, in this case, instead of enhancing the pragmatic conditions of communication in a tense relationship between facts and norms, as though, more and more, those universal pragmatic rational presuppositions were present in the lifeworld, reinforces the facts shaped by a certain conception of good. The right answer is the one that, through balancing, endorses an ethical background, not the one that ensures and strengthens communicative actions within the practices of social life. After all, the best for an *ethos* is not necessarily the right answer if rights were taken seriously: "As soon as we reduce the principle of legal equality to merely one good among others, individual rights can be sacrificed at times to collective goals."[1489]

When we analyze this rationality problem more accurately, we can conclude that it essentially stems from Alexy's fusion of facts and norms, which results in the immediate conclusion that the discourse rationality Alexy defends as the justification for his structural theory can represent a strategic rationality, one that orients action towards what is good for *us* and not what is right for *all*. Moreover, insofar as this strategic rationality may be the sign of an ideology – the *ethos*, for instance, can indicate practices oriented towards exclusion of certain individuals from exercising their communicative abilities[1490] – this fusion can imply the very impossibility of the practice of rational discourse. Naturally, this conclusion does not eliminate the fact that conceptions of *good*, particular self-understandings, an *ethos* are held by

1486 Habermas, "Reply to Symposium Participants, Benjamin N. Cardozo School of Law," 430.

1487 Habermas, *Between Facts and Norms*, 108.

1488 Habermas, "Reply to Symposium Participants, Benjamin N. Cardozo School of Law," 430.

1489 Ibid., 429.

1490 See, for instance, the *Crucifix* and *Ellwanger* cases examined in the first chapter.

adjudication, but they are so from a practice of mutual agreement that validates them as an argument. It seems, for this reason, that, despite Alexy's analysis of Habermas's communicative rationality[1491] and his defense that "balancing is as rational as discourse,"[1492] he forgot the risks of constraining the field of critical review to the context of a certain reality. The validity field, as far as constrained by the facts, can become the repetition of a certain conception of truth, deviating then from its role of regulative idea directed towards ensuring and enhancing communicative actions in reality.

The implicit metaphysical standpoint lies in the loss of the tension between facts and norms in the structure of Alexy's legal theory, for the core of his axiological concept of principles might jeopardize the intersubjective character of validity and place, instead, a binding to an *ethos* that can be ideological and lead to monologue. Four relevant outcomes can be thus visualized: (1) the construction of an axiological content that is inscribed in the legal principles, as if it were part of their structure, which implies an axiological-teleological procedure of application, whose basis, nevertheless, remains unjustified; (2) legal adjudication, so far as the decision is right by reason of its compliance with what is good for the community, is practically not distinct from that of legislation, except on account of its reference to the case, undermining then the inherent complexities of discourses of justification and discourses of application; (3) legal adjudication loses its role of protecting individuals from society, for now it is the very reality (which can be ideological and oppress an individual's exercise of her communicative skills) that validates the legal argument, and not a counterfactual validity field presumed in tension with the facts as a means to open up the space for more and more exercise by *all* individuals of their communicative potentialities; and (4) the claim to the "single right answer" becomes a nonsense, owing to the fact that the validity field is confounded with the facts and, thus, conditioned to the relativism of a definition of what is *good* for the *ethos* in the interpretation of a singular case, which, in turn, culminates in the idea that every case has multiple possible answers.

6.4.2 The First Outcome: The Construction of an Axiological Content in the Structure of Principles

The first outcome arises from a misconception about the idea of discourse rationality, at least in the way Habermas develops it. As formerly seen, at the core of Habermas's thinking, there is the validity field, characterized by a weak

[1491]See Robert Alexy, *Theorie der juristischen Argumentation: Die Theorie des rationalen Disurskes als Theorie der juristischen Begründung* (Frankfurt a.M.: Suhrkamp, 1989), 134–177.

[1492]Robert Alexy, "Balancing, Constitutional Review, and Representation," *International Journal of Constitutional Law* (Oxford University Press and New York School of Law) 3, no. 4 (2005): 577.

"transcendentalization" of some universal pragmatic presuppositions of communication. There is a tension between facts and norms, as way to make increasingly factually verifiable the practice of communicative action. As a post-metaphysical thinking, those universal pragmatic presuppositions of argumentation are, according to Habermas, "expressed in a decentered complex of pervasive, transcendentally enabling structural conditions, but [are] not a subjective capacity that would tell actors what they *ought* to do."[1493] Therefore – and this is the chief explanation why communicative reason differs from practical reason[1494] – we cannot set up material criteria at the core of the validity field. Communicative reason has a universal quality in the validity field that, in virtue of the universal aspect of communication leading to mutual understanding, cannot be thus misinterpreted as a holistic expression of an *ethos* or a social factual practice of discourse, for this would imply conditioning the pragmatic presuppositions of discourse to the way a particular community understands it. In Alexy's viewpoint, nevertheless, at the end, the legal decision relies on the idea of "unity of practical reason"[1495] and on the premise that law must necessarily be shaped by non-institutional acts. Law, in his opinion, as formerly shown,[1496] cannot provide correctness and coherence alone and must then appeal to general reflections on utility, custom and morality.[1497] The *ethos* somehow defines how the communication *in law* must be carried out. The validity field, as a consequence, has a strong content of an ethical background that constrains the universal and weak character of the presupposed pragmatic conditions of communication to be "detranscendentalized" in real practices of mutual understanding.

In the specific realm of legal adjudication, this problem gains a particular configuration, which Klaus Günther very accurately examined. He remarks that the criterion orienting balancing, as Alexy describes it, cannot rely on a material predetermined content establishing preference relations to a determinate normative standpoint over others.[1498] The axiological model Alexy ascribes to legal reasoning has the danger of conditioning the claim to appropriateness, which should orient itself by assuming beforehand the *valid* legal norms and applying them in accordance with an impartial consideration of all the circumstances of the case, to some material contents that were not subject to critical scrutiny and that will somehow shape the *valid* legal norms as a condition for correctness and coherence in adjudication: "The problem thereby indicated consists in the danger of already

[1493]Habermas, *Between Facts and Norms*, 4.

[1494]See Ibid., 3–4.

[1495]Alexy, "Jürgen Habermas's Theory of Legal Discourse," 233.

[1496]See the fourth chapter.

[1497]Robert Alexy, "Law and Correctness," in *Law and Opinion at the End of the Millenium: Current Legal Problems*, ed. Michael D. A. Freeman (Oxford: Oxford University Press, 1988): 211.

[1498]Günther, *Der Sinn für Angemessenheit: Anwendungsdiskurse in Moral und Recht*, 301.

introducing, when determining the structure of argumentation, the material criteria, which should themselves be the subject matter of an appropriateness argumentation."[1499]

Indeed, if we examine more carefully Alexy's structural theory,[1500] we can conclude that his central purpose – to provide a rational methodology for legal reasoning – lies in the belief that balancing promotes discourse rationality inasmuch as, by continuously producing preference relations between principles, arguments are necessary to justify them. For this preferential statement, which serves as a definitive rule under which a case is subsumed,[1501] there must be the justification according to the *Law of Competing Principles*. A principle itself, accordingly, is never a definitive reason but serves, as long as it is evaluated in conformity with the *Law of Competing Principles* and then selected as a dominant reason for a concrete ought-judgment, as a reason for constructing a definitive rule, which in turn serves as definitive reason.[1502] At the end, those definitive rules precede the principles in discourses of application, for they regulate in advance how, in case of conflicts of principles, a particular case should be decided. This structure of thinking is carried out when there is a collision of principles, provided that the collision of rules is resolved either by declaring one of them invalid or by inserting a clause of exception.[1503] The question, nonetheless, arises when we attempt to understand why necessarily we must follow the procedure in the way described by Alexy, i.e. why principles necessarily are related to teleological assessments, and then to a rationality that is as such considered in virtue of a procedure that furnishes axiological-teleological oriented arguments in order to establish those preferential statements. What is behind this argumentation that makes it so inevitably dependent on axiological considerations? By the same token, if it is to assume this type of argumentation as the consequence of the nature of principles, what makes us certain that we are in the face of a collision of principles or a collision of rules? Besides, why can we be certain that a definitive rule, achieved through the *Law of Competing Principles*, can already regulate how the application of law is to be carried out?

Günther saw, in this point, the primary problematic aspect of Alexy's legal theory: the distinction that Alexy formulates between principles and rules, although apparently following some of Dworkin's premises,[1504] is essentially structural and morphological.[1505] For Günther, however, we cannot, before facing the context of application itself, define whether it is a principle or a rule: "That determinate norms require appropriateness argumentation only becomes manifest in application

[1499]Ibid., 301, translation mine.

[1500]See the fourth chapter.

[1501]See Alexy, *Theorie der Grundrechte*, 86.

[1502]Ibid., 92.

[1503]Ibid., 78.

[1504]Ibid., 77.

[1505]Günther, *Der Sinn für Angemessenheit: Anwendungsdiskurse in Moral und Recht*, 268.

situations themselves."[1506] According to Günther's account, the distinction between norms must be anchored not to their concept or to their validity, but rather to discourses of application, in the "presuppositions of action on which norms are applied"[1507] – after all, it is in this realm that the problem of conflict of norms appears[1508] – which determine how norms should be handled in situations.[1509] It is only by reason of the circumstances of the case that we can establish whether the features previously foreseen in discourses of justification are the same as the ones in discourses of application[1510] and, from that, provide an adequate response to the problem of indeterminacy of law and to the distinction between rules and principles.

Rules, according to this model, are a "specific *procedure of application*"[1511] that does not require the consideration of all the features of the case and does not lead to a weighting of distinct standpoints, but rather takes as relevant only those features that are part of the semantic extension of the "if-component" embedded in the legal norm.[1512] Rules, for this reason, previously select which circumstances must be taken into account as relevant or not and disregard the changed features of the case. There are rules when the legislator, in discourses of justification, has decided the appropriateness of a norm in advance.[1513] On the other hand, principles demand an argumentative procedure of appropriateness, which will require an impartial consideration of all the features of the case. One and the other are *procedures of application,* but they, while, in any case, having the context of application as the basis for their qualification, differ in the requirement of a complementation of discourses of justification by institutional procedures that make possible the consideration of all the characteristic signs of a situation, a duty undertaken by adjudication. Principles, in conformity with this standpoint, refer to institutional procedures claiming an argumentative-interpretative endeavor by legal adjudication, whose purpose is to provide a coherent response (and, in this matter, as shown, the ideal of the *"perfect norm"* or the "*single right answer*" appears). Moreover, principles become more evident in post-conventional societies in virtue of a growing indeterminacy of law, thereby demanding more and more appropriateness argumentation.[1514] As a consequence, they are only the tip of the iceberg calling for a hermeneutical-interpretative posture within contexts of application. The consequence of this thinking is that the indeterminacy of law refers not to its structure, but to its impartial application – after all, only in virtue of the features of the case we

[1506] Ibid., 272, translation mine.

[1507] Ibid., 335, translation mine.

[1508] Ibid., 267.

[1509] Ibid., 265.

[1510] Ibid., 266.

[1511] Ibid., 336, translation mine.

[1512] Ibid., 336.

[1513] Ibid., 337.

[1514] Ibid., 339–340.

can specify whether a law is indeterminate or not.[1515] This leads to Günther's perception that every norm, in a post-conventional level,[1516] regardless of the degree of determinacy of the *prima facie* circumstances in face of the case, will require an appropriateness argumentation: "We have an obligation with every norm to enter an appropriateness argumentation, and, in fact, independently of how 'determinate' those unchanging circumstances are under which it was recognized as valid."[1517]

In consequence, since it is the context of application that will specify the procedure of application, a certain rule, as such defined based on Alexy's structural differentiation, could, actually, be a principle (now understood as a *procedure of application*) if the context demands an impartial consideration of all the features of the case: "It does not depend on the norm itself whether we apply it with or without consideration of the particular circumstances of a situation."[1518] By stressing the structure of the norm as the basis for his distinction, Alexy creates an incomprehensible premise establishing which procedure must be carried out in a particular case. In a post-conventional level, we cannot, nonetheless, define in abstract which criteria permit the definition of whether a norm is a principle or a rule and beforehand establish which procedure we must carry out: "That is why it seems easier to me to separate the manner of application of one norm from its deontological content,"[1519] says Günther. The idea of a definitive rule already regulating its situations of application, therefore, lacks any justification, for the appropriateness argumentation,[1520] instead of relying on previously defined principles and on how they are evaluated by means of the *Law of Competing Principles*, derives from selecting, among the *prima facie* valid norms, which of them is more appropriate given an exhaustive and impartial consideration of all the features of the case. It is the case, the context of application that will furnish the conditions for an impartial application of the valid norms: "If the demand of an appropriateness application is dissociated from the concept of norm structure, it can be justified only based on the idea of impartiality."[1521]

Moreover – and here Günther makes the connection with the axiological standpoint before indicated – if the case defined which procedure of application we should carry out, there would be no sense in already characterizing principles as

---

1515 Ibid., 341.

1516 Naturally, there are some rules that are artificially kept at a conventional level, but, since they violate the principle of impartial application, it is necessary a justification to do so. As he argues: "This [the impartial application] does not rule out that, for reasons demanding justification, determined norms are artificially kept at a conventional level, with the result that changes in rules, by reason of appropriateness, are only possible in exceptional cases or make necessary a decision about their validity" (Ibid., 272, translation mine).

1517 Ibid., 341–342, translation mine.

1518 Ibid., 272, translation mine.

1519 Ibid., 272, translation mine.

1520 Ibid., 274.

1521 Ibid., 274, translation mine.

optimization requirements.[1522] Indeed, there is no more direct conclusion that the distinction Alexy implements between principles and rules is morphological than his inscription of an axiological core in the structure of principles. When Alexy characterizes principles as optimization requirements, he is already making a structural parallelization of them with values.[1523] Were principles rather a procedure of application, there would be no reason to previously establish a material structure behind them. The metaphysics lies, as a consequence, in Alexy's non-submission of these material contents to critical scrutiny: they are, after all, embedded in the very structural character of principles, from which follows immediately weighting as a response, and according to which the argumentation takes place. Alexy's account, in Günther's opinion, is thus marked by the predetermination through the normative structure of particular types of justification and application,[1524] which, in other words, results from Alexy's inobservance of how dangerous it is to practically equalize – even though he remarks that this equalization would be a "wrong conceptualization of this relationship"[1525] – discourses of justification to discourses of application.

6.4.3 The Second and Third Outcomes: The Confusion between Discourses of Justification and Discourses of Application and the Loss of Protection of Minorities

The first outcome previously indicated is hence intimately connected to the second one. After all, if principles are regarded as a procedure of application, then there is no reason to assimilate them to discourses of justification. Alexy, however, by morphologically conceptualizing principles and parallelizing them to values, cannot, in essence, distinguish discourses of justification from discourses of application, except for the specification to the case. According to him, "in its logical form it [application of norms] only differs from what is generally called 'justification of norms' insofar as its object of justification is not a universal but an individual norm."[1526] More specifically, in Alexy's opinion, this differentiation is not satisfactory for a universalistic practice of decision-making nor could provide coherence, since it simply demonstrates that the relationship between the *prima facie* applicable norms and the appropriateness argumentation has an *ad hoc* character.[1527] To the extent that every new case will lead to different forms of this relationship, we cannot reach therefrom a universal parameter that will correspond

[1522]See Ibid., 274.

[1523]See Ibid., 275.

[1524]See Ibid., 276.

[1525]See Ibid., 169.

[1526]Robert Alexy, "Justification and Application of Norms," *Ratio Juris* 6, no. 2 (July 1993): 162.

[1527]Ibid., 163.

to the exigency of treating like cases alike. Moreover, inasmuch as, in Alexy's view, rationality is closely related to the capacity to achieve a universal parameter in decision-making – his structural theory is a clear example – we cannot simply rely on this distinction.

It seems that Alexy does not gather the real significance of coherence in Günther's theory (which is similar to Dworkin's and Habermas's proposals), though. Coherence, as shown, is associated with an ongoing practice of decision-making that clearly assumes the premise of observing the legal valid norms, enacted through institutional procedures of justification, and of impartially applying them in conformity with the features of the case. The concern for coherence exists in the very need to justify the meaning of each norm, and to confront them with the features of the case as a means to define the *only* appropriate one. There are, accordingly, boundaries within the practice of adjudication: its duty is not to set forth reasons that are equivalent to justification discourses, but rather to define, for that *singular* case, which is the appropriate norm. Its purpose is to establish a decision that exhausts the normative meaning of each valid principle in collision by confronting them with the characteristic signs of the case, and for that, it knows that its discourse has a completely different role and logic than the one of lawmaking. Every discourse of application is then oriented towards the "single right answer," to the "perfect norm," an ideal that, albeit unachievable, demonstrates that legal reasoning is carried out in the tense relationship between facts and norms. Alexy, nonetheless, by disregarding the complexity involved in the transition from one discourse to the other, thinks that the thesis establishing boundaries between them is wrong,[1528] since every legal interpretation and application creates a new norm that will show "an additional normative content"[1529] to the principles that were in collision. In his opinion, with respect to "discourse theory's principle of universalizability," every new norm must be then "substantiated in discourse of justification."[1530]

As shown, Alexy's claim to coherence is linked to a practice of reviewing the pertinence of a norm to other social values by way of teleological-assessments. The judge makes this evaluation by verifying whether a valid norm can be justified in a broader perspective of general interests of a determined community, an activity carried out through balancing. It is not surprising, therefore, that Alexy sustains that "producing coherence is a procedure of justification."[1531] After all, if we follow this reasoning, when the judge interprets and applies the law, she also produces a discourse of justification, which transcends the case and indeed enriches the discourses of justification themselves. In any other way, discourses of justification would become a "mere discourse of *topoi*,"[1532] which – and this is a serious

[1528]Ibid., 165.

[1529]Ibid.

[1530]Ibid.

[1531]Ibid.

[1532]Ibid., 166.

perception of how Alexy seemingly misunderstands the inherent complexities of both discourses – "would have the fatal consequence that the norms modified or newly created for the decision of a case in a discourse of application could no longer be subject matter of a discourse of justification and therefore could not be substantiated."[1533] Whereas discourses of application turns practically into discourses of justification through "substantiation," discourses of justification only have their *raison d'être* if "substantiated" by discourses of application; otherwise, they become a "mere discourse of *topoi*."[1534] If interpretation and application of a norm become a practice similar to enacting valid norms – only with the difference of its reference to a singular case – then adjudication is not discursively distinct from legislation. As a consequence, the judge can use whatever reason she finds important in a broader context of the "unity of practical reason." This explains why the judge, according to Alexy's view, is not bound to "norms already accepted."[1535] This also justifies why balancing, for it does not exclude any possible result, cannot have basic rights as binding reasons.[1536] History after all, in his opinion, gives rise to new dilemmas and demands that the existing system of rights cannot respond by itself, and therefore "has to be made more precise"[1537] by "substantiating" discourses of application into discourses of justification.

In this respect, we may think that Alexy's approach has a certain explanatory strength, because every new case can actually provide new arguments that can serve as a precedent for future cases. Naturally, no one can doubt that precedents are a very powerful source for discourses of application. However, does the fact they are relevant imply they can be practically assimilated to enacted valid norms of discourses of justification? Would there not be, in this assimilation, a serious reduction of the existing complexities between both discourses, many of them resulting from the principle of separation of powers? Habermas points out clearly his criticism of this assimilation: "Denying the difference between these two types of discourse destroys the rational basis for a functional separation of powers justified by the different possibilities of access to certain kinds of reasons."[1538]

The problem might lie precisely in this misunderstanding of the use of reasons in both discourses, an issue that is closely connected to Alexy's confusion between facts and norms. Alexy could only sustain the "substantiation" of discourses of application in discourses of justification in virtue of the material contents that are presupposed in the validity field, conditioning then the legal reasoning to an axiological reasoning that, as formerly shown, is not subject to critical scrutiny, and, as such, metaphysical. If validity is ultimately materially grounded – and not simply comprises the weak transcendental presuppositions of communication – and

1533 Ibid., 168.
1534 Ibid., 166.
1535 Ibid., 169.
1536 Alexy, "Postscript," 392.
1537 Alexy, "Justification and Application of Norms," 169.
1538 Habermas, "Reply to Symposium Participants, Benjamin N. Cardozo School of Law," 430.

these material contents are, in principle, collected and gathered from a reality represented by an *ethos,* then those material contents define how the communication *in law* should be carried out. The validity field of legal reasoning can lose its connection with the universal-reciprocal principle of impartiality and reproduce a certain reality, thereby eroding its critical review. Discourses of application and discourses of justification work with reasons that will, at the end, correspond to the satisfaction of certain facts, a certain *ethos,* which in turn furnishes the material contents that are transported to the validity field. Similarly, the criteria used in discourses of application are not distinct from the ones used in discourses of justification: they are all guided by an axiological standpoint that will shape the result according to teleological assessments of what is *good* for a particular group of people. This is why balancing, by working with the most variable reasons to reach a result that can be *accepted* by those people, can be deemed, in this standpoint, a relevant instrument for both legislation and adjudication.

It is questionable, however, that legal adjudication, by deploying balancing, is immediately concerned with accomplishing a result that can be more than the valid norms, for, otherwise, the very strength of discourses of justification would become a "mere discourse of *topoi.*" Evidently discourses of application complement discourses of justification; yet, this conclusion does not mean transforming the valid legal norms into an abstract and empty argument, if not "substantiated" by legal decisions. Although they call for complementation, legal norms are the link between adjudication and the democratic procedures of opinion- and will formation, and, as such, a condition for the validity and legitimacy of the practice of decision-making. They are not merely *topoi,* but rather, by linking ultimately to the sovereignty of people, a condition for the validation of discourses of application. In order for her discourse to be legitimate, accordingly, the judge must take seriously into account the valid legal norms. This obviously does not mean that she will disregard the new real dilemmas that emerge from case to case, which might even put in doubt the range of a particular norm. However, she understands law, in pluralistic and post-conventional societies, as just the tip of the iceberg, an opening towards the future that generates further complexities and even possibilities of abuse. If history creates new dilemmas, therefore, they do not immediately give rise to the verdict that the judge is not bound to "norms already accepted,"[1539] because a judge, by acknowledging the complexity and the opening of the law, will gather those new dilemmas exactly to confront them with an exhaustive and extensive interpretation of the system of rights. The response comes not from subverting the system of rights in favor of a history and a certain reality, but from confronting this history and this reality with the system of rights, which, at the end, has its validity anchored to democratic procedures of opinion and will-formation, and hence to the presupposition of equal rights of participation and treatment. The decision is not simply a possible result among many others, which can even not be

[1539] Alexy, "Justification and Application of Norms," 169.

bound to basic rights.[1540] Instead, it is the "single right answer" bound to basic rights, for they positively express conditions for the quest for justice as a "prior claim to an equal right to exist."[1541] The single norm originated in a legal decision is then the outcome of a procedure whose primary concern is that it can be justified by the set of accepted valid norms; it enriches the system of rights coherently insofar as it complements the discourse of justification within the boundaries of the principles thereby enacted.

With these premises, questions of justice have unconditional priority over values, for the claim to the "single right answer" cannot represent the immediate validation of a certain reality, of a singular *ethos* (even if it corresponds to the majoritarian opinion), but must assume a validity field that is marked by an intersubjective ideal of mutual understanding. Hence, the "ethical permeation of law by no means eradicates its universalistic contents."[1542] Were it otherwise, questions of justice would lose their twofold duty in pluralistic and multicultural societies to both legally neutralize value conflicts among distinct political views and prevail over "conceptions of the good that sanction authoritarian relationships within the group."[1543] Questions of justice would lose, in truth, the tension between facts, as expressing particular self-understandings, and the normative validity, as this universal claim to mutual understanding embedded in the claim to equal rights of participation and treatment. The conclusion of the second outcome thus induces the third: if questions of justice are assimilated to axiological points of view, adjudication jeopardizes its duty to protect individuals from society when it acts ideologically and arbitrarily by oppressing their exercise of communicative reason.

In addition, these conclusions explain why we must understand the judiciary as a part of a broader set of complex democratic established procedures, which create their very substantive boundaries: either discourses of justification or discourses of application have their reasons channeled into legally binding procedures, but according to distinct forms and logic of argumentation, even to guarantee that the result, in each of these procedures, is legitimately achieved not by reason of being merely bound to a particular self-understanding or form of life, but rather in virtue of being bound to legally established procedures ensuring the practice of rational discourse. Discourse rationality in law is not just the institutional use of reasons in accordance with a determined methodology, but the use of reasons in a set of complex democratic procedures that aim to safeguard, as much as possible and as a self-correcting learning process, the transcendental presuppositions of communication within the institutional framework. If legal methodology is here an issue, it is certainly not one that is teleologically oriented towards satisfying an *ethos*, but one that is concerned with expanding the communicative reason, even when it contradicts this *ethos*.

[1540] Alexy, "Postscript," 392.

[1541] Habermas, "Reply to Symposium Participants, Benjamin N. Cardozo School of Law," 393.

[1542] Ibid., 399.

[1543] Ibid., 400.

This is why, in a multicultural and post-conventional society, discourses of application must orient themselves towards reinforcing the constitutional principles that were subject to critical scrutiny in discourses of justification, while critically applying them in a way that makes justice prevail over particular conceptions of good. For this purpose, the judge, instead of ranking principles and defining axiological preference relations among them, takes as a premise that only an integral interpretation of all of them corresponds to the expectancy of a plural society that respects the other. Naturally, this is an idealization: no real judge is capable of having this extensive and exhaustive knowledge. Yet, this impossibility should not result in the loss of this ideal, in the loss of the tension between facts and norms, which is, as formerly discussed, verified in a set of legal democratic procedures that, in distinct moments (justification and application), attempt to realize the impartiality principle or the integrative interpretation of the system of rights as their very condition of legitimacy.

6.4.4 The Fourth Outcome: The Relativization and Misunderstanding of the "Single Right Answer"

By assimilating discourses of justification to discourses of application, as well as questions of justice to axiological points of view, there would be no other possible statement than the one regarding the impossibility of the "single right answer." Indeed, the claim to the "single right answer" has only sense when we understand the tension between facts and norms as an inherent characteristic of legal discourses. The skepticism about the "single right answer" is nothing but another example of how, in the structure of Alexy's legal theory, a misunderstanding of the real meaning of this tension seems to occur. Particularly, a misinterpretation of the real significance of the regulative idea of the "single right answer" seems to happen, which Günther, Dworkin and Habermas have so intensively sustained.

In his text *Legal System, Legal Principles, and Practical Reason,*[1544] while examining Dworkin's premise of the "single right answer," Alexy suggests that only a strong account of principles could prove the correction of the thesis of the "single right answer,"[1545] which, in practice, is nevertheless impossible, not only because of the limitations of human knowledge but also in virtue of some "logical reasons of broader sense."[1546] He argues that, to solve this impasse, it is necessary to appeal to a weak theory of principles on the basis of discourse rationality and the premise that legal discourse is a special case of general practical discourse.[1547] Briefly, the viable solution to a theory of principles is found in his structural theory

[1544] Alexy, "Sistema Jurídico, Principios Juridicos y Razón Práctica," 139–151.

[1545] Ibid., 145.

[1546] Ibid., 148, translation mine.

[1547] Ibid., 149.

rooted in the deployment of balancing. However, insofar as this weak version is based on the possibility of using the most variable reasons to justify a decision (after all, balancing operates in the realm of the "unity of practical reason"), there is no sense in concluding that each case will lead to the "single right answer." By assuming his point of view, only if a consensus were always guaranteed in the application of norms could we defend the possibility of achieving the "single right answer." Since, nonetheless, this consensus for each decision would require unlimited time and information, unlimited conceptual linguistic clarity, unlimited capacity and disposition to the exchange of roles among individuals, and, finally, unlimited protection against prejudices, it could only be reached approximately.[1548] For this reason, in Alexy's opinion, the thesis of the "single right answer" is not correct; it would be correct only if it could embrace "all principles, all abstract relations of priority, and, therefore, univocally [determine] the decision in each one of the cases."[1549]

After having previously examined the concept of "single right answer" in Günther's, Dworkin's and Habermas's legal theories, it seems, though, that Alexy interprets the regulative idea of the "single right answer" as the factual concretization of the "single right answer" in each case. He seems to fall into the mistake of confusing the ideal of consensus with the real achievement of consensus. Moreover, it is evident, according to those legal theories, that the "single right answer" would imply neither the need for a total factual consensus on a certain subject matter nor the conclusion that, in reality, a case could not have distinct responses.[1550] The "single right answer" is rather a counterfactual premise in tension with this reality full of disagreements and points of view, which indicates the purpose of making, within the context of indeterminacy of law in post-conventional and plural societies, a decision that is the best result of the conciliation between a consistent interpretation of the system of rights and legitimacy. For this reason, it is not necessary, contrary to Alexy's arguments, to have unlimited knowledge and a permanent consensus to think of the "single right answer." Instead, we assume the presupposition of the "single right answer" exactly because we know that, in the realm of indeterminacy of law, achieving a consensus and having unlimited knowledge is the opposite of the dynamics inscribed in the tension between facts and norms. Were consensus and total knowledge achieved, the critique would be unnecessary, thereby jeopardizing the exercise of communicative rationality. The ideal of the "single right answer" is nothing but the other facet of the realization of

---

[1548]Ibid., 151.

[1549]Ibid., 145, translation mine.

[1550]This conclusion is manifest in Dworkin's passage below:

"It is not part of this theory that any mechanical procedure exists for demonstrating what the rights of parties are in hard cases. On the contrary, the argument supposes that reasonable lawyers and judges will often disagree about legal rights, just as citizens and statesmen disagree about political rights. This chapter describes the questions that judges and lawyers must put to themselves, but it does not guarantee that they will all give these questions the same answer" (Dworkin, *Taking Rights Seriously*, 81).

justice, an ideal never entirely attainable, though, but which the judge must always presume. It requires of the judge an interpretative posture that is more than a simple deployment of a methodology to be complemented with reasons, for it takes into account that rationality has its boundaries, and thus calls for an ongoing expansion of the conditions of communication among individuals.

This is the real significance of the claim to the "single right answer": the decision must be made already acknowledging the quest for justice, which, despite being unattainable, is there as a regulative idea that has an unconditional priority over other values and transcend the context, even to legitimate the practice of adjudication: "In practical affairs, decisions must be made despite ongoing dissensus, but they should nonetheless be made in such a way that they can be considered legitimate."[1551] Were this quest for the "single right answer," however, lost, justice would become the very realization of communitarian values. This explains why, if adjudication must promote justice, in this situation, it presumably achieves it by means of axiological-teleological methodologies. On the contrary, by presupposing the "single right answer," the judge seeks justice through an interpretative posture of reconstructing the system of rights by acknowledging that its opening, complexity and indeterminacy are not simply "solved" by identifying it with a predetermined axiological point of view, but, instead, with an ongoing procedure of back-and-forward interpretations whose only safety are the discursive procedures and their boundaries. In a post-metaphysical thinking, the claim to the "single right answer" is the guarantee that discourses of application will not become a mere endorsement of a particular self-understanding, but rather will critically reflect upon this material content and confront it with the transcendental presuppositions of communication, in legal systems represented by the claim to equal rights of participation and treatment.

6.4.5 *The Final Analysis: The Problem of Rationality in Alexy's Thinking*

All these outcomes demonstrate that Alexy's legal theory, while sustaining methodologically why and how balancing provides rationality, coherence and rightness, seems to fail to explain why and how this methodology can achieve it in a post-metaphysical level. The confusion between facts and norms is at the core of all his structural theory, and this makes intelligible the reason for binding rationality to the deployment of an axiologically-based method, instead of binding rationality to a practice oriented towards mutual understanding, even when contrary to a certain axiology. His thinking seems to reveal skepticism about the democratic procedures and their boundaries being enough to respond to an ongoing construction of the contents of rights within the context of post-conventional societies. Adjudication,

[1551]Habermas, "Reply to Symposium Participants, Benjamin N. Cardozo School of Law," 396.

by way of balancing, would be then a more adequate and justified response to this impasse, and the fact that reasons are presented through this methodology would demonstrate the exercise of discourse rationality, even if it could only do it in some cases.[1552] Besides, not only would this method be more suitable for providing rationality in adjudication (after all, it is grounded in some "rational standards"[1553]), but also it would be a more adequate response to the problem of the firewall between legislation and adjudication, for the method itself could establish "firm and clear"[1554] limits to the practice of adjudication (these limits would be justified, in fact, by the law of diminishing marginal utility[1555] applied to balancing). It is thus the method – and not the procedure as the proceduralist approach presents it – that could reflectively and rationally establish boundaries for the state activity.[1556]

Yet, Alexy's confidence in his methodology seems even more radical. He not only thinks that his theory is rooted in some "rational standards" and provides rationality and limits to adjudication, but also that it better respects the "legal obligations" stemming from basic rights, which, in his opinion, does not occur in Habermas's proceduralist approach. He says, "The purely procedural model is incompatible with the legal obligation of the legislature to respect constitutional rights, since it is defined by the negation of every substantive legal obligation, including those imposed by constitutional rights."[1557] Consequently, "the legislature is left at liberty in respect of everything."[1558]

Still, all these arguments seem to go in the opposite direction of a proceduralist approach. First, rationality, according to the proceduralist model, is definitely not a problem of a method grounded in some "rational standards." It is not because the judge deploys a method with arguments that we can conclude that there is rationality, at least discourse rationality, for methods can be used to ideologically and arbitrarily undermine the exercise of communicative reason. Balancing, when axiologically justified, can result, for instance, in a decision that favors a particular *ethos* while being contrary to the premise of equal rights to participation and

1552 Alexy, "Postscript," 402.

1553 Ibid., 405.

1554 Ibid., 404.

1555 The idea that the method itself provides the *firewall* is expressed in the following passage, when Alexy explains the *Titanic* case:

"(. . .) One has reached the area in which interferences can hardly ever be justified by any strengthening of the reasons for the interference. This corresponds to the law of diminishing marginal utility, which is the firewall that Habermas misses in the theory of principles. The Titanic Case is thus not only an example of the fact that scales which can intelligently be put into relationship with each other are possible even in the case of immaterial goods such as personality and free speech, but it is also an example of the power inherent in constitutional rights as principles to set limits by way of the process of balancing, which while not right and ascertainable without balancing, are none the less firm and clear" (Ibid., 404).

1556 Ibid., 405.

1557 Ibid., 392.

1558 Ibid., 393.

treatment, which is a presupposed condition for the exercise of communicative reason.[1559] Furthermore, a proceduralist approach does not believe in the existence of "rational standards" that are not themselves subject to critical scrutiny, as though they were a conception of truth. Indeed, what are "rational standards"? Why does the use of formulas such as the *law of diminishing marginal utility, law of competing principles, etc* reveal "rational standards"? Discourse rationality in law, according to the proceduralist model, is not predetermined, but constructed within democratic procedures of mutual understanding as a condition for a post-metaphysical thinking. The only presuppositions, which are weak and exist solely to be "detranscendentalized," are the conditions for the realization of mutual agreement in reality. The tension between facts and norms is much more complex and cannot then be reduced to patterns and formulas in which arguments are placed and measured, because a proceduralist approach stems from the premise that, in complex and post-conventional societies, all knowledge is fallible. Were it otherwise, we would establish a metaphysical content – the "rational standards" themselves – behind the procedure, which, as shown, seems to happen in Alexy's legal theory when he confuses facts with norms.

Second, contrary to Alexy's words, the proceduralist legal theory never sustained the nonexistence of substantive legal rights limiting the exercise of legislation, but it simply denied that those legal rights are fixed, with a rigid content; their contents are rather constructed within those procedures and their boundaries in a tense relationship between facts and norms as a means to deliver more and more in reality the conditions for rational communication. The focus on procedures does not mean whatsoever the denial of the contents, but instead a construction of these contents within the procedures of rational dialogue. Unlike Alexy's legal theory that places some material contents in the validity field, the proceduralist approach states that "there are no source of evidence and evaluative criteria that would be given *prior* to argumentation."[1560] It is definitely not the case of absence of contents, but rather absence of contents that do not have to be validated by procedures oriented towards mutual agreement. Alexy does not seem to have grasped the interpenetration of procedures and reasons, form and content,[1561] which is at the core of the proceduralist account. The absence of contents in the validity field, for him, is the same as the "negation of every substantive legal obligation."[1562]

What really stands out, nonetheless, is that it is exactly by constructing a metaphysical standpoint in the realm of validity that the obligation to respect constitutional rights – as a "substantive legal obligation" – is in jeopardy, as he himself mentions, in the same paragraph, that balancing is not bound to basic rights

---

[1559]See the *Crucifix* and *Cannabis* cases examined in the first chapter.

[1560]Habermas, "Reply to Symposium Participants, Benjamin N. Cardozo School of Law," 408.

[1561]Ibid., 409.

[1562]Alexy, "Postscript," 392.

for they do not exclude possible results.[1563] In truth, not only is constitutional adjudication, in his opinion, exempt from respecting basic rights, but also legislation, insofar as they are "not capable of setting limits to the legislature, and cannot represent a framework."[1564] It seems that there is, accordingly, an evident contradiction and misunderstanding of the relationship between form and content in Alexy's thinking. This contradiction and misunderstanding, nevertheless, are only the outcome of a deeper and fundamental cause.

The chief problem seems to reside in the very concept of rationality Alexy holds and how he connects it to knowledge. Alexy's theory appears to stem from the belief that knowledge can be heuristically controlled and that this control can provide certainties and rationality. The morphological differentiation between rules and principles, which is the basis of his structural theory, revealed how definitive rules, reached by means of the *Law of Competing Principles*, can regulate their own application. This thinking exposed how his structural theory relies on some material contents that are not subject to critical scrutiny, and how these material contents ends in a determinate form of normative application. The outcomes of the nonexistence of practical distinction between discourses of justification and discourses of application and of the "single right answer" are other examples of how certainty, for Alexy, relies on the expectancy of following a certain methodology controlling the knowledge. Yet, this thinking might have its origins in a more serious faith, one that clearly places proceduralism and Alexy's structural theory in opposite sides. Whereas Habermas understands that the procedures oriented towards mutual agreement will lead to an ongoing critical review of law according to the boundaries those procedures create in a pragmatic fashion, Alexy seems to establish those boundaries not in the procedures, but in the capacity of texts to refuse epistemic discretion. When he states, for example, that "constitutional rights would offer more protection if the legislature were to be refused an epistemic discretion,"[1565] more than expressing the controversial conviction that texts could somehow refuse the epistemic discretion[1566] to someone, he is saying that it would be best for the protection of basic rights, if there were no uncertainties in discourses of justification. The problem of the opening of basic rights becomes accordingly a semantic problem, in a clear opposition to the developments of hermeneutics, and particularly the communicative character of the proceduralist account.

Furthermore, it is especially interesting to remark that, while attempting to semantically restrict discourses of justification by the belief in the potentiality of

---

[1563]Ibid.

[1564]Ibid.

[1565]Ibid., 416. The same happens when he states that "this would have the consequence that the legislature could only interfere in any way with constitutional rights on the basis of *empirical premises the truth of which was assured.* (Ibid., 417, emphasis mine).

[1566]According to Alexy, epistemic discretion "arises whenever knowledge of what is commanded, prohibited, or left free by constitutional rights is uncertain. Uncertainty can be caused by uncertainty about either empirical or normative premises" (Ibid., 414).

legal texts, on the one hand, by constructing his methodology rooted in balancing, he expands the space for discourses of application, on the other. This might explain why there are so many differences between Alexy's structural theory and Habermas's proceduralism: one seems to still believe in certainties,[1567] epistemic discretion controlled by texts, and methods as a condition for rationality, and places in legal adjudication the very exercise of political will (otherwise, it could undermine basic rights); the other, however, assumes that knowledge is fragile and fallible, and that its only safety – always shaky, though – does not lie in abstract methodologies and formulas, but rather is constructed, in self-correcting learning processes, in procedures oriented towards mutual understanding in all distinct institutional grounds.

6.5 Final Words

Whereas the last chapter had the purpose of examining the metaphysics in Alexy's thinking through Jacques Derrida's deconstruction, showing thereby some relevant elements that shape the concept of limited rationality,[1568] this chapter focused on another philosophical tradition, especially on Jürgen Habermas's proceduralism, in order to more directly enter, even though still searching for disclosing and undercutting metaphysics, into the realm of legal adjudication within the context of indeterminacy of law. More than the previous analysis, this chapter provided a possible *therapy* for the problem of legal adjudication in complex realities of post-conventional societies and developed a critical analysis of balancing and the concept of rationality arising therefrom. Briefly, it posed the question of procedures oriented towards mutual agreement as a direct attack on balancing, as Robert Alexy justifies it. Besides, by exposing the problems of this "rational" justification of balancing, for Robert Alexy's theory is indeed a very direct interpretation of the BVG's constitutional practice, we could say it also furnished somehow a critical review of the BVG's and, in a sense, STF's decisions.

For the debate on the indeterminacy of law relates to the claim to coherence, this chapter began by stressing how Robert Alexy understands this claim and showing how he ties it to the deployment of balancing and to the fixation of preference

[1567]According to Alexy, "epistemic discretion is only an issue in cases of uncertainty" (Ibid., 420). The same happens when he links the fact that *rights are taken seriously* to the capacity of the text to limit uncertainty:

"If the legislature were free in all cases to decide as judges in their own case what they were commanded, prohibited, and permitted to do by constitutional rights, one could no longer talk in terms of a real, reviewable, obligation to respect constitutional rights. The legal normativity of constitutional rights could no longer be taken seriously. But such an all-embracing liberty, corresponding to an unlimited normative knowledge-related discretion, is not an option" (Ibid., 420).

[1568]See the third part.

relations. By the same token, we could observe how coherence is related to his *Special Case Thesis*[1569] and to the idea that legal reasoning should embrace the totality of the legal order, in a practice that does not seem to differentiate law from morality. Accordingly, coherence, for Alexy, is in a continuous demand for a "prior supplementation," whose contents, nonetheless, seem to lead to a metaphysical standpoint, a characteristic we could already observe when we concluded, in the previous chapter, that his claim to correctness may express a *logos* of correctness.

Still, if coherence immediately results in balancing in Alexy's view, and hence in the "unity of practical reason," the study of other alternatives to legal reasoning revealed that Alexy's premises might not be entirely correct. Indeed, by examining Klaus Günther's differentiation between discourses of justification and discourses application, we could verify that coherence in adjudication can refer to a practice that does not lead to balancing with an axiological view by establishing preference relations between norms. Rather, it can refer to the search for the appropriate norm among the many *prima facie* applicable ones to a particular case, after having proceeded to an integral description of its characteristic signs. Günther's concern for the enforceable character of legal norms and the distinct procedures and discourses that take place in lawmaking and in decision-making is a serious counterargument to Alexy's theory. Indeed, while Alexy confuses discourses of justification with discourses of application in his defense of balancing (after all, for him, "producing coherence is a procedure of justification"[1570]), Günther, against balancing, remarks that adjudication, in constitutional democracies, ought to operate only in the realm of discourses of application, and thereby not reopen the discourses of justification.

Similarly, through the analysis of Ronald Dworkin's theory, we could argue that, in a similar way to the idea of appropriateness in Günther's theory, adjudication has a distinct discourse in comparison with legislation, now materialized in the differentiation between arguments of principles and arguments of policy, showing accordingly Dworkin's stress on the need to protect the individual from majoritarian points of view. More than the concern for what is good for the society in general, a judge who seeks *integrity* in law, who takes rights seriously, decides favorably for the individual, even when the majority of the population might suffer somehow. By establishing the regulative idea of the "single right answer," in the same way, we could conclude how Dworkin expresses the counterfactual premise of "equal concern and respect" as the other side of his concern for keeping consistent the system of rights. There is hence a dialectical relationship that links the judge to a community of principles through the observance of the valid legal norms, hermeneutically reconstructed according to the particularities of the case. Balancing, therefore, is not an issue here for coherence.

Finally, by focusing on Jürgen Habermas's proceduralism – which was done both by verifying how his communicative reason means an effective intervention in

---

[1569]See the fourth chapter.

[1570]Alexy, "Justification and Application of Norms," 165.

the world and by extending this debate to the dualism between constitutionalism and democracy (private and public autonomies) and justification and application, in a similar fashion as it happened with Jacques Derrida's philosophy in the last chapter – we could then develop a critical analysis of balancing through the emphasis on procedures oriented towards mutual agreement. In this respect, we were able to conclude how balancing can be a sign of a metaphysical standpoint, insofar as it might result in the fusion of facts and norms, putting thereby in jeopardy the institutional procedures to which adjudication is bound and, in the same way, the idea of the "single right answer." With proceduralism, we could envisage a response to the indeterminacy of law that is not rooted in balancing but works in this tense relationship between facts and norms in the grounds of the democratic principle, one that has the concern for the other in the idea of *intersubjectivity* and in the need to keep consistent the system of rights, or, in other words, one that satisfies simultaneously the principle of legal certainty and the legitimacy claim of law.[1571]

In all these viewpoints, it was clear that coherence does not necessarily result in balancing. Indeed, they revealed that the debate on coherence, and, hence, on rationality, demands more than abstract methods and criteria. The tense relationship between facts and norms discloses a complex reasoning that preserves the dualism between legal consistency and justice, in the very idea of equal treatment, as a necessary premise for the reconstruction of law in every new circumstance. For this reason, by focusing on the interconnections among those proposals, we could delineate a viable and robust response to the problem of indeterminacy of law and to the construction of a coherent and legitimate legal system. It was possible to conclude likewise that balancing can express metaphysics, and how this metaphysics can bring about serious outcomes for the practice of adjudication in the realm of constitutional democracy. In fact, when, in the second section of the investigation, we concluded that Alexy's premises, first, are based on an axiological point of departure, which leads to the definition of material criteria in his validity field; second, make a confusion between discourses of justification and discourses of application, which can result in the weakening of the protection of minorities; third, relativize the idea of "single right answer" by interpreting it as the factual realization of the "single right answer" in every case; and, fourth, assume rationality as the capacity to heuristically control the knowledge, it was clear that balancing, in the way Alexy justifies it as a direct reflex of the BVG's constitutional practice, might not the most adequate response to the indeterminacy of law.

Therefore, these theories complement the previous analysis of the problem of balancing through Derrida's deconstructionist premises, even though stemming from a distinct and sometimes untranslatable theoretical basis. More directly than Derrida, though, they convincingly justify, through different premises, why balancing, in the way Robert Alexy justifies and constitutional courts deploy it,[1572] can be

[1571]Habermas, *Between Facts and Norms*, 211.

[1572]See the first.

the sign of a discretionary practice that puts the principle of separation of powers in jeopardy.

Nevertheless, this step forward to the institutional ground of normative application might only have been possible because of a possible sacrifice of the very idea of justice that we can remark in Derrida's deconstructionist approach. Indeed, it is in this subject matter that we can first initiate the comprehension of how both theoretical premises – deconstruction and proceduralism – while being attempts at disclosing and undercutting metaphysics also in the institutional realm and, as such, are complementary to each other, carry with themselves some insurmountable divergences. In this case, whereas, on the one hand, we could verify that procedures oriented towards mutual agreement (in which the claim to justice refers to the premise of *equal consideration* as a regulative idea in tension with the facts) responded better than balancing to the problem of indeterminacy of law in complex, plural and post-conventional societies, on the other, we might say that the counterfactual presupposition of *equal consideration* is not enough. We can thereby reach two conclusions: first, the proceduralist account, founded upon reciprocal justice as the basis for practices of mutual understanding, is an adequate post-metaphysical response to the indeterminacy of law, which we might not achieve from balancing; and, second, the proceduralist account, however, while setting forth an institutional response to the indeterminacy of law and working with the perspective of reciprocal justice, might not yet be enough, even though paradoxically adequate.

In the realm of the debate on justice, accordingly, there is the primary conflict between the premise of *equal consideration* and *différance,* between *reciprocal* justice and a *disinterested and non-reciprocal* justice. If an idea of disinterested and non-reciprocal justice, such as Derrida's *différance,* can be applied to the institutional application of law in constitutional democracies, this is the focus of the following part of this book. The debate between deconstruction and proceduralism, therefore, in the specific question of justice and of normative application in complex, plural and post-conventional societies, might be the challenge to unfold the *concept of limited rationality* that could be directly applied to this subject matter, one that stems from a dialogue between *différance* and *intersubjectivity,* and one that might respond while being paradoxically a non-response. If this chapter confirmed the hypothesis that balancing, as a proportional evaluation of constitutional principles and values, is not the satisfactory response to the indeterminacy of law nor to the rationality in legal reasoning (at least one that acknowledges its boundaries in constitutional democracy), while revealing a possible response in this subject matter, the third part will radicalize even more this debate: it will show that only by acknowledging its boundaries can adjudication indeed grasp, although not thoroughly achieving them, the complexities and tensions of the reality and do justice to the other. It will thus confront the response we examined in this chapter with the premises of Derrida's philosophy and, from this debate, show that it is possible to think of another way to deal with cases, no longer by directly criticizing Alexy's concept of rationality – for we have already shown its metaphysical standpoint – but by directly applying the concept of limited rationality in a reconstructive manner to German and Brazilian constitutional realities.

Part III
The Concept of Limited Rationality

Chapter 7
Between *Différance* and Intersubjectivity: The Concept of Limited Rationality in Constitutional Democracy

Abstract By placing side by side Jacque's Derrida *différance* and Jürgen Habermas's *intersubjectivity,* it seems that a productive and invaluable dialogue between them is possible, for they show, through distinct views, how limited rationality is for the dilemmas of constitutional democracy and constitutional adjudication. This limited rationality, in turn, is the result of the perception of the fallible character of knowledge, of its incapacity to completely recollect and gather the complexities and tensions of history, and of its impossibility of entirely doing justice to the other's otherness. In the theme of justice in particular, these two views result in a tense but productive relationship between the idea of symmetrical justice, as equal treatment, and asymmetrical justice, as the "infinite heterogeneity". This agony, which has implications in theories such as Chantal Mouffe's agonism and Christoph Menke's paradoxical dualism at the very core of equality, leads, finally, to the perception that any resolution in this matter is a non-resolution. In this case, the resolution as a non-resolution inscribes in the practices of constitutional democracy an agony to do justice *here and now* while negotiating with a history that is always insufficiently recollected and gathered. It is in this dualism, this negotiation where the *concept of limited rationality* is unfolded, revealing how the tensions and dilemmas of constitutional democracy, either in the question of history or in the question of justice, are complex and cannot be reduced to the idea that a methodology, as long as filled with arguments, can provide rationality.

7.1 Introduction

Both Jacques Derrida and Jürgen Habermas are serious sources to grasp how the quest for a concept of limited rationality is a central issue to constitutional adjudication and, more particularly, to the problem of interpretation and application of law. From different origins and distinct outlooks, deconstruction and proceduralism have shown that they can be directly linked to the question of whether a metaphysical standpoint can lead, in the reality of constitutional courts, to the institutional

J.Z. Benvindo, *On the Limits of Constitutional Adjudication*,
DOI 10.1007/978-3-642-11434-2_7, © Springer-Verlag Berlin Heidelberg 2010

practice of judicial discretion by putting the principle of separation of powers in jeopardy. This chapter begins, for this reason, with a question that immediately stands out after having investigated the two views of the problem of legal interpretation and application when we assume balancing as a response to legal reasoning. Insofar as Derrida and Habermas aim, although in different ways, to disclose and undercut metaphysics, and, from their theories, it is possible to reveal the metaphysics embedded in balancing, particularly in the way Robert Alexy justifies and constitutional courts deploy it,[1573] how can we establish, on the other hand, a dialogue between these theories as a means to set forth a concept of limited rationality to legal reasoning?

If the concept of rationality Alexy sustains for balancing – which reflects, to a large extent, a rational response to the BVG's practice and activism[1574] (a characteristic we could extend to the STF)[1575] - already proved its metaphysical grounds, either from deconstruction or proceduralism, and a possible response to the indeterminacy of law within the context of plural and post-conventional societies was analytically discussed in the last chapter, the question now shifts to the very limits of this response and its direct application to the reality. Therefore, while in the last part of this book we focused on an influential and relevant interpretation of the BVG's practice (which could be extended somehow to the STF's) through Robert Alexy's theory, it is now time to verify that another conception of rationality is possible and apply it to those practices. For this purpose, even as a consequence of our previous debates, it is necessary to confront, in a more direct perspective, Derrida's deconstruction with Habermas's proceduralism, in order to unfold a rationality that acknowledges its boundaries in the realm of constitutional adjudication. The question is: how far is Habermas's proceduralism a viable response to the indeterminacy of law when challenged by Derrida's deconstruction? This investigation, for this reason, will concentrate on their possible connections, rather than on their possible divergences, which will lead to the following question: is it possible to argue that both theories are complementary to the analysis of constitutional democracy? Besides, how, from this complementary perspective, is it possible to extract a concept of limited rationality that best handles the dilemmas of constitutional democracy and constitutional adjudication, especially by focusing on the constitutional realities that are subject matters of this research?

This chapter will concentrate on introducing, from the tense but productive relationship between intersubjectivity and *différance*, a concept of limited rationality that can be immediately applied to the realm of constitutional adjudication. It will not yet enter, nevertheless, into the specific context of decision-making, whose investigation will be carried out in the next chapter. Rather, it will establish the premises to grasp how a concept of limited rationality can lead to a distinct look into the debates on constitutional democracy, as a necessary step to extend it later to the

[1573] See the first part.

[1574] See the second chapter.

[1575] See the third chapter.

analysis of constitutional adjudication. It is, in this chapter, that the relationship between intersubjectivity and *différance* will reveal that another way to think of *reason* is possible, one not relying on predetermined formulas and concepts, but which is instead constructed within the very democratic procedures. This reason, however, while not being able to entirely recollect and gather the history and all its tensions, cannot thoroughly do justice to the other, either. In any case, the acknowledgement of this double impossibility is what makes constitutional democracy possible, for "an event is only possible when it comes from the impossible."[1576] It is hence a reason that makes constitutional democracy possible by acknowledging its boundaries delineated both by history and by the quest for justice, whose comprehension, as we will remark, is also subject to an "irresolvable but productive tension"[1577] that enhances even more its challenges.

Since the purpose here is to unfold a concept of limited rationality by stressing the interconnections between intersubjectivity and *différance,* this chapter will begin by discussing the double bind that shapes the boundaries of reason: first, it will discuss the relationship between the quasi-transcendentalism of justice and history, in order to explore the premise that reason is limited on account of its incapacity to thoroughly recollect and gather the complexities and tensions of a determined reality (Sect. 7.2); second, it will enter into the complex theme of justice, as a means to show how reason is limited in virtue of its incapacity to entirely do justice to the other (Sect. 7.3). As long as this debate is where the connections between intersubjectivity and *différance* reveal the most relevant conflicts and where the complementary perspective can be constructed by means of a "irresolvable but productive tension"[1578] at the core of justice, we will explore more carefully some necessary aspects emerging from this concern for otherness: first, the possible relationship between the symmetry of equal treatment and the asymmetry of *différance* appears as a problem (Sect. 7.3.1); second, this question of how symmetry and asymmetry relate to each other in the realm of constitutional democracies is critically examined from Chantal Mouffe's agonism (Sect. 7.3.2); and third, the idea of equality itself emerges, which is the basis for the question of justice critically examined through Christoph Menke's stress on the individuality in an internal dualism with the symmetrical equality (Sect. 7.3.3). After this investigation, we will concentrate on the possible dialogue between intersubjectivity and *différance* in the quest for justice, in order to show that any resolution in this matter is, in fact, a non-resolution. As a non-resolution, though, it inscribes at the core of the quest for the other a tense but productive conflict that, as a self-corrective process, shows the boundaries of constitutional democracy and the boundaries of

---

[1576] Jacques Derrida, "As If It Were Possible," in *Negotiations: Interventions and Interviews, 1971–2001* (Stanford, CA: Stanford University Press, 2002), 344.

[1577] Axel Honneth, "The Other of Justice: Habermas and the Ethical Challenge of Postmodernism," in *The Cambridge Companion to Habermas*, ed. Stephen K White (Cambridge: Cambridge University Press, 1995), 319.

[1578] Ibid.

reason itself (Sect. 7.3.4). By showing the boundaries of reason and the boundaries of constitutional democracy, this chapter ends by suggesting how the concept of limited rationality, whose premises are fundamentally developed here, could be deployed in the realm of constitutional adjudication (Sect. 7.4), which is, as a matter of fact, the theme of the next and last chapter of this book.

7.2 When Proceduralism and Deconstruction Are Placed Side by Side: The First Insight into the Limits of Reason

When we place Habermas's proceduralism and Derrida's deconstruction side by side, some immediate divergences come into sight. For instance, while Habermas establishes the communicative reason as the primary issue of his philosophy, as a struggle against metaphysics and ideologies undermining the practices oriented towards mutual understanding - which he then extended to the analysis of constitutional democracies and, more specifically, to legal adjudication[1579] - Derrida places the otherness, the *différance,* as an intervenient practice against logocentrisms, and the marginalization and exclusion of the other, at the core of his thinking, then extended to the debate on democracy[1580] and on the question of justice.[1581] Habermas is more explicit in the construction of a self-reflexive society through communicative action; Derrida, on the other hand, is more emphatic in bringing the different to the fore of the debate, and in a radical asymmetrical perspective, because he says that he "[does] not believe there is a symmetry in intersubjective relations."[1582] Habermas, in any case, has in the basis of his account of reason the *intersubjectivity* and even wrote a book called *The Inclusion of the Other,*[1583] in which he treated the questions of ethnical and cultural self-understandings, minority groups in plural societies, and how constitutional democracy deals with it. He developed moreover the concept of *solidarity,* which he calls the other of justice as a reciprocal recognition of the individual as a member of a community, and thus as a concern for the well-being of the other.[1584] Derrida in turn is not unaware of the

[1579]See Jürgen Habermas, *Between Facts and Norms* (Cambridge: Polity Press, 1996).

[1580]See Jacques Derrida, *The Politics of Friendship* (London, New York: Verso, 2005).

[1581]See Jacques Derrida, "Force of Law: The 'Mystical Foundation of Authority'," *Cardozo Law Review* 11 (1990).

[1582]Jacques Derrida, "Negotiations," in *Negotiations: Interventions and Interviews, 1971–2001* (Stanford, CA: Stanford, 2002), 32.

[1583]Jürgen Habermas, *The Inclusion of the Other: Studies in Political Theory* (Cambridge, MA: The MIT Press, 1998).

[1584]See Jürgen Habermas, "Gerechtigkeit und Solidarität: Eine Stellungnahme zur Diskussion über 'Stufe 6'," in *Zur Bestimmung der Moral*, ed. Wolfgang Edelstein, Gil Noam and Fritz Oser (Frankfurt a.M.: Suhrkamp, 1986), 311.

discussions on reason,[1585] and, in his book *Rogues: Two Essays on Reason*,[1586] this link between reason and *différance* became explicit. Habermas stresses communicative action as a response to the new dilemmas of post-conventional and plural societies, as a struggle against metaphysics and ideologies undermining practices oriented towards mutual understanding; Derrida in turn draws attention to the different, to the singularity of the other, to the context as an attack on logocentrisms and the generalized thinking that historically prevailed in philosophy and in social practices. Intersubjectivity, solidarity and communicative reason, on the one hand, *différance* and simply reason, on the other, demonstrate that both philosophies might have something in common among their insoluble differences, and this "common" might reveal an interesting and relevant complementary perspective that can shape a concept of limited rationality, one that understands that its boundaries are constructed by the very history and also by the perception of the impossibility of fully doing justice, not solely in the sense of equal consideration and respect, but perhaps also in a radical, unconditional and infinite dimension of the other's otherness. In this regard, we could systematize their points of contact as follows:

1. Both inherit the philosophical linguistic turning point and stress the performative dimension of language,[1587] either by underlining the promise of the *to come*[1588] or

[1585]Indeed, Jacques Derrida is not a non-rationalist. He is rather a critic of the uses of the word reason. He sees, in the relationship between law and justice, how reason lies in the transaction between calculation and the incalculable (Jacques Derrida, *Rogues: Two Essays on Reason* (Stanford, CA: Stanford University Press, 2005), 151), "between, on the one side, the reasoned exigency of calculation or conditionality and, on the other, the intransigent, non-negotiable exigency of unconditional incalculability" (Ibid., 150). We cannot deny that reason is part of this insurmountable and incommensurable negotiation between law and justice, a negotiation that, despite the impossibility of completely undercutting metaphysics, should not be carried out with a *logos* behind it, with a *transcendental meaning* defining how this negotiation should take place. Reason lies paradoxically in the nonexistence of predetermined "rational standards," without this meaning naturally disrespect for the institutional history, the legal system, the inherited knowledge, since justice and law are paradoxically and inseparably connected, even to let deconstruction do its role. Reason lies thus in a responsible decision that, while respecting the law and its enforceability, opens them up to deconstruction, and hence to the other. He even writes about a "reason *to come*" and that he will not "[give] up on reason and on a certain "interest of reason"" (Ibid., 85). Notwithstanding that, the fact that he does not give up on reason does not mean that he embraced the Kantian rationalism, particularly the moral concept of regulative idea, since, in virtue of the "absolute and unconditional urgency of the *here and now*" (Ibid., 85), the *to come* is not "an ideal possible that is infinitely deferred" (Ibid., 84), but rather have a "structure of a promise – *and thus the memory of that which carries the future, the to-come, here and now*" (Ibid., 85–96).

[1586]Derrida, *Rogues: Two Essays on Reason.*

[1587]When Derrida writes about the negotiation between law and justice, he is there indicating that there is no justice if language is reduced to some methods and predetermined formulas. By the same token, he acknowledges the logocentric use of language as a means to destroy the singularity of the other. There is behind Derrida's deconstruction a real interest in performative acts; however, the way he treats them is distinct from Habermas's stress on communicative action.

[1588]See Jacques Derrida, "Remarks on Deconstruction and Pragmatism," in *Deconstruction and Pragmatism*, ed. Chantal Mouffe (London; New York: Routeledge, 1996), 82.

the universal pragmatic conditions of communication. In this matter, they converge on disclosing and undercutting metaphysics, for language is inherently metaphysical. They do not accept a last argument that is not subject to critical review, either through Derrida's premise that every negotiation deconstructs any assurances, dogmatic certitudes that could be placed behind negotiation, or Habermas's presupposition that every argument, every content is in interaction with a form or procedure, and hence submitted to an intersubjective practice of reciprocal deliberation.

2. While Derrida stresses the quasi-transcendentalism of the messianic character of the *to come,* the justice *to come,* the democracy *to come,* and even the reason *to* come, as if it were a "quasi-normative axis of an emancipatory, democratic politics, based in the undecontructible, context-transcendent, formal universality of justice,"[1589] Habermas underlines the quasi-transcendentalism of the counterfactual validity field of communication - expressed through weak, neutral and necessary counterfactual presuppositions arising from the very irrefutability of communication[1590] - which is projected into a procedure oriented towards mutual agreement in which all those potentially affected can freely and equally participate. Both thinkings acknowledge, in different perspectives, however, the double bind that exists between the calculable and the incalculable, between facts and norms. In this respect, deconstruction and proceduralism work with performative acts that are not limited to history, to the calculable, to the facts, but are open towards the other, thereby connecting to a quasi-transcendental premise of justice, which Derrida relates to the asymmetrical *to come,* the *impossible,* as if it were "[waving] between imperative injunction (call or performative) and the patient *perhaps* of messianicity (nonperformative exposure to what comes, to what can always not come or has already come),"[1591] and Habermas relates to the reciprocal principle of treating the other with equal consideration and respect.
3. Whereas Derrida accounts for his quasi-transcendentalism by arguing that it is not a regulative idea in the Kantian way, for it "precedes me, swoops down upon and seizes me *here and now* in a nonvirtualizable way, in actuality and not potentiality,"[1592] as "an injunction that does not simply wait on the horizon,"[1593] Habermas's association of his counterfactual premises with the concept of regulative idea, in all its expressions (for instance, in the quest for the "single right answer"), is not a deferred and passive observation of the world. It rather calls also for an intervenient attitude *here and now* by "detranscendentalizing" the counterfactual premises into the reality, in an immediate attitude towards the

[1589]Simon Critchley, "Frankfurt Improptu - Remarks on Derrida and Habermas," in *The Derrida-Habermas Reader*, ed. Lasse Thomassen (Edinburgh: Edinburgh University Press, 2006), 100.

[1590]See the last chapter.

[1591]Derrida, *Rogues: Two Essays on Reason*, 91.

[1592]Ibid., 84.

[1593]Ibid.

expansion of the practice of communication in all distinct spheres of social life, exposing then the tension between facts and norms.[1594]

4. Derrida and Habermas remark that, either in the negotiation between the calculable and the incalculable or in the procedures lying in the tension between facts and norms, there is no guarantee of success nor certainty, for this process is hermeneutical and the knowledge is fallible, always open therefore to new interpretations. Derrida develops this thinking not by assuming a validity field with some universal counterfactual presuppositions of communication, but through *iterability*, as if he were "[undermining] the possibility of a pure communication *ab initio*."[1595] It thereby inscribes the repetition in every new beginning, as if the founding moment were perpetuated in every singular situation, while opening it up to self-critique and perfectibility over time. Similarly, he stresses *undecidability*, arguing then that there are no safeties, assurances in the negotiation, but rather an asymmetrical play that remains open to interpretation and reinvention, without any origin or end, and *autoimmunity*, which is the right to self-critique and perfectibility.[1596] Habermas, on the other hand, underlines the procedures of mutual understanding, which should be carried out without any substantiality behind, transforming then any raised valid claim into a subject matter of critical scrutiny, while placing emphasis on the dimension of time through the idea of self-correcting learning process. True, one might argue that *iterability* and self-correcting learning processes are irreconcilable, for it seems that Habermas sets up a timeline of progress that is absent in the notion of iterability.[1597] If this argument seems plausible, we could say, though,

---

[1594]Indeed, the intervenient attitude *here and now* stemming from this "*detranscendentalization*" of the counterfactual premises of communication clearly exposes that Habermas's stress on regulative ideas, as we could observe in many of his developments (for instance, the "single right answer," the ideal presuppositions of communication, etc) has not the same character of Kant's moral regulative idea. It is not a deferred and passive observation of the world, but rather an effective and immediate attitude towards the expansion of communicative abilities through procedures of mutual agreement. Besides, if there are some counterfactual presuppositions transcending contexts, they exist only to be "detranscendentalized" in contexts. The validity field, therefore, is continuously subject to critical review in the practices of lifeworld; the validity of any claim thereby always "rests on shaky foundations" (Jürgen Habermas, *Moral Consciousness and Communicative Action* (Cambridge, MA: The MIT Press, 1990), 14). This is what characterizes the tension between facts and norms in all Habermas's analyses of social life: as a post-metaphysical thinking, it cannot avoid those "quasi-transcendentalizations" for a practice of mutual understanding from which communicative reason, as a response to ideologies and metaphysics, unveils itself.

[1595]Mark Devenney, *Ethics and Politics in Contemporary Theory: Between Critical Theory and Post-Marxism* (London: Routeledge, 2004), 56.

[1596]Derrida, *Rogues: Two Essays on Reason*, 86.

[1597]For Derrida, in the idea of iterability, there would be no timeline of progress, but a continuous play, without origin or end. Habermas, on the other hand, clearly announces that the procedures oriented towards mutual agreement are carried out in the dimension of time, as a self-correcting learning process. For Habermas, "citizens must see themselves as heirs to a founding generation, carrying on with the common project" (Jürgen Habermas, "On Law and Disagreement: Some that this self-correcting learning process is marked somehow by what Derrida calls *undecidability,* for there is no assurance that the present is better than the past, or that the future will be better than the present. Moreover, we could not ignore that *iterability,* while being characterized by *undecidability,* is also marked by *autoimmunity,* which is the right to self-critique and perfectibility.[1598] In other words, it seems that Derrida, at the core of the asymmetrical play where linguistic interactions take place, does not ignore the capacity of self-critique and perfectibility of a certain reality. For this reason, if the negotiation leads to *iterability* and the procedure, in turn, to a *self-correcting learning process,* we could already infer from this investigation that *iterability* (and *autoimmunity*) and *self-correcting learning process* are not so radically opposing when they are projected into practical issues: both point out the need for self-critique and also perfectibility of the negotiation or the procedure by focusing on the other, even though they might interpret this other differently. Both point to the requirement of a repetition in the very reinvention.

5. They converge on recognizing the relevance of history, institutional history and the *enforceability* or *deontology* of law in every negotiation or procedure taken place within constitutional democracies, even to maintain a minimum of security.[1599] Derrida regards enforceability of law in an intimate connection with justice as a condition for not transforming justice into a moralizing principle, and thus converting it into a sort of metaphysics. Indeed, he sees that *différance* is linked to the legitimate use of force, to the negotiation between law and justice, to the "paradoxical situations in which the greatest force and greatest weakness strangely enough exchange places"[1600] and, therefore, it refers to the "wholly history."[1601] On the other hand, Habermas's tension between facts and norms already points out how the institutional history and the deontological character of law play a primary role in his thinking, not only

Comments on 'Interpretative Pluralism'," *Ratio Juris* 16, no. 2 (June 2003), 193). While there is the possibility of ongoing disagreement, there are also, throughout history, many "stable [points] of reference" (Ibid., 193), and it is exactly in this tension between these "stable points of reference" and the possibility of disagreement that every new generation learns from the past: there is, accordingly, a movement towards perfectibility, never attained, though. Thomassen, in this respect, argues that this stress on the self-correcting learning process might point to "a *telos* of reconciliation of constitutionalism and democracy" (Thomassen, "'A Bizarre, Even Opaque Practice': Habermas on Constitutionalism and Democracy," 189), which could express a sort of "temporality, historicity and futurity understood as presence, as a modality of the present. It is a future understood as the continuation of the same, as the unfolding of a system of rights in the constitution, or as the anticipation of a future agreement" (Ibid., 184).

1598 Derrida, *Rogues: Two Essays on Reason*, 86.

1599 See Jacques Derrida, "Negotiations," in *Negotiations: Interventions and Interviews, 1971–2001* (Stanford, CA: Stanford, 2002), 17.

1600 Derrida, "Force of Law," 929.

1601 Ibid.

because they are the necessary counterpart of validity in this double bind, but also because Habermas constructs a robust theory applied to constitutional democracies centered on the democratic principle, which, more than transporting the tension between facts and norms to the legal form,[1602] demands the existence of institutionalized procedures that preserve the deontology of law.

Their thinkings, accordingly, have, as a primary focus, the perception of the very limits of reason, which does not have any assurance or certitude in the negotiation between the calculable and the incalculable or in the procedure lying in the tension between facts and norms. A second focus is on how they manage the tension between the reality and the quasi-transcendentalism calling for an immediate attitude in favor of the other's otherness. On the one hand, they understand that reason has its boundaries, which the negotiation and the procedure will construct over time; on the other, they conceive of their philosophies as strictly linked to the concern for otherness, even to transform the reality. In both cases, there is the deconstruction or the denial of some material contents or some standards establishing how to negotiate or to proceed in a determinate circumstance; in both cases, they disclose a real concern for otherness. Their theories, through distinct aspects, bring into focus two fundamental premises for the construction of a limited concept of rationality: first, there cannot be certainties, formulas, "rational standards" exempt from being submitted to critical review, either through the idea of iterability in which deconstruction appears or the accent on procedures oriented towards reciprocal agreement, and, in this matter, reason, which manifests itself in the negotiation between the calculable and the incalculable or between facts and norms, is fragile and limited by the negotiation or by the procedure; second, there cannot be the nonobservance of the complexity and the tense quality that are inscribed in the negotiation between the calculable and the incalculable, in the procedure between the facts and norms, for it is this complexity and this tension that do not transform the thinking into a simple repetition of the same structures, thereby connecting it to an intervenient attitude in the world towards the otherness.

If, nonetheless, these connections between Derrida and Habermas could, in a sense, be placed side by side and even directly indicate the boundaries of reason for its incapacity to completely gather and recollect all the tensions and complexities of the reality, the same could not be said about the quest for the other's otherness, about the quest for justice. In this aspect particularly – the justice – is where we find the most serious challenge for a possible dialogue between them. After all, how could Habermas's symmetry of equal consideration and respect and Derrida's asymmetry of the *to come* dialogue with each other in this search for a concept of limited rationality?

[1602] See Habermas, *Between Facts and Norms*, 108.

7.3 The Quest for Justice: A Dialogue Between Symmetry and Asymmetry?

7.3.1 Introduction

Not only is reason limited because of history, its tensions and complexities, which cannot be entirely gathered and recollected and must be either deconstructed or placed within procedures oriented towards mutual agreement. If the calculable or the facts already establish relevant limits to reason, the incalculable or the validity makes them even more pronounced. In this matter, reason is limited in virtue of the very impossibility of justice, not only because the idea of equal treatment and respect cannot be entirely achieved, but also because the singularity of the other, as a nonreciprocal concern for the other, can even mean, in a particular situation, the very opposite of symmetrical justice. Equal concern and respect, as the basis of symmetrical justice, might not reconcile with the idea of a justice *to come*. The regulative idea symmetrical justice expresses, rooted in the principles of equality and freedom, might not harmonize with the "unforeseeable coming of the other," with the "decision of the other, of the other in me, an other *greater and older than I am*,"[1603] that is, with an immediate demand that is more than the equal treatment an individual, by helping the other, expects for herself. The idea of reason, therefore, while encompassing the symmetrical justice that is at the core of communicative action, can become even more complex when we remark that, perhaps, in a particular circumstance, it is necessary to supplement symmetrical justice with a "principle of unilateral, entirely disinterested help."[1604] While this might be problematic within the context of a society where each individual is recognized as being equal among others[1605] and, as such, participates in public deliberations with equal rights, in the reality where reason appears, we cannot disregard this aspect *différance* sets forth. Therefore, the purpose now is to radicalize reason in its connection with the quasi-transcendentalism of justice. *Différance,* accordingly, will challenge *intersubjectivity*.

As we observed in the theories oriented towards facing the challenge of the indeterminacy of law in the last chapter (Klaus Günther, Ronald Dworkin, and Jürgen Habermas), the premise of equality and freedom is at the core of the idea of impartiality or justice, which leads decision-making to a concern for the other. In all of them, there was the modern concept of justice of treating all individuals with equal consideration and respect, an idealization that stands in a continuous struggle with a reality characterized by ideological and strategic uses of reason. More emphatically in Habermas, justice lies in the demand that every individual, when she has the chance to articulate her voice publicly, can do so freely with equal

[1603]Derrida, *Rogues: Two Essays on Reason,* 84, emphasis mine.

[1604]Honneth, "The Other of Justice: Habermas and the Ethical Challenge of Postmodernism," 319.

[1605]Ibid.

consideration and respect. There is nothing behind the idea of justice except this transposition of the idea of equal treatment to procedures oriented towards reciprocal agreement. That being so, a characteristic of these thinkings is to work with symmetry in the very diversity, which we cannot consider troublesome as long those discourses are regarded as procedures oriented towards consensus,[1606] notwithstanding its fallibilist quality. For they seek to integrate the other *discursively*, and not through any other asymmetrical form of inclusion, such as benevolence, helpfulness, and philanthropy,[1607] the reason is limited, at the end, by the very impossibility of a complete inclusion of the other in practical discourses. When these conclusions are projected into the institutional realm, the idea of the impossibility of a complete inclusion of the other in practical discourses becomes the most complex dilemma of constitutional democracy. In this case, reason is limited because of the impossibility of constitutional democracy, in all its institutional forums, to provide mechanisms that could discursively thoroughly integrate the other.

The problem arises – and this leads to the second insight into the analysis of a concept of limited rationality – when we start questioning how powerful the discourse to integrate the other, within plural and multifaceted contexts where different standpoints are placed side by side, is. In other words, the dilemma stands out when we start questioning how the singularity of the other is really a primary concern in the quest for justice, if discourse is regarded as the only post-metaphysical mechanism of integration. The asymmetrical forms of integration, in which one's responsibility and concern for the other occur regardless of any reciprocality of the other in relation to her, as we can observe in acts of benevolence, helpfulness and care, are placed in the background of this practice of mutual understanding. Habermas, in his search for a post-metaphysical thinking, needed to avoid falling into some forms of asymmetrical integration, because, if the primary parameter is the discursive procedure, any asymmetry could mean a metaphysical standpoint behind the procedure: "Every person is indeed always included in a practical discourse only as an unrepresentable individual, but the presuppositions of symmetry obtaining in practical discourse necessitate that all particular bonds be disregarded and, accordingly, that viewpoints of care recede into the background."[1608] That is why the emphasis on the particular individual and the immediate needs deriving from her own condition, which can claim more than justice in the modern sense, appears to be overcome by the need to accomplish a consensus, even though understood as a regulative idea that will be in tension with fallibilist mutual agreements. The universal claim to equal treatment, now reconfigured in a discursive basis, may not be enough to the singularity of the other's otherness.[1609]

[1606]See Ibid., 316.

[1607]Ibid.

[1608]Ibid.

[1609]Habermas knows that a sort of compensation for the recourse to the universalistic idea of justice may be necessary (See Jürgen Habermas, "Equal Treatment of Cultures and the Limits of

Yet, this ultimately symmetrical achievement of mutual agreement Habermas thematizes, albeit never entirely accomplished and thus treated solely as a regulative idea, may not culminate in the conclusion of a certain contamination of the symmetrical over the asymmetrical, but rather a tension between both. Naturally the central issue here lies in carrying out this tension within the boundaries of a democratic procedure of opinion- and will formation, which, as shown, will point to a modern concept of justice discursively remodeled.[1610] It is exactly this discursively remodeled justice that will allow a certain exercise of private autonomy in the middle of civic practices of public participation, providing thereby a protection against violations of each individual's integrity in society. Indeed, it is only in virtue of this discursively remodeled justice that positive rights can enforcedly protect the other. Yet, some doubts may still lie in how this reason, limited because of the tensions inscribed in the procedures oriented towards reciprocal understanding, can protect the singularity of the other's otherness. The final aspect that may bring something relevant to this debate refers to *différance* as a possible radicalization

Postmodern Liberalism," *The Journal of Political Philosophy* 13, no. 1 (2005): 2). Solidarity would be this compensation, as the other of justice (See Habermas, "Gerechtigkeit und Solidarität: Eine Stellungnahme zur Diskussion über 'Stufe 6'," 311), as a link that "[unites] citizens as members of a political community beyond merely legal relations" (Habermas, "Equal Treatment of Cultures and the Limits of Postmodern Liberalism," 2). But what could be a possible complementation to justice in this shift to a more emphatic consideration of the particular is treated by Habermas, nonetheless, in the same structure of rational discourse. Solidarity, in order to escape from a republican interpretation of an *ethos,* needs to be radicalized as a "solidarity among 'others'" (Ibid., 3), which connects it to democratic procedures of opinion-and will formation. Behind the idea of solidarity, therefore, lies the universal premise of equal treatment applied to procedural discourses, which must be inevitably considered if one intends to safeguard the integrity of distinct individuals: "Only the difference-sensitive egalitarian universalism of equal rights can fulfill the individualistic requirement to guarantee equally the vulnerable integrity of individuals with distinctive life histories" (Ibid., 13). That is why if, on the one hand, solidarity comprises a concern for the other's well-being, as Habermas says by quoting Schiller in the sentence "*all* men become brothers" (Habermas, "Gerechtigkeit und Solidarität: Eine Stellungnahme zur Diskussion über 'Stufe 6,'" 311, translation mine), on the other, it refers to intersubjective shared forms of life where every participant, as brothers, interact with one another as a means to achieve consensus with justice. Hence, this brotherhood, which could indicate a possible asymmetrical sentiment towards the other, is taken, however, by the symmetrical need to achieve consensus through dialogue: "Without the solidary empathy of each one in place of *all* others we could not come to a consensual solution" (Ibid., 314, translation mine). The idea of solidarity, while inserting a certain fraternal meaning into this relationship where justice comes to play, seems, ultimately, to be contaminated by the generalized and symmetrical idea of dialogue oriented towards mutual understanding. Justice, after all, "*[has] a privileged position*" (Jürgen Habermas, "Reply to Symposium Participants, Benjamin N. Cardozo School of Law," in *Habermas on Law and Democracy*, ed. Michel Rosenfeld and Andre Arato (Berkeley, CA: University of California Press, 1998)., 400.) Yet, in order to preserve this mutual equal treatment among individuals in complex societies, the dissensus are justly and solidarily articulated by the idea of tolerance, "a price for living together in an egalitarian legal community in which groups with different cultural and ethnic backgrounds must get along with one another" (Ibid., 393). Justice, therefore, refers to solidarity and tolerance through practical discourses oriented towards consensus in the middle of many and sometimes irresolvable dissensus.

[1610]See the last chapter.

of the modern concept of justice as equal concern and respect, in a complementary and irresolvable perspective (since they can, in a singular context, result in distinct actions within the empirical world). The question, in this matter is: is it possible to reconcile, in a constitutional democracy and, particularly, in constitutional adjudication, intersubjectivity with *différance*? In other words, is it possible to pursue rational consensus while leaving open the possibility of *différance*? Is there space in the democratic procedures of opinion–and will formation for asymmetrical forms of integration, where we could think of "an equality that would not be homogeneous, that would take heterogeneity, infinite singularity, infinite alterity into account?"[1611] Is it possible to bring *différance* to the nucleus of democratic procedures of normative application? Could it indeed do more justice, within the context of democratic procedures of opinion – and will formation, to the individual than the premise of equal consideration for *all*? Finally, could it bring forth a response to legal adjudication?[1612]

This theme is very intriguing and enters directly into the core of this purpose to unfold a concept of limited rationality, for it radicalizes the question of the otherness, and thus of justice. However, before bringing a final perception of this relationship between *intersubjectivity* and *différance,* we should examine two prior possible viewpoints: one that still sees, in Habermas's proceduralist approach, the asymmetrical being contaminated by symmetrical forms of discourse, and thus by the idea of consensus; and the other that states that the universal principle of equality needs to be in an internal relationship with a normative obligation towards individuality. They are both expressions of a possible dialogue between *intersubjectivity* and *différance* and, therefore, could assist us in finding a possible solution – or not – to this impasse.

7.3.2 *Is Really the Quest for Consensus Incompatible with Asymmetry? A Look Into Chantal Mouffe's "Agonist Model of Democracy"*

Chantal Mouffe is probably one of the most vehement critics of Habermas's discourse theory exactly because she sees a certain loss of asymmetry in the emphasis on consensus Habermas develops. She extends this problem to the central issues of constitutional democracy, and, by constructing what she calls an "agonistic model of democracy,"[1613] she seeks to demonstrate the importance of antagonism and the relations of power in opposition to consensus. In her book *The Democratic Paradox,* she systematizes the major premises of her thinking, while clearly

[1611]Jacques Derrida, "Politics and Friendship," in *Negotiations: Interventions and Interviews, 1971–2001* (Stanford, CA: Stanford University Press, 2005), 179.

[1612]We will examine this question in the next chapter.

[1613]Chantal Mouffe, *The Democratic Paradox* (London; New York: Verso, 2000), 98–105.

pointing out her differences from Habermas's proceduralist model. For this purpose, she demonstrates that her model is not oriented towards providing a rational justification for democracy, because she sees that privileging rationality through mechanisms of communicative aggregation "leaves aside a central element which is the crucial role played by passions and affects in securing allegiance to democratic values"[1614] as well as precludes the "conditions of existence of the democratic subject."[1615] By stressing the civic practices that disclose a "passionate commitment to a system of reference,"[1616] she argues that the proceduralist account in the way Habermas treats it loses this social dimension by setting up a "strict separation between 'procedural' and 'substantial' or between 'moral' and 'ethical."[1617] It relies hence on the belief in "purely neutral procedures"[1618] that inscribe in them the "dream of a rational consensus,"[1619] which, at the end, undermines the antagonisms and different forms of power. In Mouffe's viewpoint, the idea of a rational consensus grounded in the moral point of view of justice, as equal treatment and respect, has no justification at all, for it is unable to gather and acknowledge all the antagonisms that are at the core of plural societies.[1620] Her approach, therefore, resides in focusing on the "question of power and antagonism at its very centre,"[1621] and thus in examining the distinct exercises of power in social life, in how they can operate in a way that can be more "compatible with democratic values."[1622] In other words, Mouffe seeks to frame a model of democracy in which the "political," as this inherent antagonism of all humans, can correlate with "politics," as all institutions, discourses and practices that attempt to organize and conciliate those distinct manifestations of the 'political.'[1623]

The "agonistic model of democracy" is thus characterized by this conflictive relationship between the 'political' and the 'politics', which, unlike Habermas's proceduralist model, in her opinion, is not oriented towards consensus. Instead, its purpose is simply the "creation of unity in a context of conflict and diversity."[1624] The heterogeneity is a fundamental element of democracy, and, while being not part of the "us," it is not "an enemy to be destroyed,"[1625] either. The one that is part of this heterogeneity is simply regarded as an adversary whose rights to participate and defend her opinions are preserved. Mouffe, like Habermas, also states that, in

[1614]Ibid., 95.

[1615]Ibid.

[1616]Ibid., 97.

[1617]Ibid.

[1618]Ibid.

[1619]Ibid., 98.

[1620]Ibid., 98–99.

[1621]Ibid., 99.

[1622]Ibid., 100.

[1623]Ibid., 101.

[1624]Ibid.

[1625]Ibid., 102.

this relationship with the other, there must be tolerance, which, in any case, cannot mean indifference towards the other, but rather "treating those who defend them as legitimate opponents."[1626] Consequently, those legitimate opponents have necessarily a sentiment of justice, which, indeed, does not seem to differ from the modern concept of treating all individuals with equal concern and respect. Those adversaries, after all, share "some common ground because we have a shared adhesion to the ethico-political principles of liberal democracy: liberty and equality."[1627] The difference we could, nonetheless, remark is that Mouffe attempts to dissociate the concept of modern justice from its possible reshaping within the context of a tension between facts and norms leading to a rational resolution through consensus. As a critic of this model of rationality in Habermas's thinking, she considers more adequate to simply sustain that we must implement justice without the idea of a possible rational resolution with mutual agreement, provided that antagonisms are an inherent characteristic of democracy. We must implement justice by acknowledging the relevance of those antagonisms. Moreover, insofar as those antagonisms are part of the democratic realm, where adversaries attempt to achieve some temporary *conventions* – which she points more to Thomas Kuhn's[1628] idea of adherence to a new scientific paradigm[1629] than to the idea of rational consensus - those antagonisms become *agonisms,* in which the other is treated not as an enemy but simply as an adversary. Her model of democracy is thus characterized by this intent to "transform *antagonism* into *agonism*"[1630] whereby the passions towards democracy, and not merely rational consensus, are awakened.[1631]

Yet, except for this emphasis on the asymmetrical forms of integration that might reside in Mouffe's civic purpose of awakening and mobilizing "passions toward democratic designs"[1632] – which, in any case, will have behind the symmetrical presence of justice – and her stress on antagonism rather than on discourse, it seems that Mouffe does not radically differ from Habermas's premises. In truth, the more we examine the central arguments Mouffe brings forward against Habermas, the more it seems that she shares similar ideas, even though she continuously stresses their differences. Like Habermas, Mouffe develops her "agonistic model of democracy" with a clear interest in the possibility of including the other in democratic spaces of participation, which connects both to the modern concept of justice, notwithstanding her attempt to dissociate it from the idea of rational discourse. Both share the opinion that plural societies, while living with antagonisms, must tolerate them by bestowing on every individual, even the adversary, the right to be

[1626] Ibid.

[1627] Ibid.

[1628] See Thomas S. Kuhn, *The Structure of Scientific Revolutions* (Chicago: University of Chicago Press, 1996).

[1629] Mouffe, *The Democratic Paradox*, 102.

[1630] Ibid., 103.

[1631] Ibid.

[1632] Ibid.

treated as a legitimate opponent. For Habermas, after all, the other is certainly not treated as an enemy, but one individual with voice; otherwise, the structure of his intersubjectivity would totally lose its *raison d'être*. Besides, we could even argue that both, in a sense, have in common a certain universalism, although she states against Habermas that his philosophy transforms democracy into a "moment in the unfolding reason, linked to the emergence of universalist forms of law and morality."[1633] The concern for equal treatment and respect, which corresponds to the universal principle of justice, is behind Mouffe's "agonistic model of democracy," when she claims that every adversary has to adhere to some ethical-political principles of freedom and equality.[1634] Even her idea of temporary *conventions*, as if, in virtue of the adversary, one's position rooted in a type of political identity "[underwent] a radical change,"[1635] does not seem thoroughly distinct from the idea of rational consensus, since she assumes the dialogue as a procedure to include the other, which could raise new dilemmas for one's political identity and also promote effective changes in one's political view. On the other hand, Habermas has not understood that consensus is achieved in a permanent way. Indeed, the very character of a limited rationality stemming from his proceduralist account resides in the premise that every consensus is continuously subject to critical scrutiny, and thus submitted to procedures of revalidation throughout history.

We could read Mouffe's critique, therefore, as an attempt to set forth a concept of democracy bearing some contributions of a postmodern thinking – and here we could point out her interest in not falling into some rational paradigms of modernity – while still nevertheless being unable to release herself from some of those modern paradigms. This paradox emerges, on the one hand, from Mouffe's stress on antagonism and asymmetries, indicating then a real concern for the singularity of the other, in opposition to a possible symmetry caused by the idea of rational consensus through discourse and modern premises, and, on the other, from her still manifest maintenance of modern premises, such as the principles of freedom and equality inscribed in the idea of justice, at the core of her defense of democracy. This paradox, at a deeper level, might reveal how complex it is to refrain from these modern premises when the debate on democracy arises. Besides, it is by reason of this complexity that the critique Mouffe sustains against Habermas may be neglecting the very tension that democracy, for Habermas, unavoidably holds.

The first problem we immediately remark in Mouffe's analysis of the Habermasian model is how she treats the idea of the tension between facts and norms in his proceduralist account. When she states that Habermas leaves aside, in the quest for rational consensus, passions and affects that are part of democratic values,[1636] or

[1633]Chantal Mouffe, "Deconstruction, Pragmatism and the Politcs of Democracy," in *Deconstruction and Pragmatism*, ed. Chantal Mouffe (London; New York: Routeledge, 1996), 1.

[1634]Mouffe, *The Democratic Paradox*, 102.

[1635]Chantal Mouffe, "Deliberative Democracy or Agonistic Pluralism?," *Social Research* 66, I, no. 3 (Fall 1999): 755.

[1636]Mouffe, *The Democratic Paradox*, 95.

when she mentions that Habermas sets up a "strict separation between 'procedural' and 'substantial' or between 'moral' and 'ethical',"[1637] she may expose a certain misinterpretation of the complex relationship between form and content that is the nucleus of his philosophy. The fact that Habermas avoids placing any substantiality behind the practical discourse, for "under conditions of post-metaphysical thinking, we cannot expect a further reaching consensus that would include *substantive issues*,"[1638] in no hypothesis whatsoever signifies that he makes a strict separation between form and content. What exists is a tense relationship, since the content is constructed according to procedures whose validity is conditioned to an ongoing practice of rational dialogue. Form and content, procedures and reason interpenetrate in one another and therefore cannot be conceived as strictly separated. Indeed, for a concept of limited rationality in a post-metaphysical basis, this strict separation Mouffe advocates against Habermas would mean that his theory bears the idea of an absolute knowledge, which is exactly the reversal: "In view of the fundamental fallibility of our knowledge, neither of these two elements alone, neither form nor substance taken by itself, suffices."[1639] In Habermas's theory there is not, for this reason, "purely neutral procedures,"[1640] nor is his theory rooted in procedures without any "substantial ethical commitments,"[1641] but, instead, in procedures in tension with distinct contents (ethical, pragmatic, moral, political, etc), which are, however, not given in advance as safe arguments exempt from being submitted to critical review.

On the other hand, Habermas is not a dreamer of a consensus[1642] or of a society with "perfect harmony or transparency"[1643] where rational consensus would totally suppress the relations of power,[1644] as if the idea of democracy were to achieve a final point, as if a "final resolution of conflict [were] eventually possible."[1645] He is certainly not naïve at this point to believe that democracy would be better if there were no dissensus whatsoever. Naturally Habermas knows that antagonisms and different forms of power are necessary for the very dynamics of democracy, and this is why the core of his thinking is not characterized by a quest for the elimination of antagonisms, but rather by the search for an expansion of the institutional channels where every individual can have the right to freely and equally express her voice. In truth, this premise does not appear to be incompatible with Mouffe's purpose of "[constituting] forms of power that are compatible with democratic

[1637]Ibid., 97.

[1638]Habermas, "Reply to Symposium Praticipants, Benjamin N. Cardozo School of Law," 406.

[1639]Ibid., 408.

[1640]Mouffe, *The Democratic Paradox*, 97.

[1641]Mouffe, "Deliberative Democracy or Agonistic Pluralism? In: *Social Research,*" 749.

[1642]Mouffe, *The Democratic Paradox*, 98.

[1643]Mouffe, "Deliberative Democracy or Agonistic Pluralism?," 752.

[1644]Ibid., 753.

[1645]Mouffe, "Deconstruction, Pragmatism and the Politics of Democracy," 8.

values,"[1646] as long as there is not, in Habermas's view, a subversion of the facts and norms, but a tension between both. His counterfactual presuppositions of communication, after all, are not abstractions totally disconnected from the context but exist precisely to be "detranscendentalized" in the practices of social life. This is what configures the tension in which the other is included to the extent that she is treated with equal concern and respect, despite the impossibility of a total accomplishment of this ideal in reality. Hence, there is not, in Habermas's thinking, a "fantasy that we could escape from our human form of life"[1647] nor a denial of the ontological impossibility of ideal speech situations in the practice of social life.[1648]

Moreover, when Mouffe argues that her "agonistic model of democracy" works with the tension between "politics" and the "political," she is not distancing herself radically from Habermas, as long as discourse is unavoidably an elementary form of conciliating the distinct manifestations of the 'political' in a democracy through the participation of individuals in forums where everyone has the right to express her voice. Her aim to create a "unity in a context of conflict and diversity,"[1649] in the same way, is not antagonistic to the idea of a consensus defended by the proceduralist approach. No consensus, which is fallible, fragile, unsteady, exists without the possibility of being undermined by procedures oriented towards new consensus. No consensus, accordingly, lives democratically without the possibility of heterogeneous opinions, for this would jeopardize the tension between form and content. Antagonisms are definitely not forms of irrationalism or archaism for Habermas, nor are they eradicated in his theory, as Mouffe sustains,[1650] but rather they are, as any other content, necessary arguments for the dynamics of democracy as long as they feed the tension between facts and norms, a tension – it is necessary to repeat in this context – that always remains.

In any case, however, Mouffe introduces some relevant aspects in this discussion, and they stem from her connection to some of Derrida's premises. Her argument that Habermas's proceduralist approach establishes the idea of

[1646]Mouffe, "Deliberative Democracy or Agonistic Pluralism?," 753.

[1647]Ibid., 750.

[1648]Mouffe attempts to prove this ontological impossibility by appealing to Wittgenstein and Slavoj Zizek, through Lacan, as a critique against Habermas, as if he were sustaining the possibility of those ideal presuppositions of his theory being entirely achievable. Habermas, however, has never abandoned the tension between facts and norms in all aspects of his thinking, which exposes that he acknowledges the always exiting antagonisms, power relations, ideologies, ethical conflicts of values, etc., in the factuality. For Habermas, the obstacles for this consensus are not merely empirical or epistemological but also ontological, for the very structure of consensus carries its own dissolution. (See Mouffe, "Deliberative Democracy or Agonistic Pluralism?," 749–752).

[1649]Mouffe, *The Democratic Paradox*, 101.

[1650]According to Mouffe, in Habermas's proceduralist model, "everything that has to do with passions, with antagonisms, everything that can lead to violence is seen as archaic and irrational." Besides, his theory has, in her view, the "negation of the ineradicable character of antagonism" (Chantal Mouffe, "Decision, Deliberation, and Democratic Ethos," *Philosophy Today* 41 I, no. 1 (Spring 1997): 25).

"consensus without exclusion"[1651] in which agonism would have been "completely eliminated"[1652] is, as shown, not entirely correct, but her stress on antagonisms, exclusion, violence, passions towards democracy might reveal how asymmetry plays a central role in her thinking. She could not release herself from some modern premises, such as the idea of justice grounded in equality and freedom, but she brings forward some elements that may question those premises. One of these central elements is Derrida's concept of undecidability and the premise of the irreducibility of the other's otherness as a more adequate way to think of ethics and politics.[1653] She sees in Derrida's elegant and complex proposal the grounds for justifying that the heterogeneity and the antagonisms are the primary character of democracy and for demonstrating that consensus is impossible, for every decision necessarily produces violence to the other's otherness. Besides, she argues that deconstruction discloses the very instability of any consensus, and how this instability can be, at the same time, the risk and the chance of democracy.[1654] Democracy, by assuming these premises, would not be, accordingly, a "moment in the unfolding reason, linked to the emergence of universalist forms of law and morality,"[1655] but, rather, a moment in which deconstruction plays its role. Democracy would have conflicts and antagonisms as simultaneously "condition of impossibility of its final achievement."[1656]

However, does this contact with deconstruction make Mouffe's philosophy thoroughly distinct from the proceduralist approach? Certainly it gave her a relevant basis to sustain the emphasis on agonisms, the particular, the singularity of the other's otherness in her thinking. She could, by interpreting Derrida, escape from a certain modern tradition we can still notice in Habermas's proceduralist account and develop a theory of democracy that is not founded on the premise of a rational consensus, even though consensus is rather inevitable and necessary for democracy. She could use, in some way, *différance* against consensus. Nonetheless, as investigated, by stressing the singular, the particular, the other's otherness, Mouffe's thinking could not be released from the modern concept of justice. She could not escape from the premise of freedom and equality that is embedded in every democracy. True, she goes further than this modern concept of justice when she examines Derrida's democracy *to* come[1657] and friendship,[1658] as the infinite, the promise, the "inaccessible because it is inconceivable in its very essence, and hence in its *telos*,"[1659] which conflicts somehow with the modern concept of

---

[1651]Mouffe, "Deconstruction, Pragmatism and the Politics of Democracy," 9.

[1652]Ibid., 9.

[1653]Mouffe, *The Democratic Paradox*, 135.

[1654]Ibid., 136.

[1655]Mouffe, "Deconstruction, Pragmatism and the Politics of Democracy", 11.

[1656]Ibid., 11.

[1657]Ibid.

[1658]Mouffe, *The Democratic Paradox*, 136.

[1659]Ibid.

regulative idea, but she could not avoid the elements of the reciprocal justice at the end. Therefore, although she highlights the other's otherness and defends the impossibility of perfect democracy – an issue Habermas also remarks - it seems that there is no possibility of thinking of democracy without the premise that every individual has the right to exercise her freedom and equality. The quest for unity in the diversity, as she sustains, will inevitably fall into the dilemma of doing justice to the other, which, still, within the democratic context, may require that this treatment be carried out with the purpose of reciprocity. This is the complexity and the paradox Mouffe's thinking raises: while acknowledging the necessity of democracy to think of the singularity of the other's otherness, there is still the demand that the other shares the adherence to the ethical-political principles of liberal democracy.[1660] This dualism between liberal democracy, from which originates the principle of justice, and the singularity of the other's otherness is not resolved, accordingly, in Mouffe's thinking. This non-resolution, nonetheless, is not necessarily problematic: justice and a justice *to come* or democracy and a democracy *to come* may not indeed be reconciled.

7.3.3 *The Internal Dialects Between Modern Equality and Individuality: The Symmetry and Asymmetry in Christoph Menke's Account*

Christoph Menke investigates how the modern concept of equality, which places the premise of equal consideration for everyone in a symmetrical way, may be losing its internal relation to the inherent asymmetries the very individuality raises. His central thesis – the idea of equality should be regarded in a relationship with the commitments from individuality, no longer conceived in favor of the prevalence of the equality, but in favor of how the conflicts emerging from the individuality are faced[1661] - is a very interesting perception of how this conflict between symmetry and asymmetry is a primary concern for the contemporary debates on democracy. He clearly seeks to provide a reconfiguration of the modern idea of equality by placing at its heart an irresolvable and paradoxically conflictive dialectics with the normative commitment towards the individuality[1662], thereby leading to an ongoing questioning of its basis. By placing the individuality at the core of the idea of equality, Menke aims to avoid the modern objective comprehension of the individual as merely a person, a subject, and stress her particular qualities, intentions and plans.[1663] He adds to the modern concept of equality, which is no longer simply regarded as an objective and general understanding of person in the modern sense,

---

[1660]Ibid., 102.

[1661]Christoph Menke, *Spiegelungen der Gleichheit* (Berlin: Akademie, 2000), 7.

[1662]Ibid., 8.

[1663]Ibid., 23.

the particularity of the other. Equality must deal with the individual by embracing her "perspectively historical way of self-reference."[1664] There is a plain purpose of understanding her difference, "the difference between what they [individuals] are and want from their perspective and what they as equally treated persons may be or want."[1665] Equality must then insert in its basis the other's otherness, the normative orientation towards the individuality, which creates what he calls a "dialectical constitution of equality,"[1666] with two views that are in an irresolvable connection. On the one hand, there is what Menke calls the Justification (*Begründung*) of equality, which confirms its priority and refers to the "notion of rational essence of equality (the reason *as* its essence),"[1667] whose origins can be found in modern thinkers such as Kant, Mill and Habermas; on the other, there is the Questioning (*Befragung*) of equality, which confirms its relativity and connects to the "unfolding of the irresolvable antagonisms in which, with the other normative standpoint, the orientation towards individuality stands,"[1668] whose origins are Burke, Nietzsche and Foucault. Instead of examining the equality through either of these two sides, Menke argues that equality has inherently both sides, which raises a distinct perspective: it is no longer simply deemed a process of equal consideration, but it also considers the equal determination of the other[1669] at the core of the principle of equality.

The conflict, therefore, occurs inside the principle of equality, within its inner core, as if each one of these sides were mutually presupposing the other. While equality conflicts with other values and norms *externally*, it is also *internally* conflictive because of its twofold character. The universal character of equality, which creates a reified conception of the individual as a person or a subject, must be supplemented by the determination of the individual. Menke explains how this supplementation can be carried out by arguing that, in all processes of abstraction leading to the general idea of equal consideration, there is always, at the beginning, an act of comparison between a determinate standpoint and others. In order to establish the equality, I compare "my wishes, intents, desires and my friends, lovers, neighbors"[1670] with the others, as though they had for me no stronger weight.[1671] Hence, there is a determination of the individual at the beginning of the definition of equality, which can be rescued from that reifying abstraction. By doing this, the tension between the universality and the particularity of the other leads to a dialectal movement towards the revision of the concrete exercise of equality:

[1664] Ibid., 23, translation mine.
[1665] Ibid., 31, translation mine.
[1666] Ibid., 13, translation mine.
[1667] Ibid., 12, translation mine.
[1668] Ibid., translation mine.
[1669] Ibid., 17.
[1670] Ibid., 17, translation mine.
[1671] Ibid.

"It justifies *and* questions the equality in one."[1672] It is this justification and questioning in concert that will promote the ongoing active endeavor to do justice to the individual, which is neither merely a solidarity towards the other – since, by centralizing on the individual, there is a normative deficiency[1673] - nor the abstraction of the universal criterion of equal consideration, for it leaves aside the dimension of the other. One does justice to the individual when the insight into the normative need for equality and the repressive consequences of equality are strictly bound to each other,[1674] when justification and questioning of equality becomes an always-unavoidable practice.

Menke's proposal is thus a clear reflex of the perception of the limits of reason: equality, in the modern sense of equal treatment, cannot always do justice to the individual, thereby producing violence, which makes necessary to be supplemented by an active endeavor to practice solidarity towards the other. This manifest intent to do justice to the individual, which is more than treating her as an *equal* among others, is nevertheless marked by the continuous risk of failure.[1675] There is no safety in the paradoxical and conflictive dialectics between the idea of equal treatment for *all* and the quest for the solidarity towards the other; there is no guarantee of achieving a conciliatory solution between both forms of reflection on equality. A conciliatory solution might not be possible at the same time, and this explains why it remains aporetic and "cannot be evaded."[1676] The reason is limited because of the very limits of the practical ability[1677] to do justice to the other. On that score, his theory could be interpreted as a critique of reason[1678], and by establishing this critique, it also encouraged an ongoing practice of attempts to do justice to the individual, which does not avoid the idea of equal consideration for *all,* but is no longer presuming a success and accomplishment of the idea of doing justice to the individual, either.

Yet, it seems that his theory, by pointing out the inherent conflictive basis of the idea of equality, might be lacking the premise that the modern idea of equal treatment for *all* can already absorb the asymmetries of practical life when it is remodeled according to an intersubjective basis where citizens manifest their opinions and wills through procedures oriented towards consensus. Indeed, his theory might be misunderstanding that the abstract and universal idea of equality

---

[1672]Ibid., 35, translation mine.

[1673]Ibid., 83.

[1674]Ibid., 86.

[1675]Ibid., 33.

[1676]Christoph Menke, "Virtue and Reflection: The 'Antinomies of Moral Philosophy'," *Constellations* 12, no. 1 (2005): 45.

[1677]At the core of this thinking, we could point out Derrida's influence, when, for instance, Menke defends the thesis that Deconstruction "puts into question how philosophy (...) introduces the concept of practical ability, namely as that which makes possible successful practices" and hence encounters "faith and, with it, the limits of ability" (Christoph Menke, "Ability and Faith: On the Possibility of Justice," *Cardozo Law Review* 27 (2006): 598, 612).

[1678]See Habermas, "Equal Treatment of Cultures and the Limits of Postmodern Liberalism," 6.

is not incompatible with particular considerations of individual wills and interests in concrete, which are, nonetheless, understood, in a post-metaphysical fashion, in accordance with democratic procedures of reciprocal agreement. In brief, it might be misinterpreting the tension between form and content that is at the core of the proceduralist approach, and which will indicate, at the end, that, in democracies, there is no possibility of denying the priority of the premise of equal treatment over any other asymmetry, even to do justice to the individual.

Jürgen Habermas, when he examines Menke's account, attempts to demonstrate how this internal contradiction in the equality principle Menke suggests can be metaphysical. In his view, the modern idea of justice, discursively remodeled, cannot rely on an internal contradiction that is as such deemed because it overlooks the inherent complexity of democratic procedures aimed at achieving mutual agreement. For Habermas, "only the difference-sensitive egalitarian universalism of equal rights can fulfill the individualistic requirement to guarantee equally the vulnerable integrity of individuals with distinctive life histories."[1679] His central argument is that, at the origin of the idea of the "dialectical constitution of equality," Menke neglects the procedures taken place previously to the application of equality, particularly the procedure of legislation in which different interests are transformed into arguments of discussion.[1680] Moreover, he ignores the premise of self-correcting learning process while using the past practices of inequality as examples of an "*inconsistency* in the underlying idea of civic equality itself."[1681] Unlike Menke, Habermas argues that the fact that history has many examples of inequality does not immediately culminate in the conclusion that the modern equality principle is inconsistent or that those practices can be viewed as "conditions of impossibility,"[1682] since, by grasping it according to a procedure carried out without assurances, it is dynamically adapted to continuous reforms in concrete over the years, which is never neutral, but instead learns from the past. By learning from the past in this tension between form and content, the equality principle connects to institutional procedures of opinion – and will formation, which gives it priority over conceptions of good. Habermas, accordingly, seeks to link Menke's theory to a certain appeal to *individual goods* in the very concept of justice.

According to Habermas, "the perspective of justice and the perspective of evaluating one's own life are not equally valid perspectives in the sense that the morally required priority of impartiality can be leveled out and reversed in favor of the ethical priority of anyone's particular goals in life."[1683] When Menke, for this reason, ascribes, at the inner core of the concept of equality, the solidarity towards the individual and the premise of an unavoidable practice of violence to the other to

---

[1679]Ibid., 13.
[1680]Ibid., 6.
[1681]Ibid., 8.
[1682]Ibid., 9.
[1683]Ibid., 11.

the normative application, he rules out the priority of questions of justice over particular self-understandings, which, in Habermas's model, were already discussed in the procedures of normative justification taken place *prior* to the application of norms. In other words, he inserts an argument that was not subject to critical scrutiny within the context of procedures of legislation as a reference to the application of norms. For Habermas, the fact asymmetries occur in social life evidently makes the application of norms a complex task, but it does not transform justice into a symbiosis of two opposing perspectives[1684] that, at the end, may materialize a sort of metaphysical thinking. For this reason – and this is the main aspect of this analysis - for Habermas, "a norm can only be appropriately applied on the basis of such democratic justification"[1685] in which the citizens, bound to "procedures of *reciprocal* perspective-taking (. . .), do not want to let their individual goals be restricted in existentially unreasonable ways."[1686] This naturally brings out the "perspective of justice."[1687]

Menke's proposal, accordingly, by losing the citizen's perspective, neglects the complexity of the application of justice, which is bound to a chain of procedures in which every argument is subject to critical review – this is its post-metaphysical condition: "The non-neutral effects are a theme in the hypothetical scenarios *ex ante* within the public sphere and in the political debates of the democratic legislature and, thus, not merely in the later discourse on the application of justice."[1688] Briefly, the main problem of Menke's proposal resides in adopting an idea of modern equality that does not consider the citizen's communicative perspective in democratic procedures of opinion – and will formation and places rather a complex charge to the applicator of the norm, who will decide by herself on the meaning of a certain idea of justice rooted in a symbiosis that is, ultimately, metaphysical. It is for this reason that asymmetries, in Habermas's theory, are projected into those procedures with the priority of justice as equal treatment for *all,* and not as a final consideration the applicator of the norm, the judge, will define as an internal contradiction of the very idea of equality: "The asymmetric restrictions that are accepted on normative grounds are an expression of the principle of civic equality no less than the norm itself – and not, as Menke maintains, an indicator of its 'internal heterogeneity'."[1689]

The critiques Habermas sets forth are crucial to the investigation here, for Menke's approach seems to attempt to reconcile, paradoxically concluding that they are irreconcilable, the modern principle of justice with the need for an asymmetrical care or solidarity towards the other. He establishes, at the basis of the idea of equality, an irresolvable conflict, which demonstrates that Menke, as

[1684]Ibid.

[1685]Ibid., 12.

[1686]Ibid.

[1687]Ibid.

[1688]Ibid.

[1689]Ibid., 13.

Derrida before, is not satisfied with the idea that equal treatment for *all* is enough for the idea of equality, and consequently for the premise of justice. In comparison with Mouffe, who seems ultimately to fall into the unanswerability of reciprocal justice, despite her stress on antagonisms, undecidability,[1690] and inherent conflicts of democracies, Menke seems more radical on account of his manifest accent on the particular at the inner core of the equality principle, while, simultaneously, not abandoning the need for equal consideration for *all,* which stands in conflict with the individuality. Mouffe, while still sustaining the demand of the other to share adherence to the ethical-political principles of liberal democracy,[1691] paradoxically attempts to hold the premise, clearly influenced by her immersion in Derrida's thinking, of attacking universalism and rationalism,[1692] which, as shown, does not seem well resolved in her thinking. Menke is also an exponent of the limits of reason by stressing the other side of equality, that is, the solidarity towards the other, but his thinking seems more justified by virtue of his critical acceptance of modern rationalism while radicalizing it on the basis of a negative-dialects that has much of Adorno's thinking, but also of Derrida's deconstruction.

Menke's analysis of the *internal* conflict of the very concept of equality seems to grasp, consequently, more correctly what undecidability sets forth: it is not an anti-rationalist response to the universality inscribed at the core of the relationship between law and justice, constitutionalism and democracy, which would ultimately result in the emphasis on the particularities, the agonisms, the other. Undecidability resides instead in the tension between both sides, the calculable and the incalculable, and this is why it does not abandon, in the very premise of law, the requirement of equal treatment for *all.* On the other hand, it establishes that it is the justice *to come,* as a non-reciprocal and disinterested justice to the other in tension with the law, that will reveal the complexity of the normative application. Menke, for this reason, uses a distinct argument in comparison with Mouffe to criticize Habermas's rationalism: it is no longer an attack on consensus, rationalism and universalism, but an attack on the loss of the other facet of equality that this rationalism seems to lack, which is inherently in tension with the latter.

Nonetheless, similarly to Mouffe's approach, Menke's proposal seems to confront the boundaries of democracy. The stress on the individual, the particular, the other might reveal a certain metaphysics and indeed result in authoritarian practices of normative application. Habermas, for this reason, criticizes Menke's approach due to his lack of a more careful analysis of how the equality principle plays a fundamental role in procedures of opinion – and will formation. In sum, Menke's approach might not be adequate for democracy where the voices of all individuals are examined as arguments in a set of institutional procedures. In the

[1690]See Chantal Mouffe, "Democracy and Pluralism: a Critique of the Rationalist Approach," *Cardozo Law Review* 16 (1995): 1543–1545.

[1691]Mouffe, *The Democratic Paradox*, 102.

[1692]According to Mouffe: "Every pretension to occupy the place of the universal, or to fix its final meaning through rationality, must be rejected" (Mouffe, "Democracy and Pluralism: a Critique of the Rationalist Approach," 1544).

reality of normative application in a democratic basis, these institutional procedures are deemed indispensable, so that reciprocal justice does not become a value as any other, not conditioning thereby the equal treatment to a particular interpretation of an individual will or interest. There is a clear differentiation between discourses of justification and discourses of application that is regarded as the condition for democracy, and it is not the stress on a non-reciprocal care or solidarity that one can rule out this presupposition, even because the asymmetries are indeed interpreted as arguments in justification discourses. Consequently, by assuming this premise, the irreconcilable character of the inner conflictive principle of equality fails to face the dilemma of applying the norm democratically. This might explain why Mouffe could not avoid, lately, the premise of reciprocal justice.

On the other hand, nonetheless, we can question that Habermas's association of the non-reciprocal justice with the idea of an *ethical* value raised by an individual might be misinterpreting the complexity of this asymmetrical justice. Indeed, we could question whether the asymmetries should be merely an argument of discourses of justification, whereas discourses of application should rely solely on the presupposition of reciprocal justice. Perhaps, the stress on the other's otherness may be a pertinent concern of the applicator of the norm, while not forgetting naturally the premise of equal consideration for *all*. The question, therefore, goes directly to the core of the problem here raised: is it possible to reconcile intersubjectivity with *différance*?

7.3.4 The Resolution as a Non-Resolution: The "Irresolvable But Productive Tension"[1693] Between Différance and Intersubjectivity in the Quest for Justice

Chantal Mouffe's agonistic model of democracy poses the problem of how the focus on the particular, the singularity of the other's otherness can conflict with the premise of rational consensus of Habermas's approach. In turn, Christoph Menke's proposal attempts to justify an inherently dialectical conflictive and paradoxical relationship at the core of the equality principle. Both, in different ways, aim to establish the need for the particular, the antagonist, the other's otherness in opposition to a certain abstraction, objectivism the modern idea of justice presents within the democratic context. They are the expression of a certain discontentment with the simple application of the premise of equal consideration for *all,* and are also theories that transit between a certain presence of modern premises and strong influences of deconstruction. Even though both might raise some doubts about the way they interpret the tension between form and content that is at the core of Habermas's proceduralist approach, they, on the other hand, suggest a fundamental

[1693] Honneth, "The Other of Justice: Habermas and the Ethical Challenge of Postmodernism," 319.

question that surrounds the concept of limited rationality we can extract from the tension between facts and norms: is it possible to reconcile, in a constitutional democracy, intersubjectivity with *différance*? In other words, is it possible to seek rational consensus while keeping open the possibility of *différance*? Is there space in the democratic procedures of opinion – and will formation for asymmetrical forms of integration, in which we could think of "an equality that would not be homogeneous, that would take heterogeneity, infinite singularity, infinite alterity into account?"[1694] Is it possible to bring *différance* to the nucleus of democratic procedures of normative application? Could it indeed do more justice, within the context of democratic procedures of opinion – and will formation, to the individual than the premise of equal consideration for *all*? Finally, could it bring forth a response to legal adjudication?[1695]

These questions enter into the core of the construction of a concept of limited rationality, for they show that reason, if not able to entirely include the other in symmetrical considerations of equality and freedom, cannot thoroughly achieve, when *différance* radicalizes it, the other's otherness in the asymmetrical promise of the *to come,* of a justice *to come*. Reason, accordingly, has its limits delineated by the impossibility of inserting every individual into the procedures oriented towards mutual understanding, and mostly by its incapacity to reach, in constitutional democracies, the "*unconditional hospitality* that exposes itself without limit to the coming of the other."[1696] By the same token, if, on the one hand, tolerance appears as an inevitable price of any constitutional democracy with distinct worldviews,[1697] this price may come with an excessive cost for taking into account the "infinite alterity,"[1698] showing thereby how reason coexists with the agony of not being able to end up in a resolution to this impasse.

The concept of limited rationality thus interprets and assumes this agony as an interminable movement towards the other, even if reason cannot entirely achieve the other. It directs to the otherness in the certainty that any resolution is, in fact, a non-resolution. Whereas we must acknowledge that the idea of symmetrical justice, as equal concern and respect, might best reconcile with the characteristics of constitutional democracies, particularly in virtue of the need to institutionalize mechanisms to include the other as an equal among the others, the dissatisfaction with symmetrical justice and the emphasis on asymmetrical ways of including unconditionally every individual brings to the fore an even greater attitude to intervene in practices of social life as a means to open up the possibility of "[calling] for a consideration of a certain infinite heterogeneity,"[1699] even if it is

---

[1694]Derrida, "Politics and Friendship," 179.

[1695]We will examine this question in the next chapter.

[1696]Derrida, *Rogues: Two Essays on Reason*, 149.

[1697]See Habermas, "Reply to Symposium Participants, Benjamin N. Cardozo School of Law," 393.

[1698]Derrida, "Politics and Friendship," 179.

[1699]Ibid.

"an infinite distance."[1700] The stress on reason, due to its boundaries, becomes then an emphasis on the equality as difference, as paradoxes of the same concept, but now acknowledging that its limits might also reside in the very limits of constitutional democracy. For constitutional democracy, while calling for this *to come,* cannot even guarantee equal concern and respect, reason is constrained by the very impossibility of democracy, by this "theme of a non-presentable concept."[1701] The question Derrida ends his *Politics of Friendship* with – "when will we be ready for an experience of freedom and equality that is capable of respectfully experiencing that friendship, which would at last be just, just beyond the law, and measured up against its measurelessness?"[1702] – shapes, hence, the conclusion that reason is limited on account of the impossibility of justice, which is the other facet of the very impossibility of democracy.

For this reason, the possibility of reconciling intersubjectivity with *différance* within the context of constitutional democracies leads us to the conclusion that, despite both interacting with each other and paradoxically complementing and contradicting each other, a possible resolution of this deadlock is its very non-resolution. Still, it is this non-resolution that is a condition for its interminable attempt to resolve it *here and now* in the practices of social life. While there is a quest for achieving a consensus in the dissensus, while there are "symmetrically distributed rights and duties,"[1703] there is also the exigency of keeping open the possibility of *différance,* as a radicalization of symmetrical justice, as a relationship with the other without presupposing any reciprocal duties. Both the "affective level of sympathy and affection to my asymmetrical obligations"[1704] and the respect as "moral persons just like everyone else"[1705] enter somehow into this process. While there is the search for treating every individual as an equal among others, there is also the quest for attempting to reach the other's otherness, as an unconditional and nonreciprocal concern for the individual's well-being, as an "unlimited care."[1706]

True, by focusing on proceduralism, one could argue that this quest for the *to come,* as a promise, might lead to an appeal to an axiological point of view that was not inserted into institutional discourses of justification. However, the stress on the "exceptional singularity"[1707] might also reveal how limited constitutional democracy is, while understanding democracy as the only system that expresses the "right to self-critique and perfectibility,"[1708] whose limits "translate or call for a militant

[1700]Ibid.

[1701]Derrida, *Politics of Friendship*, 306.

[1702]Ibid.

[1703]Honneth, "The Other of Justice: Habermas and the Ethical Challenge of Postmodernism," 309.

[1704]Ibid.

[1705]Ibid.

[1706]Ibid., 313.

[1707]Derrida, *Rogues: Two Essays on Reason*, 148.

[1708]Ibid., 87.

and interminable political critique."[1709] If, with Axel Honneth, we could sustain that "this conflict is irresolvable because the idea of equal treatment necessitates a restriction of the moral perspective from where the other person in his or her particularity can become the recipient of my care"[1710] - otherwise the very idea of constitutionalism and democracy is put in jeopardy - on the other, we could conclude that "this conflict is productive because the viewpoint of care continually provides a moral ideal from which the practical attempt to gradually realize equal treatment can take its orientation – in a self-corrective manner."[1711] Naturally, it is possible to argue that, in constitutional democracies, this irresolvable conflict could, nonetheless, point to the primacy of symmetrical justice, especially when individuals can participate in procedures oriented towards mutual understanding, while reserving for those who are incapable of integrating into those discourses an obligation of care and benevolence.[1712] In any case, we could not deny that, in this debate between *différance* and intersubjectivity, the symmetrical justice received its "equally necessary counterpoint because it [supplemented] this principle of justice by the principle of unilateral, entirely disinterested help,"[1713] leading then to an "irresolvable but productive tension that prevails in the domain of moral."[1714] If, in any case, this might raise doubts whether this obligation of care and benevolence would not be found somehow in the idea of intersubjectivity itself,[1715] *différance* certainly brought it to the fore. By bringing *différance* to the fore, constitutional democracy has its tensions and complexities enhanced, while

---

[1709]Ibid.

[1710]Honneth, "The Other of Justice: Habermas and the Ethical Challenge of Postmodernism," 315.

[1711]Ibid.

[1712]According to Axel Honneth:

"An obligation to care and to be benevolent can only exist where a person is in a state of such extreme need or hardship that the moral principle of equal treatment can no longer be applied to him in a balanced manner. Thus, human beings who are either physically or mentally unable to participate in practical discourses deserve at least the selfless care of those who are close to them via emotional ties. But, conversely, the moment the other person is recognized as an equal being among all others – in that he or she can participate in practical discourses – the unilateral relation of care must come to an end; an attitude of benevolence is not permissible toward subjects who are able to articulate their beliefs and views publicly" (Ibid., 318–319).

[1713]Ibid., 319.

[1714]Ibid.

[1715]According to Thomas McCarthy:

"Though the argument for incorporating the care perspective is generally cogent, it is not clear just how and in which respects it goes beyond the bounds of a Kantian doctrine of virtue in which benevolence and beneficence figure centrally. If loving concern for concrete individuals is meant as a general but indeterminate (Kant would say "imperfect") obligation, it might be accommodated by discourse ethics along the lines that Habermas sketched in his earlier discussion of Carol Gilligan's ethics of care. If not, we will need some justification for such determinate moral obligations" (Thomas McCarthy, "The Cambridge Companion to Habermas," *Ethics* 107, no. 2 (2007), 372).

expanding its capacity to learn from this process. A learning, however, that occurs by always acknowledging the boundaries of constitutional democracy to do justice to the other in this "irresolvable but productive"[1716] tension between intersubjectivity and *différance*. A learning thus that recognizes the very limits of reason.

7.4 Final Words

After the investigation of the metaphysics embedded in the structure of balancing in the last part of this book, especially in the way Robert Alexy justifies it – which, indeed, can be extended to some of the practices of the constitutional courts examined in the first part - this chapter aimed to unfold the concept of limited rationality through the dialogue between Derrida's *différance* and Habermas's *intersubjectivity*. First, it examined the possible connections that could be found between both authors as well as their insurmountable divergences. In this regard, we could observe that, through different perspectives, they have a clear notion of the boundaries of reason, insofar as they remark that: first, there are no assurances between the reality and their quasi-transcendentalism nor can knowledge be heuristically apprehended by some abstract formulas and criteria, for it is fragile; second, history, with its complexities and tensions, while shaping reason, cannot be, nevertheless, thoroughly recollected and gathered by reason; third, it is impossible to entirely include the other or do justice to the other. These limits of reason, in any case, do not lead to a passiveness towards the world, but rather to an active attitude *here and now*, both in the quest for keeping consistent this heritage, always submitted to critical review or deconstruction, and in the quest for doing justice to the other.

In this investigation in particular – justice – as long as it is the most complex theme when *différance* and *intersubjectivity* are placed side by side this chapter examined possible theories that attempted to make this connection. First, it discussed Chantal Mouffe's "agonistic model of democracy," according to which it is necessary to establish the idea of a "consensus without exclusion,"[1717] thereby inserting the asymmetry at the core of democracy. Yet, even though Mouffe brings forward some elements from Derrida's *différance*, we concluded that her theory, contrary to her opinion, might not be so different from the stress on *intersubjectivity* that Habermas works with. Second, it discussed Christoph Menke's proposal of an internal relationship between modern equality and individuality to show that a possible irresolvable tension between symmetry and asymmetry might be necessary, for the idea of equal treatment and respect cannot always do justice to the

[1716]Honneth, "The Other of Justice: Habermas and the Ethical Challenge of Postmodernism," 319.
[1717]Mouffe, "Deconstruction, Pragmatism and the Politics of Democracy," 9.

individual. If, on the one hand, we could verify that, with Habermas, this understanding might create a risk of confusion between discourses of justification and discourses of application in constitutional democracies, on the other, however, Menke's proposal opened up the space to debate the idea of an "irresolvable but productive tension"[1718] at the core of the concept of justice. With the previous comprehension of the limits of reason in what refers to history, we could then unfold the concept of limited rationality.

The concept of limited rationality, as this chapter aimed at exploring, derives not only from the tensions and complexities of a history hermeneutically gathered in every new event, but also from the very impossibility of achieving the other's otherness, either as the premise of equal consideration and respect or the *to come,* whose conciliation, indeed, results in a resolution as a non-resolution. It is a productive tension that orients constitutional democracy, in all its institutional forums, such as the judiciary, to promoting, more and more, albeit entirely unattainable, justice to the other. The "irresolvable but productive tension"[1719] of justice and the incapacity to entirely recollect and gather the complexities and tensions of a certain reality, in an interpretative reconstruction in which there are no assurances nor certainties, shape then the two premises of the concept of limited rationality, one that acknowledges the tensions and the intricate problems of constitutional democracy as the recognition of its limits. This concept of limited rationality, which more adequately corresponds to the characteristics of institutional procedures where constitutionalism and democracy, law and justice, as double binds, realize in the practices of social life, reveals that it is not a finished product of a certain knowledge nor a way of constructing knowledge. It is rather the result of the perception of its unfinished quality, which is polished or likewise unpolished in every new event, in every new interpretation; a practice indeed that, while attempting to reconcile the quest for the other with history, is constrained by the impossibility of entirely reaching the other and the fragility of any inherited and projected certitude.

In this regard, this discussion furnishes the assumptions we must acknowledge to face the discussion of a limited rationality applied to the realm of legal adjudication. For adjudication obviously inherits many of these tensions of constitutional democracy, the conclusion is also that a concept of limited rationality in the realm of adjudication absorbs those two premises: the "irresolvable but productive tension" of justice and the fragility, notwithstanding its inexorability, of history, of institutional history. How they appear in this debate and how they connect to the institutional history of Germany and Brazil, as we examined in the first part, are the theme of the next chapter. Legal adjudication, while grasping the concept of limited rationality, radicalizes its practice as an activity designed to do justice to the other, a complex task in plural societies characterized by a multiplicity of worldviews. This complexity has implications for the collision of principles. This is

[1718]Honneth, "The Other of Justice: Habermas and the Ethical Challenge of Postmodernism," 319.
[1719]Ibid.

where the concept of limited rationality confronts the idea that a methodology, as long as filled with arguments, provides rationality to decision-making. For the concept of limited rationality refers to the tensions history brings forth and the impossibility of justice, there cannot be a silence in this matter. For the silence would mean the consent to institutional practices that neglect or erode the complexities and tensions those two premises of the concept of limited rationality reveal, the defense of limited rationality in decision-making is indeed a defense of constitutional democracy.

Chapter 8
Between *Différance* and Intersubjectivity: The Concept of Limited Rationality in the Realm of Constitutional Adjudication

Abstract If the *concept of limited rationality* revealed the complexity of constitutional democracy, for there is no possibility of completely recollecting and gathering the history and its tensions nor is it possible to entirely do justice to the other, when it is transported to constitutional adjudication, the same reasoning applies. In this specific realm, it demonstrates that a judge aware of the boundaries of reason focuses: first, on the singularity of the case, not in predetermined formulas and patterns; second, on the system of rights, keeping it coherent throughout history; and third, on the other, as the quest for doing justice to the case, in all the complexities of the resolution as a non-resolution of the dialogue between symmetrical and asymmetrical justice. As a means to demonstrate how the *concept of limited rationality* applies to the reality, the reconstruction of German and Brazilian constitutionalisms, as previously examined, appears as a relevant sign, but it is the reexamination of constitutional cases that brings it to the effective practice of decision-making. In this regard, by showing how a judge aware of the limits of reason would judge the *Crucifix, Cannabis* and *Ellwanger* cases introduced in the first chapter, the *concept of limited rationality* shows that it is necessary to think of reason in constitutional adjudication in a different perspective, one that knows that adjudication has limits, and that these limits are constructed in the very practices of this dualism between law and justice, between history and the other's otherness.

8.1 Introduction

When intersubjectivity and *différance* are presented side by side, while being irreconcilable, they open up the problematization of the other within the context of constitutional democracies, and, in this case, they complement each other. Whereas intersubjectivity indicates the need to include the other in procedures oriented towards mutual understanding, *différance* brings, as Benhabib says, the "agony of not being able to reach the ear of the other, of the inability to

J.Z. Benvindo, *On the Limits of Constitutional Adjudication*,
DOI 10.1007/978-3-642-11434-2_8,

communicate,"[1720] as a radicalization of the premise of equal consideration and respect. In both circumstances, the concern for the other is clearly announced; however, intersubjectivity works with the premise of reciprocal justice, whereas *différance* indicates a unilateral and radical nonreciprocal concern for the other. Intersubjectivity, accordingly, seems more connected to the institutional grounds of democracies where individuals are regarded as equal participants in the public sphere and can thus articulate their wills and opinions within democratic forums. *Différance,* on the other hand, relates to a care for the other, particularly those who cannot immediately articulate their opinions in practical discourses and, as such, cannot exercise their citizenship as equal members of a determinate community. As shown in the last chapter, both intersubjectivity and *différance* promote, in the realm of constitutional democracy, an "irresolvable but productive tension,"[1721] reinforcing thereby the quest for otherness within the boundaries of a reality characterized by some "stable [points] of reference"[1722] that can fail at any time. Still, how could *intersubjectivity* and *différance,* by modeling a limited concept of rationality, be articulated in the practice of legal adjudication? Besides, how could they be projected into those two constitutional realities – Germany and Brazil – that were examined in this research?

For this reason, as a counterargument to the defense of the rationality of balancing in the realm of constitutional decision-making, this chapter aims to extend the concept of limited rationality that we formerly examined within the context of constitutional democracy to the debates on constitutional adjudication. Inasmuch as the metaphysical standpoints of Robert Alexy's idea of rationality of balancing were already discussed by means of deconstruction[1723] and proceduralism,[1724] as well as explored the metaphysics embedded in the practice of constitutional courts,[1725] the problem now shifts to understanding what this limited rationality means for decision-making, and how it can be applied to the problem of constitutional adjudication. The final thesis of this book is that a dialogue between *Différance* and *Intersubjectivity,* despite their insurmountable divergences, shapes a complementary approach to constitutional decision-making in the realm of indeterminacy of law. In truth, they provide an adequate comprehension of the limited rationality within this context of interpretation and application of constitutional principles. In this respect, their points of contact can be developed by

[1720] Seyla Benhabib, "Democracy and Difference: Reflections on the Metapolitics of Lyotard and Derrida," in *The Derrida-Habermas Reader*, ed. Lasse Thomassen (Edinburgh: Edinburgh University Press, 2006), 149.

[1721] Axel Honneth, "The Other of Justice: Habermas and the Ethical Challenge of Postmodernism," in *The Cambridge Companion to Habermas*, ed. Stephen K White (Cambridge: Cambridge University Press, 1995), 319.

[1722] Jürgen Habermas, "On Law and Disagreement: Some Comments on 'Interpretative Pluralism'", *Ratio Juris* 16, no. 2 (June 2003), 193.

[1723] See the fifth chapter.

[1724] See the sixth chapter.

[1725] See the first part.

focusing on the two previous fundamental premises we could infer from the construction of the concept of limited rationality in the last chapter: the "irresolvable but productive tension"[1726] of justice, and the incapacity to entirely recollect and gather the complexities and tensions of a certain reality, in an interpretative reconstruction in which there are no assurances nor certainties.

As regards these purposes, we will develop this chapter in two central topics. First, we will develop the theoretical analysis of how the concept of limited rationality appears in the realm of constitutional adjudication (Sect. 8.2). Subsequently, the investigation will stress how this concept of limited rationality could be rebuilt in the practice of decision-making. For this end, a primary issue will be to explore it in a critical reconstruction of German and Brazilian constitutionalisms (Sect. 8.3), and then, in order to expose how the concept of limited rationality would lead to an argumentation in constitutional adjudication that differs from the idea of balancing, the emphasis will be on case study through the reanalysis of the three cases – the *Crucifix case* (Sect. 8.4.2), the *Cannabis* case (Sect. 8.4.3) and the *Ellwanger* case (Sect. 8.4.4) – that initiated this book. By exposing how a judge aware of the boundaries of reason would sustain her arguments in constitutional cases, the book ends by connecting the theoretical analysis to the effective practice of decision-making, showing thereby a real interest in elaborating a reflexive thinking that, by paraphrasing Derrida's words, indeed aims to express a militant and interminable critique[1727] of German and Brazilian constitutional realities.

8.2 The Concept of Limited Rationality in the Realm of Legal Adjudication: Intersubjectivity and *Difference* in a Complementary Fashion

In the sixth chapter, while criticizing the central premises of Robert Alexy's defense of the rationality of balancing, Habermas's proceduralist model applied to legal adjudication, as a reconstruction of Klaus Günther's and Ronald Dworkin's legal theories, provided a very robust response to the indeterminacy of law in adjudication, one that seems more adequate for the dilemmas of constitutional adjudication than balancing in the way, in some cases, constitutional courts deploy and Alexy justifies it. First and foremost, it showed that, in legal reasoning, there must be: first, a clear differentiation between discourses of justification and discourses of application; second, the quest for the "single right answer"; and third, the stress on procedures of opinion – and will formation directed towards mutual understanding. On the basis of this thinking, we could already unfold relevant aspects of the concept of limited rationality. By the same token, when we examined,

[1726]Honneth, "The Other of Justice: Habermas and the Ethical Challenge of Postmodernism," 319.

[1727]Jacques Derrida, *Rogues: Two Essays on Reason* (Stanford, CA: Stanford University Press, 2005), 87.

in the fifth chapter, Derrida's premises also applied to the problem of balancing, relevant aspects of this debate likewise appeared. Now, the purpose is to make explicit how the concept of limited rationality as worked in the previous chapter may be a more adequate response to the main dilemmas of constitutional adjudication. Specifically, the purpose is to radicalize even more the premises of Habermas's response to the indeterminacy of law in adjudication through the dialogue between him and Derrida. Furthermore, by revealing the connections between both, we will already be able to prepare the perception that, from Habermas's response to the indeterminacy of law as worked in the sixth chapter, followed by the productive tension between *intersubjectivity* and *différance* discussed in the last chapter, it is possible to envisage the application of the concept of limited rationality directly to cases.

In this regard, when we place Derrida's deconstruction and Habermas's proceduralism side by side, as we did in the last chapter, two fundamental premises gain relevance: first, the incapacity of reason to recollect and gather all the complexities and tensions of a certain reality characterized by different worldviews, leading thereby to the perception of the nonexistence of any assurance in this process, and, second, the incapacity of reason to entirely reach the other, in which the tense but productive relationship between intersubjectivity and *différance* appears as a radicalization of justice. These two premises, when transported to legal adjudication, can be examined by stressing three central propositions: first, the emphasis on the singularity of the case and the disbelief in any general methodology providing a safe way to a rational response through analytical criteria and formulas; second, the concern for the consistency of the system of rights and its enforceability in the quest for justice; and, third, the context-transcendent presupposition of justice in a tense relationship with the reality.

The first proposition refers to the disbelief in abstract methodologies that could provide some guarantees or even the criteria for achieving rationality in the negotiation between the calculable and the incalculable or in the procedures oriented towards mutual understanding, as we could remark in Alexy's defense of the rationality of balancing. In both Derrida's and Habermas's accounts, it is clear that the quest for justice in legal reasoning cannot be reduced to the deployment of a certain methodology defining beforehand some "rational" standards, analytical criteria, formulas to control the arguments in a "logical" basis. If both thinkings are taken seriously, no general rule or formula is then able to control or regulate the tensions and complexities of any practice, any action, for every context is singular, thereby producing more complexities and openings to new interpretations. Rationality, at least one that acknowledges its boundaries, does not reside, for example, in the prior definition of the general statement that, when the judge is confronted with a collision of principles, she must carry out a proportional application of legal norms and values, as a means to achieve a "rational" solution. Instead, for rationality is limited, rather than focusing on the quest for rationality in the definition of general standards, the focus transfers to the singularity of the reality, from where the complexities of decision-making arise. Therefore, it begins by centering on the case and exclusively on its features, which are integrally considered. Only after examining these features of the case, the judge can then determine which legal norm is appropriate to the case. She begins thus by acknowledging that, since rationality is limited, there cannot be any previous assurances, such as those "rational" methodological standards, in the process of decision-making. For this reason, more than the search for those standards, the concept of limited rationality orients towards the posture the judge must assume in the practice of decision-making.

In Derrida's thinking, if the context, the singular is put in jeopardy in the negotiation between law and justice, the possibility of justice in its own impossibility is undermined. In Habermas's viewpoint, if the singularity of the case is overcome by the interest in applying the law as if legal adjudication were similar to the discourse of justification – and thus oriented towards satisfying a general social interest – the risk of transforming justice, as equal consideration for *all,* into a conception of good is imminent. The direct consequence is that, in both accounts, the stress on the particular, the context results in the conclusion that there is no more justice to the case, albeit never achievable and differently considered in both theoretical perspectives, than not reducing the complexities stemming from the situation to some predetermined patterns. Rationality lies in the very negotiation between the calculable and the incalculable or in the procedure oriented towards mutual understanding, which discloses the tensions and complexities of this process while never being a guarantee of success, for reason is fragile, limited and can produce its own reversal. In other words, it means that the development of legal adjudication is intimately related, in the first instance, to iterability, which historically establishes its own limits through a process of reinterpretation and invention of the context as a way to create something new and reveals the capacity of self-critique and perfectibility[1728] of its negotiation in every new event. Or it means, in the second instance, that legal adjudication learns from history, as a self-correcting learning process, and hence reason has its boundaries in the set of democratic procedures carried out by focusing on the tension between facts and norms, that is to say, on the procedures oriented towards achieving mutual consensus, thereby doing more and more justice as equal consideration and respect.

The second proposition is concerned with the consistency of the system of rights and its enforceability in the quest for justice. Both Derrida and Habermas stress the intrinsic enforceability or deontology of law as a fundamental piece of the quest for justice. By acknowledging the limits of reason, both Derrida's and Habermas's premises carry the requirement of keeping consistent the system of rights and enforce it in every new situation while, simultaneously, reinventing it. There is, accordingly, a tension that is translated into the negotiation between law and justice (or constitutionalism and democracy) or in the procedure taken place between facts and norms, which points to the need to carry out an ongoing self-critique that will reinforce the system of rights itself. The emphasis on the enforceability or the deontology of law against possible reductions of its authoritative character in legal reasoning are clearly presented in both thinkings, which could shape, through distinct standpoints, a robust critique of balancing as examined in the previous

[1728] Ibid.

chapters. There is no possibility of thinking of justice, either in Derrida's account of the *to come* or Habermas's premise of equal consideration and respect, without acknowledging that the application of law must strengthen the law and not diminish its authoritative character by proportionally equalizing it to other axiological points of view in a teleological basis. Legal adjudication, accordingly, must observe legal principles that have been historically framed and accepted as legitimate in order to do justice to the case; it must understand, as a consequence, its position in the set of democratic procedures in which the negotiation between constitutionalism and democracy, law and justice, or, if it is to take the Habermasian language, the procedures oriented towards reaching agreement take place. We can also achieve this conclusion either by focusing on the idea of *iterability* and *autoimmunity* that is at the core of Derrida's thinking, which points to an interaction between the constative and the performative, or in Habermas's stress on the *self-correcting learning process,* which establishes the limits to reason within democratic procedures oriented towards rational consensus.

The third point of contact can be observed in the existence, in both thinkings, of a context-quasi-transcendent presupposition of justice in a tense relationship with the reality. As we examined in the last chapter, this theme is certainly one of the most complex when we think of the concept of limited rationality by placing side by side Habermas's and Derrida's theories. Habermas accounts for a symmetrical justice focused on the safeguard of equal concern and respect as the only one compatible with the institutional background of complex societies that are not restrained by a particular conception of good. In constitutional democracies, therefore, "individuals expect from one another an equal treatment that assumes that each person treats everyone else as 'one of us."[1729] Derrida in turn sets forth a concept of asymmetrical justice, as the *to come,* a justice *to come,* which is linked to the "singularity of an alterity that is not reappropriable by the ipseity of a sovereign power and a calculable knowledge."[1730] Hence, even though both underline the quest for otherness, the way they announce it is different.

As the inseparable companion to the law, as the condition for transformation, history and perfectibility of law,[1731] Derrida builds his account of justice by focusing on the three aporias *épokhè and rule,*"[1732] "*the ghost of the undecidable,*"[1733] and "*the urgency that obstructs the horizon of knowledge.*"[1734] The first leads to the requirement that any judge must make a decision that is "both regulated and without

[1729]Jürgen Habermas, "On the Cognitive Content of Morality," *Proceedings of the Aristotelian Spociety* 96 (1996): 343.

[1730]Derrida, *Rogues: Two Essays on Reason*, 148.

[1731]See Ibid., 150.

[1732]Jacques Derrida, "Force of Law: The 'Mystical Foundation of Authority'", *Cardozo Law Review* 11 (1990): 961.

[1733]Ibid., 963.

[1734]Ibid., 967.

regulation;"[1735] the second alludes to *undecidability,* as a "[deconstruction] from within [of] any assurance of presence, any certitude or any supposed criteriology that would assure us of the justice of a decision;"[1736] and the last points out to the exigency that a decision must be made *here and now*, exposing then, as Menke suggests, a "conflict *with* practice – the conflict with the attempt to bring about justice through our own conduct."[1737] Justice, while inseparable of the law,[1738] even to avoid the possibility of a decision being "governed by the caprice or partiality of the subject who decides,"[1739] shows that the judge cannot be merely attached to the law or to the calculable. There is, in the movement of interpretation and application of law, a moment of faith that guides the judge towards the construction of something new while "[appealing] to juridical determinations and to the force of law."[1740] Yet, justice has this asymmetrical character, which differs from Habermas's symmetrical perspective of equal treatment. The consequence is that, if we take Derrida's premises as our point of departure, it is possible to sustain that the judge, in different concrete cases, must complement the symmetrical concept of justice as equal treatment with the perspective of care, of a disinterested benevolence towards the other. This thinking, nevertheless, may not gather that, in the very idea of equal concern and respect, as Honneth argues, "there exists a series of special arrangements that see to it that, from within these legal relations themselves, the individual case is considered as comprehensively as possible and in a manner that Derrida can only imagine as the addition of a goodness or care perspective from without."[1741] The question, therefore, that challenges us here is exactly which are the implications of this relationship between *différance* and intersubjectivity for legal adjudication.

A serious concern is the possibility of bringing to the practice of adjudication conceptions of *good* that were not subject to critical scrutiny in the realm of discourses of justification.[1742] An axiological viewpoint, accordingly, appears in the final moment of a set of institutional procedures, allowing then the judge to temper his concern for equal treatment with a sentiment of care towards the individual. Still, by doing so, the possibility of losing the institutional grounds circumscribing the activity of adjudication can increase, thereby opening up the space for subjectivism and decisionism. Indeed, this is a serious issue, especially when we could argue that, in procedures oriented towards mutual agreement taken place in legislation, that is, in discourses of justification, this consideration

---

[1735]Derrida, "Force of Law," 961.

[1736]Ibid., 965.

[1737]Christoph Menke, "Ability and Faith: On the Possibility of Justice," *Cardozo Law Review* 27 (2006): 601.

[1738]See Derrida, *Rogues: Two Essays on Reason*, 150.

[1739]Menke, "Ability and Faith: on the Possibility of Justice," 602.

[1740]Derrida, *Rogues: Two Essays on Reason*, 150.

[1741]Honneth, "The Other of Justice: Habermas and the Ethical Challenge of Postmodernism," 315.

[1742]See the analysis of Christoph Menke's theory in the last chapter (Sect. 7.3.3).

for the other, achieving thereby those asymmetrical acts of care and benevolence, can appear as arguments for critical scrutiny. In a constitutional democracy characterized by distinct institutional deliberative procedures, the quest for symmetrical justice arrives thus as a prevalent element over any other asymmetrical perspective pointing out the relationship with the other. This Habermas's intuition may best correspond to the expectancy of individuals that intend to treat one another as equal members of a certain community, as a way to construct a self-reflexive community. Despite this understanding, we could not neglect that, by stressing *différance,* more than the emphasis on the individuals' communicative potentialities, an agony as an interminable movement towards the other comes to the fore. More than the stress on argumentative reason, a certain sensibility towards the different[1743] emerges as a new message for legal reasoning.

The practice of decision-making turns then, while acknowledging the limits of reason, into a mechanism to safeguard equality as difference, without this meaning that a possible conciliation between intersubjectivity and *différance* was achieved. Rather, both interconnect to each other without providing any resolution; it is a resolution as a non-resolution. The decision, while guaranteeing to the individual his condition of equal among others through the impartial principle of treating like cases alike, should also be aware of the need to open it up to *différance,* as a self-corrective process. As Honneth argues, "this conflict is productive because the viewpoint of care continually provides a moral ideal from which the practical attempt to gradually realize equal treatment can take its orientation – in a self-corrective manner."[1744] True, this opening to the asymmetrical acts of care every decision might carry does not mean that it will rule out the primacy of symmetrical justice, but it will also acknowledge its insufficiencies, as a reflex of the insufficiencies of constitutional democracy. By bring into focus the different, constitutional democracy learns from its own incapacity to insert every individual into procedures where she could exercise her communicative abilities. Similarly, legal adjudication learns from its incapacity to do justice, in the sense of not only equal treatment, but also of realizing the "call of the other, the arrival of the other, of an event."[1745] Constitutional democracy learns from the perception of the very limits of reason. Legal adjudication in turn learns from the consciousness of the boundaries of reason. Intersubjectivity and *différance,* accordingly, promote a "irresolvable but productive tension"[1746] that may, also in the realm of legal adjudication, foresee, perhaps as faith, what Miroslav Milovic called "a new self-reflexive community of *différance*."[1747]

[1743]See Miroslav Milovic, *Comunidade da Diferença* (Ijuí, RS; Rio de Janeiro: Unijuí; Relume Dumará, 2004), 131.

[1744]Honneth, "The Other of Justice: Habermas and the Ethical Challenge of Postmodernism," 315.

[1745]Jacques Derrida, "A Response to Simon Critchley," in *The Derrida-Habermas Reader*, ed. Lasse Thomassen (Edinburgh: Edinburgh University Press, 2006), 113.

[1746]Honneth, "The Other of Justice: Habermas and the Ethical Challenge of Postmodernism," 319.

[1747]Miroslav Milovic, *Comunidade da Diferença* (Ijuí, RS; Rio de Janeiro: Unijuí; Relume Dumará, 2004), 132, translation mine.

8.3 The Concept of Limited Rationality In German and Brazilian Constitutional Realities

With those three propositions emerging from the very limited character of rationality, and after having examined how they best handle the complexities and dilemmas of constitutional democracy and, specifically, legal adjudication, the final analysis of this book concentrates on the extension of those conclusions to reality through a brief critical reconstruction of the historical background of German and Brazilian constitutionalisms and some of their constitutional cases.[1748] If, in the last part, we explored one of the most influential concepts of rationality stemming from those characteristics of activist constitutional courts, now it is time to draw attention directly to these realities through the concept of limited rationality. It is hence a further step towards the conclusion that, while we disclosed the metaphysics embedded in Alexy's defense of the rationality of balancing, there was also, reflexively, somehow a disclosure of the existing metaphysics in the practice of the constitutional courts examined in the first part. In this respect, the purpose here is to investigate how the acknowledgment of the limits of reason would indicate a possible different scenario where constitutional courts – the *Bundesverfassungsgericht* (BVG) and the *Supremo Tribunal Federal* (STF) – rather than transforming their activisms into a protective duty towards the social values in general, would concentrate their efforts on safeguarding the constitution: (1) by focusing on the singularity of the case, not reducing then its complexities to some predetermined formulas and patterns, as if they were a sign of "rationalization" of decision-making, which, in other words, means that, if the judge deploys a general and abstract method, it is there for the case and its particularities, not the case and its particularities are there for the method; (2) by continuously keeping it consistent all through their decisions, showing thereby a real concern for the institutional history and the authoritative or deontological character of the system of rights; (3) by orienting themselves towards the quest for the other, grasping then the conflictive but productive tension that is the basis of the relationship between constitutionalism and democracy, between law and justice.

In the second chapter, the analysis focused on German constitutionalism and the BVG's development throughout history, as a means to verify how the deployment of balancing harmonized coherently with the increasing BVG's activism and casuism,[1749] indeed, with the construction of the concept of constitution as a "concrete order of values," or, as Bernhard Schlink remarks, of a concept of basic rights as objective principles of a total legal order.[1750] As seen, this BVG's approach came from a controversial historical development in the face of the vacuum of political

---

[1748] The emphasis here will be mostly on those cases we examined in the first chapter.

[1749] See Bernhard Schlink, "Abschied von der Dogmatik: Verfassungsrechtssprechung und Verfassungsrechtswissenschaft im Wandel," *Merkur* 692 (December 2006).

[1750] See Bernhard Schlink, "German Constitutional Culture in Transition," *Cardozo Law Review* 14 (1993), 711–736.

legitimacy observed after the Second World War, which led to a progressive politicization of its discourse, and whose activities reached the most distinct domains of social life. While the BVG became the most admired constitutional organ in Germany,[1751] the doubts whether its movement towards the safeguard of both the legal order and the social welfare were legitimate, as if it were its duty to centralize the social and political debates under its realm of authority, became visible. The BVG, more and more, entered into the dualism between law and politics[1752] and began interpreting many conflicts of social life as though they were problems of basic rights,[1753] using, for this purpose, instruments such as the principle of proportionality, thereby balancing. This instrument became in turn a powerful mechanism to promote an atmosphere of methodological density and rationality in decision-making. This movement towards the interest in dealing with the present and future problems of social life by interpreting them as problems of basic rights came out, though, with the consequence of the weakening of the concern for keeping consistent the system of rights.[1754] The more the legal discourse resembled a political discourse, as though arguments of principles did not differ from arguments of policy,[1755] the more the quest for keeping consistent the system of rights and its enforceable character distanced itself from constitutional decision-making, especially on account of the flexible structure (one that bears a resemblance to values, even to be balanced with them) constitutional principles should convert to when coping with political purposes. By approximating its discourses of application to discourses of justification,[1756] the second proposition above deriving from the concept of limited rationality applied to legal adjudication - continuously keep the constitution consistent all through its decisions, showing thereby a real concern for the institutional history and the deontological character of the system of rights – was put in jeopardy.

Yet, this movement is more complex. It is necessary to recall some of the characteristics of an institutional history that favored the BVG's activism and the wakening of the quest for keeping consistent the system of rights and its deontological character. The first focal point is the perception that the shift to activism in this dualism between law and politics followed a period of authoritarianism and reconstruction – in the circumstances, the National Socialism and the Second World War, respectively. While there was, accordingly, a feeling of reconstruction, there was also the need to avoid the return of authoritarianism: the first demanded a rapid engagement in dealing with the social and economic problems of the postwar

[1751]See Schlink, "Abschied von der Dogmatik: Verfassungsrechtssprechung und Verfassungsrechtswissenschaft im Wandel," 1125.

[1752]See Peter Häberle, "Grundprobleme der Verfassungsgerichtsbarkeit," in *Verfassungsgerichtsbarkeit*, ed. Peter Häberle (Darmstadt: Wissenschaftliche Buchgesellschaft, 1976).

[1753]See Schlink, "German Constitutional Culture in Transition," 722.

[1754]See Jürgen Habermas, *Between Facts and Norms* (Cambridge: Polity Press, 1996), 246.

[1755]See the sixth chapter.

[1756]See the sixth chapter.

period; the second in turn led to an extensive bill of rights that could be adapted to the circumstances of a Germany in reconstruction. In the specific realm of constitutional adjudication, these two aspects favored the erection of a powerful constitutional court, the *Bundesverfassungsgericht,* which, also guided to avoid the return of authoritarianism, and in virtue of the discredit upon traditional politics and the parliament, would not only turn into the "Guardian of the Constitution" but appear likewise as the adequate and, in a certain manner, uncontrollable[1757] institutional organ to reconstruct the social order, which was made, for instance, through positive protection of collective benefits.[1758] The need to create a constitutional court with a political intent, as a juridification of politics, was then historically justified.[1759] While the BVG increased, therefore, its domain of political activity, the parliament consolidated its secondary role in the discussion of many themes of social life. While the BVG centralized some of its activities on politics, the aim to build a consistent system of rights and enforce it in reality was relativized in favor of attending to the present and future problems of society through decision-making.

Naturally, as long as the question of legitimacy manifests itself in the BVG's shift to activism, the quest for methodological rationality appears, even to justify this practice. In this respect, the construction of a methodology as the principle of proportionality, and particularly balancing, to manage this political intent with a relativized comprehension of the constitution and its principles came out as the natural consequence of this historical development. As long as constitutional principles must be balanced with other values in order to promote this BVG's activism, not only the second proposition above of keeping consistent the system of rights and its enforceable character but also the first one regarding the reduction of the complexities of the case to some predetermined formulas and criteria or, in other words, the submission of the case and its singularities to a methodology shaping the case features according to a possible political purpose are at issue. The question now is the reduction of the singularity of the case as a means to justify the rationality, the rightness and even the legitimacy, through methodological criteria, of a political decision; that is, instead of observing the subjective rights of the individuals involved, the focus is on how, from the rights originated therefrom, they could be maximized or irradiated through a teleological analysis of possible social effects. For this purpose, the BVG took some concatenate steps: first, it interpreted

---

[1757]There was no emphatic resistance to this BVG's political role. See, for this purpose, Helmut D. Fangmann, *Justiz gegen Demokratie: Entstehung- und Funktionsbedingungen der Verfassungsjustiz in Deutschland* (Frankfurt a.M; New York: Campus Verlag, 1979), 234–235.

[1758]See Schlink, "German Constitutional Culture in Transition," 720-721; Habermas, *Between Facts and Norms,* 247.

[1759]See Heinz Laufer, "Politische Kontrolle durch Richtermacht," in *Verfassung, Verfassungsgerichtsbarkeit, Politik: Zur verfassungsrechtlichen und politischen Stellung und Funktion des Bundesverfassungsgerichts,* ed. Mehdi Tohidipur (Frankfurt a.M.: Suhrkamp, 1976), 94. See also Fangmann, Justiz gegen Demokratie: Entstehungs- und Funktionsbedingungen der Verfassungsjustiz in Deutschland, 224.

basic rights as "objective principles;"[1760] that is to say, they were no longer conceived as subjective rights but "principles of a total legal order,"[1761] which was deemed a more suitable interpretation to provide responses to the present and future problems of society;[1762] second, basic rights gained the character of optimization requirements, furnishing thereby the justification for the principle of proportionality, and particularly balancing, and losing their enforceable character insofar as they were gradually balanced according to variable degrees of satisfaction[1763] or according to the gathered social values;[1764] third, constitutional adjudication justified itself not through the interpretation of dogmatic concepts, but through a flexible methodology that could absorb these characteristics of basic rights, whose role balancing successfully undertook.

With these three steps, the BVG could not only relativize the enforceable character of principles, and thus open up the way to the disruption of a consistent interpretation of them as long as they were submitted to a gradual evaluation through balancing, but also "rationally" justify this characteristic through the deployment of some seemingly predetermined formulas and criteria. The cases we examined in the second chapter are clear examples of this movement whereby the court deployed some methodological criteria to justify an argument of policy founded on teleological analyses of basic principles as though they were objective principles embracing the totality of the legal and social order.[1765] Particularly, we could now stress the *Hochschul-Urteil*,[1766] judged in 1973, when the court held the argument that scholarship, specifically within the context of university organization, must be free from legal or any other type of state intervention; the *Schwangerschaftsabbruch I Urteil*,[1767] judged in 1975, in which the BVG stated, against the legislative definition and in agreement with what it understood as emerging from the objective axiological order, that the protection of life must prevail over the pregnant woman's self-determination in cases of abortion; and the *Schwangerschaftsabbruch II Urteil*,[1768] judged in 1995, when the court, by interpreting differently this objective axiological order, argued that it was possible to establish a temporal exception to abortion, as long as the state provided compulsory counseling in favor of the continuity of pregnancy, that is, respected the public exigency to

[1760]See Schlink, "German Constitutional Culture in Transition," 711–736.

[1761]See Habermas, *Between Facts and Norms*, 247.

[1762]Ibid., 246.

[1763]See Robert Alexy, *Theorie der Grundrechte* (Frankfurt a.M.: Suhrkamp, 1994), 76.

[1764]See Erhard Denninger, "Freiheitsordnung - Wertordnung - Pflichtordnung," in *Verfassung, Verfassungsgerichtsbarkeit, Politik: Zur verfassungsrechtlichen und politischen Stellung und Funktion des Bundesverfassungsgerichts*, ed. Mehdi Tohidipur (Frankfurt a.M.: Suhrkamp, 1976), 166–67.

[1765]See the second chapter.

[1766]BverfGE 35, 79.

[1767]BverfGE 39, 1.

[1768]BverfGE 88, 203.

create an institution responsible for these circumstances, in a clear political definition of the state's protective duty through decision-making.

These cases, among the others we examined, revealed that the court oriented its activities to laying down political functional decisions, some of them already largely discussed in the parliament, using thereby many predetermined criteria and patterns to justify its activist approach. They expressed the perception that the court, while transforming basic rights into objective principles embracing the totality of the legal order, focused more on relativizing those basic rights in favor of a political definition than really confirming their authoritative character. This deontological feature, after all, is a necessary condition for constructing a consistent system of rights not alterable by the judge's definition of the temporal axiological conscience that could lead to decisionism[1769] and to a serious violation of the principle of separation of powers. Yet, this movement not only put the construction of a consistent system of rights in jeopardy, but it also promoted a less consistent legal dogmatics[1770] and a less active scholarship's critique of the BVG's decisions,[1771] which are the immediate outcomes of the former. After all, a consistent system of rights and a consistent legal dogmatics and scholarship's critique, in this model of decision-making, become a problematic presupposition for doing politics through adjudication.

The consequences derived from disrupting the first and the second propositions of the concept of limited rationality can also, in a somewhat similar panorama to the one in Germany, be critically reconstructed when we investigate Brazilian constitutionalism and the STF's development throughout the recent history.[1772] As in Germany, this court has progressively undertaken, as its realm of responsibility, the discussion of present and future problems of society, extending thereby its authority not only to safeguard the Constitution but also to resolve many Brazilian social and political issues. In a comparable fashion, the quest for rationality in decision-making by appealing to some methodological criteria such as the principle of proportionality, and particularly balancing, has become a noticeable characteristic of its recent decisions. There is, hence, on the one hand, a growing political and intervenient attitude in the institutional ground, on the other, an expansion of justificatory mechanisms able to consolidate a seemingly rational solution that best handles this new STF's activist approach. As examined in the third chapter, this activism also arises from some relevant characteristics we could associate, in a certain and approximate way – for any resemblance in this matter must observe the historical and institutional particularities of both countries - with those from Germany: first, the process of redemocratization after a period of authoritarianism (in Brazil, the period of the military regime from 1964 to 1985), transforming then

[1769]See Ernst-Wolfgang Böckenförde, *Staat, Verfassung, Demokratie: Studien zur Verfassungstheorie und zum Verfassungsrecht* (Frankfurt a.M.: Suhrkamp, 1991), 131–133.

[1770]See Schlink, "Abschied von der Dogmatik," 1133.

[1771]See Schlink, "German Constitutional Culture in Transition," 730.

[1772]See the third chapter.

the constitutional court into the ideal institution that could catalyze the demands flowing from the emergence of a new constitutional democracy; second, the existence of a deficit of political representation in the parliament and a discredit upon the legitimacy of its activities; third, the deficient scholarship's review of the STF's decisions, which practically worships its decisions and the "rational" character of their arguments and deployed methodologies; and, fourth, the construction of mechanisms that could provide the court with the prerequisites for an activist practice justified "rationally," such as the deployment of some predetermined patterns, formulas and criteria (the principle of proportionality, and balancing in particular, for instance). These events created thus the conditions for inserting the Brazilian constitutional court into the troublesome dualism between law and politics.

By examining carefully those characteristics, we could remark that the STF changed radically its posture throughout the recent Brazilian history, from a certain conservatism and passiveness to an activism oriented towards politics through decision-making. After the promulgation of the Constitution of 1988, although it gained a vaster domain for exercising its control over governmental and parliamentary acts through the expansion and consolidation of the abstract system of judicial review (a consequence indeed of the democratic spirit of that time), the STF still remained, at the beginning of this new democratic period, very self-restrained from getting involved in political matters. This posture, nevertheless, contradicted somehow the intent to transform the STF into a court that could, in fact, act as a real legitimate protector of democracy, a "Guardian of the Constitution" and a safeguard against the return of authoritarianism, especially within a context where both the government and the parliament were strongly discredited, and the society claimed the concretization of multiple basic rights the Constitution of 1988 laid down. Not only did the STF not provide, in reality, decisions that could reinforce the new democratic model and its system of rights, creating thereby a consistent and reliable parameter for decision-making, but it also avoided transforming itself into an activist court that would consolidate, in practice, the many constitutional norms that established social benefits and provisions. By reason of a deficit of constitutional-democratic practice and constitutional-democratic knowledge, the STF was incapable of constructing a consistent system of rights in its constitutional decisions, which exposes that, right at the beginning of Brazilian constitutional democracy, the second proposition of the concept of limited rationality regarding the exigency of keeping consistent the system of rights and its enforceable character over time was not yet a real concern.

In any case, if the activism began in the lower courts,[1773] which undertook as their realm of responsibility the implementation of public policies through decision-making[1774] – a movement indeed characterized by a strong lack of concern for the

[1773]See the third chapter.

[1774]See Marcus Faro de Castro, "The Courts, Law and Democracy in Brazil," *International Social Science Journal*, no. 152 (June 1997): 241–252; Oscar Vilhena Vieira, *Supremo Tribunal Federal: Jurisprudência Política* (São Paulo: Malheiros, 2002), 135 ff.

consistency of the legal system and by the prevalence of judge's subjectivism - it also gradually appeared in the area of the Brazilian constitutional court. After the conservatism and passiveness of the initial democratic period, as if the STF were "relatively impervious to pressures for the expansion of the judicial power,"[1775] and its complacency towards the governmental and parliamentary acts, from 1993 onwards, this court started gradually but systematically to undertake a more activist role in judicial review. Many instruments, both legally and judicially, were created to concentrate the judicial review in the STF's hands,[1776] most of them clearly approximating its functions to those of the parliament while disrupting the diffuse system of judicial review, which is the only mechanism in the Brazilian Constitution whereby a common citizen can question the constitutionality of a legal statute or a governmental act. The restrictions on the ways a citizen has to file a lawsuit in this matter, creating thereby a serious encroachment on this civic practice, followed thus the concentration of judicial review in the STF's hands and its own definition of political matters. The STF's shift to politics could, accordingly, more directly intervene in governmental and parliamentary acts, but a common citizen still had a restricted space to question any of these acts.[1777]

Naturally, by distancing itself from society and expanding its activism towards politics, the STF had to transform substantially the argumentative quality of its decisions, even to justify this activism in a more legitimate way. First, as in Germany, many of the social problems became constitutional problems by transforming constitutional principles into objective principles of a total legal order. This characteristic could lead to the idea that judicial review is a political practice – as if the constitutional court were a negative and even positive legislator[1778] – oriented towards maximizing interests that are best for the whole society. As a consequence, it would result in the premise that constitutional principles have the nature of optimization requirements, and the overall constitution, in turn, the character of a

---

[1775] Castro, "The Courts, Law and Democracy in Brazil," 246.

[1776] See the third chapter.

[1777] This characteristic is more serious than the concentration of powers observed in the German BVG, if we center on the criterion that, in Germany, any individual can file the main constitutional complaint – the *Verfassungsbeschwerde* – to initiate the judicial review, which is, as a matter of fact, one important reason that makes this court the most admirable constitutional organ in that country. For this reason, the STF, while expanding its activism to politics through judicial review, became, however, a distant partner of society. This raises immediately and perhaps with greater intensity the question of the legitimacy of its decisions (even though popularity and admiration in the case of the German BVG's is not a parameter of legitimacy, as we formerly examined) and its insertion into the context of a constitutional democracy oriented by the principle of separation of powers.

[1778] See Hans Kelsen, *Wer soll de Hüter der Verfassung sein?* (Berlin: Rotschild, 1931); Enzo Bello, "Neoconstitucionalismo, Democracia Deliberativa e a Atuação do STF," in *Perspectivas da Teoria Constitucional Contemporânea*, ed. José Ribas Vieira (Rio de Janeiro: Lumen Juris, 2007), 31 ff.

"concrete order of values." Second, it held the argument that this new approach was the consequence of the irreversible evolution of Brazilian democracy, and that its decisions were the best representation of an open and pluralistic society[1779] whose values were entirely respected in decision-making. Third, it expanded the deployment of methodologies that could supply decision-making with a seemingly "rational" quality able enough to legitimately justify its rightness, while embracing more adequately this activist approach, whose methodological comprehension, such as the one observed in Germany, gained force especially through Justice Gilmar Mendes's opinions. In this matter, the principle of proportionality appeared as a technical solution able to provide answers with a rational justificatory force. Consequently, if the "constitutional court exists to make the most rational decisions"[1780] and is also engaged in political matters, what can differ it from legislation became merely the "rational" character of its decisions.

This is why, if we could identify that, in this movement towards politics, the STF relativized legal principles in order to best handle the political dilemmas emerging from a recent democratic society with many political claims, now regarded as constitutional claims, then the quest for respecting their enforceable character is put in jeopardy, in a similar fashion as that one observed in Germany. In this matter, the consequence is the risk of disrupting the consistency of the system of rights, transforming then constitutional adjudication into a malleable practice oriented towards particular interpretation of social values that are balanced with constitutional principles with no special constraints. Nevertheless, this attack on the second proposition of the concept of limited rationality is normally followed by a justificatory intent, which is, as shown, the need to provide a rational justification for this political goal and this relativization of constitutional principles, a consequence that demonstrates that the attack on the first proposition of the concept of limited rationality normally follows the second one, and vice-versa.

In Brazilian reality, this shift to activism was favored not only by those legal instruments that led to the concentration of judicial review in the STF's hands, but also by a systematic construction of legal arguments framed by methodological criteria of "rational" purposes. In the third chapter, we examined some cases that demonstrated this transition from somewhat unmethodical arguments, at least if we consider the systematization of legal reasoning, to the deployment of the principle of proportionality in an equivalent way to Germany, which temporally coincided with this movement towards activism. In the same way, we verified how the use of balancing, in different opportunities, even when not yet structured as the German model of the principle of proportionality, was used as an instrument to validate evident unconstitutional measures and guarantee the economic and political space of governability, jeopardizing thereby the normative consistency of the

---

[1779]See Peter Häberle, *Verfassung als öffentlicher Prozeß: Materialen zu einer Verfassungstheorie der offenen Gesellschaft* (Berlin: Duncker & Humblot, 1978).

[1780]Gilmar Mendes, interview by Izabela Torres, "Entrevista - Gilmar Mendes," *Correio Braziliense*, Brasília (2008, 17-August), translation mine.

Constitution. The cases of the unconstitutionality of the Provisional Measure n. 173/90 (related to the prohibition of preliminary verdicts and immediate executions of provisional sentences in economic and fiscal matters,[1781] in which the economic argument was the chief value that determined the balancing in favor of a political program), and the famous "*Apagão*" case[1782] (when the court, by holding the argument of cost/benefit, balanced the constitutional protection to the consumer in favor of a governmental program to face the energy crises) are relevant examples of this STF's political character and of the absence of a real interest in constructing a reliable legal system through decision-making. On the other hand, the progressive deployment of the principle of proportionality, with the triadic structure as observed in Germany, also pointed to this political intent. We could verify clearly these characteristics, among others,[1783] particularly in the *Disarmament Act* case,[1784] in which Justice Gilmar Mendes, against the law that incriminated the illegal possession of fire weapons and established the impossibility of parole in those circumstances, argued that the norm under scrutiny was unconstitutional, because other more serious crimes did not lead to that restriction, as if this political analysis of which crimes were subject or not to parole were the STF's responsibility; or the *Embryonic Cells*[1785] case, when the court arrived at mentioning that the STF was a "house of commons, as the parliament,"[1786] and some of its Justices' opinions established, in a similar way to that one verified in the German *Schwangerschaftsabbruch II Urteil*,[1787] the need of a Central Ethical Committee to approve the research on embryonic cells[1788] or even judicially created a crime.[1789]

In a progressive way and by recalling, by some means, certain characteristics of the German BVG's movement towards politics, these cases, among the others we previously examined,[1790] confirm the thesis that, while the constitutional court shifted to the definition of the political debates on the main themes of the social order, the quest for confirming the authoritative nature of the legal system and reinforcing its consistency through decision-making became a less serious subject of concern. More than legal certainty and legitimacy, for the respect for the impartial principle of treating like cases alike could be relativized in favor of the consolidation of a political will, the quest for decisions that would, according to the judge's temporal axiological conscience, best realize the interests of society

[1781] ADInMC 223 (published on 06.29.1990).

[1782] ADC 9 (published on 04.23.2004).

[1783] See the third chapter.

[1784] ADI 3.112 (published on 10.28.2007).

[1785] ADI 3510 (published on 05.28.2008).

[1786] Ibid., Justice Gilmar Mendes's opinion, translation mine.

[1787] BverfGE 88, 203.

[1788] ADI 3510 (published on 05.28.2008). Justice Mendes's opinion.

[1789] Ibid., Justice Menezes Direito's opinion.

[1790] See the third chapter.

by means of a "rational" justification became the main focal point. Either by conforming decision-making to these social interests or "rationally" justifying the rightness of the decision would imply a sort of "argumentative representation"[1791] legitimizing the court's activities. Moreover, this movement, similarly to Germany, has many resonances, as a worrisome deficiency of scholarship's critique of the STF's decisions and methodologies and the weakening of legal dogmatics through the growing casuism and interest in investigating Justices' personalities and idiosyncrasies. These are the consequences of a loss of the quest for legal consistency and the enforceable character of legal principles through decision-making, especially because the quest for consistency in all these contexts would make the practice of politics through adjudication even more difficult.

Therefore, we shall conclude that both the BVG and the STF can put in jeopardy, as long as they act as a "forum for the treatment of social and political problems,"[1792] the first and the second propositions of the concept of limited rationality, as we previously discussed. Not only is the quest for keeping consistent and reliable the system of rights with a real concern for the institutional history and their deontological character in danger, but also the very singularity of the case and its complexities are absorbed by the search for a rational methodological justification that could suitably assimilate this court's political intent. In this situation, it is as if the problem of rationality were a problem of deploying a certain methodology, and its legitimacy could be justified because of this rational approach and the satisfaction of people's will. In both realities, we could verify the need to deploy "rational" standards, analytical criteria to control the arguments in a "logical" basis, as though it could provide rationality in decision-making and ensure the rightness of the decision, especially in the realm of constitutional principles relativized by a political intent. This characteristic could reduce thereby the inherent tensions the relationship between constitutionalism and democracy raises, particularly the immanent and necessary connections among the distinct institutional powers and the discourses carried out therein. In short, it could reduce the intrinsic fundamental distinction between discourses of justification and discourses of application,[1793] distort the position of constitutional adjudication in the set of democratic procedures, transform the quest for justice into the realization of a particular interpretation of what is best for society, and disrupt either *iterability* or the *self-correcting learning process* – depending on how we work with these concepts. This happens because the interest in realizing social values gains primacy over the search for self-critique and perfectibility[1794] that a construction of a consistent and reliable system

[1791]See Robert Alexy, "Balancing, Constitutional Review, and Representation," *International Journal of Constitutional Law* (Oxford University Press and New York School of Law) 3, no. 4 (2005): 578 ff.

[1792]Schlink, "German Constitutional Culture in Transition," 729.

[1793]See the sixth chapter.

[1794]Derrida, *Rogues: Two Essays on Reason*, 87.

of rights, enforced in every new decision by also centering on its counterpart of justice, would engender.

Yet, if we could infer immediately the infringement of the first two propositions of the concept of limited rationality from the critical reconstruction of the German and Brazilian recent constitutionalisms, the infringement of the third proposition - the one related to the quest for the other, the incessant search for treating the other as an equal among the others in a non-resolvable but productive tension with the asymmetrical "coming of the impossible"[1795] - is more subtle. After all, one could argue that, inasmuch as the court acts in favor of the collectivity according to a political intent expressed through decision-making, it is evidently being oriented towards the otherness. In this state of affairs, it would reach the interests of society in a much more productive way than if it were merely concerned with the singularity of the case, the *prima facie* applicable norms and the application of the impartiality principle of treating like cases alike in a tense but productive relationship with the opening *différance* brings forward. Besides, one could even argue that the "rational" character of the methodologies deployed here can bring about equality, and thus justice to the case, as if some predetermined criteria associated with arguments, as long as "rationally" justified, could themselves do justice to the case. Still, on the other hand, it is possible to argue likewise that, by acting in such way, the constitutional court contradicts its function of "Guardian of the Constitution," insofar as it orients itself towards a majoritarian standpoint, i.e. one that is based on what is best for society in general, without acknowledging the existing pluralism of worldviews that characterizes complex and post-conventional societies. In a more abstract analysis, this would point to the reproduction of a substantive conception of democracy through decision-making. As long as constitutionalism or law is relativized in favor of a political intent that would conform to social values, which can be balanced with the law without any effective constraint and whose contents the judge has to express through decision-making, the quest for the particular, for the singularity of the other, is rather transformed into an affirmation of an axiological substantiality defining the future of democracy. In this circumstance, constitutional adjudication, while deciding in favor of what it understands as the best for society, can, in reality, reproduce a certain social identity that continuously threatens the possibility of including the other in the exercise of citizenship.[1796] It would guide itself thereby to the collectivity, to a certain *ethos* rather than the quest for justice, which, as we argued, ought to be worked instead in a conflictive but productive relationship between intersubjectivity

---

[1795]Jacques Derrida, "As If It Were Possible," in *Negotiations: Interventions and Interviews, 1971–2001* (Stanford, CA: Stanford University Press, 2002), 344.

[1796]See the fifth chapter, when we discussed the construction of a substantive concept of democracy through the analysis of the *logos of constitutionalism* and the *logos of democracy* based on Derrida's deconstruction.

and *différance,* as a self-corrective process that exposes the very limits of constitutional democracy and, consequently, of the reason itself.

The reconstruction of the BVG's and STF's historical developments, accordingly, while concluding on the possible violation of the first and second propositions of the concept of limited rationality, also permits addressing the question of how serious it is for constitutional democracy the conversion of the constitutional court from a "Guardian of the Constitution" into a "forum for the treatment of social and political problems,"[1797] whereby basic rights lose their authoritative character and become objective principles of a total legal order.[1798] Not only can the disruption of the set of institutional procedures that distinguish constitutional democracies from other regimes occur, especially in what refers to the principle of separation of powers, but also the quest for justice can be in jeopardy, which is the denial of the *raison d'être* of constitutional courts. As we formerly mentioned, the practice of decision-making, by acknowledging the limits of reason, must be an institutional instrument to safeguard equality as difference. This means, in other words, that while not resolving the tense but productive relationship between intersubjectivity and *différance,* there must be the quest for the impartial principle of treating like cases alike and the awareness to open up the decision to *différance,* as a self-corrective process that shows the boundaries of reason and the boundaries of constitutional democracy. More than stressing this search for rationality as if it derived from some predetermined patterns and criteria, it is important to understand that reason lies in this relationship between constitutionalism and democracy, between law and justice, which requires of the judge an effective interest in bringing into focus the different, the other. By acting in this way, she frees herself from a belief that methods, predetermined criteria, for they provide "rationality," also do justice to the case, and acknowledges rather that, from a "logical" belief, she might be disrespecting the multiplicity of worldviews that characterizes constitutional democracy. Furthermore, the judge realizes that only by acting towards the other, and not towards what is good for the whole society in a political behavior, will the sought after legitimacy and rationality reveal themselves. Only by grasping that constitutional adjudication has a fundamental role in constitutional democracies insofar as it is an essential institutional instrument to protect the equality as different, in a tense but constructive relationship between constitutionalism and democracy, between law and justice, will it be possible to articulate reason, a limited reason, that is not a simple reproduction of a determined *ethos,* of a majoritarian point of view, but rather a reason whose limits disclose a self-reflexive practice towards the other.

[1797]Schlink, "German Constitutional Culture in Transition," 729.

[1798]See Schlink "German Constitutional Culture in Transition," 711–736.

8.4 When the Concept of Limited Rationality Meets Constitutional Cases

8.4.1 Introduction

Having said that, it is still necessary to address this historical reconstruction of the German and Brazilian recent constitutionalisms by concentrating upon the question of how the concept of limited rationality would indicate another manner of working with the tense but productive dilemmas of constitutional democracy in the realm of constitutional adjudication. More specifically, the analysis now will focus on investigating, by acknowledging that reason lies in this non-resolvable tension between the quasi-transcendentalism of justice and the complexity of a reality marked by a pluralism of worldviews, how the judge should face the challenging activity of decision-making, without this meaning paradoxically the construction of some general formulas or criteria that would indeed diminish the singularity of the case and reduce the complexities originated therefrom. In reality, by assuming that reason has its boundaries erected by the impossibility of entirely recollecting and gathering the history and its complexities and by the impossibility of justice, now grasped in this non-resolvable relationship between intersubjectivity and *différance,* the judge will uphold those three propositions of the concept of limited rationality. For those propositions are, in truth, consequences of a theoretical construction that centers primarily on the other as the focal point of adjudication within the complex context of constitutional democracy, a practice of decision-making that endorses those propositions is not oriented towards achieving rationality by means of general predetermined formulas and criteria, but rather is oriented towards affirming, as much as possible, the otherness, for it knows that reason has its limits, both from history and justice, demanding thereby of the judge a posture, and not a belief in "logical" patterns as the condition for rationality. It requires of the judge a posture that: (1) exhaustively focuses on the singularity of the case, not reducing then its complexities to some predetermined formulas and criteria; (2) hermeneutically reconstructs the institutional history, as if the founding moment of the law were somehow perpetuated in every reinvention carried out in accordance with the particularities of the context, keeping thereby consistent and enforceable the system of rights; and (3) orients itself to the affirmation of the otherness, for decision-making is not a "dogmatic and irresponsible mechanics that would drown decision in the environment of dogmatic generality,"[1799] but rather should address itself to an "irresolvable but productive tension"[1800] of intersubjectivity and *différance* in the quest for justice.

[1799]Jacques Derrida, "Politics and Friendship," in *Negotiations: Interventions and Interviews, 1971–2001* (Stanford, CA: Stanford University Press, 2005), 182.

[1800]Honneth, "The Other of Justice: Habermas and the Ethical Challenge of Postmodernism," 319.

The question remains, nevertheless, in how we could visualize these three propositions in the concrete practice of decision-making. After having critically reconstructed both German and Brazilian constitutionalisms through the investigation of the BVG's and STF's developments in the recent history and revealed the metaphysical standpoint of the prevailing concept of rationality stemming therefrom in the last part of this book, it seems that there is a call for bringing this discussion directly to the complex task of decision-making. In other words, there is a demand for linking those propositions of the concept of limited rationality to cases. It is time, accordingly, to recall the three constitutional cases that initiated this research: the *Crucifix case,*[1801] the *Cannabis case*,[1802] and the *Ellwanger case.*[1803] If they served as instigation for this study, at this time, they will show that, were the concept of limited rationality taken seriously, perhaps another viewpoint would be possible, one that, more than pointing to the general interests of society with a flexible and objective structure of basic rights, would stress the "irresolvable but productive tension" between the quest for equal treatment and the differences that really 'make' difference,[1804] strengthening thereby the constitution in this new context the case gives rise to.

8.4.2 The Crucifix Case

In the *Crucifix case* in Bavaria,[1805] the discussion of the collision between the freedom of faith and the state's duty of religious and philosophical neutrality,[1806] on the one hand, and the prevalent traditional values in Bavaria, on the other, is a very interesting theme to verify how some of the arguments therein presented could be immediately deemed incompatible with the concept of limited rationality. The question hence is how a judge, in those circumstances, ought to examine this discussion. In this regard, a judge aware of the limits of reason would, at first, attempt to gather, as much as possible, the distinct features of the case. She would thus verify the specific condition of the complainants, that is, that they are school-age minor children from a family of followers of Rudolf Steiner's anthroposophical philosophy of life and as such do not profess Christian beliefs. As non Christians, she would see that the children's parents are concerned with the fact that the affixation of crosses could influence their children in a Christian direction, against, therefore, the beliefs of their family and their right to bring up their children without

1801 BVerfGE 93,1.

1802 BVergGE 90, 145.

1803 HC 82.424-2/RS.

1804 See Richard Bernstein, "Introduction," in *Habermas and Modernity*, ed. Richard Bernstein (Cambridge, MA: The MIT Press, 1988), 88.

1805 BVerfGE 93,1.

1806 BVerfGE 93,1. Translation: Institute for Transnational Law.

direct state interference in religion or philosophical matters. On the other hand, she would observe that the school, even though questioned by the complainants and having exchanged the crucifix for smaller ones without Christ's body in some classes, could not commit itself to cooperating with the children's parents by changing the crucifixes in every classroom where their children attended classes. In the social perspective, she would infer that the Bavarian State is characterized by a majority of Christians, who traditionally profess this belief in multiple spheres of social life, including public offices and schools. After having exhaustively described all the features of the case, the next step would be to explore the arguments of the lower courts. Here, she would remark that there is an intent to relieve the burden the children are suffering by arguing, first, that the crucifixes serve merely "for constitutionally unobjectionable support to parents in the religious upbringing of their children,"[1807] or that "in the educational sphere religious and philosophical conceptions had always been of importance,"[1808] and, second, that this confrontation would take place elsewhere too.[1809] By the same token, she would understand that the lower courts held the argument of tolerance in the tension between positive and negative religious freedom, as a means to conclude that, in those circumstances, the complainants could not demand absolute priority of their negative confessional freedom over the prevailing religious convictions of others.[1810]

Once exhaustively explored the particular features of the case and the lower courts' arguments, she would then focus on which legal norms are *prima facie* applicable to the case and, likewise, how this debate has been carried out within the legal institutions throughout history. Therefore, she would, first, examine the §13 (I) 3 of the School Law for Fundamental School in Bavaria (VSO), whose constitutionality is at issue and whose contents say: "the school shall support those having parental power in the religious upbringing of children. School prayer, school services and school worship are possibilities for such support. In every classroom a cross shall be affixed. Teachers and pupils are obliged to respect the religious feelings of all."[1811] Second, she would interpret it in conformity with the constitution. Particularly, she would do it grounded in article 4 (1) of the German Basic Law (*Grundgesetz*), which defines that "freedom of faith and of conscience, and freedom of creed religious or ideological, are inviolable," and with it, article 7, whose contents say that "the entire education system is under the supervision of the state," and "the persons entitled to bring up a child have the right to decide whether they shall receive religious instruction." Similarly, this article establishes that "religious instruction forms part of the ordinary curriculum in state and municipal schools, excepting secular schools. Without prejudice to the state's right of supervision, religious instruction is given according to the tenets of the religious

[1807] BVerfGE 93,1. Translation: Institute for Transnational Law.

[1808] BVerfGE 93,1. Translation: Institute for Transnational Law.

[1809] BVerfGE 93,1. Translation: Institute for Transnational Law.

[1810] See Ibid.

[1811] BVerfGE 93,1. Translation: Institute for Transnational Law.

communities. No teacher may be obliged against his will to give religious instructions." In the state level, she would also verify that the Bavarian State Constitution, in its article 135, establishes that "the public elementary schools shall be joint schools for all children of elementary-school age. In them pupils shall be taught and brought up in conformity with the principles of the Christian confessions."[1812]

The investigation of the distinct *prima facie* applicable norms would be, in the sequence, followed by the analysis of possible collisions between them. Consistent with this analysis, she would observe that articles 4 (1) and 7 of the Basic Law establish the freedom of faith, the state's duty of religious and philosophical neutrality and the parents' right to bring up their children in religious matters and to decide whether their children should receive religious instruction. On the other hand, she would verify that education is under state's authority and that, while being the religious instruction part of the ordinary curriculum, it is also given based on the tenets of the religious community. What she could, in principle, envisage as a collision of constitutional norms leading to balancing, however, would be treated as a search for the appropriate norm to the particular circumstances of the case. Indeed, she would conclude that, if the Basic Law establishes that religious instruction is given in accord with the tenets of the religious community, and the Bavarian State Constitution connects this norm to the Christian confession, it does not mean at all that the state can, on account of these norms, disregard the freedom of faith and its duty of religious and philosophical neutrality, or even the parents' right to bring up their children in religious matters without direct state interference. In a society characterized by the pluralism of worldviews and which respects the other as a member of this community, regardless of his or her religious or philosophical beliefs, she would be certain that there would be a collision of constitutional norms leading to balancing, only if she interpreted those norms in conformity not with the singularity of the case, but instead with the political effects of her decision. She would be sure that the subjective right to freedom of faith has an enforceable character that cannot be simply transported to an objective structure maximizing its content according to the factual and legal possibilities. For she would be really concerned with keeping consistent the system of rights and its deontological character, she would not balance the freedom of faith as a means to gradually harmonize its contents with the constitutional norm establishing that religious instruction is given in accordance with the tenets of the religious community. Rather, she would verify which of these norms is the appropriate one given the special features of the case and *only* the case.

A closer look into the original decision, nonetheless, reveals that both the BVG's majority and dissenting opinions interpreted this issue as a collision of principles leading to balancing. When we examined this case in the first chapter, we could comment that the BVG's majority opinion, even though stressing that the affixation of crucifixes would infringe state's duty of religious and philosophical neutrality, proceeded to balancing using, for this purpose, the concept of practical concordance

---

[1812] BVerfGE 93,1. Translation: Institute for Transnational Law.

(*praktische Konkordanz*). According to this opinion, insofar as the "the Land legislature is not utterly barred from introducing Christian references in designing the public elementary schools, even if those with parental power who cannot avoid these schools in their children's education may not desire any religious upbringing,"[1813] the resolution of this conflict lay thus in the reasonability of the compulsion, i.e., "there is a requirement (...) that this be associated with only the indispensable minimum of elements of compulsion."[1814] In other words, as means to solve the collision of principles, those constitutional norms establishing that religious instruction is part of the ordinary curriculum and is given in conformity with the tenets of the religious community would not prevail within that context as long as "the affixing of crosses in classrooms goes beyond the boundary thereby drawn to the religious and philosophical orientation of schools."[1815] The problem, therefore, transferred to the definition of what exactly was the maximal burden an individual could suffer. The resolution to this balancing was, above all, founded on the graduation of the possible encroachment an individual would experience in this situation. To all appearances, the solution was a balanced solution grounded, therefore, in the existence of a collision of constitutional norms.

This argumentation founded on balancing and on the concern for the political effects of the decision was even more radical in the dissenting opinion. Here, the court, first, connected the Christian beliefs to the contents of the Western cultural heritage;[1816] second, introduced a teleological-utilitarian argument: "the state, which through compulsory schooling is deeply involved in the upbringing of children by the parental household, is largely dependent on acceptance by parents of the school system it organizes;"[1817] and third, argued that neutrality does not mean indifference or secularism.[1818] Yet, what is particularly remarkable in this decision is exactly how the dissenting opinion interpreted the same burden in another direction. In this matter, it stated that "the psychic impairment and mental burden that non-Christian pupils have to endure from the enforced perception of the cross in class is of only relatively slight weight,"[1819] especially when children do not suffer the risk of being discriminated thereby. This conclusion is what led to the construction of the idea of tolerance. For this purpose, the main question, as it was in the BVG's majority opinion, was to verify which religious freedom, either positive or negative, should prevail. Again, the solution was a balanced solution grounded in the collision of principles.

Yet, a judge conscious of the boundaries of reason would not carry out this debate founded upon how much burden an individual could tolerate in this

[1813]BVerfGE 93,1. Translation: Institute for Transnational Law.

[1814]BVerfGE 93,1. Translation: Institute for Transnational Law.

[1815]BVerfGE 93,1. Translation: Institute for Transnational Law.

[1816]BVerfGE 93,1. Translation: Institute for Transnational Law.

[1817]BVerfGE 93,1. Translation: Institute for Transnational Law.

[1818]BVerfGE 93,1. Translation: Institute for Transnational Law.

[1819]BVerfGE 93,1. Translation: Institute for Transnational Law.

circumstance, for, by centering on the particularities of the case, she would know that this case does not require a balanced solution, but rather the simple definition of which constitutional norm is more appropriate to the circumstance. She would know that connecting the right to freedom of faith to a graduation based on what the court understands as the "indispensable minimum of elements of compulsion" weakens the authoritative character of this subjective right, opening up thereby the space for the judge's subjective evaluation of how much burden an individual could tolerate. Therefore, her concern would not be whether the affixation of crucifixes is a missionary spread of Christianity that goes beyond the "indispensable minimum of elements of compulsion"[1820] or, on the contrary, a tolerable practice that makes positive religious freedom prevails over the negative one. Instead, in order to define which is the appropriate norm to the case, she would concentrate, first, on the institutional history of this discussion, on how it has been considered in the distinct institutional forums and, second, on how it connects to the quest for justice.

The institutional history, how this theme has been developed in the distinct democratic procedures, appears as a primary source, but how this material connects to the quest for justice is what leads it to the tense or, as Derrida says, mad[1821] moment of decision-making. With a clear perception of the limits of reason, she would acknowledge that, in the relationship between constitutionalism and democracy, there is no tradition that is exempt from critical review or, in Derrida's words, deconstruction. As a consequence, she would be aware that all reconstruction of this history developed in decision-making, while context dependent, must also be open to the quest for justice. The right solution is hence the one that interprets those norms and this institutional history focusing on the "irresolvable but productive tension"[1822] that the quest for justice brings forth. Accordingly, a decision that overvalues the Western heritage or the Christianity of a certain community, of a determined *ethos*, and sustains that the constitutional norm associating religious instruction with the tenets of the religious community is the appropriate one fails to pass through the double bind of law and justice. First, because a simple investigation of the institutional history that led to the Basic Law's definition of freedom of faith and state's duty of religious and philosophical neutrality shows that the coexistence of distinct worldviews and the expansion of the channels where the individuals can exercise their rights as citizens of a complex and plural society is a concern of German democracy. This concern has been reproduced, in different ways, in the debates taken place in Germany's distinct institutional spheres, even as a primary characteristic to avoid the return of the past authoritarianism. In the context of reinforcing the constitution as a means to expand the exercise of democracy by all citizens, the inclusion of the other becomes a primary concern.

[1820]See Sonja M. Esser, *Das Kreuz - ein Symbol Kultureller Identität? Der Diskurs über das Kruzifix-Urteil (1995) aus kulturwissenschaftlicher Perspektive* (Münster, New York, München, Berlin: Waxmann, 2000), 33.

[1821]See Derrida, "Force of Law," 967.

[1822]Honneth, "The Other of Justice: Habermas and the Ethical Challenge of Postmodernism," 319.

Even though the religious instruction according to the tenets of the community still plays a relevant role in German democracy – and this is the reason for the existence of article 7(3) of the Basic Law – it cannot be used as a means to endanger the freedom of faith which is a subjective right of all individuals living in a society that respects distinct worldviews. The history proves that, by placing a majoritarian view at the center of state's activity regardless of distinct points of view, the German democracy may be in jeopardy.

By the same token, the quest for justice, which immediately flows from the previous debate, also leads the judge to the conclusion that there would be no justice to the case, if her decision upheld the argument that the negative religious freedom was not as absolute as to avoid the expression, with the affixation of crucifixes in school classrooms, of the Christian beliefs. Insofar as her concern is the case and *only* the case, she would know that the question does not refer to the establishment of an acceptable *degree* of religious freedom, for this is only a matter for whom issues a decision grounded in balancing, not for a judge conscious of the limits of reason. Her concern, therefore, would not be to verify how much freedom of faith one individual owns, but rather if the freedom of faith, as a subjective right against the state encroachment on the private sphere, and the state's duty of religious and philosophical neutrality are the appropriate norms to the case, and, if they are, how to confirm their authoritative character in the case. In this regard, she would conclude that there are some individuals being effectively encroached by reason of the state's activity in favor of the majoritarian religious profession of the Bavarian population. By centering on the "irresolvable but productive tension"[1823] of intersubjectivity and *différance,* she would infer that there is nothing more undemocratic than compelling individuals to share, regardless of the *degree* of the burden they suffer, a majoritarian religious belief, for constitutional democracy, while treating the other as an equal among others, ought to also instigate *différance* as a self-corrective process, as a "negotiation [that] must readjusts itself each day in relation to differing places."[1824]

For this reason, in a state that endeavors all efforts to include the other in the different domains of society, as a sign of its commitment to constitutionalism and democracy, the right decision is the one that would conclude that the norms establishing the freedom of faith, the state's duty of religious and philosophical neutrality, and the parents' right to bring up their children in religious matters and to decide whether their children should receive religious instruction are the applicable norms to the circumstances. It would thus declare the unconstitutionality of the §13 (I) 3 of the School Law for Fundamental School in Bavaria (VSO). This is a solution that: first, exhaustively focused on the singularity of the case, and not on the political intent to obtain the "assent of the majority of the population,"[1825] as the original BVG's dissenting opinion argued, nor reduced its complexities to some

[1823]Ibid.

[1824]Derrida, "Politics and Friendship," 180.

[1825]BVerfGE 93,1. Translation: Institute for Transnational Law.

predetermined formulas and criteria, such as the idea of balancing; second, hermeneutically reconstructed the institutional history, keeping thereby consistent and enforceable the system of rights, for the norms were not balanced, but applied in all their enforceable character; and, third, oriented itself to the affirmation of the otherness, grounded in the "irresolvable but productive tension"[1826] of intersubjectivity and *différance* in the quest for justice.

8.4.3 The Cannabis Case

If the *Crucifix case* is an interesting example to explore the dualism between law and justice in the realm of constitutional adjudication, the *Cannabis case,* even though the BVG, in different passages, made an allusion to the legislator's discretionary margin in this matter, is particularly relevant on account of its clear debate on the juridification of politics; that is, how political the constitutional courts' exercise of judicial review is. When we look into the *Cannabis case,* as we did in the first chapter, the immediate conclusion is that it is a remarkable example of how the dualism between law and politics enters into decision-making. It demonstrates how the previous conclusions on the BVG's shift to politics are correct,[1827] and how it, to sustain an evident political decision, falls into the need to provide methods and criteria, such as the principle of proportionality, and particularly balancing, to "rationally" justify this incursion into political matters. By deploying it, the BVG could argue that its decision was expressing the best possible solution to *society,* given that it could rationally examine whether the *Narcotic Act* was not disproportionate regarding the right to free development of personality combined with the right to freedom as well as infer whether it was compatible with the equality principle. As a decision of manifest political contents, what prevailed in the argumentation was the discussion of collective values in terms of social consequences as a way to affirm, through balancing and criteria of efficiency (what could promote the best social results), the proportional extension of the right to freedom in the circumstances. The court undertook the role of assessing social values as well as attempted to justify its conclusions in this matter through a dogmatic system (the principle of proportionality) based on criteria of efficiency and intensity. However, as formerly showed, the quest for making a decision that is grounded primarily in arguments of policy[1828] normally comes with side effect of the relativization of basic rights. The transformation of subjective rights into objective principles of a total legal order[1829] is the immediate outcome of this process, and the weakening of the authoritative character of those constitutional

[1826]Honneth, "The Other of Justice: Habermas and the Ethical Challenge of Postmodernism," 319.

[1827]See the second chapter.

[1828]See the fifth chapter.

[1829]See Schlink, "German Constitutional Culture in Transition," 711–736.

norms, leading thereby to the loss of the concern for keeping consistent the system of rights, appears as the checkmate of a troublesome reality where the principle of separation of powers plays a fundamental role.

For this reason, when we focus on an imaginary scenario where a judge conscious of the limits of reason would face the challenge to judge the constitutionality of some *Narcotic Act* provisions (§29, I, BtMG), the primary aspect we would discover is her integral consideration of her role in the set of democratic procedures in which she is part of. More than placing emphasis on criteria of efficiency and on a possible assessment of social values related to the issue under her scrutiny, as a means to attend to them as much as possible, she would concentrate first on interpreting coherently the norms at issue based on an exhaustive description of all the features of the case. In this respect, she would verify that the case relates to some *Narcotic Act* provisions that incriminated, among others, the possession of *cannabis* products, and afterwards place them in the face of the constitutional principles that were indicated in the lawsuits, both in the various *Richtervorlage* and the constitutional complaint (*Verfassungsbeschwerde*). Particularly, she would confront the *Narcotic Act* provisions with articles 2 (I) (free development of personality), 2 (II) 2 (right to freedom), and 3(I) (equality principle) of the Basic Law.

After having examined how these constitutional principles have been discussed in the distinct institutional forums, and particularly how they have been associated with those *Narcotic Act* provisions at issue, she would conclude that the equality principle of article 3 (I) of the Basic Law is not an appropriate norm to the case, especially when the argument those suits raised was that the *cannabis* products bring about less harmful effects than other legal intoxicating substances, such as alcohol and nicotine, or that, by banning the consumption of *cannabis* products, the potential consumers of this drug would consume other drugs that would cause more social damages, even when these other drugs were legal. Unlike the BVG's original decision, which, despite its recognition of the legislator's discretionary margin in this field,[1830] provided, nevertheless, some problematic arguments such as the one sustaining that nicotine is not a narcotic, for it could not lead someone to "get high" (*Rausch*), or the one defining alcohol as a substance with cultural and historical qualities that justify its different treatment,[1831] she would simply argue that the definition of which substance is a narcotic or not - and, as such, able to initiate a criminal prosecution – is not of her competence, for her duty is not political. It is not of her authority to evaluate whether a substance can be qualified or not as narcotic, and much less whether it is similar or not to others not legally considered narcotic. This is far beyond the scope of her activity, which should not be oriented towards balancing the quality of a certain product with cultural goods or how the society conceives of certain products.

[1830] See BVerfGE 90, 145 (III).

[1831] See the first chapter.

Specifically in this case, she would acknowledge that a consistent interpretation of the equality principle does not allow extending it to a political definition of the quality of a certain substance, such as those *cannabis* products, in comparison with others. For the equality principle has an enforceable character, she would know that it could not be relativized as an objective structure embracing any theme of social life, as though any social claim could be deemed an offense to the equality principle, demanding thus the constitutional court's assessment. By acting in this way, she would avoid the risk of transforming the equality principle into a value as any other, able to be balanced with arguments of policy, resulting then in serious damages on its enforceable nature. Indeed, when the BVG interpreted the case using the equality principle as one of the main reasons, we could immediately verify how arbitrary its conclusions were, using, for instance, scientific references to the possibility of "getting high" (*Rausch*) and the cultural tradition of alcohol in the European circle.[1832] This is a serious example of how an axiological standpoint emerges in the concept of the equality principle when its enforceable character is relativized thanks to its extension to any theme of social life, as if it were under court's authority the definition of the scientific effects of a drug in comparison with others, and the specification of which is the cultural heritage in the drugs area. This entire axiological viewpoint shaped somehow, notwithstanding the result showing that the equality principle was thereby not infringed, the political character of the decision.

However, it was in the analysis of the right to free development of personality (article 2 (I) of the Basic Law) in combination with the right to freedom (article 2 (II) 2 of the Basic Law) that the political focus of the decision expressed itself more intensively. In this regard, the BVG's original position took some relevant steps: first, it delineated the "inner core" of the right to free development of personality, which did not embrace the right to "get high" (*Rausch*); and, second, it deployed the principle of proportionality grounded, first and foremost, in the possible effects the use of drugs causes in the society. In this particular subject matter, after the examination of the suitability and necessity of the measure, balancing was deployed as a way to conclude that the consequences of unbanning the dealing of *cannabis* products were more serious or weightier than the entire protection of the free development of personality in combination with the right to freedom in these circumstances.[1833] The restriction on these personal freedoms was justified as long as they could be proportionally evaluated according to the social effects the consumption and handling of *cannabis* products promoted. Considering this analysis of social effects, the court understood that the measure was disproportionate in what refers to the occasional private possession and consumption in small amounts of *cannabis* products, because the effects, in these circumstances, would be minimal, even if they could motivate in some way the illegal drug market. Therefore, the court's decision oriented primarily to understanding that the criminal

[1832]See the first chapter.

[1833]BVerfGE 90, 145, translation mine.

prosecution in these situations and a possible condemnation based on the general provision of penalties were disproportionate, for the effects on the individual offender could be inadequate and, from the point of criminal special prevention, disadvantageous.[1834] Yet, inasmuch as there were legal mechanisms that would refrain the authorities from criminal prosecution and imposition of a penalty in these situations, the court finally decided favorably to the constitutionality of the *Narcotic Act* provisions at issue, although some other judges interpreted this evaluation of the proportionality of the measure differently, leading them to defend another solution.[1835]

The doubt is nevertheless how far this political discourse, founded upon evaluations of social effects and the consequent proportionate graduation of those rights to personal freedom, is not revealing a serious encroachment on legislature's activity. Indeed, by deploying the principle of proportionality, and balancing in particular, it seems that most of the reasoning was composed of arguments of justification.[1836] The question becomes then how a judge aware of the limits of reason would examine this last point differently. In this regard, the judge would be carefully concerned with how her decision could be coherently inserted into the set of democratic procedures she, as a member of the judiciary, is part of. For this reason, she would examine how the right to free development of personality and the right to freedom have been framed in the distinct institutional forums, and how the constitutional scholarship has historically interpreted it. In what refers to the free development of personality (article 2 (I) of the Basic Law), she would acknowledge that it is a residuary fundamental right (*Auffanggrundrecht*),[1837] demanding thereby the combination with another special basic right,[1838] such as the right to freedom. In the same way, she would verify that the right to freedom refers to the physical freedom of movement (*körperliche Bewegungsfreiheit*), that is, the right everyone has to move oneself to a closer or more distant place and to decide not to go to a certain place.[1839] In other words, it means that where someone does not intend to stay, he or she has not to stay, either, a content that has been historically coined.[1840] As a consequence, this constitutional principle can reach any arrest an individual may suffer, regardless of whether it is a short detention or an imprisonment for life.[1841] Likewise, it can be associated with the examination of the proportionality of the condemnation one individual may suffer, based on the special features of the singular and *concrete* case. Indeed, in this respect, she would observe that the BVG has a solid

[1834]BVerfGE 90, 145, translation mine.

[1835]See the first chapter.

[1836]See the fifth chapter.

[1837]See BVerfGE 6, 32 (*Elfes-Urteil*).

[1838]See Bodo Pieroth and Bernhard Schlink, *Grundrechte: Staatsrecht II* (Heidelberg: C.F. Müller, 2006), 86.

[1839]See Ibid., 98.

[1840]See Ibid.

[1841]See Ibid., 99.

understanding that, the longer the penalty is, the more material and procedural conditions for its justification the judge must fulfill in her sentence.[1842]

Once examined the usual institutional discussion of the extension of these subjective rights, she would conclude that the combination of articles 2 (I) and 2 (II) 2 of the Basic Law, if not deployed carefully, would open up the possibility for any social theme related to a possible encroachment an individual might suffer to become a constitutional issue. Particularly, it would open up the possibility of politically reexamining the contents of any political measure. For she would be concerned with the enforceable character of basic rights, she would not transform the right to freedom combined with the right to free development of personality into a malleable argument, an objective principle embracing the totality of the legal order.[1843] After all, a judge conscious of the limits of reason would not envisage her activity as a "forum for the treatment of social and political problems,"[1844] whose decisions could be made by balancing those constitutional principles in order to proportionally adapt them to a certain expected social outcome. Therefore, inasmuch as those two principles combined open up the possibility of arguments of policy in decision-making, replacing then the parliament's political evaluation, they make even more necessary the careful examination of the special features of the case.

In the analysis of both principles combined, she would conclude that the debate on the possibility of incriminating an individual by reason of consumption of small amounts of *cannabis* products, as specified in the case, could not immediately become a problem of infringement of the physical freedom of movement, at least not *in abstract*. If she understood this otherwise, indeed any other political measure that could *potentially* cause any imaginable limitation on this right, even on account of a practice of a crime, would become a problem of basic rights. Every political program destined to the repression of criminality and the definition of the possible penalties an individual might suffer in virtue of committing a crime would be under the BVG's subjective interpretation of the reasons for that political program and its social repercussions, in a typical political function. For this reason, this problem could at most center on a discussion of the proportionality of the punishment in a singular and individual case, with a determined subject involved, since the proportional graduation of a punishment according to the facts is a judicial activity *par excellence*. This was not, nevertheless, the issue raised before the BVG – and actually this would demand other relevant factual and legal aspects that were not even brought to discussion - but rather a *general* analysis of the constitutionality of those *Narcotic Act* provisions at issue. The BVG undertook this *abstract* and political analysis by balancing those principles with the social effects: how each Justice interpreted the proportionality of the measure in the circumstances of an

---

[1842]See Ibid., 101.

[1843]See Schlink, "German Constitutional Culture in Transition." In: *Cardozo Law Review*, 711–736.

[1844]Ibid., 729.

individual who sporadically consumes *cannabis* products in small amounts varied according to each one's axiological interpretation of this particular reality, in an evident discourse that did not differ in truth from the legislator's.

For this reason, the question relating to this political criminal program could be evaluated with those two principles by shaping their contents according to the possibilities of a balanced solution. However, since a balanced solution is not one that reinforces those principles in the practice of decision-making, but instead relativizes them based on analyses of social costs and efficiency, a judge conscious of the limits of reason would remark that it is not her duty to proceed to balancing, because, in doing so, she would put in jeopardy the authoritative character of those constitutional principles and indeed replace legislator's interpretation of social values by hers.[1845] After all, she knows that basic rights should not be interpreted as optimization requirements of a total legal order, for this weakens their enforceable character, transforming them into malleable arguments the judge could use to sustain her subjective interpretation of social values. Inasmuch as the *Narcotic Act* provisions could not directly infringe both principles but only relatively and obliquely make a connection with them, she would conclude that the question is no longer under her authority. The right solution, therefore, would be the one that points out that articles 2 (I) and 2 (II) 2 of the Basic Law cannot justify the judicial review of the legal measure at stake.

Moreover, this solution is the one that more coherently connects to the quest for otherness, although it might not come out immediately as it occurred in the *Crucifix case*. For the judge does not occupy a political position in the definition of a proportional comprehension of those constitutional principles under discussion, she immediately avoids transforming decision-making into a "forum for the treatment of social and political problems,"[1846] in which she, rather than acting as an essential institutional instrument to protect the equality as a counter-majoritarian voice, would uphold the argument of what is best for the entirely society. As long as law and justice are a co-original double bind, the more decision-making becomes an activity oriented towards the definition and revision of political programs through the deployment of political arguments – as balancing leads to - the more law is put in jeopardy, and consequently justice, as the "irresolvable but productive tension"[1847] between intersubjectivity and *différance*, is also in danger. The singularity of the case, the context with all its features, which is a fundamental task of a judge aware of the limits of reason, becomes a secondary concern when the investigation of the social repercussions of a certain measure, balanced with the constitutional principle under scrutiny, comes to the fore. This is an immediate outcome of the deprivation of the tense but productive relationship between law and justice: whereas the law loses its enforceable character and the system of rights becomes

[1845]For a very interesting analysis of the political character of balancing, see Bernhard Schlink, *Abwägung im Verfassungsrecht* (Berlin: Duncker & Humblot, 1976).

[1846]Schlink, "German Constitutional Culture in Transition," 729.

[1847]Honneth, "The Other of Justice: Habermas and the Ethical Challenge of Postmodernism," 319.

a malleable structure that can be balanced with any judge's political interpretation of social effects, as we could observe in this case, the concern for the singularity of the case by treating it not only as an equal case among others alike, but even by radicalizing the possible insufficiencies of this symmetrical justice, is disregarded. As the *Crucifix case,* not only may the disruption of the set of institutional procedures that distinguish constitutional democracies from other regimes happen, now clearly demonstrated by the evident confusion between discourses of justification and discourses of application in decision-making,[1848] but also the quest for justice is in peril, for, the more political the decision is, the more distanced it is from the particular features of the case, weakening thereby its potentiality as a counter-majoritarian utterance.

For this reason, the right decision, in the *Cannabis case,* would be the one that: first, avoids entering into a political debate on social effects through the deployment of predetermined formulas, such as balancing, and centers primarily on the singularity of case, for only by doing so is the quest for justice not put in peril; and, second, reconstructs the dogmatic and institutional debates on every one of the legal rights at issue, as a means to verify which norm should prevail in this context, strengthening thereby their contents and leading to a consistent interpretation of the system of rights. Briefly, it would be a decision that would materialize the tense but productive relationship between law and justice. From the previous examination, we could conclude that the right decision would not be the one whose arguments would lead to a political definition of the social effects of the consumption, even in small amount and occasionally, of *cannabis* products, but rather the one that would grasp how adjudication is inserted into the set of democratic procedures in a society characterized by the principle of separation of powers. In this respect, the right decision would be the one that would point out that the *Narcotic Act* provisions are constitutional, for articles 2 (I), 2 (II) 2, 3 of the Basic Law are not able to incite the judicial review in the matter brought to the court.

8.4.4 The Ellwanger Case

Finally, the *Ellwanger case* makes the link of the application of the concept of limited rationality to the context of Brazilian constitutionalism. As the *Crucifix* and *Cannabis* cases, the question here lies in this dualism between the quasi-transcendentalism of justice, now grasped in its "irresolvable but productive tension" between intersubjectivity and *différance*, and the reality characterized by multiple worldviews, whose complexities and tensions cannot be entirely recollected and gathered, especially because every new interpretation relies on shaky grounds. The *Ellwanger* case is particularly remarkable, because it enters into the core of the debate on otherness, and how methodologies, when not deployed with the concern

[1848]See the fifth chapter.

for the quest for justice, on the one hand, and the consistent and enforceable character of the system of rights, on the other, can conceal the tensions of this debate. Indeed, they can culminate in arbitrary axiological points of view, transforming thereby the practice of decision-making into an activity that shapes the system of rights according to "broad discretionary powers."[1849] If we could observe, in this case, the use of the principle of proportionality, and particularly balancing, with a certain responsibility towards the propositions of the limited concept of rationality in Justice Gilmar Mendes's opinion, even though still characterized by this need to place the conflict in some formulas and predetermined concepts, there was likewise, especially in Justice Marco Aurélio's opinion, the deployment of this seemingly rational framework exactly to uphold an axiological standpoint that, while apparently leaving aside the system of rights as it has been historically developed, also defended a serious position whose concern for otherness could be severely criticized.

In the first chapter, we could remark this dual perspective in the deployment of the principle of proportionality. While both opinions revealed a clear intent to examine the question through the eyes of the principle of proportionality, and particularly balancing, the way they worked with this methodology was completely different, showing thereby that the problem lies much more in the *posture* of whom acts as a judge than, in reality, in the systematization of the arguments. Evidently, as we formerly and exhaustively discussed, the idea that rationality comes from methods such as balancing contradicts the premises that orient the concept of limited rationality, above all the one that asks the judge not to reduce the complexities of the case and its features to some predetermined formulas and patterns, as if they expressed the "rationalization" of decision-making. For this reason, a judge conscious of the limits of reason, if she intends to deploy a general and abstract method, must understand that the method is there for the case and not the case is there merely for the method itself. A generic and abstract purpose of "rationalization" cannot reduce the singularity of the case, of the context, and much less subvert those propositions of the concept of limited rationality as a means to justify the use of discretionary power. If we recall Justice Gilmar Mendes's and Marco Aurélio's opinions, we could conclude that the first used the method for the case, with the risks this process carries, while the second concentrated on deploying this methodology to justify an axiological understanding that was rather both disconnected from the institutional history, notwithstanding that he attempted to motivate his opinion historically, and even more dissociated from the discussion of the otherness.

Indeed, Justice Gilmar Mendes, although we could infer that he oriented his opinion to providing a "rational" and systematic solution to the collision of principles under his scrutiny, as if this structural framework could "rationalize" his

[1849] Jürgen Habermas, "Reply to Symposium Participants, Benjamin N. Cardozo School of Law," in *Habermas on Law and Democracy*, ed. Michel Rosenfeld and Andre Arato (Berkeley, CA: University of California Press, 1998), 430.

opinion – contradicting, therefore, the concept of limited rationality - he carefully examined the features of the case, connected this factual examination to the discussion of the legal norms at stake, and verified how they are part of a society marked by a pluralism of worldviews. When he employed balancing, he remarked that the proportion between the pursued goal – the preservation of the inherent values of a pluralistic society, the human dignity – and the burden imposed on the defendant's freedom of speech would lead to the conclusion that this freedom does not embrace racial intolerance and the incitation to violence, especially in a democratic society.[1850] He, accordingly, used the principle of proportionality and followed some of the propositions of the concept of limited rationality, except for his need to transform this debate into a proportional analysis of constitutional principles. The concern for the consistency of the legal system and the quest for the other were rather evident in his decision, which demonstrates that the problem is much more of a posture in decision-making than really the methodology itself.

Nonetheless, if we examine Justice Marco Aurélio's opinion, we can conclude that he deployed exactly the same methodology to reinforce a naturalistic (in Brazil, no one cultivates any repulsive feeling against the Jewish community,[1851] the "Brazilian society is [not] predisposed to practice discrimination against the Jewish people),"[1852] semantic (the restriction on legal principles demands "an almost literal" interpretation), originalist (the appeal to the Constitution's Framers, whose purpose was not to create a constitution for the Germans, French, Italians, Polish or Europeans in general), and discriminatory (there is only racism in Brazil against black people) argumentation. He balanced the legal principles at issue with a peculiar and criticizable interpretation of the facts by placing a naturalistic and originalist axiological point of view as weightier in the circumstances, relativizing consequently those principles according to a discretionary power. Yet, what is more serious is that he interpreted history to build a discourse of majority – which is the predisposition of Brazilian people towards Jews? – as an argument to exclude the minority from the possibility of being victims of racism. The crime of racism, which has a constitutional status in Brazil and is in truth an instrument to be used especially to defend the minority from discriminatory acts, became, insofar as the Brazilian people have a predisposition only to discriminate certain groups, but not Jews, an instrument that just some individuals – the "black people" – could make use of. The question here is thus far beyond the discussion of equal treatment for *all*, and much more distant from a possible interest in asymmetrical forms of care. Justice Marco Aurélio interpreted democracy according to a criticizable interpretation of Brazilian history, naturalizing its contents as if they were a static reality and, even in a synchronic perspective, built a questionable understanding of history by using it against the other's otherness.

[1850] See the first chapter.

[1851] HC 82.424-2/RS (Justice Marco Aurélio's opinion).

[1852] HC 82.424-2/RS (Justice Marco Aurélio's opinion).

Hence, a judge aware of the boundaries of reason would examine this question through a distinct perspective, one that, rather than focusing on predetermined methods and formulas, would stress the dualism of law and justice in order to reach the right decision. By following the first proposition of the concept of limited rationality, the judge's first concern would be to exhaustively describe the features of the case as well as connect them to the *prima facie* applicable norms. In this respect, she would verify that the case refers to a *Habeas Corpus*[1853] filed by the author and editor Siegfried Ellwanger, a Brazilian resident in the state of Rio Grande do Sul, who was condemned to prison in virtue of the racist contents of the books he wrote or edited. The condemnation was grounded in article 20 of the Law 7.716/89, which establishes that to "practice, induce or incite, by means of social communication or publication of any nature, the discrimination or prejudice of race, religion, ethnics, or national precedence" is a crime of racism, whose punishment is the confinement from two to five years. In the same way, the decision was also justified by the equality principle (article 5, *caput,* of the Constitution) and the incrimination of racism (article 5, XLII, of the Constitution). She would, accordingly, examine the lower courts' description of the contents of those books, as a way to grasp whether their contents could be embraced by the principle of freedom of speech (article 5, IV, of the Constitution), or whether they caused serious offense to a determined social group, in this situation, the Jewish community. She would verify, for instance, that many of his written or edited books were indeed directed to attacking the Jews, either through a questionable reconstruction of history or through an emphasis on possible conspiracies of the Jewish people. Books such as the *Jewish or German Holocaust – Behind the Lie of the Century, The Conquers of the World – The Real War Criminals,* and the famous *The Protocols of the Elders of Zion* were in the list of his publications. On the other hand, she would verify that the defendant's arguments were mostly directed to defending the freedom of speech and the fact that his books were not racist, but instead had an intellectual and scientific purpose that the Brazilian Constitution protects.

When examining the possible collision between the freedom of speech and the equality principle combined with the constitutional crime of racism, the judge would not, nevertheless, proceed to balancing, as it occurred in the STF's original decision, especially in Justices Gilmar Mendes's and Marco Aurélio's opinions. For she knows that there is no possibility of balancing an illegal practice, constitutionally defined as a crime, with the freedom of speech without losing both their deontological force, as if there were a more or less crime or a more or less constitutional right. For this reason, she would rather stress the debate on the institutional development of those constitutional rights at stake, their historical implications and connections with the characteristics of Brazilian constitutional democracy. In this matter, the first concern would be to verify that one of the most fundamental principles of the process of democratization in Brazil is the freedom of speech, especially after the military regime that inflicted an effective censorship,

[1853]See HC 82.424-2/RS (published on 03.19.2004).

with its deleterious effects, on Brazilian society. She would remark that many decisions in the distinct judicial branches have strongly upheld the freedom of speech as an indispensable constitutional principle to the healthy maintenance of the sought after democracy, reinforcing it in distinct opportunities. By the same token, she would acknowledge that this principle, albeit its democratic relevance, has not always prevailed when in collision with the equality principle, particularly when the question lies in a possible incrimination by reason of racism. Indeed, if the freedom of speech is a fundamental constitutional principle for Brazilian democracy, so it is the equality principle, for it refers to the need to include the other in all the different democratic forums of society. There is not, after all, democracy if individuals have denied their rights to participate and express their voices in distinct areas of Brazilian social life. Furthermore, the equality principle is the one that closely connects to the quest for justice, to the interminable but necessary quest for the other, characterizing then the other side of the double bind of law.

For this reason, she would acknowledge the boundaries of freedom of speech, when the discourse, while free, is directly oriented towards excluding the other, for, in this situation, freedom of speech is no longer being exercised in favor of democracy, but rather as its contrary. If freedom of speech, which is a crucial instrument for democracy, is used instead to exclude the other, to attack the freedom and equality of others, its exercise in that particular circumstance is not sheltered by the Constitution. Especially in a society characterized by the agony to overcome the historical social inequalities, and a society that constitutionalized, as a basic right, the crime of racism, demonstrating thereby the seriousness with which Brazilian society treats this issue, the freedom of speech cannot signify the possibility of expressing any content, regardless of its discriminatory nature, as if it were a guarantee of democracy. After all, the freedom of speech can be used against democracy, against the possibility of including the other. A judge aware of the limits of reason would infer that, if it is to keep consistent and enforceable the system of rights, there is no more justice to the case than considering the freedom of speech paradoxically a condition and a risk for democracy, without therefore any assurances in this matter. If the risk, nonetheless, is necessary, even to the perfectibility of democracy, it can also be reduced, if the social practice and, particularly, the democratic institutions avoid practices whose immediately intent is to disrupt constitutional democracy by excluding the other, as the exhaustive investigation of the features of this case could reveal.

The examination of how these principles in collision are envisaged in the multiple institutional forums and in their connections with a history marked by the intent to open it up to the different is, anyhow, complemented by the third proposition of the concept of limited rationality. The question here lies in how those principles can be interpreted in a way that they could orient Brazilian society to become a self-reflexive society that, while open to new possibilities in history, demonstrates an effective sensibility towards the different.[1854] Heterogeneity is,

[1854]See Milovic, *Comunidade da Diferença*, 131.

after all, indispensable or, as Derrida remarks, "heterogeneity is not depoliticizing, it is rather the condition of politicization: it is the way of broaching the question of the political, of the history and genealogy of this concept, with the most concrete consequences."[1855] Insofar as an individual acts to exclude the other, which could certainly be done by encouraging a racist practice through books, the freedom of speech is used against the premise of an "irresolvable but productive tension"[1856] between intersubjectivity and *différance* in the quest for justice. By examining the particularities of the case, the contents of those books, and the real purpose behind those books, it is clear that the hypothesis is of the use of freedom of speech against the other, and hence against the quest for justice.

The solution, accordingly, would concentrate on the dualism between law and justice, in order to show that either by focusing on the characteristics of Brazilian constitutionalism and its institutional history or by examining the boundaries of any constitutional principle with reference to the debate on otherness, the argument sustaining the revisionist content of those books based on the right to freedom of speech does not convince a judge conscious of the limits of reason. In fact, by, (1) examining carefully the singularity of the case and all its distinct and relevant features, such as the contents of the books and all other facts involved in this case; (2) hermeneutically reconstructing the contents of the constitutional principles at issue, as a means to reinforce their authoritative character in the case, which is done by consistently interpreting them, both by showing the boundaries of the freedom of speech and indicating the appropriateness of the equality principle and the right against racism to the case; and (3) concentrating on the quest for justice as a real concern for the other, in a close connection with a reality characterized by a pluralism of worldviews – which, in the situation, is clearly verified by the need to avoid practices that intend to exclude the other as a citizen in the democratic procedures – the right decision would be the one dismissing the *habeas corpus*. It would consider that the appropriate norms, in the circumstances, which were not proportionally balanced to adapt to a particular context, but examined in their very enforceable character, would be the equality principle and the right against racism.

For a judge conscious of the limits of reason works in the dualism between law and justice, the dismissal of the *habeas corpus*, according to the singularities of this case, is the right answer, and, indeed, the confirmation that her activity is not a representation of the interests of a prevailing *ethos;* it is not a political representation of the will of people, as it happened, even though questionable in what refers to the interpretation of this predisposition of Brazilian society, in Justice Marco Aurélio's opinion. Rather, it can be oriented against majoritarian points of view, as long as the quest for the other and the reconstruction of history by stressing the other are assumed as the primary concern of a responsible practice of decision-making, one that grasps the tense but productive relationship of intersubjectivity

[1855]Jacques Derrida, "Remarks on Deconstruction and Pragmatism," in *Deconstruction and Pragmatism*, ed. Chantal Mouffe (London; New York: Routeledge, 1996), 81.

[1856]Honneth, "The Other of Justice: Habermas and the Ethical Challenge of Postmodernism," 319.

and *différance* in the realm of the negotiation between constitutionalism and democracy, between law and justice.

8.5 Final Words

If the last chapter aimed to unfold the concept of limited rationality by revealing the productive tension between *intersubjectivity* and *différance*, this chapter targeted to enter into the intricate practice of decision-making in order to reveal that the concept of limited rationality, as this book has been focusing on, is not merely an abstract and disconnected discourse from the real and vivid practices of constitutional adjudication. Indeed, by stressing both the impossibility of entirely recollecting and gathering the complexities of reality, on the one hand, and the impossibility of thoroughly doing justice, on the other, a challenge sprung from this perception, which appeared as a strong message to verify how, from different aspects, it is possible to envisage another way to face constitutional dilemmas from case analysis. The three propositions that followed this research – first, exhaustively focus on the singularity of the case, not reducing then its complexities to some predetermined formulas and criteria; second, hermeneutically reconstruct the institutional history, keeping thereby consistent and enforceable the system of rights; and, third, directs itself towards the affirmation of the otherness, the "irresolvable but productive tension"[1857] of intersubjectivity and *différance* in the quest for justice – opened up the possibility to reconstruct the German and Brazilian recent constitutional realities, and to verify how, in a sense, their shift to activism and the possible encroachment on the principle of separation of powers could go in the opposite direction of the concept of limited rationality.

This chapter, for this reason, recalled the German and Brazilian constitutionalisms and showed that, the more their courts provide a legal discourse resembling a political discourse, the less the quest for keeping consistent the system of rights becomes a serious concern, disrupting then the second proposition above. In the same way, while this political activism becomes a reality, the reduction of the complexities of the case through their submission to some predetermined formulas - shaping hence basic rights as though they were objective principles, and, as such, more adapted for this activism - becomes a continuous practice, affecting directly the first proposition above. Finally, insofar as this activism relativizes the law in favor of a political goal, the quest for the other, for the singularity of the other, transforms into an axiological substantiality that can place a general value, a majoritarian standpoint, over the individual, infringing thereby the third proposition of the concept of limited rationality.

By the same token, through case analysis, by creating an imaginary judge to rejudge those real and existent cases from Germany and Brazil, those propositions

[1857]Ibid.

arising from the concept of limited rationality brought to light that it is possible, in the *here and now* of constitutional adjudication, to put in practice a discourse that is not political, for it is not oriented towards satisfying a majoritarian view, but is rather inherently concerned with the constitution, with its enforceability in every new case, for it knows that only by protecting the constitution will it protect the other. In the investigation of the *Crucifix* case, we could observe this perspective by verifying that the right answer is the one that points out the enforceable character of the state's duty of religious and philosophical neutrality and the freedom of faith, whose arguments would not lead to balancing as a way to evaluate the minimum of elements of compulsion the children could suffer, but rather to a discourse that is really concerned with keeping consistent the system of rights and, by the same token, protecting the other – in the case, the members of a minority – from a prevailing *ethos*. In the *Cannabis* case, we could remark that the right answer would clearly avoid entering into the convoluted space of policy and, for this purpose, would not interpret the equality principle and the principle of free development of personality combined with the right to freedom, as though they could embrace any theme of social life. Finally, in the *Ellwanger case,* we could conclude that, in Brazilian constitutional culture, the right answer would point out that the crime of racism was committed, without this meaning that the judge would balance a crime with the freedom of speech, for this would make them lose their deontological force.

This chapter, therefore, ends by unfolding the concept of limited rationality to adjudication, one that, by emerging from the tense but productive relationship between *différance* and *intersubjectivity,* reveals its potential to face constitutional dilemmas and, in this respect, to confront balancing and the belief in the rationality of balancing. If balancing has appeared as a characteristic intimately related to the constitutional courts' shift to political activism, perhaps the problem lies in which activism is a democratic activism. In this regard, the concept of limited rationality can be the sign of a practice that knows that the most democratic activism is the one whose actions are oriented towards the other, and, as a consequence of its double bind, towards the law and constitution.

Conclusion

When Bernhard Schlink suggested in his text *German Constitutional Culture in Transition* the need to rethink constitutional scholarship, one that would confront more directly the *Bundesverfassungsgericht's* political character, releasing thereby a "significant critical potential,"[1858] a relevant message appeared as an encouragement for this research. Indeed, after having concluded that important part of German constitutional scholarship "canonizes the decisions of the Bundesverfassungsgericht" and "interprets these decisions, and their reasoning, as though they were a codified law,"[1859] a similar perception could also be directly applied to Brazilian constitutional culture and, in particular, to Brazilian constitutional scholarship. Those words stemming from the critical investigation of a distinct context, with a historical tradition of untranslatable nuances, strongly stimulated the perception that, despite the singularities of both constitutional realities, it seemed it would be possible to outline a connection between Germany and Brazil. In this respect, by attempting to follow this quest for a "significant critical potential," the question of rationality in decision-making came out as a stimulating theme to initiate this dialogue between both constitutional cultures.

The discussion of which rationality is behind the practice of constitutional courts within the context of a historical movement towards political activism, and how constitutional scholarship aims to justify this movement rationally, revealed many possible associations between these two realities, particularly because of the increasing influence of German constitutionalism, with its reasoning and methodologies, in Brazilian constitutional life. This comparative study became them a relevant subject matter of this research, not only because of the empirical connections it unveiled, but mostly as a result of the outcomes these empirical connections have brought about in the debates on rationality and the convictions and beliefs this word "reason," within this background, has intensively promoted. While these

[1858] Bernhard Schlink, "German Constitutional Culture in Transition," *Cardozo Law Review* 14 (1993): 735.

[1859] Ibid.

convictions and beliefs were been disclosed and realized the "critical potential" as a means to reveal their insufficiencies in the realm of constitutional democracies, in a typical exercise of deconstruction or critical review of their basis, the reconstruction of the idea of reason was being unfolded. From the critique of the prevailing conception of rationality behind this movement of constitutional courts towards political activism, and the main methodological instrument constitutional courts deploy in this dualism between law and politics, now reinforced by a possible rational justification – balancing, specifically – the concept of limited rationality gradually came out. Therefore, the book, which began as an investigation and critique of constitutional cultures and the concept of rationality originated therefrom, ended by stressing the concept of limited rationality as a response to the indeterminacy of law in constitutional democracies.

This research, accordingly, in order to realize this "critical potential," had to be both empirically grounded and theoretically justified. Insofar as it is from the singularities of the everyday practice of decision-making and from its factual historical features that the question of rationality originates not as an entity detached from the real and vivid problems of social life, the first part of this investigation concentrated on initially bringing forward, more as an instigation, three constitutional cases, two from the German *Bundesverfassungsgericht* (*Crucifix case* and *Cannabis case*) and one from the Brazilian *Supremo Tribunal Federal* (*Ellwanger case*), which exposed how constitutional courts deal with methodologies, particularly balancing, in complex circumstances of an apparent collision of constitutional principles. After the case analysis, the investigation examined the constitutional backgrounds of those cases, primarily by focusing on how history has converged upon the erection of constitutional courts with the characteristics of a strong dualism between law and politics, and how this dualism has created a serious debate on the possible encroachment on the principle of separation of powers. In this respect, the discussion of the institutional history from which the German *Bundesverfassungsgericht* (second chapter) and the Brazilian *Supremo Tribunal Federal* (third chapter) arose as "[forums] for the treatment of social and political problems"[1860] brought relevant links between both constitutional realities.

First, it revealed how the increasing constitutional courts' activism is related to the emergence of the intent to reestablish a constitutional democracy after an authoritarian period and to the discredit upon the governmental and parliamentary institutions, in a context therefore where the constitutional court undertakes this role of exercising, through decision-making, a political activism and of protecting society from a possible reemergence of the authoritarian past. Second, for the political character of their decisions increases in this reality of activism, the consequence is that there is the loss of consistency in the interpretation and application of the system of rights, now directed to solving the present and future dilemmas of social life through the idea that basic rights being objective principles

[1860]Schlink, "German Constitutional Culture in Transition," 729.

of a total legal order.[1861] Third, as an attempt to methodologically justify this political character of constitutional adjudication, balancing, which appears as a suitable and efficient instrument for this activist nature of constitutional courts, turns up as a technique of decision-making and, with support of relevant part of constitutional scholarship (as we could observe in Robert Alexy's thinking), a rational technique for this new constitutionalism. From this direct contact with both realities, hence, the question of rationality and, more immediately, the rationality of balancing, emerged naturally as a relevant subject of analysis, showing thereby that the *praxis* and *theory* are not two independent worlds.[1862]

The second part was then, above all, a debate on rationality and, more specifically, the rationality that is behind this movement of constitutional courts towards activism. After having examined the constitutional courts' shift to activism and political matters, leading thereby to a serious discussion of the possible encroachment on the principle of separation of powers, the question that needed to be posed was how this emphasis on the rationality of balancing handles the tensions and complexities of constitutional democracy. In this respect, the immediate subject of analysis was to verify how the practice of decision-making is linked, particularly in those two constitutional realities at issue, to a concept of rationality that flows from the belief in methodologies and predetermined concepts that could solve the most complex dilemmas of constitutional adjudication, especially within the context of indeterminacy of law. In other words, it examined how it is connected to the belief that abstract methods and formulas are responsible, as long as filled with arguments, for achieving the rationality, the rightness and even the legitimacy of the activity of decision-making.

As a theoretical source for this investigation, given that he is one of the most influential and incisive defenders of the rationality of balancing and, basically, one of the main representatives of this new constitutionalism we observe in Germany (with direct outcomes in Brazil), the fourth chapter, which inaugurated the second part, examined Robert Alexy's thinking, particularly his *Special Case Thesis* (*Sonderfallthese*) and his *Theory of Constitutional Rights* (*Theorie der Grundrechte*). The central premises of his defense of balancing, such as the idea that legal discourse is a special case of the general practical discourse, the characterization of principles as optimization requirements, or the definition of the Weight Formula to prove how balancing can be rationally inferred, were directly discussed. But, insofar as Alexy sustains, against his critiques, that balancing is "not a danger for rights but, on the contrary, a necessary means of lending them protection"[1863] or that it is "not an alternative to argumentation but an indispensable form of rational practical discourse,"[1864] it became necessary to examine his premises and verify how

[1861]See Ibid., 711–736.

[1862]See Jürgen Habermas, *Nachmetaphysisches Denken: philosphische Aufsätze* (Frankfurt a.M.: Suhrkamp, 1988), 41.

[1863]Robert Alexy, "Constitutional Rights, Balancing, and Rationality," *Ratio Juris* 16, no. 2 (June 2003): 131.

[1864]Ibid.

metaphysical, and as such, unjustified in their own basis, they are. The fifth chapter initiated this direct confrontation with Alexy's premises through Jacques Derrida's deconstruction, which, as complex as his philosophy is, is also a powerful thinking to disclose the metaphysics embedded in Alexy's claim to rationality, correctness and legitimacy. Indeed, by stressing the *to come*, as "a militant and interminable critique"[1865] that is, as a "weapon aimed at the enemies of democracy"[1866] and, as such, a weapon against the use of a discourse on reason that could work against the law and justice in the realm of decision-making, Derrida's deconstruction proved a valuable contribution to this research. It also raised the question of the necessary dualism, the double bind that occurs in the negotiation between constitutionalism and democracy and between law and justice, which opened up the premises to foresee the concept of limited rationality.

Yet, this confrontation should also occur by introducing possible alternatives to balancing that revealed to be more adequate responses to the indeterminacy of law. The quest for disclosing and undercutting metaphysics we examined through Derrida's deconstruction, accordingly, would now gain different contours and a more direct visualization in the practice of decision-making. A distinct but complementary language would add a very robust perception of how problematic Alexy's defense of balancing might be as well as its troublesome consequences in the reality of constitutional democracy. In this respect, the sixth chapter provided this critical analysis by focusing on Habermas's proceduralism, from which we discussed Klaus Günther's and Ronald Dworkin's views on the question of decision-making. There, we could explore the construction of a post-metaphysical thinking clearly concerned with the consistency of the legal system and the inclusion of the other, in a dialectical movement whereby the tensions between facts and norms unfold themselves. While exposing the fragilities of Alexy's standpoint, it gradually opened up, in a similar fashion to that of Derrida's deconstruction in the fifth chapter, the premises for grasping the concept of limited rationality.

The message of the third part immediately arose from the debates of the second part. The question that remained was how, from Derrida's deconstruction and Habermas's proceduralism, we could foresee a concept of limited rationality that could serve as a relevant counterargument to this incessant defense of the rationality of balancing and, more emphatically, to the very characteristics of this new constitutionalism flowing from the historical and effective practices of the German and Brazilian constitutional courts. If the second part proved that Alexy's theory sustaining the rationality of balancing, indeed presented as a viable justification for this constitutional courts' shift to activism, is metaphysical and can generate problematic effects when operated in reality, it did so by confronting it with the characteristics that gradually unfold the concept of limited rationality. In this regard, it demonstrated that, either because it reduces the complexities and tensions

---

[1865]Jacques Derrida, *Rogues: Two Essays on Reason* (Stanford, CA: Stanford University Press, 2005), 86.
[1866]Ibid.

that characterize constitutional democracies, whose contents are continuously hermeneutically reconstructed in inevitable shaky foundations, or because it does not place, as its primary concern and as the counterpart of the quest for a consistent system of rights, the quest for the other, the deployment of balancing seems to go in the opposite direction of the concept of limited rationality.

The last two chapters of this book aimed to enter more directly into this debate on the concept of limited rationality. First, the question was how to investigate this theme in the grounds of a constitutional democracy characterized by multiple worldviews and by the quest for including the other. The seventh chapter, which introduced the third part, had the purpose of unfolding, from the relevant conclusions already outlined in the previous chapters, this concept of limited rationality. In this respect, while the very complexity of history and its tensions shaped the first challenge of this rationality, incapable of entirely recollecting and gathering all its features, either through the stress on Derrida's *iterability* or Habermas's *self-correcting learning process,* the question of the other came out as the most intricate subject matter in a possible dialogue between both authors. In this double bind of constitutionalism and democracy, as well as of law and justice, the question of the other became a primary concern, as long as, without the quest for the other, there is no law, and thus institutional history, the system of rights loses in consistency. Therefore, for this aspect – the loss of consistency – is intimately related to the previous debate on balancing (second part) and on constitutional courts' shift to activism (first part), it was necessary to further investigate which otherness is the one that more adequately corresponds to the characteristics of constitutional democracy, and thus to constitutional adjudication. It was necessary to place side by side *intersubjectivity* and *différance.*

As a conclusion of a possible dialogue between these two perspectives of justice, the symmetrical equal concern and respect of Habermas's intersubjectivity and Derrida's asymmetrical justice *to come,* the question of the other became a discussion of the "irresolvable but productive tension"[1867] between intersubjectivity and *différance.* This revealed that constitutional democracy learns from the perception of the very limits of reason and vice-versa. In this matter, the sought after possibility of thinking of a "*new self-reflexive community of différance*"[1868] became a primary justification for a new posture regarding the otherness, one that seeks to thematize and problematize *here and now* in practice the search for the other. This is where the connection with constitutional adjudication appears, as an instrument to operationalize, in reality, the quest for the other. The eighth and last chapter aimed to explore this complex extension of the debate on democracy to decision-making and, more specifically, reveal how it is possible to verify the concept of limited rationality, initially, in the critical reconstruction of German and Brazilian constitutionalisms.

[1867]Axel Honneth, "The Other of Justice: Habermas and the Ethical Challenge of Postmodernism," in *The Cambridge Companion to Habermas*, ed. Stephen K White (Cambridge: Cambridge University Press, 1995), 319.

[1868]Miroslav Milovic, *Comunidade da Diferença* (Ijuí, RS; Rio de Janeiro: Unijuí; Relume Dumará, 2004), 132, translation mine.

This investigation exposed how the movement towards politics in constitutional adjudication is problematic in a society that aims to preserve, as one of its main guarantees, the principle of separation of powers and, above all, the counter-majoritarian protection that constitutional adjudication ought to represent in reality. When this analysis directed to cases, we could see that that another practice of decision-making is possible, one that centers carefully on the singularity of the case, hermeneutically reconstructs its contents by acknowledging the consistent and enforceable character of the system of rights, and, finally, is intimately concerned with the quest for the other.

The connection of the empirical world with the concept of limited rationality appeared thus as a final message, one that might point out this "*new self-reflexive community of différance*."[1869] If it is possible to achieve it, it might be best to recall the Derridian words, that "for an event to take place, for an event to be possible, it must be, as event, as invention, the coming of the impossible."[1870] Still, it might be best to see that all this debate indicates that the tense but productive relationship between *intersubjectivity* and *différance* is indeed a resolution as a non-resolution that makes this *here and now* of the other even more urgent. It is, by the same reason, the perception that Schlink's words towards constructing a "significant critical potential"[1871] in constitutional scholarship translate into an incessant revision of our beliefs and practices. This is the reason why this research should end by remembering Theodor Adorno, when he remarked that "the greatness of Freud [consisted] in that, like all great bourgeois, he left standing undissolved these contradictions and disdained the statement of pretended harmony in which the thing itself is contradictory. He revealed the contradictory character of social reality."[1872] In the same way, the greatness of the concept of limited rationality is that it leaves standing undissolved its contradictions and does not attempt to provide insurmountable truths.

[1869]Ibid., translation mine.

[1870]Jacques Derrida, "As If It Were Possible," in *Negotiations: Interventions and Interviews, 1971–2001* (Stanford, CA: Stanford University Press, 2002), 361.

[1871]Schlink, "German Constitutional Culture in Transition," 735.

[1872]Theodor W. Adorno, *Die revidierte Psychoanalyze*, Vol. 8, in *Gesammelte Werke* (Frankfurt a.M: Suhrkamp, 1972), translation mine.

Bibliography

Ávila, Humberto. *Teoria dos Princípios: da Definição à Aplicação dos Princípios Jurídicos*. São Paulo: Malheiros, 2008

Adorno, Theodor W. Die revidierte Psychoanalyze. Vol. 8, in *Gesammelte Werke*, by Theodor W. Adorno. Frankfurt a.M: Suhrkamp, 1972

Aleinikoff, T. Alexander. "Constitutional Law in the Age of Balancing." *Yale Law Journal* 96, no. 5 (April 1987): 943–1005

Alexy, Robert. "Balancing, Constitutional Review, and Representation." *International Journal of Constitutional Law* (Oxford University Press and New York School of Law) 3, no. 4 (2005)

Alexy, Robert. "Coherence and Argumentation or the Genuine Twin Criterialess Super Criterion." In *On Coherence Theory of Law*, edited by Aulis et al. Aarnio. Lund: Juristförlaget I Lund, 1998

Alexy, Robert. "Constitutional Rights, Balancing, and Rationality." *Ratio Juris* 16, no. 2 (June 2003)

Alexy, Robert. "Die Gewichtsformel." In *Gedächtnisschrift für Jürgen Sonnenschein*, edited by Joachim Jickely, Peter Kreutz and Dieter Reuter. Berlin: de Gruyter, 2003

Alexy, Robert. "Discourse Theory and Fundamental Rights." In *Arguing Fundamental Rights*, edited by Agustín José Menéndez and Erik Oddvar Eriksen. Dordrecht: Springer, 2006

Alexy, Robert. "Discourse Theory and Human Rights." *Ratio Juris* 9, no. 3 (August 2007): 209–235

Alexy, Robert. "Jürgen Habermas's Theory of Legal Discourse." In *Habermas on Law and Democracy: Critical Exchanges*, edited by Michel Rosenfeld and Andrew Arato. Berkeley, CA: University of California Press, 1998

Alexy, Robert. "Justification and Application of Norms." *Ratio Juris* 6, no. 2 (July 1993)

Alexy, Robert. "Law and Correctness." In *Law and Opinion at the End of the Millenium: Current Legal Problems*, edited by Michael D. A. Freeman. Oxford: Oxford University Press, 1988

Alexy, Robert. "On Balancing and Subsumption A Structural Comparison." *Ratio Juris* 16, no. 4 (December 2003)

Alexy, Robert. "On the Structure of Legal Principles." *Ratio Juris* 12, no. 4 (September 2000)

Alexy, Robert. "Postscript." In *A Theory of Constitutional Rights*, by Robert Alexy. Oxford: Oxford University Press, 2002

Alexy, Robert. *Recht, Vernunft, Diskurs*. Frankfurt a.M.: Suhrkamp, 1995

Alexy, Robert. "Sistema Jurídico, Princípios Jurídicos y Razón Práctica." *Doxa* 5 (1998)

Alexy, Robert. *The Argument from Injustice: A Reply to Legal Positivism*. Oxford: Clarendon Press, 2002

Alexy, Robert. "The Special Case Thesis." *Ratio Juris* 12, no. 4 (December 1999)

Alexy, Robert. *Theorie der Grundrechte*. Frankfurt a.M: Suhrkamp, 1994

Alexy, Robert. *Theorie der juristischen Argumentation: Die Theorie des rationalen Disurskes als Theorie der juristischen Begründung*. Frankfurt a.M.: Suhrkamp, 1989

Apel, Karl-Otto. *Auseinandersetzungen in Erprobung des transzendentalpragmatischen Ansatzes*. Frankfurt a.M.: Suhrkamp, 1998

Apel, Karl-Otto. *Transformation der Philosophie: Das Apriori der Kommunikationsgemeinschaft*. Vol. 2. Frankfurt a.M.: Suhrkamp, 1976

Arendt, Hannah. *Eichmann in Jerusalem: ein Bericht von der Banalität des Bösen*. Leipzig: Reclam, 1990

Arendt, Hannah. *Was ist Politik?: Fragmente aus dem Nachlaß*. München: Piper, 1993

Böckenförde, Ernst-Wolfgang. *Staat, Verfassung, Demokratie: Studien zur Verfassungstheorie und zum Verfassungsrecht*. Frankfurt a.M.: Suhrkamp, 1991

Baracho Júnior, José Alfredo de Oliveira. "O Supremo Tribunal Federal e a Teoria Constitucional." In *15 Anos de Constituição: História e Vicissitudes*, edited by José Adércio Leite Sampaio. Belo Horizonte: Del Rey, 2004

Barnstedt, Elke Luise. "Judicial Activism in the Practice of German Federal Constitutional Court: Is the GFCC an Activist Court?" *Juridica International II* (2007)

Barros, Suzana de Toledo. *O Princípio da Proporcionalidade e o Controle de Constitucionalidade das Leis Restritivas de Direitos Fundamentais*. Brasília: Brasília Jurídica, 2003

Barroso, Luis Roberto. "Os Princípios da Razoabilidade e da Proporcionalidade no Direito Constitucional." *Revista dos Tribunais Cadernos de Direito Constitucional e Ciência Pollítica* 23 (1998): 65–79

Bello, Enzo. "Neoconstitucionalismo, Democracia Deliberativa e a Atuação do STF." In *Perspectivas da Teoria Constitucional Contemporânea*, edited by José Ribas Vieira. Rio de Janeiro: Lumen Juris, 2007

Benhabib, Seyla. "Democracy and Difference: Reflections on the Metapolitics of Lyotard and Derrida." In *The Derrida-Habermas Reader*, edited by Lasse Thomassen. Edinburgh: Edinburgh University Press, 2006

Benjamin, Walter. "Zur Kritik der Gewalt." In *Zur Kritik der Gewalt und andere Aufsätze*, by Walter Benjamin. Frankfurt a.M.: Suhrkamp, 1965

Bennington, Geoffrey. "Derridabase." In *Jacques Derrida*, by Geoffrey Bennington and Jacques Derrida. Paris: Éditions de Seuil, 1991

Benvindo, Juliano Z. *Racionalidade Jurídica e Validade Normativa: da Metafísica à Reflexão Democrática*. Belo Horizonte: Argvmentvm, 2008

Bernstein, Richard J. "An Allegory of Modernity/Postmodernity: Habermas and Derrida." In *The Derrida-Habermas Reader*, edited by Lasse Thomassen. Edinburgh: Edinburgh University Press, 2006

Bernstein, Richard. "Introduction." In *Habermas and Modernity*, edited by Richard Bernstein. Cambridge, MA: The MIT Press, 1988

Boechat, Lêda Rodrigues. *História do Supremo Tribunal Federal*. Rio de Janeiro: Civilização Brasileira, 1991

Bonavides, Paulo. *Curso de Direito Constitucional*. São Paulo: Malheiros, 1994

Cannizaro, Enzo. *Il Principio della Proporzionalità nell' Ordinamento Internazionale*. Milano: Giuffrè, 2000

Caputo, John D. "L'Idée Même de L'à Venir." In *La Démocratie à Venir: Autour de Jacques Derrida*, by Marie-Louise (ed.) Mallet, edited by Marie-Louise Mallet. Paris: Galigée, 2004

Carbonell, Miguel. *El Princípio de Proporcionalidad en el Estado Constitucional*. Bogotá: Universidad Externado de Colombia, 2007

Carvalho Netto, Menelick de. "A Contribuição do Direito Administrativo Enfocado da Ótica do Administrado para uma Reflexão acerca dos Fundamentos do Controle de Constitucionalidade das Leis no Brasil: Um Pequeno Exercício de Teoria da Constituição." *Forum*, March 2001

Carvalho Netto, Menelick de. "A Hermenêutica Constitucional e os Desafios Postos aos Direitos Fundamentais." In *Jurisdição Constitucional e Direitos Fundamentais*, edited by José Adércio Leite Sampaio. Belo Horizonte: Del Rey, 2003

Carvalho Netto, Menelick de. "A Revisão Constitucional e a Cidadania: a Legitimidade do Poder Constituinte que deu Origem à Constituição da República Federativa do Brasil de 1988 e as

Potencialidades do Poder Revisional nela Previsto." *Revista do Ministério Público Estadual do Maranhão*, no. 9 (2002)
Carvalho Netto, Menelick de. Pragmatic Requirements of Legal Interpretation under the Paradigm of the Democratic Rule of Law. Athens: Paper presented at the VII World Congress of the International Association of Constitutional Law, 2007, 11–15 June
Castro, Marcus Faro de. "The Courts, Law and Democracy in Brazil." *International Social Science Journal*, no. 152 (June 1997): 241–252
Cattoni de Oliveira, Marcelo Andrade. "Devido Processo Legislativo e Controje Jurisdicional de Constitucionalidade no Brasil." In *Jurisdição Constitucional e Direitos Fundamentais*, edited by José Adércio Leite Sampaio. Belo Horizonte: Del Rey, 2003
Cattoni de Oliveira, Marcelo Andrade. *Direito Processual Constitucional*. Belo Horizonte: Mandamentos, 2001
Cattoni de Oliveira, Marcelo Andrade. "O Projeto Constituinte de um Estado Democrático de Direito." *In 15 Anos de Constituição: História e Vicissitudes*, edited by José Adércio Leite Sampaio. Belo Horizonte: Del Rey, 2004
Chassard, Pierre. Derrida: *La Destruction du Monde*. Brussel: Mengal, 2004
Clemens, Thomas. "Das Bundesverfassungsgericht im Rechtsund Verfassungsstaat: Sein Verhältnis zur Politik und zum einfachen Recht; Entwicklungslinien seiner Rechtssprechung." In *Das Bundesverfassungsgericht: ein Gericht im Schnittpunkt von Recht und Politik*, edited by Michael Piazolo. Mainz-München: Hase & Koehler, 1995
Clérico, Laura. *Die Struktur der Verhältnismäßigkeit*. Baden-Baden: Nomos, 2001
Corrêa, Oscar Dias. "O 160o. Aniversário do STF e o Novo Texto Constitucional." *Arquivos do Ministério da Justiça*, no. 173 (1988)
Corrêa, Oscar Dias. *O Supremo Tribunal Federal, Corte Constitucional do Brasil*. Rio de Janeiro: Forense, 1987
Critchley, Simon. "Frankfurt Improptu Remarks on Derrida and Habermas." In *The Derrida-Habermas Reader*, edited by Lasse Thomassen. Edinburgh: Edinburgh University Press, 2006
Cruz, Álvaro Ricardo de Souza. "Habermas, Ação Estratégica e Controle de Constitucionalidade." In *15 Anos de Constituição: História e Vicissitudes*, edited by José Adércio Leite Sampaio. Belo Horizonte: Del Rey, 2004
Dastur, Françoise. *Philosophie et Différence*. Paris: Les Éditions de la Transparence, 2004
Denninger, Erhard. "Freiheitsordnung Wertordnung Pflichtordnung." In *Verfassung, Verfassungsgerichtsbarkeit, Politik: Zur verfassungsrechtlichen und politischen Stellung und Funktion des Bundesverfassungsgerichts*, edited by Mehdi Tohidipur. Frankfurt a.M.: Suhrkamp, 1976
Derrida, Jacques. "A Response to Simon Critchley." In *The Derrida-Habermas Reader*, by Lasse Thomassen, edited by Lasse Thomassen. Edinburgh: Edinburgh University Press, 2006
Derrida, Jacques. "As If It Were Possible." In *Negotiations: Interventions and Interviews, 1971–2001*, by Jacques Derrida. Stanford, CA: Stanford University Press, 2002
Derrida, Jacques. *De la Grammatologie*. Paris: Les Éditions de Minuit, 1967
Derrida, Jacques. "Declarations of Independence." In *Negotiations: Interventions and Interviews, 1971–2001*, by Jacques Derrida. Stanford, CA: Stanford University Press, 2002
Derrida, Jacques. "Deconstruction and the Possibility of Justice." *Cardozo Law Review II*, no. 5–6 (July-August 1990)
Derrida, Jacques. "Entretien par Jerôme-Alexandre Nielsberg Penseur de l'événement." *L'Humanité* 28 (January 2004)
Derrida, Jacques. "Force of Law: The 'Mystical Foundation of Authority'." *Cardozo Law Review* 11 (1990)
Derrida, Jacques. *L'écriture et la Différence* . Paris: Éditions de Seuil, 1979
Derrida, Jacques. *Marges de la Philosophie*. Paris: Les Éditions de Minuit, 1972
Derrida, Jacques. "Negotiations." In *Negotiations: Interventions and Interviews, 1971–2001*, by Jacques Derrida. Stanford, CA: Stanford, 2002
Derrida, Jacques. "Performative Powerlessness A Response to Simon Critchley." In *The Derrida-Habermas Reader*, by Lasse Thomassen, edited by Lasse Thomassen. Edinburgh: Edinburgh University Press, 2006

Derrida, Jacques. "Politics and Friendship." In *Negotiations: Interventions and Interviews, 1971–2001*, by Jacques Derrida. Stanford, CA: Stanford University Press, 2005
Derrida, Jacques. *Positions*. Paris: Minuit, 1972
Derrida, Jacques. *Psyché: Inventions de l'Autre*. Paris: Galilée, 1987
Derrida, Jacques. "Remarks on Deconstruction and Pragmatism." In *Deconstruction and Pragmatism*, by Chantal Mouffe, edited by Chantal Mouffe. London; New York: Routeledge, 1996
Derrida, Jacques. *Rogues: Two Essays on Reason*. Stanford, CA: Stanford University Press, 2005
Derrida, Jacques. "The Deconstruction of Actuality." In *Negotiations: Interventions and Interviews, 1971–2001*, by Jacques Derrida. Stanford, CA: Stanford University Press, 2002
Derrida, Jacques. *The Politics of Friendship*. London, New York: Verso, 2005
Devenney, Mark. *Ethics and Politics in Contemporary Theory: Between Critical Theory and Post-Marxism*. London: Routeledge, 2004
Dietze, Gottfried. "Unconstitutional Constitutional Norms? Constitutional Develpment in Postwar Germany." *Virginia Law Review* 42, no. 1 (January 1956)
Dooley, Mark, and Liam Kavanagh. *The Philosophy of Derrida*. Stocksfield: Acumen, 2007
Dworkin, Ronald. *A Matter of Principle*. Cambridge, MA: Harvard University Press, 1985
Dworkin, Ronald. *Law's Empire*. Cambridge, MA: Harvard University Press, 1987
Dworkin, Ronald. *Taking Rights Seriously*. Cambridge, MA: Harvard University Press, 1978
Ellis, Evelyn. *The Principle of Proportionality in the Laws of Europe*. Oxford: Hart, 1999
Emiliou, Nicholas. *The Principle of Proportionality in European Law: A Comparative Study*. London: Kluwer Law Internat, 1996
Ericksen, Erik Oddvar. "Democratic or Jurist-Made Law?" In *Arguing Fundamental Rights*, edited by Agustín José Menéndez and Erik Oddvar Eriksen. Dordrecht: Springer, 2006
Esser, Sonja M. *Das Kreuz ein Symbol Kultureller Identität? Der Diskurs über das 'Kruzifix-Urteil (1995) aus kulturwissenschaftlicher Perspektive*. Münster, New York, München, Berlin: Waxmann, 2000
Fangmann, Helmut D. *Justiz gegen Demokratie: Entstehungund Funktionsbedingungen der Verfassungsjustiz in Deutschland*. Frankfurt a.M; New York: Campus Verlag, 1979
Günther, Klaus. *Der Sinn für Angemessenheit: Anwendungsdiskurse in Moral und Recht*. Frankfurt a.M.: Suhrkamp, 1988
Günther, Klaus. Ein normativer Begriff der Kohärenz für eine Theorie der juristischen Argumentation. *Rechtstheorie* Vol. 20. Berlin: Duncker & Humblot, 1989
Günther, Klaus. "Legal Adjudication and Democracy: Some Remarks on Dworkin and Habermas." *European Journal of Philosophy* 3, no. 1 (April 1995)
Günther, Klaus. "The Idea of Impartiality and the Functional Determinacy of Law." *Northwestern University Law Review* 83, no. 1 & 2 (1989)
Gadamer, Hans-Georg. "Rethorik, Hermeneutik und Ideologiekritk: Metakritische Erorterungen zu 'Wahrheit und Methode'." In *Hermeneutik und Ideologiekritik*, edited by Karl-Otto Apel. Frankfurt a.M: Suhrkamp, 1971
Gaffney, Paul. *Ronald Dworkin on Law as Integrity: Rights as Principles of Adjudication*. Lewiston, NY: The Edwin Mellen Press, 1996
Gasché, Rodolphe. *Inventions of Difference: on Jacques Derrida*. Cambridge, MA: Harvard University Press, 1994
Gasché, Rodolphe. "L'Étrange Concept de Responsabilité." In *La Démocratie à Venir: Autour de Jacques Derrida*, by Marie-Louise Mallet, edited by Marie-Louise Mallet. Paris: Galilée, 2004
Grimm, Dieter. "Die Entfältung der Freiheitsrechte." In *Das Bundesverfassungsgericht: Geschichte, Aufgabe, Rechtssprechung*, edited by Jutta Limbach. Heidelberg: C. F. Müller, 2000
Gusy, Christoph. *Richteliches Prüfungsrecht: Eine verfassungsgeschchtliche Untersuchung*. Berlin: Duncker & Humblot, 1985
Häberle, Peter. *Die Wesengehaltsgarantie des Artikel 19 Abs. 2 Grundgesetz: zugleich ein Beitrag zum Institutionellen Verständnis der Grundrechte und zur Lehre vom Gesetzesvorbehalt*. Heidelberg: C. F. Müller, 1983

Häberle, Peter. "Grundprobleme der Verfassungsgerichtsbarkeit." In *Verfassungsgerichtsbarkeit*, edited by Peter Häberle. Darmstadt: Wissenschaftliche Buchgesellschaft, 1976

Häberle, Peter. *Verfassung als öffentlicher Prozeß: Materialen zu einer Verfassungstheorie der offenen Gesellschaft*. Berlin: Duncker & Humblot, 1978

Habermas, Jürgen. "A Short Reply." *Ratio Juris* 12, no. 4 (December 1999)

Habermas, Jürgen. *Between Facts and Norms*. Cambridge: Polity Press, 1996

Habermas, Jürgen. "Constitutional Democracy: a Paradoxical Union of Contradictory Principles." *Political Theory* 29, no. 6 (December 2001)

Habermas, Jürgen. *Der philosophische Diskurs der Moderne: Zwölf Vorlesungen*. Frankfurt a.M: Suhrkamp, 1985

Habermas, Jürgen. "Equal Treatment of Cultures and the Limits of Postmodern Liberalism." *The Journal of Political Philosophy* 13, no. 1 (2005)

Habermas, Jürgen. "Gerechtigkeit und Solidarität: Eine Stellungnahme zur Diskussion über 'Stufe 6'." In *Zur Bestimmung der Moral*, edited by Wolfgang Edelstein, Gil Noam and Fritz Oser. Frankfurt a.M.: Suhrkamp, 1986

Habermas, Jürgen. "How to Respond the Ethical Question." In *The Derrida-Habermas Reader*, by Lasse Thomassen, edited by Lasse Thomassen. Edinburgh: Edinburgh University Press, 2006

Habermas, Jürgen. *Kommunikatives Handeln und detranszendentalisierte Vernunft*. Stuttgart: Philipp Reclam, 2001

Habermas, Jürgen. *Moral Consciousness and Communicative Action*. Cambridge, MA: The MIT Press, 1990

Habermas, Jürgen. *Nachmetaphysisches Denken: philosphische Aufsätze*. Frankfurt a.M.: Suhrkamp, 1988

Habermas, Jürgen. "On Law and Disagreement: Some Comments on 'Interpretative Pluralism'." *Ratio Juris* 16, no. 2 (June 2003)

Habermas, Jürgen. "On the Cognitive Content of Morality." *Proceedings of the Aristotelian Society* 96 (1996)

Habermas, Jürgen. "Reply to Symposium Participants, Benjamin N. Cardozo School of Law." In *Habermas on Law and Democracy*, edited by Michel Rosenfeld and Andre Arato. Berkeley, CA: University of California Press, 1998

Habermas, Jürgen. *The Inclusion of the Other: Studies in Political Theory*. Cambridge, MA: The MIT Press, 1998

Habermas, Jürgen. *The Theory of Communicative Action: Reason and the Rationalization of Society*. Vol. I. Boston: Beacon Press, 1985

Habermas, Jürgen. "Zu Gadamers 'Wahrheit und Methode'." In *Hermeneutik und Ideologiekritik*, edited by Karl-Otto Apel. Frankfurt a.M.: Suhrkamp, 1971

Halberstan, Daniel. "Desperately Seeking Europe: On Comparative Methodoloy and the Conception of Rights." *International Journal of Constitutional Law* 5, no. 1 (2007): 166–182

Hanau, Hans. *Der Grundsatz der Verhältnismäßigkeit als Schranke privater Gestaltungsmacht*. Tübingen: Mohr-Siebeck, 2004

Heidegger, Martin. *Identität und Differenz*. Pfullingen: Neske, 1957

Heidegger, Martin. *Sein und Zeit*. Tübingen, : Neomarius, 1949

Hillgruber, Christian, and Christophh Goos. *Verfassungsprozesrecht*. Heidelberg: C. F. Müller, 2004

Hirschberg, Lothar. *Der Grundsatz der Verhältnismäßigkeit*. Göttingen: Otto Schwartz & CO, 1981

Honneth, Axel. "The Other of Justice: Habermas and the Ethical Challenge of Postmodernism." In *The Cambridge Companion to Habermas*, edited by Stephen K White. Cambridge: Cambridge University Press, 1995

Honnig, Bonig. "Dead Rights, Live Futures: On Habermas's Attempt to Reconcile Constitutionalism and Democracy." In *The Derrida-Habermas Reader*, edited by Lasse Thomassen. Edinburgh: Edinburgh University Press, 2006

Jackson, Vicki C. "Ambivalent Resistance and Comparative Constitutionalism: Opening Up the Conversation on 'Proportionality", Rights and Federalism." *University of Pennsylvania Journal of Constitutional Law* 1 (1999)

Jansen, Nils. *Die Struktur der Gerechtigkeit*. Baden-Baden: Nomos , 1998

Kates, Joshua. *Essential History: Jacques Derrida and the Development of Deconstruction*. Evanston, Il: Northwestern University Press, 2005

Kelsen, Hans. "Wesen und Entwicklung der Staatsgerichtsbarkeit". *Berichte [der] Verhandlungen der Tagung der Deutschen Staatsrechtslehrer zu Wien am 23. und 24. April 1928." Veröffentlichungen der Vereinigung der deutschen Staatsrechtslehrer 5 (1929)*: 30–88

Kelsen, Hans. *Wer soll de Hüter der Verfassung sein?* Berlin: Rotschild, 1931

Kirk, Jeremy. "Constitutional Guarantees, Characterization and the Concept of Proportionality." *Melbourne University Law Review* 21, no. 1 (1997)

Koch, Oliver. *Der Grundsatz der Verhältnismäßigkeit in der Rechtsprechung des Gerichtshofs der Europäischen Gemeinschaften*. Berlin: Duncker & Humblot, 2003

Kohlberg, Lawrence. "The Claim to Moral Adequacy of a Highest Stage of Moral Judgment." *The Journal of Philosophy* 70, no. 8 (1973)

Krauss, Rupprecht von. *Der Grundsatz der Verhältnismäßigkeit in seiner Bedeutung für die Notwendigkeit des Mittels in Verwaltungsrecht*. Hamburg: Appel, 1955

Kuhn, Thomas S. *The Structure of Scientific Revolutions*. Chicago: University of Chicago Press, 1996

La Torre, Massimo. "Nine Critiques to Alexy's Theory of Fundamental Rights." In *Arguing Fundamental Rights*, edited by Agustín José Menéndez and Erik Oddvar Eriksen. Dordrecht: Springer, 2006

Laufer, Heinz. "Politische Kontrolle durch Richtermacht." In *Verfassung, Verfassungsgerichtsbarkeit, Politik: Zur verfassungsrechtlichen und politischen Stellung und Funktion des Bundesverfassungsgerichts*, edited by Mehdi Tohidipur. Frankfurt a.M.: Suhrkamp, 1976

Leisner, Walter. *Der Abwägungsstaat: Verhältnismäßigkeit als Gerechtigkeit?* Berlin: Duncker & Humblot, 1997

Lenkner, Theodor. *Der rechtfertigende Notstand: Zur Problematik der Notstandregelung im Entwurf eins Strafgestzbuches*. Tübingen: Mohr-Siebeck, 1965

Lenoble, Jacques. "Law and Undecidability: Toward a New Vision of the Proceduralization of Law." In *Habermas on Law and Democracy*, edited by Michel Rosenfeld and Andrew Arato. Berkeley, CA: University of California Press, 1998

Lerche, Peter. *Übermaß und Verfassungsrecht: zur Bindung des Gesetzgebers an die Grundsätze der Verhältnismäßigkeit und der Erforderlichkeit*. Goldbach: Keip, 1999

Limbach, Jutta. "*The Effects of the Jurisdiction of the German Federal Constitutional Court*." European University Institute. EUI Working Paper Law. 99/5. http://cadmus.iue.it/dspace/bitstream/ 1814/150/1/law99_5.pdf (accessed 2009, 14-July)

Luhmann, Niklas. "Kausaliät im Süden." *Soziale Systeme* 1 1 (1995): 7–28

Möller, Kai. "Balancing and the Structure of Constitutional Rights." *International Journal of Constitutional Law* 5, no. 3 (2007)

Müller, Friedrich. *Juristische Methodik*. Berlin: Duncker & Humblot, 1995

Maus, Ingeborg. "Justiz als gesellschaftliches Über-Ich: Zur Funktion von Rechtsprecung in der 'vaterlosen Gesellschaft." In Sturtz der Göter? *Vaterbildeer im 20. Jahrhundert*, edited by Werner Faulstisch and Gunter Grimm. Frankfurt a.M.: Suhrkamp, 1989

McCarthy, Thomas. "The Cambridge Companion to Habermas." *Ethics* 107, no. 2 (2007)

McCormick, John P. "Habermas' Discourse Theory of Law: Bridging Anglo-American and Continental Legal Traditions?" *The Modern Law Review* 60, no. 5 (September 1997)

Mendes, Gilmar. *Controlling Constitutionality in Brazil*. Lecture presented at Harvard Law School. : http://www.stf.jus.br/arquivo/cms/noticiaArtigoDiscurso/anexo/Controle_ de_Con stitu cionalidade_v__Ing.pdf (accessed 2009, 14-July)

Mendes, Gilmar. *Constitutional Jurisdiction in Brazil: The Prbolem of Unconstitutional LegislativeOmission*.http://www.stf.jus.br/arquivo/cms/noticiaArtigoDiscurso/anexo/Omisao_Legislativa_v__Ing.pdf (accessed 2009, 7-July)

Mendes, Gilmar. *Controle de Constitucionalidade: Aspectos Jurídicos e Políticos*. São Paulo: Saraiva, 2004

Mendes, Gilmar. *Die abstrakte Normenkontrolle vor dem Bundesverfassungsgericht und vor dem brasilianischen Supremo Tribunal Federal*. Berlin: Duncker & Humblot, 1991

Mendes, Gilmar. *Direitos Fundamentais e Controle de Constitucionalidade*. São Paulo: Saraiva, 2004

Mendes, Gilmar. Interview by Izabela Torres. "Entrevista Gilmar Mendes." *Correio Braziliense*. Brasília, (2008, 17-August)

Mendes, Gilmar. *New Challenges of Constitutional Adjudication in 21st Century: A Brazilian Perspective*. Lecture presented in Washington (US). 2008 10-October. http://www.stf.jus.br/arquivo/cms/noticiaArtigoDiscurso/anexo/Jurisdicao_Constitucional_no_Seculo_XXI_v__Ing.pdf (accessed 2009, 14-July)

Mendes, Gilmar. *Judicial Reform as a Fundamental Element to Ensure Legal Security to Foreign Investments in Brazil*. New York (US) Lecture presented at the Council of Americas. http://www.stf.jus.br/arquivo/cms/noticiaArtigoDiscurso/anexo/Reforma_do_Sistema_Judiciario_no_Brasil_v_Ing.pdf (accessed 2009, 14-July)

Mendes, Gilmar. *Jurisdição Constitucional: Controle Abstrato de Normas no Brasil e na Alemanha*.São Paulo: Saraiva, 2004

Mendes, Gilmar. "O papel do Senado Federal no Controle de Constitucionalidade: um Caso Clássico de Mutação Constitucional." *Revista de Informação Legislativa*, no. 162 (April, June 2004): 149–168

Mendes, Gilmar. "O Princípio da Proporcionalidade na Jurisprudência do Supremo Tribunal Federal." *Revista Diálogo Jurídico* I, no. 5 (2001)

Mendes, Gilmar. *Recent Evolution of the Brazilian Judicial System*. Washington (US) Lecture Presented in the Library of Congress. 2008. http://www.stf.jus.br/arquivo/ cms /noticia Artigo Discurso/anexo/Evolucao_Recente_do_Sistema_Judiciario_Brasileiro_Ing.pdf (accessed 2009, 14-July)

Mendes, Gilmar, and Ives Gandra da Silva Martins. *Controle Concentrado de Constitucionalidade*. São Paulo: Saraiva, 2007

Menéndez, Agustín José, and Erik Oddvar Eriksen. "Introduction." In *Arguing Fundamental Rights*, by Agustín José Menéndez and Erik Oddvar Eriksen, edited by Agustín José Menéndez and Erik Oddvar Eriksen. Dordrecht: Springer, 2006

Menke, Christoph. "Ability and Faith: On the Possibility of Justice." *Cardozo Law Review* 27 (2006)

Menke, Christoph. *Spiegelungen der Gleichheit*. Berlin: Akademie, 2000

Menke, Christoph. "Virtue and Reflection: The 'Antinomies of Moral Philosophy'." *Constellations* 12, no. 1 (2005)

Michelman, Frank. "Democracy and Positive Liberty." Boston *Review*. 1996, November. http://bostonreview.net/BR21.5/michelman.html (accessed 2009, 15-July)

Michelman, Frank. "The Supreme Court 1985 Term Foreword: Traces of Self-Government." *Harvard Law Review* 100 (1987)

Milovic, Miroslav. A Impossibilidade da Democracia," *Anais do Congresso Nacional do Compedi* 14 (Florianópolis: Fundação Boiteux, 2005)

Milovic, Miroslav. *Comunidade da Diferença*. Ijuí, RS; Rio de Janeiro: Unijuí; Relume Dumará, 2004

Milovic, Miroslav. *Filosofia da Comunicação*: Para uma Crítica da Modernidade. Brasília: Plano, 2002

Morris, Martin. "Deliberation and Deconstruction: The Condition of Post-National Democracy." In *The Derrida-Habermas Reader*, edited by Lasse Thomassen. Edinburgh: Edinburgh University Press, 2006

Mouffe, Chantal. "Decision, Deliberation, and Democratic Ethos." *Philosophy Today* 41 I, no. 1 (Spring 1997)

Mouffe, Chantal. "Deconstruction, Pragmatism and the Politcs of Democracy." In *Deconstruction and Pragmatism*, edited by Chantal Mouffe. London; New York: Routeledge, 1996

Mouffe, Chantal. "Deliberative Democracy or Agonistic Pluralism?" *Social Research* 66, I, no. 3 (Fall 1999)

Mouffe, Chantal. "Democracy and Pluralism: a Critique of the Rationalist Approach." *Cardozo Law Review* 16 (1995)

Mouffe, Chantal. *The Democratic Paradox*. London; New York: Verso, 2000

Neuman, Gerald L. *Constitutional Conception of the Rule of Law and the Rechtsstaatsprinzip of the Grundgesetz.* http//papers.ssrn.com/paper.taf?abstract_id=195368 (accessed 2009, 14-July)

Niquet, Marcel. *Moralität und Befolgunsgültigeit: Prolegomena zu einer realistischen Diskurstheorie der Moral.* Würzburg: Königshausen & Neumann, 2002

Paixão, Cristiano, and Paulo Henrique Blair Oliveira. "O Julgamento das Células-Tronco: Ponderação contra a Constituição." *Constituição e Democracia*, 2008, June

Percesepe, Gary John. Future(s) of Philosophy: *The Marginal Thinking of Jacques Derrida.* New York: Lang, 1989

Piazolo, Michael. "Zur Mittlerrolle des Bundesverfassungsgerichts in der deutschen Verfassungsordnung eine Einleitung." In *Das Bundesverfassungsgericht: ein Gericht im Schnittpunkt von Recht und Politik*, edited by Michael Piazolo. Mainz-München: Hase & Koehler, 1995

Pieroth, Bodo. *Verfassungsbeschwerde: Einführung, Verfahren, Grundrechte.* Münster: ZAP Verlag, 2008

Pieroth, Bodo, and Bernhard Schlink. *Grundrechte: Staatsrecht II.* Heidelberg: C. F. Müller, 2006

Pulido, Carlos Bernal. *El Principio de Proporcionalidad y los Derechos Humanos.* Madrid: Centro de Estúdios Políticos y Constitucionales, 2003

Pulido, Carlos Bernal. "On Alexy's Weight Formula." In *Arguing Fundamental Rights*, edited by Agustín José Menéndez and Erik Oddvar Eriksen. Dordrecht: Springer, 2006

Pulido, Carlos Bernal. "The Rationality of Balancing." *Archiv für Rechtsund Sozialphilosophie* 92, no. 2 (2006)

Ramp, Ulrich. *Hüter der Verfassung oder Lenker der Politik.* München: Grin Verlag, 2003

Regh, William. "Translator's Introduction." In *Between Facts and Norms*, by Jürgen Habermas. Cambridge

Rorty, Richard. *Truth and Progress: Philosophical Papers.* Vol. 3. Cambridge: Cabridge University Press, 1998

Rosenfeld, Michel. "Can Rights, Democracy, and Justice be Reconciled through Discourse Theory? Reflections on Habermas's Proceduralist Paradigm of Law." In *Habermas on Law and Democracy*, edited by Michel Rosenfeld and Andrew Arato. Berkeley, CA: University of California Press, 1998

Rothemberg, Elisabeth. "Introduction." In *Negotiations: Interventions and Interviews, 1971–2001*, by Jacques Derrida. Stanford, CA: Stanford University Press, 2002

Santos, Marcelo Paiva dos. *A História Não Contada do Supremo Tribunal Federal.* Porto Alegre: Sergio Antonio Fabris, 2009

Sarmento, Daniel. "Eficácia Temporal do Controle de Constitucionalidade: O Princípio da Proporcionalidade e a Ponderação de Interesses das Leis." *Revista do Direito Administrativo* (Renovar), no. 212 (April-June 1998): 27–40

Schlink, Bernhard. "Abschied von der Dogmatik: Verfassungsrechtssprechung und Verfassungsrechtswissenschaft im Wandel." *Merkur* 692 (December 2006)

Schlink, Bernhard. *Abwägung im Verfassungsrecht.* Berlin: Duncker & Humblot, 1976

Schlink, Bernhard. "Der Grundsatz der Verhältnismäßigkeit." In *Festschrift 50 Jahre Bundesverfassungsgericht*, edited by Peter Badura and Horst Dreier. Tübingen: Mohr Siebeck, 2001

Schlink, Bernhard. "Freiheit durch Eingriffsabwehr Rekonstruktion der klassischen Grundrechtsfunktion." *Europäische Grundrechte Zeitschrift* (N.P. Engel Verlag), 1984

Schlink, Bernhard. "German Constitutional Culture in Transition." *Cardozo Law Review* 14 (1993)

Schlink, Bernhard. "Open Justice in a Closed Legal System?" *Cardozo Law Review* 13 (1992)

Schlink, Bernhard. "The Dynamics of Constitutional Adjudication." In *Habermas on Law and Democracy*, edited by Michel Rosenfeld and Andrew Arato. Berkeley, LA: University of California Press, 1998

Schlink, Bernhard. "The Journey into Activism." *Cardozo Law Review* 17 (1996)

Schlink, Bernhard, and Arthur J. Jacobson. "Introduction - Constitutional Crises: The German and the American Experience." In *Weimar: a Jurisprudence of Crisis*, edited by Arthur J Jacobson and Bernhard Schlink. Berkeley, CA: Univeristy of California Press, 2000

Schmitt, Carl. *Der Hüter der Verfassung*. Tübingen: Mohr, 1931

Schmitt, Carl. *Verfassungslehre*. Berlin: Duncker & Humblot, 1957

Seifert, Jürgen. "Verfassungsgerichtliche Selbsbescrhänkung." In *Verfassung, Verfassungsgerichtsbarkeit, Politik: Zur verfassungsrechtlichen und politischen Stellung und Funktion des Bundesverfassungsgerichts*, edited by Mehdi Tohidipur. Frankfurt a.M.: Suhrkamp, 1976

Silva, Luís Virgílio Afonso da. "O Proporcional e o Razoável." *Revista dos Tribunais*, no. 798 (April 1992): 23–50

Silva, Luís Virgílio Afonso da. *Grundrechte und gesetzgeberische Spielräume*. Baden-Baden: Nomos, 2003

Smend, Rudolf. *Verfassung und Verfassungsrecht*. Berlin: Duncker & Humblot, 1928

Smith, Jeffery. "Justifying and Applying Moral Principles." *The Journal of Value Inquiry* 40 (2006)

Steffani, Winfrieed. "Verfassungsgerichtsbarkeit und Demokratischer Entscheidungsprozess." In *Verfassungsgerichtsbarkeit*, edited by Peter Häberle. Darmstadt: Wissenschaftliche Buchgesellschaft, 1976

Stelzer, Manfred. *Das Wesesgehaltsargument und der Grundsatz der Verhältnismäßigkeit*. Wien; New York: Springer, 1991

Stolleis, Michael. "Judicial Review, Administrative Review, and Constitutional Review in the Weimar Republic." *Ratio Juris* 16, no. 2 (June 2003): 266–280

Streck, Lenio Luiz, Marcelo Andrade Cattoni de Oliveira, and Martonio Mon'alverne Barreto Lima. *A Nova Perspectiva do Supremo Tribunal Federal sobre o Controle Difuso: Mutação Constitucional e Limites da Legitimidade da Jurisdição Constitucional*. http://www.mundojuridico.adv.br/sis_artigos/artigos.asp?codigo=912 (accessed 2009, 7-July)

Streck, Lênio Luiz. *Entrevista ao Conjur: Lênio Streck Fala sobre o STF*. http://www.conjur.com.br/2009-mar-15/entrevista-lenio-streck-procurador-justica-rio-grande-sul (accessed 2009, 14-July)

Streck, Lênio Luiz. *O Fahrenheit Sumular no Brasil: o Controle Panóptico da Justiça*. http://leniostreck.com.br/index2.php?option=com_docman&task=doc_view&gid=17&Itemid=40 (accessed 2009, 14-July)

Sweet, Alec Stone. *Constitutionality, Balancing and Global Constitutionalism*. http://www.law.columbia.edu/null/Stone-Sweet+-+Proportionality+Balancing?exclusive =filemg r.download &file_id=101159&showthumb=0 (accessed 2009, 14-July)

Tholen, Toni. *Erfahrung und Interpretation: Der Streit zwischen Hermeneutik und Dekonstruktion*. Heidelberg: Universitätsverlag C. Winter, 1999

Thomassen, Lasse. "'A Bizarre, Even Opaque Practice': Habermas on Constitutionalism and Democracy." In *The Derrida-Habermas Reader*, edited by Lasse Thomassen. Edinburgh: Edinburgh University Press, 2006

Vieira, Oscar Vilhena. *Supremo Tribunal Federal: Jurisprudência Política*. São Paulo: Malheiros, 2002

Wellmer, Albrecht. "Reason, Utopia, and the Dialectic of Enlightenment." In *Habermas and Modernity*, edited by Richard J Bernstein. Cambridge, MA: The MIT Press, 1988

Wojciech, Sadursky. *Rights Before Courts: a Study of Constitutional Courts in the Post-Communist States of Central and Eastern Europe*. Dordrecht: Springer, 2005

FSC
www.fsc.org
MIX
Papier | Fördert
gute Waldnutzung
FSC® C083411

Kolibri 360 GmbH
Ferdinand-Jühlke-Straße 7
99095 Erfurt, Deutschland
produktsicherheit@kolibri360.de